Programming Excel with VBA

A Practical Real-World Guide

Flavio Morgado

Apress®

Programming Excel with VBA: A Practical Real-World Guide

Flavio Morgado
Rio de Janeiro, Brazil

ISBN-13 (pbk): 978-1-4842-2204-1 ISBN-13 (electronic): 978-1-4842-2205-8
DOI 10.1007/978-1-4842-2205-8

Library of Congress Control Number: 2016957877

Cover image designed by FreePik

Managing Director: Welmoed Spahr
Lead Editor: Gwenan Spearing
Technical Reviewer: Fabio Claudio Ferracchiati
Editorial Board: Steve Anglin, Pramila Balan, Laura Berendson, Aaron Black, Louise Corrigan, Jonathan Gennick, Robert Hutchinson, Celestin Suresh John, Nikhil Karkal, James Markham, Susan McDermott, Matthew Moodie, Natalie Pao, Gwenan Spearing
Coordinating Editor: Mark Powers
Copy Editor: Kim Wimpsett
Compositor: SPi Global
Indexer: SPi Global
Artist: SPi Global

Distributed to the book trade worldwide by Springer Science+Business Media New York, 233 Spring Street, 6th Floor, New York, NY 10013. Phone 1-800-SPRINGER, fax (201) 348-4505, e-mail orders-ny@springer -sbm.com, or visit www.springer.com. Apress Media, LLC is a California LLC and the sole member (owner) is Springer Science + Business Media Finance Inc (SSBM Finance Inc). SSBM Finance Inc is a Delaware corporation.

For information on translations, please e-mail rights@apress.com, or visit www.apress.com.

Apress and friends of ED books may be purchased in bulk for academic, corporate, or promotional use. eBook versions and licenses are also available for most titles. For more information, reference our Special Bulk Sales–eBook Licensing web page at www.apress.com/bulk-sales.

Any source code or other supplementary materials referenced by the author in this text are available to readers at www.apress.com/9781484222041. For detailed information about how to locate your book's source code, go to www.apress.com/source-code/. Readers can also access source code at SpringerLink in the Supplementary Material section for each chapter.

To my beloved sons, Georgia and Diego.

You are the light of my life!

Contents at a Glance

Contents

About the Author

Flavio Morgado is a food engineer with a master's degree in food science and technology. He is also a VBA professional developer, a technical writer, an English to Brazilian Portuguese technical translator, and a professor at UNIFESO - Centro Universitário Serra dos Órgãos in the city of Teresopolis in Rio de Janeiro, Brasil.

He has written more than 30 books (and translated an equal number) — all published just in Brazil.

Flavio lives in Teresopolis; when he is not teaching, writing, or developing a love of something (or somebody...), he is running or riding his mountain bike throughout the Teresopolis Mountains, followed by his 11 dogs (2016's dog count).

About the Technical Reviewer

Fabio Claudio Ferracchiati is a senior consultant and a senior analyst/developer using Microsoft technologies. He works for BluArancio (`www.bluarancio.com`). He is a Microsoft Certified Solution Developer for .NET, a Microsoft Certified Application Developer for .NET, a Microsoft Certified Professional, and a prolific author and technical reviewer. Over the past ten years, he's written articles for Italian and international magazines and coauthored more than ten books on a variety of computer topics.

Acknowledgments

This book shows how to improve worksheet applications using Visual Basic for Applications (VBA). A great part of it was inspired or is based on content and knowledge available for free on the Internet.

So, I like to thank the existence of these VBA Internet sites (in alphabetical order):

- Better Solutions: www.bettersolutions.com/
- Excel Matters: http://excelmatters.com/
- ExcelVBA: http://excelevba.com.br/
- Mr. Excel: www.mrexcel.com
- Microsoft System Development Network (MSDN): https://msdn.microsoft.com/
- Ozgrid: www.ozgrid.com/VBA/find-method.htm
- Pearson Software Consulting: www.cpearson.com/
- Ron de Bruin Excel Automation: www.rondebruin.nl/win/section2.htm
- StackExchange: http://stackexchange.com/
- StackOverflow em Português: http://pt.stackoverflow
- StackOverflow: http://stackoverflow
- VB-fun.de: www.vb-fun.de/

And thanks for the existence of these nutrient Internet sites (in alphabetical order):

- Agricultural Research Serving of United States: www.ars.usda.gov
- EatingWell: www.eatingwell.com/
- Food and Agriculture Organization for the United Nations: www.fao.org
- National Heart, Lung, and Blood Institute: www.nhlbi.nih.gov
- The Dash Diet Eating Plan: dashdiet.org
- Wikipédia: www.wikipedia.com

I also want to thank Microsoft and all the people on the Microsoft Excel development team for giving us Excel—a superb piece of software that is versatile and powerful. As a professional developer, I know how difficult it is to produce something so good, even though it is not perfect, but what is? Please Microsoft, receive my most profound respect and compliments.

Introduction

This book was created to teach you how to use Visual Basic for Applications (VBA) to automate worksheet applications.

It is a code book that was written to teach everyone, including people with zero experience in programming and people with decent programming knowledge, how to use and apply programming techniques to better interact with the users of your Microsoft Excel solutions in a more professional way.

It uses some ready-made worksheet applications as practical examples of how you can produce solid, precise, and reliable worksheet applications based on VBA programming.

First I'll introduce you to the VBA environment and the language structure and show you some basic examples that will take you on a consistent journey through the Excel object model. All the proposed examples use VBA to teach you how to interact with the Excel object model and its many properties, methods, and events.

As a basic strategy to teaching Excel VBA programming, this book uses the VBA Immediate window to first test each proposed Excel feature before showing how to use it in a code procedure. It is full of programming examples whose complexity grows from the book's beginning to its end. This means it was written to be read one chapter at a time, with each chapter using the knowledge of previous chapters to provide a jump on your programmer skills.

My teaching strategy is to show a `UserForm` interface or code procedure step-by-step and to comment on all its instructions, one at a time, to give you a better understanding of how the VBA code can use good programming techniques to produce the desired result, with clear, concise, and reusable code.

What's in the Book

This book is divided into 11 chapters, each one approaching Excel VBA programming with a new complexity. To get a big picture of what you will find inside it, here is a summary of each chapter:

- Chapter 1, "Understanding Visual Basic for Applications (VBA)," is intended to show you the VBA metaphor, including how to use VBA integrated development environment (IDE) and the VBA language structure. It is a basic chapter to show you a first approach to VBA and a programming language.
- Chapter 2, "Programming the Microsoft Excel Application Object," touches on the first programmable object in the Microsoft Excel object model hierarchy: the `Application` object, which represents the Microsoft Excel application window, with some of its main properties, methods, and events. In this chapter, you will learn about the `FileDialog`, `GetOpenFileName`, `GetSaveAsFileName`, and `OnTime` methods; when the `Application` object events fire; and how to use a `Class` module to watch and/or control whether a worksheet tab name can be changed, as a first approach to producing VBA objects.

- Chapter 3, "Programming the Microsoft Excel Workbook Object," talks about the second layer of the Microsoft Excel object model, which represents the Workbooks collection and contains all the open Workbook objects inside the main Excel window. You will learn about how and when the Workbook object events fire and will see the VBA UserForm and its controls, learning how to use the ListBox control and its interface (properties, methods, and events) to interact with the Workbook object using VBA code.
- Chapter 4, "Programming the Microsoft Excel Worksheet Object," touches on the third object level in the Excel object model hierarchy: the Worksheet object, which represents the sheet tab in a Excel workbook file. It also shows how its programmable interface is composed (it properties, methods, and events). In this chapter, you will use again a VBA UserForm to learn how to add, delete, copy, move, rename, sort, and change sheet tab visibility inside a workbook file. You will also learn the many ways (and the preferable one) to reference sheet tabs in the VBA code and how to control the "cascade events" phenomenon that happens with VBA objects.
- Chapter 5, "Programming the Microsoft Excel Range Object," talks about the deepest object inside the Excel object model hierarchy: the Range object, which can represent any number of cells inside a Worksheet object. Using another UserForm interface, you will learn how to programmatically define the cell addresses that compose any selected range, how to use VBA to name a range (using the Names collection), how to use the VBA Collection object, and how to produce a similar interface to the Excel Name Manager interface using VBA, improving the Excel interface by allowing you to change the range name visibility inside the workbook file.
- Chapter 6, "Special Range Object Properties and Methods," expands your knowledge about the Range object by covering the Cells and CurrentRegion properties and the End, OffSet, Find, AutoFilter, Sort Copy, and PasteSpecial methods, using again the UserForm approach. At the end of this chapter you are presented with the frmFindFoodItems UserForm that uses most of this knowledge to find food items in the USDA food table using different search criteria (by food item name or nutrient content).
- Chapter 7, "Using Excel as a Database Repository," presents you with a programmable approach to implementing a database storage system to store worksheet data as database records in unused worksheet rows, based on a data validation list cell and range names. This chapter uses all the knowledge gathered so far to produce a standard code module full of procedures that use module-level constant values to define the database parameters, allowing you to adapt it to any worksheet application design.
- Chapter 8, "Creating and Setting a Worksheet Database Class," expands the database code module to a Class module that uses range names to store the database properties inside unused worksheet rows. It defines the SheetDBEngine class interface, showing how to implement its properties, methods, and events. To allow you to easily use the SheetDBEngine class, this chapter also presents the frmDBPRoperties UserForm, which was produced as a database wizard, to help implement the database storage system on any worksheet application.

- Chapter 9, "Exchanging Data Between Excel Applications," answers important questions regarding how to programmatically update worksheet-based tables from where the worksheet application gathers its data. It teaches how to update the USDA worksheet to any version released by the ARS-USDA web page, using either a simple, silent procedural approach or a `UserForm` that reacts to the updating process, finding food item name inconsistencies between two releases of `SRxx.accdb` or `SRxx.mdb` Microsoft Access database nutrient tables. You will be also presented with the `frmManage UserForm` to allow mass operation on the worksheet database so you can delete and save recipe nutrient data and export and import recipes, using VBA automation.
- Chapter 10, "Using the Windows API," takes your VBA knowledge gathered so far in this book to the next level by teaching you how to understand and use the Windows application programmable interface (API), based on dynamic linked libraries (DLLs) and C++ procedures. The API can be declared and called from within a VBA project. In this chapter, you will learn how to implement a `Timer` class to create programmable timer objects and how to change the `UserForm` appearance by removing its title bar; adding a resizable border and minimize, restore, and maximize buttons; and adding transparency and creating a fade effect. You'll also learn how to animate a `UserForm` when it loads and how to apply a skin effect to change the `UserForm` shape.
- Chapter 11, "Producing a Personal Ribbon Using RibbonEditor.xlam," uses the RibbonX VBA add-in developed by Andy Pope as free software to produce a personalized ribbon with tools that interact with your worksheet application, giving it a professional appearance.
- The afterword is a brief testimony of the path used to write this book, with my insights about how I imagined it and created it step-by-step, with trial-and-error experimentation as I learned the Excel object model.

This Book's Special Features

Programming Excel 2016 with VBA was designed to give you all the information you need to understand how to replicate a behavior, insert a formula, define an interface, and so on, without making you wade through ponderous explanations and interminable background. To make your life easier, this book includes various features and conventions to help you get the most of the book and Excel itself.

- *Steps*: Throughout the book each Excel task is enumerated in step-by-step procedures.
- *Commands*: I used the following style for Excel commands: Conditional Formatting button in Styles area on the Home tab of the ribbon (this mean you must click the Conditional Formatting button that you find in the Styles group of the Home tab).
- *Menus*: To indicate that you click the File menu and then select Save, I use File ➤ Save.

- *Functions*: Excel worksheet functions appears in capital letters and are followed by parenthesis: `IF( )`. When I list the arguments you can use with a function, they will appear in a bullet list of items, using the same order that they appear inside the function arguments list to explain the meaning of each one.
- *Rows, columns, cell address, ranges, and sheet names*: Excel rows, columns, cell address, range names, and sheet names appear using an Arial font to detach them from the text.

This book also uses the following box to call attention to important (or merely interesting) information:

Attention The Attention box presents asides that give you more information about the topic under discussion. These tidbits provide extra insights that give you a better understanding of the task at hand, offer complementary information about the issue that is being currently discussed, or even talk about an unexpected Excel behavior regarding a given task.

Web Site Extras

All the examples presented in this book are available at `www.apress.com/9781484222041` as ZIP files for each book chapter. In addition to the example workbooks or worksheet applications, I'll post any updates, corrections, and other useful information related to this book.

Your Feedback Is (Very) Important!

Before you continue reading, I would like to say that your opinion is important to me. I really don't know how many of you will write to me to give any feedback, but I hope I can answer everyone and, whenever possible, resolve any questions or problems that arise. Since I have many other duties, sometimes it may take a little while so I can answer you, but I promise to do my best. Please feel free to write to me at the following e-mail: `flaviomorgado@gmail.com`.

CHAPTER 1

Understanding Visual Basic for Applications (VBA)

This chapter will teach you how to program Microsoft Excel using Visual Basic for Applications code procedures. It will show you how to interact with the VBA integrated development environment (IDE), show how VBA programming code works, and give lots of examples. This introduction is for people who do not know anything about VBA and want to get started. If you already have some experience, you can skip to Chapter 2.

You can obtain all the procedure code in this chapter by downloading the `Chapter01.zip` file from the book's page at Apress.com, located at `www.apress.com/9781484222041`, or from `http://ProgrammingExcelWithVBA.4shared.com`.

What Is Macro Code?

There was a time in computer usage that if you needed to accomplish a series of successive commands in software, you could use the appropriate program syntax to type the menu commands to follow, one command per row, and then make the program execute them, row by row. In the DOS age of the early 1980s, the spreadsheet program called Lotus 123 called this method a *macro*, and since then, every time a program tries to allow the user to implement such a resource to execute any desired number of successive menu commands, the convention is to say that the program is following a *macro code*.

The first versions of Windows and all Microsoft Office applications (including Microsoft Excel) could implement this sort of automation by recording successive series of actions performed on the interface using the keyboard or the mouse.

Until 1994 when Microsoft Excel was on version 3.0, each Microsoft Office application had its own way to store its macro codes, meaning that these codes were not interchangeable between Office applications. In other words, Word has its own set of commands, PowerPoint had another set, and Excel had yet another one. At this time you needed to be very specialized to read and understand each Microsoft Office application macro code.

Everything changed in 1995, when Microsoft released Microsoft Visual Basic 4; its IDE used the VBA language and made it part of all Microsoft Office 6 applications. From this time on, all Microsoft Office applications were capable of being programmed using a single language, allowing all applications to interact with each other. Even though this code was produced with VBA as a true programming language, the term *macro code* stuck to this sort of application automation and continues to be used today.

Electronic supplementary material The online version of this chapter (doi:10.1007/978-1-4842-2205-8_1) contains supplementary material, which is available to authorized users.

F. Morgado, *Programming Excel with VBA*, DOI 10.1007/978-1-4842-2205-8_1

Although you can consider Excel macro code to be any kind of code automation created to interact with Excel and its environment, you can divide the code into two types of VBA code.

- *Excel macros*: The VBA code created by Excel itself when you activate Excel macro recording and execute any successive actions on its interface
- *VBA programming code*: The code you create by hand to implement the same automation, but with more control and many times with more efficiency

In this book I will teach you how to understand and create your own VBA code procedures so you can take greater control of your worksheet applications.

The VBA Environment

Before you begin to learn the Visual Basic for Applications language, you must acquaint yourself with the VBA IDE; this is the place where you can create and test the VBA code of any Microsoft Office application.

There are two ways to open the VBA IDE.

- By going to the Microsoft Excel Developer tab and then clicking the Visual Basic button, as shown in Figure 1-1
- By pressing Alt+F11 in any Microsoft Office application

Figure 1-1 shows the Developer tab on the Excel ribbon with the Visual Basic command button.

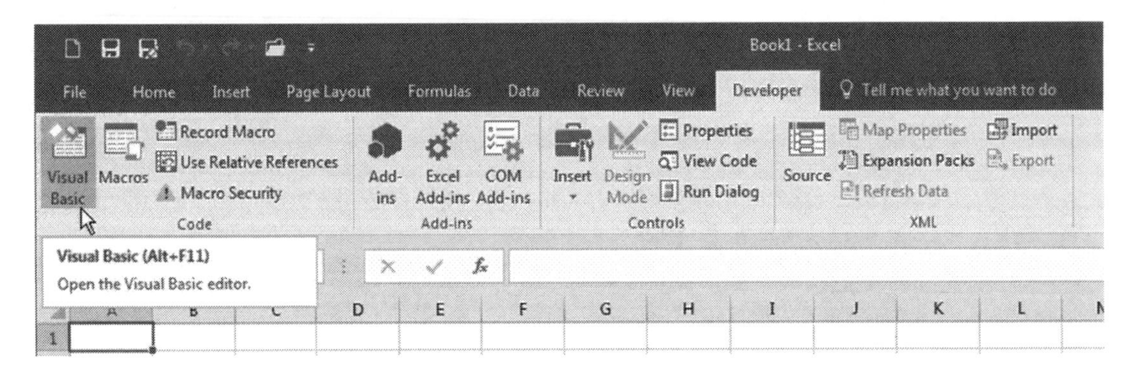

Figure 1-1. *This is the Excel Developer tab, where you will find, among other tools, the Visual Basic command button to activate the Visual Basic IDE*

To show the Excel Developer tab, follow these steps:

1. Click the Excel File tab and select Options to show the Excel Options dialog box.
2. Select the Customize Ribbon option.
3. In the right list box, check the Developer tab (Figure 1-2).
4. Click OK to close the Excel Options dialog box.

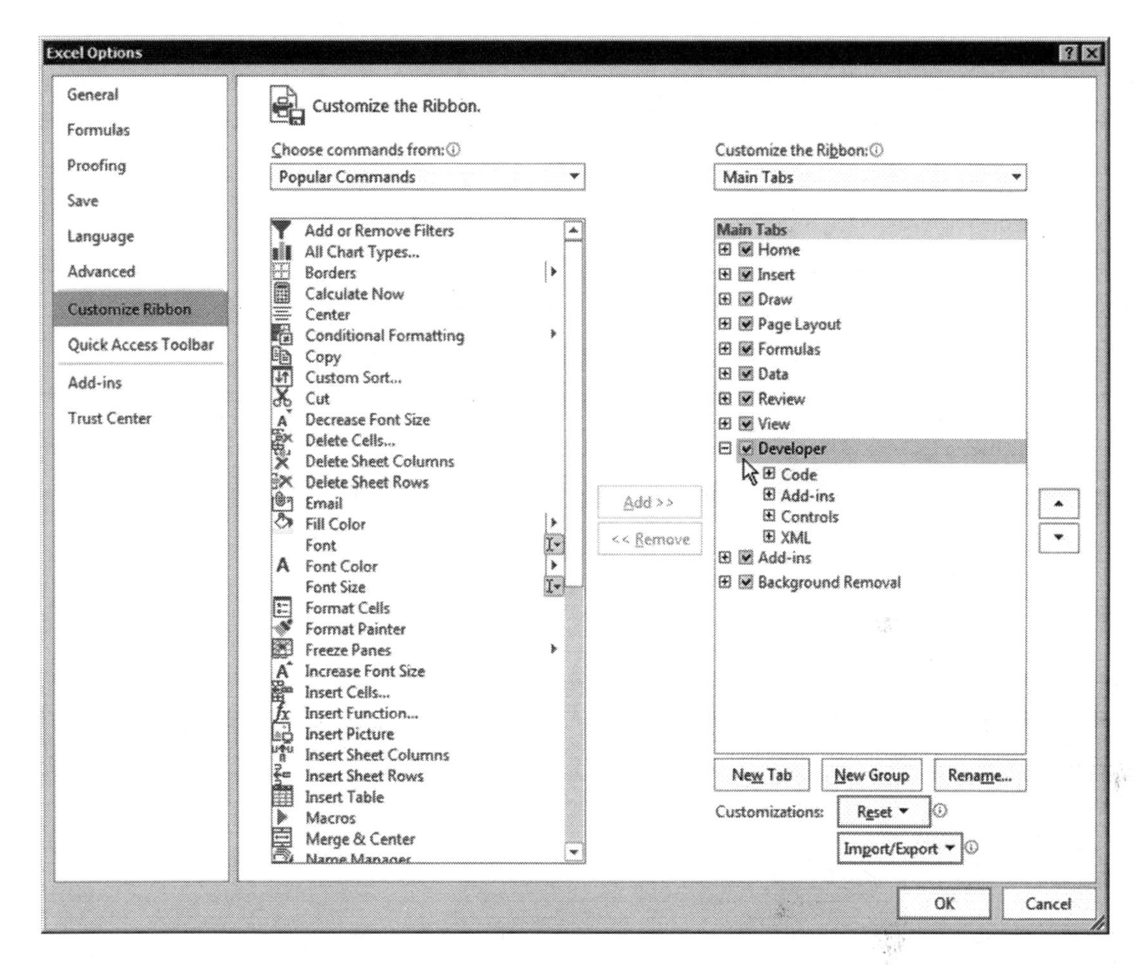

***Figure 1-2.** To show the Developer tab on the Excel ribbon, select File ➤ Options, click Customize Ribbon, and check the Developer option in the right list box*

Figure 1-3 shows the Visual Basic IDE that you can activate by pressing Alt+F11 or by clicking the Visual Basic command on the Excel Developer tab.

■ **Attention** You can turn on the visibility of the Project Explorer, the Properties window, and the Immediate window using the appropriate commands in the Visual Basic View menu item.

Modules: The VBA Documents

VBA works this way on Office documents: each document created by any Microsoft Office application has one specific VBA object attached to it called a *code module*. It is almost impossible to dissociate the object from the code module, and if this happens (which is rare but possible), the file becomes corrupted and unusable until you fix it by deleting the offending object.

Excel has the ability to open workbook files, which are a special kind of file made up of one or more worksheets (called Sheet1, Sheet2, and Sheet3). Indeed, each new file created by Microsoft Office 2003 or later versions has by default three blank worksheets (although you can configure Excel to create new workbooks with 1 to 100 blank worksheets).

■ **Attention** To configure the number of blank worksheets a new Excel workbook (or Excel file) has, select File ➤ Option, select "Include this many sheets" option, and indicate how many worksheets you desire. Note, however, that every workbook must have at least one worksheet inside it.

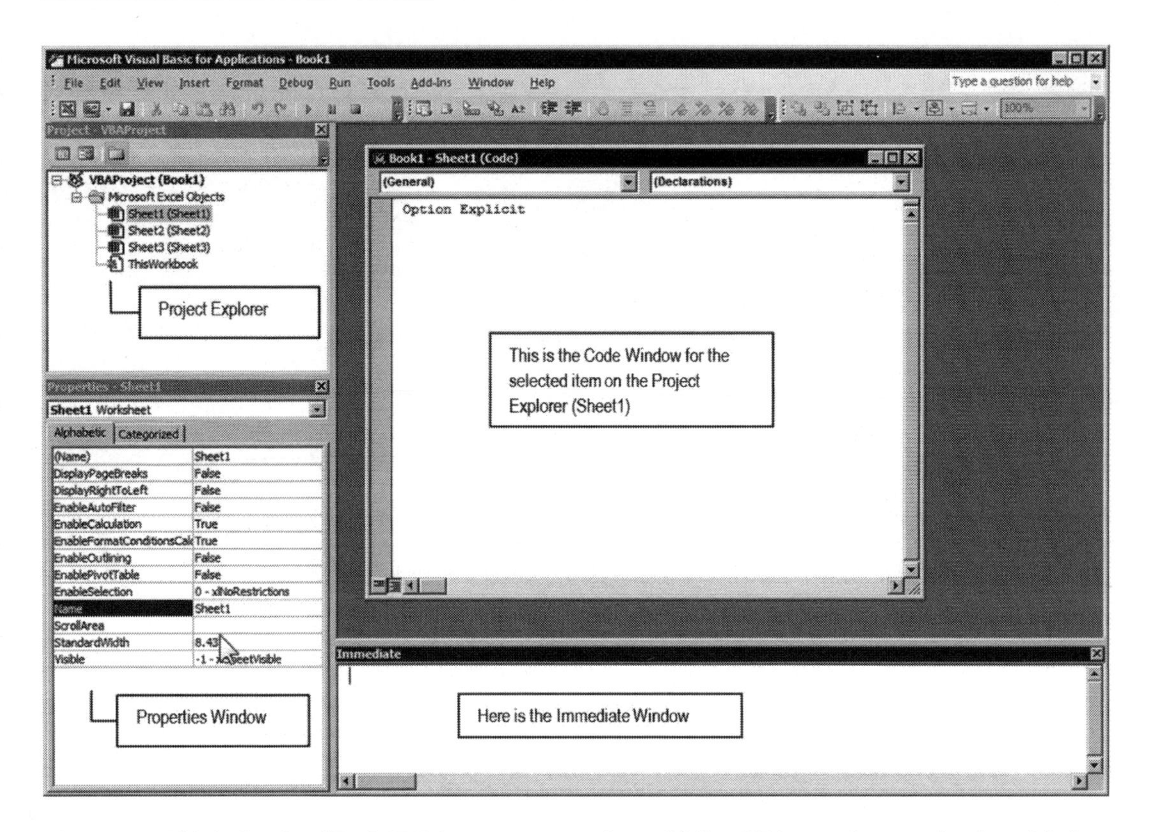

Figure 1-3. *This is the Visual Basic IDE that you use to write and follow VBA procedures and codes, with the Project Explorer, the Properties window, the Immediate window, and the Code window (for Book1, Sheet1) visible*

From VBA's point of view, each Excel object has its own code module: one for the entire workbook (it is the last object on the Microsoft Excel Objects branch of the Project Explorer tree, named `ThisWorkbook`) and one for each worksheet inserted by default on the workbook, named with the same worksheet name (`Sheet1` to `Sheet3` on a default workbook with three blank worksheets). They are always positioned at the beginning of the Microsoft Excel Objects branch of the Project Explorer tree (Figure 1-4).

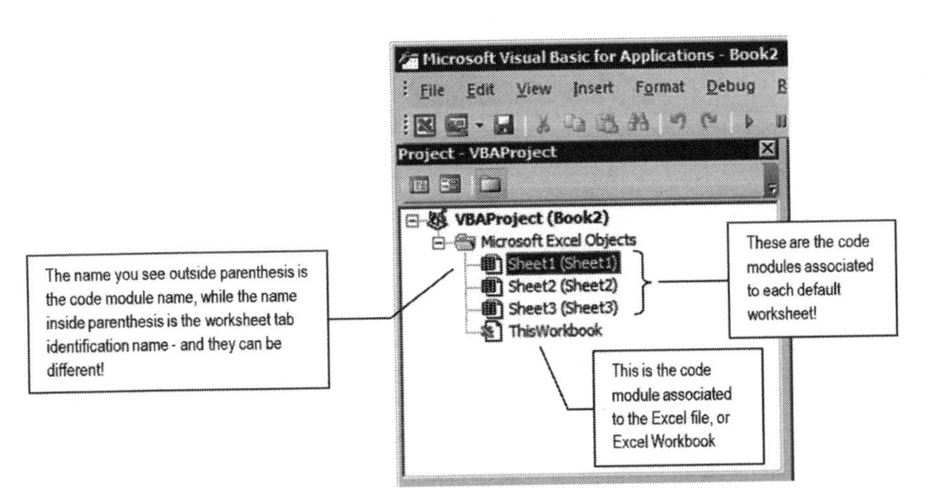

Figure 1-4. *Every Excel file, or Excel workbook, has at least one code module associated with the workbook (named ThisWorkbook) and one code module for each worksheet, having the same name of the sheet tab. They are all shown on the Project Explorer tree*

To see the VBA code module associated with each default Excel object, double-click the code module in the Project Explorer tree. Figure 1-3 shows the `Sheet1` code module, while Figure 1-5 shows the `ThisWorkbook` code module opened in the VBA IDE.

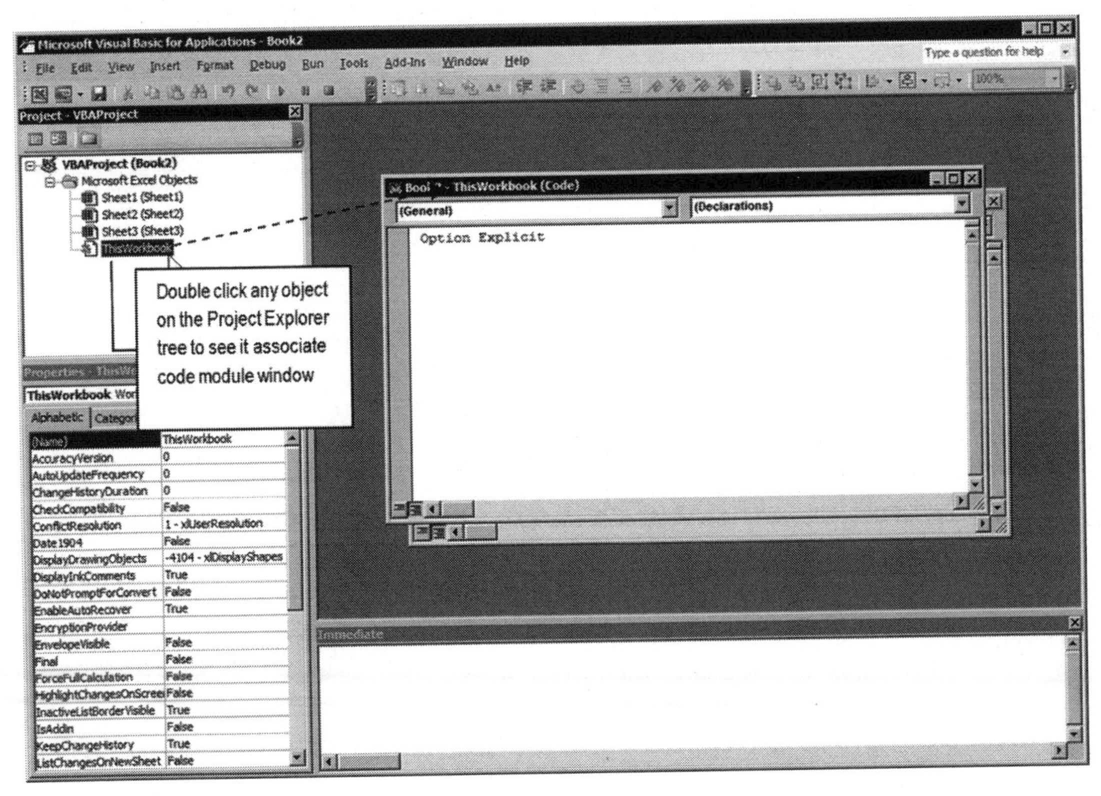

Figure 1-5. *This is the VBA code module associated with the Excel workbook file. By default it is called ThisWorkbook, and every Excel file has just one of them*

Standard and Class Modules

Although no Excel object can be dissociated from its code module, you can create independent code modules that are not associated with any object using the VBA Insert ➤ Module and Insert ➤ Class Module menu commands. Figure 1-6 shows that after you execute VBA Insert ➤ Module, a new Module branch of independent code modules is created inside the Project Explorer tree, with the new `Module1` code module already selected in the VBA interface.

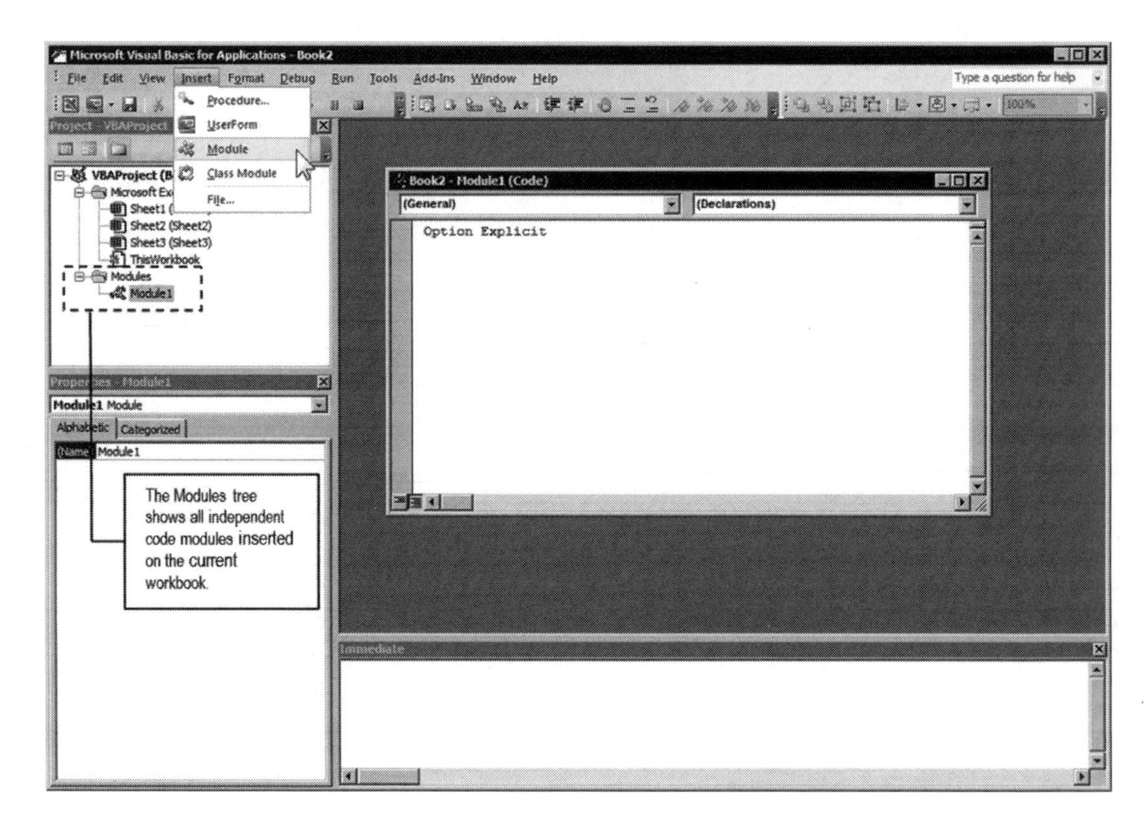

Figure 1-6. *This is the VBA IDE after a new independent code module called Module1 was created using the VBA Insert ➤ Module menu command*

■ **Attention** You can change the focus from one code module to another by double-clicking the Project Explorer tree or by using VBA Windows menu command to select it from among any open code module windows.

Note that whenever you select any Excel object in the VBA Project Explorer tree, all its properties are shown in the Properties window, even if the code module is closed. Later in this book I will talk about some interesting code module properties associated with just the VBA IDE or with the object itself (properties associated with the worksheet or workbook object).

This VBA structure of different code module windows is part of any Excel macro-enabled workbook, even if you don't use them! They are the documents that can be manipulated just by the VBA IDE, and it is where all the action will occur to control the automation of your Excel solution.

■ **Attention** To delete any module or class module inserted in the VBA structure of an Excel file, right-click its name on the Project Explorer tree and select Remove <Module Name> in the pop-up menu. You will be prompted to export the code module code before you delete it from the workbook structure.

Class Modules

`Class` modules are a special type of code module that allows the creation of a programmable "object." Using `Class` modules, you can trap other existing VBA objects (like the Excel `Application` object, `UserForm`, and so on), create your own objects, define behavior using `Property` procedures and methods represented by `Public` and `Sub` procedures, and raise events (automatically respond to user actions). Consider a `Class` module as a code repository that can be reused, creating many instances of the class for the same purpose, where all instances share the same base code. The `Class` module is the core module of object-oriented programming in VBA, and you will see it in action later in this book.

The VBA Language

Like any other programming language, VBA executes individual statements one at a time from the first row after the procedure declaration to the last row.

You should strive to write procedure code using VBA strict syntax rules with one statement per row. The shorter the procedure code is (the fewer rows it has), the faster it executes and the easier it is to understand.

To know how to write VBA code, you must understand its syntax for variable declarations, its instructions, and its functions.

Procedures: The VBA Code

Procedures are the heart of any programming language. They are responsible for executing the actions you need to grant a given functionality to your application.

You use VBA code modules to write the procedures that must be executed; there are three types of procedures:

- *Function procedures*: These can return a value when finishing execution.
- *Sub procedures*: These can't return a value.
- *Property procedures*: These are a special pair of procedures with the same name that can implement an object property.

For now we will stay with the concept of `Function` and `Sub` procedures. Property procedures will be addressed later in this chapter.

Using Function and Sub Procedures

You may ask why `Function` and `Sub` procedures exist, and my answer is that they are a holdover from the very first programming language. They allow a programmer to divide the executing code into reusable parts. The main code is executed by one procedure (`Function` or `Sub`), while any reusable can is executed by a `Sub` procedure, thus reusing its code.

For our usage, let's just stay with this simple definition: use a `Function` procedure whenever you want to return a value, and use a `Sub` procedure whenever you want to execute any necessary and successive programming steps without returning any value (and this `Sub` procedure eventually can be reused by other procedures).

Both `Function` and `Sub` procedures have specific syntax rules for any VBA code module: a begin statement that indicates its scope (`Public` or `Private`), its name, its argument list (if any), the type of returned value for `Function` procedures, and an `End` statement.

The following VBA code fragment declares a simple `Function` procedure named `FindFoodITem( )` (note that it has a return value declared as `Boolean`):

```
Function FindFoodItem( ) as Boolean
    [Statements]
    FindFoodItem = <FunctionResult>
End Function
```

■ **Attention** In this book, everything that appears inside the < > characters means the name you must give to the procedure or argument it represents.

The following is another VBA code fragment declaring a `Sub` procedure named `FindFoodItem( )` (note that it *does not* return a value):

```
Sub FindFoodItem( )
        ...
End Sub
```

■ **Attention** In the same VBA code module there can be only one `FindFoodItem( )` procedure, declared as a `Function` or `Sub` procedure.

The following is the complete VBA syntax necessary to declare a `Function` procedure (optional parts of the `Function` declaration are between brackets):

```
[Public | Private | Friend | Static] Function <FunctionName> ([Optional] [Arg1[ [as
Type],...) [As<Type>]
    [Statements]
    <FunctionName> = <FunctionResult>
End Function
```

In this code:

> `[Public | Private | Friend | Static]`: These are VBA keywords to specify the scope of the `Function` procedure. Note that they are between brackets, meaning that they are optional. If you don't use a VBA keyword, the `Function` will be considered `Private`, meaning that it can be accessed only by the procedures inside the code module where it belongs. If you use the keyword `Public` to declare a `Function` procedure, it will be capable of being accessed from other procedures in any other code module of your workbook application.

■ **Attention** The Friend keyword is used only on Class modules to create procedures that are public to all Excel objects on the same workbook while it is private to the code. The Static keyword indicates that all variables declared inside the Function procedure will keep their values between function calls.

<FunctionName>: This is the name you want to associate with the Function procedure. It must begin with a letter and have no spaces or reserved characters. A good programming practice for Function names is to use a clear name that indicates what it executes. It must be made up of concatenated names, where each name must begin with a capital letter like Function FindFoodItem( ).

[Optional] [Arg1]: Both Function and Sub procedures must be followed by a pair of parentheses, which may contain one or more declared arguments that the function can receive. These arguments are the values upon which the function will operate. VBA allows you to pass optional arguments if you precede them with the Optional keyword (meaning that you are not obliged to pass them to the procedure). Once one argument is declared as Optional, all others that follow it must also be declared as Optional.

[As <Type>]: Arguments for Function and Sub procedures, as the values returned by a Function procedure, can be of a specific type. They can be an integer or real number, a text string, a date, a logical value, or an object. If you do not specify the type expected to be received by an argument or returned by a Function, the type will be considered as the Variant data type, which is a special kind of VBA data that is polymorphic, meaning it can represent any data type.

End Function: This must be the last function statement indicating to VBA where the code procedure ends.

To declare a Sub procedure, you use the following syntax. Note that it is almost identical to the Function procedure declaration, except it doesn't return a value (it doesn't have the [As Type] part of the Function declaration) and has an End Sub statement to indicate its finish.

```
[Public, Private] Sub <SubName> ([Optional] [Arg1[ [as Type],…)
    [Statements]
End Sub
```

Calculating Age in Years

Let's try to write a simple VBA procedure to calculate the age in years of any person. How we can do that?

We need simple basic information: the person's date of birth! Since your computer always has the current date, we can use it to count the number of days between these two dates (current date minus date of birth) and do some math to express the result in years.

Before we try to create this simple procedure, you need to know a basic concept regarding how VBA treats date values.

To Windows—and all its applications—any date is just a consecutive counting of the number of days since 01/01/1900. The first day of the 20th century is considered as day 1, and any later date is a consecutive day counting from it (earlier dates are negative values). I am writing these words on 02/17/2014, and this date is considered by Windows as day 41,687, meaning that it has been 41,687 days since January, 1, 1900.

Attention You can easily get the number for any date by typing it in any Excel cell and changing the cell format from Date to General number.

By subtracting two different valid dates, you will receive the number of days between the two dates. Dividing this result by 365 days per year, you will get the difference in years, right?

Attention For your information, a valid Date value is one that has no more 30 days in April, June, September, and November; has no more than 29 days in February on lap years (28 days in normal years); has no more than 31 days in January, March, May, July, August, October, and December; and has a month value between 1 and 12.

The next `Function` procedure code calculates any person's age in years by receiving the person's date of birth as an argument to the procedure:

```
Function AgeInYears(DateOfBirth)
    AgeInYears = (Date - DateOfBirth) / 365
End Function
```

Attention Extract the `AgeInYears.xlsm` macro-enabled workbook from the `Chapter01.zip` file to test this function procedure.

To build this `AgeInYears` function, follow these steps:

1. In the VBA interface of any Excel worksheet, choose the Insert ➤ Module menu command to create a new standard code module.
2. Make the VBA Properties window visible by executing the View ➤ Properties Window menu command.
3. In the Property window, type `basAgeInYears` for the `Name` property of this new code module.
4. Insert in the `basAgeInYears` code module a new `Public Function` procedure by executing the Insert ➤ Procedure menu command. In the Add Procedure dialog box, type `AgeInYears` as the procedure name and select Function on Type and Public on Scope.
5. Click OK to create the `Function AgeInYears( )` procedure structure in the `basAgeInYear` module (Figure 1-7).

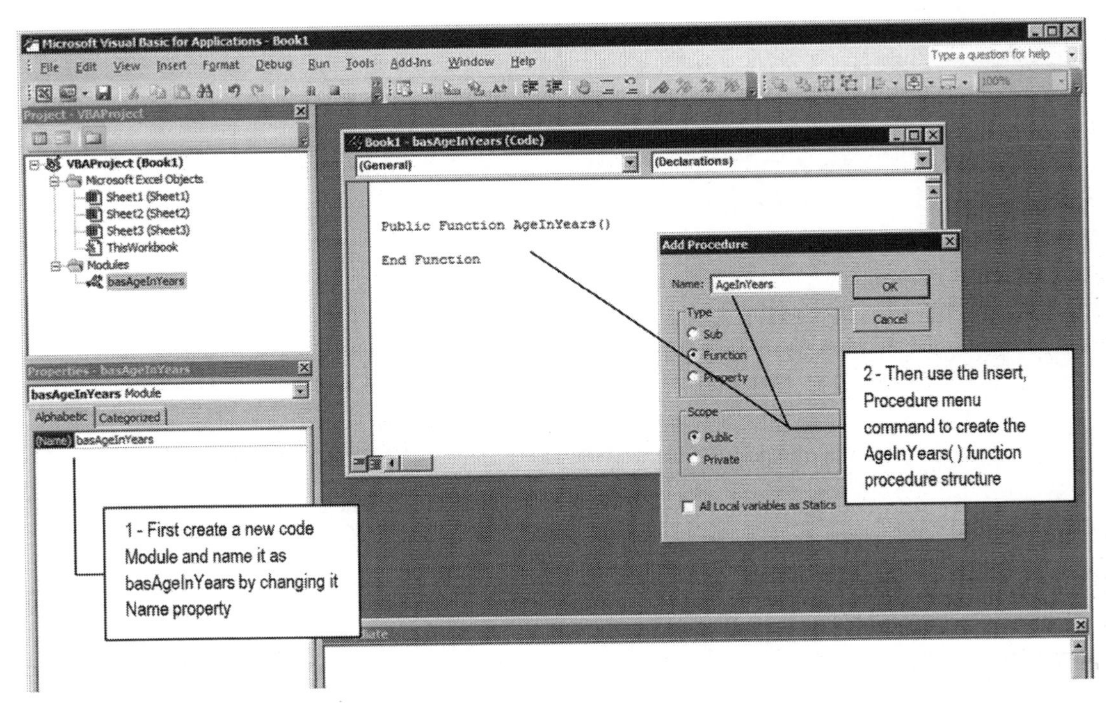

Figure 1-7. *To create the AgeInYears function procedure, first create a new code module and change its Name property to basAgeInYears. Then use the Insert ➤ Procedure menu command to create a Public Function procedure structure and name it as the AgeInYears function*

6. Insert the `DateOfBirth` argument between the `Function` procedure parentheses to tell VBA that this function must receive an argument to operate on. Your function must become like this:

```
Function AgeInYears(DateOfBirth)

End Function
```

7. Inside the procedure code (between the `Function` declaration and the `End Function` instructions), press Tab to indent the code, type the function name exactly as declared, and enter the = sign to tell VBA that the function value will be attributed by the right side of the equation.

```
Function AgeInYears(DateOfBirth)
    AgeInYears =
End Function
```

8. Calculate the number of days between the `System` date and the `DateOfBirth` argument by subtracting from the VBA `Date( )` function the `DateOfBirth` value (evolve this subtraction between parentheses to assure that this subtraction operation will be executed first, before any other math). Your function code will look like this:

■ **Attention** You may be wondering why you are naming the code module with the bas prefix. This is good programming practice. "Bas" comes from "Basic" and as you will see later is the recommended prefix for all normal code modules inserted on any VBA project.

```
Function AgeInYears(DateOfBirth)
AgeInYears = (Date - DateOfBirth)
End Function
```

■ **Attention** VBA will automatically take any parenthesis pair you type before the Date( ) function.

9. Now that you have calculated the days past between these two dates, divide this result by 365 (assuming a constant 365 days per year value). Your function will look like this:

```
Function AgeInYears(DateOfBirth)
    AgeInYears = (Date - DateOfBirth) / 365
End Function
```

Test Procedure Codes Using VBA Immediate Window

Since the first Basic programming language, every time you want to return the value of a Function procedure or variable on the programming environment (or computer screen), you must use a question mark (?) as the print character just before the function/variable name.

Use the VBA Immediate window to print Function or Variable values or to execute any Sub or Function procedure, following these two simple rules:

- To return Function procedures or Variable values, type ? and the Function procedure name in the Immediate window, followed by an opening parenthesis and a closing parenthesis (if it does not require arguments). If the function requires one or more arguments, type them between the parentheses, separating arguments with a comma (,).
- To execute Sub of Function procedures, discarding a return value, do not type the ? character. Type just the Sub or Function procedure name, *without* the opening and closing parentheses. If the Sub or Function procedure requires one or more arguments, type them after the Sub of Function name, without parentheses, separating the arguments with comma characters.

Since the AgeInYears( ) code procedure is a Function procedure (which returns a value) and requires the DateOfBith as an argument to the function to calculate and return a value, you can test it this way in the VBA Immediate window (note that I typed #4/25/1961# for the DateOfBirth argument, as shown in Figure 1-8):

```
?AgeInYears ( #4-25-1961# )
```

■ **Attention** You have to put the date argument between hash marks (#) so Visual Basic can understand that you are passing a valid date to the DateOfBirth argument of the Function procedure. If you do not do this, VBA will interpret that you want to first calculate the result of the expression 4-25-1961 = -1982 and will pass it to the Function argument. Since this is a negative value, when you subtract it from the system date, VBA will effectively add 1,982 days to your computer's current date, miscalculating the person's age.

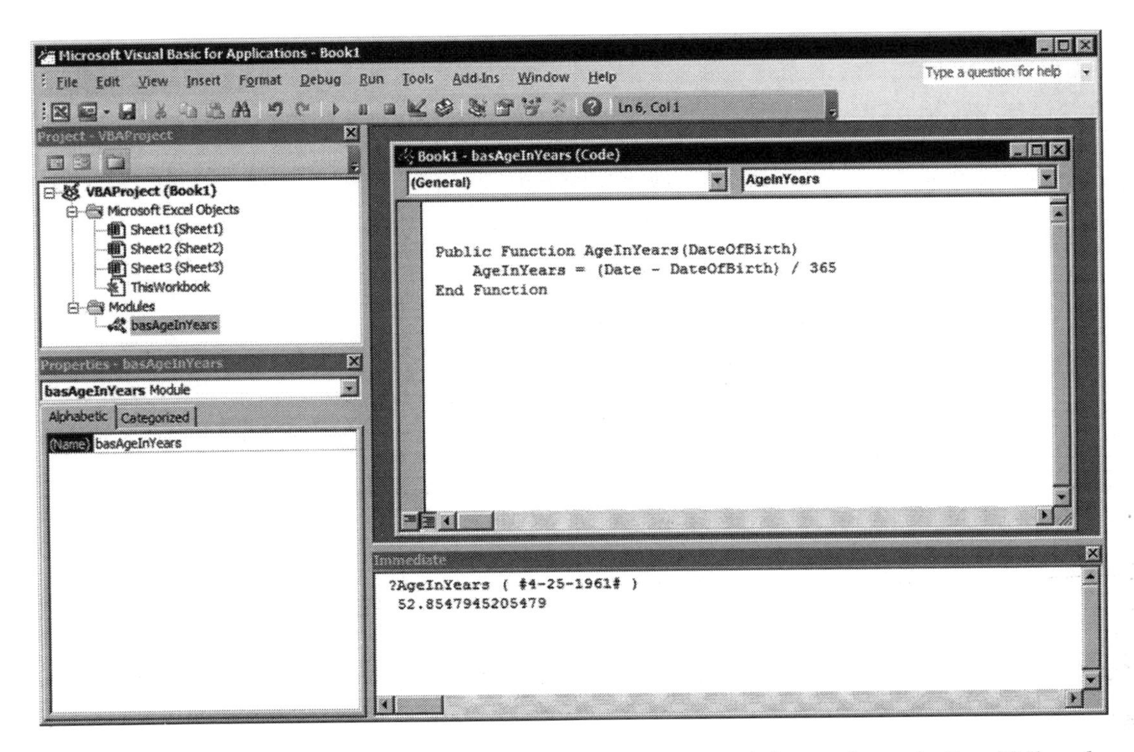

Figure 1-8. *Use the VBA Immediate window to execute any Function or Sub procedure code. Type ? followed by the Function procedure or Variable name followed by parentheses and press Enter. VBA will execute the procedure and return a value*

The number you see behind the `?AgeInYears( #4-25-1961#)` procedure call in the VBA Immediate window of Figure 1-8 is the age in years returned by the `AgeInYears( )` function procedure for my age at February 17, 2014—around 52.85 years!

■ **Attention** When you type the opening parenthesis after the `Function` procedure name in the Immediate window, VBA immediately shows the `DateOfBirth` argument to tell you that this function expects to receive some information (any optional argument will be shown between brackets, as shown in Figure 1-9).

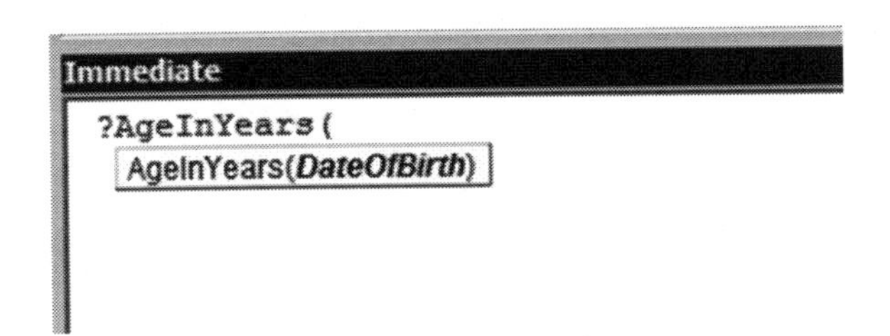

Figure 1-9. *Whenever you type the opening parenthesis after any procedure name in the VBA environment, all arguments required by the code procedure will be shown to you*

■ **Attention** Although a `Function` procedure always returns a value, VBA allows you to discard it by calling it as a `Sub` procedure (which never returns value) by simply calling its name inside other procedure code without attributing its value to a test condition or variable declaration. Alternatively, you can use the VBA `Call` keyword to explicitly indicate that a procedure code is being called, discarding its return value (if any).

In this book I will use the `Call` keyword to call `Sub` procedures in order to follow good programming practices.

Using Your Function Procedure Inside Excel

Now that you have created the `Function AgeInYears( )` procedure, let's try to see how it performs in Excel by following these steps:

1. Select any worksheet in the same workbook that you wrote the `AgeInYears( )` code procedure.
2. In cell `A1`, type **Date of Birth**; in cell `B1`, type **Age in Years**.
3. In cell `A2`, type any valid date (such as your date of birth). If you want, continue to type more dates of birth below cell `A2`.
4. In cell `B2`, type this formula:

```
=AgeInYears(A2)
```

5. Press Enter to calculate in cell `B2` the age in years for the date of birth typed in cell `A2`. If you typed any other dates of birth below cell `A2`, drag the cell `B2` selector down to propagate its formula.

Figure 1-10 shows `Sheet1` of the `AgeInYears.xlsm` workbook that you can extract from the `Chapter01.zip` file with some calculated ages in years using the `AgeInYears( )` function procedure.

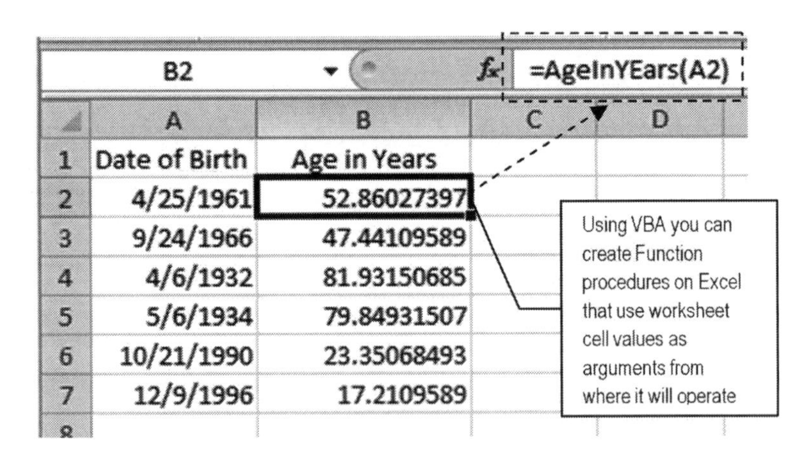

Figure 1-10. *You can test any function procedure code that returns a value using Excel worksheet cells. Sheet1 from the AgeInYears.xlsm workbook calculates many ages in years using the Function AgeInYear( ) procedure created on module basAgeInYear of this workbook*

■ **Attention** I really don't know why Excel capitalized the *E* in the =AgeInYEars(A2) formula...

Executing Code Procedures Step-by-Step

VBA executes code procedures at lightning speeds using your computer processor. To force VBA to execute any procedure code one step at a time, use VBA breakpoints to stop the code execution and follow it step-by-step. Let's do it on the `AgeInYears( )` function procedure to see how it performs. Just follow these steps:

1. Activate VBA by pressing the Alt+F11 function key (or click the VBA button of the Developer tab).
2. Insert a breakpoint on any procedure code by clicking the gray bar at the left of the instruction where you want the code to stop.
3. VBA will put a maroon dot on the left of the code instruction.

Figure 1-11 shows that I clicked at the left of the `AgeInYears( )` function procedure declaration instruction to insert a breakpoint when the function starts.

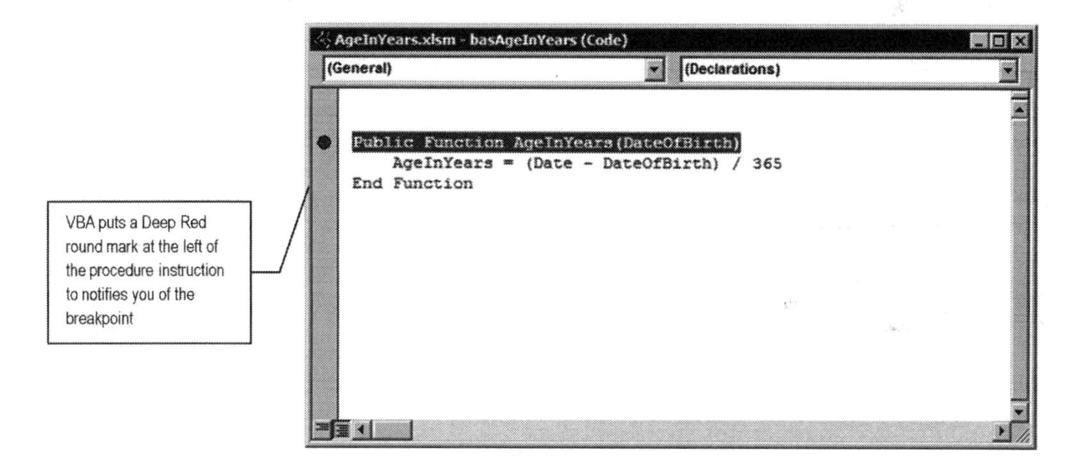

Figure 1-11. *Use the gray bar at the left of the procedure code instruction where you want to insert a breakpoint to force VBA to stop the code*

You can insert as many breakpoints as you want on any code procedure. To remove a breakpoint, just click it again.

■ **Attention** To remove all breakpoints on every code procedure of your entire VBA project, execute the Debug ➤ Clear All Breakpoints menu command (or press Ctrl+Shift+F9).

By putting a breakpoint on the `Function AgeInYears( )` procedure declaration, every time Excel tries to execute it, VBA will immediately stop the code execution at this point.

You can test the breakpoint you put on the `AgeInYears( )` function procedure in Excel using one of these two approaches:

- By executing again the `Function` procedure code from the Excel Immediate window (you can just put the blinking VBA cursor on this procedure call inside the Immediate window and press Enter)
- By editing any Excel formula cell that uses the `AgeInYears( )` function and pressing Enter

Whenever you try one of these two methods, the `AgeInYears( )` function will be executed again, and VBA will stop the execution on the breakpoint inserted on the procedure declaration, putting the entire row instruction in yellow. It is said that VBA has entered *Break mode* (Figure 1-12).

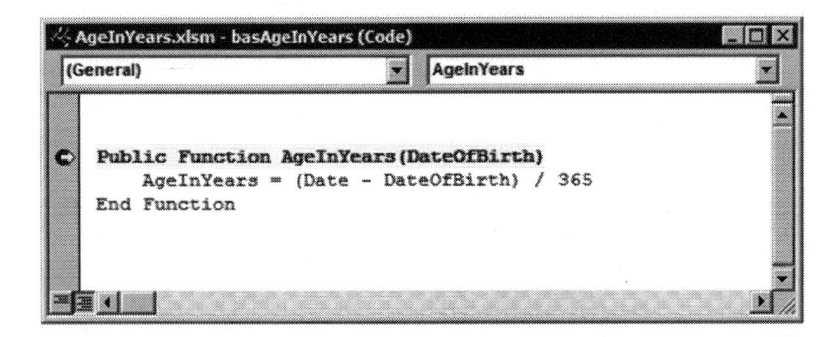
AgeInYears.xlsm - basAgeInYears (Code)
(General)
AgeInYears
Public Function AgeInYears(DateOfBirth)
AgeInYears = (Date - DateOfBirth) / 365
End Function

Figure 1-12. *When you try to execute any VBA procedure code that has a breakpoint, VBA will stop the code execution on the breakpoint and format the entire instruction row in yellow*

Whenever VBA enters Break mode, you can use the code module or use the Immediate window to inspect any variable value.

I called the `AgeInYear( )` function procedure by editing the cell `B2` formula of the `Sheet1` worksheet and pressing Enter, meaning that according to the formula inserted on this cell, Excel has passed to the `DateOfBirth` argument the value typed in cell `A2` of the `Sheet1` worksheet.

While the procedure is executing in Break mode, point the mouse to the `DateOfBirth` argument and VBA will show you the value that the argument had received (Figure 1-13).

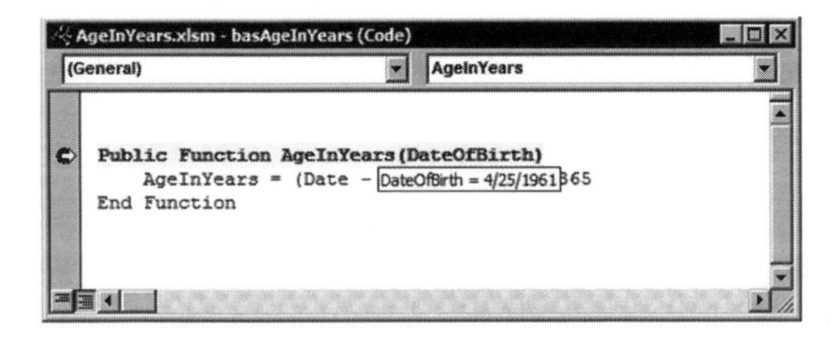
AgeInYears.xlsm - basAgeInYears (Code)
(General)
AgeInYears
Public Function AgeInYears(DateOfBirth)
AgeInYears = (Date - DateOfBirth = 4/25/1961 365
End Function

Figure 1-13. *When VBA enters Break mode, point the mouse cursor to any procedure argument (or value used inside the procedure) to see how VBA is evaluating it. The DateOfBirth argument had received the 4/25/2014 date value*

When VBA has entered Break mode, you can continue executing the procedure code in two different ways:

- Press F8 to execute just the instruction in yellow. Press F8 continuously to execute one instruction at a time until the code ends.
- Press F5 to exit VBA Break mode and execute all the procedure code from the currently detached yellow instruction to the end of the code.

Try to press F8 to initiate the `AgeInYear( )` function procedure and put the VBA focus on its first (and only) instruction. You can continue to inspect any procedure value by just pointing the mouse to it or by selecting part of the code and watching its evaluation. Figure 1-14 shows what happens when you point the mouse to the `Date` instruction, when you drag the mouse over the `(Date - DateOfBirth)` expression and point the mouse to this selection, and when you drag the mouse to select the right-side equation expression and point the mouse to it.

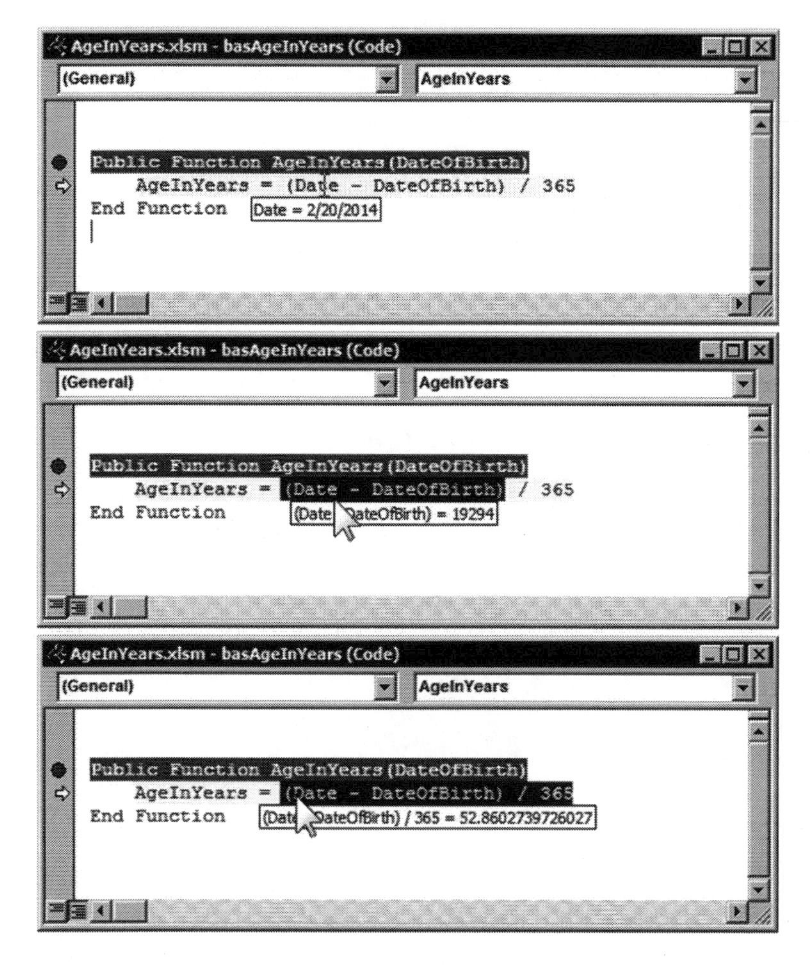
AgeInYears.xlsm - basAgeInYears (Code)
(General)
AgeInYears
Public Function AgeInYears(DateOfBirth)
AgeInYears = (Date - DateOfBirth) / 365
End Function
Date = 2/20/2014
AgeInYears.xlsm - basAgeInYears (Code)
(General)
AgeInYears
Public Function AgeInYears(DateOfBirth)
AgeInYears = (Date - DateOfBirth) / 365
End Function
(Date - DateOfBirth) = 19294
AgeInYears.xlsm - basAgeInYears (Code)
(General)
AgeInYears
Public Function AgeInYears(DateOfBirth)
AgeInYears = (Date - DateOfBirth) / 365
End Function
(Date - DateOfBirth) / 365 = 52.8602739726027

Figure 1-14. *When VBA enters Break mode, you can point to any function or variable or select any complete expression by dragging and pointing the mouse over it to see how VBA will evaluate it*

Press F8 one more time to attribute the right side of the equation to the value of the `AgeInYear` function procedure. VBA will select the `End Function` instruction, ending the code execution when you press F8 one more time (Figure 1-15).

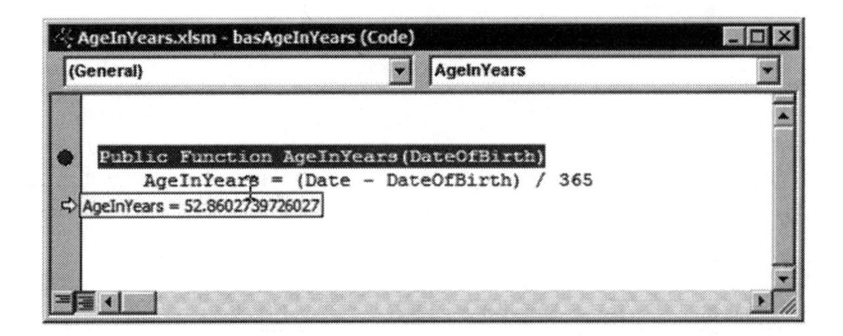
AgeInYears.xlsm - basAgeInYears (Code)
(General)
AgeInYears
Public Function AgeInYears(DateOfBirth)
AgeInYears = (Date - DateOfBirth) / 365
AgeInYears = 52.8602739726027

Figure 1-15. *You can inspect the value that any Function procedure will return by executing the code attribution instruction in Break mode and pointing the mouse to the function name. This is one of the best ways to know whether a function code procedure is working as expected*

■ **Attention** Don't forget to remove the breakpoints in your code after you have inspected your VBA code. Use the Debug ➤ Remove All Breakpoints VBA menu command to do it quickly!

Variable Declaration

A *variable* is a programming expression, meaning a place in the computer memory where a value is stored. Since this value can change, its content can vary.

Every time you write a `Function` or `Sub` procedure that receives an argument, you are indeed creating a variable with the name of the argument (such as the `DateOfBirth` argument of the `AgeInYears( )` function procedure).

To take a better control of the values used inside the code procedure, you can use the recommended strategy of making partial calculations, storing them in variables, and then using the variable results in further calculations, effectively making the code clear and easy to follow in VBA Break mode.

Besides the variables automatically created on the procedure code declaration, you can use the VBA `Dim` declaration statement syntax to create new variables, as shown here:

```
[Dim | Private | Public] <VariableName> <as Type>
```

In this code:

> `Dim`: Can be used inside or outside code procedures to create private variables to the procedure or code module.
>
> `Private`: Can be used just outside code procedures to create private code module variables (in the `Declaration` section of any code module).
>
> `Public`: Can be used just outside code procedures to create `Public` code module variables (in the `Declaration` section of any code module).

<VariableName>: Is the name associated to the variable being created. It must contain only alphabetic characters, decimal digits, and underscores (_); it must begin with an alphabetic character or an underscore; and it must contain at least one alphabetic character or decimal digit if it begins with an underscore.

As Type: Specifies the variable data type.

Inside any code procedure you can use the Dim statement only to declare a variable name (Dim comes from "Dimension"). The next procedure code uses the Dim statement to create the procedure-level variable named Years. It then receives the number of years that have passed between the system date (returned by the VBA Date( ) function) and the DateOfBirth argument. The Years variable value is then attributed as the AgeInYears1( ) function procedure value.

```
Function AgeInYears1 (DateOfBirth)
    Dim Years
    Years = (Date - DateOfBirth) / 365
    AgeInYears1 = Years
End Function
```

Implicit vs. Explicit Variable Declaration

By default VBA has the bad habit of allowing you to use a variable without declaring it. This is called an *implicit* declaration. When this happens, you do not need to declare the variable you want to use. Just write its name on any operation inside a procedure code and VBA will automatically create it for you.

Carefully look at the next code for the Function AgeInYears2( ) procedure.

Attention Both Function AgeInYears1( ) and AgeInYear2( ) procedures can be found inside the AgeInYears_Implicit Declaration.xlsm macro-enabled workbook that you can extract from the Chaper01.Zip file.

```
Function AgeInYears2 (DateOfBirth)
    Dim DaysCount
    DaysCount =Date - DateOfBirth
    AgeInYears2 = DayCount / 365
End Function
```

Looking at the AgeInYears2( ) function, you can see that it uses a different strategy to calculate the age in years from any date of birth. It first uses the VBA Dim statement to declare the DaysCount variable on the first procedure instruction and then uses this variable to store the number of days between the system date and the DateOfBirth argument.

```
Dim DaysCount
DaysCount =Date - DateOfBirth
```

In the last procedure row, the DaysCount variable value is then divided by 365 to calculate the age in years, and this result is defined as the Function AgeInYears2( ) procedure return value.

```
AgeInYears2 = DayCount / 365
```

Did you notice something wrong with the last code row of this function procedure? Look again...

Yes, it has an error: the `DaysCount` variable was misspelled as `DayCount` (without an *s* after `Day`). Since VBA is operating in its *implicit* variable declaration mode, it automatically creates the `DayCount` variable at this moment, which has no value, and the `AgeInYears( )` function returns... zero!

Figure 1-16 shows `Sheet1` worksheet and the `basAgeInYears` code module from the `AgeInYears_Implicit Declaration.xlsm` macro-enabled workbook. `Sheet1` uses the functions `AgeInYears1( )` and `AgeInYears2( )` to calculate the age in years in columns `B` (Age in Years 1) and `C` (Age in Years 2) for each date of birth in column `A`. Note that the Age in Years 2 column shows zero for the calculated ages!

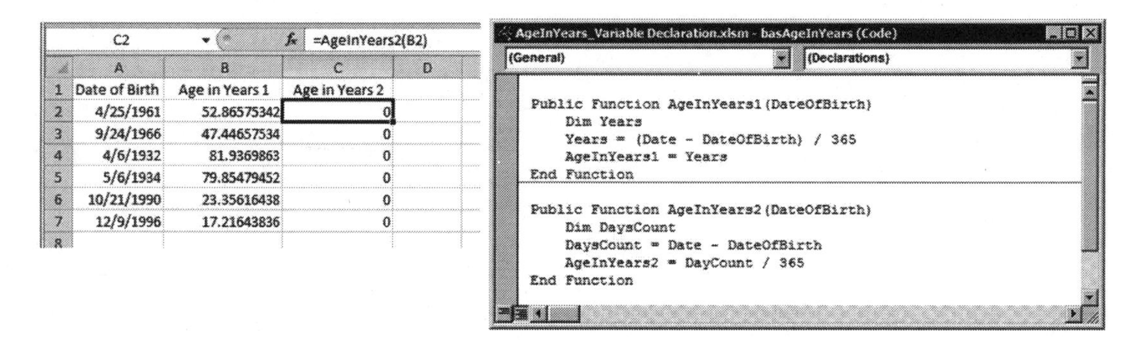

***Figure 1-16.** The Function AgeInYears2( ) procedure of the AgeInYears_Implicit Declaration.xlsm macro-enabled workbook has incorrectly misspelled the DaysCount variable on the last procedure row, leading to a "bug" in the procedure result*

The lesson is quite simple: besides declaring variable names, you must always tell VBA to avoid using implicit declaration mode.

Using Option Explicit

You can force VBA to require the explicit declaration of all variables used inside a code module by typing `Option Explicit` as the first code module instruction. Alternatively, you can ask VBA to always insert this instruction on every new code module by following these steps:

1. In the VBA interface, execute the menu command Tools ➤ Options.
2. Select the Require Variable Declaration option.
3. Click OK to update the changes.

From now on, every time you open or create a new module, it will insert on its very first row the instruction `Option Explicit`, meaning that all variables must be explicitly declared (Figure 1-17).

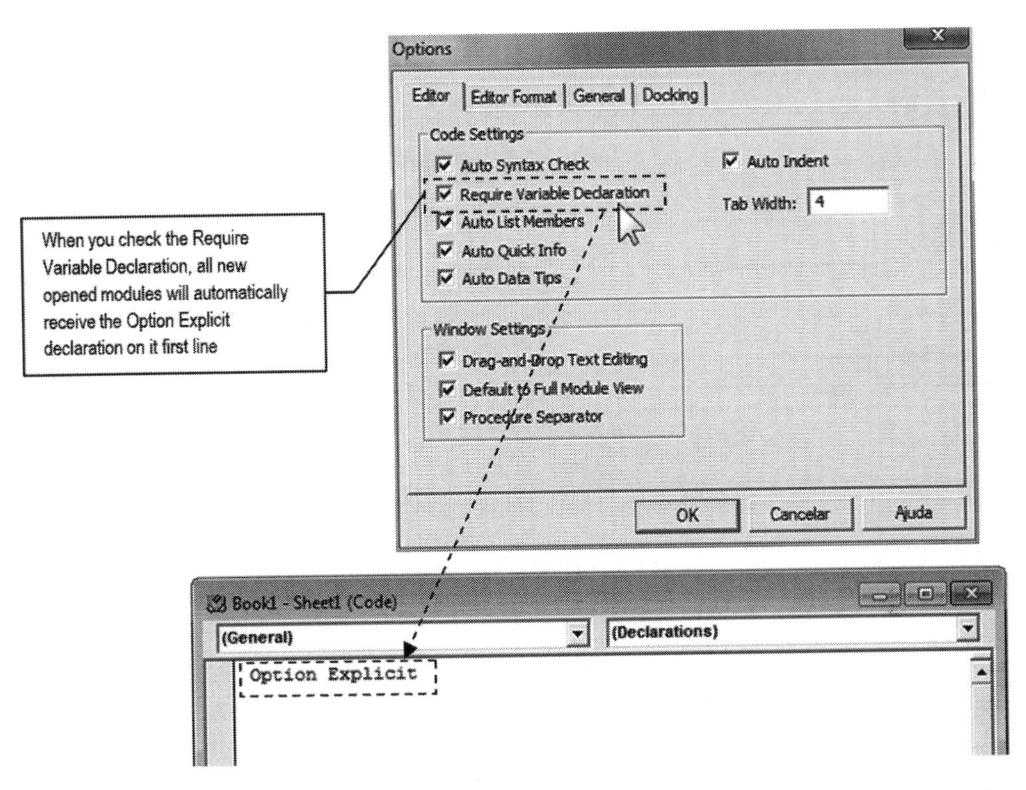

Figure 1-17. *Select the Require Variable Declaration check box in the VBA Option dialog box to make new code modules automatically receive the Option Explicit declaration when it begins. All code modules that you had open before this action must manually receive on its first line of code the Option Explicit declaration*

■ **Attention** All code modules that had already been opened inside VBA *before* you checked the Require Variable Declaration option *will not* be automatically updated with the `Option Explicit` statement. You must manually insert it as the first module instruction, before any variable or procedure declaration.

Try to insert the `Option Explicit` statement on the first row of `basAgeInYears` of the **AgeInYears_Implicit Declaration.xlsm** macro-enabled workbook to force VBA to catch the variable misspelling error of the `AgeInYears2( )` function procedure.

Once you have typed the `Option Explicit` statement, you can make VBA search for any misspelled variable name by using the VBA Debug ➤ Compile VBA Project menu command or you can try to edit any formula cell of column `C` (Age In Years 2) of the `Sheet1` worksheet. VBA will immediately find and detach the misspelled `DayCount` variable, stopping the code execution until you fix it (Figure 1-18)!

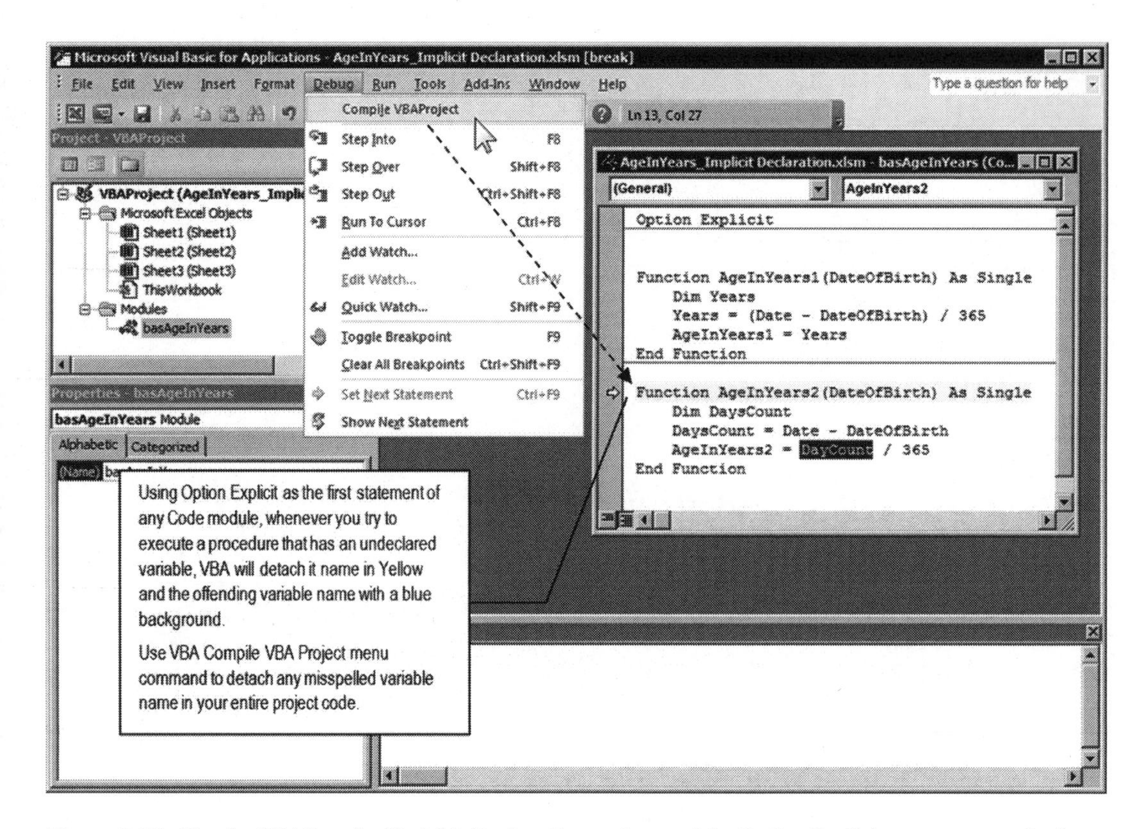

Figure 1-18. *Use the VBA Require Variable Declaration option and the Option Explicit statement on the first line of every code module to make VBA search and find any misspelled variable name inserted on your code*

Variable Types

When you declare a variable name as an argument to any procedure or inside the procedure code using the VBA `Dim` statement, you are reserving a memory place to the value it will hold.

VBA has a default data type for variables called the `Variant` type; this is a special type of variable that can hold any kind of data such as numbers, text, dates, arrays, and objects. Any time you declare a variable and do not specify its type (or declare a `Function` procedure and do not declare the data type it must return), you are telling VBA that it must create a `Variant` type variable (or return a `Variant` value).

The `Variant` type is great! But since it can hold any kind of data, it reserves additional memory, meaning that for long operations, your processor will do more effort to manipulate its data. Variables with the `Variant` data type are the only ones that can represent some special values like `Empty`, `Error`, `Nothing`, and `Null`.

Good programming practices recommend that you must declare the variables with the exact type they will hold (or even better, with the smaller possible data type, so it can use the minimum amount of memory necessary to store it content).

Table 1-1 lists the VBA data types, the range of values they can hold, and the number of bytes each one takes in computer memory.

Table 1-1. *VBA Variable Data Types (in Alphabetical Order)*

Data Type	Range of Values	Memory Usage
`Boolean`	`True` or `False` where 0 = `False`, -1 = `True`.	1 byte
`Byte`	0 through 255 (unsigned).	1 byte
`Currency`	-922,337,203,685,477.5808 to 922,337,203,685,477.5807 (4 decimals precision).	2 bytes
`Date`	0:00:00 (midnight) on January 1, 0001, through 11:59:59 PM on December 31, 9999.	8 bytes
`Double` (double-precision floating-point)	-1.79769313486231570E+308 through -4.94065645841246544E-324 † for negative values. 4.94065645841246544E-324 through 1.79769313486231570E+308 † for positive values.	8 bytes
`Integer`	-32,768 through 32,767 (signed).	1 byte
`Long` (long integer)	- 2,147,483,648 through 2,147,483,647 (signed).	2 bytes
`Object`	Any type can be stored in a variable of type `Object`.	4 bytes on 32-bit CPU 8 bytes on 64-bit CPU
`Single` (single-precision floating-point)	-3.4028235E+38 through -1.401298E-45 † for negative values. 1.401298E-45 through 3.4028235E+38 † for positive values.	4 bytes
`String` (variable-length)	0 to approximately 2 billion Unicode characters.	Varies
`User-Defined` (structure)	Each member of the structure has a range determined by its data type and independent of the ranges of the other members.	Varies
`Variant`	0 through 65,535 (unsigned).	Varies

† In scientific notation, E refers to a power of 10. So, 9.23E+2 signifies 9.23 x 10^2 or 923, and 9.23E-2 signifies 9,23/10^2 or 0.0923.

Table 1-2 classifies all numeric VBA data types by its footprint on computer memory or by its size.

Table 1-2. *VBA Numeric Data Types Ordered by Its Size*

Data Type	Number Type	Range of Values
`Boolean`	Integer	-1 = `True`, 0 = `False`
`Byte`	Integer	0 to 255
`Integer`	Integer	-32,768 to 32,767
`Long`	integer	- 2,147,483,648 through 2,147,483,647
`Currency`	Real	Ideal for monetary calculations using up to 4 decimals precision; -922,337,203,685,477.5808 to 922,337,203,685,477.5807
`Date`	REAL	Date/time calculations, where the `Integer` part holds the number of days past from 1/1/1900 and the `Decimal` part, with up to 15 digits precision, holds the hour regarding the number of milliseconds past from midnight
`Single`	Real	-3.4028235E+38 through -1.401298E-45 for negative values; 1.401298E-45 through 3.4028235E+38 for positive values
`Double`	Real	-1.79769313486231570E+308 through -4.94065645841246544E-324 for negative values; 4.94065645841246544E-324 through 1.79769313486231570E+308 for positive values

As you can see from Table 1-2, the smallest possible numeric integer data type is `Boolean` (which can hold just 0 or -1), followed by `Byte` (0 to 256), `Integer` (maximum integer value = 32,767), and `Long`, which is the greatest integer data type (maximum value = 2,147,483,647).

On the real side of the numerical data types, the one with less precision is `Currency`, using just four digits for its decimal part (using all other digits for an enormous integer part), followed by `Date` (yes, the `Date` type is a real number that uses a precision with up to 15 decimal digits), `Single`, and `Double`. This last one is the greatest real number that you can manipulate with VBA.

Let's look now to the new versions of the `Function AgeInYears1( )` and `Function AgeInYears2( )` procedures from the `basAgeInYears` code module of the `AgeInYears_Explicit Declaration.xlsm` macro-enabled workbook (also inside the `Chapter01.zip` file), which now has all variable names declared with its adequate data types.

```
Function AgeInYears1 (DateOfBirth as Date) as Single
    Dim Years as Single
    Years = (Date - DateOfBirth) / 365
    AgeInYears1 = Years
End Function

Function AgeInYears2 (DateOfBirth as Date) as Single
    Dim DaysCount as Long
    DaysCount =Date - DateOfBirth
    AgeInYears2 = DayCount / 365
End Function
```

Note the following:

- Both functions declare just one argument, `DateOfBirth as Date`. This guarantees that just valid dates can be received by these functions. Any Excel formula cell that uses an invalid date will receive error code `#VALUE!`. Any function call from the VBA Immediate window will receive a VBA message error ("Compile Error: Expected expression," as shown in Figure 1-19).
- Both functions declare a return value `as Single`, meaning that they return a single precision floating-point number (the integer part is the number of years; the decimal part is proportional to 12 months).
- `AgeInYears1( )` declares `Years as Single`, so the `Years` variable can store floating-point values (real numbers with an integer and a decimal part).
- `AgeInYears2( )` declares `DaysCount as Long`, so `DaysCount` can store a number that is long enough to represent any person's age in days.

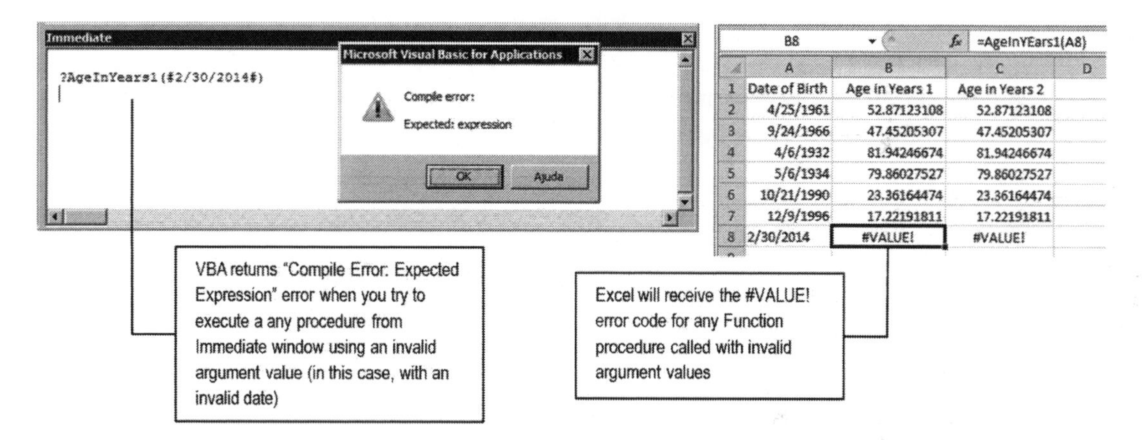

Figure 1-19. *When any procedure argument has a specific data type, VBA will raise different errors in Excel. If you try to use an invalid date in the VBA Immediate window, you will receive a Compile error, but when you use the invalid date from an Excel formula, Excel will receive a #VALUE! error code*

I will now discuss each of these two `Function` procedures in consideration of each data type, so you can understand the reason to use them on each declared variable.

Begin paying attention that both functions return a `Single` data type as specified on its declaration instruction; this is the smallest possible real number that uses up to six decimal floating-point precision.

`Function AgeInYears1( )` uses as a strategy the calculation of the years between two consecutive dates. Since this value is a real number with an integer part representing the years portion of the age and the decimal part representing the percentage of 12 months of age, it must be declared with any data type that is capable of representing a real number.

Looking at Table 1-1, you will notice that the `Single` data type takes 4 bytes of computer memory and is the smaller `Real` number you can use with a good decimal floating-point precision. That's why `AgeInYears1 ( )` declares the `Years` variable with the `Single` data type: it is the smallest possible real number that is also in accordance with the `AgeInYears1( )` return data type.

■ **Attention** You can use the `Currency` data type to declare the `Years` variable using just 2 bytes of computer memory. But as a rule of thumb, programmers use the `Currency` data type just for monetary values.

Whenever you need a `Real` value, try to use the `Single` data type. Reserve the `Double` data type to such calculations that need greater precision or can result in very big numbers.

Function `AgeInYears2( )`, besides being declared as using the `Single` data type as its return value, uses another strategy: it first calculates the number of days between two consecutive days. Since this number is an integer value, the `DaysCount` variable was declared with the `Long` data type, which is one that can hold a large number of ages using a days counting.

On the last procedure instruction, the `DaysCount as Long` integer value is divided by 365, leading to a `Real` value that is converted to `Single` precision—the data type used to declare the procedure!

■ **Attention** If you are wondering why I do not use the `Integer` data type, it is because its maximum value can't be greater than 32,767 (or 32,767 days of age). Dividing this number by 365 days per year, you can achieve just about 89.8 years of age. Any people older than these ages will generate a days count greater than this value, leading to a VBA error called an *overflow* error.

Overflow is an error that happens when you try to attribute to a numeric variable a value that is greater than it can contain.

Array Declares

You can declare arrays in VBA using the `Dim` statement and adding to the variable name a pair of parentheses that may include the array dimensions. The next statement declares the `Ages( ) as Variant` array that can contain up to ten elements:

```
Dim Ages(0 to 9) as Variant
```

By default, VBA arrays are zero-based, meaning that the first array item has index = 0. `Ages( )` can use indexes `Ages(0)` to `Ages(9)`, storing up to ten different `Variant` values.

The next statement declares a bidimensional array that can receives just the `String` data type:

```
Dim Peoples(1,9) as String
```

VBA does not allow a fixed-size array to change its dimensions, but you can declare a dynamic array using just parentheses to indicate that it represents an array.

```
Dim varAges( ) as String
```

Use the VBA `Redim( )` statement to change the first array dimension at run time (you cannot change the array data type):

```
Redim varAges(200)
```

■ **Attention** Since a `Variant` variable can receive anything, it can also be considered as an array (or be associated with another array variable) and be redimensioned using the `Redim` statement.

Alternatively, you can use a VBA `Array( )` function to create an array based on some values.

```
Dim MyValues as Variant
MyValues = Array(10, 20, 30)
```

■ **Attention** I will use arrays in many examples of this book. For now this basic knowledge is enough.

Variable Scope and Lifetime

Besides the data type, each variable has a *scope*—or a lifetime in your code procedure.

Variables declared inside any code procedure with a `Dim` statement have just the procedure scope, being created with it declaration instruction and being destroyed when the procedure ends. Both the `Years` and `DaysCount` variables of the functions `AgeInYears1( )` and `AgeInYears2( )` are created inside its code procedures and destroyed as the procedure ends.

■ **Attention** To extend the lifetime of any variable to the time your application is active, use the `Static` statement to declare it inside any code procedure (or declare a `Static` procedure).

Using the Static Statement to Hold Any Variable Value

The next `Function AgeInYears3( )` procedure declares the `Static Years as Single` variable so it can hold its last value between procedure calls.

■ **Attention** The function `AgeInYears3( )` procedure can also be found in the `basAgeInYears` module of the `AgeInYears_Explicit Declaration.xlsm` macro-enabled workbook.

Note that the `Function AgeInYears3( )` procedure has a totally different approach. By using the `Optional` keyword to declare the `DateOfBirth` argument as the `Variant` data type (instead of the `Date` data type), it can now be avoided—the user of this function is not obliged to pass any value to the function.

```
Function AgeInYears3 ( Optional DateOfBirth as Variant) as Single
    Static Years as Single
    If Not IsEmpty(DateOfBirth) then
        Years = (Date - DateOfBirth) / 365
    End If
   AgeInYears3 = Years
End Function
```

Note that after the declaration of the `Static Years` variable, the procedure code uses the `IF` statement to verify with the VBA `IsEmpty( )` function if the `DateOfBirth` argument has any value (or is empty).

```
Static Years as Single
If Not IsEmpty (DateOfBirth) then
```

Attention Just the `Variant` data type can receive the `Empty` value (in other words, can return `True` or `False` to the `IsEmpty( )` VBA function).

The `IsEmpty (DateOfBirth)` function will return `True` whenever the optional `DateOfBirth` argument is missing and `False` otherwise. The `IF` statement then makes a logical test using `Not IsEmpty (DateOfBirth)` to verify that the argument is missing. `DateOfBirth` is *not missing*; it executes the code right after the `IF` statement, and the `Years` variable is recalculated using the same method discussed for the `Function AgeInYears1( )` procedure.

```
If Not IsEmpty ( DateOfBirth ) then
    Years = ( Date - DateOfBirth ) / 365
End If
```

This row will be executed just if IsEmpty (DateOfBirth) = False, and Not IsEmpty ((DateOfBirth) =True

On the last procedure row, the `Years` variable is returned as the `Function` result, being calculated or not on the last procedure call!

```
    AgeInYears3 = Years
End Function
```

The `Function AgeInYears3( )` procedure usage can be demonstrated using the `Sheet2` worksheet of the `AgeInYears_Explicit Declaration.xlsm` macro-enabled workbook (Figure 1-20). Note that column `B`, Age in Years 3, calculates the age in years using this formula in cell `B2`.

```
=AgeInYears3(A2)
```

You must edit the cell `B2` formula and press Enter so the `Static Years` variable of `AgeInYears3( )` has a value defined. Then drag cell `B2` down to cell `B3`. Since `A3` is empty, the `AgeInYears3( )` function procedure uses the `Static Years` variable value to return the last calculated value for cell `B3`.

Attention If you drag cell B3 selector down, every other row will receive the last calculated age in years because of the static nature of the `Years` variable.

Microsoft Excel does not execute your `Function` procedures when it opens the workbook where it resides unless you use the `Application.Volatile` method to force its recalculation. You must edit any formula cell that uses your `Function` procedure so the `Static` value of any variable is determined again.

Figure 1-20. *This is Sheet2 of the AgeInYears_Explicit Declaration.xlsm macro-enabled workbook. It uses the Function AgeInYears3( ) procedure that has declared as Optional the DateOfBirth argument, with the Variant data type, and as Static the Years variable, meaning that it does not lose its value between function calls*

■ **Attention** You can use the `Static` keyword on any procedure declaration to indicate to VBA that every procedure variable has a static (or persistent) value between function calls. The next example of the `AgeInYears4( )` function procedure follows this strategy, indicating that both the `DaysCount` and `Years` variables holds its values between function calls:

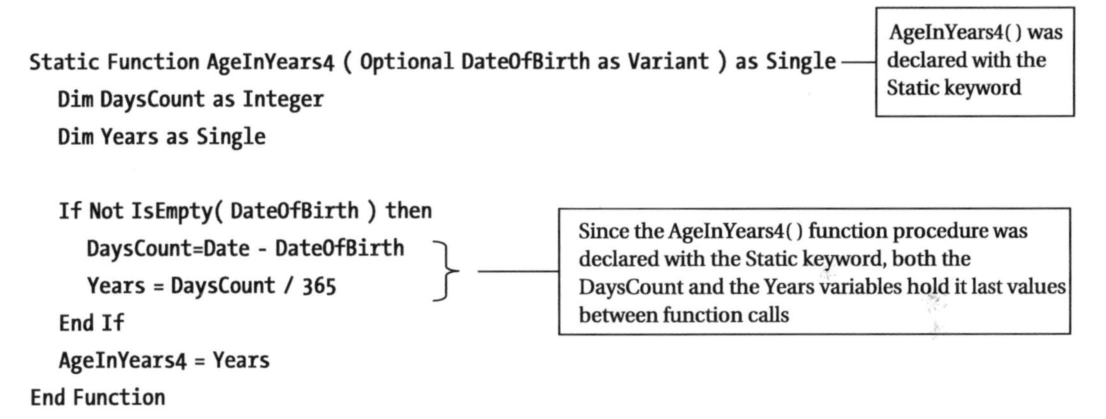
AgeInYears4( ) was declared with the Static keyword
Since the AgeInYears4( ) function procedure was declared with the Static keyword, both the DaysCount and the Years variables hold it last values between function calls

```
Static Function AgeInYears4 ( Optional DateOfBirth as Variant ) as Single
   Dim DaysCount as Integer
   Dim Years as Single

   If Not IsEmpty( DateOfBirth ) then
      DaysCount=Date - DateOfBirth
      Years = DaysCount / 365
   End If
   AgeInYears4 = Years
End Function
```

You can test the `Function AgeInYears4( )` procedure on the `Sheet3` worksheet of the `AgeInYears_Explicit Declaration.xlsm` macro-enabled workbook.

Using Code Module Variables

You already know that a `Static` variable holds its values between procedure calls; in other words, its values persist while your workbook is open.

But there is a problem with `Static` variables: you can use just one inside the procedure where it was declared. What about using the variable value with more than one procedure code? This is where the code module variables enter the action!

Code module variables are the ones declared at the top of any code module, outside any procedure code. They can be of these two types:

- *Private to the code module*: All variables declared with the `Dim` or `Private` statement on the `Declarations` section of any code module. The value can be used by all procedure codes of the same module.
- *Public to all code modules*: If it is declared with the `Public` statement and its value is available for every code module of your VBA project, you can access it from anywhere!

As a programmer you will use code mode variables when you want to make intermediate calculations in one procedure and store them outside the procedure code so they can be used with other procedures of your application.

Using Private Code Module Variables

Suppose for a moment that the `DaysCount` and `Years` values used to calculate the age in years were derived from complex calculations that can made only once for a given data of birth (this is not the case for such simple values). In this case, you must declare both the `DaysCount` and `Years` variables as private module variables so their calculated value can be easily accessed from other procedures of the same module.

The next example of the `AgeInYears5( )` function procedure from the module `basAgeInYears` of the `AgeInYears_Explicit Declaration_PrivateModuleVariables.xlsm` Excel macro-enabled workbook (also inside `Chapter01.zip`) shows this concept for you.

```
Option Explicit

Dim DaysCount As Integer
Dim Years As Single

Public Function AgeInYears5(DateOfBirth As Variant) As Single
    DaysCount = Date - DateOfBirth
    Years = DaysCount / 365
    AgeInYears5 = Years
End Function
```

Do you see the difference? Now since both the `DaysCount` and `Years` variables have been declared in the `Declaration` section of the `basAgeInYears` module, their values will remain on your computer memory while the workbook is open and can be accessed from any other code procedure inserted in `basAgeInYears`.

This is the case for the `AgeInWeeks( )` procedure of the `basAgeInYears` module, which uses the `DaysCount` variable value to return the number of weeks lived so far for anyone whose date of birth was already processed by the `AgeInYears5( )` procedure.

```
Public Function AgeInWeeks(Optional DateOfBirth As Variant) As Single
    If Not IsMissing(DateOfBirth) Then
        DaysCount = Date - DateOfBirth
    End If
    AgeInWeeks = DaysCount / 7
End Function
```

■ **Attention** Both the `DaysCount` and `Years` module variables can be declared using the `Private VBA` statement (instead of `Dim`) as an indication that they can be accessed from procedures created inside the `basAgeInYears` module.

You may wondering why VBA has this ambiguous way to declare a private module variable. The answer is because up to Visual Basic 4, VBA uses just `Dim` to declare local variables and the retired statement `Global` to declare public variables. When Visual Basic 4 arrived, Microsoft discarded the `Global` statement and used just `Private` and `Public` as ways to declare module variables. `Dim` still works, though, because it has been traditionally used by all basic languages.

Note that the `Function AgeInWeeks( )` procedure declares an `Optional DateOfBirth as Variant` argument so you can calculate the age in weeks without needing to first calculate the age in years.

■ **Attention** I want to call your attention to the fact that this is just a didactic example of a bad use of code module variables. If you try to use the `Function AgeInWeeks( )` procedure on any worksheet to calculate the age in weeks for more than one date of birth without passing it the `DateOfBirth` argument, it eventually may fail to calculate, returning the last calculated `AgeInYears( )` or `AgeInWeek( )` values. `Sheet1` of the `AgeInYears_Explicit Declaration_PrivateModuleVariables.xlsm` Excel macro-enabled workbook shows how this can happen (Figure 1-21).

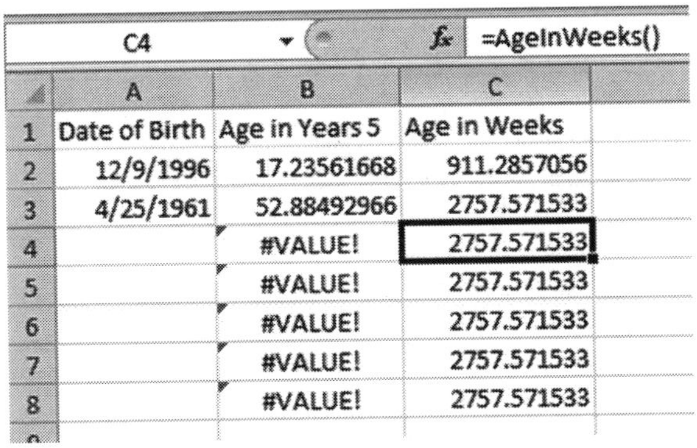

C4 =AgeInWeeks()

	A	B	C
1	Date of Birth	Age in Years 5	Age in Weeks
2	12/9/1996	17.23561668	911.2857056
3	4/25/1961	52.88492966	2757.571533
4		#VALUE!	2757.571533
5		#VALUE!	2757.571533
6		#VALUE!	2757.571533
7		#VALUE!	2757.571533
8		#VALUE!	2757.571533

Figure 1-21. *This is Sheet1 from AgeInYears_Explicit Declaration_PrivateModuleVariables.xlsm Excel macro-enabled workbook using the Function AgeInWeeks( ) procedure from basAgeInYears. Since both DaysCount and Years variables were declared as code module variables, their values remain while the workbook is open. Note that column B uses the Function AgeInYears5( ) procedure to calculate the age in years for the associated date of birth in column A, returning constant error #VALUE! on cells B4:B8 because these rows do not receive the DateOfBirth argument. But the Function AgeInWeeks( ) procedure continues to return the last calculated because of the last value stored on the DaysCount code module variable*

Create a Flow Chart for the Algorithm of Complex Procedures

Whenever you need to create more complex procedures, a good idea is to create a flow chart to represent the instructions that your code must follow before trying to implement it on VBA.

Let's suppose you want to express the age as a literal expression in years and months based on the `DaysCount` and `Years` code module variables from `basAgeInYears` of the `AgeInYears_Explicit Declaration_PrivateModuleVariables.xlsm` Excel macro-enabled workbook. The idea is that based on any date of birth, you express the age for a person with 52 years and 10 months this way: `52y, 10m`.

How can you do that?

Here's an example using a `Years = 52.88492966` value:

1. Take the integer part of the `Years` variable (`52`) and concatenate it the y suffix for the literal age expression.

```
52y;
```

2. Take the decimal part of the `Years` variable (`0.88492966`) and multiply it by 12 to find the months part of the age.

```
52.88492966 - 52 = 0.88492966 x 12 = 10.61916
```

3. Verify whether the integer part of the month is greater than zero. If it is, concatenate a comma to the literal age in years (`52y`), the integer part of the months, and the `m` suffix.

```
52y, 10m
```

4. Return this result as the literal expression for the age (return value of the `Function` procedure).

Figure 1-22 shows the flow chart that represents all the operations that must be followed by any procedure code to return the literal expression of the age in years and months. You read this flow chart by following the arrows, the same way the procedure code will do.

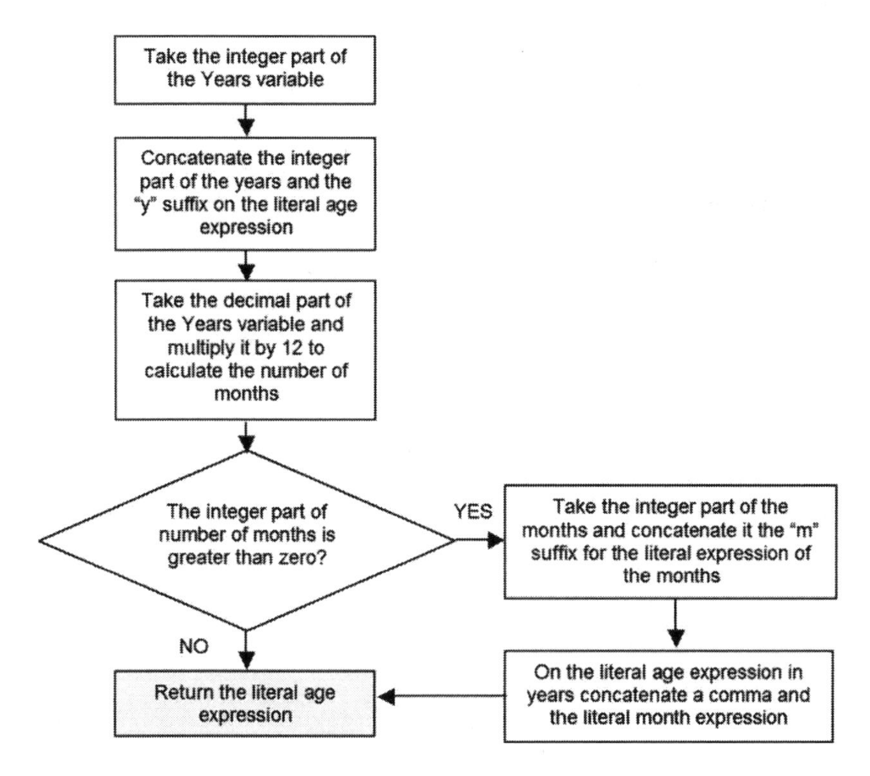

Figure 1-22. *This is the flow chart for all steps needed to calculate the literal expression of the age, adding the suffix y to the years part and the suffix m to the month part (if any)*

Note on the flow chart that the diamond represents the logical test needed to verify whether the months part is greater than zero. If it is, the months literal expression is calculated and concatenated to the years part of the age. If it is not, the literal age part is returned.

The next Function AgeInYearMonth( ) procedure does the same operations to return the literal expression of the age based on the Years Module variable already calculated using the AgeInYears5( ) function procedure:

```
Option Explicit

Dim DaysCount As Integer
Dim Years as Single
```

Years is a Module-level variable.

```
Public Function AgeInYearMonth(Optional DateOfBirth As Variant) As String
   Dim YearsInteger As Integer
   Dim Age As String
   Dim Months As Single

   If Not IsMissing(DateOfBirth) Then
      DaysCount = Date - DateOfBirth
      Years = DaysCount / 365
   End If

   YearsInteger = Int( Years )
   Age = YearsInteger & "y"

   Months = Years - YearsInteger
   Months = CInt( Months * 12 )

   If Months > 0 Then
      Age = Age & ", " & Months & "m"
   End If

   AgeInYearMonth = Age
End Function
```

If you follow the AgeInYearMonth( ) function procedure, you will see all the steps pictured on the flow chart of Figure 1-22. But it has some programming techniques that deserve consideration, mentioned here:

- The AgeInYearMonth( ) function procedure returns a String value and declares an optional argument DateOfBirth as Variant, which will be tested inside the procedure using the VBA Not IsMissing( ) expression to verify whether it was received.
- If IsMissing(DateOfBirth) is false, Not IsMissing(DateOfBirth) is true (DateOfBirth *is not* missing), and it recalculates the DaysCount and Years code module variables.

- If this is true, it will use the last calculated value already stored on these code module variables:

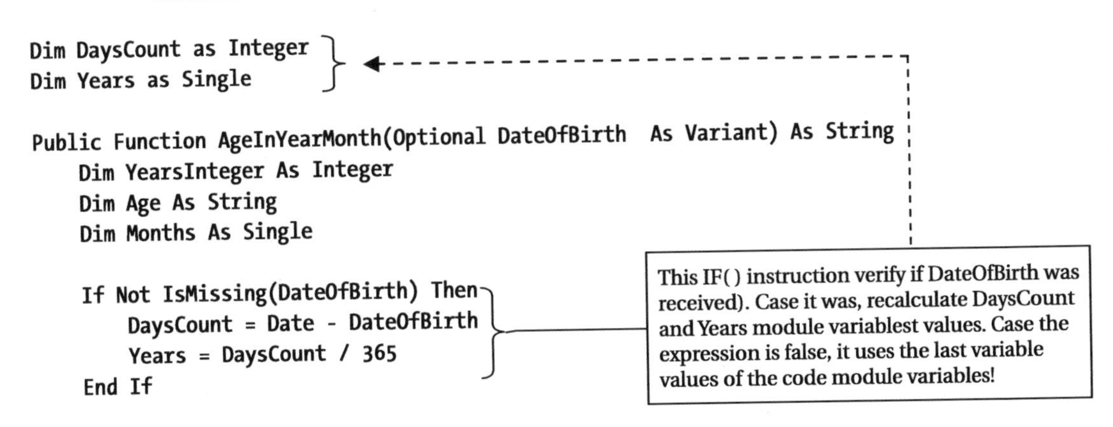
Dim DaysCount as Integer
Dim Years as Single
Public Function AgeInYearMonth(Optional DateOfBirth As Variant) As String
Dim YearsInteger As Integer
Dim Age As String
Dim Months As Single
If Not IsMissing(DateOfBirth) Then
DaysCount = Date - DateOfBirth
Years = DaysCount / 365
End If
This IF( ) instruction verify if DateOfBirth was received). Case it was, recalculate DaysCount and Years module variablest values. Case the expression is false, it uses the last variable values of the code module variables!

- Also note that `AgeInYearMonth( )` declares the variables as they appear in the code. The first variable used is `YearsInteger as Integer`, followed by `Age as String` and `Months as Single`.
- The procedure then uses the VBA `Int( )` function to take the integer part of the `Years` variable without rounding it and attributes it to the `YearsInteger` variable.

```
YearsInteger = Int(Years)
```

■ **Attention** The VBA `Int( )` function never rounds results when it takes off the integer part of any real number. To round the result, just attribute any real number to an `Integer` variable or use the VBA `CInt( )` function.

`CInt( )` means "convert to integer."

- Now that `YearsInteger` holds the integer part (not rounded) of the `Years` variable, the procedure concatenate the y suffix to the `Age as String` variable value to create the literal age in years expression. Note that VBA uses the & character as the concatenation operator.

```
Age = YearsInteger & "y"
```

- The Age string variable now holds the literal age in the years expression, and the procedure will calculate the number of months of the age. It does this by subtracting from `Years` its integer part (the `YearsInteger` variable value) and attributing the result to the `Months` variable, declared as `Single`, which is a type that can hold decimal values.

```
Months = Years - YearsInteger
```

Since the Months variable was declared as Single, it will now hold the decimal part of Years variable.

- Now that the Months variable holds the decimal value of the age, it calculates the age in months by multiplying Months * 12 and uses the VBA CInt( ) function to round the result to the next greatest integer.

```
Months = CInt( Months * 12 )
```

- The Months variable now has the months expression of the age, and the code uses another IF( ) structure to verify that it is greater than zero. If it is, it concatenates a comma, the Months variable, and the m suffix to the Age variable (note that it concatenates to the Age variable the value that Age has stored from previous operations).

```
If Months > 0 Then
    Age = Age & ", " & Months & "m"
End If
```

- The procedure then uses the Age string variable as the function return result.

```
    AgeInYearMonth = Age
End Function
```

Comment Your Code!

All the explanations that I'm giving for the inner workings of the Function AgeInYearMonth( ) procedure can be done inside the procedure code by using VBA comments.

To insert a comment on any procedure row, just begin the row (or the comment) with a single quote character. Everything you type before the single quote on the same row will be considered as a comment and will be ignored by VBA, which automatically displays comments in green.

Inside basAgeInYears from the AgeInYears_Explicit Declaration_PrivateModuleVariables.xlsm Excel macro-enabled workbook you will find the Function AgeInYearMonth( ) procedure, which has many comments to explain how it works (Figure 1-23).

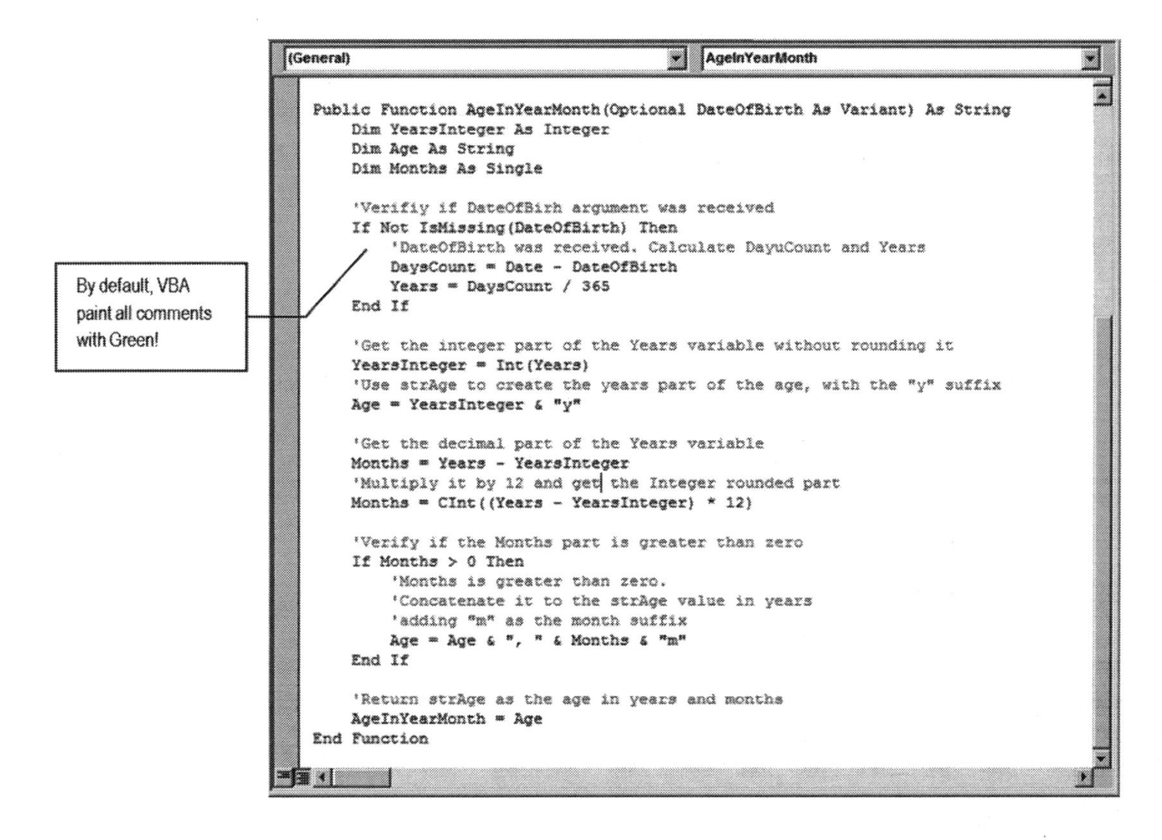

(General)
AgeInYearMonth
Public Function AgeInYearMonth(Optional DateOfBirth As Variant) As String
Dim YearsInteger As Integer
Dim Age As String
Dim Months As Single
'Verifiy if DateOfBirh argument was received
If Not IsMissing(DateOfBirth) Then
'DateOfBirth was received. Calculate DayuCount and Years
DaysCount = Date - DateOfBirth
Years = DaysCount / 365
End If
'Get the integer part of the Years variable without rounding it
YearsInteger = Int(Years)
'Use strAge to create the years part of the age, with the "y" suffix
Age = YearsInteger & "y"
'Get the decimal part of the Years variable
Months = Years - YearsInteger
'Multiply it by 12 and get the Integer rounded part
Months = CInt((Years - YearsInteger) * 12)
'Verify if the Months part is greater than zero
If Months > 0 Then
'Months is greater than zero.
'Concatenate it to the strAge value in years
'adding "m" as the month suffix
Age = Age & ", " & Months & "m"
End If
'Return strAge as the age in years and months
AgeInYearMonth = Age
End Function
By default, VBA paint all comments with Green!

Figure 1-23. *Use the single quote character to insert comments on your code procedures so you can easily understand and maintain it in the future*

When you comment your code, you can document the steps you use to achieve a solution, which will make it easier for you or any others in the future to follow the logic behind the code. Yes, the procedure will become bigger, but the steps will become clearer!

■ **Attention** By commenting your procedure code, you end up validating it. Think about comments as the procedure quality control system.

You can see the `Function AgeInYearMonth( )` procedure in action by selecting the `Sheet2` tab worksheet from the `AgeInYears_Explicit Declaration_PrivateModuleVariables.xlsm` Excel macro-enabled workbook. Note that since its `DateOfBirth` argument is optional, whenever it is missing on the `Function` procedure call, it will use the last calculated value of the `Years` variable, which may return wrong results to the worksheet (Figure 1-24).

D4 =AgeInYearMonth()

	A	B	C	D	E
1	Date of Birth	Age in Years 5	Age in Weeks	Age in Years and Months	
2	12/9/1996	17.23835564	2757.714355	17y, 3m	
3	4/25/1961	52.88767242	2757.714355	52y, 11m	
4		#VALUE!	2757.714355	52y, 11m	
5		#VALUE!	2757.714355	52y, 11m	
6		#VALUE!	2757.714355	52y, 11m	
7		#VALUE!	2757.714355	52y, 11m	
8		#VALUE!	2757.714355	52y, 11m	
9					
10					

When there is no date of birth, the AgeInYearMonth( ) Function will return a wrong result based on the last values of the Years Private Module variables

Figure 1-24. *This is Sheet2 worksheet from the AgeInYears_Explicit Declaration_PrivateModuleVariables.xlsm Excel macro-enabled workbook that uses the Function AgeInYearMonth( ) procedure to return a literal expression of the age in years and months (if any). Note that whenever there is no DateOfBirth argument, the procedure continues to return the last calculated value (stored in the Years code module variable)*

Using Public Code Module Variables

Use `Public` module variables when you want to store values that can be accessed by any procedure of your VBA project. The `basAgeInYear` module from the `AgeInYears_Explicit Declaration_PublicModuleVariables.xlsm` Excel macro-enabled workbook uses this strategy to declare the `DaysCount`, `Years`, `Weeks`, and `Age` as `Public` variables (Figure 1-25).

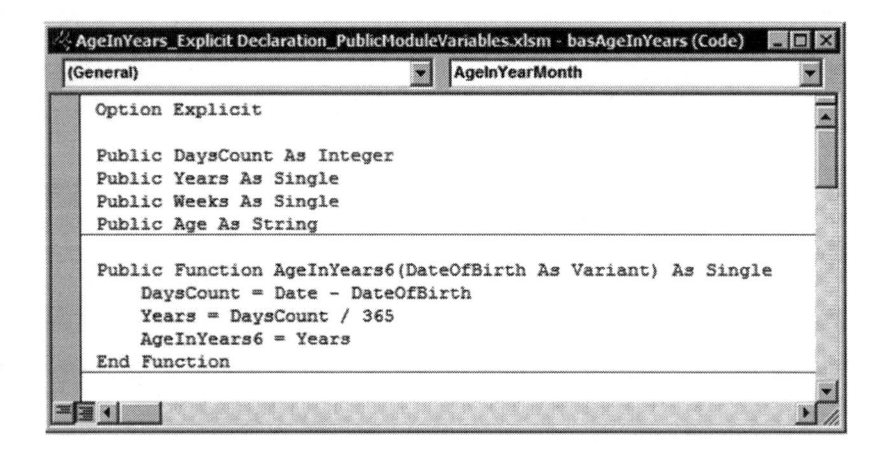

AgeInYears_Explicit Declaration_PublicModuleVariables.xlsm - basAgeInYears (Code)
(General)
AgeInYearMonth
Option Explicit
Public DaysCount As Integer
Public Years As Single
Public Weeks As Single
Public Age As String
Public Function AgeInYears6(DateOfBirth As Variant) As Single
DaysCount = Date - DateOfBirth
Years = DaysCount / 365
AgeInYears6 = Years
End Function

Figure 1-25. *When you declare Public variables in any module, their value can be accessed and changed from anywhere in your VBA project*

When you declare a `Public` variable on a standard module, you can verify or change its value from the VBA Immediate window. Take a look at the `Sheet1` worksheet from the `AgeInYears_Explicit Declaration_PublicModuleVariables.xlsm` Excel macro-enabled workbook, which uses the `AgeInYears6( )`, `AgeInWeeks2( )`, and `Function AgeInYearMonth2( )` procedures from `basAgeInYears`. It has just two rows of formulas, with two different dates of birth. Since it uses the `Public` module variables `Age`, `DaysCount`, `Weeks`, and `Years` to store its values, they will hold the last calculation regarding the date of birth of cell `A3` (4/25/1961, as shown in Figure 1-26).

D3 =AgeInYearMonth2(A3)

	A	B	C	D
1	Date of Birth	AgeInYears6	AgeInWeeks2	AgeInYearMonth2
2	12/9/1996	17.25479507	899.7142944	17y, 3m
3	4/25/1961	52.90410995	2758.571533	52y, 11m
4				

***Figure 1-26.** Sheet1 from the AgeInYears_Explicit Declaration_PublicModuleVariables.xlsm Excel macro-enabled workbook uses AgeInYears6( ), AgeInWeeks2( ), and Function AgeInYearMonth2( ) procedures from basAgeInYears, which base their calculations on Public module variables, whose values will hold the last calculated Excel formula*

You can verify the values of any `Public` variable declared on a standard module by typing ? followed by the variable name in the VBA Immediate window. Use the comma character to separate different variable names and print them on the same row of the VBA Immediate window (Figure 1-27).

■ **Attention** Note that Excel *does not* execute any `Function` procedure when it opens your workbook, unless it has the `Application.Volatile` instruction. In such cases you must edit any formula cell that uses it so all public variables will be updated by VBA.

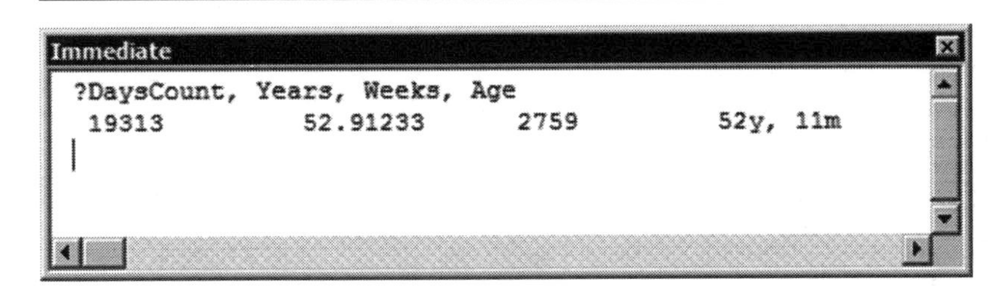
Immediate
?DaysCount, Years, Weeks, Age
19313 52.91233 2759 52y, 11m

***Figure 1-27.** Use the VBA Immediate window to inspect the value of any Public variable while your workbook is open. You can inspect more than one variable on a single row, separating them with comma characters*

Since you can inspect any public variable value, you can also change its value using the VBA Immediate window. Just type the variable name followed by a = character, then type its new value, and press Enter to update it (Figure 1-28).

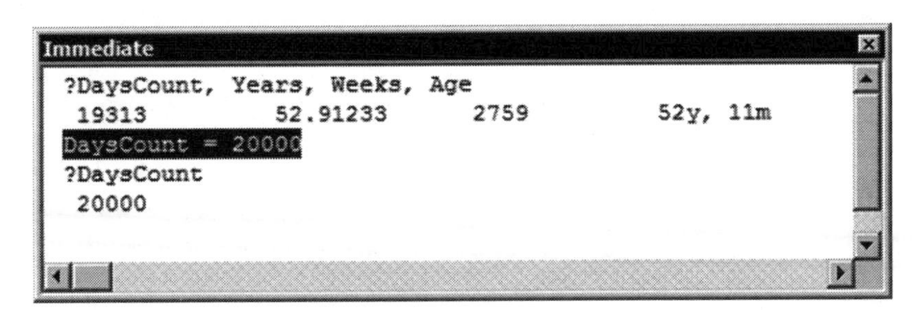
Immediate
?DaysCount, Years, Weeks, Age
19313 52.91233 2759 52y, 11m
DaysCount = 20000
?DaysCount
20000

***Figure 1-28.** You can also use the VBA Immediate window to change the value of any public variable declared on a standard module (or to change the value of any variable of a running procedure using a VBA breakpoint). Use the = character to attribute it with the desired value and check the value to confirm the change*

■ **Attention** As you will see later in this chapter, you can protect the value of the variables whose value must be accessible by any code module of your VBA project by implementing a `Property` procedure associated with a `Private` code module variable.

Public Procedures and Variables Constitute the Module Interface

All `Public` variables (and `Public` procedures) can be accessed in the VBA Immediate window by using this syntax: code module name, a dot, and the public variable name:

```
?<CodeModuleName>.<VariableName>
```

You can verify that this is true by typing `?basAgeInYears.` (note the dot after the module name) in the VBA Immediate window; you will see all public declarations appear in alphabetical order as interface members of the `basAgeInYears` code module (Figure 1-29).

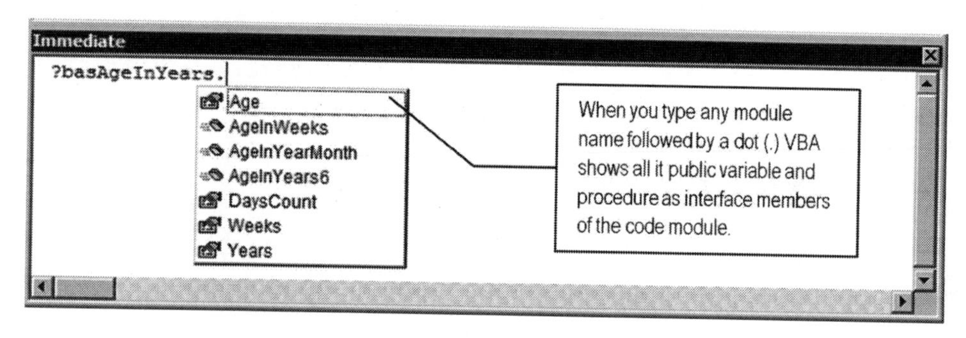

Figure 1-29. *When you declare public variables or procedures, they will appear as interface members of the code module in the VBA Immediate window*

By object rule convention, a *property* is just a value, while a *method* is an action performed by the object. Looking at the `basAgeInYears` module interface on Figure 1-29, you can note that `Age`, `DaysCount`, `Weeks`, and `Years` are considered properties (using the hand property icon), while `AgeInWeeks`, `AgeInYearMonth`, and `AgeInYears6` are considered methods (using the running green method icon).

■ **Attention** Any `Public` variable is considered a property, even though it is not associated to a `Property` procedure.

Using Enumerators

Many VBA functions require that you use appropriate numeric values as arguments, and they are already coded in the VBA language. For example, there is a constant value associated with each day of the week on the enumerator called `vbDayOfWeek`. If you type `vbDayofWeek.` in the VBA Immediate window (note the dot after the `vbDayOfWeek` enumerator name), VBA will promptly show all the members of this enumerator, representing each weekday in its quick list information (Figure 1-30).

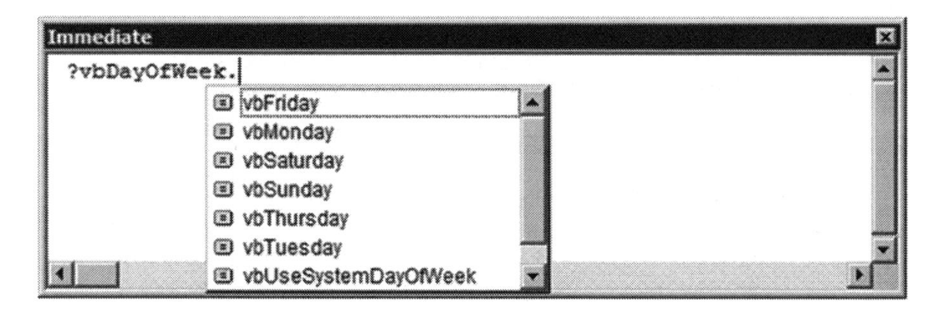

Figure 1-30. *Type vbDayOfWeek. (note the dot after the enumerator) in the VBA Immediate window to see all its members, representing each weekday*

To create such a structure of numeric constant values associated to declared names, you must use a VBA `Enum` instruction, which has this syntax:

```
||Public| |Private| Enum <EnumerationName> [ As DataType ]
    Member_1
...
    Member_n
End Enum
```

In this code:

`EnumerationName`: This is the name you want to give to the enumerator, following the same rules for variable names.

`DataType`: This is optional; it is the data type returned by any enumerator member. By default it receives the `Integer` data type but can be specified as `Byte`, `Integer`, or `Long`.

`Member1 to n`: These are the names of each member of the enumerator.

Any enumerator created by you must be declared on the `Declaration` section of the code module, as public or private.

When you declare an enumerator, you must type the `End Enum` instruction to indicate to VBA where it finishes or VBA will raise an error code. You can assign any numerical value to each enumeration member, or you can assign just the value of the first member and VBA will assign increments of 1 to each successive member. The next declaration creates the `Gender` enumerator, assigning 1 to the Male member, meaning that `Female` will automatically receive value = 2:

```
Enum Gender
    Male = 1
    Female
End Enum
```

In this book, I will call your attention to this important and simple programming technique whenever an enumerator can be used to make the code easier to read, follow, or use.

Passing Arguments by Reference or by Value

Procedure arguments are implicit declared *by reference* but can be explicit declared *by value*, using this syntax:

```
Sub MyProcedure(arg1 as variant, ByVal arg2 as variant)
...
End Sub
```

In this example code, the `arg1` argument is declared *by reference*, while the `arg2` argument is declared *by value*. Declaring procedure arguments by reference means that the memory address of the argument is received by the procedure, which can then change the memory address value. Procedure arguments declared using the `ByVal` keyword are converted to the current value before being passed to the calling procedure and do not suffer any change when the code executes.

In this book most procedure arguments will be implicitly declared *by reference*, and I will bring it to your attention whenever a *by value* reference can be used.

■ **Attention** You can use the VBA `ByRef` keyword to explicitly indicate that a procedure argument is declared by reference, which is not necessary.

Most VBA functions declare their arguments *by value*, meaning that they will not be changed by the calling function.

Using a Naming Convention

If you take a good look at all the `DaysCount` and `Years` variables used in the procedure examples described so far, you will note that you can't identify the data type or scope just by the variable name. You need to look at where and how it was declared to see if the variables have a specific data type (or received the default, polymorphic `Variant` data type) or, if it is a procedure, are a private or public module variable.

By using a naming convention, anyone can identify the scope and type of a variable just by reading its name. This is a recommended programming practice, and the most widespread naming convention for programming objects used on VBA procedures are the Reddick name conventions (earlier known as Reddick-Lezinsky naming convention), which are based on the Hungarian conventions invented by Charles Simonyi, a programmer who worked at Xerox PARC circa 1972–1981 and who later became chief architect at Microsoft.

The Reddick VBA naming conventions (RVBA) for programming objects (variables, controls, forms, reports, and so on) use a prefix to qualify the scope of a variable, a tag to qualify the type of the object, followed by the object name that uses one or more capitalized words, without spaces or underscores to separate them. The basic structure of the RVBA is as follows:

```
<prefix><tag><ObjectName>
```

Tables 1-3 to 1-6 show the prefix and tags used by the RVBA rules to define the variable scope, variable type, some Excel programmable object tags, and `UserForm` control tags.

■ **Attention** You can find all the latest Reddick naming conventions by searching Google or by extracting `Reddick Naming Conventions 6.01.pdf` from the `Chapter01.zip` file.

Table 1-3. *Reddick Naming Conventions for Variable Scope*

Variable Scope	Prefix
Procedure variable	No prefix
Local static variable	s
Module-level variable	m
Public variable in a UserForm or Sheet module	p
Public variable in a standard module	g

Table 1-4. *Reddick Naming Convention Tags for Variable Data Type*

Variable Type	Tag
Byte	byt
Boolean	bln
Currency	cur
Date	dat
Decimal	dec
Double	dbl
Integer	int
Long	lng
Object	obj
Single	sng
String	str
Variable Type	**Tag**
Type (user-defined)	typ
Variant	var

Table 1-5. *Reddick Naming Convention Tags for Some Excel Programmable Objects*

Excel Object	Tag
Class module	cls
Standard module	bas
UserForm	frm
Range	Rng or rg
Worksheet	Wks or ws
Workbook	Wkb or wb

Table 1-6. Reddick Naming Convention Tags for UserForm Controls

UserForm Control Type	Tag
Control (generic)	ctl
Checkbox	chk
ComboBox	cbo
CommandButton	cmd
Frame	fra
Image	img
Label	lbl
ListBox	lst
OptionButton	opt
Shape	shp
Tab control	tab
TextBox	txt

By using the prefixes and tags in Tables 1-3 to 1-6, you can rewrite some Function procedure codes to acquaint yourself with good programming practices for writing clear, concise, and professional code structures.

Let's start with the Function AgeInYears2( ) procedure that uses just two variables: DateOfBirth and DaysCount. Using the RVBA naming conventions, its variables should be declared this way:

```
Function AgeInYears2 (datDateOfBirth as Date) as Single
    Dim lngDaysCount as Long

    lngDaysCount =Date - datDateOfBirth
    AgeInYears2 = lngDayCount / 365
End Function
```

Note that now the procedure argument is named datDateOfBirth, receiving the dat tag, so you can easily see that it was declared with the Date type. The same happened to lngDaysCount: it receives the lng tag because it was declared with the Long type. Since both variables are local to the procedure code, none of them received a scope prefix.

Now look at Function AgeInYears3( ), which declared the Years variable as Static:

```
Function AgeInYears3 (Optional datDateOfBirth as Variant) as Single
    Static ssngYears as Single

    If Not IsEmpty(datDateOfBirth) then
        ssngYears = (Date - datDateOfBirth) / 365
    End If
    vAgeInYears3 = ssngYears
End Function
```

Note that now you use the s prefix plus the sng tag to indicate that the ssngYear variable was declared as Static with the Single data type!

The Function AgeInYears5( ) procedure was the first to use private module-level variables, and according to the RVBA naming conventions, they must be declared this way:

```
Option Explicit

Dim mintDaysCount As Integer
Dim msngYears As Single

Public Function AgeInYears5(datDateOfBirth As Variant) As Single
    mintDaysCount = Date - datDateOfBirth
    msngYears = mintDaysCount / 365
    AgeInYears5 = msngYears
End Function
```

Now the private module variables received the m prefix (to indicate *m*odule scope) plus the int or sng tag to indicate the data type. When you look at the procedure code, you can easily recognize that mintDaysCount is a module with an Integer data type variable, while the msngYears is a module with a Single data type variable.

To end this RVBA naming conventions exercise, note how the Function AgeInYears6( ) procedure from Figure 1-25 should be rewritten to adapt to best programming practices:

```
Option Explicit

Public gintDaysCount As Integer
Public gsngYears As Single
Public gsngWeeks As Single
Public gstrAge As String

Public Function AgeInYears6(datDateOfBirth As Variant) As Single
    gintDaysCount = Date - datDateOfBirth
    gsngYears = gintDaysCount / 365
     AgeInYears6 = gsngYears
End Function
```

This time all Public standard module variables were declared with the g prefix followed by a specific data type tag. Now, every time you face such variable names in code, like gsngYears, you can easily spot that they are global (Public variables from a standard module) Single data type variables!

From now on, all procedures in this book will use such naming conventions to name variables, modules, forms, controls, and variables.

■ **Attention** As an Excel developer, whenever you need to create a public Function procedure that will be used directly on Excel worksheet formulas, you must avoid adding tag data types to name the procedure arguments. Microsoft Excel functions do not use tagged data types to identify the type of its arguments, and it would be strange to most Excel users to find tagged arguments when they use your Function procedures on their formulas. For example, the Function AgeinYears( ) procedure will be better accepted by a wider user audience if it asks for a DateOfBirth argument instead of datDateOfBirth argument.

Using Property Procedures

Although you can access any `Public` variable from Excel interface formulas, any procedure of your VBA project can eventually change its value when it shouldn't.

To avoid the direct manipulation of `Public` variables in any code module, VBA offers a third procedure type called `Property`. This is the one that is used to define and return the value of a private module variable.

To implement a property, VBA uses a pair of different procedure types: a `Property Let` procedure to define the property value and a `Property Get` procedure to return the property value. You can insert this pair of `Property` procedures by executing the VBA Insert ➤ Procedure menu command and selecting the property type. Figure 1-31 shows how to implement a public property named `Years` with assistance from the VBA Add Procedure dialog box.

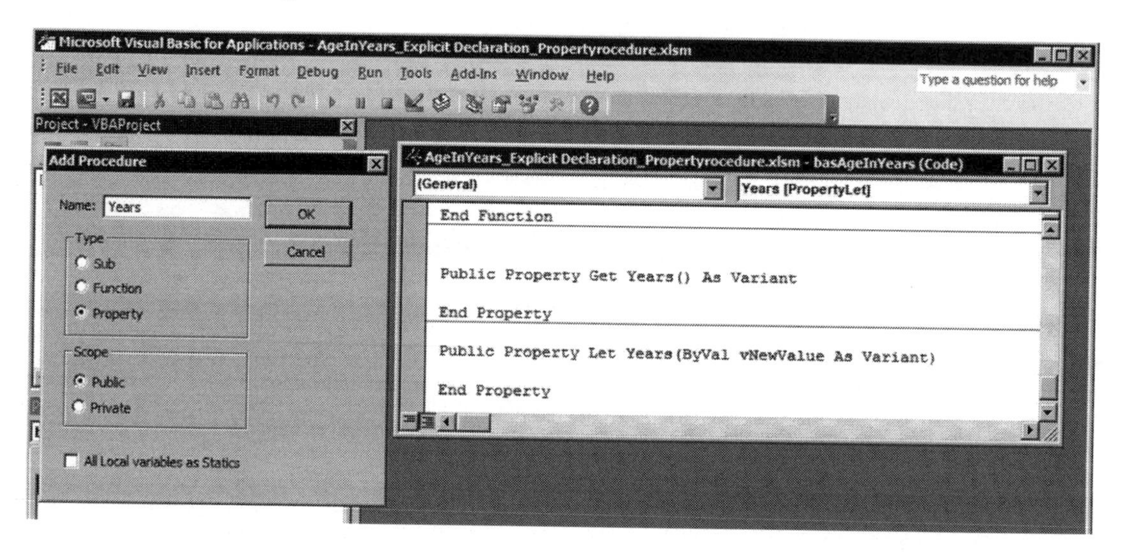

Figure 1-31. *Use the VBA Insert ➤ Procedure dialog box to create the pair of Property Let and Get procedures to implement a property on any code module*

Property procedures work like this:

- Use the `Propery Let` procedure (the one that receives the `vNewValue` argument) to store the property value.
- Use the `Property Get` procedure (with no arguments) to return the property value.

Since you must store the property value using the `Property Let` procedure, declare a `Private` module variable to hold the property value, and use the `Property Get` procedure to return the `Private` variable value. The next listing shows how to implement such a `Years Property` procedure that manipulates the `Private msngYears as Single` module-level variable:

```
Private msngYears as Single

Public Property Get Years() As Single
    Years = msgnYears
End Property

Public Property Let Years(ByVal vNewValue As Single)
    msngYears = vNewValue
End Property
```

Please note that the msngYears variable is a Private variable and as such can be accessed from inside the module where it is declared. Also note that the Property Let Years( ) procedure vNewValue as Single argument must have the same data type returned by the Property Get Years( ) procedure (also declared as Single) or VBA will generate a compile error.

Once the Years Property procedures have been implemented on the basAgeInYears code module, you can change or verify the value using the VBA Immediate window. You can find such an example by using the AgeInYears_Explicit Declaration_PropertyProcedure.xlsm macro-enabled workbook also found inside the Chapter01.zip file. Use this syntax to set the Years property by executing the Property Let procedure (Figure 1-32):

```
basAgeInYears.Years = 50
```

And use this syntax from the VBA Immediate window to verify the value of the Years property by executing the Property Get procedure:

```
?basAgeInYears.Years
50
```

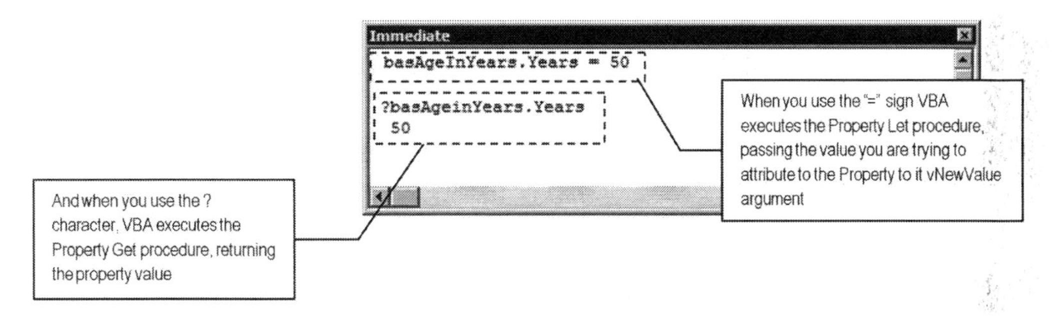

Figure 1-32. *You can set or verify any Public property that implements both the Property Let and Property Get procedures using the VBA Immediate window. Note that the code module name followed by a dot must precede the property name*

Note that basAgeInYears from the AgeInYears_Explicit Declaration_PropertyProcedure.xlsm macro-enabled workbook implements the Years property using Public procedures for both Property Get Years( ) and Property Let Years( ), allowing you to recover or set the property value using the VBA Immediate window.

But the properties Age, DaysCount, and Week implement just the Property Get procedure as a Public procedure, keeping the Property Let procedure as Private. This way they can be changed only from any code procedure inside the basAgeInYears value or from the VBA Immediate window; you cannot change them from another code module.

■ **Attention** Here is a secure way to create a Read Only Property procedure: implement the Property Let procedure as Private to the code module. You can also implement just the Property Get( ) procedure and let your code directly interact with the private variable to set the property value.

Property Procedures Allow Greater Control of Private Variables

You are probably wondering what the difference is between using a `Private` or `Public` variable and a `Property` procedure to do the same task to hold and return a value.

The answer to this question is quite simple. By using `Property` procedures, you can take great control of what is stored on your private module variable or even validate the value before you store it.

For example, let's suppose that you want to avoid invalid values being attributed to the `Years` variable, such as negative years or years greater than 130. You can do this by refining the code of the `Property Let Years( )` procedure, like this:

```
Public Property Let Years(ByVal vNewValue As Single)
    If vNewValue > 0 and vNewValue <=130 then
        msngYears = vNewValue
    End If
End Property
```

Or you can allow the `Years` property to be set just once for your entire project, using a `Property Let Years( )` procedure like this:

```
Public Property Let Years(ByVal vNewValue As Single)
    If msngYears <> 0 then
        If vNewValue > 0 and vNewValue <=130 then
            msngYears = vNewValue
        End If
    End If
End Property
```

Note that this last code procedure still validate the `vNewValue` argument to be greater than zero and lower than 130. But this time, it will be set just once.

In the next chapter of this book, we will use such approaches to create some `Property` procedures, and I will call your attention whenever I feel that it will be useful to your full comprehension of the code strategy used to program Excel with VBA.

VBA Statements, Functions, and Instructions

VBA like any other programming language has a lot of defined statements and a large array of ready-made functions and instructions that you will use to program your procedure code. The differences between them are as follows:

- A *statement* is a code structure that you can strictly follow to execute a given task one or many times, such the `If.. Then... Else... End If` set of instructions that you saw in previous examples.
- A VBA *function* is a ready-made `Function Procedure` implemented by VBA, which may receive one or more arguments, process them, and return a value. They are grouped by many different categories, such as Math, String, Conversion, Date and Time, Financial, File and Folder Manipulation, and so on.
- A VBA *instruction* is a ready-made `Sub Procedure` implemented by VBA that performs an action without returning a value.

In this book, we will use specific VBA statements, functions, and instructions to make special code operations, and as such I will call your attention to what is the function's purpose, it syntax, and why I used it to create the code.

Let's see some special cases of VBA instructions that will be used to interact with your code procedures.

Using VBA Instructions

VBA statements are the core of the VBA programming language, allowing you to declare variables and procedure types, implement logical decisions in your code, and execute loops that repeatedly execute a code section to perform a given action.

These are the two types of VBA instructions used inside code procedures:

- Instructions used to perform a logical decision
- Instructions used to execute a loop

VBA Logical Decision Instructions

The most common VBA logical decision instructions are as follows:

- `If... End If`
- `Select Case... End Select`

Making Decisions with If...End If Instructions

The `If... End If` VBA instructions have these basic instructions (elements inside brackets are optional):

```
If <Condition1> Then
    [Code Statements1]
[ElseIf <Condition2> Then]
    [Code Statements2]
[Else]
    [Code Statments3]
End If
```

The `If... End If` instruction basically executes a logical test on the `Condition1` statement and, if it returns `True`, branches your code to execute the `[Code Statements1]` code part.

If `Condition1` is evaluated as `False`, you can either do nothing by just closing the `If` branch with the `End If` statement, execute an alternative `Else` clause (and its `[Code Statements3]` code part), or even make another set of tests using a successive `ElseIf <Condition2> Then` statement.

The `If` instruction has an obligatory syntax of making a valid logical test on `Condition1` and then closes the branch with the `End If` statement. It is advised to indent your code inside the `If`... instruction to never forget to type the `End If` of the last `If` instruction.

You can nest an If instruction inside the Then, Else, or ElseIf clause of an outer If instruction. The next code statement shows two nested If ...End If instructions. Note that the code uses indentation to visually set where each If Instruction ends with its associated End If instruction.

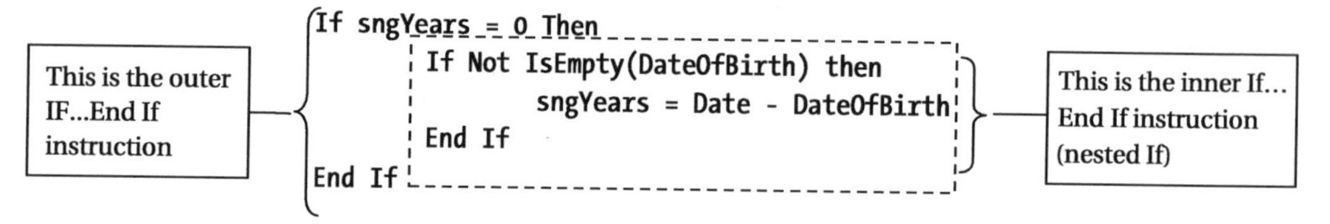

You can use the If... End If instruction to make as many logical tests as you want by using the optional ElseIf <Condition n> Then clause, and if any logical test is evaluated as True, you can also plan an emergency action using the Else clause that will always be executed whenever all previous tests fail.

The next code fragment verifies whether the KeyPress variable contains the key code for the F1, F2, F3, and F4 function keys, and if this is true, it calls the Procedure1 procedure. If the KeyPress variable has any other keyboard key code, it executes the Procedure2 procedure on the Else clause.

```
If KeyPress = vbKeyF1 then
    Call Procedure1
ElseIf KeyPress = vbKeyF2 then
    Call Procedure1
ElseIf KeyPress = vbKeyF3 then
    Call Procedure1
ElseIf KeyPress = vbKeyF4 then
    Call Procedure1
Else
    Call Procedure2
End if
```

This last procedure can be rewritten using just one, extent logical test to verify whether the KeyPress variable is either F1, F2, F3, or F4, using the OR operator:

```
If KeyPress = vbKeyF1 OR KeyPress = vbKeyF2 OR KeyPress = vbKeyF3 OR KeyPress = vbKeyF4 then
    Call Procedure1
Else
    Call Procedure2
Endif
```

Note that you must compare the KeyPress variable value with the desired VBA function key constant. Every time you need to do this kind of long logical test, it may be better to use the Select Case instruction.

Making Decisions with the Select Case...End Select Instruction

The Select Case instruction allows you to use a simple test structure by making many different tests in one code row. It has this general form:

```
Select Case <condition>
    Case <value1>
        [Code Statments1]
     [Case <value2>]
         [Code Statments2]
```

```
    [Case Else]
        [Code Statments3]
End Select
```

The Select Case instruction allows you to make an inference about the <condition> just once, then use the Case option to verify whether Condition fits the case, and finally execute a different set of code statements. Note that it also has a [Case Else] instruction that will be executed when every other Case instruction fails (or returns False).

The last procedure used to test the KeyPress variable against the F1, F2, F3, or F4 function keys can be rewritten this way using the Select Case instruction:

```
Select Case KeyPress
    Case vbKeyF1, vbKeyF2, vbKeyF3, vbKeyF4
        Call Procedure1
    Case Else
        Call Procedure2
End Select
```

Shorter, isn't it? It does the same test and makes the same decision. It is up to you to select the If... End If or Select Case instructions.

Note that the Select Case instruction has an obligatory syntax of verifying a variable value or making a logical test on Condition and then closes the branch with the End Select statement. For every Select Case instruction there must exist an End Select statement, and as such, you may indent your code to never forget to type it.

Using the same technique described for the If.. End If instruction, you can also nest a Select Case instruction inside any Case instruction of an outer Select Case. The next code statement shows a nested Select Case instruction inside an outer Select Case instruction. Note that the code uses indentation to visually set where each Select instruction ends with the associated End Select instruction.

```
Select Case KeyPress
    Case >=112, <=123
        'Function key pressed. Test KeyPress again!
        Select Case KeyPress
            Case vbKeyF1, vbKeyF2, vbKeyF3, vbKeyF4
                Call Procedure1
            Case Else
                Call Procedure2
        End Select
Case >=48, <=47
    'Numeric key pressed
    Call Procedure 3
End Select
```

This is the nested Select case instruction

VBA Looping Statements

A loop statement is a programmable structure that allows code to be executed until a condition is met, such as processing all the cells of a given Excel range.

VBA uses four types of loop statements: For...Next, For Each...Next, While...End, and Do...Loop. These will be used throughout this book.

The For...Next Statement

The For...Next loop statement iterates through the code a maximum number of times. It has this syntax:

```
For counter [ As datatype ] = start To end [ Step step ]
    ' Statement block to be executed for each value of counter.
    [ Exit For ]
Next [ counter ]
```

In this code:

Counter: This must be an integer data type that supports the greater-than or equal (>=), less-than or equal (<=), and addition (+) operators.

start, end, step: These must be integer data types that set the iteration values of the start, end, and step counter defining the loop length. The optional step can be positive or negative. If it is omitted, it is taken to be 1.;

Exit [For]: You use this instruction to exit the For...Next loop at any moment. The Exit instruction alone (without For) will also work.

Whenever VBA finds a For...Next loop, it will process the loop incrementing the counter argument by the step value. When the counter passes the end value (either positive or negatively), it will terminate the loop, passing control to the statement following the Next statement.

To end the loop before the counter passes the end argument, use an Exit For statement inside the loop.

The next code processes a For...Next loop 100 times, using the Cells method to process the first 100 column A cells and exiting the loop if it finds a cell value that is greater than 1000:

```
For intI = 1 to 100
    If Cells(intI, 1) > 1000 then
        Exit For
    End If
Next
```

The For Each...Next Statement

The For Each...Next loop differs from the For...Next statement by processing each element in a collection, instead of a specified number of times. It has this syntax:

```
For Each elementvariable [ As datatype ] In collection
  ' Statement block to be executed for each value of elementvariable.
Next [ elementvariable ]
```

Where

elementvariable: This must be such that each element of the collection can be converted to a data type. You can always declare it as Variant.

Exit For: You use this instruction to exit the For...Next loop at any moment.

The For Each... Next iteration loop sets the variable elementvariable to each element in the collection (beginning from the first element), executing the statement block until all collection elements have been processed. Then it terminates and passes control to the statement following the Next statement.

The next code examples uses a For Each...Next instruction to loop through all Range.Cells collections of the MyRange range name:

```
Dim rg as Range
For Each rg in Range("MyRange").Cells
...
Next
```

The While...End and Do...Loop Statements

The While...End and Do...Loop statements differ from the For...Next statements by not determining a maximum number of interactions; also, they cannot define a step and may or may not use a conditional test when the loop begins or ends to define the duration. They have this syntax:

```
While condition
    [ Exit While ]
    [ statements ]
End

Do { While | Until } condition
    [ statements ]
    [ Exit Do ]
Loop { While | Until } condition
```

In this code:

> While, Until: These are instructions that indicate whether the loop must last until While condition = True (the loop ends *when* condition = False) or Until condition is true (the loop last *while* condition = False). They can be used on the beginning or end of the Do...Loop statements.
>
> condition: This is a logical test, a Boolean variable, or a number that can be evaluated to True or False (just 0 = False).
>
> Exit [While [, Exit [Do]: These are used to exit the Loop at any moment. The Exit instruction alone (without While or Do) will also work.

Be careful to guarantee that the condition becomes True or False, according to the test made, to not enter on a perpetual loop, which will freeze your computer until you press Ctrl+Break to put VBA in Break mode and do a step-by-step code iteration with the F8 function key to solve the problem.

These simple While...End and Do...Loop statements have no duration specified. The conditional test is made inside the loop in this way:

```
While
    If condition then Exit While
End
Do
    If condition then Exit Do
Loop
```

To guarantee that the Do...Loop statements will be executed at least one time, use the While or Until instruction at the end of the loop.

```
Do
    [statements]
    [ Exit Do ]
Loop Until Condition = True
```

By putting the test condition on the beginning of the Do...Loop statement, it behaves like a While...End statement, where the inner instructions may never be executed.

```
Do While Condition = False
    [statements]
    [ Exit Do ]
Loop
```

You will see plenty of examples of all these loop instructions in later chapters.

Using Event Procedures

Events are special VBA Sub procedures that you can use to take action whenever events happen. Any Microsoft Office application fires events in response of user actions. For example, when you open an Excel workbook, the Workbook_Open( ) event fires, and if you create any code inside the Workbook_Open( ) event code, you can make something happen whenever the workbook is opened.

The same thing happens when you close the workbook (the Workbook_BeforeClose event fires), when you select another sheet tab (the Worksheet_Activate event fires), or even when you select another cell (the Worksheet_SelectionChange event fires) and change the cell value (the Worksheet_Change event fires).

You cannot control *when* the event fires, but as a programmer, being aware that a given Event procedure exists allows you to take control of your worksheet application by writing VBA code inside the event procedure associated with the user action performed on the interface.

To see the Workbook_Open event procedure, open a new Excel workbook and double-click the ThisWorkbook object in the Project Explorer tree to show its code module. Click the ComboBox located at the top-left corner of the code module window (where you read (General)) and select the only object it has: Workbook. When you do this, it will automatically select and create the default event procedure structure inside the ThisWorkbook code module: Workbook_Open( ) (Figure 1-33).

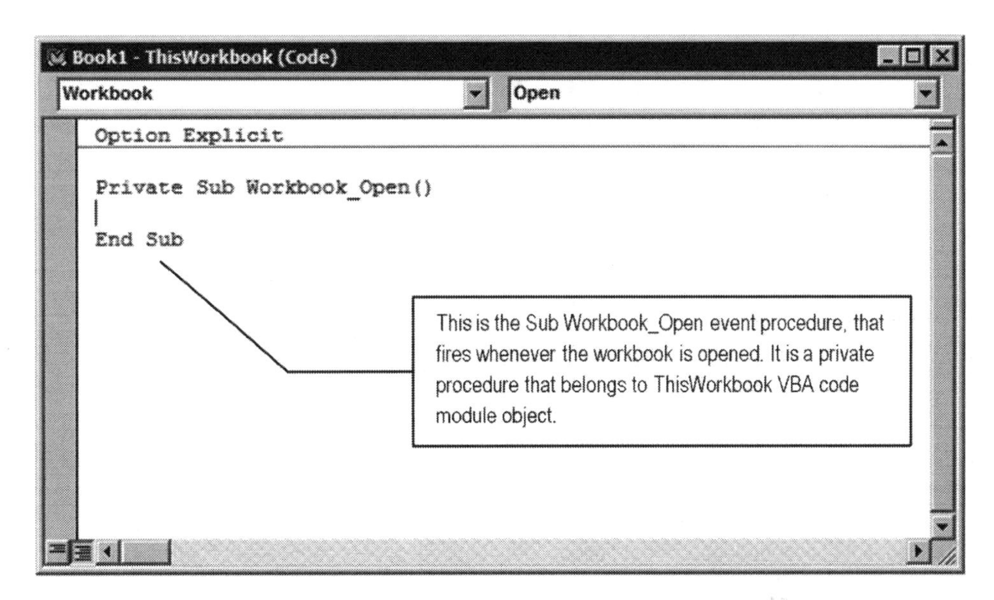

Figure 1-33. *In the ThisWorksheet code module, select the Workbook object on the top-left list box to automatically create the object default event procedure:* Sub *Workbook_Open. This event fires whenever the workbook is opened and is up to you insert any code on it*

Note that the `Workbook_Open` event has the structure of a common `Sub` procedure and is declared as a `Private` procedure, meaning that it can be accessed only by the code inserted inside the `ThisWorkbook` code module. Also note that this event has no argument.

The `Workbook` object of the `ThisWorbook` code module has a lot of different events that fire against the user action. They can be found by keeping the `Workbook` object selected on the left `ComboBox` of the code module window while expanding the right `ComboBox` list.

Try to select the `BeforeClose` event of any the `Workbook` object to make VBA automatically create the structure of the `Sub Workbook_BeforeClose( )` event procedure. This will fire every time the user tries to close the workbook or the Excel window (Figure 1-34).

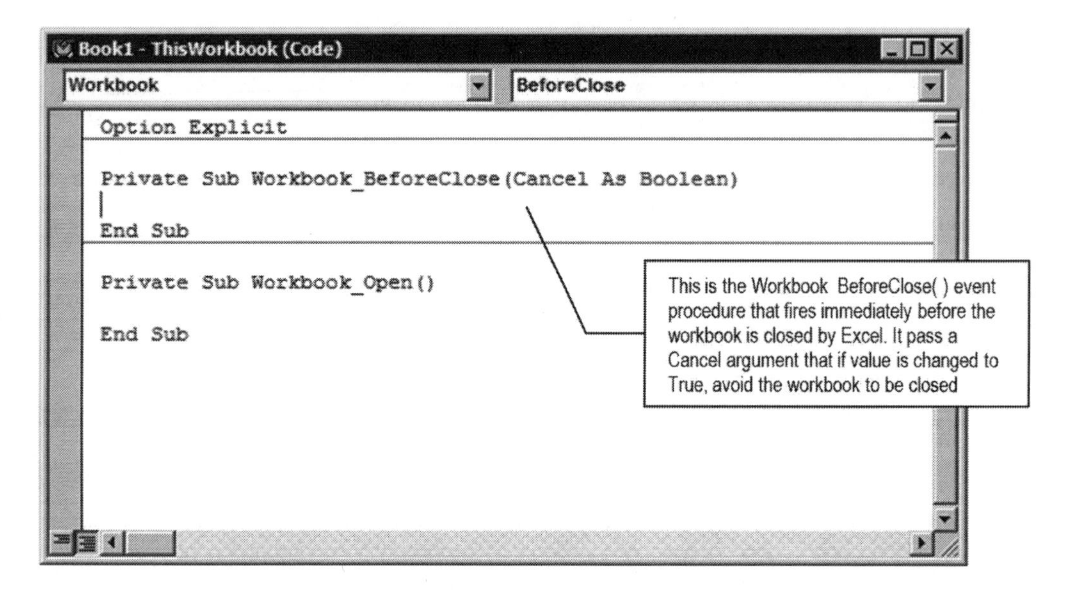

Figure 1-34. *While the ThisWorkbook code module has the Workbook object selected in the top-left list box, click the right list box to show all events fired by the workbook object and select the BeforeClose event to force VBA to create the code procedure structure*

■ **Attention** You *cannot* modify (add or delete) the arguments of any object event procedure automatically created by VBA, but you can use or change any argument value inside the event procedure to make the Excel environment react to the proposed change.

Note that the Sub `Workbook_BeforeClose( )` event procedure has a `Cancel` argument, declared as `Boolean`. For your knowledge, on the VBA environment every event procedure that has a `Cancel` argument means that the event can be canceled if you change the `Cancel` argument to `True` (`Cancel = True`) inside the event procedure code.

Since the `Workbook_BeforeClose( )` event procedure fires whenever you try to close the workbook (which will happen when you try to close the file or close Excel), this means that if anywhere inside the `Workbook_BeforeClose( )` event procedure you set `Cancel = True`, the workbook (or Excel) will not be closed. I will use this trick later in this book.

The next code fragment inserted on the `ThisWorkbook` code module to the `Workbook_BeforeClose( )` event procedure changes the `Cancel` argument to `True`, avoiding the workbook to be closed by any means:

```
Private Sub Workbook_BeforeClose(Cancel As Boolean)
    Cancel = True
End Sub
```

■ **Attention** An event procedure `Argument`, like the `Cancel` argument from the `Workbook_BeforeClose( )` event, is a value that is received by the code associated with the event. The `Cancel` argument was declared as `Boolean`, meaning that it can be either `True` or `False`. The default value is `False` (or zero), and since you can change the value (and Excel can perceive this change), programmers always refer to this argument as "received" by the Excel interface to the procedure.

Some event procedures can pass different types of arguments, related to what happens to the worksheet. To show the Sheet1 object code module, double-click it in the VBA Project Explorer tree. Select the Worksheet object on the top-left ComboBox of the code module window to create the default event procedure: the Sub Worksheet_SelectionChange( ) event (Figure 1-35).

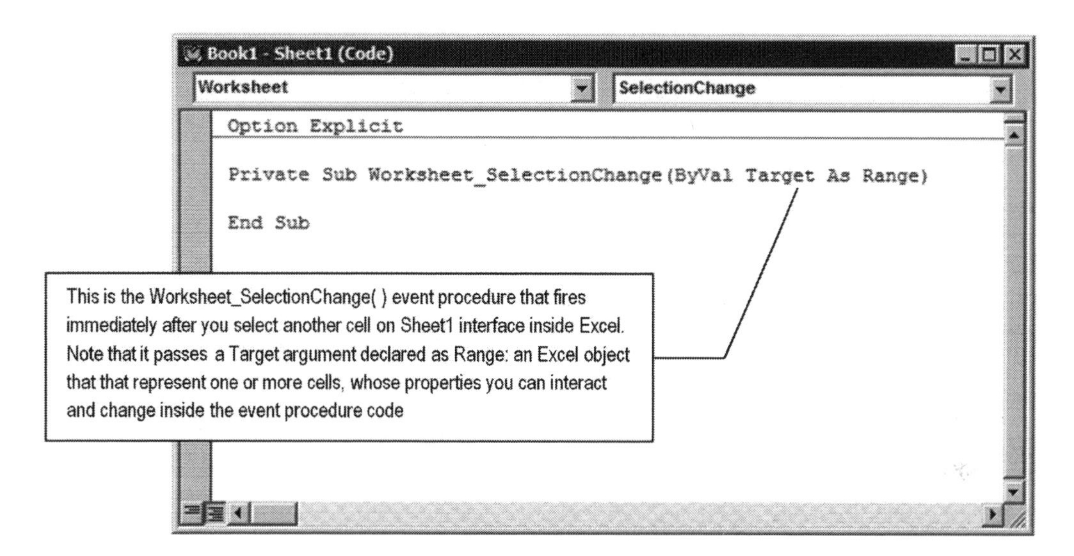

Figure 1-35. *In the Sheet1 code module, select the Worksheet object on the top-left list box to force VBA to create the code structure of it default event procedure: the Worksheet_SelectionChange( ) event*

Note that this time VBA passes to the Sub Worksheet_SelectionChange( ) event procedure an Excel object declared as Target as Range. This time you do not have a value by itself but an entire object interface that you can manipulate inside the event procedure (like the range value, cells font, color, and so on).

In the following chapters, you will interact with Excel in a deeper way, using the Application, Workbook, and Worksheet events as examples so you can understand how to use them to automate your Excel solutions with VBA code.

Using Class Modules

A Class module is like any other module on appearance, but it can create a programmable object that can be reused when it is declared as an object variable inside other code modules.

The public Sub and Function procedures declared inside a Class module will become the object methods, while the public variables or public Property procedures will become the properties. The Class module's Name property will be used as the object name, so you can reference it on other code modules by its name.

Let's suppose you created a class whose property Name = Timer, which allows you to create and set any timers you want. If this class has a public variable named Interval and a public Sub Enabled( ) procedure, it has an Interval property (in miliseconds) and an Enabled method. The Class module code will be like this (Sub Enabled( ) still has no instructions):

```
Option Explicit

Public Interval as Integer
```

```
Public Sub Enabled( )
...
End Sub
```

Supposing that all the code inside `Public Sub Enabled( )` creates a timer that will fire at each `Interval` value (defined in milliseconds), you can create a new `Timer` object on any other code module by declaring an object variable whose type is the `Timer` calls it on the module `Declaration` section:

```
Option Explicit
Dim mTimer1 as Timer
```

You use another procedure to instantiate the class using VBA `Set` and `New` keywords.

```
Public Sub BeginTimer
    Set mTimer = New Timer
    mTimer.Interval = 3000
    mTimer.Enabled
End Sub
```

This last instruction will be enough to create the `mTimer1` timer that will fire every three seconds (3000 miliseconds). In later chapters, you will be presented with different `Class` modules that create interesting programmable objects.

Declaring and Raising Events on Object Code Modules

Events are code procedures that fire when a condition is met. As explained in the section "Using Event Procedures" earlier in this chapter, most Excel programmable objects have a rich set of events that you can program to react to a given situation or user action.

By using the VBA `Event` keyword, you can declare `Private` and `Public` events on any object code module (such as an Excel object like `Thisworkbook` or `Sheet1` code modules or inside any `UserForm` or `Class` module), like you declare a `Sub` or `Function` procedure, with all the arguments it needs (if any), using this syntax:

```
[ Private | Public ] Event <EventName> [( [ Arg1 [ as Type ] ], ..., [Argn [as Type ] ] )
```

In this code:

> `Arg1, ..., Arg n`: These are the optional event argument names and types.

To raise any event declared inside an object module, use the VBA `RaiseEvent` instruction that has this syntax:

```
RaiseEvent <EventName>
```

In this code:

> `EventName`: This is an event declared with the `Event` keyword on the object or class module that you want to raise on the object interface.

To use the events exposed by any `Class` module, you must declare an object variable to represent it using the VBA `WithEvents` keyword, like so:

```
Dim WithEvents <VariableName> as <ClassName>
```

In this code:

`VariableName`: This is the variable name that represents an instance of the `Class` module.

`ClassName`: This is the `Name` property of the `Class` module that identifies the class object.

■ **Attention** In Chapter 2, you will find good examples about how to expose `Class` module events.

Using VBA UserForms

The last VBA object that I will use on this book is the `UserForm` object. `UserForm` is the only VBA structure that may be visible on Excel by allowing you to create forms that interact with your solution.

Many Excel add-ins use the VBA `UserForm` object to collect and execute VBA code on the selected worksheet, like the Data Analysis ToolPak that you can install in Excel to execute some interesting statistical operations, such as Descriptive Statistical analysis of your data, ANOVA with single and Two-factor, with or without replication, Correlation, Covariance, Histogram, Moving average, T-Test, and so on. All these operations use a VBA `UserForm` to interact with Excel and your data to calculate and return the desired results. Figure 1-36 shows the Data Analysis `UserForm`, where you can select among the many statistical analysis available. Figure 1-37 shows the Descriptive Statistics `UserForm` interface as implemented by the Analysis ToolPak add-in.

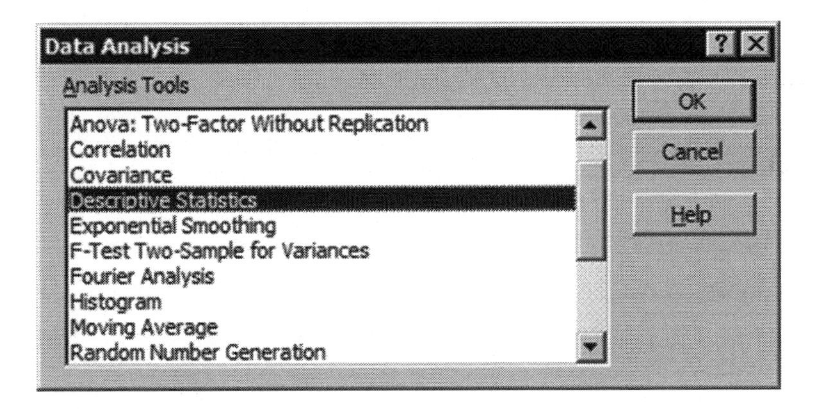

***Figure 1-36.** This is the Analysis ToolPak UserForm interface where you can select many different statistical analysis to perform on your Excel worksheet data*

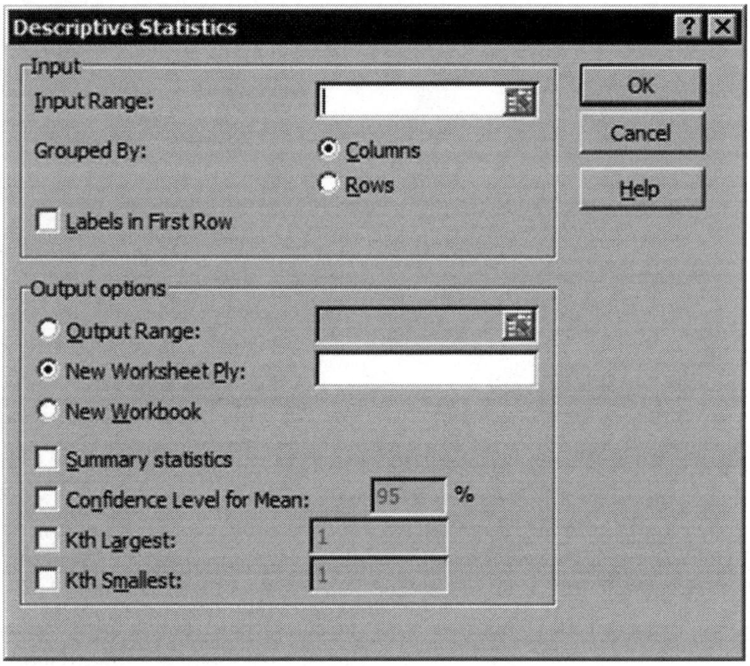

Figure 1-37. *This is the Descriptive Statistics interface, implemented on a VBA UserForm object by the Analysis ToolPak Excel add-in*

Such dialog box interfaces are first created on your workbook by using the VBA Insert ➤ UserForm command. When you do this, another branch will appear on the VBA Project Explorer Tree called Forms. The `UserForm1` object window will be selected in the VBA IDE, which will also show the Toolbox, which is where you can select many different controls to compose your `UserForm` interface (Figure 1-38).

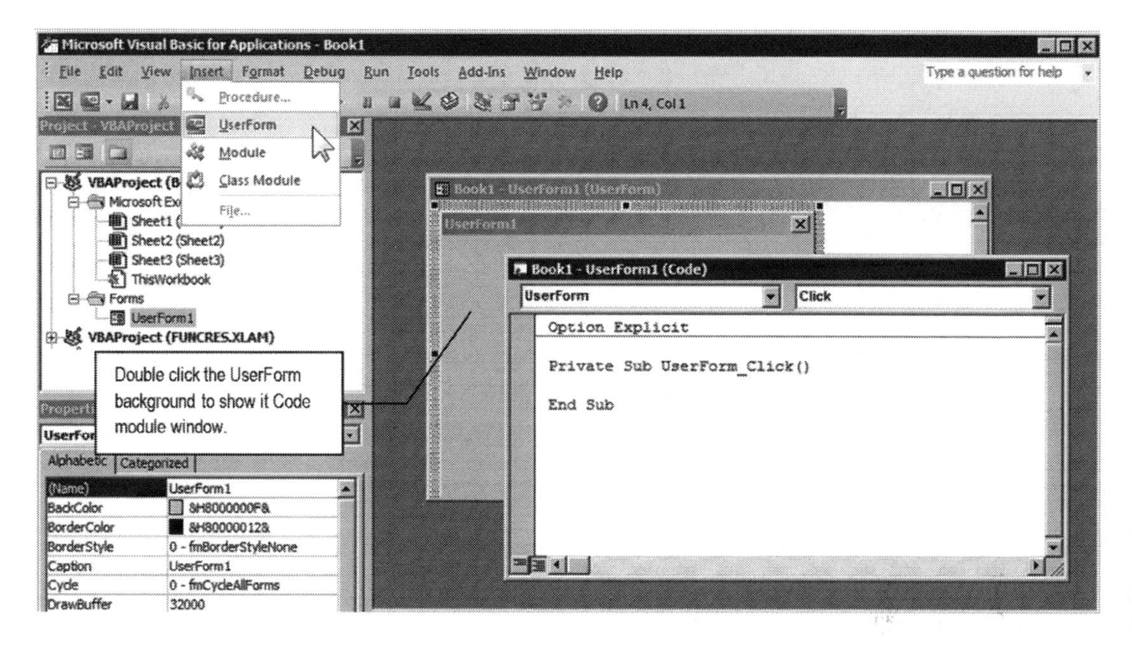

Figure 1-38. *When you use VBA Insert ➤ UserForm menu command, VBA will insert a new Project Explorer branch called Forms, with the new UserForm selected. It will show the standard UserForm interface and will make the VBA Toolbox visible so you can select the controls you want to use to create your UserForm interface*

Every `UserForm` object has its own code module that you can show by double-clicking the `UserForm` background. Like any other code module window, you can use it to create `Function`, `Sub`, and `Event` procedures, and like every VBA object, the `UserForm` also has its own set of `Event` procedures, like `UserForm_Initialize( )`, `UserForm_Activate( )`, and so on, from where you can control how it will work to interact with the users of your application.

To load any UserForm using VBA code, you must use the UserForm object's Show method in this way (note the dot concatenating the UserForm1 object with the Show method):

```
UserForm1.Show
```

To unload any UserForm, use the VBA Unload method in this way (where UserForm1 is the name of the UserForm object you want to unload):

```
Unload UserForm1
```

Attention To unload any UserForm using VBA code inside the UserForm code module, you can use the VBA Me keyword to make the UserForm code module refer to itself. Such code is commonly created inside a CommandButton control that you put on the form to close it when the user clicks in this way:

```
Unload Me
```

Since you need to reference the Show method of the UserForm object via VBA to load it, it must be made from an external procedure, usually inserted on an independent code module, like the ShowUserForm( ) Sub procedure inserted on the basUserForm code module of the UserForm.xlsm Excel macro-enabled workbook that you can extract from the Chapter01.zip file. It has this simple code:

```
Public Sub ShowUserForm1()
    UserForm1.Show
End Sub
```

You can test this procedure by typing its name in the VBA Immediate window and pressing Enter, making the UserForm1 window appear in the Excel interface. Since every UserForm is by default a modal window, you can't use the VBA Immediate window to close it using the Unload method. The user of your application needs to manually close the UserForm by clicking the Close button, pressing Alt+F4, or clicking some CommandButton control such as the one placed on the UserForm1 interface shown in Figure 1-39, which executes this simple code:

```
Private Sub CommandButton1_Click()
    Unload Me
End Sub
```

Figure 1-39 shows the VBA code modules for the UserForm1 with the CommandButton1_Click event procedure that fires whenever the button is clicked to unload the UserForm1, as well as the ShowUserForm1( ) Sub procedure from basUserForm from the UserForm.xlsm Excel macro-enabled workbook.

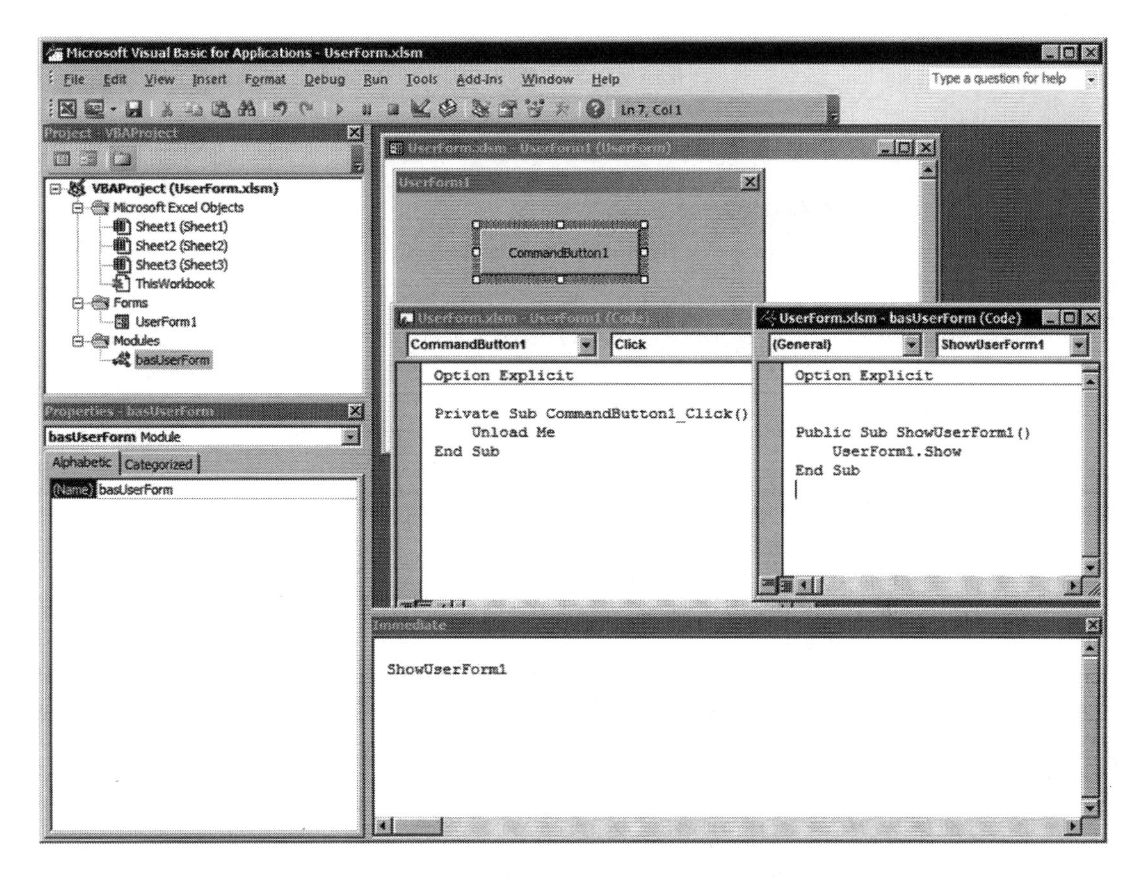

Figure 1-39. *This is the UserForm1 code module with the CommandButton1_Click event procedure and the ShowUserForm1 Sub procedure from the basUserForm code module used to load the UserForm1 in the Excel interface*

For now I just want you to understand that the heart of the VBA environment resides in code modules where you can write your own `Function` or `Sub` procedure codes, or you can use the structure of the Excel event procedures and `UserForm` objects to execute any code regarding the actions the users perform in your application interface.

In the next chapters, you will make plenty use of VBA `UserForm` objects to create user interfaces to Excel objects, exploring the properties, methods, and events.

The VBA Me Keyword

The VBA `Me` keyword represents the code module where it is declared. In the previous section, it was used to indicate a `UserForm` to unload itself.

Use the `Me` keyword in an object code module to make the code refer to itself (the object), whether it be the `ThisWorkbook`, `Sheet1`, `UserForm1`, or any standard module or the `Class` module.

There is a great advantage to using `Me` in VBA. When you type `Me` and a dot (`Me.`), VBA will show all object interfaces, including the properties, methods, and events, allowing you to gain time writing the code.

And when using `Me` to refer to the `UserForm` from the code module, it will also show you all the control names on the VBA list interface, making it easy to correctly reference them in the code. That is why you will see so many `Me.`-prefixed control names throughout this book!

Evoking a VBA Procedure from an Excel Worksheet

To evoke a VBA procedure from any Excel worksheet, you must call it from any Excel event procedure or allow it to be executed by a user action by putting a `Button` control on the worksheet that must execute the procedure. To insert a `Button` control on any worksheet, just follow these steps (I will use the `Sub ShowUserForm1( )` procedure from `basUserForm` of the `UserForm.xlsm` workbook as an example):

- On the Developer tab, select Insert ➤ Button (Form control), as shown in Figure 1-40.
- Drag the mouse over the worksheet to define the size of the Command button you are inserting.

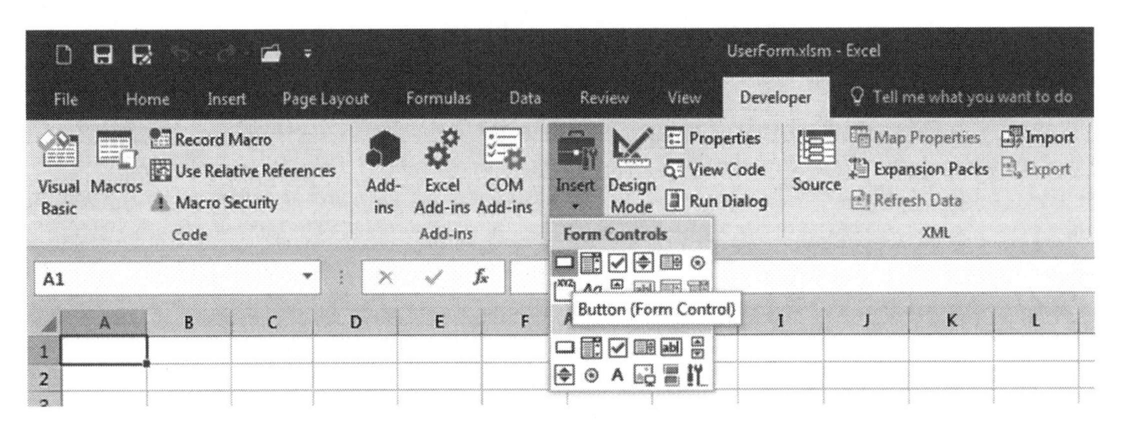

Figure 1-40. *To allow the user to execute specific Public procedures from your Excel solution, on the Developer tab select the Insert ➤ Button (Form control) command and drag the button over the worksheet*

- Excel will open the Assign Macro dialog box, which is where you choose `Public Sub` procedures available on your VBA project.
- Select the `Public` procedure you want to associate to the `Button` control's `Click` event (in this case `ShowUserForm1`, as shown in Figure 1-41).

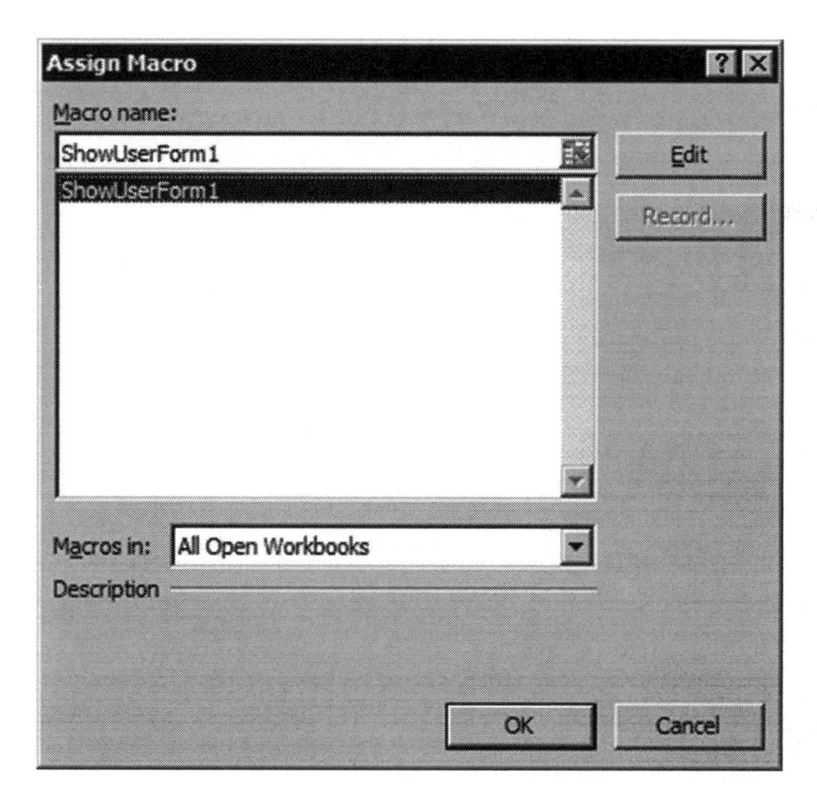

Figure 1-41. *In the Assign Macro dialog box, select the Public Sub procedure you want to associate with the Click event of the Button control*

- Click OK to terminate the association. Excel will name the `Button` control as `Button1`.
- Once the `Button` control is associated with the desired `Public Sub` procedure, it is inserted on Edit mode over you worksheet so you can change the caption and size (Figure 1-42).

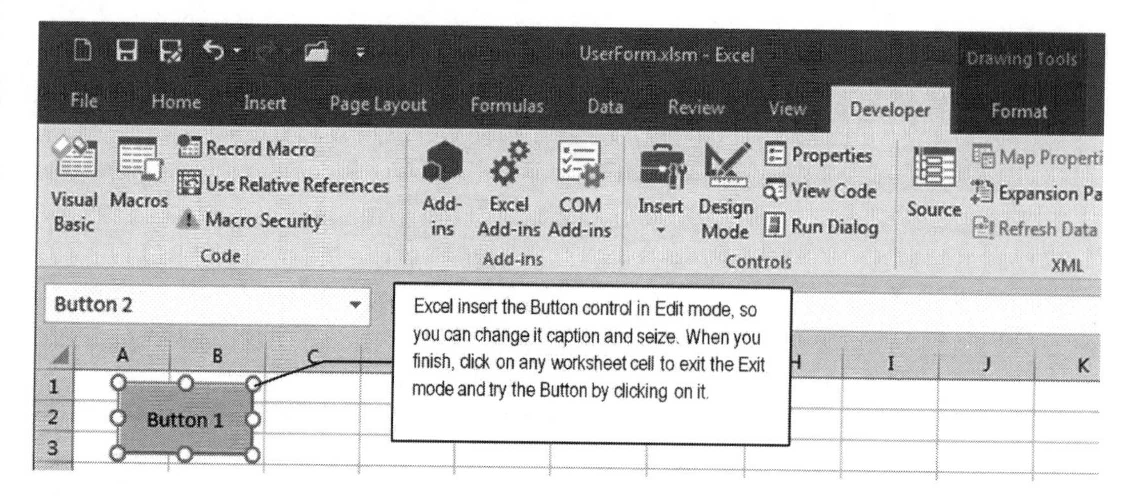

Figure 1-42. *To allow the user to execute a specific Public procedure from your Excel solution, on the Developer tab select the Insert ➤ Button (Form control) command and drag the button over the worksheet*

- To exit the Edit mode, just click any worksheet cell and then click the Button control to try the associated code (in this case, it must open `UserForm1`, as shown in Figure 1-43).

■ **Attention** Just `Public Sub` procedures from Standard modules are automatically shown by the Assign Macro dialog box. To associate any other `Public Function` procedure from `ThisWorkbook`, any `Sheet` object, or any standard module, type the object name followed by a dot and the `Public` procedure name in the "Macro name" text box of the Assign Macro dialog box and press Enter. Excel will check the procedure existence before associating it to the `Button` control.

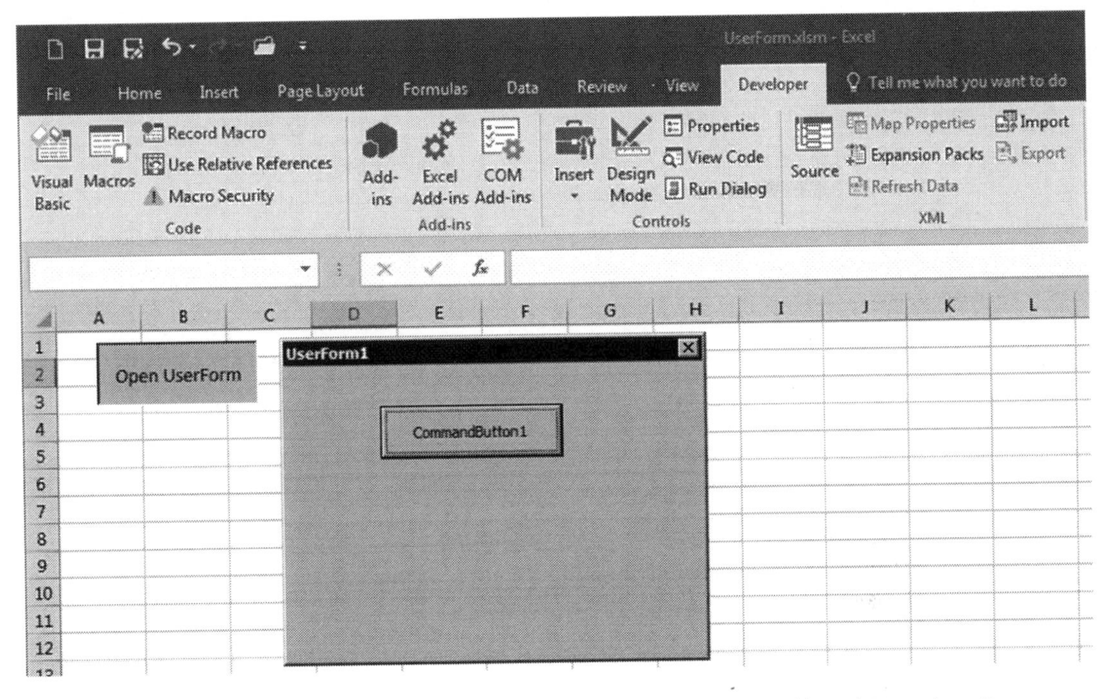

Figure 1-43. *Since the Button was intended to execute the ShowUserForm1 Public Sub Procedure, it was named as Open UserForm, and when you click it, you will see the UserForm1 interface modal floating over the selected worksheet*

Two Special VBA Functions: MsgBox and InputBox

VBA has two special functions that you will use often to interact with the user of your application: `MsgBox( )` and `InputBox( )`. The first one is used to grant messages and capture user decisions, while the second is used to receive user input.

Using MsgBox( )

Use the VBA `MsgBox( )` function to show an information dialog box to the user and eventually let the user decide what to do by pressing a button. The `MsgBox( )` function syntax is as follows:

```
MsgBox(Prompt as String, Buttons as Long, Title as String) as Long
```

In this code:

`Prompt`: This is required; it is the message that will be shown by the dialog box, with a maximum length of about 1,024 characters depending on the width of the characters used. To break the `Prompt` message in lines, concatenate it with the VBA constant `VBCrLf`, which consists of a carriage return and line feed characters.

`Buttons`: This is optional; it is a numeric expression that is the sum of up to four different values that specify the buttons to display, the icon style to use, the default button when the user presses Enter, and the modality of the dialog box. When omitting the button's value, a default value of zero will show just the OK button.

`Title`: This is optional. This is the expression displayed in the title bar of the dialog box. If you omit the title, the application name will be placed in the title bar.

The VBA `MsgBox( )` function will return the code of the button selected by the user, according to the Table 1-7 values.

Table 1-7. *Button Constants and Values Clicked on the Dialog Box Displayed by the VBA MsgBox( ) Function*

Bottom Pressed	Constant Name	Constant Value
OK	`vbOk`	1
Cancel	`vbCancel`	2
Abort	`vbAbort`	3
Retry	`vbRetry`	4
Ignore	`vbIgnore`	5
Yes	`vbYes`	6
No	`vbNo`	7

Tables 1-8 to 1-11 describe the VBA constant and values used by the `Buttons` argument of the `MsgBox( )` function.

Table 1-8. *Buttons Constant and Values Displayed by the VBA MsgBox( ) Function*

Buttons Displayed	Constant Name	Constant Value
OK button only	`vbOKOnly`	0
OK and Cancel	`vbOKCance`	1
Abort, Retry, and Ignore	`vbAbortRetryIgnore`	2
Yes, No, and Cancel	`vbYesNoCancel`	3
Yes and No	`vbYesNo`	4
Retry and Cancel	`vbRetryCancel`	5

Table 1-9. *Icon Constants and Values Displayed by the VBA MsgBox( ) Function*

Icon Displayed	Constant Name	Constant Value
⊗	`vbCritical`	16
?	`vbQuestion`	32
!	`vbExclamation`	48
i	`vbInformation`	64

Table 1-10. *Default Selected Button (from Left to Right) Constants and Values on the Dialog Box Displayed the VBA MsgBox( ) Function*

Default Selected Button	Buttons Displayed	Constant Value
First button is default	`vbDefaultButton1`	0
Second button is default	`vbDefaultButton2`	256
Third button is default	`vbDefaultButton3`	512

Table 1-11. *Application Modal Constants and Values for the Dialog Box Displayed by the VBA MsgBox( ) Function*

Modality of the Dialog Box	Constant Name	Constant Value
Modal dialog box. The user must respond to the message box before continuing work in the current application.	`vbApplicationModal`	0
System modal dialog box. All applications are suspended until the user responds to the message box.	`vbSystemModal`	4096

Let's try a simple `MsgBox( )` function with just a "Hello World" message and the OK button. In the VBA Immediate window, type this code:

```
MsgBox "Hello World"
```

When you press Enter, VBA will show the dialog box you see in Figure 1-44 and will discard the code returned by the `MsgBox( )` function, because you did not type ? (the print character) in the Immediate window before the instruction.

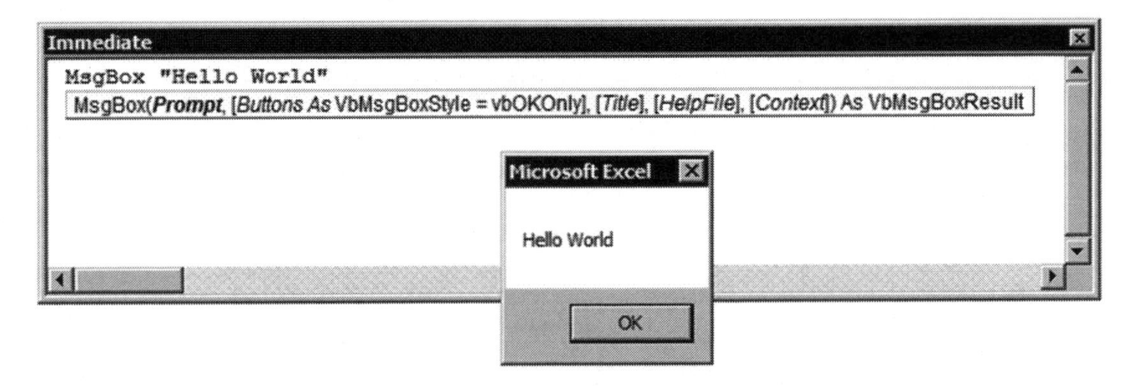

Figure 1-44. *Use the VBA Immediate window to test the MsgBox( ) function. If you do not type the ? character before the function name or surround the arguments with parentheses, the return value will be discarded*

Although it seems too difficult to dominate the VBA `MsgBox( )` function with the many different possible constants, it isn't. You can take advantage of VBA constant lists and easily select the buttons, icons, default button, and modality of the message box to be displayed while writing your code. Do not forget that you must select just one constant of each possible type (button, icon, default button, and modality).

Try the next exercise in the VBA Immediate window to show a dialog box with the Critical icon and three buttons: Abort, Retry, and Ignore:

1. Type `?MsgBox(` in the Immediate window (VBA will display the argument list for the function).

2. Type the `Prompt` argument between double quotes (like "Message string"), press the comma key, and watch VBA offer all possible constants for the `Buttons` argument. Select the first constant, `vbAbortRetryIgnore`, meaning that the dialog box will show three buttons: Abort, Retry, and Ignore. The `MsgBox( )` function call will become the following:

   ```
   ?MsgBox("Message string", vbAbortRetryIgnore)
   ```

3. After selecting the `vbAbortRetryIgnore` constant, press the + character and select the `vbCritical` constant to show the associated Critical icon. The `MsgBox( )` function call will become the following:

   ```
   ?MsgBox("Message string", vbAbortRetryIgnore+vbCritical)
   ```

4. Having select `vbAbortRetryIgnore+vbCritical` constants, press again the + character and select the `vbDefaultButton2` constant, meaning that the default button will be Retry.

   ```
   ?MsgBox("Message string", vbAbortRetryIgnore+vbCritical+vbDefaultButton2
   ```

5. Press another comma key and type the `Title` argument string between double quotes (the title of the dialog box). Press a closing parenthesis to finish the `MsgBox( )` function.

   ```
   ?MsgBox("Message string", vbAbortRetryIgnore+vbCritical+vbDefaultButton2,
   "Dialog box title")
   ```

6. Press Enter to generate the desired dialog box.
7. Note that the Ignore button (second dialog box button) is selected by default. Press Enter or click the Abort or Ignore button and watch the constant value returned by the VBA `MsgBox( )` function in the Immediate window (Figure 1-45).

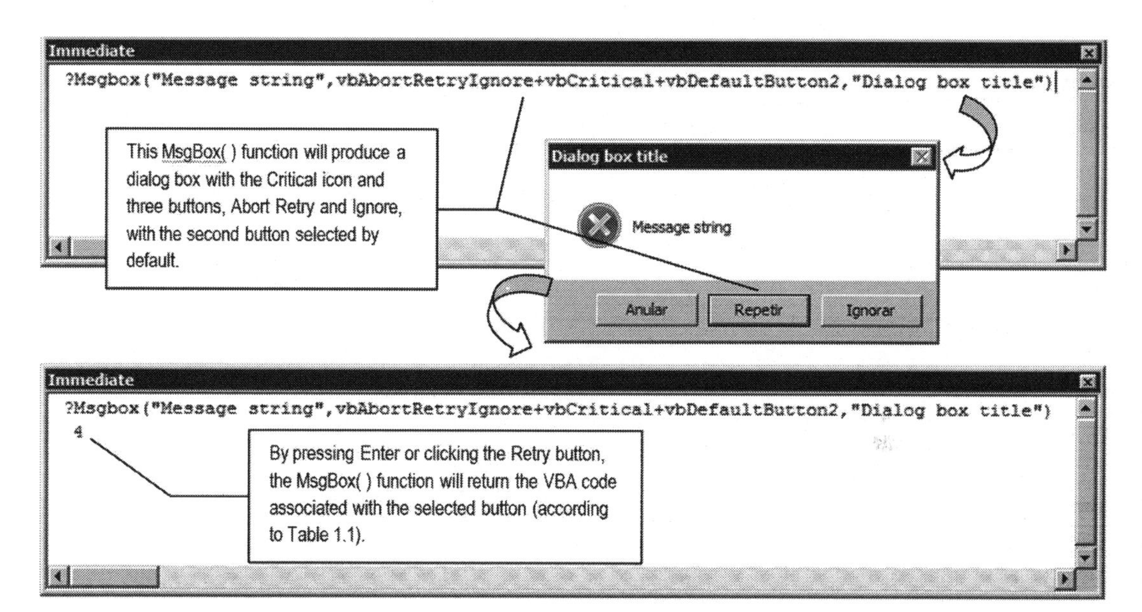

Figure 1-45. *Using the VBA Immediate window to test the* `MsgBox( )` *function, you can take advantage of the VBA quick list feature to select the constant you want for the button, icon, default button, and modality of the dialog box that will be displayed*

To control the message string format granted by the `MsgBox( ) Prompt` argument, you must use the VBA constant `vbCrLf` (meaning carriage return and line feeding) to break the message into different lines.

The file `MsgBox.xlsm` Excel macro-enabled workbook that you can extract from the `Chapter01.zip` file has this code inserted on the `Workbook_Open` event procedure of the `ThisWorkbook` object, meaning that it will fire every time the workbook is opened:

```
Private Sub Workbook_Open()
    Dim strMsg As String
    Dim strTitle As String

    strMsg = "File MsgBox.xlsm" & vbCrLf
    strMsg = strMsg & "ThisWorkbook object" & vbCrLf
    strMsg = strMsg & "Workbook_Open Event has fired"
    strTitle = "MsgBox( ) function test"
    MsgBox strMsg, vbInformation, strTitle
End Sub
```

Note that the Workbook_Open event code declares two string variables: strMsg to hold the message box string and strTitle to hold the title string. It then uses the VBA constant vbCrLf to insert two line breaks on the strMsg variable containing the message string. Also note that it successively concatenates strMsg to the previous value to compose the three-line string shown by MsgBox( ) (bold in the next code):

```
strMsg = "File MsgBox.xlsm" & vbCrLf
strMsg = strMsg & "ThisWorkbook object" & vbCrLf
strMsg = strMsg & "Workbook_Open Event has fired"
```

The dialog box you receive every time the MsgBox.xlsm workbook is opened is shown in Figure 1-46.

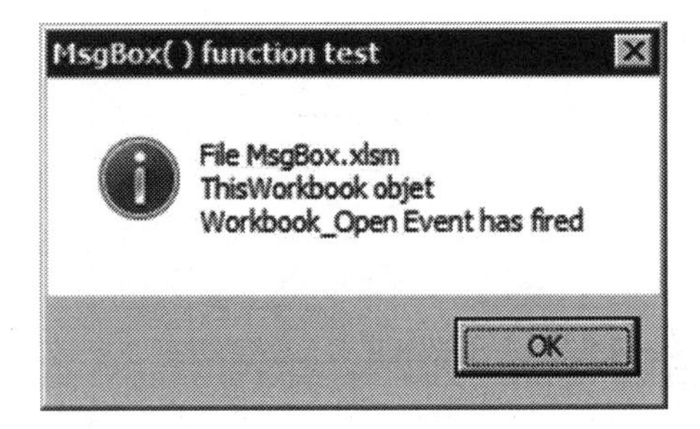

Figure 1-46. *This is the dialog box created by the MsgBox( ) function inserted on the Workbook_Open event procedure of the ThisWorkbook object, which will appear every time the workbook is opened*

Using InputBox

The VBA InputBox( ) function is used to collect user data, allowing the user to type a value that will be used on your code procedure. It has this syntax:

```
InputBox(prompt[, title] [, default] [, xpos] [, ypos] [, helpfile, context])
```

In this code:

Prompt: This is required; it is the message that will be shown by the dialog box, with a maximum length of about 1,024 characters, depending on the width of the characters used. To break the Prompt message into lines, use the VBA constant VbCrLf, which consists of a carriage return character and line feed character.

Title: This is optional; it is the expression displayed on the title bar of the dialog box. If you omit the title, the application name will be placed on the title bar.

DefaultResponse: This is optional; it is a string expression representing the default value that the dialog box will show, if no other input is provided. By omitting this value, the displayed text box will be empty.

Xpos: This is optional; it is a numeric expression that specifies the distance (in pixels), of the left edge of the dialog box from the left edge of the screen.

Ypos: This is optional; it is a numeric expression that specifies the distance (in pixels), of the top edge of the dialog box from the top edge of the screen.

HelpFileContext: This is the number of the Help Context page that will be shown if the user presses the F1 key (if any).

■ **Attention** If you omit XPos and YPos, the dialog box will be centered on the screen.

The InputBox( ) function always returns a string value. It will show a dialog box with just a text box with two buttons, OK and Cancel, where the user must type the required value or press Enter to accept the DefaultResponse argument (if supplied). If the user presses Cancel or if the text box is empty when the user presses Enter (or clicks OK), InputBox( ) will return an empty string (represented by two successive double quotes). Any other value will be returned as a string and must be treated as well.

As with the MsgBox( ) function, you can use the VBA Immediate window to test how InputBox( ) performs. Since InputBox( ) is used to return a value, you must always type the ? character on the Immediate window to watch the value it returns. The next code will exhibit an InputBox( ) asking for the user to type a numeric value (Figure 1-47):

```
?InputBox("Please type a numeric value", "Amount required")
```

Note in Figure 1-47 that 2300 was typed in the InputBox( ) text box, and this value was returned by the function to the Immediate window.

When you ask the user to input using the InputBox( ) function, you must validate whether something has been typed (since the user can click the OK button without typing anything or click the Cancel button; both situations returning an empty string), whether the data has the expected type (like a date or numeric value), or whether it is inside the expected range (for numeric values).

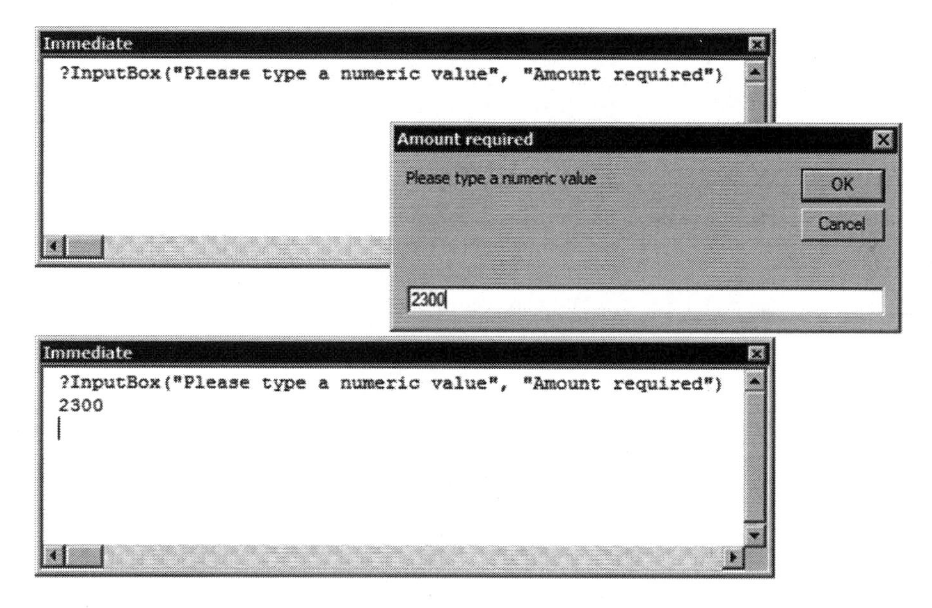

Figure 1-47. *Use the VBA Immediate window to test the InputBox( ) function. The value you type in the text box will be returned by the function*

The InputBox.xlsm Excel macro-enabled workbook that you can extract from the Chapter01.zip file has the following code inserted on the Workbook_Open( ) event procedure of the ThisWorkbook object (meaning that it will fire every time the workbook is opened) to ask the user to type his birthdate):

```
Private Sub Workbook_Open()
    Dim strMsg As String
    Dim strTitle As String
    Dim strDate As String

    strTitle = "Birthday date needed!"
    strMsg = "Please, type your Birhday date:"
    strDate = InputBox(strMsg, strTitle)
    If Len(strDate) Then
        If IsDate(strDate) Then
            strMsg = "You have now " & AgeInYearMonth(CVDate(strDate)) & "!"
            strTitle = "Welcome to MyApplication!"
            MsgBox strMsg, vbInformation, strTitle
        End If
    End If
End Sub
```

Whenever the `InpuBox.xlsm` workbook is opened, the `Workbook_Open` event fires and a `InputBox( )` asks for the user's birthdate.

```
    strTitle = "Birthday date needed!"
    strMsg = "Please, type your Birhday date:"
    strDate = InputBox(strMsg, strTitle)
```

The user answer to the input box and the next line of code verify whether something has been typed by the user, using the VBA `Len( )` function to verify the length of the string returned:

```
    If Len(strDate) Then
```

The VBA `Len( )` function returns the number of characters any string variable has. Since the VBA zero means `False` and anything else means `True`, if nothing had been typed or if the user clicked the Cancel button, `strDate` will be a zero-length string, `Len(strDate)` will return zero (0) or `False`, and the `If Len(strDate)… End If` structure will end the `Workbook_Open` event.

```
        If Len(strDate) Then
            ...
        End If
    End If
End Sub
```

If strDate is a zero length string, Len(strDate) will return zero, which also means false. In this case, the code inside the If… End If statement will not be executed

But if the user types anything, `Len(strDate)` will return the length of the string typed: the `Len(strDate)` test will evaluate to `True`, and the code inside the `If Len(strDate)… End If` structure will be executed, using the VBA `IsDate( )` function to verify whether what the user typed is a valid date of birth.

```
If Len(strDate) Then
    If IsDate(strDate) Then
```

Again, if what the user type is not a valid date, the `If IsDate(strDate)… End If` structure will end the `Workbook_Open` event, but if it is a valid date, the code will use the `strMsg` variable to create the message string that will be returned by the `MsgBox( )` function, using the `AgeInYearMonth( )` function (created earlier in this chapter) from the basAgeInYears module to calculate the age in year and months for the date inserted in the `strDate` variable, taking care to first convert the `strDate` string to a valid date value using the VBA `CVDate( )` function.

```
If IsDate(strDate) Then
    strMsg = "You have now " & AgeInYearMonth(CVDate(strDate)) & "!"
    strTitle = "Welcome to MyApplication!"
    MsgBox strMsg, vbInformation, strTitle
```

Figure 1-48 shows what happens when you open the InputBox.xlsm workbook and type a valid date of birth in request to the InputBox( ) function. Please note that although all VBA variables had been correctly initiated by the AgeInYearMonth( ) function procedure, cell A2, B2, and C2 formulas will not be automatically updated. You must edit each one so their true value will be returned.

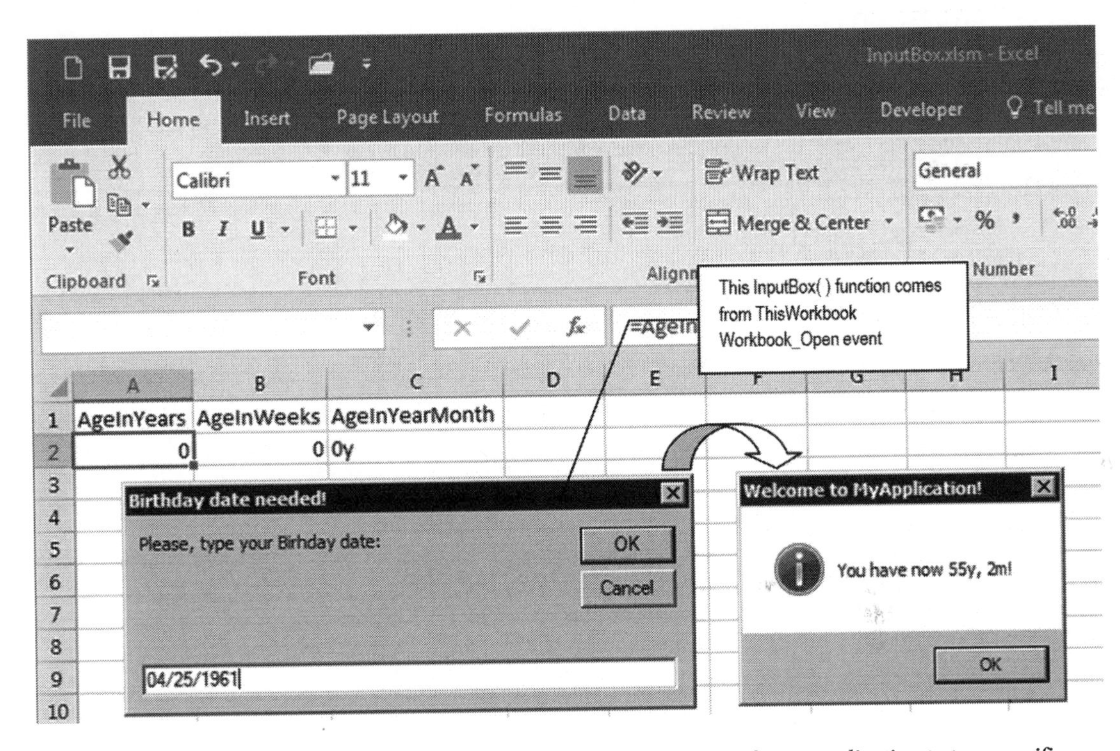

Figure 1-48. *Using the VBA InputBox( ) function, you may ask the user of your application to type specific data that must be validated by the procedure code, before it gets used inside your workbook application. Note that although Sheet1 has formulas using AgeInYears( ), AgeInWeeks( ), and AgeInYearMonth( ) functions, they are not updated after you type the values. You must edit each cell so the values are updated by Excel*

Dealing with VBA Errors

Let's talk about errors that may happen on your code. They may come from many sources: bad code logic, impossible operations (like zero division), and a handful of unexpected errors.

The question here is that VBA will also break the code whenever an executable error is found on your code, selecting the instruction that generates the error on the code module.

Since this is an issue that has its own chapter in most VBA programming books, let's review what you can do to avoid unexpected errors:

1. Use the `On Error Resume Next` instruction.
2. Create an error trap.

The On Error Resume Next Instruction

Whenever an unexpected error happens in your code, VBA will issue a `MsgBox( )` warning indicating the error number and the message associated with it. The last generated error will be associated with the `Err` object, and the error message can be retrieved by the `Error(Err)` function, were `Err` represents the last error code.

Whenever you put the `On Error Resume Next` instruction in your code, all instructions after it may generate an error, but VBA will ignore the errors, giving no message to the user. As the `On Error Resume Next` instructions imply, when an error is encountered by VBA, it will ignore and resume the code on the next instruction.

This seems great, doesn't it? But it is not. You have to use this instruction carefully, especially on such occasions that intermediate calculations are expected inside the code. If some intermediate calculation step were missed because of an `On Error Resume Next` instruction, the result may completely fail, returning no result. Worse, it may return a wrong result constituting what is called a *code bug*.

Sometimes you need to make a code instruction that may generate an error (such as trying to access a range name that should exist but doesn't) and verify whether the code generates an error, like `Function CheckRange( )` code:

```
Function CheckRange(strRange as string) as Boolean
    Dim varValue as Variant

    On Error Resume Next

    'Try to get Range value
    varValue = Range(sstrRange).Value
    CheckRange = (Err = 0)
End Function
```

This simple function returns `True` or `False` according to the `strRange` existence. If it does not exist, VBA will generate an error that will be ignored while the error code is stored on the `Err` object. `CheckRange = (Err = 0)` is a logical test, which returns `True` whenever the range exists (`Err = 0`, no error found) and `False` otherwise.

▒ **Attention** To reenable VBA errors issued inside a code procedure after an `On Error Resume Next` instruction is executed, use the `On Error Goto 0` (zero) statement.

Setting an Error Trap

Instead of disabling VBA errors in your code, you may want to set up an error trap to specify where the code should continue whenever an unexpected error is raised.

The error trap is defined by the `On Error GoTo <label>` instruction, where `<label>` is a code line that has a word with no spaces and ends with a ":" character, which specifies the beginning of the error trap. The error is then analyzed and eventually treated, and the code must continue on the same line code where it was generated using a `Resume` instruction, on the next line of code using a `Resume Next` instruction, or may be sent to another label using a `Resume <label>` instruction.

The next Sub TestErrorTrap( ) procedure implements an error trap using a Select Case statement to control any possible error code that may arise (it just implements the Else clause for all possible errors):

```
Sub TestErrorTrap( )
    On Error GoTo TestErrorTrap_Error
    ...
    <Procedure instructions>
    ...
TestErrorTrap_End: 'label that sets the procedure exit door
    Exit Sub
TestErrorTrap_Error: 'label that sets the Error trap
    Select Case Err
        Case Else
            MsgBox "Error "" & Err & ": "& Error(Err)
    End Select
    Resume TestErrorTrap_End
End Sub
```

The error trap sends any error to the Test_Error label instruction

A MsgBox( ) warns about the error code and text, and send the code to the Test_End label

This error trap will deviate the code to the TestErrorTrap_Error: label whenever an error is raised. The error will be shown by a VBA MsgBox( ) function using the "Error <error code>: Error message" style and then will continue on the procedure TestErrorTrap_End: label, which will exit the code. If no error happens, the code will end by executing the Exit Sub statement right before the error trap.

The error trap will be used in some procedures of this book, and you will be guided step-by-step, whenever it happens.

Protecting Your VBA Code

Once you have produced solid, trustworthy code, sometimes you may want to protect your intellectual property by keeping the VBA code of your application from being accessed and eventually changed by other people.

To protect your VBA code, use the Tools ➤ VBAProject Properties menu command, which will show the VBAProject - Project Properties dialog box. This dialog box has two tabs: General and Protection (Figure 1-49).

By default, every application that implements VBA attributes the name VBAProject to the project code (hence the menu command VBAProject Properties). You can give another name to your VBA project and provide a description (this information will appear in the VBA Object Browser window, and if the Excel application is saved as an Excel add-in, it will also appear on Excel Add-In dialog box). Once you change the VBA project name, it will appear on top of the Project Explorer tree and change the menu command inside the Tools menu.

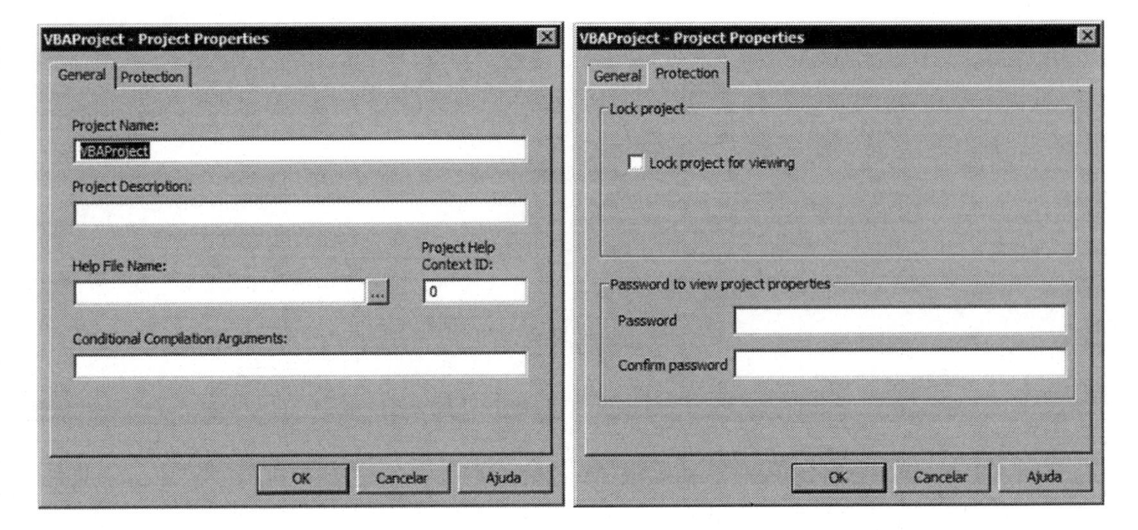

Figure 1-49. *This is the VBAProject - Project Properties dialog box, where you can give a name and description to your VBA code, lock a project for viewing, and assign it password protection*

To protect your VBA code, just click the Protection tab, check the "Lock project for viewing" option, and type and confirm password protection.

From now on, the protected VBA project will ask for this password to allow access to the content.

■ **Attention** There is no absolute protection on the digital word, meaning that breaking password code protection is just a matter of time and persistence. A simple Google search will show many tutorials about how to break a VBA project password.

Conclusion

Visual Basic for Applications, despite the word Basic in its name, is not basic in any sense. It is a complete computer programming language that touches on some important object-oriented programming principles, offering a full set of data types, instructions, statements, and functions that you can use to control the Microsoft Excel object model and any other program that adopts it.

It has a steep learning curve for first-time users. It is based on code modules that may be independent or attached to an object, where you write `Private` and `Public Sub`, `Function`, and `Properties` procedures to execute a defined set of instructions in successive order. It uses declared variable names of many different data types to store intermediate results that are manipulated by the code and has a full set of language structures to allow you to make decisions inside the code procedure.

Summary

In this chapter, you learned about the following:

- How VBA code works and why you should learn it to create great interfaces to interact with the users of your Excel application
- What VBA procedures are and the difference between `Function` and Sub procedures

- That VBA procedures can receive one or more arguments that are the values from where the code will operate
- That just VBA `Function` procedures can return a value
- That every procedure argument is a variable that can receive different types of values
- That some procedure arguments can be declared as `Optional` and, as such, are not obliged to be passed to the procedure
- That you must declare any VBA variable using a specific data type to make better use of your computer memory
- That just the `Variant` variable data type can receive `Null` or `Empty` values
- That variables have a scope and a lifetime (`Static`, `Private`, and `Public` variables will keep its values between procedure calls)
- That you must use a naming convention as good programming practice to easily recognize any variable type by just reading it name inside the procedure code
- That you can use `Property` procedures to validate `Private` variable values
- That VBA objects expose `Event` procedures, a special type of procedure that fires against some user actions on an Excel interface
- That you can use VBA `UserForm` objects to create a dialog box to interact with the user of your application data
- How you can insert a `Button` control on any worksheet to execute any `Public` procedure
- How to use the VBA `MsgBox( )` function to show a message to the user of your application and eventually ask the user to make a decision by clicking a button
- How you can control some specifications of the `MsgBox( )` function, like the text string, the icon, the buttons, and which button is selected by default
- How to use the VBA `InputBox( )` function to ask the user to type specific information needed by your application
- How to validate the information returned by the `InputBox( )` function before it gets used by other procedures of your VBA project
- How to protect your VBA code with a password

In the next chapter, you will learn how to use VBA to interact with the top-level object on a Microsoft Excel object model: the `Application` object.

CHAPTER 2

Programming the Microsoft Excel Application Object

Microsoft Excel exposes a set of objects that you must master to interact with the Excel application interface. In this chapter, you will start to master VBA programming by seeing practical examples. You can obtain all the procedure code in this chapter by downloading the `Chapter02.zip` file from the book's Apress.com product page, located at `www.apress.com/9781484222041`, or from `http://ProgrammingExcelWithVBA.4shared.com`.

The Microsoft Excel Object Model

Excel offers a lot of different objects in its application programmable interface; it is a large object model. Figure 2-1 shows one of the most simplified views of the Excel object model found on the Office Dev Center of the Social MSDN web site; it illustrates the main Excel objects. (You can find images with much more complexity and precision by searching Google using the words *Excel Object Model Diagram*.)

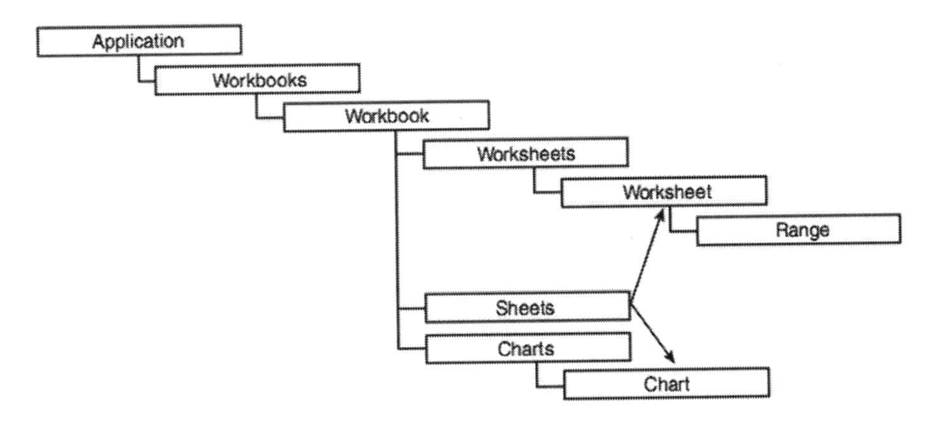

Figure 2-1. *This is a simplified vision of the Microsoft Excel object model diagram, showing the main objects and the interdependent relationships*

F. Morgado, *Programming Excel with VBA*, DOI 10.1007/978-1-4842-2205-8_2

This object model shows that everything starts with the Application object, which represents Excel itself. The next object is the Workbooks collection, where you can find a programmable reference for each opened Excel file as a single Workbook object.

Since each Excel workbook consists of different worksheets, each Workbook object has its own Worksheets collection, which keeps a programmable reference for each sheet tab the Excel file contains using independent Worksheet objects to represent them. Since each worksheet has a collection of cells, they are represented by the Range object.

Every Microsoft Excel chart can be displayed in two different ways: floating over any worksheet cells or inside its own Sheet tab. That is why the Excel object model also shows the Sheets and Charts collections, which contain references to both the Sheets and Charts tabs.

Confusing, isn't it? Welcome to the world of object model programming!

The Application Object

As mentioned, the main Excel object is the Application object, which represents the Microsoft Excel interface window. You use the Application object to perform some specific programmable operations in the Excel interface, such as the following:

- *Get access to some important Excel objects*: Access the active workbook, active worksheet, and what is currently selected in the Excel interface.
- *Interact with Excel behavior*: Set Excel screen updating on/off; fire events; set calculation options; manipulate the Excel status bar text; set window visibility, size, and position; and hide or show certain objects of the Excel interface, like the Excel ribbon.
- *Use file access operations*: Open, save, and close workbooks, and change the file name and folder where a workbook must be saved.

The Excel Application object has a large programmable interface that, like any other programmable object, can be inspected using the VBA Immediate window or the VBA Object Browser. You can use the VBA window to inspect the properties, methods, and events of every programmable object you can access through VBA programming code.

To inspect the Application object interface, just type ?Application. (note the dot after Application) in the VBA Immediate window and scroll through the object programmable interface. To use the VBA Object Browser, press the F2 function key in the VBA environment and search for the Application object. The icons at the left of any interface member show whether the item is a property (hand pointing), a method (running green), or an event (yellow lightning) (Figure 2-2).

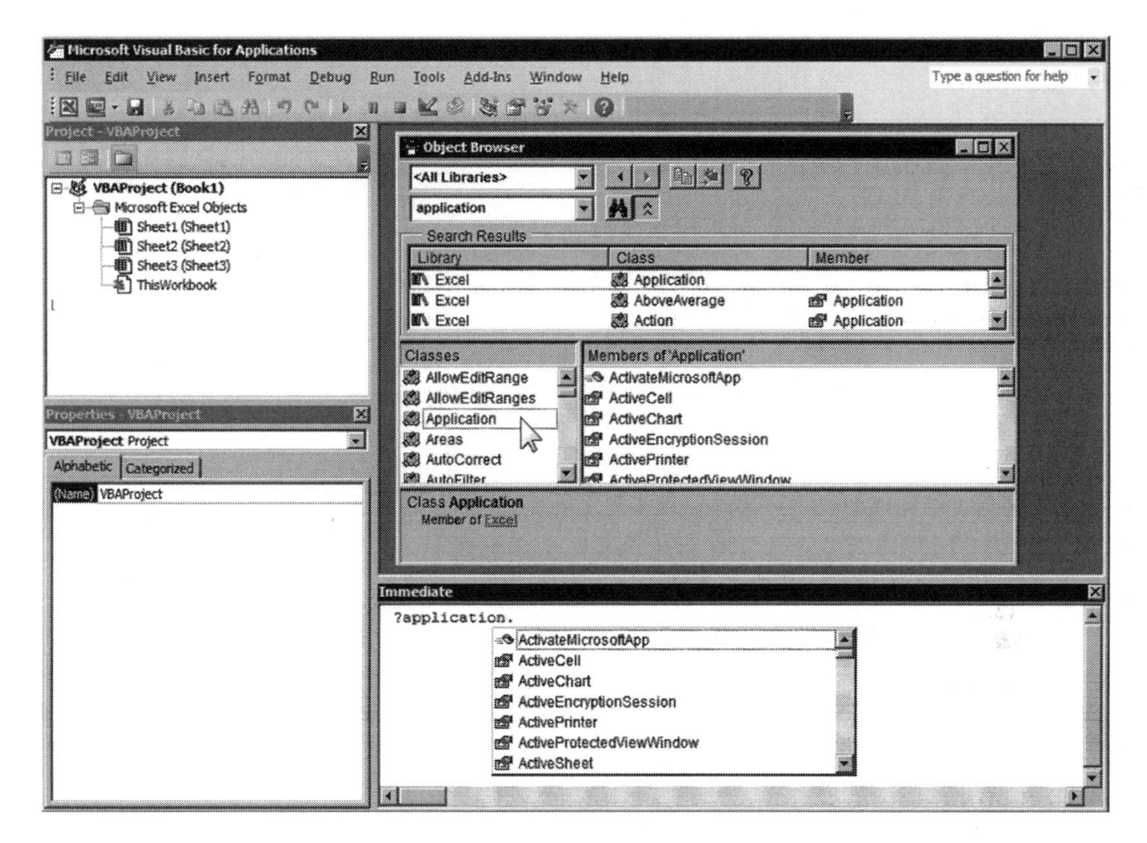

Figure 2-2. *Use the VBA Immediate window or the Object Browser (press F2 in the VBA interface) to inspect the programmable interface of any available object of the Excel object model*

To avoid wading through the vast Excel `Application` object interface, I will talk about some of the important objects, methods, and events that we will use in this book.

Table 2-1 lists some important `Application` object properties that return important Excel objects. Table 2-2 lists some important `Application` object properties that control the way the Excel window behaves while your VBA code executes. Table 2-3 lists some important `Application` object methods that you can use to perform specific Excel actions.

Table 2-1. *Some Important Microsoft Excel Application Object Properties That Return Important Workbook Objects*

Application Object Property	Value Returned	Represents
`ActiveCell`	Cell value	Value of the selected cell on the selected worksheet
`ActiveChart`	Chart object	Reference to the selected chart on the selected worksheet
`ActivePrinter`	String text	Name of the active printer
`ActiveSheet`	Sheet object	Reference to the active sheet on the selected workbook
`ActiveWindow`	Window object	Reference to the window of the current Excel application
`ActiveWorkbook`	Workbook object	Reference to the selected workbook
`Selection`	Object	Refers to the selected object in the active window, usually the value of the selected cell
`ThisCell`	Range object	The cell from which a user-defined function is called
`ThisWorkbook`	Workbook object	The workbook containing the current VBA code

Table 2-2. *Some Important Microsoft Excel Application Object Properties That Control the Excel Workbook*

Application Object Property	Value	Used To
`DisplayAlerts`	`Boolean`	Controls whether Excel displays alerts and messages while a macro is running
`Calculation`	`Numeric`	Returns or sets an `XlCalculation` value that represents the calculation mode
`DisplayScrollBars`	`Boolean`	Displays or hides Excel scroll bars for all workbooks
`DisplayStatusBar`	`Boolean`	Displays or hides the Excel status bar
`EnableEvents`	`Boolean`	Controls whether Excel events are enabled for the specified object
`ScreenUpdating`	`Boolean`	Turns on/off Excel screen updating

***Table 2-3.** Some Important Microsoft Excel Application Object Methods and the Actions They Perform*

Application Object Method	Action Performed
`ConvertFormula`	Converts an Excel formula format from R1C1 style to A1 style, and vice versa
`FileDialog`	Opens a File Open or Save As dialog box
`GetOpenFileName`	Shows the Excel Open dialog box where you select the folder and files to be opened
`GetSaveAsFileName`	Shows the Excel Save As dialog box to select the folder and file to save
`InputBox`	Shows Excel `InputBox`
`Intersect`	Returns a range object that is the intersection of one or more ranges
`OnKey`	Performs a specific `Public Function` procedure of a Standard module when a given key is pressed
`OnTime`	Performs a specific `Public Function` procedure on a specified time of the day
`SendKeys`	Send keys to the Excel interface as if they have been typed by the user
`Quit`	Quits the Excel interface
`Volatile`	When used inside a `Function` procedure, forces it to always recalculate

Since the `Application` object is the top member object in the Excel interface, to use any Excel `Application` object property or method, you don't need to explicitly reference the `Application` object in your code. Although if you reference it and insert the dot character, VBA will open a quick list with all the object interfaces, and your code will become easier to understand.

For example, Table 2-1 shows that the `Application` object `ActivePrinter` property returns some string text with the name of your default Windows printer. By using the VBA Immediate window, you can verify your default printer name using two different syntaxes (Figure 2-3), as shown here:

```
?ActivePrinter
?Application.ActivePrinter
```

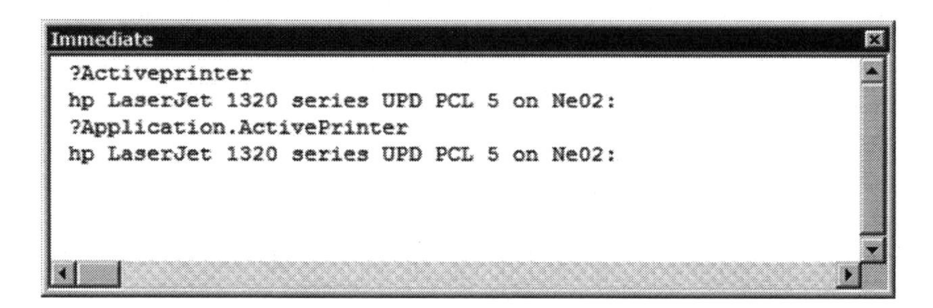

***Figure 2-3.** Since the Application object is the top-level object on the Excel object model, you don't need to reference it when you want to refer to any of its properties or methods. You can access any property or method with or without preceding it by Application*

Since many Application object properties cited in Table 2-1 do not return a value (they return specific Excel object pointers), you must explicit reference some property of this returned object to print its value in the VBA Immediate window. For example, the Application object ActiveCell property returns a Range object. Since a Range object returns a value, you can inspect the ActiveCell value in the VBA Immediate window, or you can specify any Range object property that returns a value to be printed. Since a Range object has an Address property, you can print the address of the current selected cells in the VBA Immediate window using these two different syntaxes:

```
?Application.ActiveCell.Address
```

or

```
?ActiveCell.Address
```

You can also navigate through the object model separating objects with a dot character (.) to print some useful properties you need to investigate while your code is running. For example, you can get the number of sheet tabs on the active workbook by exploring the ThisWorkbook object Worksheets collection Count property, as follows:

```
?ThisWorkbook.Worksheets.Count
```

Or you can print the number of rows of the third sheet tab (which has the same number of rows of every other sheet tab) using this syntax (Figure 2-4):

```
?ThisWorkbook.Worksheets(3).Rows.Count
```

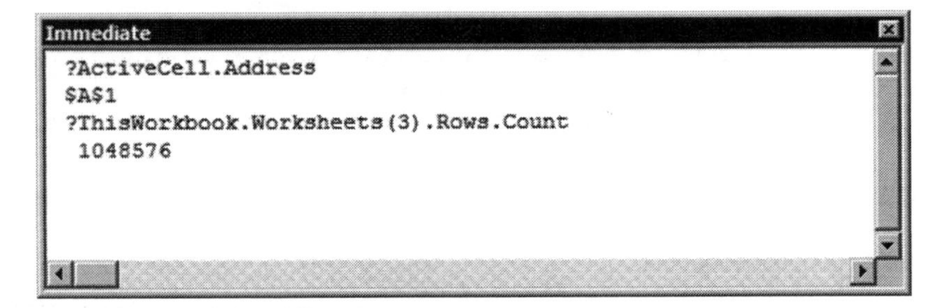

Figure 2-4. *If the Application object properties return an object reference instead of a value, you must use some of the object properties to inspect it in the VBA Immediate window, like the Rows collection Count property for the Worksheets collection of the Application object. You can avoid preceding the Worksheets collection by the Application object*

Using Application Properties to Control the Way the Excel Interface Behaves

Some Application object properties control the way Excel behaves, and you will use them in your VBA code to make your procedures run smoother, without disturbing the Excel environment while your code runs.

Table 2-2 shows some of these properties, like the Application.DisplayScrollBars property, which you can use to hide Excel scroll bars from every sheet tab, enabling the user to avoid scrolling your application interface through all the possible worksheet cells. To set any Table 2-2 properties, you must not use the VBA print character (?) in the Immediate window. You use the equal sign to define the property value.

For example, try to use this syntax in the VBA Immediate window to hide the Excel scroll bars:

```
Application.DisplayScrollBar = False
```

Just the Application.Calculation method needs to be strictly defined to a given calculation state to guarantee that Excel will not try to calculate the workbook while your code change cell values. To do this, you must set the Calculation property to the xlCalculationManual constant, execute your worksheet changes, and then set it again to xlCalculationAutomatic.

```
Application.Calculation =xlCalculationManual
 ...  'Code here
Application.Calculation = xlCalculationAutomatic
```

■ **Attention** Note that the Excel Application object also has a Calculate method, which forces the entire workbook to calculate, no matter which value the Calculation property has.

Most times when you execute the VBA code behind Excel, you notice that the Excel environment seems to do nothing while the code is running and doing different operations over one or more worksheets. Three different Excel properties must be disabled to guarantee that the Excel window remains unchangeable while your code is running: Calculation, EnableEvents, and ScreenUpdating.

To turn on/off such properties, create a simple VBA procedure that receives a Boolean argument and change these three properties at one time, turning them off before your code runs and turning them on again after your code finishes. The next code procedure can be used to *blind* the Excel interface for every change made by your code:

```
Public Sub CalculationEventsScreenUpdating ( bolEnabled As Boolean )
    Application.Calculation = IIf(bolEnable, xlCalculationAutomatic, xlCalculationManual)
    Application.EnableEvents = bolEnable
    Application.ScreenUpdating = bolEnable
End Sub
```

Note on the Sub CalculationEventsScreenUpdating( ) procedure that the bolEnabled argument (declared As Boolean as its three-letter prefix implies) is used to alternate the Calculation property using the VBA IIf( ) function (also called Immediate If) according to the value it receives. If bolEnabled is True, IIf( ) will return xlCalculationAutomatic; if it is False, it will return xlCalculationManual.

```
Application.Calculation = IIf(bolEnable, xlCalculationAutomatic, xlCalculationManual)
```

To use the Sub CalculationEventsScreenUpdating( ) procedure, you simply make a call for it at the beginning and at the end of your code procedure, in this way:

```
Sub Test( )
   Call CalculationEventsScreenUptdating(False)  'Turn off Calculation, Events and Screen
   updating
   ...  'Your code will run here
   Call CalculationEventsScreenUptdating(True)   'Turn on Calculation, Events and Screen
   updating
End Sub
```

▒ **Attention** As you will see later, if you set the Application.EnableEvents property to False, Excel will not fire any event while your code is running. If by any means your code breaks because of errors before you set the EnableEvents property to True, your Excel application will stop reacting to Excel events, as if it were frozen.

Using Application Methods to Show Excel File Dialogs

The Excel Application object also has useful methods to allow you to interact with the Windows and Office dialog boxes used to load and save files, with the Windows timer, and with your keyboard.

Three Excel Application object methods can be used to interact with folder and files. There is one low-level method, FileDialog, and two high-level methods, GetOpenFileName and GetSaveAsFileName, which are the simplest to use. They are a great way to introduce you to good VBA programming techniques.

Using the FileDialog Method

Among the three possible methods to show an Excel File Open or Save As dialog box, the Application.FileDialog method is the most complex. But it allows you to take more control of the dialog box, with this syntax:

```
Application.FileDialog(FileDialogType)
```

In this code:

- FileDialogType: This is an Excel VBA constant that can be set to any of the following:
  - msoFileDialogFilePicker: This allows the user to select a file.
  - msoFileDialogFolderPicker: This allows the user to select a folder.
  - msoFileDialogOpen: This creates an Open File dialog box.
  - msoFileDialogSaveAs: This creates a Save As file dialog box.

The Application.FileDialog method returns a FileDialog object, which has a defined interface of properties and methods, as stated in Table 2-4.

Table 2-4. *FileDialog Object Properties and Method*

Property	Type	Description
AllowMultiSelect	Property	Determines whether the user is allowed to select multiple files from a file dialog box
Application	Property	Returns an Application object that represents the container application for the object
ButtonName	Property	Returns or sets a String representing the text that is displayed on the action button of a file dialog box, after you select a file
Creator	Property	Returns a 32-bit integer that indicates the application in which the specified object was created
DialogType	Property	Returns an MsoFileDialogType constant representing the type of file dialog box that the FileDialog object is set to display
FilterIndex	Property	Returns or sets an Integer indicating the default file filter of a file dialog box
Filters	Property	Returns a FileDialogFilters collection
InitialFileName	Property	Returns or sets a String representing the path and/or file name that is initially displayed in a file dialog box
InitialView	Property	Returns or sets a MsoFileDialogView constant representing the initial presentation of files and folders in a file dialog box
Item	Property	Returns the text associated with an object
Parent	Property	Returns the Parent object for the specified object
SelectedItems	Property	Returns a SelectedItems collection with all selected files
Title	Property	Returns or sets the title of a file dialog box displayed using the FileDialog object
Show	Method	Show the desired dialog box with all set properties; returns True if any file/folder is selected and returns Cancel otherwise

▒ **Attention** There is no effective difference between the msoFileDialogFilePicker and msoFileDialogOpen constants. Both will show a dialog box to open one or more files, according to the value defined for the AllowMultiSelect property. The first constant will by default show "Browse" in the dialog box title bar, while the second will show "File Open" (if you do not change the Title property).

You can test the Application.FileDialog method by typing these commands in the VBA Immediate window, passing one of the FileDialogType constants to the method.

```
?Application.FileDialog(msoFileDialogOpen).Show
```

Figure 2-5 shows how the FileDialog object properties relate to interface items of the evoked dialog box.

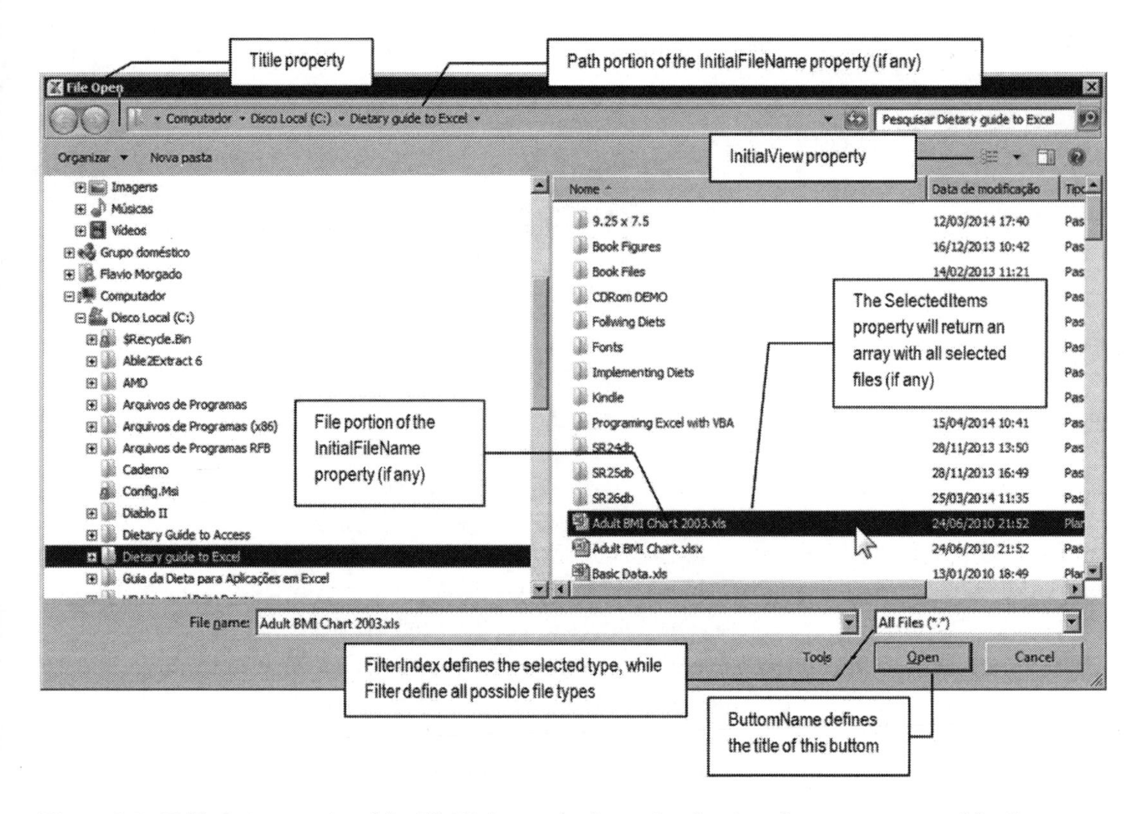

Figure 2-5. *Table 2-4 properties of the FileDialog method associated to interface components of the Open dialog box*

To use the `FileDialog` object, you must use a `With... End With` VBA construct to reference the `FileDialog` object just once, set any of it properties, and then evoke the Show method, as follows (note that inside the `With...End With` statement, you just use the dot before the Show method):

```
Public Sub ShowFileDialog()
   Dim fd as FileDialog
   Dim strFile as string

   Set fd = Application.FileDialog(msoFileDialogOpen)

   With fd
       '... Set the FileDialog properties here
      If .Show Then    'Evoke the dialog box Open file
         '.Show returns True
         'User selected a file.
         strFile  = .Selecteditems(1)
         'Process the file here
      Else
         ' .Show returns False
         'User click Cancel or close the dialog box selecting anything
      End If
   End With
End Sub
```

Note in the previous code that if the user select one file or folder, its name will be returned to the strFile string variable by the first item (item 1) of its SelectedItems() collection.

By setting the AllowMultiSelect property to True, the user can select multiple files in the Open dialog box. You will process all selected files by looping through the SelectedItems( ) collection using a VBA For Each statement (you must declare a Variant variable since the For Each loop just works with Variant and Object variables).

```
Public Sub ShowFileDialog()
   Dim fd as FileDialog
   Dim varFile as Variant

   Set fd = Application.FileDialog(msoFileDialogOpen)

   With fd
       '... Set the FileDialog to allow the selection of multiple files
       . AllowMultiSelect = True
      If .Show Then '.Show evokes the dialog box Open file
         '.Show method returns True
         'User selected one or more files
         For Each varFile in .SelectedItems
            Debug.Print varFile 'Print in the Immediate window all selected files and its path
            'Process the selected file here
         Next
      Else
         ' .Show method returns False
         'User click Cancel or close the dialog box selecting anything
      End If
   End With
End Sub
```

To effectively use the Application object FileDialog method in Excel, your best bet is to implement a Private function that takes care of all Application.FileDialog arguments and creates Public Sub procedures that call these functions and can be easily associated to a ControlButton inserted in the Excel worksheet application interface.

The file Application FileDialog method.xslm that you can extract from the Chapter02.zip file uses this strategy. It implements on its basFileDialog standard module the Private Function ShowDialogBox( ) procedure, which is accessed from different Public Sub procedures of this module associated to ControlButton objects of the Sheet1 interface. Right-click any ControlButton and select the Assign Macro command in the context menu to show the Assign Macro dialog box, which displays the procedures associated to the selected button (if any) and all other Public Sub procedures available in your VBA project (Figure 2-6).

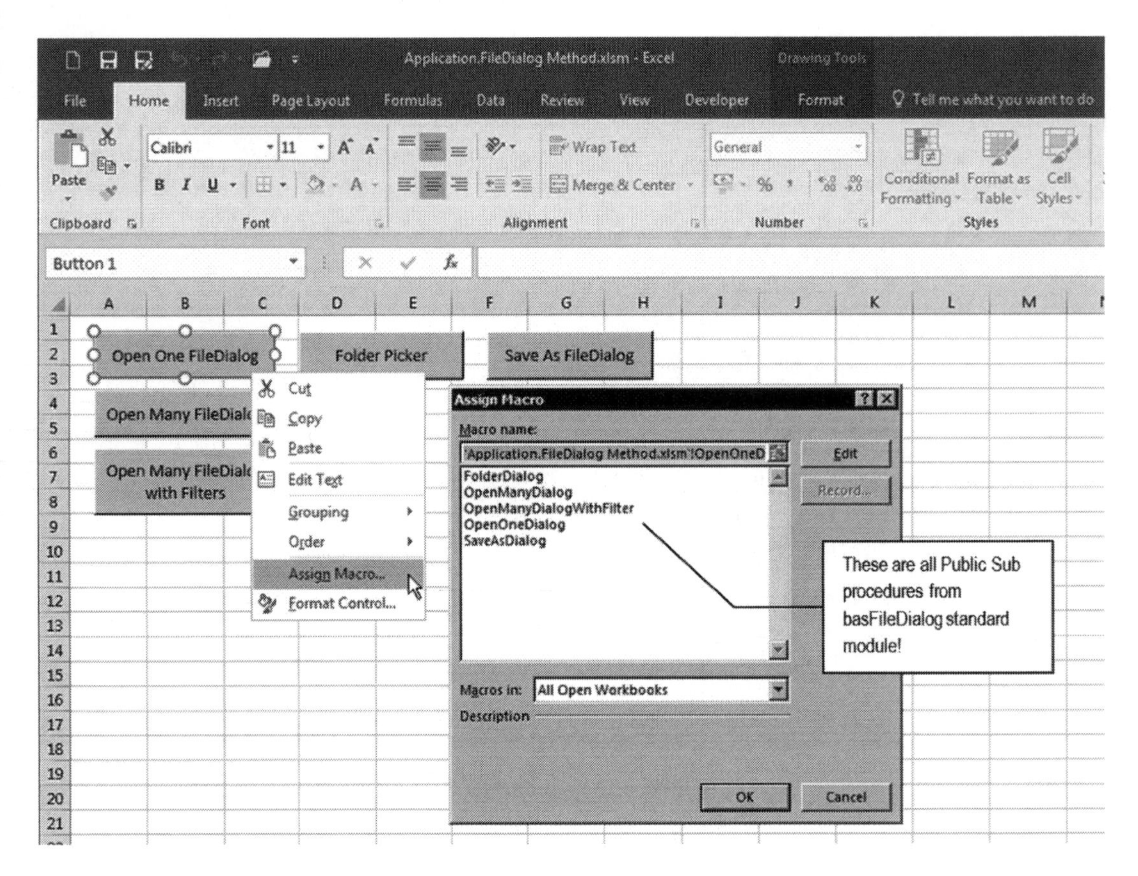

Figure 2-6. *The file Application FileDialog method.xlsm has Button controls associated with Public Sub procedures from basFileDialog, which makes a call to its Private Function ShowDialogBox( ) that uses the Application.FileDialog method*

Note that the Open One FileDialog `ControlButton` is associated with the `Public Sub OpenOneDialog( )` procedure, and if you click the Edit button in the Assign Macro dialog box, the VBA editor will promptly show this procedure code:

```
Public Sub OpenOneDialog()
    Dim strFile As String

    strFile = ShowDialogBox() & ""

    If Len(strFile) Then
        MsgBox "File selected: " & strFile, vbInformation, "1 File Selected!"
    End If
End Sub
```

Open One Single File

The Sub OpenOneDialog( ) procedure makes a single call to the ShowDialogBox( ) function without passing it any argument and associates the selected file returned by the function (if any) to the strFile string variable. The value returned by the ShowDialogBox( ) function is concatenated to an empty string ("") because it returns a Variant, which by default is set to Empty (the value returned if the user clicks the Cancel or Close button of the dialog box):

```
strFile = ShowDialogBox() & ""
```

The Private Function ShowDialogBox( ) from basFileDialog of the Application FileDialog method.xlsm Excel file executes this code:

```
Private Function ShowDialogBox(Optional DialogType As DialogType = OpenFile, _
                              Optional MultiSelect As Boolean, _
                              Optional Title As String, _
                              Optional ButtonName As String, _
                              Optional FileFilters As String, _
                              Optional FilterIndex As Integer = 2) As Variant

    'Purpose: Show a dialog box Open or Folder Picker
    'Optional arguments: DialogType = specify File or Folder selection
    '                                 MultiSelect = specify multi file selection
    '                                 Title = Title off the dialog box
    '                                 ButtonName = Title off the action button
    '                                 Filters = string with one filter or Array of filter strings
    '                                              Each file filter must contain filter
                                                   name, a colon, and filter extension
    '                                              Separate file filters with colon
    '                                 FilterIndex: file filter to be used
    'Returns: string with selected file/folder or array with selected files/folders

    Dim fd As FileDialog
    Dim varItem As Variant
    Dim varFiles() As Variant
    Dim strFllter As String
    Dim strExtension As String
    Dim intPos As Integer
    Dim intPos2 As Integer
    Dim intI As Integer

    'Define dialog type to Open dialog
    Set fd = Application.FileDialog(DialogType)

    With fd
        'Define some dialog properties
        .AllowMultiSelect = MultiSelect

        If Len(Title) Then
            .Title = Title
        End If
```

```
        If Len(ButtonName) Then
            .ButtonName = ButtonName
        End If

        If DialogType <> OpenFolder Then
            'Verifiy if FileFilters argument was passed to the procedure
            If Len(FileFilters) Then
                'Clear default Excel filters
                .Filters.Clear

                'FileFilters must be a string with one or more file filter
                ' using Comma as separating character
                Do
                    intPos2 = InStr(intPos + 1, FileFilters, ",")
                    strFllter = Mid(FileFilters, intPos + 1, intPos2 - intPos - 1)
                    intPos = InStr(intPos2 + 1, FileFilters, ",")
                    If intPos > 0 Then
                        strExtension = Mid(FileFilters, intPos2 + 1, intPos - intPos2 - 1)
                    Else
                        strExtension = Mid(FileFilters, intPos2 + 1)
                    End If
                    .Filters.Add strFllter, strExtension
                Loop Until intPos = 0
            End If

            If FilterIndex > .Filters.Count Then
                FilterIndex = 1
            End If
            .FilterIndex = FilterIndex
        End If

        If fd.Show Then
            If .SelectedItems.Count = 1 Then
                'Just one file was selected. Return it name!
                ShowDialogBox = .SelectedItems(1)
            Else
                'More than one file was selected
                'Fullfill an array of file names and return it!
                ReDim varFiles(.SelectedItems.Count - 1)
                For Each varItem In .SelectedItems
                    varFiles(intI) = varItem
                    intI = intI + 1
                Next
                ShowDialogBox = varFiles
            End If
        End If
    End With
End Function
```

Note that the Function ShowDialogBox( ) procedure may receive six optional arguments (DialogType, MultiSelect, Title, ButtonName, FileFilters, and FilterIndex) on the declaration.

```
Private Function ShowDialogBox(Optional DialogType As DialogType =
                              OpenFile, _
                              Optional MultiSelect As Boolean, _
                              Optional Title As String, _
                              Optional ButtonName As String, _
                              Optional FileFilters As String, _
                              Optional FilterIndex As Integer = 2) As Variant
```

The DialogType optional argument is declared as DialogType, a Private Enum declaration that uses the standard VBA constants associated with the FileDialogType values of the Application.FileDialog method, receiving by default the OpenFile value (which is associated to the msoFileDialogOpen constant).

```
Option Explicit

Private Enum DialogType
    OpenFile = msoFileDialogOpen
    OpenFolder = msoFileDialogFolderPicker
    SaveAsFile = msoFileDialogSaveAs
End Enum
```

Since the DialogType argument is defined by default as msoFileDialogOpen, when the function is called without any arguments, it shows an Open dialog box, associating it to the fd variable, declared as FileDialog type:

```
Private Function ShowDialogBox(Optional DialogType As DialogType = OpenFile, _
    ...
    Dim fd As FileDialog
    ...
    Set fd = Application.FileDialog(DialogType)
```

Once the desired FileDialog is associated to the fd variable, a With fd... End With structure is created so the code can make easier use of the fd object variable properties, verifying whether any value has passed to the MultiSelect, Title, and ButtonName procedure arguments and associating the value to the appropriate FileDialog arguments (note that there is a dot at the left of each property to indicate the association with the procedure argument, if received). The existence of the Title and ButtonName arguments is tested by the VBA Len( ) function, which returns the length of the argument text if it has been received. When the argument is omitted, Len( ) will return 0, which means False.

```
With fd
    'Define some dialog properties
    .AllowMultiSelect = MultiSelect

    If Len(Title) Then
        .Title = Title
    End If

    If Len(ButtonName) Then
        .ButtonName = ButtonName
    End If
```

■ **Attention** The ButtonName argument changes the OK button text when you select a file.

The ShowDialogProcedure( ) function then tests whether the FileFilters argument has been received, by verifying its length with the VBA Len( ) function. If it is true, it first clears the default Excel filters used by the FileDialog method.

```
'Verifiy if FileFilters argument was passed to the procedure
If Len(FileFilters) Then
    'Clear default Excel filters
    .Filters.Clear
```

The ShowDialogBox( ) function expects to receive on its FileFilters argument a string of file filters, where the filter names and filter extensions are separated by comma characters. Let's say you are trying to pass two file filters: one for .xlsx files and another to .xls files. In this case, you should create a string like this:

```
"Excel 2007-2010 files, *.xlsx, Excel 97-2003 files, *.xls"
```

The procedure must search the FileFilters string argument by each comma character, selecting the filter name (strFilter) and filter extension (strExtension) and adding them to the FileDialog.Filters collection, which requires this syntax:

```
fd.Filters.Add strFilter, strExtension
```

It does this by executing a VBA Do...Loop structure that persistently uses the VBA InStr( ) function to search for two successive commas inside the FileFilters string. The first comma defines the end of the file filter name, and the second comma defines the end of the file filter extension. To extract the filter name and filter extension from inside the FileFilters string, you must use the VBA Mid( ) function.

The VBA InStr( ) function performs a string search inside any string variable from an initial set position. It has this syntax:

```
InStr( [Start], String, Substring, [Compare] )
```

In this code:

Start: This is optional; this is a numeric expression, 1-based, that sets the starting point for the search. If omitted, the search begins at the first character.

String: This is required; it refers to the string variable or expression being searched.

Substring: This is required; it refers to the string expression sought.

Compare: This is optional; it specifies the type of string comparison. If omitted, the Option Compare setting of the code module determines the type of comparison. You can use the following:

Binary: This performs a binary comparison (taking into account capitalization).

Text: This is the default type, which performs a text comparison (disregarding capitalization).

The InStr( ) function returns the position of the first character of the Substring argument if Substring is found inside the String searched. It returns zero otherwise.

The VBA Mid( ) function extracts any substring from a string, starting at a defined position. It has this syntax:

```
Mid ( String, Start, Length )
```

In this code:

> String: This is required; it is the string variable or expression from which the characters must be returned.
>
> Start: This is required; it is a 1-based integer expression defining the starting position of the characters to return. If Start is greater than the number of characters in String, Mid( ) will return a zero-length string ("").
>
> Length: This is optional; it is an integer expression that indicates the number of characters to be returned. If omitted (or if there are fewer characters than Length characters in the text), all characters from the start position to the end of the string are returned.

The strategy used to consistently extract the file filter name and file extension from the FileFilters string argument is to use two integer variables (intPos and intPos2) that point to two successive commas inside the argument. If the FileFilters string has just one pair of filter name and file extension, there will be just one comma. If it has two pairs of filter names and file extensions, it will have three commas. The number of commas will be equal to n-1 regarding the count of each pair of filter name and filter extension.

As soon as the Do... Loop structure begins, the first comma is used to separate the first filter name, and its filter extension is defined by intPos2. Note that when the Do... Loop structure begins, intPos has the default zero value attributed to any numeric variable, so the search begins on intPos+1 = 1 or on the first character of the FileFilters variable.

```
'FileFilters must be a string with one or more file filter
' using Comma as separating character
Do
    intPos2 = InStr(intPos + 1, FileFilters, ",")
```

Supposing that the FileFilters string has at least one comma, intPos2 will find its position, which then will be used by the VBA Mid( ) function to extract the filter name from the FileFilters string and attribute it to the strFilter string.

```
    strFllter = Mid(FileFilters, intPos + 1, intPos2 - intPos - 1)
```

Note that Mid( ) begins its search on the position defined by the intPos+1 variable and extracts intPos2 - intPos1 - 1 characters from the FileFilters string. Figure 2-7 shows how the proposed FileFilters string will be operated on by these two last instructions.

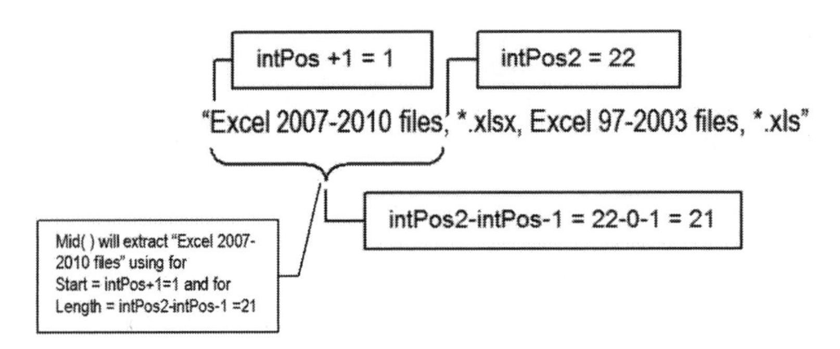

Figure 2-7. *When the Do...Loop structure begins, the first comma is found on character 22 and attributed to the intPos2 variable. The VBA Mid( ) function then uses intPos+1 as the first extraction point and intPos2-intPos-1 as the number of characters to be extracted from the FileFilters string variable*

To extract the file extension for the proposed file filter, the Do...Loop structure performs another `InStr( )` function to search for the next comma character, beginning at intPos2 +1 (the next character after the first found comma).

```
intPos = InStr(intPos2 + 1, FileFilters, ",")
```

If it finds the position of another comma character, it means that the `FileFilters` string probably has another pair of file filters and file extensions, and the `strExtension` variable receives the desired filter extension by using the `Mid( )` function again with the new `intPos` value (Figure 2-8).

```
If intPos > 0 Then
    strExtension = Mid(FileFilters, intPos2 + 1, intPos - intPos2 - 1)
```

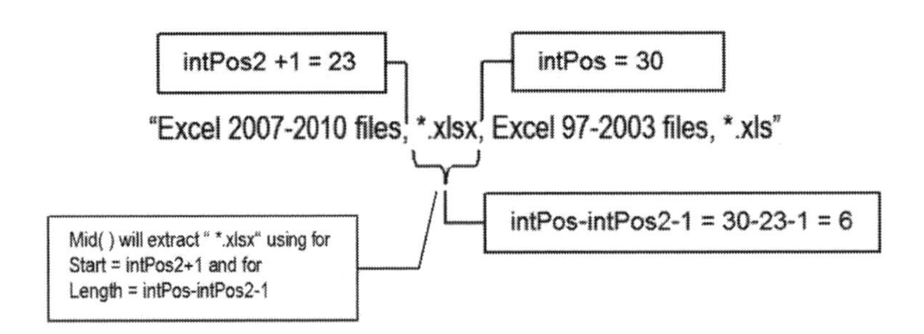

Figure 2-8. *By redefining the search starting point to intPos2 +1, the Do...Loop structure uses InStr( ) to search for the next comma character. If it is found, the Mid( ) function is used to extract the filter extension that is inside the two comma characters defined by the intPos2 and intPos variables*

Since both the `strFilter` and `strExtension` variables have now the filter name and filter extension, they are used to compose the first filter using the Add method of the `Filters` collection.

```
.Filters.Add strFllter, strExtension
```

The Do...Loop structure then verifies whether intPos =0 will define the end of the loop.

```
Loop Until intPos = 0
```

With intPos = 30, the loop begins again, now searching for the next comma beginning at intPos +1 = 31. This time intPos2 will receive character position 51, extracting the filter name from the next 19 characters (in this case, "Excel 97-2003 files," as shown in Figure 2-9).

```
Do
    intPos2 = InStr(intPos + 1, FileFilters, ",")
    strFllter = Mid(FileFilters, intPos + 1, intPos2 - intPos - 1)
```

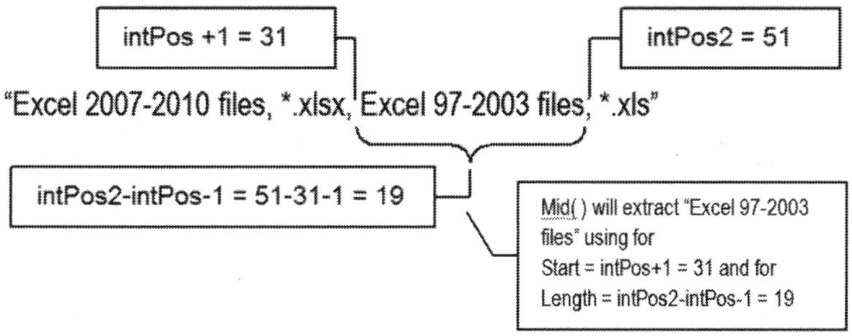

intPos +1 = 31

intPos2 = 51

"Excel 2007-2010 files, *.xlsx, Excel 97-2003 files, *.xls"

intPos2-intPos-1 = 51-31-1 = 19

Mid( ) will extract "Excel 97-2003 files" using for Start = intPos+1 = 31 and for Length = intPos2-intPos-1 = 19

Figure 2-9. *When the Do...Loop structure begins again, the third comma is found on character 51 and attributed again to the intPos2 variable. The VBA Mid( ) function then uses intPos + 1 = 31 as the first extraction point and intPos2 - intPos - 1 = 19 as the number of characters to be extracted from the FileFilters string variable*

But now, there is no other comma to find, and the next InStr( ) instruction will return zero to intPos. In this case, the If intPos>0 test will return False. The Then clause will be executed, using Mid( ) without the third argument (Length), meaning that all the text beginning from Start (inPos2+1=52) to the end of the FileFilters string must be extracted (the last file extension). Both strFile and strExtension will be added again to the Filters collection.

```
    intPos = InStr(intPos2 + 1, FileFilters, ",")
    If intPos > 0 Then
        strExtension = Mid(FileFilters, intPos2 + 1, intPos - intPos2 - 1)
    Else
        strExtension = Mid(FileFilters, intPos2 + 1)
    End If
    .Filters.Add strFllter, strExtension
```

Since intPos = 0, the Do...Loop test will return True, effectively ending the loop:

```
Loop Until intPos = 0
```

When the Do...Loop ends, the Filters collection Count property will have the number of file filters added by the loop. Since the ShowDialogBox( ) function procedure can receive the default file filter on the FilterIndex argument, the procedure verifies that the FilterIndex is greater than Filter.Count. If it is, it will be redefined to FilterIndex = 1 (the first filter added to the Filters collection).

```
            If FilterIndex > .Filters.Count Then
                FilterIndex = 1
            End If
            .FilterIndex = FilterIndex
        End If
```

▪ **Attention** By default, the Application.FileDialog method will show a lot of filters in the Open or Save dialog box, being that the first filter is All Files, with file extensions defined to *.*. The second default filter is Excel .xlsx files, and as such, it is defined by default to the FilterIndex argument on the ShowDialogBox( ) declaration.

The dialog box is then shown to the user using the Show method. Since the dialog box is modal, the procedure code will stop, waiting the user response.

```
        If fd.Show Then
```

When the dialog box is closed, the Show method will return True when a file is selected and will return False otherwise. If the user clicked Cancel or clicked the dialog box's Close button (at the top-right corner), the function ends, returning Empty (the default value for the Variant data type).

```
Private Function ShowDialogBox(Optional DialogType As DialogType = OpenFile, …) as Variant
    ...
        If fd.Show Then
         ...
        End If
    End With
End Function
```

If the user selected one or more files, you must explore the SelectedItems.Count property to verify how many files were selected. If the AllowMultiSelect property is defined to False or the user selected just one file, SelectedFiles.Count = 1, the ShowDialogBox( ) function must return the selected file name by referencing the first item of the SelectedFiles( ) collection.

```
            If .SelectedItems.Count = 1 Then
                'Just one file was selected. Return it name!
                ShowDialogBox = .SelectedItems(1)
            Else
                ...
            End If
        End If
    End With
End Function
```

If more than one file was selected, the ShowDialogBox( ) function will dimension the one-dimensional varFiles( ) array to be capable of receiving all file names and will fulfill it with the selected files. Since VBA arrays are zero-based (their first index is zero), it is dimensioned using the SelectedFiles.Count - 1 value.

```
If .SelectedItems.Count = 1 Then
    ...
Else
    'More than one file was selected
    'Fullfill an array of file names and return it!
    ReDim varFiles(.SelectedItems.Count - 1)
```

The procedure makes a For Each... Next loop through all files on the SelectedItems collection, adding each selected file name to the varFiles( ) array and using the intI variable to indicate the array index (note that intI has the default value of 0, being incremented on each loop passage).

```
For Each varItem In .SelectedItems
    varFiles(intI) = varItem
    intI = intI + 1
Next
```

When the array is fulfilled with all the selected file names, it is returned by the ShowDialogBox( ) procedure.

```
ShowDialogBox = varFiles
```

Open Many Files

You can now use VBA Break mode on the Public Sub OpenManyDialogWithFilter( ) procedure from basFileDialog of the Application FileDialog method.xlsm macro-enabled workbook file to go step-by-step through the procedure code, verifying how the ShowDialogBox( ) function procedure deals with the FileFilters argument and selecting many files. It has this code:

```
Public Sub OpenManyDialogWithFilter()
    Dim varFiles As Variant
    Dim varItem As Variant
    Dim strFilters As String
    Dim strTitle As String
    Dim strMsg As String

    strFilters = "Microsoft Excel 2007-2010,*.XLS?"
    strFilters = strFilters & ",Microsoft Excel 97-2003,*.XL?"

    varFiles = ShowDialogBox(OpenFile, True, "Open My Excel Files", "Which File?", strFilters, 1)

    If Not IsEmpty(varFiles) Then
        If IsArray(varFiles) Then
            strTitle = UBound(varFiles, 1) + 1 & " Files selected"
            strMsg = strTitle & vbCrLf
            For Each varItem In varFiles
                strMsg = strMsg & varItem & vbCrLf
```

```
            Next
        Else
            strTitle = "File selected"
            strMsg = "1 File selected:" & vbCrLf & varFiles
        End If
        MsgBox strMsg, vbInformation, strTitle
    End If
End Sub
```

When you click the Open Many FileDialog with Filters ControlButton of the Application FileDialog method.xlsm macro-enabled workbook, the procedure declares the variables, defines the filters to be used on the strFilters string variable, and returns the ShowDialogBox( ) function result to the varFiles Variant variable. Note its arguments DialogType = OpenFile, AllowMultiSelect = True, Title = "Open My Excel Files", ButtonTitle = "Which Files?", FileFilters = strFilters, and FilterIndex = 1):

```
Public Sub OpenManyDialogWithFilter()
    Dim varFiles As Variant
     ...
    strFilters = "Microsoft Excel 2007-2010,*.XLS?"
    strFilters = strFilters & ",Microsoft Excel 97-2003,*.XL?"

    varFiles = ShowDialogBox(OpenFile, True, "Open My Excel Files", "Which File?",
               strFilters, 1) & ""
```

When the dialog box closes, the user (you!) did one of these two actions:

- Clicked Cancel or clicked the Close dialog box button. In this case, the varFiles variable will continue with its default Empty value, IsEmpty(varFiles) = True and Not IsEmpty(varFiles) = False, and nothing happens:

```
    If Not IsEmpty(varFiles) Then
       ...
    End If
End Sub
```

- Selected one or more files and clicked the Which File? button. In this case, Not IsEmpty(varFiles) = True, and the procedure verifies how many files have been selected:
    - If two or more files have been selected, varFile has an array of file names, IsArray(varFiles) = True, and the procedure must process all files in the array returned by the VBA Ubound( ) function, which has this syntax:

      ```
      UBound ( Array, [Dimension])
      ```

      In this code:

      Array: This is required; it is the array variable in which you want to find the highest possible subscript of a dimension.

      Dimension: This is optional; it is an Integer 1-based dimension for which the highest possible subscript is to be returned (1 means the first dimension, 2 means the second, and so on). If omitted, 1 is assumed.

The VBA UBound( ) function returns the item count for the desired array dimension considering the array base count, which obeys the Option Base statement that defines the lower bound for array subscripts. If Option Base 1 is not specified, arrays declared are 0-based, and the value returned by Ubound( ) must be added to 1 to produce the effective item count and compose the message box title (for example, an array that Ubound( ) returns as 5 has six items, with array indexes from zero to five). A For Each... Next loop processes each array item (selected file).

```
If IsArray(varFiles) Then
    strTitle = UBound(varFiles, 1) + 1 & " Files selected"
    strMsg = strTitle & vbCrLf
    For Each varItem In varFiles
        strMsg = strMsg & varItem & vbCrLf
    Next
```

- If just one file was selected, varFiles holds just the string regarding the selected file. Note that the strMsg variable breaks the message into two lines using the VBA vbCrLf constant (Cr means "carriage return," and Lf means "line feed"):

```
If IsArray(varFiles) Then
    ...
Else
    strTitle = "File selected"
    strMsg = "1 File selected:" & vbCrLf & varFiles
End If
```

The MsgBox( ) function displays the file names selected ending with the Sub OpenManyDialogWithFilter() procedure (Figure 2-10):

```
        MsgBox strMsg, vbInformation, strTitle
    End If
End Sub
```

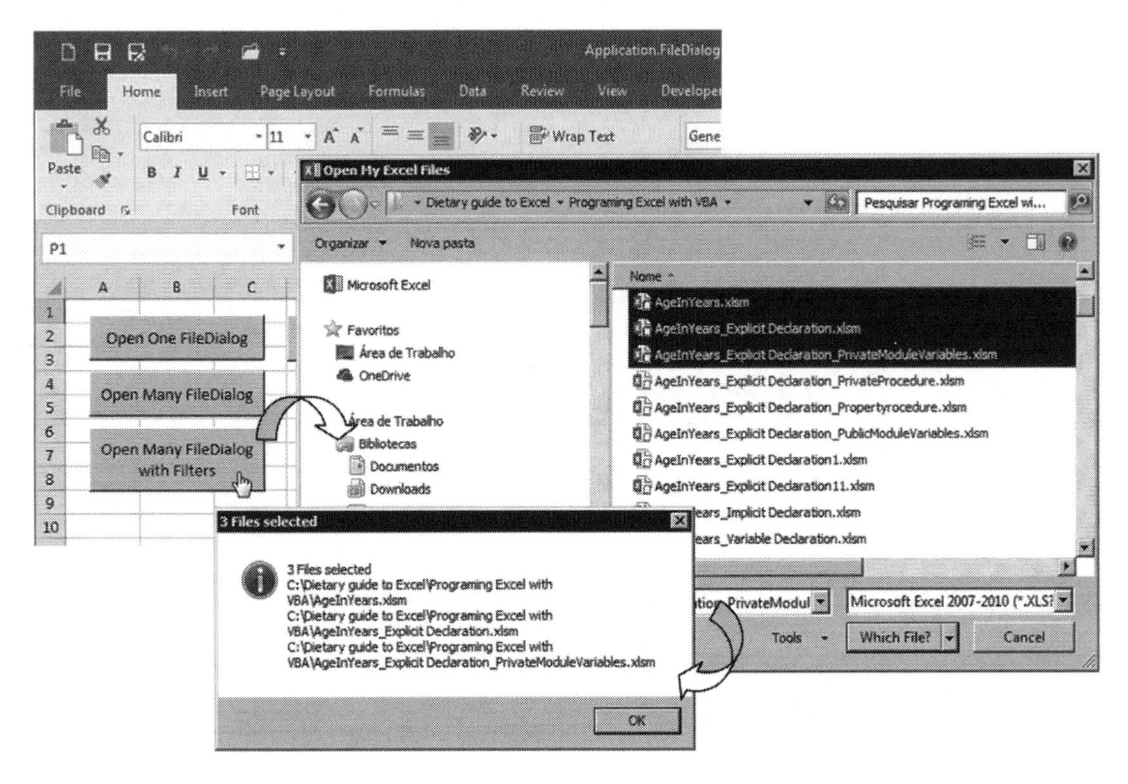

Figure 2-10. *When you click the Open Many FileDialog with the Filters Button control, the Public Sub OpenManyDialogWithFilter( ) procedure is executed and calls the Private ShowDialogBox( ) function, allowing the selection of one or more files, which are exhibited by a MsgBox( ) function*

You must study all other `Public Sub` procedures from `basFileDialog` of the `Application FileDialog method.xlsm` macro-enabled workbook to understand how they interact with the `ShowDialogBox( )` function to allow the selection of files and folders.

■ **Attention** By using the `OpenFolder` enumerator on the `DialogType` argument of the `ShowDialogBox( )` method, Excel will show a dialog box where just folders are shown and can be selected (try the Folder Picker control button and the associated `Sub FolderDialog( )` procedure).

Using the `SaveAsFile` enumerator, the `ShowDialogBox( )` function will show the Save As dialog box, which is essentially the same as the Open dialog box. The main difference between them is that if you select an existing file name on the Save As dialog box, the system will automatically warn you that the file already exists, and you will have an extra chance to cancel the operation.

Using the GetOpenFileName and GetSaveAsFileName Methods

The Application object's GetOpenFileName and GetSaveAsFileName methods work just like the ShowDialogBox( ) procedure, both returning a Variant value with the selected file or array of files, using a different order of optional arguments:

```
Application.GetOpenFilenName(FileFilter, FilterIndex, Title, ButtonText, MultiSelect)
Application.GetSaveAsFileName(InitialFileName, FileFilter, FilterIndex, Title, ButtonText)
```

In this code:

InitialFilename: This is optional. It is used just to the GetSaveAsFileName method and is an optional Variant data type to the default name of the file that will be saved.

FileFilter: This is optional. It is a string that specifies the pair of filter name and filter extension, separated by commas, to be used by the dialog box.

FilterIndex: This is optional. It is the index number (1-based) of the filter name to be used by default.

Title: This is optional. It is a string that specifies the title of the dialog box.

ButtonText: This is optional; it works just on the Mac operating system.

MultiSelect: This is optional; it is used just by the GetOpenFileName method, which when set to True allows the selection of multiple files.

There are no essential differences using the Application object's GetOpenFileName, GetSaveAsFileName, or FileDialog method, besides the fact the latter method is the only one that allows you to browse and return a folder's name by using the msoFileDialogFolderPicker constant on the FileDialogType argument.

Actually, these two methods are easier to use than the FileDialog method, requiring small pieces of VBA code interaction, considering the following:

- Both GetOpenFileName and GetSaveAsFileName return False if the user cancels the associated dialog box shown by each method.
- GetSaveFileName and GetOpenFileName always return the typed or selected file name whenever the MultiSelect argument is defined as False (the default value).
- If the GetOpenFileName MultiSelect argument is defined as True, it returns a 1-based one-dimensional array of file names, even if just one file will be selected.

To see how you can program both methods, open the file Application GetOpenFileName and GetSaveAsFileName methods.xlsm macro-enabled workbook (also available in the Chapter02.zip file). Right-click any ControlButton and then select the Assign Macro menu command to show the Assign Macros dialog box to show all the Public Sub procedures available in this workbook (Figure 2-11).

As Figure 2-11 implies, the Open One File ControlButton is associated with the Public Sub OpenOneFile( ) procedure, which has this VBA code:

```
Public Sub OpenOneFile()
    Dim varFile As Variant
    Dim strFile As String
    Dim strFolder As String
    Dim strMsg As String
    Dim intPos As Integer
```

```
    varFile = Application.GetOpenFilename
    If varFile <> False Then
        intPos = InStrRev(varFile, "\")
        strFolder = Left(varFile, intPos - 1)
        strFile = Mid(varFile, intPos + 1)
        strMsg = "Folder: " & strFolder & vbCrLf
        strMsg = strMsg & "File: " & strFile

        MsgBox strMsg, vbInformation, "1 file selected"
    End If
End Sub
```

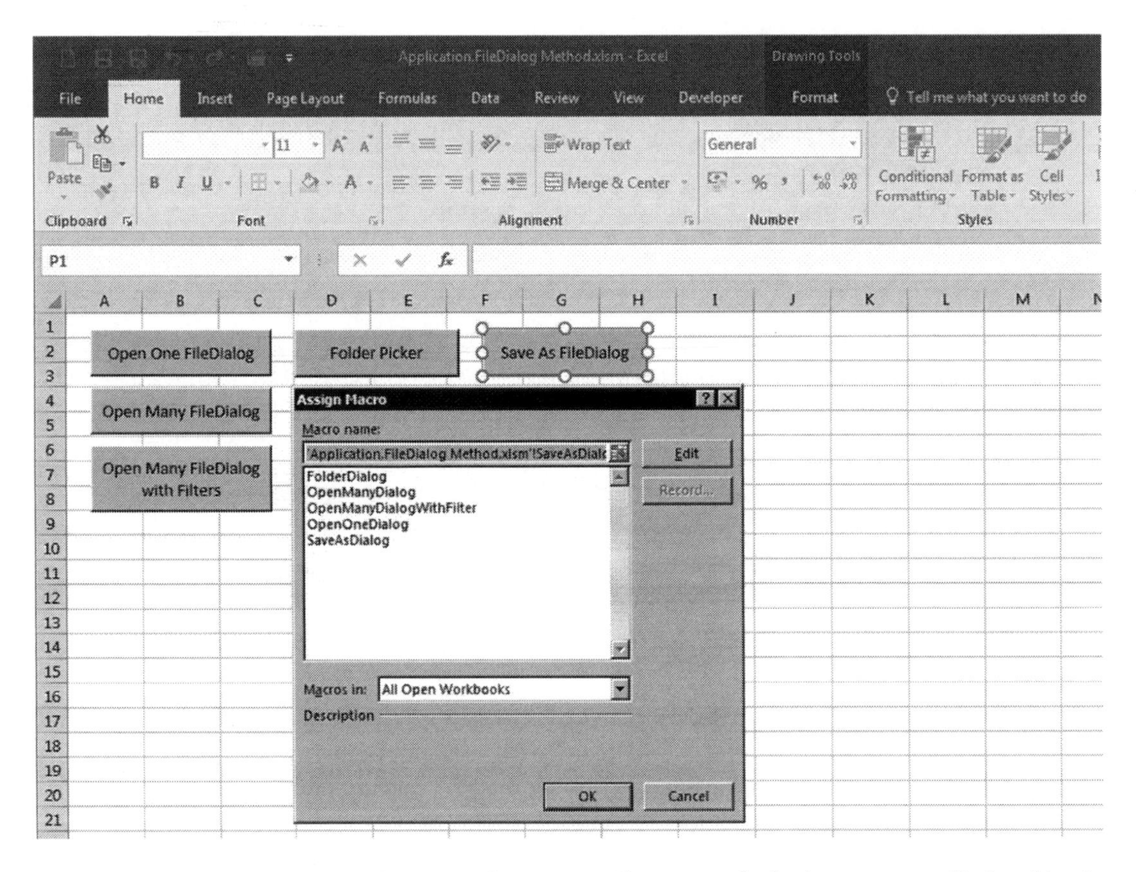

Figure 2-11. *This is the GetOpenFileName and GetSaveAsFileName methods.xlsm macro-enabled workbook, with Button controls associated to Public Sub procedures that use Application.GetOpenFileName and GetSaveAsFileName to interact with files and folders*

Open One Single File

When you click the Open One File Button control, it makes a simple call to the Application.GetOpenFileName method without passing it any argument and waits for the dialog box to be closed to attribute a value to the varFile variable.

```
varFile = Application.GetOpenFilename
```

If the user clicks the Cancel or Close button, the varFile variable will receive False, and the procedure will end. If a file is selected, the procedure will break the file path and file name into two different pieces and show it using a MsgBox( ) function (a common issue regarding Excel programming on Internet sites).

Note that to get the file name from the returned selected file path, the procedure makes use of the VBA InStrRev( ) function.

```
varFile = Application.GetOpenFilename
If varFile <> False Then
    intPos = InStrRev(varFile, "\")
```

The InStrRev( ) VBA function searches any string backward for a given string match and has this syntax:

```
InStrRev ( String, StringMatch [, Start [ , Compare] ] )
```

In this code:

String: This is required; it is the string expression being searched.

StringMatch: This is required; it is the string expression being searched for.

Start: This is optional; it is a numeric expression setting the 1-based starting position for each search, starting from the left side of the string. If Start is omitted, then -1 is used, meaning the search begins at the last character position. Search then proceeds from right to left.

Compare: This is optional; it is a numeric value indicating the kind of comparison to use when evaluating substrings. If omitted, a binary comparison is performed (meaning that upper/lowercase are different letters).

Note that the InStrRev( ) function searches backward the varFile variable for the last backslash that separates the folder name from the file name and stores it when found in the intPos variable.

The folder path where the file is stored is then retrieved using the VBA Left( ) function, from the beginning of the varFile variable to the character immediately before the last backslash:

```
strFolder = Left(varFile, intPos - 1)
```

The VBA Left( ) function has this syntax:

```
Left ( String, Length )
```

In this code:

String: This is required; it is the string expression from which the leftmost characters are returned.

Length: This is required; it is an Integer expression indicating how many characters to return. If zero, a zero-length string ("") is returned. If greater than or equal to the number of characters in String, the complete string is returned.

To extract the file name from the varFile variable, the procedure uses the VBA Mid( ) function, from the next character after the last backslash to the end of the varFile string:

```
strFile = Mid(varFile, intPos + 1)
```

Note that the Mid( ) function uses just the Start argument, meaning it must extract from varFile, beginning on intPos+1 position, all characters to the end of the string, literally extracting the file name varFile.

■ **Attention** VBA also has the Right( ) function, which extracts characters from right to left of any string and has this syntax:

```
Right ( String, Length )
```

In this code:

String: This is the string that you want to extract from.

Length: This indicates the number of characters that you want to extract starting from the rightmost character.

Using Right( ): You can achieve the same Mid( ) result to extract the file name from varFile using this syntax:

```
strFile = Right(varFile, Len(varFile) - intPos)
```

The folder name and file name are then presented to the user using a MsgBox( ) function, which breaks the message in two lines by concatenating the vbCrLf constant after the folder name:

```
strMsg = "Folder: " & strFolder & vbCrLf
strMsg = strMsg & "File: " & strFile

MsgBox strMsg, vbInformation, "1 file selected"
```

Open Many Files

The Open Many Files and Open Many Files with Filter buttons use the same programming technique to extract the files from the array returned by the Application.GetOpenFileName method, when define the last argument MultiSelect = True. This is the Sub OpenManyFilesWithFilter procedure code associated to the Open Many Files with Filter Button control:

```
Public Sub OpenManyFilesWithFilter()
    Dim varFiles As Variant
    Dim varItem As Variant
    Dim strFilters As String
    Dim strTitle As String
    Dim strMsg As String
```

```
    strFilters = "Microsoft Excel 2007-2010,*.XLS?"
    strFilters = strFilters & ",Microsoft Excel 97-2003,*.XL?"

    varFiles = Application.GetOpenFilename(strFilters, 1, "Open My Excel File!", , True)
    If IsArray(varFiles) Then
        strTitle = UBound(varFiles) & " File(s) was selected!"
        strMsg = "File(s) selected:" & vbCrLf
        For Each varItem In varFiles
            strMsg = strMsg & varItem & vbCrLf
        Next
        MsgBox strMsg, vbInformation, strTitle
    End If
End Sub
```

After all variable declarations (note that they are grouped by type), the procedure first mounts the desired file filters on the strFilters string variable.

```
strFilters = "Microsoft Excel 2007-2010,*.XLS?"
strFilters = strFilters & ",Microsoft Excel 97-2003,*.XL?"
```

It then makes a call to the Application.GetOpenFileName method, passing it the strFilter argument (1 for the default FilterIndex), sets "Open My Excel File!" as the dialog title, and defines the last argument as MultiSelect = True.

```
varFiles = Application.GetOpenFilename(strFilters, 1, "Open My Excel File!", , True)
```

The procedure will stop at this point waiting the modal dialog box to be closed. Once again, if the user clicks Cancel or clicks the dialog box's Close button, varFiles will receive False, and the procedure will end. But since the Multiselect argument is defined as True, if the user selects one or more files, varFiles will receive an array with the complete path to each selected file. So, the next instruction verifies with the VBA IsArray( ) function if varFiles is an array structure:

```
If IsArray(varFiles) Then
```

To create the message box title, the procedure uses the VBA UBound( ) function to count how many file names have been inserted on the varFiles array.

```
    strTitle = UBound(varFiles) & " File(s) was selected!"
```

Note that the procedure does not use the UBound( ) Dimension argument, meaning that it must retrieve the first (and only) dimension of the returned array of selected file names. And since the UBound( ) function is 1-based, the value returned means the total files selected.

The strMsg variable is then initiated with the "File(s) selected:" string, using a vbCrLf constant to break the string message.

```
    strMsg = "File(s) selected:" & vbCrLf
```

Inside a `For Each... Next` loop, a file name on the `varFiles` array is added to the `strMsg` string, breaking the string with a `vbCrLf` for each file, and at the end of the loop the `MsgBox( )` function shows the selected files.

```
    For Each varItem In varFiles
        strMsg = strMsg & varItem & vbCrLf
    Next
    MsgBox strMsg, vbInformation, strTitle
```

The Save as File `ControlButton` is associated to the `Sub SaveFile( )` procedure, which uses the `Application.GetSaveAsFileName` method to allow the user to select or type the file name that will be used to save a file.

```
Public Sub SaveFile()
    Dim varFile As Variant

    varFile = Application.GetSaveAsFileName("Default file name", , , "Save My Excel File!")
    If varFile <> False Then
        MsgBox "Selected File to save: " & varFile, vbInformation, "File will be saved"
    End If
End Sub
```

Using Application InputBox Method

The Excel `Application.InputBox` method differs from the VBA `InputBox( )` function because VBA raises a modal dialog box and always returns a text value, while the Excel `Application.Inputbox` raises a semi-modal dialog box that allows the user to select cells on the workbook; it also can validate the user input and can capture Excel object errors and formula values. The `Application.InputBox` method has this syntax:

```
Application.InputBox(Prompt, Title, Default, Left, Top, HelpFile, HelpContextID, Type)
```

In this code:

`Prompt`: This is required; it is a string representing the message displayed in the dialog box.

`Title`: This is optional; it is a string for the title of the dialog box.

`Default`: This is optional; it specifies the value that will appear in the text box when the dialog box is initially displayed. If omitted, the text box is left empty (it can be defined to a Range object).

`Left`: This is optional; it specifies an *x* position for the dialog box in relation to the upper-left corner of the screen, in points.

`Top`: This is optional; it specifies a *y* position for the dialog box in relation to the upper-left corner of the screen, in points.

`HelpFile`: This is optional; it is the name of the Help file associated to the input box. If the `HelpFile` and `HelpContextID` arguments are present, a Help button will appear in the dialog box.

`HelpContextID`: This is optional; it is the context ID number of the Help topic in `HelpFile`.

`Type`: This is optional; it specifies the return data type. If this argument is omitted, the dialog box returns text.

Table 2-5 specifies the values that can be passed to the Type argument of the Application.InputBox method.

Table 2-5. *Values Allowed to the Application.Inputbox Type Argument and the Value Expected or Returned by the InputBox*

Type Value	Return
0	Formula (string beginning with =)
1	Number
2	Text (default)
4	Logical (True or False)
8	Range object
16	Error object
64	Array of values

In Table 2-5, the Return column indicates that besides returning False whenever the user clicks the Cancel button, the Application.InputBox method can also return different types of values, an array, or even object pointers (Range or Error).

You can try the Application.InputBox( ) method in the VBA Immediate window, typing a command line that has at least the Prompt argument (which by default uses Type = Text).

```
?Application.InputBox("Select a cell")
```

By clicking any cell value, the Application.InputBox method will return the value as text in the Immediate window (Figure 2-12).

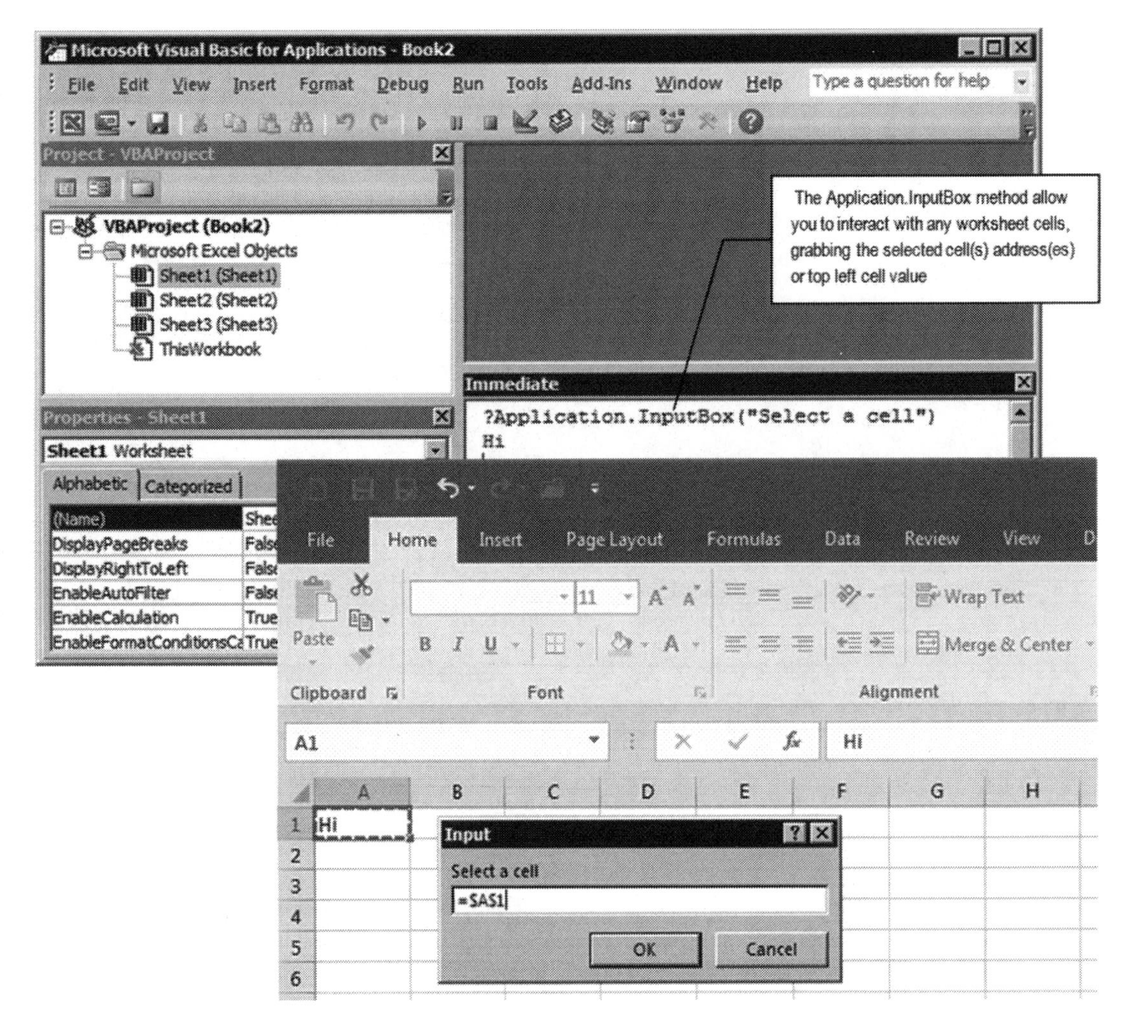

***Figure 2-12.** Use the VBA Immediate window to test the Application.Inputbox method, passing it first the required argument prompt. Excel will show a InputBox dialog over the current worksheet ready to grab any text value from any worksheet. Click any cell to see its address in the InputBox dialog, and press Enter to return the selected cell value (or top-left cell value of any selected range) to the VBA Immediate window*

If you define the Type argument to 1 (`Number`) and select a cell with a text value, the `InputBox` will validate the selected data and show a dialog box warning indicating that a number is expected.

To select both `Number` and `Text` values, you can pass 1 + 2 = 3 to Type (the sum of arguments `Text + Number`). To allow the selection of `Number`, `Text`, and `Logical` values, pass 1 + 2 + 4 = 7 to Type.

The `Application.InputBox` method does its best when you want to recover any range of addresses selected by the user on the worksheet and pass the selected range address to some VBA procedure. Whenever you need such user interaction, use Type `= 0` (Formula).

Let's see some practical examples. The `Application InputBox method.xlsm` macro-enabled workbook that you can extract from the `Chapter02.zip` file has command buttons associated to `Public Sub` procedures of the `basInputBox` method to allow you to test each possible `Application.Inputbox Type` argument value. It also has some number, text, and error values in the cell range `D2:F4`.

Figure 2-13 shows that the Application InputBox method.xlsm workbook has one Public Sub procedure for each possible Type argument value (according to Table 2-5). To easily select the Type argument, basInputBox also has the Public Enum InputBoxType.

```
Public Enum InputBoxType
    Formula = 0
    Number = 1
    Text = 2
    Logical = 4
    Range = 8
    ExcelError = 16
    ArrayOfValues = 64
End Enum
```

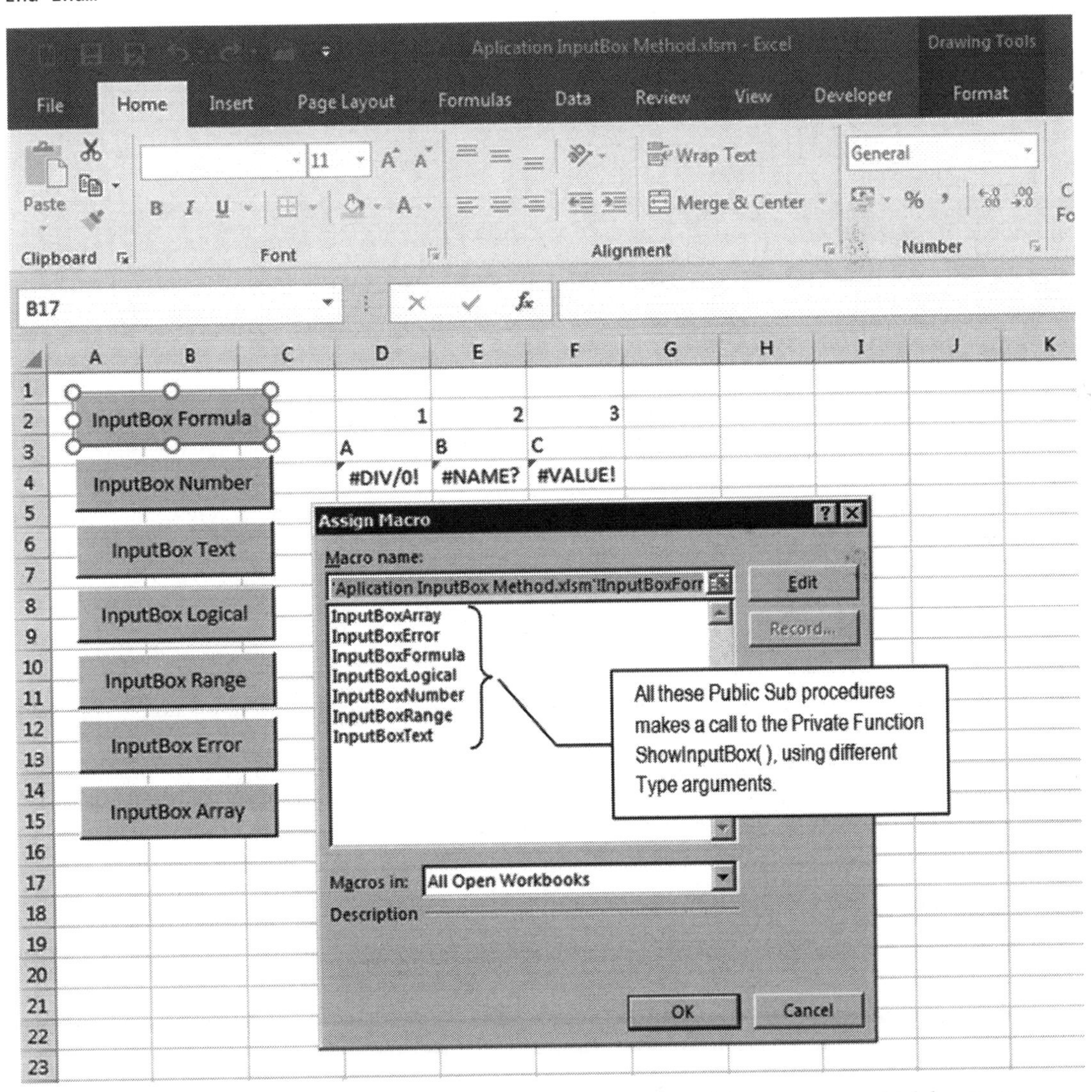

Figure 2-13. *This is the Application InputBox method.xlsm macro-enabled workbook, which has Button controls associated to Public Sub procedures that make use of a centralized Private Function to make calls to the Application.Inputbox method using different Type arguments*

All these Public Sub procedures make a call to the generic Private Function ShowInputBox( ) procedure from basInputBox, using different values for the SelectionType argument.

```
Private Function ShowInputBox(Optional Prompt As String, _
                              Optional Title As String, _
                              Optional DefaultValue As Variant, _
                              Optional SelectionType As InputBoxType = Formula) As Variant
    Dim varValue As Variant

    Select Case SelectionType
        Case Range
            On Error Resume Next
            Set varValue = Application.InputBox(Prompt, Title, DefaultValue, , , , ,
            SelectionType)
            If IsObject(varValue) Then
                Set ShowInputBox = varValue
            End If
        Case Else
            varValue = Application.InputBox(Prompt, Title, DefaultValue, , , , ,
            SelectionType)
            If SelectionType = Formula Then
                varValue = Application.ConvertFormula(varValue, xlR1C1, xlA1)
            End If
            ShowInputBox = varValue
    End Select
End Function
```

Note that the Function ShowInputBox( ) function has been declared with four optional arguments and returns a Variant value, allowing it to return all possible Application.InputBox returned values: number, text, array of values, or object reference. Also note that to take advantage of the InputBoxType enumerator, the last optional argument, SelectionType, is declared as InputBoxType, receiving by default the Type = Formula value.

```
Private Function ShowInputBox(Optional Prompt As String, _
                              Optional Title As String, _
                              Optional DefaultValue As Variant, _
                              Optional SelectionType As InputBoxType = Formula) As Variant
```

Besides being capable of returning False whenever the user clicks the dialog box's Cancel button, according to Table 2-5, when Type = 8, Application.InputBox returns a Range object. Every other Type argument returns a variant value, an array of variant values, or eventually a VBA Error object (Type = 16). To take care of these different possible values, ShowInputBox( ) uses a Select Case instruction to first verify which value has been attributed to the SelectionType argument.

```
    Dim varValue As Variant

    Select Case SelectionType
```

If SelectionType = Range, the procedure uses a VBA Set instruction to associate the varValue Variant variable to the Range object returned by the Application.InputBox method. It is necessary to disable VBA errors using an On Error Resume Next instruction because if the user clicks Cancel or the Close button of the InputBox, False will be returned, and the Set instruction will generate an undesirable runtime error (Set works just to object references).

```
        Case Range
            On Error Resume Next
            Set varValue = Application.InputBox(Prompt, Title, DefaultValue, , , , , SelectionType)
```

Since we actually don't know whether there is or not a reference to a Range object on varValue, the procedure uses the VBA IsObject(varValue) function to verify this possibility. If IsObject(varValue) = False (the user clicked the Cancel button), the function runs until the end, and varValue returns Empty. If IsObject(varValue) = True, the selected Range object is then returned by the ShowInputBox( ) function, using again the VBA Set statement.

```
            If IsObject(varValue) Then
                Set ShowInputBox = varValue
            End If
```

Every other possible Type value fits on the varValue Variant variable without the need to use a VBA Set statement.

```
        Case Else
            varValue = Application.InputBox(Prompt, Title, DefaultValue, , , , , SelectionType)
```

But if SelectionType = Formula, then varValue will receive the formula using the Excel R1C1 format. To change the formula to Excel A1 format, the procedure uses the Excel Application.ConvertFormula method to change the default R1C1 to the customary A1 format, and ShowInputBox( ) returns whatever the user selected on the worksheet.

```
            If SelectionType = Formula Then
                varValue = Application.ConvertFormula(varValue, xlR1C1, xlA1)
            End If
            ShowInputBox = varValue
```

Since every ControlButton of the worksheet uses a specific Type value when calling the ShowInputBox( ) function, let's see how each one works, beginning with Sub InputBoxFormula( ), which has this code:

```
Public Sub InputBoxFormula()
    Dim varValue As Variant
    Dim strMsg As String

    varValue = ShowInputBox("Select any cells range or type a formula:", "Application.
    InputBox Type = Formula", , Formula)
    If varValue <> False Then
        strMsg = "Formula created by Application.InputBox:" & vbCrLf
        strMsg = strMsg & varValue
        MsgBox strMsg, vbInformation, "Formula created with Application.InputBox Type = Formula"
    End If
End Sub
```

When you click the InputBox Formula Button control, the dialog box will allow you to type or create a formula with any cell references you select on the worksheet. As mentioned, these references are returned using the R1C1 style, and the ShowInputBox( ) function uses the Application.ConvertFormula method to change them to the A1 style. Figure 2-14 shows the formula returned after creating a formula with cells D2, E2, and F2.

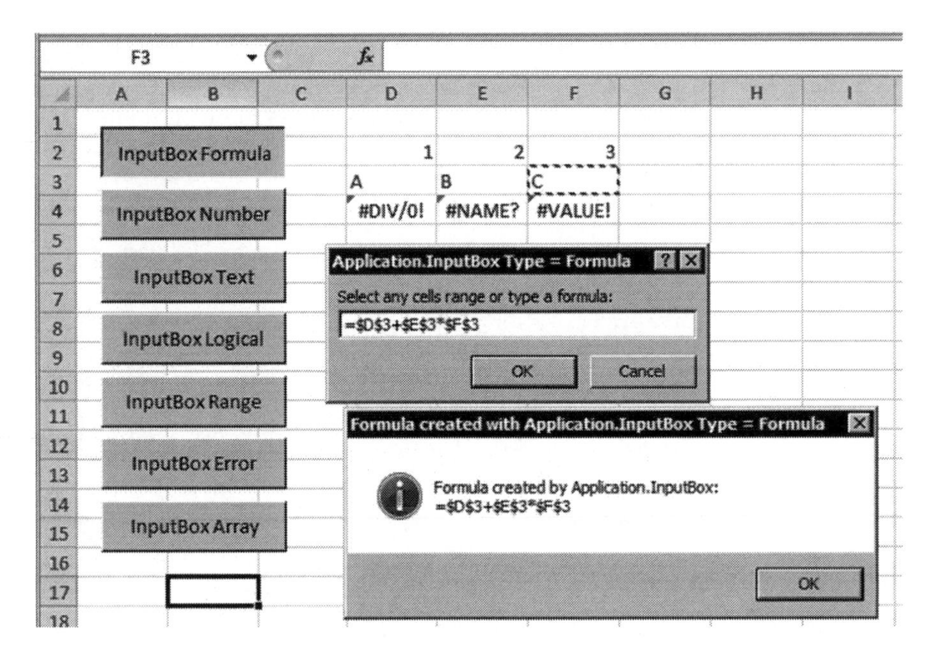

Figure 2-14. *Using Type = Formula with the Application.Inputbox method, you can click any cells of the worksheet, use any Excel mathematical operators, and return an Excel formula to your procedure*

■ **Attention** You can see the R1C1 style returned by the Application.InputBox method setting a breakpoint on the ShowInputBox( ) function row where the Application.ConvertFormula method is inserted and inspecting the varValue variable (Figure 2-15).

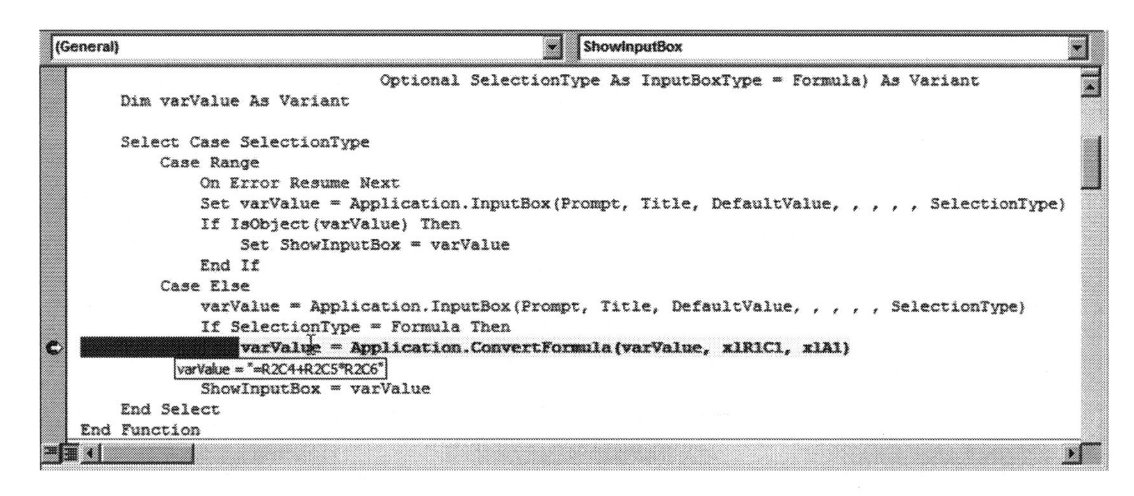

Figure 2-15. *To see the R1C1 formula style returned by the Application.InputBox method when Type = 0 (Formula), place a breakpoint on the row of Application.ConvertFormula and inspect the varValue variable content*

All other control buttons used to test `Application.InputBox` (the selection of `Number`, `Text`, `Logical`, and `Error` values) use a `Sub` procedure similar to that used to create a formula. The InputBox Range `ControlButton` uses the `Address` property of the `Range` object to return the selected range. The InputBox Array `ControlButton` is the only one that needs to be mentioned because it must verify some array properties, such as row and column counting, before showing with the `MsgBox( )` function each selected array item. Here is the code procedure:

```
Public Sub InputBoxArray()
    Dim varValue As Variant
    Dim varItem As Variant
    Dim strMsg As String
    Dim intRows As Integer
    Dim intCols As Integer
    Dim intI As Integer
    Dim intJ As Integer

    varValue = ShowInputBox("Select any cell values:", "Application.InputBox Type =
ArrayOfValues", , ArrayOfValues)
    If IsArray(varValue) Then
        intRows = UBound(varValue, 1)

        On Error Resume Next
        intCols = UBound(varValue, 2)
        On Error GoTo 0

        If intCols = 0 Then
            strMsg = "Array has " & intRows & " column(s):" & vbCrLf
            For intI = 1 To intRows
                varItem = varValue(intI)
                If IsError(varItem) Then
                    varItem = CStr(varItem)
```

```
                End If
                strMsg = strMsg & "A(" & intI & ")=" & varItem & " "
            Next
        Else
            strMsg = "Array has " & intRows & " row(s) and " & intCols & " column(s):" & vbCrLf
            For intI = 1 To intRows
                For intJ = 1 To intCols
                    varItem = varValue(intI, intJ)
                    If IsError(varItem) Then
                        varItem = CStr(varItem)
                    End If
                    strMsg = strMsg & "A(" & intI & "," & intJ & ")=" & varItem & " "
                Next
                strMsg = strMsg & vbCrLf
            Next
        End If
        MsgBox strMsg, vbInformation, "Array returned by Application.InputBox Type = ArrayOfValues"
    End If
End Sub
```

Note that this time the procedure uses the VBA `IsArray( )` function to verify whether `varValue` contains an array of items.

```
    varValue = ShowInputBox("Select any cell values:", "Application.InputBox Type =
    ArrayOfValues", , ArrayOfValues)
    If IsArray(varValue) Then
```

If this is true, the procedure tries to recover the array dimensions. There is no VBA function that you can use to return how many dimensions an array has, but the `Application.InputBox` method can just return one-dimensional arrays (when you select cells on the same row) or bidimensional arrays (when you select cells in different rows, even if you use just one column). The procedure tries to recover how many rows the array has using `UBound(varValue, 1)` to return it to the `intRows` variable.

```
    If IsArray(varValue) Then
        intRows = UBound(varValue, 1)
```

It then tries to recover the second array dimension, if any. If there is no second dimension on the `varValue` array, VBA will raise an error, and to avoid this error, you must use `On Error Resume Next`. Since this instruction disables error raising until the procedure ends, you must turn on again the error catch using an `On Error GoTo 0` statement.

```
        On Error Resume Next
        intCols = UBound(varValue, 2)
        On Error GoTo 0
```

Now you have the array dimensions: `intRows` contains the rows count, and `intCols` contains the column count. Both are 1-based for the `varValue` array returned by the `Application.Inputbox` method. If `intCols = 0`, then `varValue` is a *row array* with just one row and many columns. The code will process just the `intRows` items, and a string message will show how many columns have been selected.

```
        If intCols = 0 Then
            strMsg = "Array has " & intRows & " column(s):" & vbCrLf
```

To process all array items, the procedure uses a `For... Next` loop to loop through all array items, extracting each array item to the `varItem Variant` variable (note that it is 1-based):

```
For intI = 1 To intRows
    varItem = varValue(intI)
```

Since there is a possibility to select cell error values on the worksheet, the procedure uses the VBA `IsError(varItem)` function to verify whether the selected cell has any errors. If it does, the captured error is changed to a string.

```
        If IsError(varItem) Then
            varItem = CStr(varItem)
        End If
```

At this point, `varItem` has a value that can be added to the `strMsg` string text, indicating the array index and the value.

```
        strMsg = strMsg & "A(" & intI & ")=" & varItem & " "
    Next
```

The loop will continue to process all other array items, and when finished, a `MsgBox( )` function shows the items selected.

```
        MsgBox strMsg, vbInformation, "Array returned by Application.InputBox Type =
        ArrayOfValues"
    End If
End Sub
```

Note that when there is a bidimensional array, you need to perform a double `For...Next` loop: an outer loop for each array row (using the `intI` Integer variable) and an inner loop for every array column of each row (using the `intJ Integer` variable). At each inner loop passage, the array item is returned to the `varItem` variable and tested with the `IsError( )` function as the `strMsg` string is created, item by item.

```
strMsg = "Array has " & intRows & " row(s) and " & intCols & " column(s):" & vbCrLf
For intI = 1 To intRows
  For intJ = 1 To intCols
    varItem = varValue(intI, intJ)
    If IsError(varItem) Then
        varItem = CStr(varItem)
    End If
    strMsg = strMsg & "A(" & intI & "," & intJ & ")
           =" & varItem & " "
  Next
  strMsg = strMsg & vbCrLf
Next
```

This is the For...Next outer loop to grab array row items
This is the For...Next inner loop to grab array column items

Note that every time the inner loop (column items) ends, a line break is inserted on the `strMsg` variable using the `vbCrLf` constant.

```
        Next
        strMsg = strMsg & vbCrLf
    Next
```

You must click the InputBox Array `ControlButton` and try to select different array types: a single-row array, a single-column array, and a multirow column array (the last two types lead to a bidimensional array). Using cells in the range `D2:F4`, which has number, text, and error values, you will receive the `MsgBox( )` message shown in Figure 2-16.

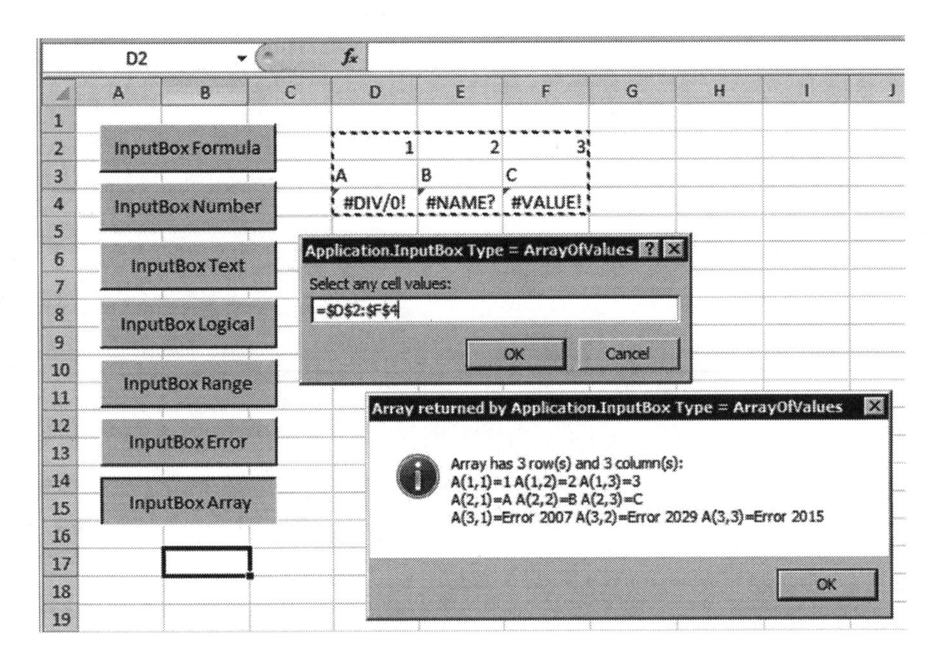

Figure 2-16. *When you use Type=64 (ArrayOfValues), the Application.InputBox method returns an array of items. You must use a For... Next loop to extract, analyze, and use each returned array value. Note that there is a special case when you select one or more cells with errors: InputBox will return an Error object that must be converted to a string (or treated with Excel CvError( ) function) so you can verify the error returned*

Attention Whenever you use a VBA `CStr( )` function to convert cell error values to a string, you will receive the associated error code. Figure 2-16 shows that Error 2007 is associated with a #DIV/0! error, Error 2029 with a #NAME! error, and Error 205 with #VALUE! error. To deal with cell formula errors in code, use an Excel CvError( ) function, which is covered in Chapter 5.

Using Application OnTime Method

Microsoft Excel exposes the Application.OnTime method as a way to create a timer to execute any public procedure of a standard code module at a specified moment (public procedures from ThisWorkbook, Sheet, or UserForm modules won't work). It has the next syntax:

```
Application.OnTime(EarliestTime, Procedure, LatestTime, Schedule)
```

In this code:

> EarliestTime: This is the exact time the procedure must be executed (using the format hh:mm:ss). Use the TimeValue( ) VBA function to convert any time string to a VBA valid time.
>
> Procedure: This is a string with the procedure name (must be a public procedure of a standard module).
>
> LatestTime: This is the maximum time to the procedure executing.
>
> Schedule: This is a Boolean (True/False) to set or cancel the OnTime method for a given procedure. The default is True.

The next example shows how to execute the Public Sub Procedure1 procedure exactly at 12:00:

```
Application.OnTime TimeValue("12:00"), "Procedure1"
```

It must be mentioned that once the Application.OnTime method is set for a given procedure, if you want to cancel it, you need to call it again with the exact set time using Schedule = False. So, you need to store the set time on a public variable to cancel the timer. The next example shows how you can cancel the timer created to execute Procedure1 at 12:00 with the last code instruction before it fires:

```
Application.OnTime TimeValue("12:00"), "Procedure1", , False
```

To create a timer ahead of time, you must define it, adding to the current system time a specific time value. The next examples execute the Procedure1 public procedure five seconds ahead, using VBA Now( ) or Time( ) functions to get the system time and adding five seconds using the VBA TimeValue( ) function that converts any time string with the "hh:mm:ss" format to the associated time value:

```
Application.OnTime Now +TimeValue("00:00:05"), "Procedure1"
Application.OnTime Time +TimeValue("00:00:05"), "Procedure1"
```

If Excel is in Edit mode or has any procedure running when the Application.OnTime time comes, the defined procedure will not run. So, you can use the LatestTime argument to state the maximum time of day that Application.OnTime should try to run the procedure again.

But once the Application.OnTime is set for a given time, if you close the workbook and keep the Excel window opened, it will open again the workbook and run the procedure when the time has come.

Although a timer's usefulness on a worksheet application may be very task specific, one of its applications is to allow a UserForm to close after some exhibited time, like the common splash screen shown when most professional applications are opened.

The Application OnTime method.xlsm macro-enabled workbook that you can also find inside the Chapter02.zip file allows you to verify how the Application.OnTime method can be used to automatically close a UserForm. It has two UserForms (UserForm1 and UserForm2) and the basOnTime standard module, where you can find the Public Sub procedures UnloadUserForm1( ) and UnloadUserForm2( ) that close the associate UserForm using the Unload instruction.

```
Option Explicit

Public Sub UnloadUserForm1()
    Unload UserForm1
    Debug.Print "UnloadUserForm1 fired"
End Sub

Public Sub UnloadUserForm2()
    Unload UserForm2
    Debug.Print "UnloadUserForm2 fired"
End Sub
```

Note that both code procedures use the VBA Debug.Print instruction to print in the VBA Immediate window a string expression identifying when each method is fired. Both UserForm1 and UserForm2 define the Application.OnTime method on the UserForm_Initialize event, which fires every time a UserForm is opened. This is the UserForm1_Initialize event procedure, which sets the UnloadUserForm1 procedure to fire five seconds before the UserForm is opened:

```
Private Sub UserForm_Initialize()
    Application.OnTime Time + TimeValue("0:0:5"), "UnloadUserForm1"
End Sub
```

Note that UserForm1 has no other code to cancel the Application.OnTime method if you close the form before it fires the UnloadUserForm1 procedure.

UserForm2 sets the Application.OnTime method on the UserForm_Initialize event and unsets it on the UserForm_Terminate event (using Schedule=False), to guarantee that when the UserForm is closed, the Application.OnTime event is canceled. Note that UserForm2 uses the mvarTime module-level variable to store the exact time the OnTime method fires, reusing this value on the UserForm_Terminate event. An On Error Resume Next instruction is used on the UserForm_Terminate event to avoid any errors that may arise if the Application.OnTime event had been fired before the form is closed.

```
Dim mvarTime As Variant

Private Sub UserForm_Initialize()
    mvarTime = Time + TimeValue("0:0:5")
    Application.OnTime mvarTime, "UnloadUserForm2"
End Sub

Private Sub UserForm_Terminate()
    On Error Resume Next
    Application.OnTime mvarTime, "UnloadUserForm2", , False
End Sub
```

Let's see the action. The file Application OnTime method.xlsm has two Button controls associated to Public Sub procedures of the Sheet1 worksheet, used to open UserForm1 and UserForm2 (Figure 2-17).

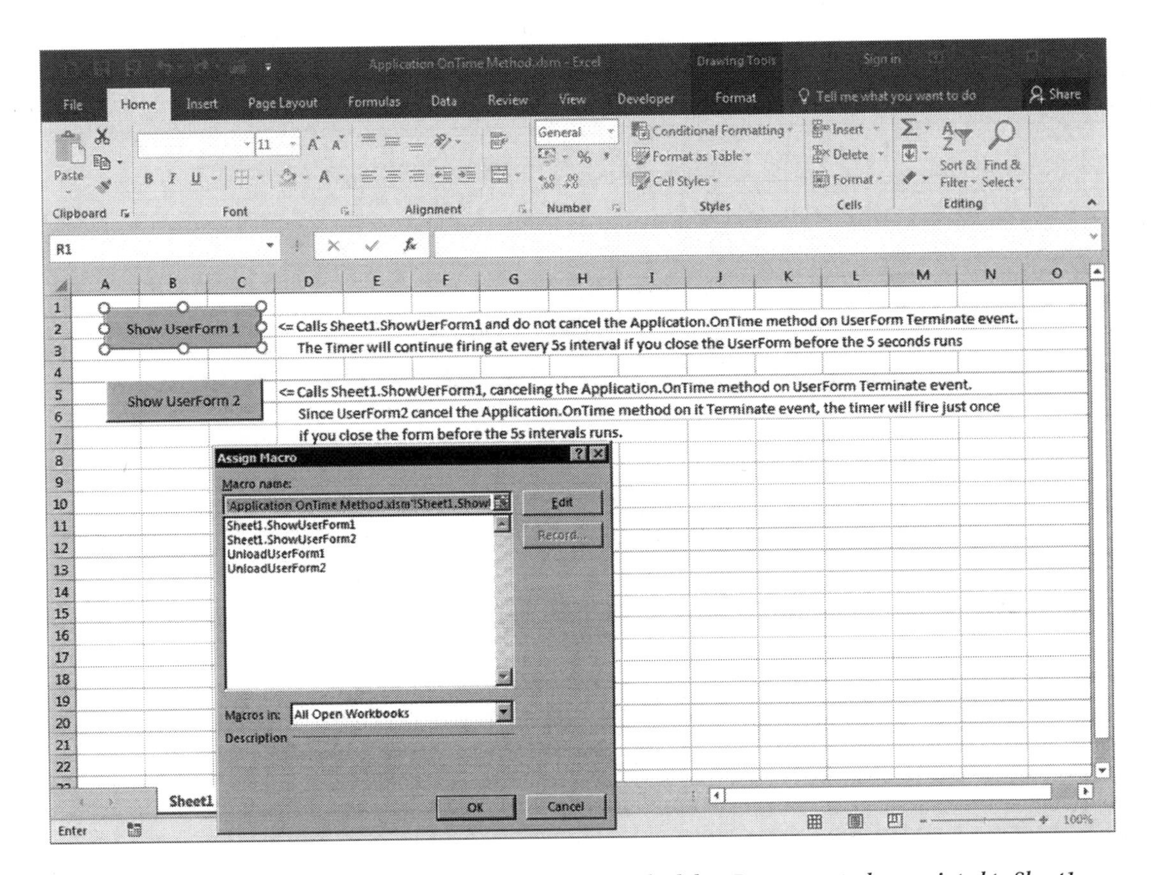

Figure 2-17. *This is file Application OnTime method.xlsm, which has Button controls associated to Sheet1 Public Sub procedures that simply load UserForm1 and UserForm2. When each UserForm is loaded, they set the Application.OnTime method to unload it five seconds ahead*

Try to click the Show UserForm1 `ControlButton` and see that the `UserForm1` will be loaded. Wait five seconds and it will close, leaving a message in the VBA Immediate window when the `Application.OnTime` event fires (Figure 2-18).

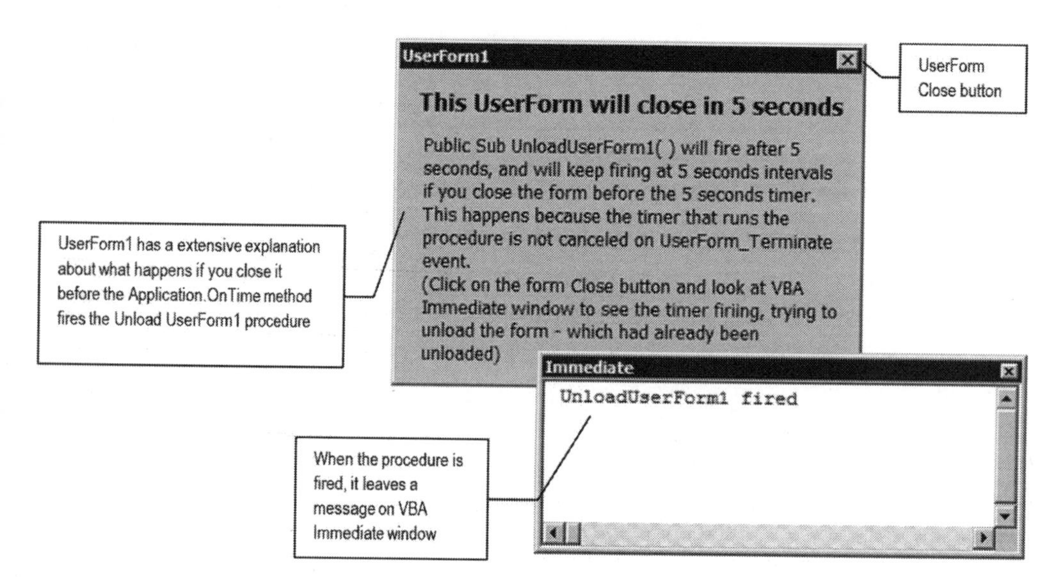

Figure 2-18. *Click Show UserForm1 ControlButton to load UserForm1 and wait five seconds for the Application.OnTime method to fire the Public Sub UnloadUserForm1( ) procedure and unload the UserForm*

After the `UserForm1` had been automatically unloaded by the `Application.OnTime` method, click the Show UserForm1 `ControlButton` again, but this time *do not* wait five seconds: click the `UserForm` Close button to close it and watch the VBA Immediate window. At each five seconds you will see that the VBA window blinks and another "Unload UserForm1 fired" message appears in the Immediate window, meaning that the `Application.OnTime` method continuously fires, again and again, trying to unload the *not loaded* `UserForm`.

Try to close the workbook leaving Excel opened (use the File ➤ Close command), and when `Application.OnTime` runs again, Excel will reopen the workbook, firing the timer again and again. I don't have a decent explanation for this odd Excel behavior, but you should be aware of it (Figure 2-19)!

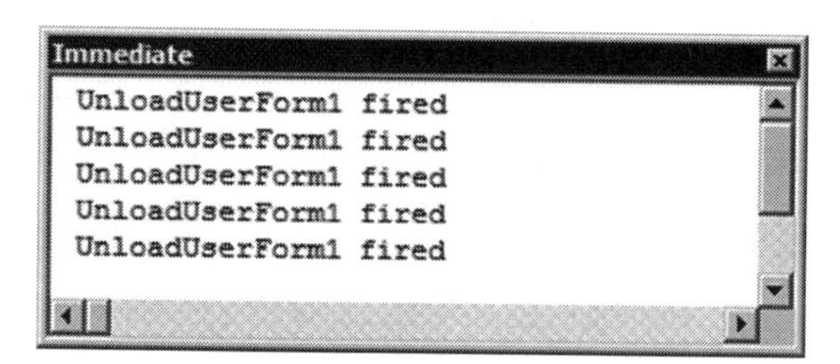

Figure 2-19. *If you click the Show UserForm1 ControlButton of the Application InputBox method.xlsm workbook and close UserForm1 before the five-second timer runs out, Excel will keep firing the UnloadUserForm1( ) public procedure again and again. And if you try to close the workbook leaving Excel opened, it will reopen the workbook to keep firing the OnTime event. This is a clear Excel bug!*

The easiest way to cancel this odd Excel behavior is to close the Excel window, reopen it, and open again the `Application OnTime method.xlsm` macro-enabled workbook.

Do the same exercise by clicking Show UserForm2 and you will see that this time, you can wait five seconds for `UserForm2` to be automatically closed, or you can click the Close button, without needing to wait until the timer runs to completion. The code inserted on the `UserForm2_Terminate` event kills the timer set, and Excel behaves as expected (Figure 2-20).

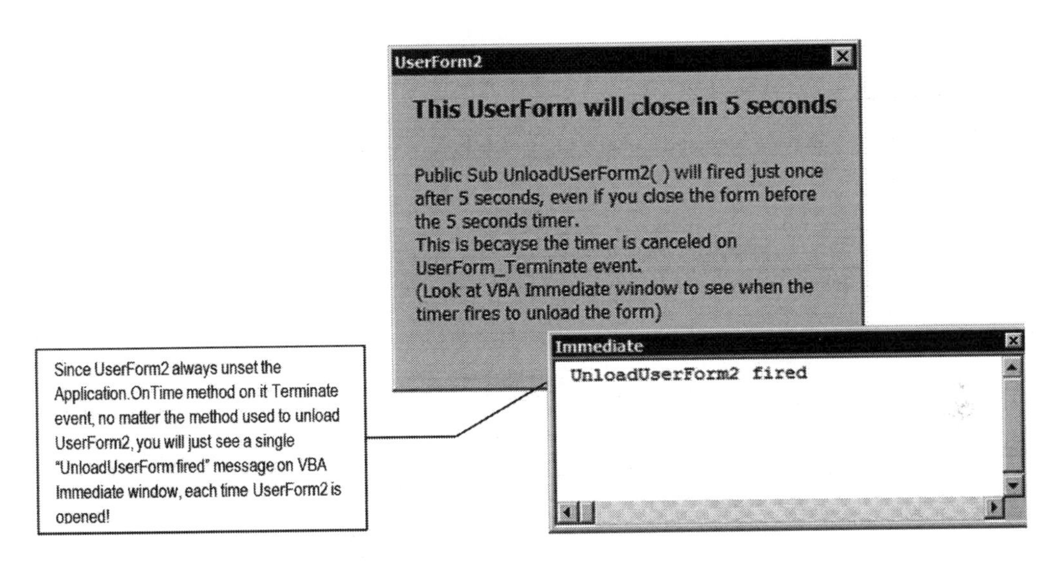

Figure 2-20. *UserForm2 stores the Application.OnTime time on a module-level variable and uses the UserForm_Terminate event to unset the timer, by calling again Application.OnTime with at the same time but using Schedule = False (last method argument). You will see just one "UnloadUserForm2 Fired" message in the VBA Immediate window whenever UserForm2 is closed*

The lesson here is quite simple. You should always disable each `Application.OnTime` method before the workbook is closed or when it is used to close any `UserForm`.

■ **Attention** If you do not use a `On Error Resume Next` instruction before you try to disable any `Application.OnTime` method, VBA will raise an expected error. Another Excel bug?

Using Application Events to React to User Actions

Looking to the Microsoft Excel object model depicted in Figure 2-1, you must understand that there is an object hierarchy where the higher-level object is the `Application` object (first level), followed by the `Workbook` object (second level) and the `Worksheet` object (third level).

■ **Attention** Although the `Workbooks` and `Worksheets` collections stay in between the `Workbook` and `Worksheet` objects, they do not represent an independent level in the Excel object hierarchy. They just indicate where these object references are kept.

Each of these objects has its own set of *events*, which are code procedures that fire according to user action, and all three objects (Application, Workbook, and Worksheet) have a group of the similar event procedures with cascading firing: from the top-level object (Application object) to the bottom (Worksheet object) when the workbook is opened or from the bottom-level object (Worksheet object) to the higher one (Application object) whenever the user acts on your worksheet application data.

For example, the Worksheet object (represented by the Sheet1 object, for example) has a Change event that will fire whenever any Sheet1 worksheet cell value changes. Both the Workbook object (represented by the ThisWorkbook object in the VBA Project Explorer tree) and the Application object have a SheetChange event that cascade-fires when any cell changes its values on the Sheet1 worksheet. The order of this event firing is as follows:

```
Worksheet_Change( )  →  Workbook_SheetChange( )  →  Application_SheetChange( )
```

It is up to you where you will place code to control the change of any cell in the Excel environment. If you place the code in the Sheet1_Change event, you will just control any Sheet1 worksheet cell changes. If you place code on the ThisWorkbook_SheetChange event, you may control the changes in any cell of every worksheet of this workbook. And if you place code in the Application_SheetChange event, you can control changes in any cell of every workbook opened in the Excel interface. If every one of these events has a code, they will be cascading, from the bottom level (Worksheet_Change) to the higher level (Application_SheetChange).

You need to study this event order with more attention to absolutely understand how to use them on the behalf of your application usability.

Creating an Excel.Application Object Reference

You already know by now that your application code will run by programming the code module of the ThisWorkbook object, any Sheet object code module, Standard or Class modules, and UserForms code modules of a given Excel workbook, and that all these code modules can be easily selected in the VBA Project Explorer tree.

But when you search for the Application code module in the VBA Project Explorer tree, it is simply not there. So, how you can catch and program the Application object events?

The answer to this question is quite simple: you need to create an object variable that represents the Excel.Application object and propagate its events. And you do this using the Dim WithEvents VBA instruction, like so:

```
Dim WithEvents app as Excel.Application
```

Note that the expression Excel.Application is the one you *must* use when you want to have access to the Microsoft Excel object model from any outside application, such as Microsoft Access code modules (given that the VBA project of the Microsoft Access database has a reference to the correct Microsoft Office x.x Object Library, where x.x is the version number; x.x = 14.0 refers to Microsoft Office 2010 version).

When referencing the Excel.Application object from inside any Microsoft Excel code module, you can simply use this short syntax declaration because of the Application object being the higher programming object on the Excel object model:

```
Dim WithEvents app as Application
```

Whenever you do any of the last two object variable declarations, your code module will be ready to react to any Excel.Application object event procedure. Figure 2-21 shows a Microsoft Excel VBA ThisWorkbook code module where such an object reference is declared. After the declaration of the app object variable using the WithEvents instruction, you can select it in the left Object combo box of the code module window and all the Application events will be shown in the right Procedure combo box of the code module.

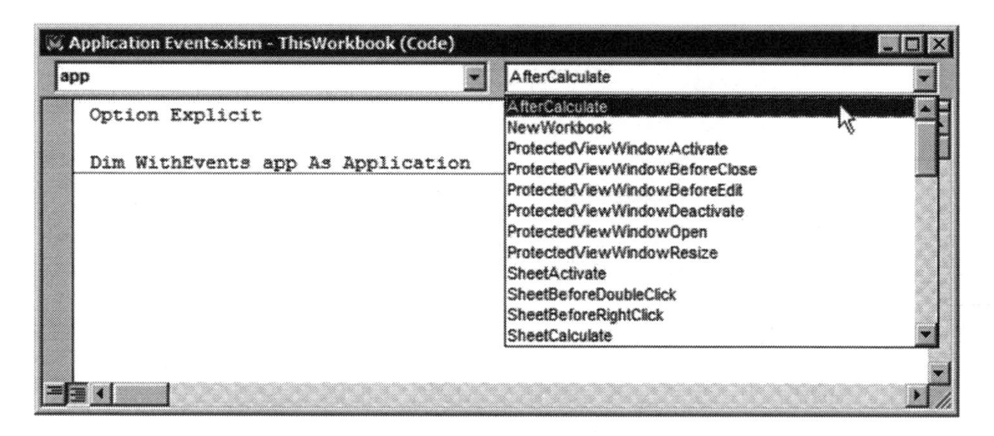

Figure 2-21. *Whenever you declare a variable object using WithEvents and the object fires events, all the events can be selected in the right Procedure combo box of the code module window*

The `Dim WithEvents app as Excel.Application` instruction defines the app object variable that is capable of reacting to all `Application` object events. But it is still not pointing to any Microsoft Excel window application. To do this, I need to put some code on any worksheet application to set the app object variable to the current Excel application window, which is made using such simple code.

```
Set app = Application
```

This last instruction uses the `Application` object that, when used without any prefix, means a pointer to the current Excel window and all its events.

Table 2-6 shows some of the main `Excel.Application` events fired by such an app object variable, when they trigger, and the event declaration code.

■ **Attention** Search the Internet with the keywords *Excel Application Events* to find a complete list of Excel `Application` object events. Table 2-6 shows some of the results returned by this Internet address:

`http://msdn.microsoft.com/en-us/library/microsoft.office.interop.excel.application_events(v=office.15).aspx`

Table 2-6. *Application Object Events and Their Occurrence*

Event name	Occurrence	Event Declaration and Arguments
`AfterCalculate`	After all calculation activities have been completed	`Private Sub app_NewWorkbook(ByVal Wb As Workbook)`
`NewWorkbook`	When a new workbook is created	`Private Sub app_NewWorkbook(ByVal Wb As Workbook)`
`SheetActivate`	When any sheet is activated	`Private Sub app_SheetActivate(ByVal Sh As Object)`
`SheetBeforeDoubleClick`	When any worksheet is double-clicked	`Private Sub app_SheetBeforeDoubleClick(ByVal Sh As Object, ByVal Target As Range, Cancel As Boolean)`
`SheetBeforeRightClick`	When any worksheet is right-clicked	`Private Sub app_SheetBeforeRightClick(ByVal Sh As Object, ByVal Target As Range, Cancel As Boolean)`
`SheetCalculate`	After any worksheet is recalculated	`Private Sub app_SheetCalculate(ByVal Sh As Object)`
`SheetChange`	When cells in any worksheet changed	`Private Sub app_SheetChange(ByVal Sh As Object, ByVal Target As Range)`
`SheetDeactivate`	When any sheet is deactivated	`Private Sub app_SheetDeactivate(ByVal Sh As Object)`
`SheetSelectionChange`	When the selection changes on any worksheet	`Private Sub app_SheetSelectionChange(ByVal Sh As Object, ByVal Target As Range)`
`WindowActivate`	When any workbook window is activated	`Private Sub app_WindowActivate(ByVal Wb As Workbook, ByVal Wn As Window)`
`WindowDeactivate`	When any workbook window is deactivated	`Private Sub app_WindowDeactivate(ByVal Wb As Workbook, ByVal Wn As Window)`
`WindowResize`	When any workbook window is resized	`Private Sub app_WindowResize(ByVal Wb As Workbook, ByVal Wn As Window)`
`WorkbookActivate`	When any workbook is activated	`Private Sub app_WorkbookActivate(ByVal Wb As Workbook)`
`WorkbookAfterSave`	After the workbook is saved	`Private Sub app_WorkbookAfterSave(ByVal Wb As Workbook, ByVal Success As Boolean)`
`WorkbookBeforeClose`	Before any open workbook closes	`Private Sub app_WorkbookBeforeClose(ByVal Wb As Workbook, Cancel As Boolean)`
`WorkbookBeforePrint`	Before any open workbook is printed	`Private Sub app_WorkbookBeforePrint(ByVal Wb As Workbook, Cancel As Boolean)`

(continued)

Table 2-6. (*continued*)

Event name	Occurrence	Event Declaration and Arguments
`WorkbookBeforeSave`	Before any open workbook is saved	`Private Sub app_ WorkbookBeforeSave(ByVal Wb As Workbook, ByVal SaveAsUI As Boolean, Cancel As Boolean)`
`WorkbookDeactivate`	When any open workbook is deactivated	`Private Sub app_WorkbookDeactivate (ByVal Wb As Workbook)`
`WorkbookNewChart`	When a new chart is created in any open workbook	`Private Sub app_ WorkbookNewChart(ByVal Wb As Workbook, ByVal Ch As Chart)`
`WorkbookNewSheet`	When a new sheet is created in any open workbook	`Private Sub app_ WorkbookNewSheet(ByVal Wb As Workbook, ByVal Sh As Object)`
`WorkbookOpen`	When a workbook is opened	`Private Sub app_WorkbookOpen(ByVal Wb As Workbook)`

You can see when the `Application` events in Table 2-6 fire by placing some code on each event procedure, using a VBA `MsgBox( )` function to show a message that identifies it. This a good starting point to see the `Application` event order and how your VBA code can react to them.

Firing Application Events

The `Application SheetChange Event.xlsm` macro-enabled workbook, which you can also extract from the `Chapter2.zip` file, shows how this is done: it declares the app variable on the `ThisWorkbook` code module and uses the `Workbook_Open( )` event procedure (the event that runs only once when the workbook is opened) to set the app object variable to reference the current Excel window.

```
Option Explicit

Dim WithEvents app As Application

Private Sub Workbook_Open()
    Set app = Application
End Sub
```

And once this reference is set, it uses the `app_SheetChange( )` event to catch any changes on any opened workbook cell using this code procedure:

```
Private Sub app_SheetChange(ByVal Sh As Object, ByVal Target As Range)
    Dim strMsg As String
    Dim strTitle As String

    strTitle = Sh.Parent.Name & " workbook changed!"
    strMsg = "File is " & Sh.Parent.Name & vbCrLf
    strMsg = strMsg & "Sheet name is: " & Sh.Name & vbCrLf
    strMsg = strMsg & "Cell affected: " & Target.Address & " = " & Target.Value
    MsgBox strMsg, vbInformation, strTitle
End Sub
```

Note that the Application_SheetChange( ) event passes the Sh as Object and Target as Range arguments, which are used to grab the workbook name, sheet name, and cell address where the change happened.

The first two code rows use the Name property of the Parent property (which represent the workbook where the sheet resides) of the Sh argument (which represents the worksheet where the change happened) to grab the workbook name using the syntax Sh.Parent.Name.

```
strTitle = Sh.Parent.Name & " workbook changed!"
strMsg = "File is " & Sh.Parent.Name & vbCrLf
```

■ **Attention** On most object models, whenever a given object has dependent objects (like Workbook has Worksheets or Worksheets has Ranges), the dependent object frequently exposes a Parent property, which returns an object reference to the object it belongs to, according to the object model hierarchy. Workbook.Parent returns a reference to the Application object, and Worksheet.Parent returns a reference to the Workbook object.

Since the Sh argument represents the worksheet where the change happened, Sh.Parent returns a pointer to the Workbook object that it belongs to, and Sh.Parent.Name returns the workbook name represented by the Workbook.Name property!

Next the code gets the worksheet name where the change happens, using the Name property of the Sh object.

```
strMsg = strMsg & "Sheet name is: " & Sh.Name & vbCrLf
```

And finally it gets the cell address and the new value, using the Address and Value properties of the Target object passed to the event procedure.

```
strMsg = strMsg & "Cell affected: " & Target.Address & " = " & Target.Value
```

The code then shows the changed cell using the VBA MsgBox( ) function.

Try to change any cell value of the Application SheetChange Event.xlsm workbook and you will receive a message as shown in Figure 2-22, stating the workbook name, sheet name, cell address, and value changed.

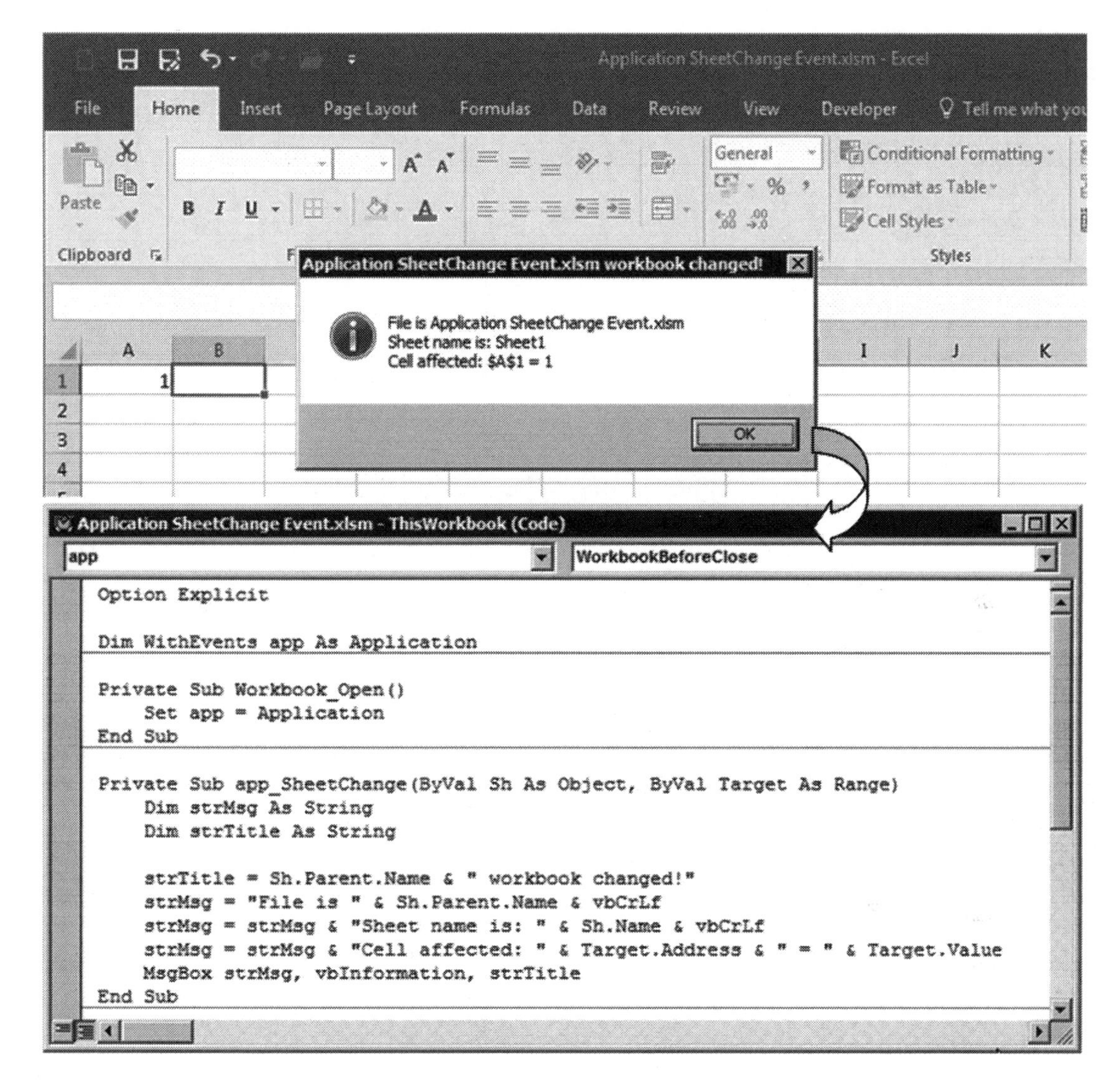

Figure 2-22. *The Application SheetChange Event.xlsm macro-enabled workbook declares Dim WithEvents app as Excel.Application and uses the ThisWorkbook Workbook_Open( ) event to set a reference to the current Excel application, allowing the app variable to react to any event raised by the Application object. By programming the app_SheetChange( ) event, the code can catch any cell change on every workbook opened in this Excel window*

And if you create a new blank workbook or open any existing workbook and change any cell value, this Excel window will continue to fire `app_SheetChange( )` events for every cell changed on any opened workbook. Figure 2-23 shows what happened when a new workbook named `Book3` was created and the `Sheet1.A2` cell had changed.

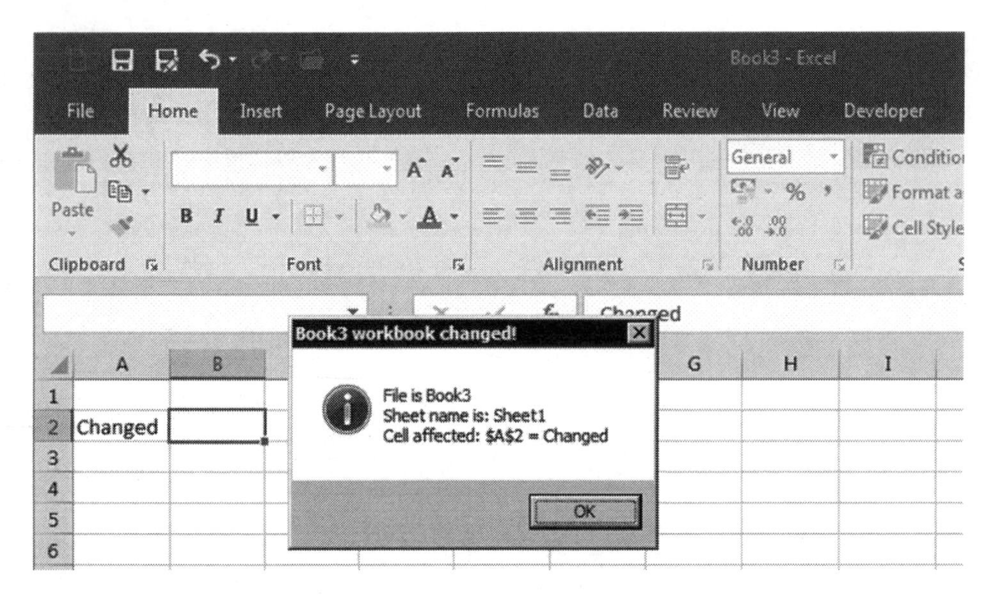

Figure 2-23. Since the Application object takes care of every workbook that is opened inside the Excel window, when you open another workbook, those events are also captured by the app object variable, declared on ThisWorkbook code module

---

■ **Attention** Any object variable reference you set to the Excel `Application` object is very sensible to disturbances on the code module where it is declared. If the code module raises any unpredictable error, you declare another module-level variable, insert another procedure on the code module, modify any code procedure, or simply change the procedure presentation order, the reference to the `Application` object set to the app object variable will be lost, and it will stop raising the events it should.

Using Class Modules to Control Application Object Events

You can control the `Application` object methods and properties from any code module, but if you want to have a distinct code module to represent the `Application` object and its events, you will need to use a `Class` module to program it because the `Dim WithEvents` instruction can be used only in object code modules (the `Thisworkbook`, `Worksheet`, and `UserForm` code modules).

The `Application Events.xlsm` macro-enabled workbook, which you can extract from the `Chapter02.zip` file, has such a `Class` module inside the VBA project, and when you open it, having no other Excel workbook opened (neither the default Blank Workbook), Excel will fire four successive `Application` object events: `WorkbookOpen( )`, `WorkbookActivate( )`, `WindowActivate( )`, and `Application_AfterCalculate( )` (Figure 2-24).

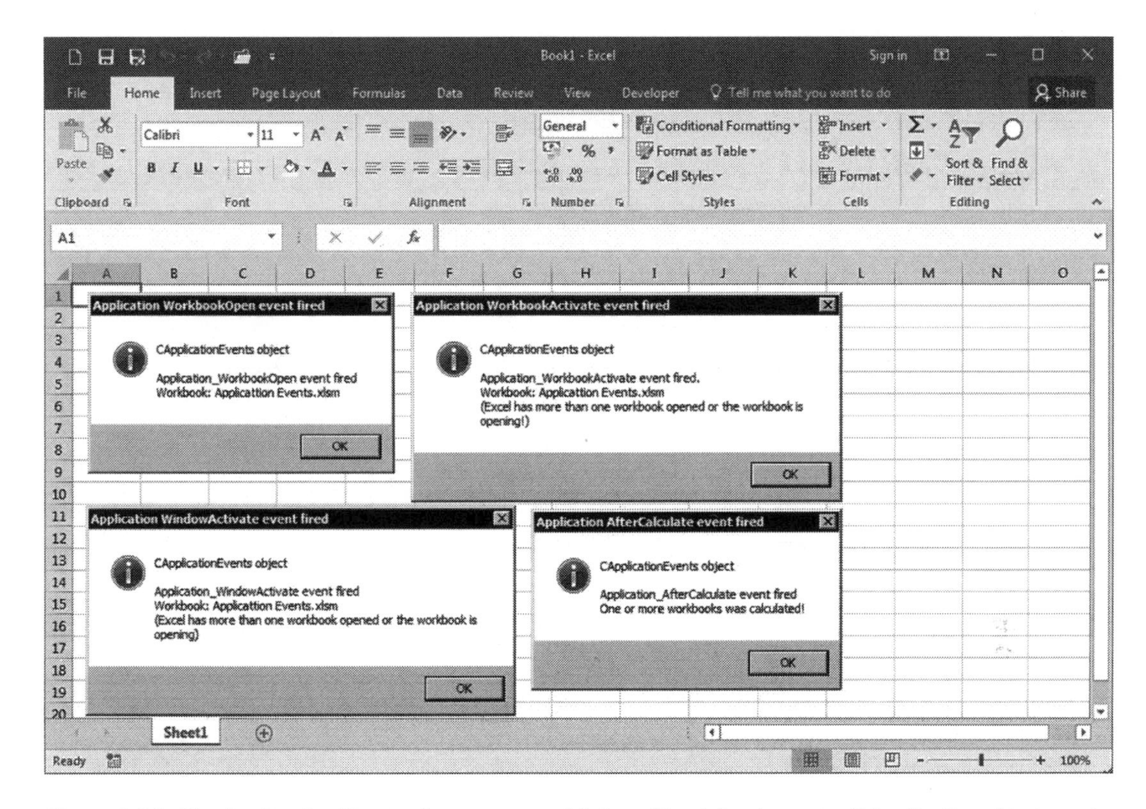

Figure 2-24. *The Application Events.xlsm macro-enabled workbook implements all Application object events cited in Table 2-6 using a VBA Class module named CApplicationEvents*

The `Class` module was named `CApplicationEvents` (as a good programming practice, every `Class` module must begin with a *C*), and the `ThisWorkbook` code module has a module-level object variable declared as `CApplicationEvents`.

```
Option Explicit
Dim app As New CApplicationEvents
```

When the `Application Events.xlsm` macro-enabled workbook is opened, the `WorkbookOpen( )` event of `ThisWorkbook` object fires, creating a new instance of this class.

```
Private Sub Workbook_Open()
    Set app = New CApplicationEvents
End Sub
```

This is the beauty of `Class` modules: they live by themselves once they are instantiated on any part of your code. Whenever a new instance of `CApplicationEvents` is created, the `Class_Initialize( )` event fires, creating a reference to the current Excel window using the same technique described later for the `Application SheetChange Event.xlsm` macro-enabled workbook, as you can see in Figure 2-25, which shows the `CApplicationEvent` code module:

```
Option Explicit

Dim WithEvent.cel.Application
```

```
Private Sub Class_Initialize()
    Set app = Application
End Sub
```

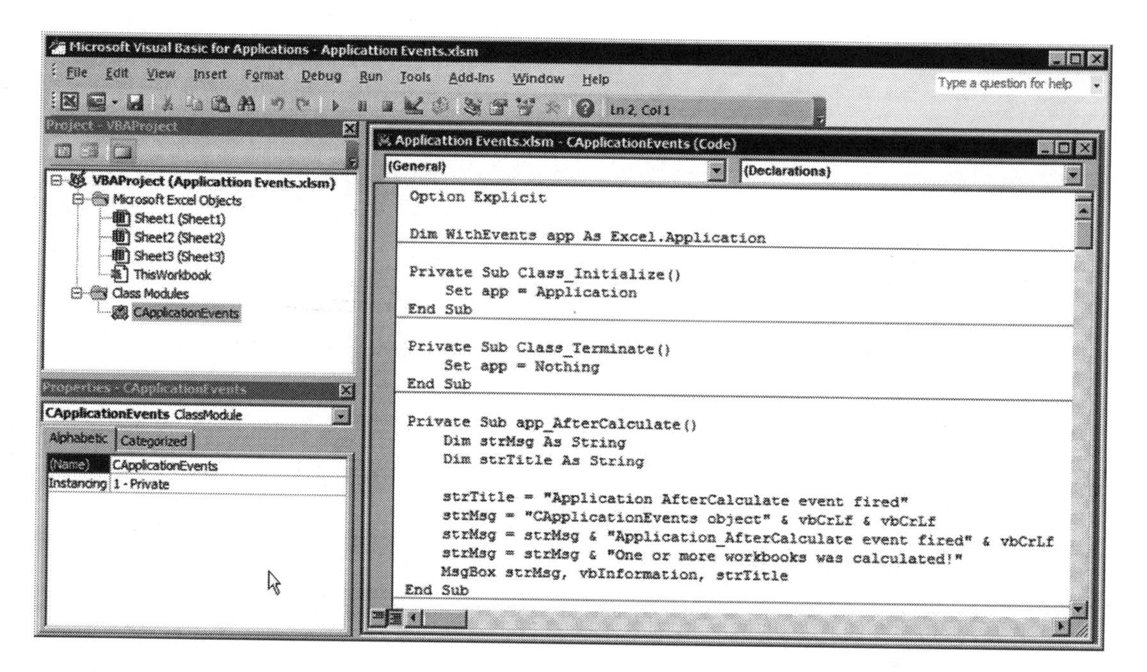

Figure 2-25. *This is the CApplicationEvents Class module from the Application Events.xlsm macro-enabled workbook. It declares a Dim WithEvents app as an Application variable object on the Class module declaration section, uses the Class_Initialize event to set the reference of the current Excel window to the app variable, and codes each Application object event cited in Table 2-6 indicating when they fire*

Now, all `Application` events are propagated to the new instance of the `CApplicationEvents Class` module and are ready to be programmed whenever they fire.

The `CApplicationEvents Class` module has a code procedure for each event cited in Table 2-6, using the VBA `MsgBox( )` function to show the event name that was fired.

Note that the `WorkbookOpen( )` and `WorkbookActivate( )` events have an `Wb as Workbook` argument, which is used to extract the name of the workbook that was opened. This is the code used for the `app_WorkbookOpen( )` event on the `CApplicationEvents Class` module:

```
Private Sub app_WorkbookOpen(ByVal Wb As Workbook)
    Dim strMsg As String
    Dim strTitle As String

    strTitle = "Application WorkbookOpen event fired"
    strMsg = "CApplicationEvents object" & vbCrLf & vbCrLf
    strMsg = strMsg & "Application_WorkbookOpen event fired" & vbCrLf
    strMsg = strMsg & "Workbook: " & Wb.Name & vbCrLf
    MsgBox strMsg, vbInformation, strTitle
End Sub
```

Note that the code identifies the event fired in the title bar of the MsgBox( ) function and then shows in the message dialog the object from where it fired (CApplicationEvents), the event fired, and the workbook name from where it fired. Similar code was used to program the app_WorkbookActivate( ) event.

```
Private Sub app_WorkbookActivate(ByVal Wb As Workbook)
    Dim strMsg As String
    Dim strTitle As String

    strTitle = "Application WorkbookActivate event fired"
    strMsg = "CApplicationEvents object" & vbCrLf & vbCrLf
    strMsg = strMsg & "Application_WorkbookActivate event fired." & vbCrLf
    strMsg = strMsg & "Workbook: " & Wb.Name & vbCrLf
    strMsg = strMsg & "(Excel has more than one workbook opened or the workbook is opening!)"
    MsgBox strMsg, vbInformation, strTitle
End Sub
```

The app_WindowActivate( ) event has two arguments: Wb as Workbook and Wn as Window. The first represents the workbook and is used to identify it in the MsgBox( ) function, and the second represents the current Excel window.

```
Private Sub app_WindowActivate(ByVal Wb As Workbook, ByVal Wn As Window)
    Dim strMsg As String
    Dim strTitle As String

    strTitle = "Application WindowActivate event fired"
    strMsg = "CApplicationEvents object" & vbCrLf & vbCrLf
    strMsg = strMsg & "Application_WindowActivate event fired" & vbCrLf
    strMsg = strMsg & "Workbook: " & Wb.Name & vbCrLf
    strMsg = strMsg & "(Excel has more than one workbook opened or the workbook is opening)"
    MsgBox strMsg, vbInformation, strTitle
End Sub
```

Finally, the app_AfterCalculate( ) event fires, with no arguments.

```
Private Sub app_AfterCalculate()
    Dim strMsg As String
    Dim strTitle As String

    strTitle = "Application AfterCalculate event fired"
    strMsg = "CApplicationEvents object" & vbCrLf & vbCrLf
    strMsg = strMsg & "Application_AfterCalculate event fired" & vbCrLf
    strMsg = strMsg & "One or more workbooks was calculated!"
    MsgBox strMsg, vbInformation, strTitle
End Sub
```

This is the Application object event's sequence order when Excel is opened:

```
WorkbookOpen → WorkbookActivate → WindowActivate → AfterCalculate
```

■ **Attention** If you are wondering why Excel fires both the `WorkbookActivate` and `WindowActivate` events, here is the explanation: the `WindowActivate` event returns an object pointer to the Excel window on its `Wn as Window` argument, which has specific Microsoft Office methods, properties, and events that you can use to control the Microsoft Excel window. Since a single Excel window can open many workbooks at a time (where each one will fire its own events), use the `Workbook_WindowActivate` event to control the Excel window appearance for different workbook applications.

If you try to open another workbook, there will be a change in the event order (considering that the opened workbook will not fire its own events).

WorkbookOpen → **WindowDeActivate** → **WorkbookDeActivate** → WorkbookActivate → WindowActivate → AfterCalculate

This time, the workbook that was first opened (the one that has the `CApplicationEvents Class`) fires the `app_WindowDeactivate( )` and `app_WorkbookDeactivate( )` events (the `MsgBox( )` function will show the name of the workbook that is firing the event). Try it for yourself!

And when you try to close the last opened workbook, the `app_WorkbookBeforeClose( )` event will fire, asking you for a confirmation to close it (Figure 2-26).

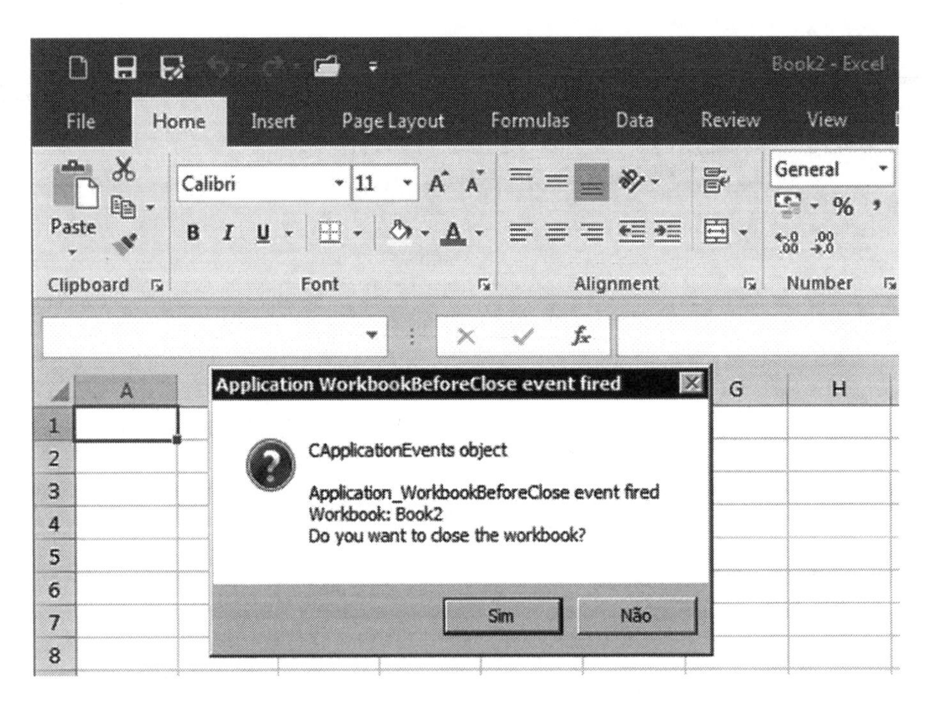

Figure 2-26. *Whenever you try to close a opened workbook, the app_WorkbookBeforeClose( ) event from the CApplicationEvents Class will fire, asking for a confirmation with the MsgBox( ) function*

This is the code used on the app_WorkbookBeforeClose( ) event. Note that it receives two arguments: Wb as Workbook and Cancel as Boolean.

```
Private Sub app_WorkbookBeforeClose(ByVal Wb As Workbook, Cancel As Boolean)
    Dim strMsg As String
    Dim strTitle As String

    strTitle = "Application WorkbookBeforeClose event fired"
    strMsg = "CApplicationEvents object" & vbCrLf & vbCrLf
    strMsg = strMsg & "Application_WorkbookBeforeClose event fired" & vbCrLf
    strMsg = strMsg & "Workbook: " & Wb.Name & vbCrLf
    strMsg = strMsg & "Do you want to close the workbook?"
    If MsgBox(strMsg, vbYesNo + vbQuestion, strTitle) = vbNo Then
        Cancel = True
    End If
End Sub
```

Every event procedure that has Before in its name has a Cancel argument. If you make Cancel=True inside the event procedure, the event is canceled.

Whenever the user clicks the MsgBox( ) No button on the app_WorkbookBeforeClose( ) event, the procedure will make Cancel = True, and the WorkbookBeforeClose( ) event will not fire, meaning that the workbook will be kept open by Excel (try it!).

```
    If MsgBox(strMsg, vbYesNo + vbQuestion, strTitle) = vbNo Then
        Cancel = True
    End If
```

But if you click the MsgBox( ) Yes button, the event sequence will be fired by the CApplicationEvents Class object to close a opened workbook.

WorkbookBeforeClose → WindowDeActivate → WorkbookDeActivate → WorkbookActivate → WindowActivate

There are other Application object events that have Before as part of their names, as cited in Table 2-6. This is the case for the SheetBeforeDoubleClick, SheetBeforeRightClick, WorkbookBeforeClose, WorkbookBeforePrint, and WorkbookBeforeSave events, and the VBA code inserted on each event procedure will always ask if you want to continue with the event processing. Try to double-click any worksheet cell and you will fire the app_SheetBeforeDoubleClick event, receiving a MsgBox( ) that asks if you want to double-click the cell and put it in Edit mode. By answering No, the event will be canceled, and the double-click will not happen (Figure 2-27).

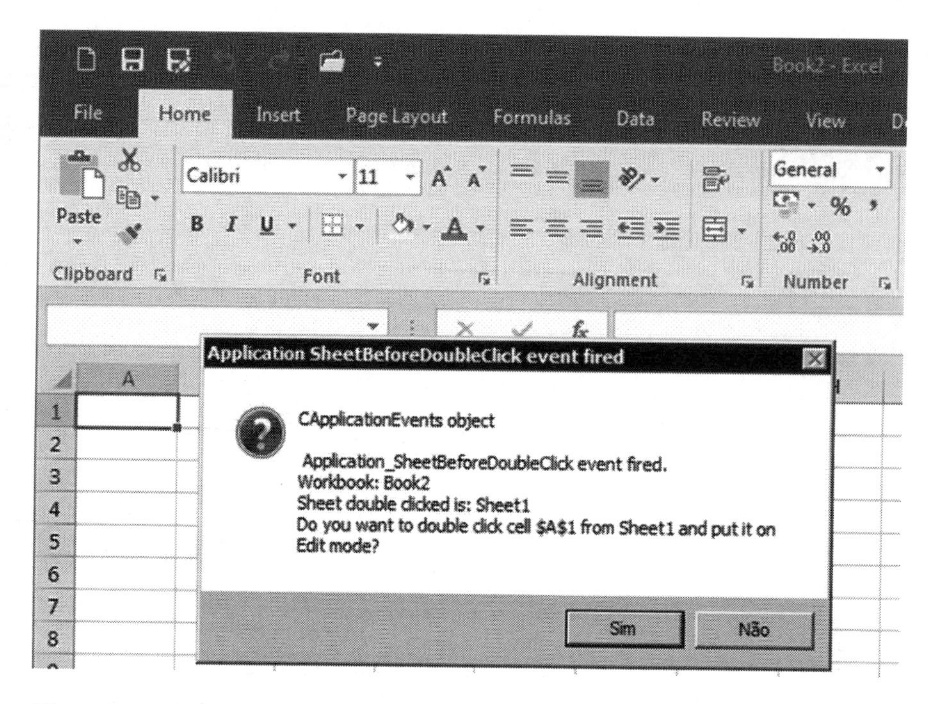

Figure 2-27. *Every event procedure that has Before in its name receives a Cancel argument that allows you to cancel the event. This is the case with the app_SheetBeforeDoubleClick event, which receives three arguments: Sh as Object (which represents the sheet tab), Target as Range (which represents the cell that was double-clicked), and Cancel as Boolean (which can be used to cancel the event)*

This is the code inserted on the app_SheetBeforeDoubleClick( ) event procedure of the CApplicationEvents class. Note how the Cancel argument is turned True when the user clicks the No button of the MsgBox( ) function box displayed by it.

```
Private Sub app_SheetBeforeDoubleClick(ByVal Sh As Object, ByVal Target As Range, Cancel As
Boolean)
    Dim strMsg As String
    Dim strTitle As String

    strTitle = "Application SheetBeforeDoubleClick event fired"
    strMsg = "CApplicationEvents object" & vbCrLf & vbCrLf
    strMsg = strMsg & " Application_SheetBeforeDoubleClick event fired." & vbCrLf
    strMsg = strMsg & "Workbook: " & Sh.Parent.Name & vbCrLf
    strMsg = strMsg & "Sheet double clicked is: " & Sh.Name & vbCrLf
    strMsg = strMsg & "Do you want to double click cell " _
            & Target.Address & " from " & Sh.Name & " and put it on Edit mode?"
    If MsgBox(strMsg, vbYesNo + vbQuestion, strTitle) = vbNo Then
        Cancel = True
    End If
End Sub
```

You will see similar code when you try to right-click any cell or cell range, try to print, or save any opened workbook using the Application Events.xlsm workbook.

To finish the event procedure exercise, try to select any cell range on any worksheet and watch which cells' address range were selected by reading the MsgBox( ) function inserted on the app_SheetSelectionChange( ) event (Figure 2-28).

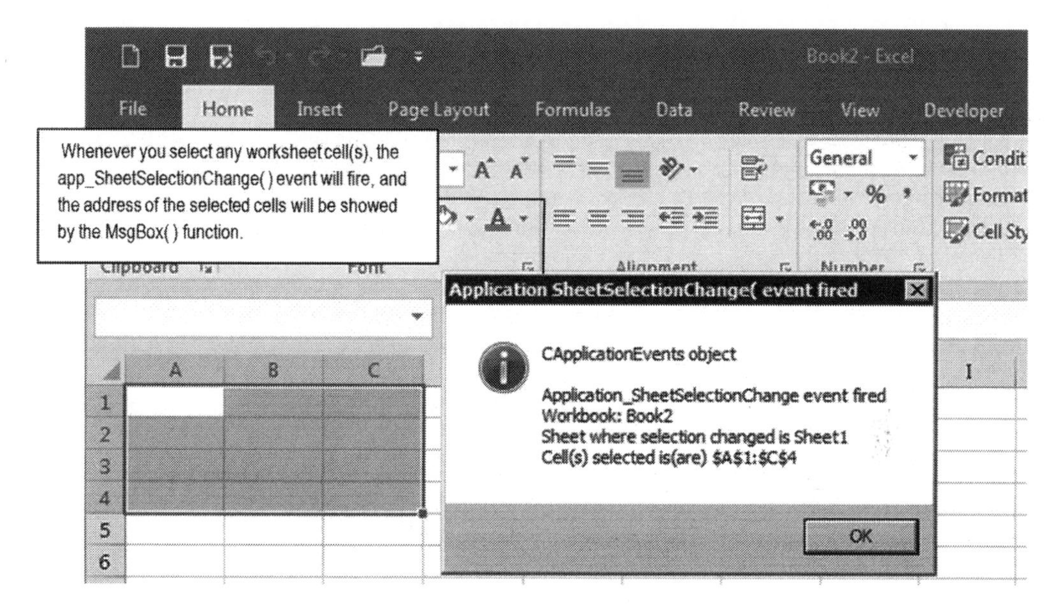

Figure 2-28. *If you select any cell range on any worksheet of the Application Events.xslm workbook, the app_SheetSelectionChange( ) event of the CApplicationEvents Class will fire, and the MsgBox( ) function inserted on it will show the address of the cell(s) selected*

Note in Table 2-6 that the app_SheetSelectionChange( ) event receives two arguments: ByVal Sh As Object and ByVal Target As Range. The first argument, Sh as Object, is an acronym for "Sheet" and has information about the worksheet tab where the cell selection change happened, while the second argument, Target as Range, represents the cell or cell range that was selected on the Sh object. The word ByVal used on both argument declarations means that the objects received by the event have no link with the true object they represent (the sheet tab and selected cells). If you make changes on these object variables inside the event procedure, they will not be propagated to the true object they represent.

```
Private Sub app_SheetSelectionChange(ByVal Sh As Object, ByVal Target As Range)
    Dim strMsg As String
    Dim strTitle As String

    strTitle = "Application SheetSelectionChange( event fired"
    strMsg = "CApplicationEvents object" & vbCrLf & vbCrLf
    strMsg = strMsg & "Application_SheetSelectionChange event fired" & vbCrLf
    strMsg = strMsg & "Workbook: " & Sh.Parent.Name & vbCrLf
    strMsg = strMsg & "Sheet where selection changed is " & Sh.Name & vbCrLf
    strMsg = strMsg & "Cell(s) selected is(are) " & Target.Address
    MsgBox strMsg, vbInformation, strTitle
```

The app_SheetSelectionChange( ) event uses the Parent.Name property of the Sh object to identify the workbook name (parent of the sheet) where the cells have been selected.

```
strMsg = strMsg & "Workbook: " & Sh.Parent.Name & vbCrLf
```

It then uses the Name property of the Sh object to identify the sheet tab name and the .Address property of the Target object to identify the address of the selected cell(s).

```
strMsg = strMsg & "Sheet where selection changed is " & Sh.Name & vbCrLf
strMsg = strMsg & "Cell(s) selected is(are) " & Target.Address
```

If you continue to use the Application Events.xlsm workbook and click the sheet tabs (the app_SheetActivate event fires), change cell values on any worksheet (the app_AfterCalculate, app_SheetChange, and app_SheetSelectionChange events fire), insert a formula on any cell (the app_SheetCalculate, app_AfterCalculate, app_SheetChange, and app_SheetSelectionChange events fire), insert a new worksheet (app_WorkbookNewSheet, app_SheetDeactivate, and app_SheetActivate events fire), delete an existing sheet tab (the app_SheetDeactivate, app_SheetActivate, and app_AfterCalculate events fire), and so on, you will receive a MsgBox( ) function warning you of the event name that has been fired by the user action.

Trying it is the best way to learn the order that Microsoft Excel Application events fire.

■ **Attention** Whenever you close the Application Events.xlsm macro-enabled workbook, the Dim WithEvents app as CApplicationEvents variable will be destroyed, and the Class Terminate event will fire. It is a good programming practice to release any object references used by a Class module on the Terminate event so the CApplicationEvents Class module executes this code to release the object module-level variable app from the association to the Application object (you can see this event procedure in Figure 2-24):

```
Private Sub Class_Terminate( )
    Set App = Nothing
End Sub
```

Inside the Chapter02.zip file you will also find the CApplicationEvents.cls file that has all the code used by the CApplicationEvents Class module. You can use the VBA File ➤ Import menu command to add it to any other Microsoft Excel workbook.

Using a Class Module to Control Sheet Tab Name Changes

Many people ask for an event that fires when the user changes any sheet tab name. Unhappily, Microsoft did not code such an event on any Microsoft Excel object, so you must improvise, using the Application object events in a Class module to control when any sheet tab name changes and let each workbook decide what to do by itself.

You must base this code on a mechanism that is based on Application object events that will work in this way:

1. Declare two Class module variables: one object variable to store a pointer to the current Worksheet object (called mWks) and another String variable to store the current worksheet sheet tab name (called mstrWksName).

2. Whenever the Application Initialize( ) and WorkbookActivate( ) events fire, you will use the Application.ActiveSheet property to set a pointer to the active sheet to the mwks object variable and the Worksheet.Name property to store the active sheet tab name on the mstrWksName string variable, like so:

```
Set mWks = Application.ActiveSheet
mstrWks = mWks.Name
```

3. If the user of your application changes the current sheet tab name, nothing will happen. But if the user changes the selected cell (the Application_SheetChange( ) event fires), tries to select another cell (the Application_SheetSelectionChange( ) event fires), tries to select another sheet tab (the Application_SheetDeactivate( ) event fires), or tries to save the workbook (the Application_WorkbookBeforeSave ( ) event fires), you will use a Sub procedure to compare the current worksheet name with the name stored on the string variable, like so:

```
If mWks.Name <> mstrWksName then
```

4. If the module-level variable name (mWks.Name) has a different name than the current sheet tab name, it indicates that the sheet tab name changed, and the Class module will raise the SheetNameChange( ) event, passing it two arguments: the mWks object variable pointer to the worksheet that suffers the name change and a Cancel argument that allows the code to cancel the change.

Let's see this in action before analyzing the Class module code. Open the Application SheetNameChange Event.xlsm macro-enabled workbook, which you can extract from the Chapter02.zip file and change any sheet tab name: double-click any sheet tab and type a new name (such as Sheet1 to Sheet10).

If you try to change the current cell value, select another cell on the same worksheet, select another sheet tab, or close the workbook, you will receive a MsgBox( ) warning indicating that the sheet tab name can't be changed, and it will be returned to its previous name (see Figure 2-29).

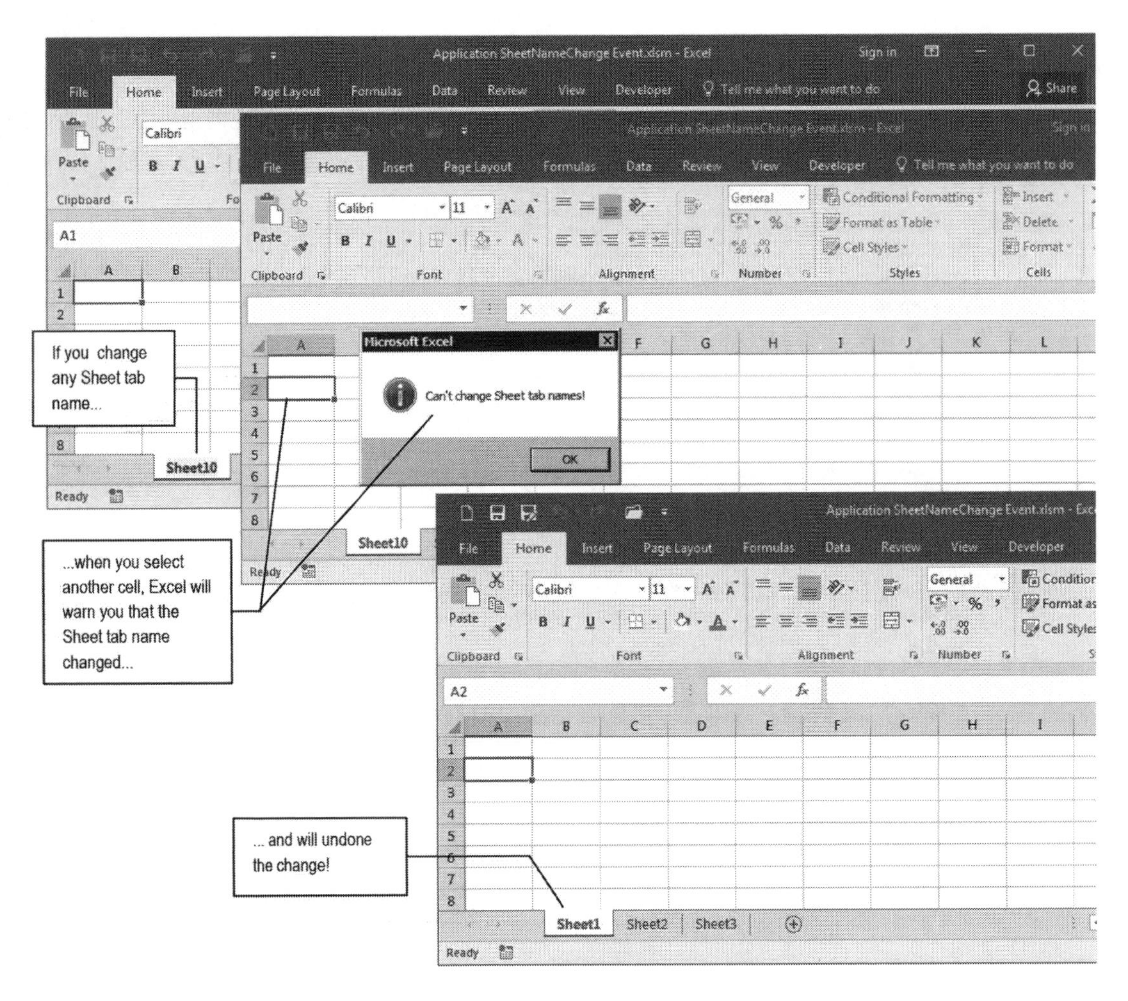

Figure 2-29. *The Application SheetNameChange Event.xlsm macro-enabled workbook has a Class module that uses the Application object events to track when any sheet tab name changes. The change will be perceived if you try to change the current cell value, select another cell, select another sheet tab, or close the workbook, and the name change will be undone*

It works this way: the `Application SheetNameChange Event.xlsm` macro-enabled workbook has the `CSheetNameChange Class` module, which declares in the `Declaration` section the event procedure `SheetNameChange( )` and three module-level variables: `app`, `mwks`, and `mstrWksName`.

```
Option Explicit

Event SheetNameChange(ByVal Sh As Object, ByRef Cancel As Boolean)

Private WithEvents app As Excel.Application
Private mWks As Worksheet
Private mstrWksName As String
```

Note that the event SheetNameChange( ) passes two arguments: ByVal Sh as Object and ByRef Cancel as Boolean, meaning that whenever this event is raised, the code cannot make any change to the Sh object (which is passed *by value* to represent the worksheet) but will allow you to make changes to the Cancel argument. Also note that it uses the same object names employed by some Application object events to keep tuned with the Microsoft events syntax.

Whenever a new instance of the CSheetNameChange Class module is created, the Class_Initialize( ) event fires, the app object variable is set to the current Excel window, and the mWks object variable is set to the active worksheet.

```
Private Sub Class_Initialize()
    Set app = Application
    Set mWks = ActiveSheet
    mstrWksName = mWks.Name
End Sub
```

And as a good programming practice, whenever the CSheetNameChange Class is destroyed (which will happen when the Application SheetNameChange Event.xlsm workbook is closed), the Class_Terminate( ) event fires, releasing the object variables.

```
Private Sub Class_Terminate()
    Set mWks = Nothing
    Set app = Nothing
End Sub
```

Although the Class Initialize( ) event correctly sets the mWks object variable to the Application. ActiveSheet property and the mstrWksName string variable to the active worksheet name, if you open another workbook and then close it, the mWks reference will be lost. To avoid future VBA automation errors on the code, you also need to code the Application object WorkbookActivate( ) event, resetting the mWks and mstrWksName variables every time another workbook is activated to reflect the new active worksheet properties.

```
Private Sub app_WorkbookActivate(ByVal Wb As Workbook)
    Set mWks = ActiveSheet
    mstrWksName = mWks.Name
End Sub
```

The CSheetNameChange Class uses the Private Sub NameChange( ) procedure to control what it should do whenever any sheet tab name changes.

```
Private Sub NameChange(Optional wks As Worksheet)
    Dim bolCancel As Boolean

    If wks Is Nothing Then
        Set wks = app.ActiveSheet
    End If

    If mstrWksName <> mWks.Name Then
        RaiseEvent SheetNameChange(mWks, bolCancel)
        If bolCancel Then
            mWks.Name = mstrWksName
        End If
    End If
```

```
    Set mWks = wks
    mstrWksName = mWks.Name
End Sub
```

Note that Sub NameChange( ) declares the Optional wks as Worksheet argument, and inside the procedure it declares the bolCancel as Boolean variable.

```
Private Sub NameChange(Optional wks As Worksheet)
    Dim bolCancel As Boolean
```

The first procedure action verifies that the Optional wks as Worksheet argument was received. If it was not received, it is defined to the current worksheet using the Application.ActiveSheet property. This is necessary because when any different sheet tab is selected, you will need to grab the reference and name to keep following sheet tab name changes, as you see here:

```
    If wks Is Nothing Then
        Set wks = app.ActiveSheet
    End If
```

The Sub NameChange( ) procedure then verifies if the worksheet name stored on the mstrWksName string module-level variable differs from the current worksheet name.

```
    If mstrWksName <> mWks.Name Then
```

If this is true, it means that the worksheet sheet tab name changed, and the SheetNameChange( ) event is raised, passing as arguments the mWks object variable (pointer to current worksheet) and the bolCancel Boolean variable.

```
        RaiseEvent SheetNameChange(mWks, bolCancel)
```

▪ **Attention** Compare this last row to the SheetNameChange( ) Event declaration and you will note that mWks is passed to the Sh argument and bolCancel is passed to the Cancel event argument.

When an event is raised by an object, it doesn't mean that it will be coded. But if it is coded by the application that uses an instance of the CSheeNameChange class, the code can define the Cancel SheetNameChange( ) event argument to True, as an indication that the event (sheet name change) *must be canceled*. So, the next line of code will verify if bolCancel = True, meaning that the sheet name change must be undone. If it is, it will change the mWks object variable Name property to its previous name, stored on the mstrWksName string variable.

```
        If bolCancel Then
            mWks.Name = mstrWksName
        End If
```

And the last procedure code associates again the mWks object variable to the optional Wks argument (which points to the active sheet) and the mstrWksName string variable to worksheet name.

```
    Set mWks = wks
    mstrWksName = mWks.Name
```

The action happens on the CSheetNameChange Class when some Application event fires. These are the codes for the App_SheetActivate( ), app_SheetChange( ), app_SheetDeactivate( ), and app_SheetSelectionChange( ) events. Note that all event procedures make a simple call to the Sub NameChange( ) procedure, passing it the Sh argument, which means the activated sheet that is passed by the Application object event procedure.

```
Private Sub app_SheetActivate(ByVal Sh As Object)
    Call NameChange(Sh)
End Sub

Private Sub app_SheetActivate(ByVal Sh As Object)
    Call NameChange(Sh)
End Sub

Private Sub app_SheetChange(ByVal Sh As Object, ByVal Target As Range)
    Call NameChange(Sh)
End Sub
Private Sub app_SheetDeactivate(ByVal Sh As Object)
    Call NameChange(Sh)
End Sub

Private Sub app_SheetSelectionChange(ByVal Sh As Object, ByVal Target As Range)
    Call NameChange(Sh)
End Sub
```

▒ **Attention** There is really no need to use the VBA Call keyword when you want to call a Sub procedure from another procedure. But I recommend you do so as good programming practice to make your code clear.

The only Application event that does not pass the Sh argument is the App_WorkbookBeforeSave event. This event calls the Sub NameChange( ) procedure without passing it any argument (and that is why NameChange( ) declares the wks as Optional argument).

```
Private Sub app_WorkbookBeforeSave(ByVal Wb As Workbook, ByVal SaveAsUI As Boolean, Cancel
As Boolean)
    Call NameChange
End Sub
```

And this is all the code that the CSheeNameChange Class module has. Now you must look at how Class is used in the Application SheetNameChange Event.xlsm workbook. The ThisWorkbook code module begins declaring an object variable of the CSheetNameChange Class using the VBA WithEvents keyword so their events can be raised on this module.

```
Option Explicit

Dim WithEvents mSheetName As CSheetNameChange
```

Now Class must be instantiated on the ThisWorkbook Workbook_Open( ) event, the one that fires whenever the workbook is opened.

```
Sub Workbook_Open()
    Set mSheetName = New CSheetNameChange
End Sub
```

And that is all! Now, whenever any sheet tab name changes in this workbook, this change will be perceived if you try to change any cell, select another cell, select another sheet tab, or try to save the workbook. And this perception is made by programming the msheetName object variable's SheetNameChange( ) event, which passes the Sh as Object (representing the active worksheet) and Cancel arguments.

```
Private Sub mSheetName_SheetNameChange(ByVal Sh As Object, Cancel As Boolean)
    MsgBox "Can't change Sheet tab names!", vbInformation
    Cancel = True
End Sub
```

Figure 2-30 explains how this happens in the ThisWorkbook VBA code.

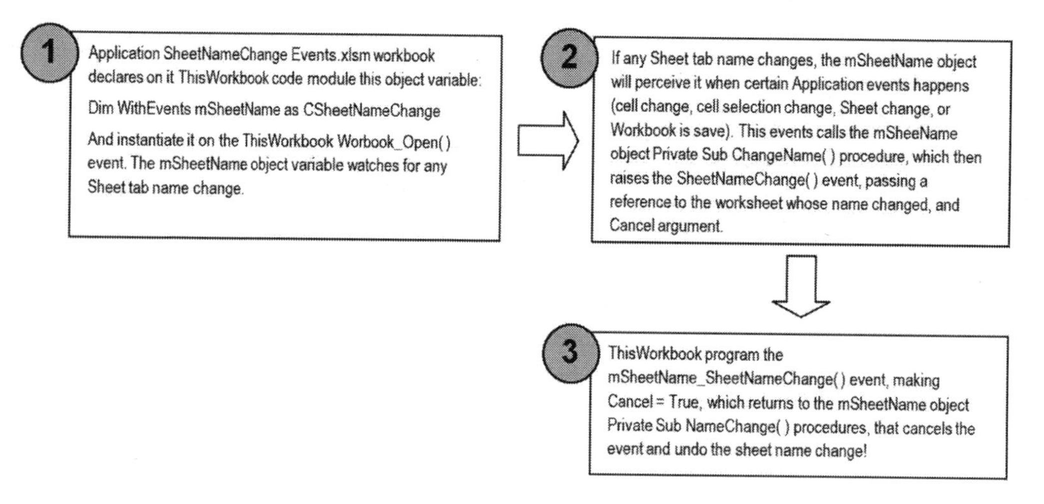

Figure 2-30. *This is a view of how the CSheetNameChange Class is used to control any sheet name changes in the Application SheetNameChange Event.xlsm macro-enabled workbook*

■ **Attention** In the Chapter02.zip file you will also find the CSheetNameChange.cls file, which has all CSheetNameChange Class module code. Use the VBA File ➤ Import menu command to add this class module to any Excel workbook VBA project and take control of the sheet name change.

Chapter Summary

In this chapter, you learned the following:

- The hierarchy of the Microsoft Excel object model
- How to access some workbook objects using important `Application` object properties
- Which are the main `Application` object methods, what they do, and how to use them
- How to use the `Application` object `FileDialog` method
- How to use the `Application` object `GetOpenFileName` and `GetSaveFileAs` methods to select the path and file name to load and save files
- How to use the `Application` object's `InputBox` method to grab Microsoft Excel data and formulas
- How to use the `Application` object's `OnTime` method to create a timer on your applications
- How to declare and instantiate an object variable to program `Application` events
- How to use a `Class` module to easily propagate `Application` object events
- How to create a `CSheetNameChange` `Class` to control any sheet tab name changes

In the next chapter, you will learn about the next object on Microsoft Excel object model hierarchy: the `Workbook` object (and `Workbooks` collection).

CHAPTER 3

Programming the Microsoft Excel Workbook Object

In this chapter you will learn about the second object in the Microsoft Excel object model hierarchy: the `Workbook` object. Like its parent (the `Application` object), it has a rich interface with many properties, methods, and events that you should be aware of to program your application with VBA. You can obtain all the procedure code in this chapter by downloading the `Chapter03.zip` file from the book's Apress.com product page, located at `www.apress.com/9781484222041`, or from `http://ProgrammingExcelWithVBA.4shared.com`.

The Workbook Object

The Microsoft Excel `Workbook` object represents each individual Excel file and has a lot of properties, methods, and events that you can interact with using VBA code to take control of your application. Table 3-1 lists some important `Workbook` object properties, their values, and what they mean.

Attention Search the Internet with the keywords *workbook properties* to find a complete list of Excel `Workbook` object properties. Table 3-1's brief list comes from the following location on the Microsoft MSDN web site:

`http://msdn.microsoft.com/en-us/library/microsoft.office.tools.excel.workbook_properties.aspx`

F. Morgado, *Programming Excel with VBA*, DOI 10.1007/978-1-4842-2205-8_3

Table 3-1. *Some Important Microsoft Excel Workbook Object Properties*

Workbook Object Property	Value Returned	Represents
ActiveChart	Chart	The active Microsoft Excel Chart object. When no chart is active, this property returns Nothing.
ActiveSheet	Worksheet	A reference to the active Worksheet (the sheet on top).
Application	Application	A reference to the current Microsoft Excel Application (the one that opened this workbook).
Parent	Application	A reference to the current Microsoft Excel Application (the one that opened this workbook).
CodeName	String	Name of the workbook code module (usually ThisWorkbook).
CommandBars	Object	Gets a CommandBars object that represents the Microsoft Office Excel command bars.
FullName	String	File name of the associated Workbook.
HasPassword	Boolean	Indicates whether the workbook has a protection password.
Names	Object	Returns a Collection object with all range names in the workbook (including worksheet-specific names).
Password	String	Gets or sets the password that must be supplied to open the workbook.
Path	String	Gets the complete path to the associated Workbook.
ProtectStructure	Boolean	Indicates whether the order of the sheets in the workbook is protected.
ProtectWindows	Boolean	Indicates whether the windows of the workbook are protected.
Saved	Boolean	Indicates whether any changes have been made to the workbook since it was last saved.
Sheets	Object	Returns a Collection that represents all the sheets in the workbook.
Worksheets	Object	Same as Sheets; returns a Collection that represents all the sheets in the workbook.

Table 3-2 lists some important Workbook object methods and the actions they perform when evoked by your VBA code.

■ **Attention** Search the Internet with the keywords *workbook methods* to find a complete list of Excel Workbook object properties. Table 3-2's brief list comes from the following location on the Microsoft MSDN web site:

http://msdn.microsoft.com/en-us/library/microsoft.office.tools.excel.workbook_methods.aspx

Table 3-2. *Some Important Microsoft Excel Workbook Object Methods*

Workbook Object Method	Action Performed
`Activate`	Activates the first window associated with the workbook
`Close`	Closes the workbook
`PrintOut`	Prints the workbook (all worksheets)
`Protect`	Protects a workbook so it cannot be modified, with or without a password
`ProtectSharing`	Saves the workbook and protects it for sharing, with or without a password
`Save`	Saves the workbook
`SaveAs`	Saves the workbook in a different file or folder
`SaveCopyAs`	Saves a copy of the workbook to another file, keeping intact the open workbook in memory
`Unprotect`	Removes workbook protection if it was set; may require a password
`UnprotectSharing`	Turns off protection for sharing and saves the workbook; may require a password

Using Workbook Object Events

The `Workbook` object has a subset of the `Application` object events that you can use to react to the actions made by the user of your application. Since you now understand that the Microsoft Excel `Application` object has a set of event procedures that will fire for every workbook and worksheet opened inside the same Excel window, you must also understand that the `Workbook` object events must be used to control just one single workbook and its worksheets—the ones associated with the `ThisWorkbook` code module of the VBA project (if there is more than one workbook opened, each one will have its own `ThisWorkbook` object on the VBA Explorer tree).

Note, however, that the event order of all these events (`Application` object events and `Workbook` object events) will first fire on the dependent object and then on the parent object, according to the Microsoft Excel object model cited in Figure 2-1. In other words, the `Workbook_Activate` event of the `ThisWorkbook` object will fire before the `Application_WorkbookActivate` event of the `Application` object, if both are programmed on the VBA project.

Table 3-3 shows some of the main `Workbook` object events, when they trigger, and their event declaration code. Note that they are quite similar in name to the related `Application` object events preceded by the `Workbook_` prefix, and since they refer to the `ThisWorkbook` object, they don't need to pass the `Wb as Workbook` object as an argument for any of the event procedures.

■ **Attention** Search the Internet with the keywords *workbook events* to find a complete list of Excel `Workbook` object events. Table 3-3's brief list comes from the following location on the Microsoft MSDN web site:

`http://msdn.microsoft.com/en-us/library/microsoft.office.tools.excel.workbook_events.aspx`

***Table 3-3.** Workbook Object Events*

Event Name	Occurrence	Event Declaration and Arguments
`Activate`	When Excel has more than one workbook opened and the workbook is activated.	`Private Sub Workbook_Activate()`
`AfterSave`	After the workbook is saved.	`Private Sub Workbook_AfterSave(ByVal Success As Boolean)`
`BeforeClose`	Before the workbook closes. If the workbook has been changed, this event occurs before the user is asked to save changes.	`Private Sub Workbook_BeforeClose(Cancel As Boolean)`
`BeforePrint`	Before the workbook (or anything in it) is printed.	`Private Sub Workbook_BeforePrint(Cancel As Boolean)`
`BeforeSave`	Before the workbook is saved.	`Private Sub Workbook_BeforeSave(ByVal SaveAsUI As Boolean, Cancel As Boolean)`
`Deactivate`	When the workbook is deactivated.	`Private Sub Workbook_Deactivate()`
`NewChart`	When a new chart is created in the workbook.	`Private Sub Workbook_NewChart(ByVal Ch As Chart)`
`NewSheet`	When a new sheet is created in the workbook.	`Private Sub Workbook_NewSheet(ByVal Sh As Object)`
`Open`	When the workbook is opened.	`Private Sub Workbook_Open()`
`SheetActivate`	When any sheet is activated.	`Private Sub Workbook_SheetActivate(ByVal Sh As Object)`
`SheetBeforeDoubleClick`	When any worksheet is double-clicked, before the default double-click action.	`Private Sub Workbook_SheetBeforeDoubleClick( ByVal Sh As Object, ByVal Target As`
`SheetBeforeRightClick`	When any worksheet is right-clicked, before the default right-click action.	`Private Sub Workbook_SheetBeforeRightClick( ByVal Sh As Object, ByVal Target As`
`SheetCalculate`	After any worksheet is recalculated or after any changed data is plotted on a chart.	`Private Sub Workbook_SheetCalculate(ByVal Sh As Object)`
`SheetChange`	When cells in any worksheet are changed by the user or by an external link.	`Private Sub Workbook_SheetChange(ByVal Sh As Object, ByVal Target As Range)`
`SheetDeactivate`	When any sheet is deactivated.	`Private Sub Workbook_SheetDeactivate(ByVal Sh As Object)`
`SheetSelectionChange`	When another cell (or cells) is selected on any worksheet.	`Private Sub Workbook_SheetSelectionChange( ByVal Sh As Object, ByVal Target As Range)`

(*continued*)

Table 3-3. (*continued*)

Event Name	Occurrence	Event Declaration and Arguments
`WindowActivate`	When any workbook window is activated.	`Private Sub Workbook_WindowActivate(ByVal Wn As Window)`
`WindowDeactivate`	When any workbook window is deactivated.	`Private Sub Workbook_WindowDeactivate(ByVal Wn As Window)`
`WindowResize`	When any workbook window is resized.	`Private Sub Workbook_WindowResize(ByVal Wn As Window)`

Once again, you can see when these workbook events fire by placing some code on each event procedure that uses a VBA `MsgBox( )` function to show a message that identifies the event. The `Workbook Events.xlsm` Excel macro-enabled workbook that you can extract from the `Chapter03.zip` file has such code for all events cited in Table 3-3. When you open the `Workbook Events.xlsm` Excel macro-enabled workbook, you will receive three successive `MsgBox( )` dialogs saying that Excel successively fired the `Workbook_Open`, `Workbook_Activate`, and `Workbook_WindowActivate` events (Figure 3-1).

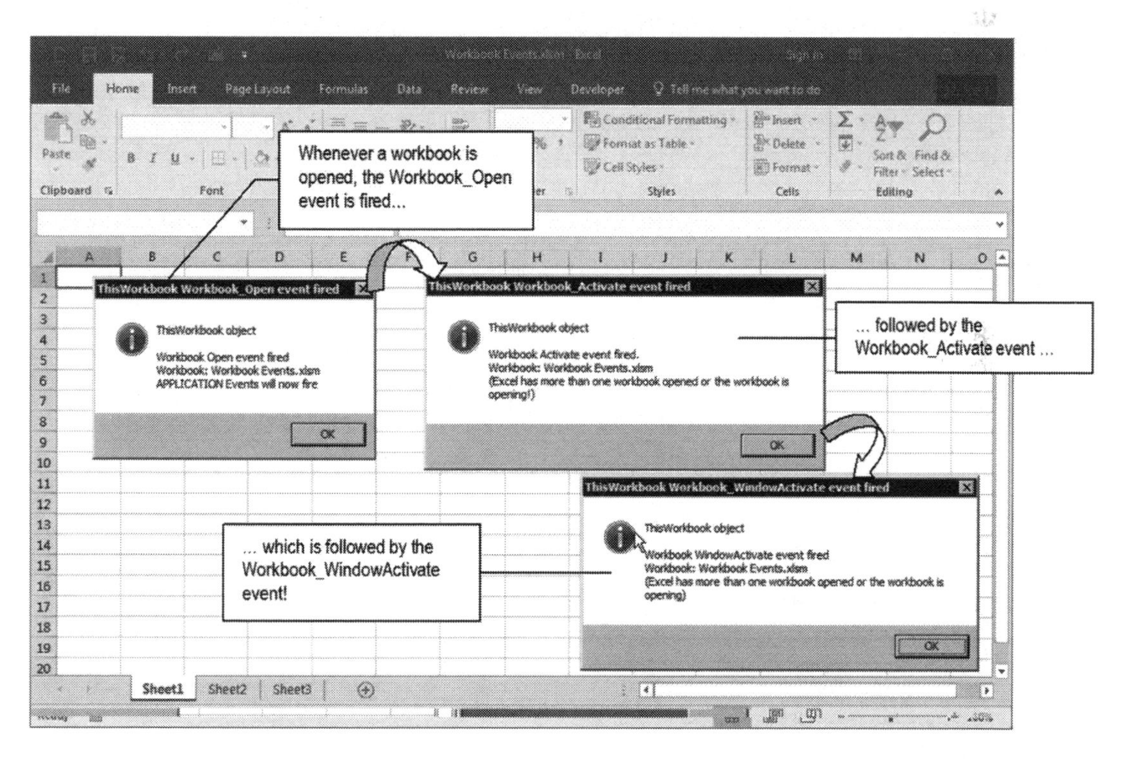

Figure 3-1. *The Workbook Events.xlsm macro-enabled workbook has VBA MsgBox( ) functions inserted on all event procedures cited in Table 3-1, so you can watch when they fire. Whenever you open any workbook, you will see that Excel successively fires the Workbook_Open, Workbook_Activate, and Workbook_WindowActivate events*

The next lines of code show how each one of these event procedures was coded on the ThisWorkbook code module of the Workbook Events.xlsm macro-enabled workbook.

```
Private Sub Workbook_Open()
    Dim strMsg As String
    Dim strTitle As String

    strTitle = "ThisWorkbook Workbook_Open event fired"
    strMsg = "ThisWorkbook object" & vbCrLf & vbCrLf
    strMsg = strMsg & "Workbook Open event fired" & vbCrLf
    strMsg = strMsg & "Workbook: " & ThisWorkbook.Name & vbCrLf
    strMsg = strMsg & "APPLICATION Events will now fire"
    MsgBox strMsg, vbInformation, strTitle
End Sub

Private Sub Workbook_Activate()
    Dim strMsg As String
    Dim strTitle As String

    strTitle = "ThisWorkbook Workbook_Activate event fired"
    strMsg = "ThisWorkbook object" & vbCrLf & vbCrLf
    strMsg = strMsg & "Workbook Activate event fired." & vbCrLf
    strMsg = strMsg & "Workbook: " & ThisWorkbook.Name & vbCrLf
    strMsg = strMsg & "(Excel has more than one workbook opened or the workbook is
    opening!)"
    MsgBox strMsg, vbInformation, strTitle
End Sub

Private Sub Workbook_WindowActivate(ByVal Wn As Window)
    Dim strMsg As String
    Dim strTitle As String

    strTitle = "ThisWorkbook Workbook_WindowActivate event fired"
    strMsg = "ThisWorkbook object" & vbCrLf & vbCrLf
    strMsg = strMsg & "Workbook WindowActivate event fired" & vbCrLf
    strMsg = strMsg & "Workbook: " & ThisWorkbook.Name & vbCrLf
    strMsg = strMsg & "(Excel has more than one workbook opened or the workbook is opening)"
    MsgBox strMsg, vbInformation, strTitle
End Sub
```

By coding the same event procedures using the Application object, as covered in Chapter, your code will successively fire events for both objects (Workbook and Application, in this order), which can eventually turn your application into a mess.

It is easy to implement all Application object events on the Workbook Events.xlsm macro-enabled workbook! Just follow these steps:

1. Extract from the Chapter02.zip the CApplicationEvents.cls file (or open the Application Events.xlsm macro-enabled workbook, show the VBA project, click the CApplicationEvents class module in the VBA Explorer tree, and export it using the VBA File > Export menu command).

2. In the VBA Explorer tree, be sure to select the `Workbook Events.xlsm` project (in case more than one workbook is opened inside Excel) and execute the VBA File ➤ Import menu command.

3. Select the `CApplicationEvents.cls` class module file and click OK to import it to the `Workbook Events.xlsm` macro-enabled VBA project.

4. Double-click the `ThisWorkbook` object in the VBA Explorer tree to show its code module and declare an object variable to represent an instance of the `CApplicationEvents` class, in this way:

```
Dim mapp as CApplicationEvents
```

5. In the `ThisWorkbook` class module, select the `Workbook_Open` event procedure and add this line of code right below the `Dim` variable declaration instructions to create the instance of the `CApplicationEvents` class:

```
Set mapp = New CApplicationEvents
```

6. Save the `Workbook Events.xlsm` file with a new name, close the workbook, and open it again to see it successively fire both the `Workbook` and `Application` object events (Figure 3-2).

▩ **Attention** To make the file `Workbook Events.xlsm` fire the `Application` object events, you can call the `ThisWorkbook Workbook_Open` event procedure from the VBA Immediate window by typing the following code instruction (without using the `?` print character):

```
Thisworkbook.Workbook_Open
```

Since the workbook is already opened, you will not see that it fires the associated `app_WorkbookOpen( )` events.

The `Application` and `Workbook Events.xlsm` macro-enabled workbook that you can extract from the `Chapter02.zip` file already has the `CApplicationEvents Class` module instantiated on its `ThisWorkbook Workbook_Open( )` event procedure.

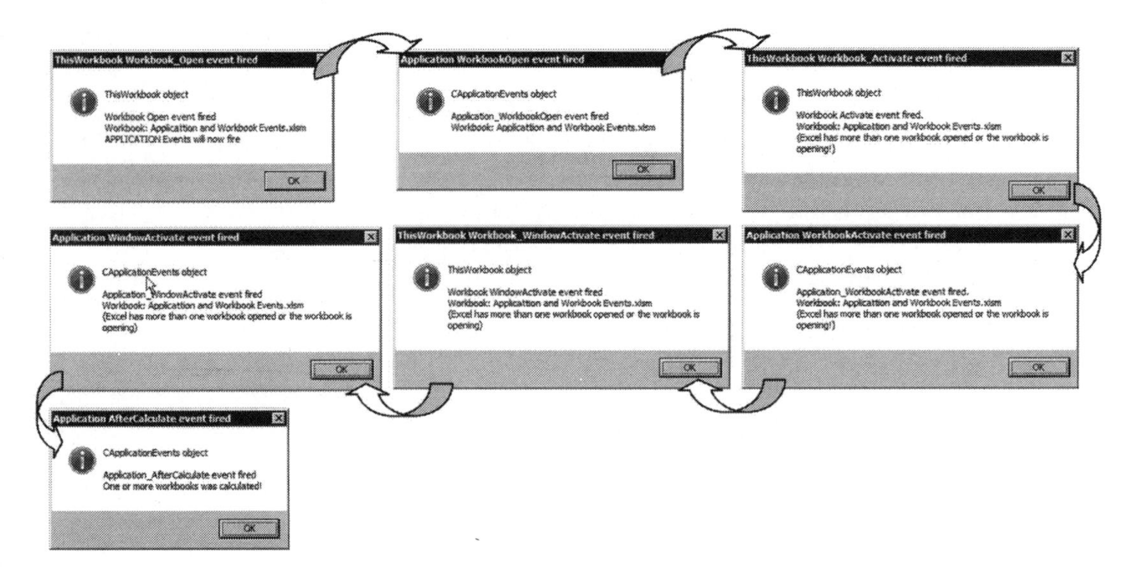

Figure 3-2. *If you add a copy of the CApplicationEvents Class module to the Workbook Events.xlsm macro-enabled workbook, it will fire both the Workbook and Application object events (in this order), which eventually may make a mess in your worksheet application code. The Application and Workbook Events. xlsm macro-enabled workbook has a CApplicationEvents Class module and fires both the Workbook and Application events. Note the arrows indicating the event procedure order*

Note in Figure 3-2 that Excel now fires this event order whenever the `Application` and `Workbook Events.xlsm` macro-enabled workbook is opened:

```
ThisWorkbook.Workbook_Open → Application.WorkbookOpen → ThisWorkbook.Workbook_Activate →
→ Applicaton.WorkbookActivate → ThisWorkbook.Workbook_WindowActivate → Application.
WindowActivate → Application_AfterCalculate
```

As you can see, both `Workbook` and `Application` object events fire in pairs. First `ThisWorkbook.Workbook_Open` fires, followed by its parent event `Application_WorkbookOpen`, and so on.

Workbook Open Event and the frmSplashScreen UserForm

Let's try some practical VBA code procedures using the `Workbook_Open` event, which fires whenever a workbook is opened.

You must use the `Workbook_Open` event to show the application splash screen to initialize `Class` modules or values you must use on your application and/or to redirect the user to a specific workbook point of your application (select a specific worksheet cell).

The main use of the `Workbook.Open` event is to show your Excel application *splash screen*, a VBA `UserForm` that automatically appears and disappears after a few seconds to identify your application and yourself as the creator or owner of the solution, similar to the way Excel does when you open it. Thus, this simple example is a good one to improve your VBA knowledge regarding basic programming steps.

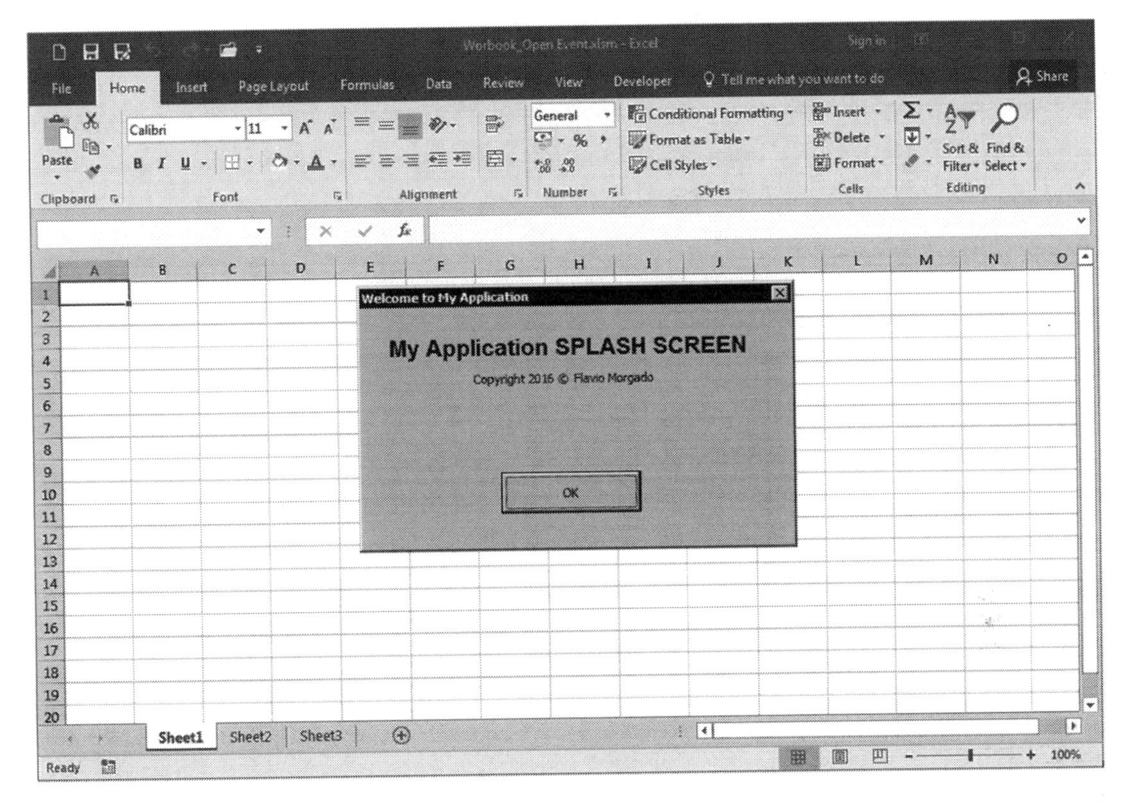

Figure 3-3. *This is the frmSplashScreen UserForm from the Workbook_Open Event.xlsm Excel macro-enabled workbook that is opened by the code inserted in the Workbook_Open event of the ThisWorkbook object*

Open the `Workbook_Open Event.xlsm` Excel macro-enabled workbook that you can find inside the `Chapter03.zip` file and note that it shows a VBA `UserForm` over the `Sheet1` interface to welcome the user to the (supposed) application (Figure 3-3).

Since any `UserForm` is opened as a modal window, you need to click the OK button of the `UserForm` to close it and then get access to the workbook itself.

If you activate the VBA interface by pressing Alt+F11 in Excel, you will see that the Project Explorer tree has just one object inserted on its `Forms` branch, named `frmSplashScreen`, while the `ThisWorkbook` code module is opened to show the next simple code that does this trick inside its `Workbook_Open` event procedure (Figure 3-4):

```
Private Sub Workbook_Open()
    frmSplashScreen.Show
End Sub
```

When the VBA project is opened, double-click the `frmSplashScreen` object in the `Forms` tree and note the next properties of `frmSplashScreen` by inspecting the VBA Properties window:

- `Name = frmSplashScreen` (to change the default `UserForm1` name)
- `Caption = "Welcome to My Application"`, the text that appears in the title bar of the `frmSplashScreen` window

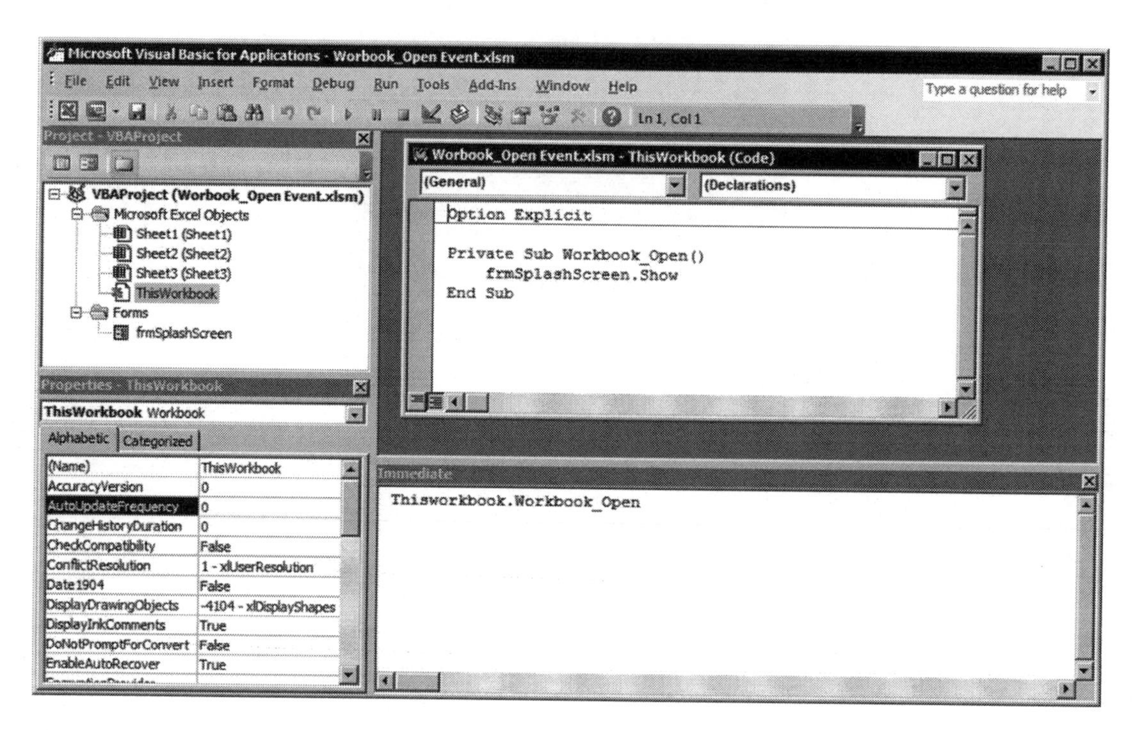

Figure 3-4. This is the Workbook_Open event procedure from the Workbook_Open Event.xlsm Excel macro-enabled workbook that uses the frmSplashScreen.Show method to show it in the Excel interface every time the workbook is opened

Also note that the `frmSplashScreen UserForm` has two `Label` controls (one for the welcome message and another for the copyright information) and one `CommandButton` with an "OK" caption. Click the `CommandButton` and inspect the VBA Properties window, namely, these properties:

- `Name = cmdOK` (to change the default `CommandButton1` name)
- `Caption = "OK"`, the text that appears on the `CommandButton` caption

Implementing a UserForm Timer

To automatically close the `frmSplashScreen UserForm` object the way professional applications do (like Excel), you need to implement a timer using the `UserForm Activate` event. This timer will fire every time the `UserForm` is opened and show its interface on the screen.

The timer can be implemented with different programming techniques. The simplest technique is to use the `Application.OnTime` method, but you can also use the VBA `Timer( )` function or base it on Windows DLL calls (covered in Chapter 10). Let's see how this can be done using the first two programming techniques.

Using the Application.OnTime Method

The `Workbook_Open( )` event in the `Application.OnTime method.xlsm` Excel macro-enabled workbook, which you can extract from the `Chapter03.zip` file, has a `UserForm` object implemented as the application splash screen. It is automatically shown when the workbook is opened and is closed after five seconds (Figure 3-5).

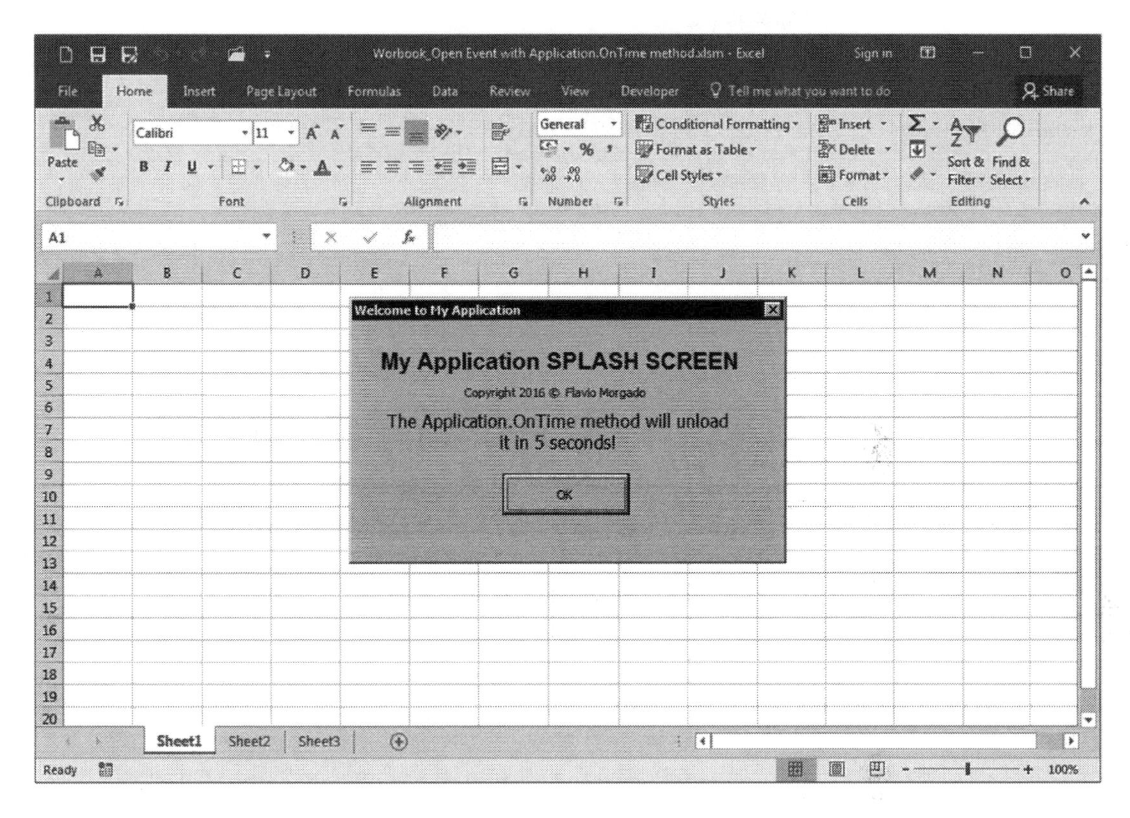

Figure 3-5. *The Workbook_Open( ) event in Application.OnTime method.xlsm uses the Application.OnTime method to implement a timer on the frmSplashScreen UserForm and automatically closes it after five seconds*

After you open the workbook, press Alt+F11 to show the VBA IDE, and select the `ThisWorkbook` object, note that the `Workbook_Open( )` event shows the `frmSplashScreen UserForm` using its `Show` method.

```
Private Sub Workbook_Open()
    frmSplashScreen.Show
End Sub
```

Inspect the `frmSplashScreen UserForm` code module and note that the timer is implemented using a technique described in Chapter 2: the timer is set on the `UserForm_Initialize( )` event and unset on the `UserForm_Terminate( )` event, as follows:

```
Private Sub UserForm_Initialize()
    mvarTime = Now + TimeValue("00:00:05")
```

```
    Application.OnTime mvarTime, "ClosefrmSplashScreen"
End Sub

Private Sub UserForm_Terminate()
    On Error Resume Next
    Application.OnTime mvarTime, "ClosefrmSplashScreen", , False
End Sub
```

The mvarTime module-level variable holds the exact time used to set the timer, and the same value is used again to unset it. If the timer runs to the end, the Application.OnTime method will call the Public Sub ClosefrmSplashScreen( ) procedure of the basUserForm module, which simply closes the frmSplashScreen using the VBA Unload method.

```
Public Sub ClosefrmSplashScreen()
    Unload frmSplashScreen
End Sub
```

If the frmSplashScreen UserForm is closed by the user action by clicking the OK CommandButton, the Click( ) event will fire, and the form will be unloaded.

```
Private Sub cmdOK_Click()
    Unload Me
End Sub
```

■ **Attention** The frmSplashScreen UserForm can be also closed by pressing the Esc key or clicking the UserForm Close button (the *X* in the top-right corner). The CommandButton control has two properties called Default and Cancel; when they are set to True, they map the Click( ) event to the Enter and Esc keyboard keys, respectively. This means that whenever one of these keys is pressed, the associated CommandButton Click( ) event fires. The frmSplashScreen UserForm cmdOK CommandButton has both its Default and Cancel properties set to True, which is why the user can close it by pressing either the Esc or Enter key (Figure 3-6).

Only one UserForm CommandButton can have its Default and/or Cancel properties set to True.

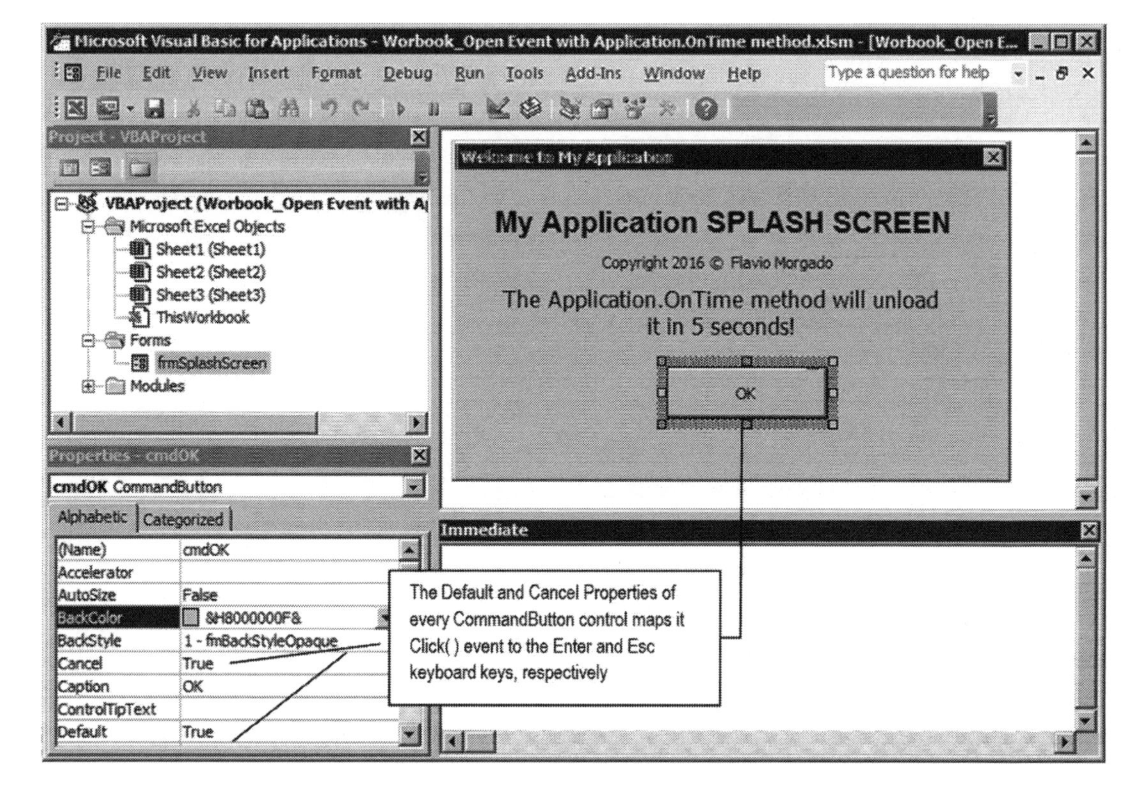

Figure 3-6. *Use the Cancel and Default properties of the CommandButton control to associate the Click( ) event to the Esc and Enter keyboard keys, respectively. Only one CommandButton on the UserForm can have each property set to True*

Using the VBA Timer( ) Function

You can also create a timer using a simple VBA programming technique. You will need to code the `UserForm_Activate( )` event using the VBA `Timer( )` function (which returns the number of seconds past since midnight, or 0:00) and using a simple loop that will last for the desired amount of seconds. When the loop finishes, the VBA `Unload` method will unload the `UserForm` by itself. Supposing that you want to create a five-second timer, you just store the VBA `Timer( )` returned value on the `lngSeconds` variable, and inside a `Do While... Loop` you subtract this value from the value returned by successive `Timer( )` function calls. The loop must end when the subtraction is greater than five seconds, as follows:

```
Private Sub UserForm_Activate()
    Dim lngSeconds As Long

    lngSeconds = Timer
    Do
    Loop Until (Timer - lngSeconds) > 5
    Unload Me
End Sub
```

Such code was inserted in the `frmSplashScreen Activate( )` event of the `Workbook_Open` event with the `Timer.xlsm` Excel macro-enabled workbook that you will find in the `Chapter 03.zip` file.

Since the `Workbook_Open` event with the `Timer.xlsm` workbook also uses the `ThisWorkbook.Open` event to show the `frmSplashScreen UserForm`, the form will be automatically opened when the workbook is opened, and although it will be kept open in the Excel interface for five seconds, you will not be able to see its interface; you will see just the `UserForm` title bar with a white form background (Figure 3-7).

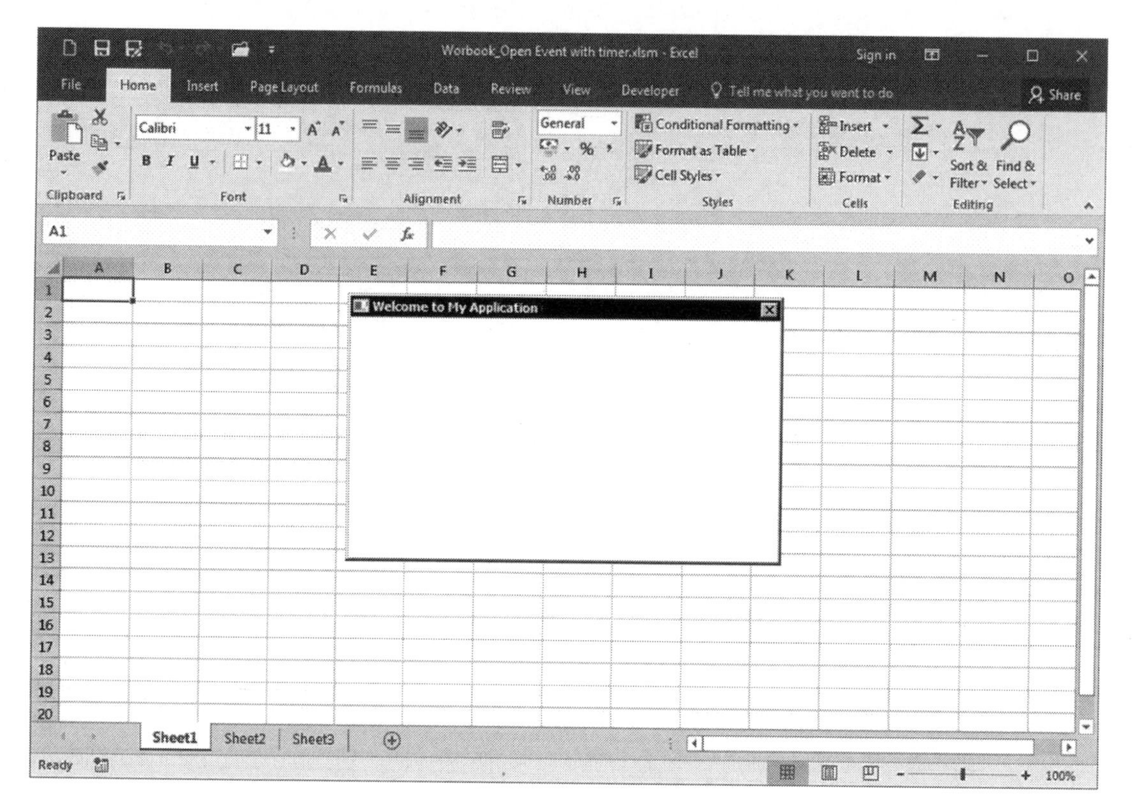

Figure 3-7. *If you try to implement a timer using a Do...Loop statement of the UserForm Activate( ) event, the code will not allow the form to repaint its interface onscreen, and you will see just its title bar and window border*

This strange behavior happens because of the VBA code of the `ThisWorkbook_Open` and `UserForm_Activate( )` events firing so fast that they do not give enough processor time to `frmSplashScreen` to repaint its interface. So, you must force it to repaint itself using the `UserForm Repaint` method.

Using the UserForm Repaint Method

You can see this new version of the `UserForm_Activate( )` event by opening the `Workbook_Open` event with the `Timer Repaint.xlsm` Excel macro-enabled workbook that uses this code on its `UserForm_Activate( )` event.

```
Private Sub UserForm_Activate()
    Dim lngSeconds As Long

    Me.Repaint
    lngSeconds = Timer
    Do
    Loop Until (Timer - lngSeconds) > 5

    Unload Me
End Sub
```

The Me.Repaint instruction executes the Repaint Method of the UserForm object to force it actualization on the computer screen

Now, before the five-second loop begins, the form is repainted on the screen, and you can see its interface while the loop is running. But now you will have another problem. You can't close the form while the five-second loop is executing because the VBA running code does not allow you to click the OK `CommandButton`! You need to add another VBA programming trick in the code to allow the `cmdOK_Click` event to fire while the `Do While... Loop` code is executing inside the `UserForm_Activate( )` event.

Using the VBA DoEvents Instruction

This is where the VBA `DoEvents` instruction enters into action: every time you have a loop in your code that should be interrupted by any `UserForm CommandButton` (such as the common Cancel `CommandButton` used to stop any longer process), you must put a `DoEvents` instruction *inside the loop* to tell VBA that it must *watch* if another interface event is fired while the loop is running.

Open the `Workbook_Open` event in the `Timer Repaint DoEvents.xlsm` Excel macro-enabled workbook and you will see that this time the `frmSplashScreen UserForm` will remain in the Excel interface for five seconds or will be closed and vanish from the screen whenever you click the OK or Close button. Now it executes the next code to implement its programmable timer on its `UserForm_Activate( )` event:

Since the `cmdOK_Click( )` event uses the VBA `Unload` method to unload the `UserForm`, the loop is instantly stopped and the form is closed!

```
Private Sub UserForm_Activate()
    Dim lngSeconds As Long

    Me.Repaint
    lngSeconds = Timer
    Do
        DoEvents
    Loop Until (Timer - lngSeconds) > 5
    Unload Me
End Sub

Private Sub cmdOK_Click()
    Unload Me
End Sub
```

The DoEvents VBA instruction allow the VBA code to watch if another concurrent event fires and process it while this event code is running, such as the cmdOK_Cick event, which executes the Unload Me instruction, literally closing the UserForm and interrupting the loop.

■ **Attention** To create a `UserForm SplashScreen` that is kept on the screen for a given amount of time without user interference, don't put the OK button on it. This will let the user appreciate the information you want to give about your application.

Setting Workbook Object References

Since each Microsoft Excel window can open an undefined number of workbooks, you must take care when your application deals with more than one workbook opened at the same time and the way you reference them.

If you need to refer to *your workbook application* from any code module of your application, always use the `Application` object's `ThisWorkbook` property with one of these two syntaxes:

```
Dim wkb as Workbook
Set wkb = Application.ThisWorkbook
Debug.Print wkb.Name
```

or

```
Set wkb = ThisWorkbook
Debug.Print wkb.Name
```

But if you need to reference *your application* from the `ThisWorkbook` object code module, you can simply use the VBA keyword `Me`, without setting any variable to reference it. The next instruction will print your application name in the VBA Immediate window if you put it in any procedure of the `ThisWorkbook` object code module:

```
Debug.Print Me.Name
```

Looking again to the Microsoft object model in Figure 2-1 of Chapter 2, you will note that the `Workbooks` collection is the second main object behind the `Application` object.

Every time you open a workbook inside the same Microsoft Excel window (represented by the `Application` object), Excel puts a reference to it in the `Workbooks` collection, which has two main properties to identify its members: the `Index` property (an `Integer` value) and the `Name` property (a `String` value).

For the `Application.Workbooks` collection, the `Index` property sets the unpredictable order of each opened workbook inside Excel, which can be inadvertently changed as the user closes and opens another workbook, while the `Name` property sets the stable workbook name, which can also be changed when you use Excel's Save As menu command or the `Workbook` object's `SaveAs` method.

Suppose you have a worksheet application named `Book1` and it is the only workbook open inside a single Microsoft Excel window. The next two instructions will set a reference to it on the `wkb as Workbook` object variable:

```
Dim wkb as Workbook
Set wkb = Workbooks(1)        '< Reference by Index
Set wkb = Workbooks("Book1")  '< Reference by Name
```

The first thing you should never forget is that you simply can't consider that your application will ever receive `Index = 1` because this will just happen if your application is the first opened inside Excel; this is a great cause of error flaws when programming Excel applications.

Don't ever use the `Index` property to refer to the workbooks opened by your application because you can't anticipate which value it will be. But you can trust in the `Name` property, because if any opened workbook should be saved with another name, this will be reflected in the `Application.Workbooks` collection.

Let's see some practical examples about workbook referencing while we also extend your VBA knowledge to deal with the `UserForm` object and its `ListBox` control. Close all opened workbooks and open the `Workbook Referencing.xlsm` macro-enabled workbook, which you can extract from the file `Chapter03.zip`. You will receive the `frmOpenWorkbooks UserForm` from where you can open another Excel workbook and look at the `Index` and `Name` properties (Figure 3-8).

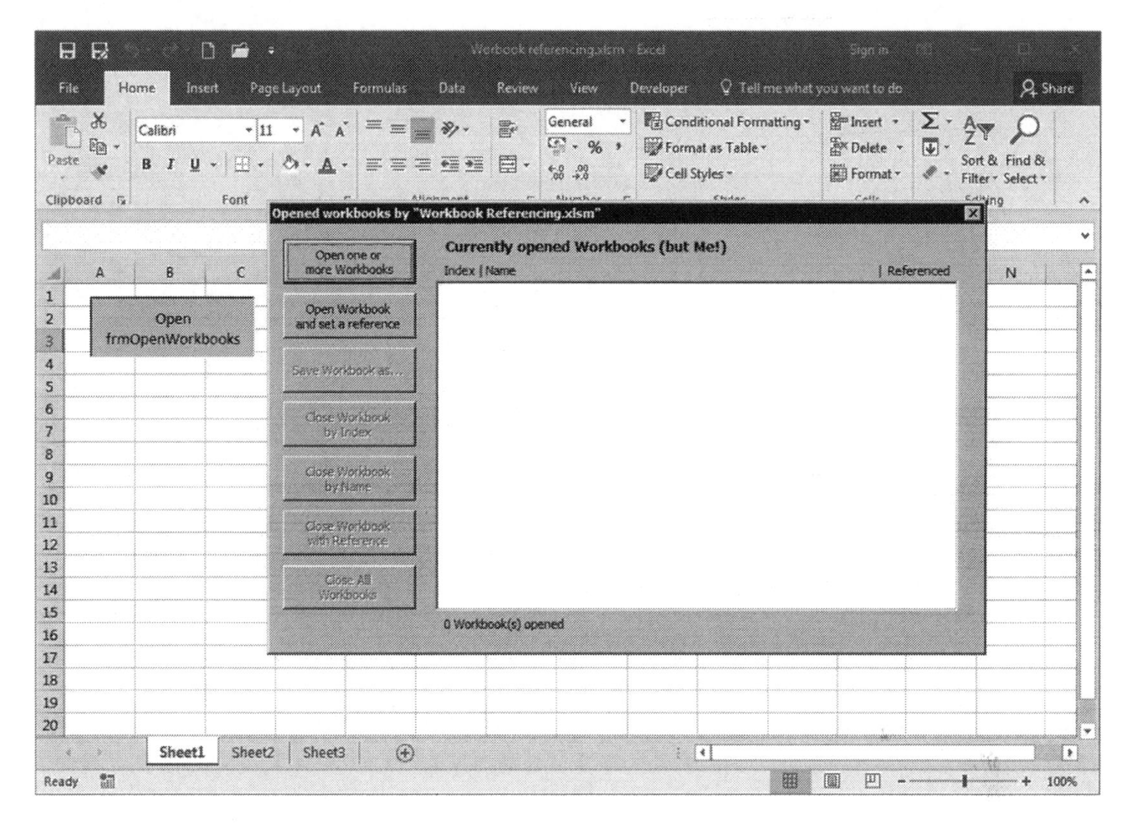

Figure 3-8. *The Workbook Referencing.xlsm macro-enabled workbook shows the frmOpenWorkbooks, which will show every other workbook opened in the same Microsoft Excel application window. This figure shows an empty list of open workbooks, meaning that only this application is opened at this moment*

■ **Attention** If you open Excel and leave the default empty workbook `Book1` opened before opening `Workbook Referencing.xlsm`, you will see it in `frmOpenWorkbooks UserForm ListBox` as the first workbook opened (having `Index = 1`).

Click the Open one or more Workbooks button (`cmdOpenWkbs`), select one or more Excel files, and click OK to see them be opened behind the `UserForm` while their `Index` and `Name` properties appear inside the `lstWorkbooks ListBox`. Figure 3-9 shows what happened after I selected and opened four workbooks already explored in this book: their `Index` and `Name` properties, as well as the information if the workbook is referenced by any object variable, are shown on the `lstWorkbooks ListBox`. Also note that the number of workbooks opened is displayed at the bottom of `lstWorkbooks`, while the Close All Workbooks `CommandButton` becomes enabled.

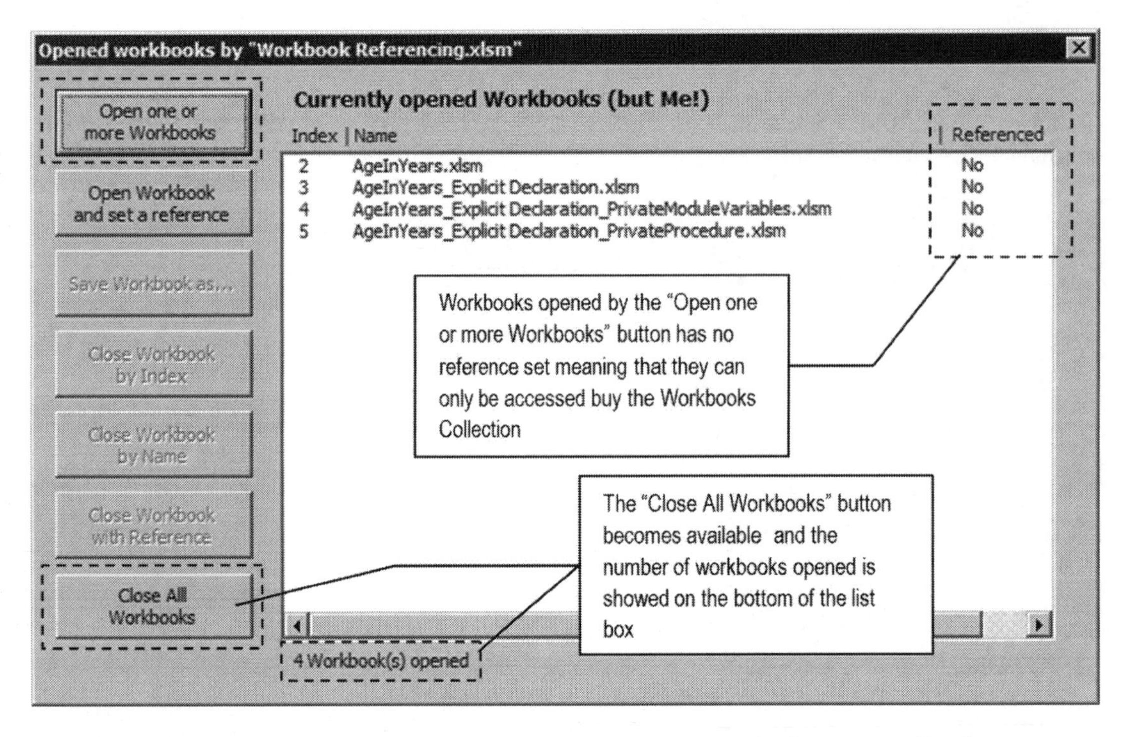

Figure 3-9. *Use the Open one or more Workbooks CommandButton from the frmOpenWorkbooks UserForm to select and open workbooks inside this Excel application window. The lstWorkbooks ListBox will show each workbook index and whether they are referenced by any variable*

■ **Attention** Note in Figure 3-9 that there is no `Index = 1` on this list because this index belongs to the application itself (`Workbook Referencing.xlsm`) since it was the only open application before I selected four other workbooks to be opened. Note that above the `lstWorkbooks ListBox` a `Label` control states "Currently opened Workbooks (but Me!)."

Any workbook event will normally fire for the workbooks you select. For example, if you select the `Workbook Events.xlsm` workbook from the `frmOpenedWorkbooks UserForm`, all the initializing events (the `Workbook_Open`, `Workbook_Activate`, and `WindowActivate` events) will fire, and the associated `MsgBox( )` will pop up on the screen. Although all workbook projects run in the same VBA interface (under the same Excel `Application` object), each workbook runs on its own process, and you cannot share workbook code inside the same Excel application window.

This is the code that executes when you click the `cmdOpenWkbs CommandButton` and fire its `Click( )` event on the `frmOpenWorkbooks UserForm` code module:

```
Private Sub cmdOpenWkbs_Click()
    Dim varFiles As Variant
    Dim varItem As Variant
```

```
    varFiles = ShowDialogBox(OpenFile, True)
    If Not IsEmpty(varFiles) Then
        If IsArray(varFiles) Then
            For Each varItem In varFiles
                Application.Workbooks.Open (varItem)
            Next
        Else
            Application.Workbooks.Open (varFiles)
        End If
        Call FilllstWorkbooks
        Call DefineButtons(False)
    End If
End Sub
```

You can see that the cmdOpenWkbs_Click( ) event uses the ShowDialogBox( ) procedure from basFileDialog explored in Chapter 2.

After the Open dialog box appears, the user can select one or more files (because of the True value used in the second procedure argument) or click Cancel, which will leave the varFiles Variant variable with the default Empty value. That is why the procedure uses the VBA IsEmpty( ) function to test varFiles.

```
varFiles = ShowDialogBox(OpenFile, True)
If Not IsEmpty(varFiles) Then
```

The procedure then uses the VBA IsArray( ) function to verify whether you select more than one file, and if this is true, it uses a For Each ... Next loop to open all selected files using the Application.Workbooks.Open method. Or it opens the only selected file.

```
If IsArray(varFiles) Then
    For Each varItem In varFiles
        Application.Workbooks.Open (varItem)
    Next
Else
    Application.Workbooks.Open (varFiles)
End If
```

After all files have been opened, the Sub FillsWorkbooks( ) procedure is called to fill the ListBox, the Sub DefineButtons( ) procedure is called to synchronize the enabled state of all CommandButton objects of the UserForm (note in Figure 3-9 that the Close all Workbooks CommandButton becomes available), and the procedure ends.

```
        Call FilllstWorkbooks
        Call DefineButtons(False)
    End If
End Sub
```

If you want to close all opened workbooks, just click the cmdCloseAll CommandButton to execute its Click( ) event. Since we are closing all opened workbooks except the Workbooks Events.xlsm workbook, where the code is running, you just need to loop through the Workbooks collection and apply its Close method, executing this code:

```
Private Sub cmdCloseAll_Click()
    Dim varItem As Variant

    For Each varItem In Workbooks
        If Not varItem Is ThisWorkbook Then
            varItem.Close
        End If
    Next
    Set mWkb = Nothing
    Call FilllstWorkbooks
    Call DefineButtons(False)
End Sub
```

Note in the previous code that to avoid closing Me (Workbook Referencing.xlsm), the code verifies inside the For Each... Next loop if the referenced workbook is equal to the object returned by the ThisWorkbook method of the Application object.

```
If Not varItem Is ThisWorkbook Then
```

If this is true, the workbook referenced by the varItem variable is closed, using the Workbook object Close method.

```
varItem.Close
```

After all workbooks have been closed, the mWbk module-level variable is set to Nothing, and the Sub FillWorkbooks( ) and DefineButtons( ) procedures are called to synchronize the UserForm interface.

■ **Attention** The VBA For Each... Next loop needs to use a variable declared with the same type of the collection it belongs to or a Variant variable because this is the only data type that can hold any kind of value or object reference. Since we are looping through the Workbooks collection, the varItem variable holds a reference to a Workbook object and can execute its Close method.

Since varItem is declared as Variant, VBA can't anticipate at compile time if the object it will receive has a Close method. This will be verified when the code is running, using the object association known as *late binding*.

Now click Open Workbook and set a reference CommandButton (cmdOpenWkbReference) to select and open another workbook inside Excel. The Index and Name properties will appear in the ListBox along with "Yes" in the Referenced column (Figure 3-10). Note that now both Close Workbook with Reference and Close all Workbooks CommandButtons become available and that the label displaying the number of workbooks opened by the UserForm below the ListBox is updated.

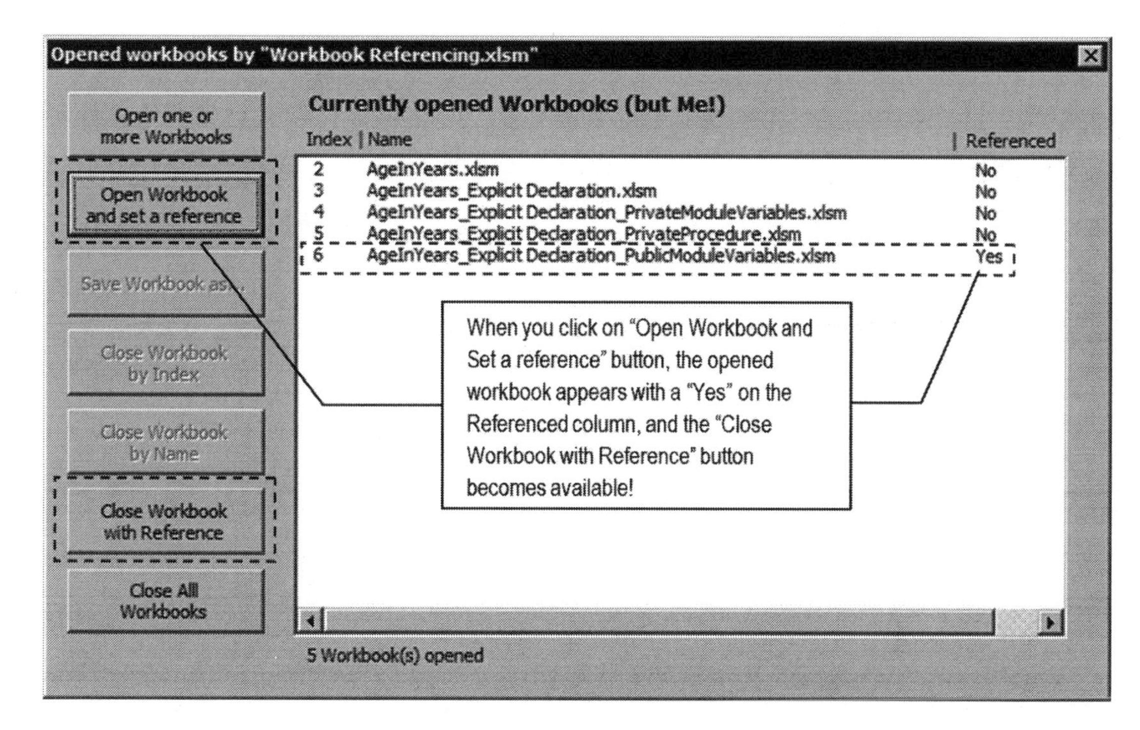

Figure 3-10. *When you click the "Open Workbook and set a reference" button, the selected workbook will be opened, and the referenced value will show "Yes" in the ListBox, as a confirmation that there is a module variable referencing this workbook*

Whenever you click the cmdOpenWkbReference CommandButton, VBA executes this code:

```
Option Explicit
Dim mWkb As Workbook

Private Sub cmdOpenWkbReference_Click()
    Dim strFile As String

    strFile = ShowDialogBox() & ""
        If Len(strFile) Then
        'Workbook was selected. Open it!
        Set mWkb = Application.Workbooks.Open(strFile)
        Call FilllstWorkbooks
        Call DefineButtons(False)
    End If
End Sub
```

Did you notice that this time the mWkb module-level variable holds a reference to the opened workbook? This reference is important because you can close the referenced workbook without using the Workbooks collection's Index or Name property. In fact, if you click the Close Workbook with Reference button (cmdCloseWkbByRef), you will execute this code:

```
Private Sub cmdCloseWkbByRef_Click()
    mWkb.Close
    Set mWkb = Nothing
    Call FilllstWorkbooks
    Call DefineButtons(False)
End Sub
```

Once again, after the referenced workbook is closed, both the Sub FillsWorkbook( ) and DefineButtons( ) procedures are called to synchronize the UserForm interface.

Using the ListBox Control

Let's take a look at how you can use the ListBox control to show information in the UserForm interface. Close the frmOpenedWorkbooks UserForm, select it in the VBA environment, and click the lstWorkbooks ListBox to show its properties in the VBA Properties window (Figure 3-11).

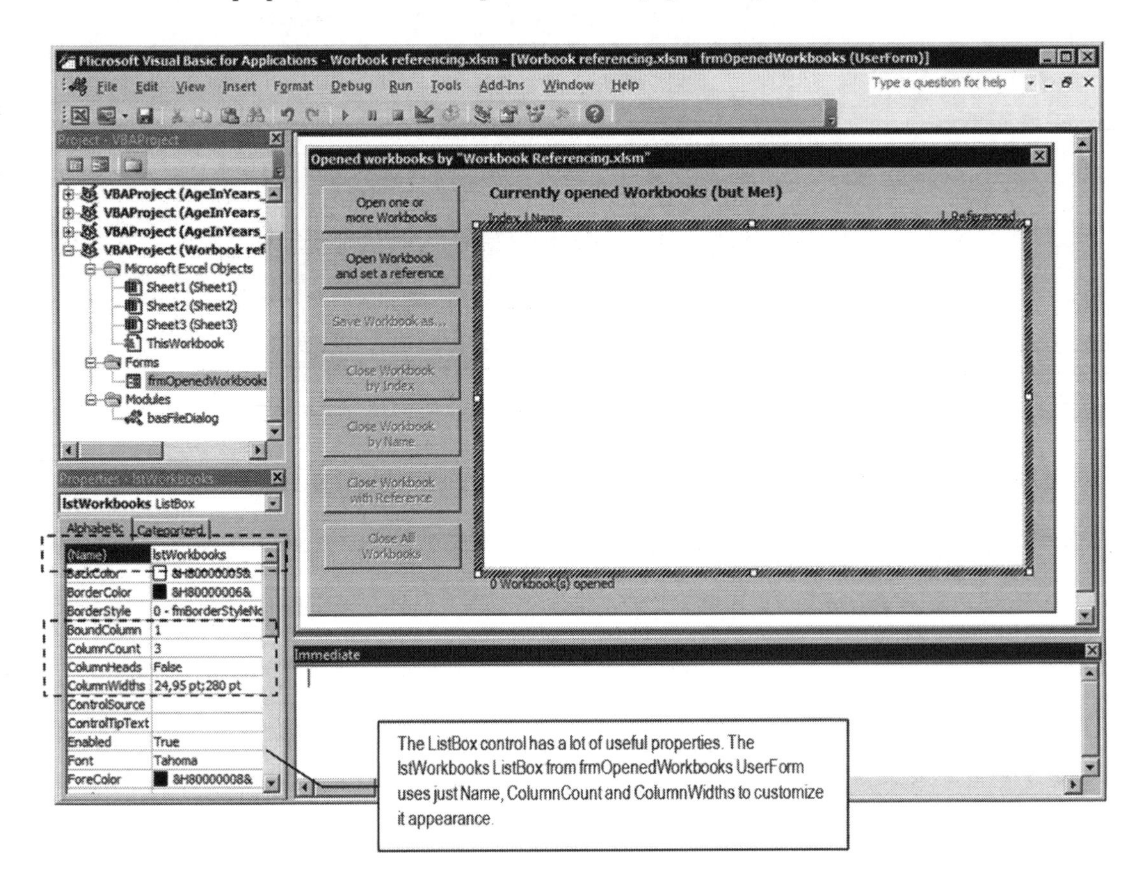

Figure 3-11. *This is the lstWorkbooks ListBox and its properties in the VBA environment. Note that the ListBox's Name, BoundColumn, ColumnCount, ColumnHeads, and ColumnWidths have been set*

The UserForm ListBox control has a lot of useful properties. Table 3-4 shows some of the most important properties for this control.

Table 3-4. *Some Important ListBox Control Properties*

ListBox Control Property	Value Returned	Represents
Name	String	The control name inside the UserForm
BackColor	Long integer	Sets the back color of the ListBox control
BoundColumn	Integer	Sets the column number (1-based) that returns the ListBox value
Column	Integer	Sets the column number (0-based) for the ListBox selected item information
ColumnCount	Integer	Indicates the number of data columns (1-based) that the ListBox has
ColumnHead	Boolean	Indicates whether the ListBox will show a heard with the name of the column fields used to populate it; works just with SQL data
ColumnWidths	String	Sets the column width (in points) of each ListBox column, separated by semicolons
ControlTipText	String	Sets the text that will appears in a yellow window when the mouse rests over the control
Enabled	Boolean	Enables and disables the ListBox control
Font	String	Sets the font name used to display text on the ListBox
ForeColor	Long Integer	Sets the font color of the ListBox control
ListCount	Integer	Returns the number of items (1-based) inserted on the ListBox
Listindex	Integer	Returns the index (order, 0-based) of the selected item on the ListBox
MultiSelect	Boolean	Indicates whether the user can select more than one item of the ListBox
TabIndex	Integer	Sets the tab order of the ListBox on the UserForm
TabStop	Boolean	Indicates whether the ListBox will receive the focus by pressing the Tab key
Text	String	Used to insert fixed values on the ListBox from the Properties window
TextAlign	Integer	Sets the text alignment used on all ListBox columns
Value	String	Sets the value of the ListBox (selected item)
Visible	Boolean	Indicates whether the ListBox is visible on the UserForm interface

The ListBox control also has some important methods that you must be aware of to deal with the ListBox items. Table 3-5 shows the most important for you.

***Table 3-5.** Some Important ListBox Control Methods*

ListBox Control Method	Action Performed
AddItem	Adds a new item to the ListBox control
Clear	Removes all items inserted on the ListBox control
RemoveItem	Removes a given item from the ListBox control
SetFocus	Sets the focus to the ListBox control

The lstWorkbooks ListBox has been defined to show three columns (ColumnCount = 3), where each column has a different width, as defined on the ColumnWidths property. The first column (Index value) has 24,95 pt (we tried to insert 25 points, but VBA converted it), the second column (Name value) has 280 pt, and the third column (Referenced value) has no specification, meaning that it goes from the second column to the width of the ListBox control. To indicate which value will return the ListBox value whenever a list item is selected, use the BoundColumn column to indicate the desired column number (1-based).

```
BoundColumn =1
ColumnCount = 3
ColumnWidths = 24,95 pt;280 pt
```

Adding Items to the ListBox

To add items to any ListBox control, you must use the AddItem method, which has this syntax:

```
expression.AddItem(Item, [Index])
```

In this code:

expression: This is required; it is the ListBox name property.

Item: This is required; it is a String value with the display text for the new item.

Index: This is optional; it is an Integer value indicating the position (0-based) of the new item in the list. If omitted, the item is added to the end of the list.

The ListBox AddItem method always inserts the information on the ListBox's first column. To add the Workbook.Index property as a new item to the bottom of the lstWorkbooks ListBox control, you can use code like this (where wkb represents a variable declared as Workbook):

```
lstWorkbooks.AddItem wkb.Name
```

Using the ListBox Column Property

Every ListBox control that uses more than one column has a Column property that you can use to set or get any column value for a given ListBox item. It has this syntax:

```
ListBox.Column( column, row ) [= String]
```

column: This is optional; it is a 0-based Integer from 0 to one less than the total number of columns.

row: This is optional; it is a 0-based Integer from 0 to one less than the total number of rows.

Since both the column and row arguments of the ListBox.Column property are 0-based, the first ListBox column receives the 0 index, while the last ListBox row receives the ListBox.ListCount - 1 value. At any moment, the selected ListBox item receives the ListBox.ListIndex property (which is also 0-based).

Supposing the ListBox has three columns, the first column is used to show the Workbook Index property on the Workbooks collection, and the second column is used to show the Workbook.Name property. Supposing that intI has the Workbook Index property inside the Workbooks collection, to add the workbook Index on the first lstWorkbooks column and the Workbook.Name property on the second lstWorkbooks column, use syntax like this:

```
lstWorkbooks.AddItem intI
lstWorkbooks.Column(1, lstWorkbooks.ListCount - 1) = wkb.Name
```

Note that lstWorkbooks.ListCount - 1 returns the last ListBox item—the one that had been inserted by the last lstWorkbooks.AddItem method operation.

This is all the information you need in order to understand what happens with the Sub FilllstWorkbooks( ) procedure that is called from every CommandButton of the frmOpenedWorkbooks UserForm. It has this code:

```
Private Sub FilllstWorkbooks()
    Dim varItem As Variant
    Dim intI As Integer
    Dim intIndex As Integer

    Me.lstWorkbooks.Clear
    If Workbooks.Count > 1 Then
        For intI = 1 To Workbooks.Count
            If Not (Workbooks(intI) Is ThisWorkbook) Then
                With Me.lstWorkbooks
                    .AddItem intI
                    .Column(1, .ListCount - 1) = Workbooks(intI).Name
                    If Workbooks(intI) Is mWkb Then
                        .Column(2, .ListCount - 1) = "Yes"
                    Else
                        .Column(2, .ListCount - 1) = "No"
                    End If
                End With
            End If
        Next
    End If
    Me.lblWorkbooksCount.Caption = Me.lstWorkbooks.ListCount & " Workbook(s) opened"
End Sub
```

Note that the first action performed by the Sub FilllstWorkbooks( ) procedure is to clean the lstWorkbooks ListBox, removing all its items.

```
Me.lstWorkbooks.Clear
```

It then verifies whether the Workbooks collection has more than one workbook (because it must have at least one for the Workbook Referencing.xlsm application, which is running the code).

```
If Workbooks.Count > 1 Then
```

If Excel has two or more workbooks open, the code performs a For...Next loop through the Workbooks collection using the Index property to reference each opened workbook. Note that the intI variable is used to set the desired index, which goes from 1 (the first opened workbook) to the Workbooks.Count property:

```
For intI = 1 To Workbooks.Count
```

The code now verifies whether the referenced workbook is not ThisWorkbook to add it to the list.

```
If Not (Workbooks(intI) Is ThisWorkbook) Then
```

If this is true (it is another opened workbook), the code uses a With...End With construction to reference just once the lstWorkbooks ListBox and creates more concise code.

```
With Me.lstWorkbooks
...
End With
```

The Index and Name properties of the referenced workbook are then added to the first and second columns of the lstWorkbooks ListBox. Note that the intI variable holds the Index property and that .ListCount - 1 (note the dot before the property name) indicates the item where the second column information will be changed.

```
With Me.lstWorkbooks
    .AddItem intI
    .Column(1, .ListCount - 1) = Workbooks(intI).Name
```

To confirm that the workbook has a variable reference set, the code verifies whether the referenced workbook is the same one referenced by the mWkb module-level variable. If this is true, it adds "Yes" to the third lstWorkbooks column. If not, it adds "No."

```
If Workbooks(intI) Is mWkb Then
    .Column(2, .ListCount - 1) = "Yes"
Else
    .Column(2, .ListCount - 1) = "No"
End If
```

■ **Attention** You can use the VBA IIF( ) function to do this last task using just one instruction, in this way:

```
.Column(2, ListCount - 1) = Iif( Workbooks(intI) Is mWkb, "Yes", "No")
```

And when all opened workbook information has been added to the lstWorkbooks ListBox, the code updates the lblWorkbooksCount label information, changing the Caption property to reflect how many items the lstWorkbooks has and using the ListCount property.

```
Me.lblWorkbooksCount.Caption = Me.lstWorkbooks.ListCount & " Workbook(s) opened"
```

Note that the Me keyword used on the Me.lstWorkbooks.ListCount instruction refers to the UserForm—the owner of the code module.

I also want to call your attention to the programming technique used by the UserForm: whenever another workbook is opened or closed, the lstWorkbooks ListBox is cleared and filled again. It does not use the ListBox AddItem or RemoveItem to manage the list items. They are all removed and re-inserted.

Referencing ListBox Items

We are now able to talk about what happens when you click any lstWorkbooks item and the CommandButtons named Save Workbook as..., Close Workbook by Index, and Close Workbook by Name. All these three buttons get a reference from the selected workbook on the ListBox either by its Index on the Workbooks collection or by its Name property. This is the code that runs when you click the Close Workbook by Index button (cmdCloseWbkByIndex):

```
Private Sub cmdCloseWkbIndex_Click()
    Dim intIndex As Integer

    intIndex = Me.lstWorkbooks.Value
    Workbooks(intIndex).Close
    Call FilllstWorkbooks
    Call DefineButtons(False)
End Sub
```

Since the workbook index was added by the lstWorkbooks.AddItem method *and* the lstWorkbooks BoundColumn property is defined to 1 (default value), you can trust that the lstWorbooks value is the selected workbook Index on the Workbooks collection and can be easily closed by calling the Workbooks collection's Close method.

```
intIndex = Me.lstWorkbooks.Value
Workbooks(intIndex).Close
```

The procedure calls the FilllstWorkbooks( ) procedure to clear the ListBox, fills it again with the new Workbooks collection indexes, and then calls the DefineButtons( ) procedure to synchronize the Enabled property of the UserForm CommandButtons.

Select the first opened workbook on the list and click the Close Workbook by Index button. Note that when you do this, all other opened workbooks change their Index values inside the Workbooks collection (Figure 3-12).

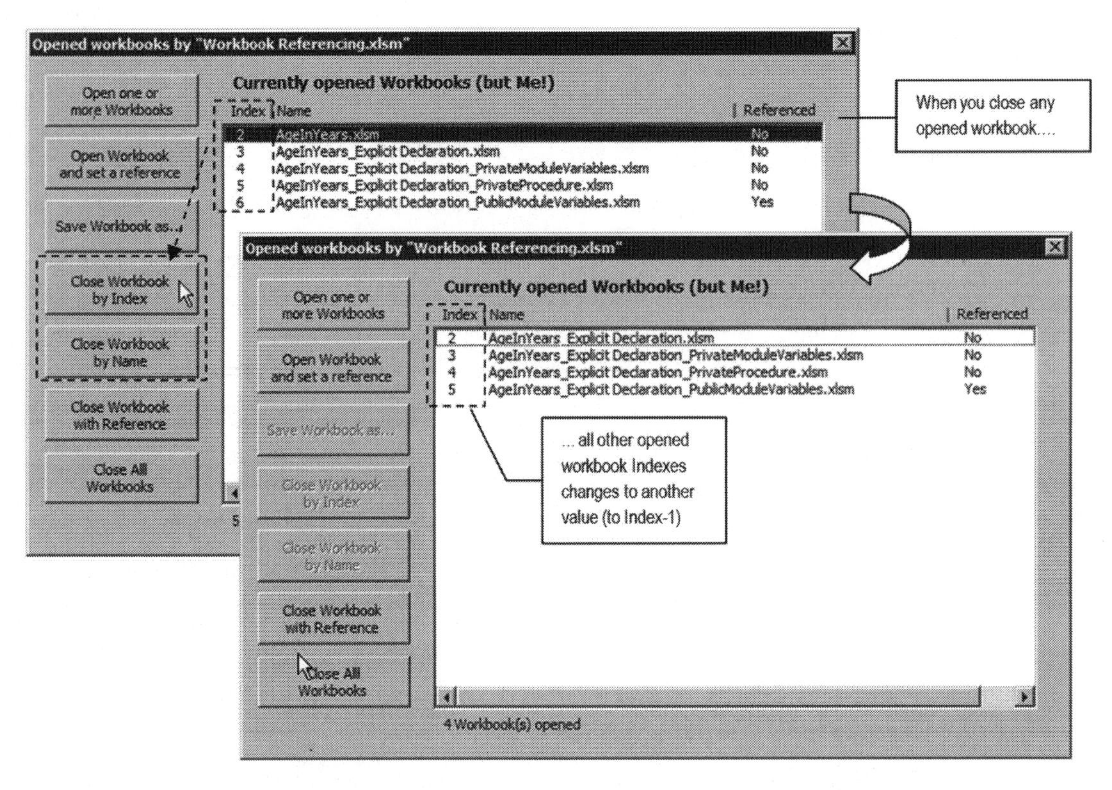

Figure 3-12. *When you click any lstWorkbooks item and click the Close Workbook by Index or Close Workbook By Name button, the selected workbook is closed, and all remaining workbook Index values inside the Workbooks collection change*

You can continue to select workbooks on `lstWorkbooks` and close them by its `Index` or by its `Name`, because the `FilllstWorkbooks( )` procedure always cleans the `ListBox` and fills it again using its new `Workbooks` collection indexes. When you click the Close Workbook by Name button, the `cmdCloseWkbName_Click( )` event fires and executes this code:

```
Private Sub cmdCloseWkbName_Click()
    Dim strName As String

    strName = Me.lstWorkbooks.Column(1)
    Workbooks(strName).Close
    Call FilllstWorkbooks
    Call DefineButtons(False)
End Sub
```

This time, the `strName` string variable receives the second `ListBox` column value, where the workbook name is displayed, and executes the `Workbook` collection's `Close` method, referencing it by its `Name` property:

```
strName = Me.lstWorkbooks.Column(1)
Workbooks(strName).Close
```

Now see what happens when you select any workbook, click the "Save workbook as" button, and give it another name. Although the workbook name changes, the Index remains the same inside the Workbooks collection (Figure 3-13).

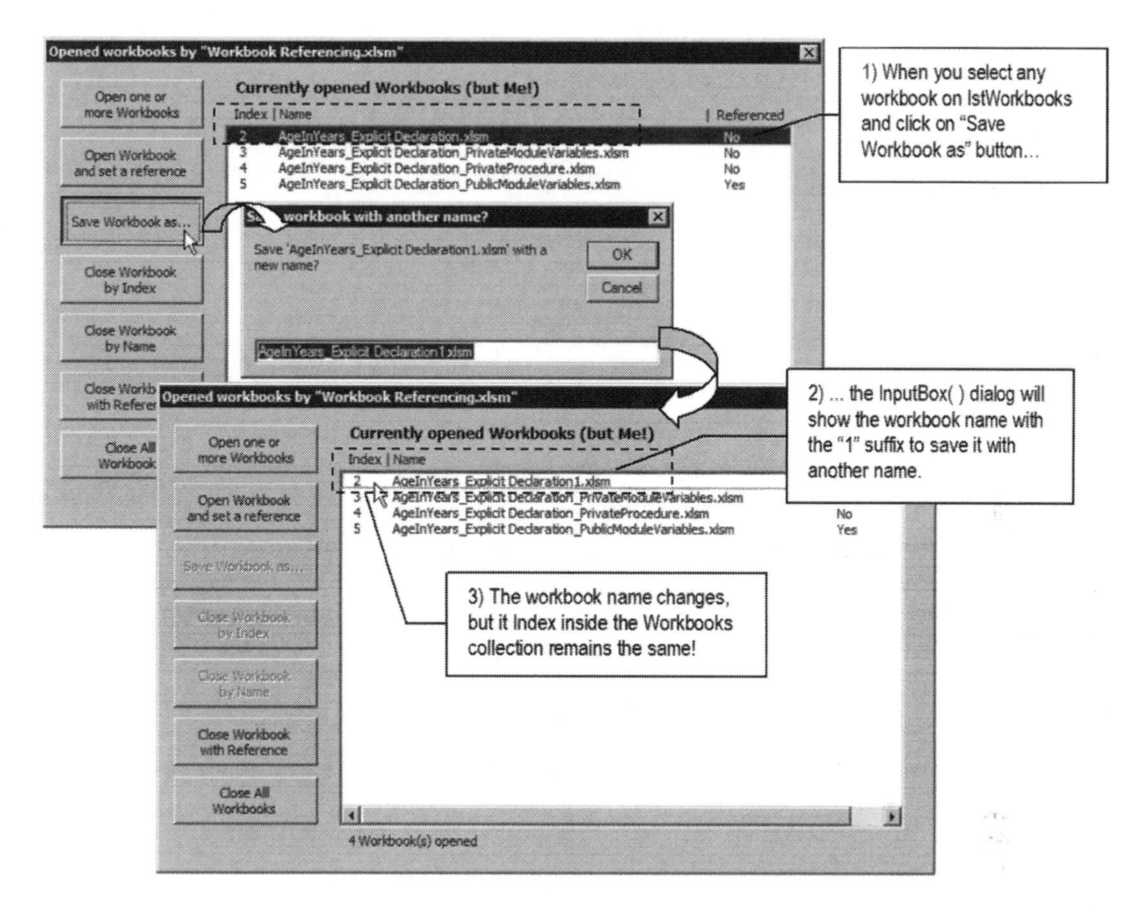

Figure 3-13. *When you click the "Save Workbook as" button of frmOpenedWorkbooks, the workbook name appears in the Inputbox( ) dialog with the 1 suffix added. Click OK to save the workbook with a new name and watch that it doesn't change its Index value inside the Workbooks collection*

■ **Attention** If the workbook you are trying to change the name of is already saved on your disk drive, the Workbook.SaveAs method will show a warning dialog box asking your permission to overwrite the file. If you click the No or Cancel button, the SaveAs method will raise a unexpected error on your code (Figure 3-14).

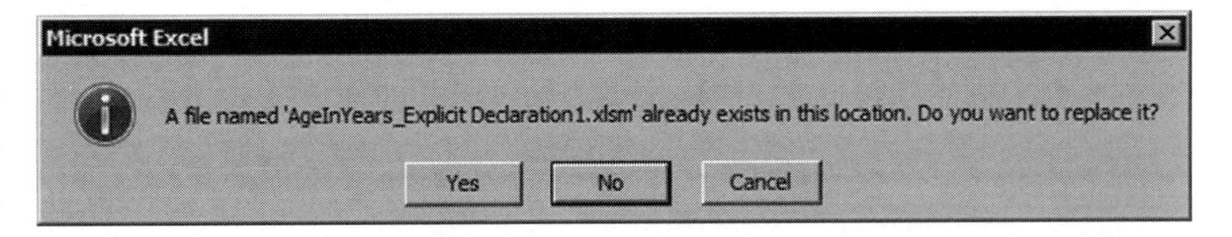

Figure 3-14. *The Workbook.SaveAs method will show a warning dialog before overwriting any existing workbook. If you cancel the saving operations by clicking the No or Cancel button, the Workbook.SaveAs method will raise an unexpected error on your code*

This is the code executed when you click the "Save Workbook as" (cmdSaveAs) Command button:

```
Private Sub cmdSaveAs_Click()
    Dim varWkb As Variant
    Dim strName As String
    Dim intIndex As Integer
    Dim intPos As Integer
    Const conErrSaveAsFailed = 1004

    On Error GoTo cmdSaveAs_Err

    intIndex = Me.lstWorkbooks.Value
    strName = Me.lstWorkbooks.Column(1)
    intPos = InStrRev(1, strName, ".")
    strName = Left(strName, intPos - 1) & "1" & Mid(strName, intPos)

    strName = InputBox("Save '" & strName & "' with a new name?", "Save workbook with
    another name?", strName)
    If Len(strName) Then
        Workbooks(intIndex).SaveAs strName
        Call FilllstWorkbooks
        Call DefineButtons(False)
    End If

cmdSaveAs_End:
    Exit Sub
cmdSaveAs_Err:
    Select Case Err
        Case conErrSaveAsFailed
        Case Else
            MsgBox "Error " & Err & ": " & Error(Err), vbCritical, "Error on Save As
workbook"
    End Select
    Resume cmdSaveAs_End
End Sub
```

Setting an Error Trap

Besides the variables needed in the code, the cmdSaveAs_Click( ) event procedure also declares the constant conErrSaveAsFailed = 1004 to represent the error code raised whenever the Workbooks collection.SaveAs method is canceled and sets the error trapping to catch this on any other unexpected errors to the cmdSaveAs_Err: label.

```
Private Sub cmdSaveAs_Click()
    ...
    Const conErrSaveAsFailed = 1004

    On Error GoTo cmdSaveAs_Err ---┐
                                   ▼
```

Now, whenever any error is raised inside the event procedure, the code stream will be transferred to the cmdSaveAs_Err: label and treated by a Selected Case statement. If the error code is equal to the conErrSaveAsFailed constant, it means that the user just canceled the Workbooks collection's SaveAs method, and nothing will happen. If a MsgBox( ) function will show the error, the procedure will continue on the cmdSaveAs_End: label (its exit door) and end as usual.

```
    cmdSaveAs_End:
┌--►   Exit Sub
¦   cmdSaveAs_Err:      ◄---┘
¦       Select Case Err
¦           Case conErrSaveAsFailed
¦           Case Else
¦               MsgBox "Error " & Err & ": " & Error(Err), vbCritical,
¦               "Error on Save As workbook"
¦       End Select
¦       Resume cmdSaveAs_End
└---End Sub
```

Whenever an error happened on this procedure, the code will be sent to the cmdSaverAs_Err label to be treated by a Select Case statement...

... and the code will be transferred to the cmdSaveAs_End label: the procedure exit door !

Saving the Workbook with a New Name

The procedure begins storing the Index and Name values of the selected workbook in the Workbooks collection in the intIndex and strName variables, while the intPos Integer variable receives the position of the dot (.) that divides the workbook file name from its extension using the VBA InStrRev( ) function (search string on reverse order), because the Workbook.Name property used to fill the second column of the lstWorkbooks ListBox always returns the workbook file extension:

```
intIndex = Me.lstWorkbooks.Value
strName = Me.lstWorkbooks.Column(1)
intPos = InStrRev(1, strName, ".")
```

The next operation uses the intPos value to get the workbook file name (without the file extension) using the VBA Left( ) function to extract it from the first character to the point immediately before intPos (the dot position), concatenate it with the suffix 1 to create another file name, and then concatenate it again the file extension, using the VBA Mid( ) function to extract everything from the dot position to end of the strName string so the file can be saved with another name but with the same type used to open it.

```
strName = Left(strName, intPos - 1) & "1" & Mid(strName, intPos)
```

The strName variable (which actually holds the proposed new workbook file name) will then receive the value returned by the VBA InputBox( ) function.

```
strName = InputBox("Save '" & strName & "' with a new name?", "Save workbook with another
name?", strName)
```

If the user clicks the Cancel button of the InputBox( ) function shown in Figure 3-13, strName will receive an empty string. Otherwise, a file name was proposed by the user, and it is now stored in the strName variable. The next two lines of code verify with the VBA Length( ) function whether the strName variable has any characters inside it. If it does, it calls the Workbooks collection.SaveAs method to try to save the workbook with the proposed name. Note that the code refers to the selected workbook on the Workbooks collection using its Index, while the proposed name stored on the strName variable is used by the SaveAs method.

```
If Len(strName) Then
    Workbooks(intIndex).SaveAs strName
```

If no error occurs, the Workbooks.SaveAs method saves the workbook with another name and updates it into the Workbooks collection without changing its Index. Once more, the Sub FilllstWorkbooks( ) and DefineButtons( ) procedures are called to clear and fill the lstWorkbooks ListBox and synchronize the frmOpenedWorkbooks UserForm interface, and the procedure ends on its exit door, which is the Exit Sub instruction inside the cmdSaveAs_End: label.

```
        Call FilllstWorkbooks
        Call DefineButtons(False)
    End If

cmdSaveAs_End:
    Exit Sub
    ...
End Sub
```

Synchronizing the UserForm Interface

All CommandButton procedures commented on in the previous sections have a call to the DefineButtons( ) procedure, which is used to synchronize the Enabled property of the CommandButtons. The lesson here is the *synchronization* of the interface elements with the user operations.

Look again to Figures 3-11, 3-12, and 3-13 and note how the CommandButton availability changes with the select state of the lstWorkbooks ListBox. If lstWorkbooks has no selected items with no other workbook open in the interface, just the first two buttons used to open workbooks (with or without a variable reference) are available. When any workbook is opened by the UserForm action, the Close All Workbooks or Close Workbook by Reference button (if there is any referenced workbook) becomes enabled, and if the user clicks any ListBox item, both Close Workbook by Index and Close Workbook by Name become enabled.

Whenever you select any workbook, save it with another name, or close one or more workbooks, the Close... CommandButtons become disabled. This is called *interface synchronization*, and you must be aware of this simple, necessary technique to create great, solid, professional-looking interfaces to your application. Let's see what happens behind the curtains by looking at the DefineButtons( ) code procedure:

```
Public Sub DefineButtons(bolEnabled As Boolean)
    Me.cmdCloseWkbIndex.Enabled = bolEnabled
    Me.cmdCloseWkbName.Enabled = bolEnabled
    Me.cmdSaveAs.Enabled = bolEnabled
    Me.cmdCloseAll.Enabled = bolEnabled Or (Workbooks.Count > 1)
    Me.cmdCloseWkbByRef.Enabled = (Not (mWkb Is Nothing))
End Sub
```

The procedure is quite simple: it receives the bolEnabled as Boolean argument, which must be True or False. The first action is to set the focus to the lstWorkbooks ListBox and then set the Enabled properties of the Close Workbook by Index, Close Workbook by Name, and Save Workbook As CommandButtons, according to the value received.

```
Me.cmdCloseWkbIndex.Enabled = bolEnabled
Me.cmdCloseWkbName.Enabled = bolEnabled
Me.cmdSaveAs.Enabled = bolEnabled
```

Note that two buttons must not obey just the bolEnabled argument. cmdCloseAll CommandButton must be active whenever bolEnabled = True *or* if there are two or more workbooks open, which can be verified using the Workbooks collection Count property.

```
Me.cmdCloseAll.Enabled = bolEnabled Or (Workbooks.Count > 1)
```

The last case is the cmdCloseWbkByRef CommandButton (Close Workbook by Reference). This button must be always enabled when there is a reference set on the mWkb object variable. And you can confirm that verifying if mWkb is Nothing (it must be enabled whenever this is False):

```
Me.cmdCloseWkbByRef.Enabled = (Not (mWkb Is Nothing))
```

That is why when the Close All Workbooks (cmdCloseAll) and Close Workbook by Reference (cmdCloseWkbByRef) Click events fire, their code sets the mWkb variable to nothing.

The only place that calls the DefineButtons( ) procedure using a True argument is the lstWorkbooks_Click( ) event, which has this code to enable all relevant CommandButtons:

```
Private Sub lstWorkbooks_Click()
    Call DefineButtons(True)
End Sub
```

There is also a simple lesson here: synchronize your UserForm interface using a single Sub procedure and call it after every user action (using Click or Change events) that is coded on your interface.

Disabling Screen Updating

The Application object has the ScreenUpdating method, which can be used by your applications to turn on and off all screen-updating operations made by Excel. Chances are that you need to turn screen updating off while you are opening another workbook inside your applications, making any necessary operations, and closing it, before the user of your application even notices what is happening behind the scenes.

The `Workbook Referencing ScreenUpdating.xlsm` macro-enabled workbook that you can also extract from the `Chapter03.zip` file has such a feature on its `frmOpenedWorkbooks UserForm`, implemented by a simple CheckBox control below its `lstWorkbooks ListBox` (Figure 3-15).

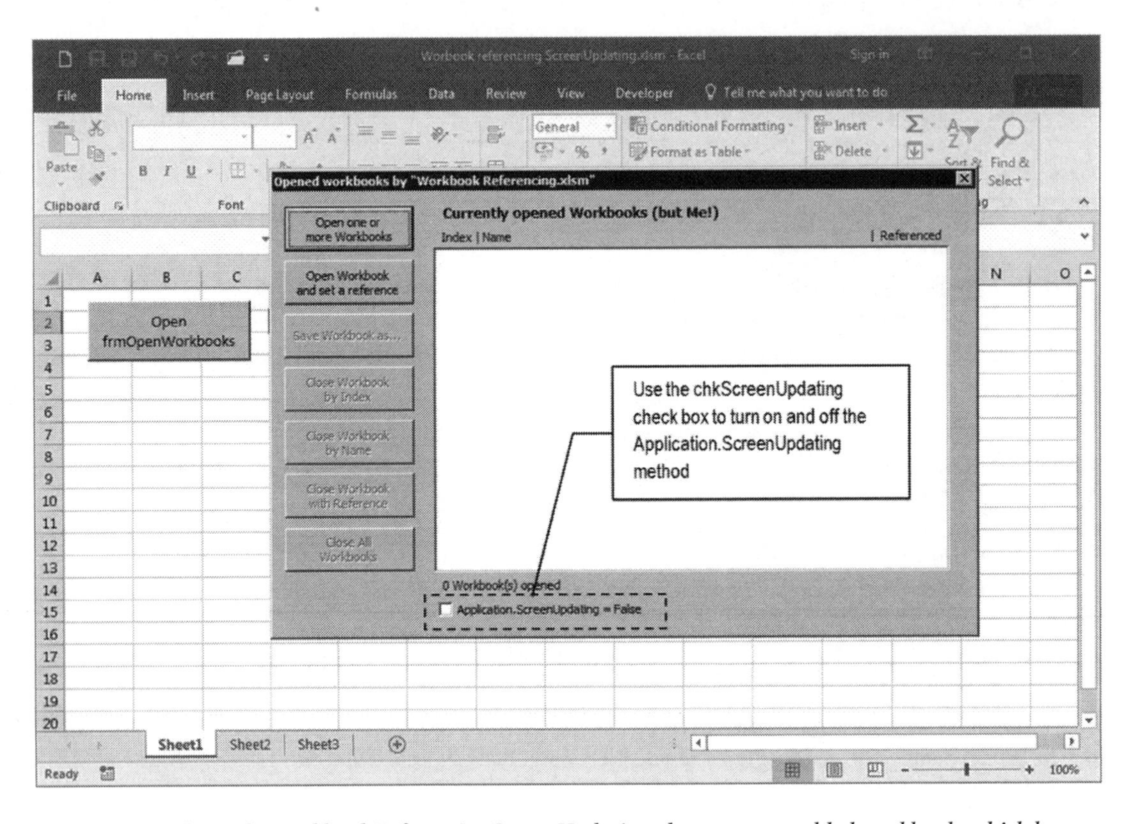

Figure 3-15. *This is the Workbook Referencing ScreenUpdating.xlsm macro-enabled workbook, which has the chkCheckBox below its lstWorkbooks ListBox that allows you to control the Application ScreenUpdating method. When chkCheckBox is unchecked, you can open and close workbooks without notice of any operation in the Excel interface*

This `UserForm` makes the same operations already commented on in the previous section but can open and close workbooks without updating the Excel user interface while `chkScreenUpdating Checkbox` is unchecked: you can open any number of workbooks without notice in their worksheet interfaces, while the `lstWorkbooks ListBox` shows what is opened.

■ **Attention** If you open some workbooks and check the `chkScreenUpdating CheckBox`, Excel will update its interface and immediately show the active `worksheet` of the last-opened workbook.

To turn on and off the `Application.ScreenUpdating` method, just one line of code is needed on both `cmdOpenWkbs_Click( )` and `cmdOpenWbkWithReference_Click( )` events.

```
Private Sub cmdOpenWkbs_Click()
    Dim varFiles As Variant
    Dim varItem As Variant

    varFiles = ShowDialogBox(OpenFile, True)

    If Not IsEmpty(varFiles) Then
        Application.ScreenUpdating = Me.chkScreenUpdating
        If IsArray(varFiles) Then
            For Each varItem In varFiles
                Application.Workbooks.Open (varItem)
            Next
        Else
            Application.Workbooks.Open (varFiles)
        End If
        Call FilllstWorkbooks
        Call DefineButtons(False)
    End If
End Sub

Private Sub cmdOpenWkbReference_Click()
    Dim strFile As String

    strFile = ShowDialogBox() & ""

    If Len(strFile) Then
        'Workbook was selected. Open it!
        Application.ScreenUpdating = Me.chkScreenUpdating
        Set mWkb = Application.Workbooks.Open(strFile)
        Me.cmdCloseWkbByRef.Enabled = True
        Call FilllstWorkbooks
    End If
End Sub
```

Note that the `Application.ScreenUpdating` method is turned on/off by the state of the `chkScreenUpdating` CheckBox, which returns `True` when checked and `False` when unchecked.

```
Application.ScreenUpdating = Me.chkScreenUpdating
```

When you click the `chkScreenUpdating` CheckBox, the `chkScreenUpdating_Click( )` event fires, executing this code:

```
Private Sub chkScreenUpdating_Click()
    Application.ScreenUpdating = Me.chkScreenUpdating
End Sub
```

And whenever you close `frmOpenedWorkbooks`, the `UserForm_Terminate( )` event fires, turning on again the `Application.ScreenUpdating` method.

```
Private Sub UserForm_Terminate()
    Application.ScreenUpdating = True
End Sub
```

■ **Attention** You should take care to always turn on again the `Application.ScreenUpdating` method or the Excel screen will seem to be freeze. If this ever happens in your code, press Alt+F11 to show the VBA interface and type `Application.ScreenUpdating=True` in the VBA Immediate window to turn the screen updating on again.

Through this entire book we will deal with the `Workbook` object and its interface, using some of its events, properties, and methods to improve the application usability.

Chapter Summary

In this chapter, you learned about the following:

- The sequence of Excel `Workbook` object events
- How to pop up a splash screen to your application using a VBA `UserForm` and the `Workbook.Open` event
- How you can implement a `Timer` using the `Application.OnTime` method or VBA `Timer( )` function
- How you can use the VBA `DoEvents` statement inside a `Do...Loop` structure
- How to set `Workbook` object references
- How you can create a `UserForm` interface to open, save, and close workbooks
- How to deal with the `UserForm` object and `ListBox` control properties and methods

In the next chapter, you will learn about the next object in the Microsoft Excel object model where all the action of your worksheet application really happens!

CHAPTER 4

Programming the Microsoft Excel Worksheet Object

In this chapter you will learn about the third object in the Microsoft Excel object model hierarchy, which is the `Worksheet` object. This is where the real action of most of your worksheet applications will happen. Like its parent and grandparent objects (the `Workbook` and `Application` objects, respectively), it has a rich interface with many properties, methods, and events that you should be aware of to program your application with VBA. You can obtain all the procedure code in this chapter by downloading the `Chapter04.zip` file from the book's Apress.com product page, located at `www.apress.com/9781484222041`, or from `http://ProgrammingExcelWithVBA.4shared.com`.

The Worksheet Object

The Microsoft Excel `Worksheet` object is the core of any worksheet application and as such is the focus of many Excel VBA applications. Similar to its parent, the `Workbook` object, and grandparent, the `Application` object, the `Worksheet` object is full of properties, methods, and events that you can interact with using VBA code to take absolute control of your application.

Table 4-1 shows some important `Worksheet` object properties.

Attention Search the Internet with the keywords *worksheet properties*, *worksheet methods*, or *worksheet events* to find a complete list of Excel `Worksheet` object properties, methods, and events, respectively. Tables 4-1, 4-2, and 4-3 come from the following location on the Microsoft MSDN web site:

`http://msdn.microsoft.com/en-us/library/microsoft.office.tools.excel.worksheet_properties.aspx`

F. Morgado, *Programming Excel with VBA*, DOI 10.1007/978-1-4842-2205-8_4

Table 4-1. *Excel Worksheet Object Properties That Control the Way the Excel Window Behaves*

Worksheet Object Property	Value	Used To
`Application`	`Object`	Gets a reference to the `Application` object.
`AutoFilter`	`Object`	Gets an `AutoFilter` that provides information about filtered lists on the worksheet if filtering is enabled. Gets `Nothing` if filtering is off.
`AutoFilterMode`	`Boolean`	Gets or sets a value that indicates whether filtering is currently enabled on the worksheet (that is, whether the filter drop-down arrows are currently displayed).
`Cells`	`Range`	Gets a `Range` object that represents all the cells on the worksheet (not just the cells that are currently in use).
`CircularReference`	`Range`	Gets a `Range` object that represents the range containing the first circular reference on the sheet. Gets `Nothing` if there is no circular reference on the sheet.
`CodeName`	`String`	Sets the sheet tab code module name on the VBA project.
`Columns`	`Range`	Gets a `Range` object that represents one or more columns on the worksheet.
`Comments`	`Collection`	Gets a `Comments` collection that represents all the comments.
`DisplayPageBreaks`	`Boolean`	Gets or sets a value that indicates whether page breaks (both automatic and manual) on the worksheet are displayed.
`EnableAutoFilter`	`Boolean`	Gets or sets a value that indicates whether `AutoFilter` arrows are enabled when user-interface-only protection is turned on.
`EnableCalculation`	`Boolean`	Gets or set a value that indicates whether Microsoft Office Excel automatically recalculates the worksheet when necessary.
`EnableSelection`	`Boolean`	Gets or sets a value indicating which cells can be selected on the sheet.
`Name`	`String`	Gets or sets the name of the worksheet.
`Names`	`Collection`	Gets a `Names` collection that represents all the worksheet-specific names (names defined with the `WorksheetName!` prefix).
`Next`	`Worksheet`	Gets a `Microsoft.Office.Interop.Excel.Worksheet` that represents the next sheet.
`Parent`	`Workbook`	Gets the parent object for the worksheet.
`Previous`	`Worksheet`	Gets a `Microsoft.Office.Interop.Excel.Worksheet` that represents the previous sheet.
`ProtectContents`	`Boolean`	Gets a value that indicates whether the contents of the worksheet (the individual cells) are protected.

(*continued*)

Table 4-1. *(continued)*

Worksheet Object Property	Value	Used To
Protection	Object	Gets a Protection object that represents the protection options of the worksheet.
Range	Range	Gets a Range object that represents a cell or a range of cells.
Rows	Range	Gets a Range object that represents one or more rows on the worksheet.
ScrollArea	Range	Gets or sets the range where scrolling is allowed, as an A1-style range reference.
Sort	Object	Gets the sorted values in the current worksheet.
Tab	Object	Gets a tab for the worksheet.
Type	Object	Gets the worksheet type.
UsedRange	Range	Gets a Range object that represents all the cells that have contained a value at any time.
Visible	Integer	Gets or sets a value that determines whether the object is visible (xlSheetVisible), hidden (xlSheetHidden), or very hidden (xlSheetVeryHidden).

Table 4-2 shows some important Worksheet object methods and the actions they perform when evoked by your VBA code.

Table 4-2. *Some Important Excel Worksheet Object Methods*

Worksheet Object Method	Action Performed
Activate	Makes the underlying Worksheet object the active sheet
Delete	Deletes the underlying Worksheet object
Move	Moves the worksheet to another location in the workbook
Paste	Pastes the contents of the clipboard onto the worksheet
PasteSpecial	Pastes the contents of the clipboard onto the worksheet, using a specified format
PrintOut	Prints the worksheet
PrintPreview	Shows a preview of the worksheet
Protect	Protects a worksheet
SaveAs	Saves changes to the worksheet in a different file
Select	Selects the worksheet
Unprotect	Removes protection if the worksheet is protected

The Excel Worksheet object has a small set of events when compared to the Excel Application and Workbook objects. You use these events to control the user actions on each sheet tab.

Table 4-3 shows the most important Worksheet object events, when they fire, and the event declarations and arguments (if any).

Table 4-3. *Worksheet Object Events and Their Occurrence*

Event Name	Occurrence	Event Declaration and Arguments
Activate	When the worksheet is activated	Private Sub Worksheet_Activate()
BeforeDoubleClick	When the worksheet is double-clicked, before the default double-click action	Private Sub Worksheet_ BeforeDoubleClick(ByVal Target As Range, Cancel As Boolean)
BeforeRightClick	When the worksheet is right-clicked, before the default right-click action	Private Sub Worksheet_ BeforeRightClick(ByVal Target As Range, Cancel As Boolean)
Calculate	After the worksheet is recalculated or after any changed data is plotted on a chart	Private Sub Worksheet_Calculate()
Change	When cells in the worksheet are changed by the user or by an external link	Private Sub Worksheet_Change(ByVal Target As Range)
Deactivate	When the sheet is deactivated	Private Sub Worksheet_Deactivate()
SelectionChange	When another cell (or cells) is selected on the worksheet	Private Sub Worksheet_ SelectionChange(ByVal Target As Range)

Note that all Worksheet object events are preceded by the word Worksheet_ in their procedure declaration. Once again, for each associated event in the Workbook and Application objects, Excel will first fire the Worksheet object event, followed by the associated Workbook object event and then by the Application object event (the event order always fires from the bottom to the higher object level).

Using Worksheet Object Events

By now you must already know when and why these events fire. The Worksheet Events.xlsm Excel macro-enabled workbook that you can extract from the Chapter03.zip file has all these events coded on the Sheet1 object code module.

It is important to note that the Worksheet object events *do not* fire when the workbook is opened. For example, the Sheet1_Activate( ) event does not fired when the workbook is opened and Sheet1 is activated. To see the Worksheet Events.xlsm macro-enabled workbook Sheet1 object events fire, you must select another sheet tab (the Worksheet_Deactivate event fires), select again the Sheet1 tab (the Worksheet_Activate event fires), select another cell on the Sheet1 worksheet (the Worksheet_ SelectionChange fires), change any cell value (the Worksheet_Change and Worksheet_SelectionChange events fire), insert a formula on any cell (the Worksheet_Calculate, Worksheet_Change and Worksheet_ SelectionChange events fire), double-click any cell (the Worksheet_BeforeDoubleClick event fires), or right-click any cell (the Worksheet_BeforeRightClick event fires).

Figure 4-1 shows the event sequence that fires after typing the formula =1 on cell A1 of the Sheet1 worksheet of the Worksheet Events.xlsm file and pressing Enter.

Worksheet_Calculate → Worksheet_Change → Worksheet_SelectionChange

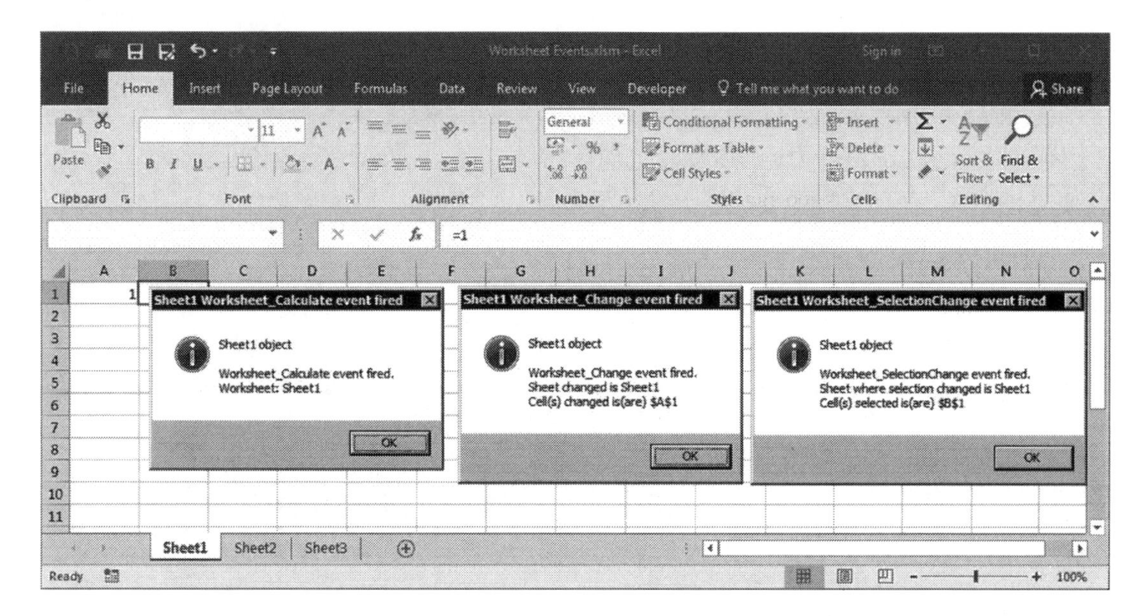

Figure 4-1. *The Worksheet Events.xlsm code shows all Worksheet object events in Table 4-3. When you insert a formula on any Sheet1 cell (cell A1 received formula =1) and press Enter, Excel will change cell A1's value, recalculate the worksheet, and select cell A2, firing three successive events: Worksheet_Calculate, Worksheet_Change, and Worksheet_SelectionChange*

Please note that every `Sheet1` object event has a slightly different code strategy to print the `Sheet1` object name and `Sheet1` tab name, which are different names. This is the `Sheet1 Worksheet_Change( )` event procedure:

```
Private Sub Worksheet_Change(ByVal Target As Range)
    Dim strMsg As String
    Dim strTitle As String
    Dim strCodeName As String

    strCodeName = Application.ActiveSheet.CodeName
    strTitle = strCodeName & " Worksheet_Change event fired"
    strMsg = strCodeName & " object" & vbCrLf & vbCrLf
    strMsg = strMsg & "Worksheet_Change event fired." & vbCrLf
    strMsg = strMsg & "Sheet changed is " & Application.ActiveSheet.Name & vbCrLf
    strMsg = strMsg & "Cell(s) changed is(are) " & Target.Address
    MsgBox strMsg, vbInformation, strTitle
End Sub
```

To get the `Worksheet` object name, the code uses the `Worksheet.CodeName` property (returned by the `Application.ActiveSheet` property, which represents the active sheet) and stores it on the `strCodeName` string variable.

```
strCodeName = Application.ActiveSheet.CodeName
```

And to get the worksheet sheet tab name, the code uses the `Worksheet.Name` property (returned by the `Application.ActiveSheet` property).

```
strMsg = strMsg & "Sheet changed is " & Application.ActiveSheet.Name & vbCrLf
```

Now you know that every worksheet tab has two names: one that can be easily changed by the user of your application (the Sheet1 tab name, returned by the Worksheet.Name property) and another that belongs to the Worksheet object code module and represents it in the VBA Explorer tree (the Worksheet.CodeName property).

Although this is quite simple, it has a tremendous impact on your worksheet applications. In fact, you should never code the Worksheet.Name property into your VBA procedures (like the Sheet1 tab name). Always refer to the Worksheet.CodeName property in VBA, because it can be changed in the VBA environment by first selecting the desired Worksheet object in the Project Explorer tree and then changing its Name property in the VBA Properties window (Figure 4-2), even though you can trust the technique described in Chapter 2 to avoid any sheet tab name changes.

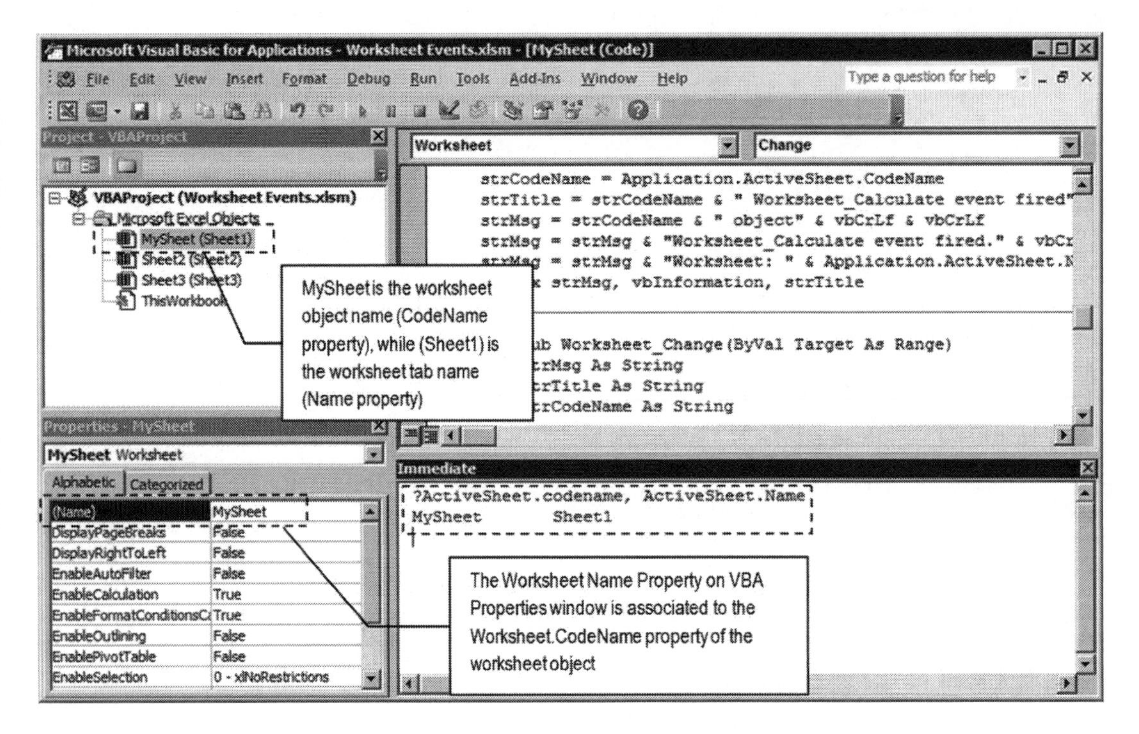

Figure 4-2. *The VBA Project interface allows you to change any worksheet code name using the Properties window. This figure changes the Sheet1 worksheet tab Name property from Sheet1 to MySheet, even though the worksheet tab continues to show Sheet1 as the worksheet name. Note how the Name and CodeName properties of the ActiveSheet object are printed in the VBA Immediate window*

■ **Attention** The Name property of any selected worksheet object in the VBA Project Explorer tree is associated to the CodeName property of the Worksheet object. Its name must obey the variable name declaration. In other words, it must begin with a letter or underscore and cannot contain spaces. The VBA Project Explorer tree always shows the object Name property (Worksheet.CodeName), followed by the current sheet tab name (the Worksheet.Name property, between parentheses).

To avoid any Worksheet object's Name property value from being changed in the VBA interface, protect the VBA project of your application! See Chapter 1 for more information.

After the VBA Name property for the Sheet1 worksheet was changed to MySheet, try to change any cell value and note how the event code procedure will always use the worksheet object's Name property (using Application.ActiveSheet.CodeName) to correctly identify the worksheet object name (Figure 4-3).

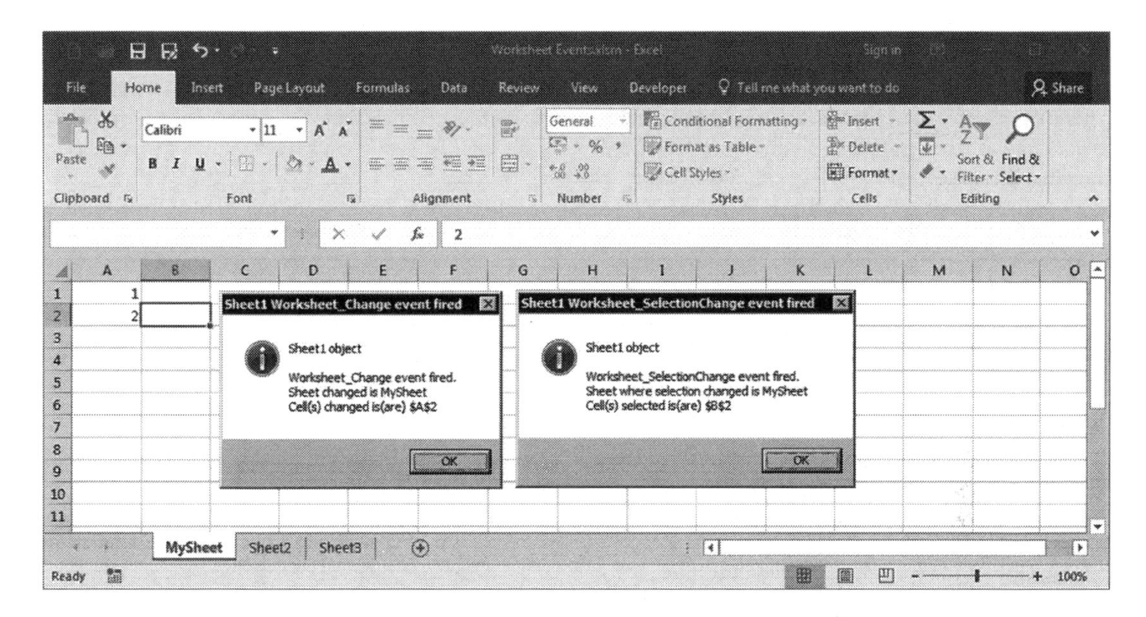

Figure 4-3. *After changing the VBA Name property of the Sheet1 object, try to change any cell value and note how the event procedure's code captures the current object name using the Application.ActiveSheet.CodeName property*

The file Workbook and Worksheet Events.xlsm fires all Worksheet and Workbook object events (in this order), and the files Application, Workbook, and Worksheet Events.xlsm fire all three object events. You must extract them from the Chapter04.zip file to see how all object-associated events fire from the bottom-level object (the Worksheet object) to the higher-level object (the Application object).

Referring to Worksheets

To refer to any worksheet that your application needs to access, you must set a reference to it. If the code needs to refers to the worksheet that owns the code module (for example, you are coding the Sheet1 code module and need to refer to Sheet1 itself), use the VBA keyword Me.

Note, however, that Me can't be used in the VBA Immediate window unless your code is running and you place an interruption point to make the code break. So, this *will not work* in the Immediate window if no code or event is running:

```
?Me.Name
```

To refer to the active sheet (the one whose sheet tab is already selected in the Excel interface), you can use the ActiveSheet property of the Application object, as follows:

```
?Application.ActiveSheet.Name
```

But since the Application object is the top-level object on the Microsoft Excel object model hierarchy, you don't need to refer it when using the ActiveSheet property (or any other workbook object cited in Table 2-1 in Chapter 2). This will also work:

```
?ActiveSheet.Name
```

There will be times that you will need to set a reference to another worksheet of the same workbook (or from another opened workbook in the Excel interface) from the current Worksheet object code module (the one that is running your VBA code). For all these cases, the easiest way to get a reference to the desired worksheet is to use the Worksheets or Sheets collection of the Workbook object, referencing the worksheet by its Index (sheet order) or its Name.

To refer to any worksheet from *your workbook* application, use the ThisWorkbook object's Worksheets or Sheets collection. The next code fragment sets a reference to the first sheet tab of the application workbook to a Worksheet object variable using either the sheet index number or the sheet tab name string.

```
Dim ws as Worksheet
Set ws = ThisWorkbook.Worksheets(1)
```

Or use the following:

```
Set ws = ThisWorkbook.Worksheets("Sheet1")
```

■ **Attention** As Table 3-1 implies, Microsoft Excel exposes two different collections for all worksheets of any workbook: the Worksheets and Sheets collections. So, you can also use the Sheets collection, which is used in a lot of Microsoft Excel VBA code you find on the Internet.

```
Dim ws as Worksheet
Set ws = ThisWorkbook.Sheets(1)
```

Here's another example:

```
Set ws = ThisWorkbook.Sheets("Sheet1")
```

To set a reference to any worksheet of any opened workbook inside the same Microsoft Excel interface, you must first use the Application object's Workbooks collection to set a reference to the desired Workbook object using an object variable declared as Workbook, using either the workbook Index or the Name properties. Here's an example:

```
Dim wb as Workbook
Set wb = Application.Workbooks(1)    'Reference by Index
```

Here's another example:

```
Set wb = Workbooks("Book2")     'Reference by Name
```

Once the reference to the desired Workbook object has been set, use its Worksheets or Sheets collection to get a reference to the desired worksheet (either by Index or by Name), in this way:

```
Dim wb as Workbook
Dim ws as Worksheet
```

```
Set wb = Workbooks("Book2")
Set ws = wb.Worksheets("Shee1")
```

If you do not need to reference again the desired opened workbook in the Excel interface, you can set a reference to any of its worksheets using a simplified syntax using the .Name property of both collections (Workbooks and Worksheets), in this way:

```
Dim ws as Worksheet
Set ws = Workbooks("Book2").Worksheets("Shee1")
```

Or you can use the Workbook.Index property of both collections (Workbooks and Worksheets):

```
Dim ws as Worksheet
Set ws = Workbooks(2).Worksheets(1)
```

■ **Attention** Do never forget that you can't trust that Workbook.Index refers to any workbook opened by your application using VBA. To guarantee that you are using the right workbook, get the Name property of the supposed workbook inside the Workbooks collection and then test it to verify whether it is the right workbook.

```
strName = Workbooks(2).Name
If strName = <workbookname> then
    ...
End If
```

Setting the Worksheet Object Reference

Let's try a simple exercise. Open a new Excel workbook, press Alt+F11 to show the VBA IDE, and try this simple experience using the VBA Immediate window to print the sheet name based on its index:

```
?Thisworkbook.Worksheets(1).Name
```

If the first sheet tab is Sheet1, the Immediate window must print Sheet1 for Thisworkbook.Worksheets(1).Name. Now, drag the Sheet1 sheet tab to the right, placing it between Sheet2 and Sheet3. If Sheet2 becomes the first sheet tab, when you execute the Thisworkbook.Worksheets(1).Name command again in the Immediate window, it will print Sheet2 (Figure 4-4).

The lesson is quite simple: your application should never refer to the supposed Worksheets collection's Index or Name properties in its VBA code because they are prone to changes. This is the main cause of so many people asking for a way to avoid the sheet tab name change: they inadvertently coded these values inside the VBA code modules.

Use the VBA Me keyword when you want to set a self-reference to the current worksheet object (the Worksheet.CodeName property) to refer to any worksheet of your application or set one to the ActiveSheet property to refer to the worksheet that has the focus in the Excel interface. Here's an example:

```
?Me.Name
?Sheet1.Name   'Sheet1 is the Worksheet.CodeName property
?ActiveSheet.Name
```

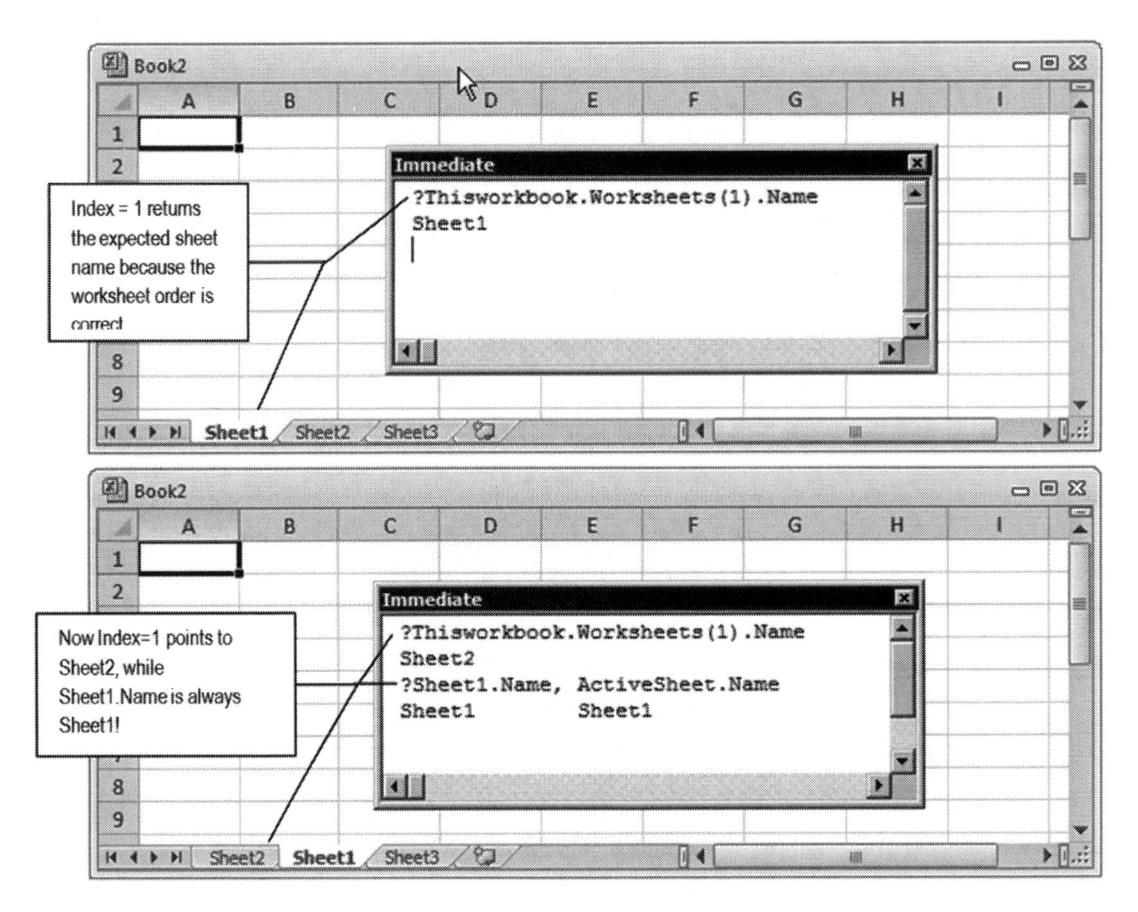

Figure 4-4. *Beware of referring to any worksheet using its Index on the Worksheets collection because if you change the sheet order, the worksheet index also changes, and you may refer to the wrong sheet tab. Use the Worksheet object's Codename property, the ActiveSheet property of the Application object, or the keyword Me to reference it (Me means the worksheet where the code is running)*

■ **Attention** To avoid that the `Worksheet` object's `Name` property (associated to the `Worksheet.CodeName` property) in the VBA Properties window can be changed, protect your VBA project. See Chapter 1 for more information.

Let's see some practical examples of worksheet references and methods.

Using the CSheetNameChange Class to Avoid a Single Sheet Change Name

Let's see a practical example of the `Me` keyword to avoid just a single sheet change name, using the beauty of the `Class` module's object programming.

The `Worksheet Events with CSheetNameChange Class.xlsm` macro-enabled workbook has a copy of the `CSheetNameChange` class (developed in section "Using a Class Module to Control Sheet Tab Name Changes" in Chapter 2) that is used to watch for any sheet tab name changes. Figure 4-5 shows the VBA environment window with the `Sheet1` object code module and its coded events.

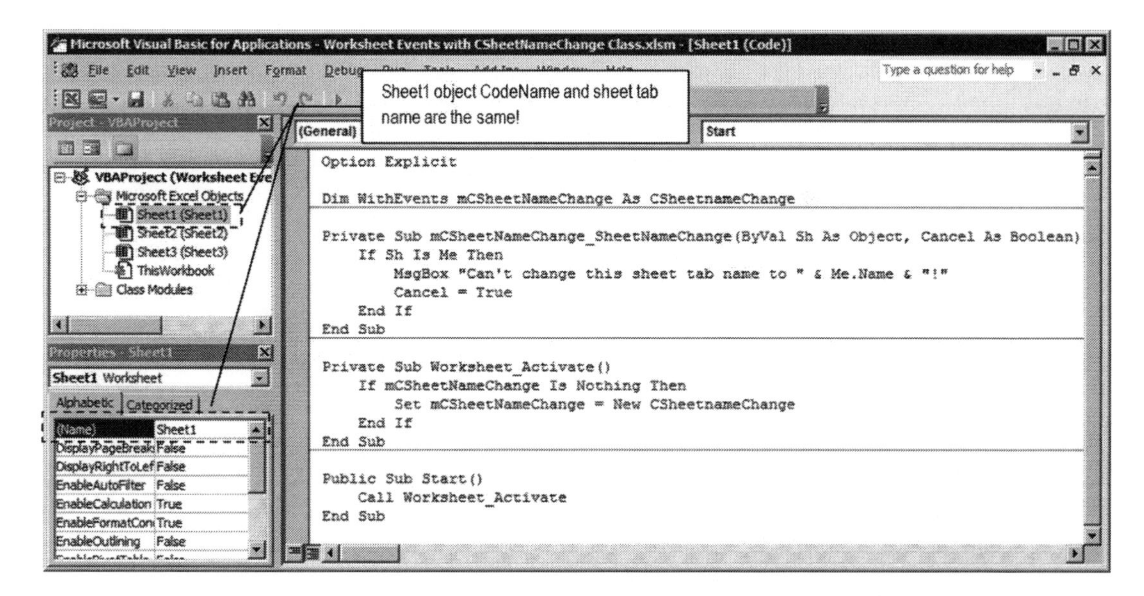

***Figure 4-5.** This is the macro-enabled workbook in the VBA Project Explorer, showing its CSheetNameChange Class module and the Sheet1 object code module that do not allow the Sheet1 tab name to be changed*

You can't change the `Worksheet Events with CSheetNameChange Class.xlsm` file's `Sheet1` tab name because the `Sheet1` object uses the `CSheetNameChange` class to watch and fire its `SheetChangeName( )` event whenever this happens. It works this way:

1. The `Sheet1` object uses `WithEvents` to declare the module-level `mCSheetNameChange as CSheetNameChange` object variable on its module `Declaration` section.

```
Option Explicit
Dim WithEvents mCSheetNameChange As CSheetNameChange
```

2. When the worksheet is activated, the `Worksheet_Activate( )` event fires and verifies whether the `mCSheetNameChange` variable has already been instantiated, comparing it to the VBA `Is Nothing` value (object variables not instantiated have the default `Nothing` value). If this is true, a new instance of the `CSheetNameChange` class is set to the `mCSheetNameChange` variable.

```
Private Sub Worksheet_Activate()
    If mCSheetNameChange Is Nothing Then
        Set mCSheetNameChange = New CSheetnameChange
    End If
End Sub
```

3. But since the `Worksheet_Activate` does not fire when the workbook is opened, we need to create a `Public Sub` procedure on the `Sheet1` code module that makes a call to the `Sheet1.Worksheet_Activate` event, so this procedure can be called from the `Thisworkbook` code module. The `Sheet1` object has the `Public Sub Start( )` procedure, which has this code:

```
Public Sub Start()
    Call Worksheet_Activate
End Sub
```

4. And when the workbook is opened, you force the Sheet1 object Activate( ) event to fire by making a call to the Sheet1.Start( ) public procedure from the ThisWorkbook object's Workbook_Open event, which has this code (note that Sheet1 object means the Worksheet.CodeName property for the Sheet1 worksheet tab):

```
Private Sub Workbook_Open()
    Call Sheet1.Start
End Sub
```

5. Since Sheet1 set an instance of the CSheetNameChange class, whenever *any* sheet tab name changes, the CSheetNameChange object's SheetNameChange( ) event will fire, passing the Sh argument to identify the Worksheet object whose name has been changed. To verify whether the Sh argument refers to the current worksheet, you must compare it to the Me keyword (*this* code module object). If they are the same, *this sheet tab name* (Me) is changed, the event is canceled, and a MsgBox( ) warns that the Sheet1 tab name can't be changed to the desired new name.

```
Private Sub mCSheetNameChange_SheetNameChange(ByVal Sh As Object, Cancel As
Boolean)
    If Sh Is Me Then
        MsgBox "Can't change this sheet tab name to " & Me.Name & "!"
        Cancel = True
    End If
End Sub
```

■ **Attention** Please note that Me.Name indicates the new name typed for this sheet tab because the event has not finished yet and the previous worksheet name was not restored.

Try to change the Sheet1 tab name to anything and see for yourself. After changing the Sheet1 tab name, if you try to save the workbook, select another cell, change any selected cell value, double-click or right-click any cell, or select another worksheet tab, the mCSheetNameChange_SheetNameChange( ) event will fire, and the sheet tab name will be turned again to Sheet1, although you can still change the Sheet2 or Sheet3 tab name (Figure 4-6)! Isn't it beautiful?

Exercise!

How can you avoid that just two sheet tab names (like the Sheet1 and Sheet2 worksheets) from the current workbook can't be changed? Where must you declare and instantiate the object variable that represents the CSheetNameChange class?

■ **Attention** The answer is in the file `Workbook Events with CSheetNameChange Class.xlsm` that you can extract from the `Chapter04.zip` file.

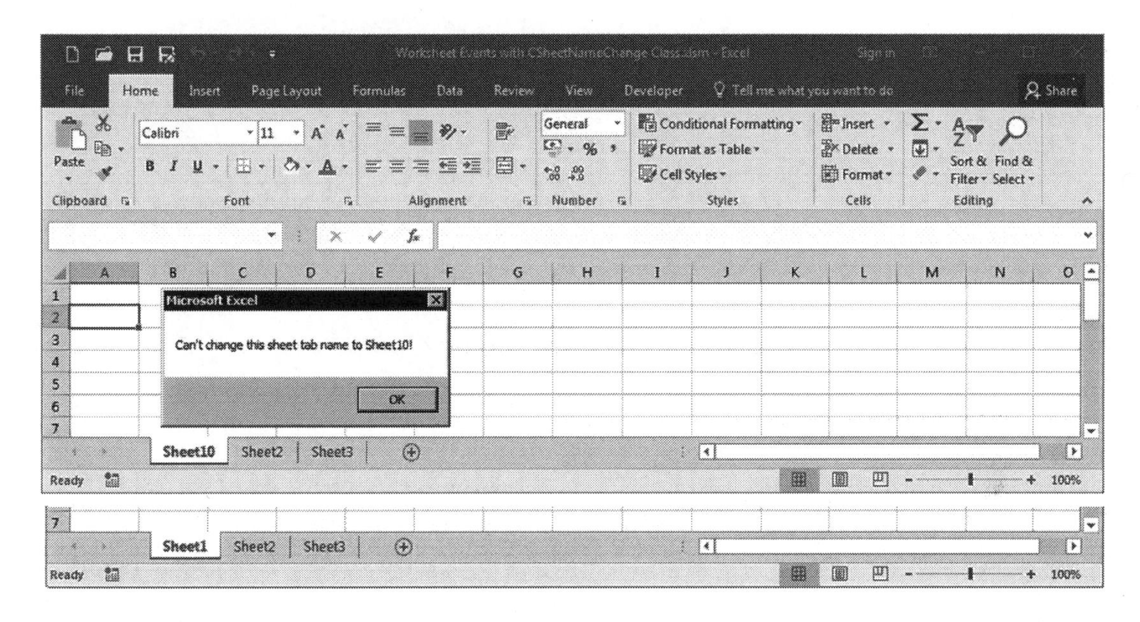

Figure 4-6. *Using the CSheetNameChange class and its SheetChangeName( ) event procedure just on the Sheet1 code module, you can avoid that just this worksheet changes its name*

Using Worksheet Object Properties and Methods

Most worksheet application actions will happen inside individual `Worksheet` object code modules, and you must be aware of how to deal with the `Worksheets` or `Sheets` collection and some very popular `Worksheet` object methods, like `Move`, `Copy`, `Protect`, and `Delete`, as well as the `Worksheet` object's `Visible` property, to really control the behavior of your VBA Excel applications.

The `Worksheet Referencing.xlsm` macro-enabled workbook that you can extract from the `Chapter04.zip` file has the `frmWorksheets UserForm` that is automatically opened by the `ThisWorkbook.Workbook_Open` event and that allows you to deal with the `Worksheet` object (Figure 4-7).

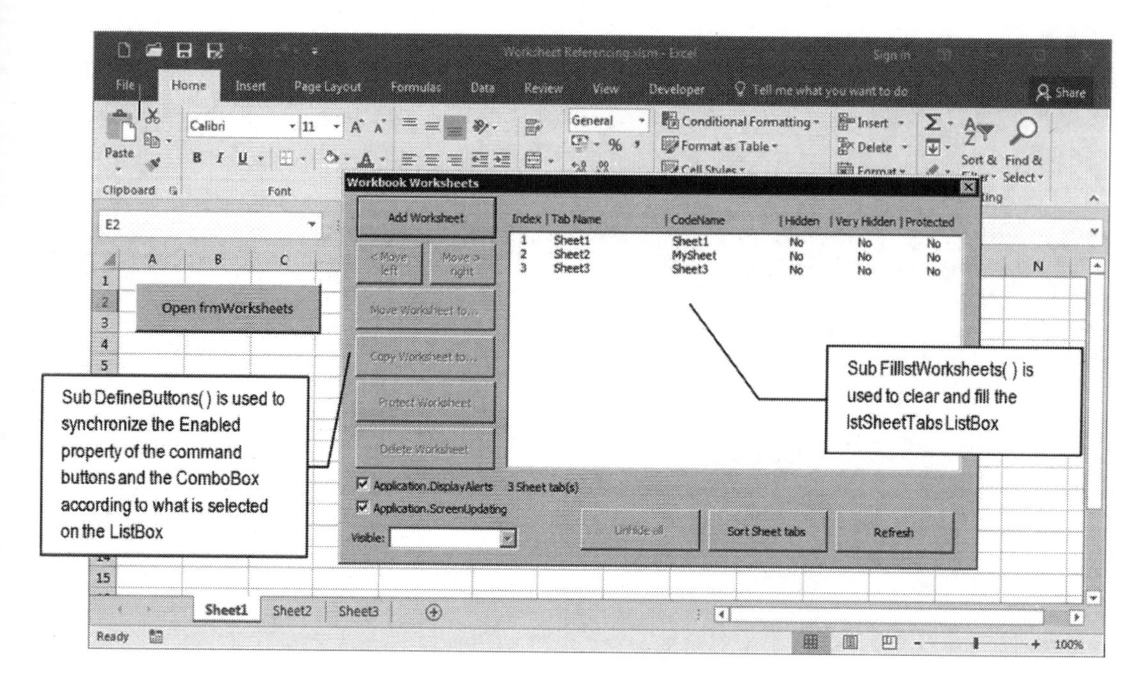

Figure 4-7. *This is the frmWorksheets UserForm from Worksheet Referencing.xlms macro-enabled workbook that allows you to learn how to do some simple operations with VBA using the Worksheets collection and the Worksheet object*

■ **Attention** Before we begin to explore the `frmOpenedWorkbooks` interface and its code strategy, I would like you to know that this `UserForm` code, interface, and controls synchronization were not made by chance. This took a lot of work to get perfect, at least for me! If your `UserForms` do not behave like this the first time, be patient and try again and again until they work the way you expect.

Using the same strategy employed by `frmOpenedWorkbooks`, discussed in the section "Setting Workbook Object References" earlier in this chapter, `frmWorksheets` also bases its interface synchronization on two main `Sub` procedures: `FilllstSheetTabs( )` to fill the `lstSheetTabs ListBox` with information about all workbook worksheets and `DefineButtons( )` to synchronize the `Enabled` property of the `UserForm` controls according to the `lstSheetTabs` selection state.

When the `Worksheet Referencing.xlsm` workbook is opened (or when the Open frmWorksheets command button is clicked), the `ThisWorkbook.Workbook_Open( )` event fires and executes this code to open the `frmWorksheets UserForm` in a modeless state (passing `False` to its `Modal` argument):

```
Private Sub Workbook_Open()
    frmWorksheets.Show False
End Sub
```

And when `frmWorksheets` loads, the `Initialize( )` event fires, executing this code:

```
Option Explicit

Dim mbolCancelEvent As Boolean
```

```
Dim mintHidden As Integer

Private Sub UserForm_Initialize()
    mbolCancelEvent = True

    'Fill ComboBox cboVisible
    Me.cboVisible.AddItem xlSheetVisible
    Me.cboVisible.Column(1, cboVisible.ListCount - 1) = "xlSheetVisible"
    Me.cboVisible.AddItem xlSheetHidden
    Me.cboVisible.Column(1, cboVisible.ListCount - 1) = "xlSheetHidden"
    Me.cboVisible.AddItem xlSheetVeryHidden
    Me.cboVisible.Column(1, cboVisible.ListCount - 1) = "xlSheetVeryHidden"

    'Update ListBox lstSheetTabs
    Call FilllstSheetTabs
End Sub
```

You may noticed that the `UserForm` declares two module-level variables in its `Declaration` section (`mbolCancelEvent` and `mintHidden`), and the first action of the `UserForm_Initialize( )` event is setting `mbolCancelEvent = True`.

If you look at the VBA Properties window, you will see that the `cboVisible` ComboBox has two columns (`ColumnCount = 2`), that the value returned by the control refers to its first column (`BoundColumn=1`), and that its first column is invisible (`ColumnWidths= 0 pt`, as shown in Figure 4-8).

So, the next action of the `UserForm_Initialize( )` event is to use the `AddItem` method of the ComboBox control to fill the two columns of the `cboVisible` ComboBox with the three possible worksheet visible states: `xlSheetVisible`, `xlSheetHidden` or `xlSheetVeryHidden`.

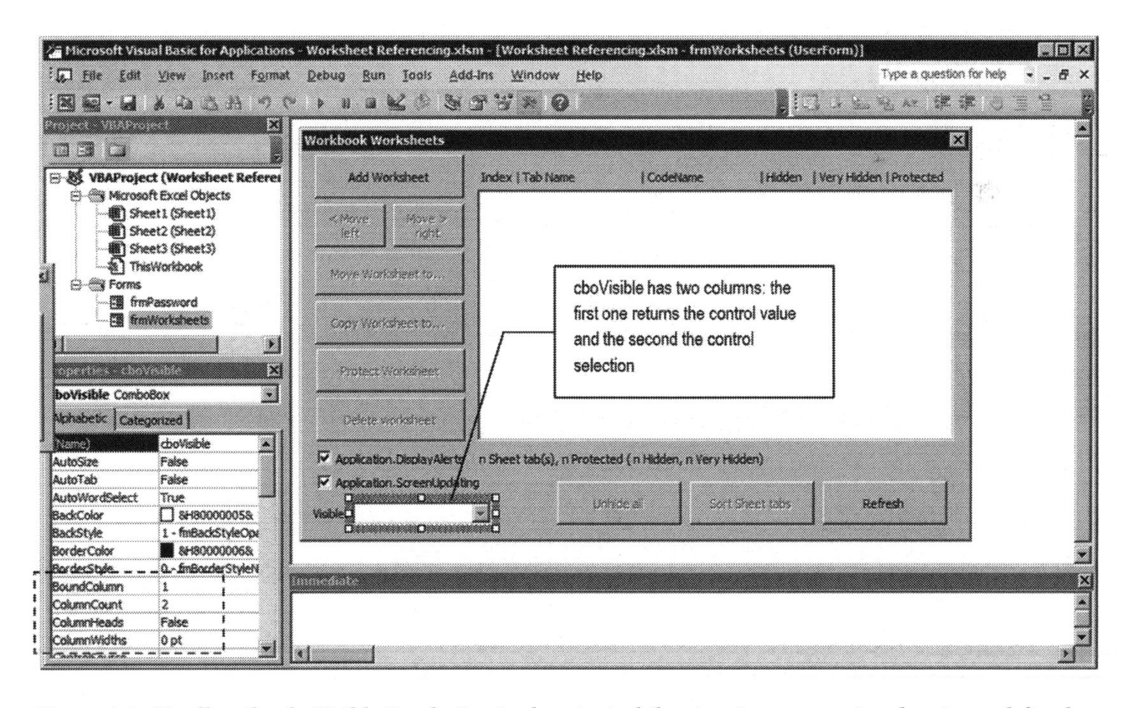

Figure 4-8. *To allow the cboVisible ComboBox to show text while returning a numeric value, it was defined with two columns (ColumnCount=2), with the first column being the control value (BoundColumn=1) even though it is invisible (ColumnWidth=0)*

The cboVisible.AddItem method sets the first column value, and the cboVisible.Column property sets the second column value, as follows:

```
Me.cboVisible.AddItem xlSheetVisible
Me.cboVisible.Column(1, cboVisible.ListCount - 1) = "xlSheetVisible"
```

And as soon as cboVisible is filled, the UserForm_Initialize( ) event calls Sub FilllstSheetTabs( ) to fill the lstSheetTabs ListBox with information about all the workbook sheet tabs.

```
Public Sub FilllstSheetTabs()
    Dim varItem As Variant
    Dim intI As Integer
    Dim intProtected  As Integer
    Dim intVeryHidden As Integer

    mintHidden = 0
    With Me.lstSheetTabs
        .Clear
        For Each varItem In Worksheets
            intI = intI + 1
            .AddItem intI
            .Column(1, .ListCount - 1) = varItem.Name
            .Column(2, .ListCount - 1) = varItem.CodeName

            Select Case varItem.Visible
                Case xlSheetVisible
                    .Column(3, .ListCount - 1) = "No"
                    .Column(4, .ListCount - 1) = "No"
                Case xlSheetHidden
                    .Column(3, .ListCount - 1) = "Yes"
                    .Column(4, .ListCount - 1) = "No"
                    mintHidden = mintHidden + 1
                Case xlSheetVeryHidden
                    .Column(3, .ListCount - 1) = "Yes"
                    .Column(4, .ListCount - 1) = "Yes"
                    mintHidden = mintHidden + 1
                    intVeryHidden = intVeryHidden + 1
            End Select

            If varItem.ProtectContents Then
                .Column(5, .ListCount - 1) = "Yes"
                intProtected = intProtected + 1
            Else
                .Column(5, .ListCount - 1) = "No"
            End If
        Next
    End With

    Me.lblSheetTabs.Caption = Sheets.Count & " Sheet tab(s)" & _
    IIf(intProtected > 0, ", " & intProtected & " protected", "") & _
    IIf(mintHidden > 0, " (" & mintHidden & " hidden", "") & _
    IIf(intVeryHidden > 0, ", " & intVeryHidden & " very hidden", "") & _
    IIf(mintHidden, ")", "")
```

```
    Call DefineButtons(False)
End Sub
```

The programming technique should now be familiar to you: after you declare all the variables it needs, the FilllstSheetTabs( ) procedure begins by resetting the mintHidden form-level variable (you will see later why), a With Me.lstSheetTabs... End With loop is initiated (so you can just press the dot character to access all the lstSheetTabs ListBox control interfaces), and the lstSheetTabs ListBox is cleared using the Clear method to remove all the previous items (if any).

```
    mintHidden = 0
    With Me.lstSheetTabs
        .Clear
```

A For Each...Next loop is then initiated to access all items of the Worksheets collection and fill the lstSheetTabs ComboBox with the information. The intI Integer variable is used to count the collection items and generate the Index property of each Worksheet object inside the Worksheets collection, and this value is used by the lstSheetTabs ListBox AddItem method to define the ListBox value.

```
        For Each varItem In Worksheets
            intI = intI + 1
            .AddItem intI
```

The lstSheetTabs ListBox has six columns (defined on its ColumnCount property). The first column (which is also the BoundColumn) holds the Worksheet object's Index value inside the Worksheets collection, the second column holds the Worksheet object's Name property (tab name), the third column holds the Worksheet.CodeName, the fourth and fifth columns hold the Visible property, and the sixth column holds the Protect property. All columns are visible and have specific widths (as you can see in their ColumnWidths properties) determined by trial and error (Figure 4-9).

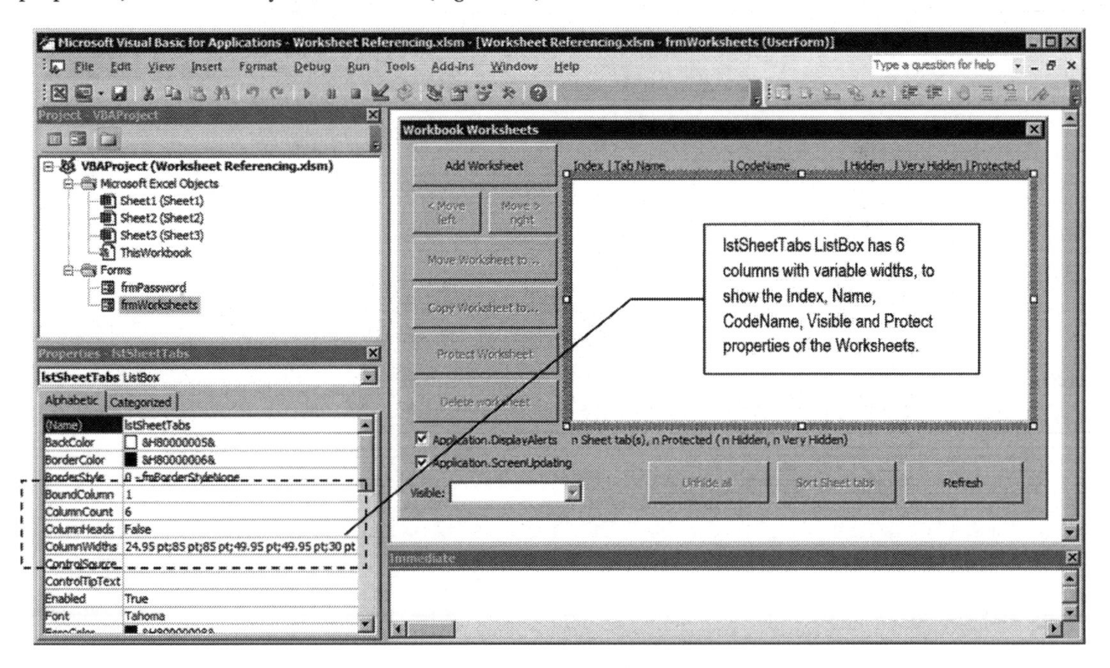

Figure 4-9. This is the lstSheetTabs ListBox used to show some worksheet properties, like Index, Name, CodeName, Visible, and Protect. It uses six columns, with all visible and with different column widths determined by trial and error

After determining the `lstSheetTabs` value with the `AddItem` method, the current `Worksheet` object's `Name` and `CodeName` are added to the second and third columns of the same item of the `ListBox`, using `ListCount - 1` to determine the position.

```
.Column(1, .ListCount - 1) = varItem.Name
.Column(2, .ListCount - 1) = varItem.CodeName
```

Attention The position of any item inside a `ListBox` control is determined by the `ListIndex` property, which is a 0-based value. The `ListCount` property, a 1-based value, indicates the number of items of any `ListBox` control. Since the `AddItem` method by default adds new items to the end of the list (unless you specify the desired position), you must use `ListCount-1` to reference the last added item.

Any Excel sheet tab of every Excel workbook has three possible visible states: visible (associated to the `xlSheetVisible` constant, the default value), hidden (associated to the `xlSheetHidden` constant, which you can set by right-clicking the sheet tab and selecting Hidden), and very hidden (associated to the `xlSheetVeryHidden` constant). Very hidden is a state that you can set only by selecting the desired `Worksheet` object in the VBA Project Explorer tree and using the `Visible` property in the VBA Properties window (Figure 4-10).

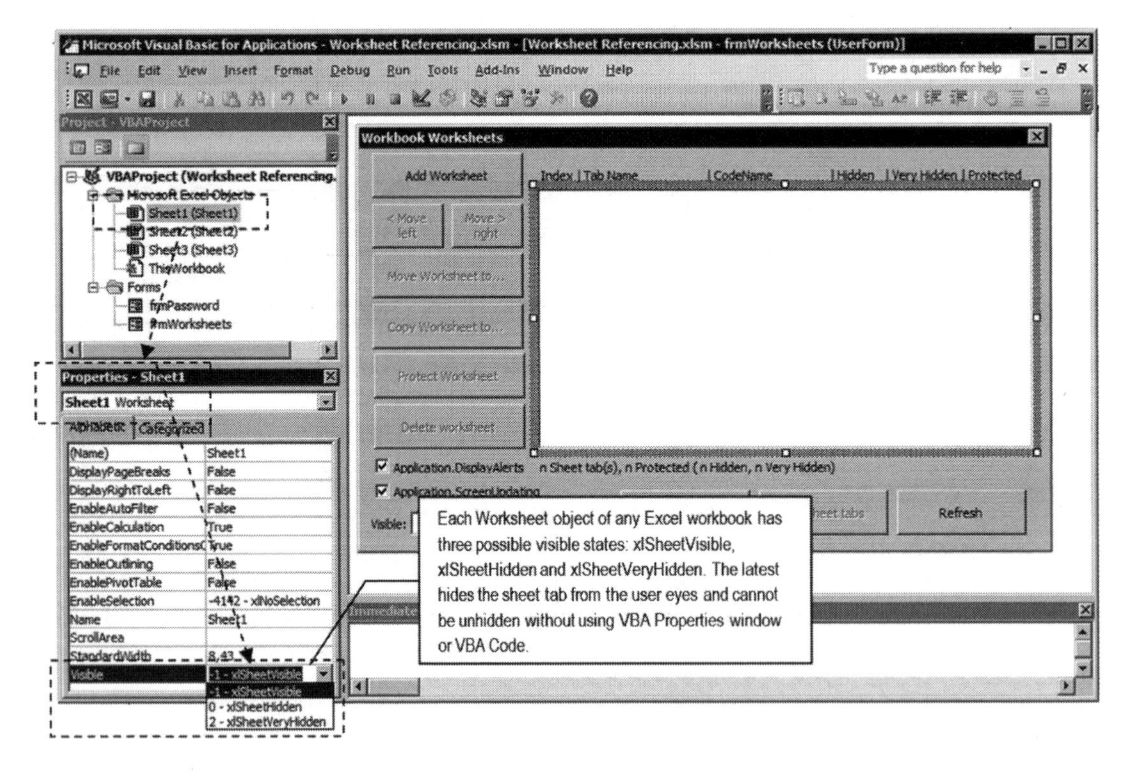

Figure 4-10. *All Worksheet objects of any Excel workbook file have three possible Visible properties represented by the constants xlSheetVisible, xlSheetHidden, and xlSheetVeryHidden. The last one hides the sheet tab so deep that it cannot be turned visible again by the user action right-clicking any sheet tab and selecting Unhide. You must use the VBA Properties window to turn it visible again or VBA code*

So, the next instructions of the `FilllstSheetTabs( )` procedure uses a `Select Case` instruction to verify the `Visible` property of the current worksheet and determine the fourth and fifth `ListBox` column values. Note that it compares the `Visible` property with the possible Excel constants. If the sheet tab is visible, both columns receive the word "No" to the Hidden (fourth) and Very Hidden (fifth) `ListBox` columns.

```
Select Case varItem.Visible
    Case xlSheetVisible
        .Column(3, .ListCount - 1) = "No"
        .Column(4, .ListCount - 1) = "No"
```

If the sheet is just hidden, the fourth `ListBox` column receives the word "Yes," and the form-level `mintHidden Integer` variable is incremented.

```
Case xlSheetHidden
    .Column(3, .ListCount - 1) = "Yes"
    .Column(4, .ListCount - 1) = "No"
    mintHidden = mintHidden + 1
```

Bu if the sheet is defined as very hidden, both the fourth and fifth columns receive the word "Yes" and the `mintHidden` and `intVeryHidden` variables are incremented.

```
Case xlSheetVeryHidden
    .Column(3, .ListCount - 1) = "Yes"
    .Column(4, .ListCount - 1) = "Yes"
    mintHidden = mintHidden + 1
    intVeryHidden = intVeryHidden + 1
End Select
```

■ **Attention** `mintHidden` was declared as a module-level variable because you need its value to synchronize the enabled state of the `cmdUnHideAll CommandButton` on the `DefineButtons( )` procedure, as you will see.

Next, the `FilllstSheetTabs( )` procedure defines the value of the sixth `ListBox` column according to the `Protect` property of the current sheet tab. If it is protected, the column receives the "Yes" value, and the `intProtected` variable is incremented.

```
        If varItem.ProtectContents Then
            .Column(5, .ListCount - 1) = "Yes"
            intProtected = intProtected + 1
        Else
            .Column(5, .ListCount - 1) = "No"
        End If
    Next
End With
```

When the `With lstSheetTabs... End With` structure ends, the procedure has all the information it needs to define the `Caption` property of `lblSheetTabs`: the `Label` control at the bottom of the `lstSheetTabs ListBox`. It must resume how many sheet tabs the workbook has (how many are protected, hidden, and very hidden using the `intI`, `mintHidden`, `intVeryHidden`, and `intProtected` variables), and the `DefineButtons( )` procedure is called to synchronize the `UserForm` interface.

```
    Me.lblSheetTabs.Caption = Sheets.Count & " Sheet tab(s)" & _
                              IIf(intProtected > 0, ", " & intProtected & " protected", "") & _
                              IIf(mintHidden > 0, " (" & mintHidden & " hidden", "") & _
                              IIf(intVeryHidden > 0, ", " & intVeryHidden &
                              " very hidden", "") & _
                              IIf(mintHidden, ")", "")
    Call DefineButtons(False)
End Sub
```

Avoiding Cascading Events

Before you read about how the `DefineButtons( )` procedure of `frmWorksheets` works, we must talk about an important programming issue regarding the unpredictable behavior of some object events that unexpectedly fire from inside other events, creating a phenomenon called *cascading events*.

This is the case of the ComboBox and `ListBox` controls' `Change( )` events that you expect to fire just when the user changes the control value but can fire in surprisingly moments. For example, both controls fire the `Change( )` event whenever the control value is changed from inside another code procedure or eventually when the first `AddItem` method is used in another procedure to populate it. This last case is what happens when the `frmWorksheets` `UserForm_Initialize( )` event fires!

The mechanism by which you can avoid any object `Change( )` event from being inadvertently executed when not fired by the user action is quite simple:

- Declare a `Boolean` module-level variable (like `mbolCancelEvent`).
- Make the variable `True` at the beginning of any procedure event that will cascade another object event.
- On the first instruction of the cascading object event, verify whether the module-level variable is `True`. If it is, turn it `False` again and exit the event procedure, doing nothing! The next code procedure examples express how this can be done:

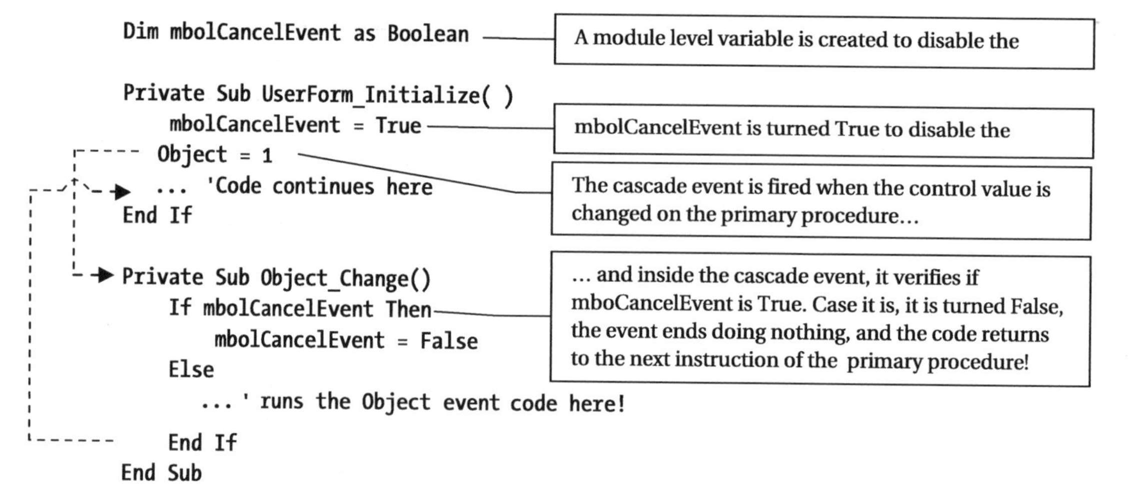

Try this: close `frmWorksheets` if it is open, press Alt+F11 to open the VBA interface, select the `frmWorksheets` code module, and put an interruption point on the first instruction of the `UserForm_Initialize( )` event (the `mbolCancelEvent=True` instruction).

■ **Attention** You set an interruption point on any VBA code module by clicking the gray bar at the left of the instruction where you want the code to stop, as shown in Figure 4-11.

Press the F5 function key to force `frmWorksheets` to open, and when the code stops at the interruption point, press F8 to run the code in step-by-step mode (the Step Into command), executing one code line at a time. Note that the first instruction of the `UserForm_Initialize( )` event sets `mbolCancelEvent = True`. Immediately after the second event instruction is executed to insert the first `ComboBox` list item (`Me.cboVisible.Additem 1`), it will unexpectedly jump to the `cboVisible_Change( )` event (Figure 4-11).

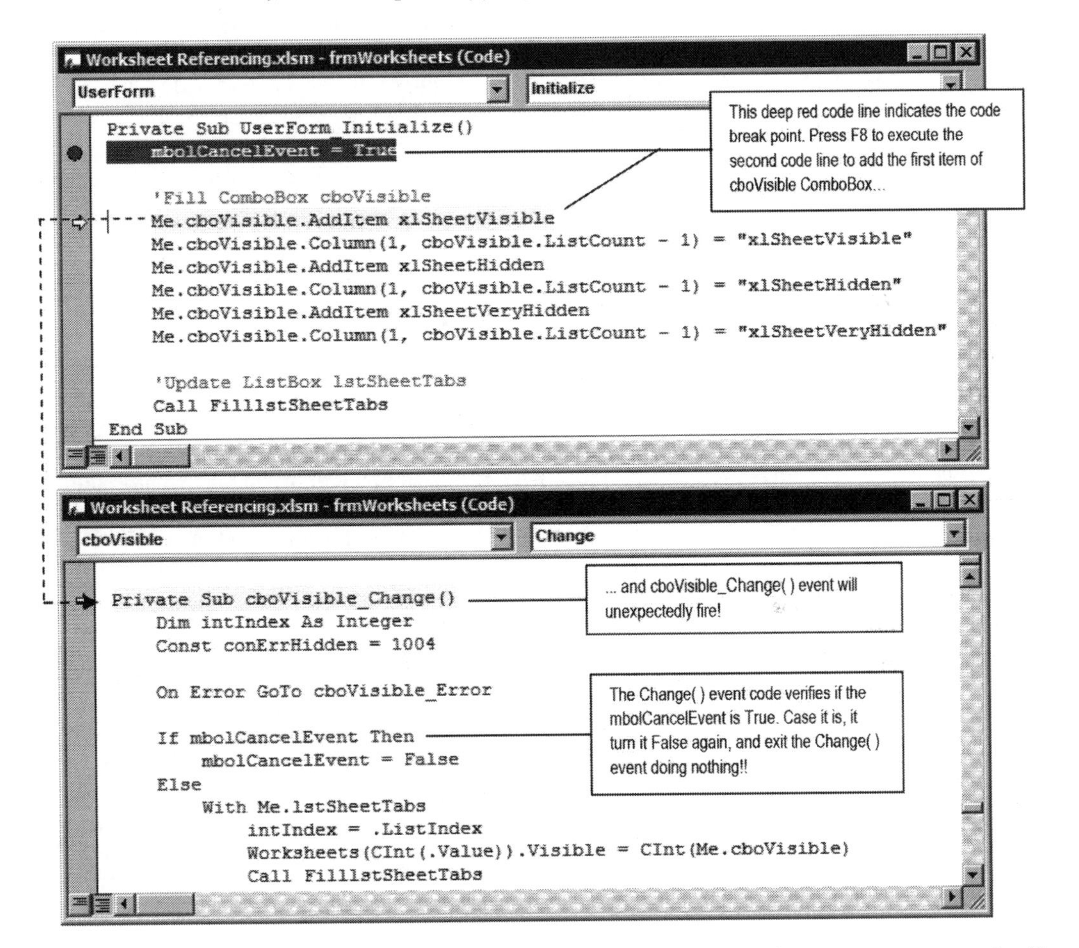

Figure 4-11. *If you put a code break on the first instruction of the frmWorksheets UserForm_Initialize( ) event and run through the code step by step by pressing the F8 function key, you will see that when the first item is added to the cboVisible ListBox using the AddItem method, the cboVisible_Change( ) event unexpectedly fires*

```
Private Sub UserForm_Initialize()
    mbolCancelEvent = True
```

Inside the `cboVisible_Change( )` event, `mbolCancelEvent` is tested, and if it is `True`, it is set to `False` and the event exits, doing nothing.

```
Private Sub cboVisible_Change()
    ...
    If mbolCancelEvent Then
        mbolCancelEvent = False
    Else
    ...
End Sub
```

You should be aware that the `frmWorksheets` interface must synchronize the `cboVisible ComboBox` value with the selected sheet tab item on the `lstSheetTabs ListBox`. When you click and select any item of `lstSheetTabs`, the `cboVisible` value must be changed to reflect the visible state of the selected sheet tab.

In other words, the `lstSheetTabs_Change( )` event must change the `cboVisible Value` property, which will cascade the `cboVisible_Change( )` event. And since the `frmWorksheets` interface synchronization is made from the `Sub DefineButtons( )` procedure, this procedure must also turn the module-level variable `mbolCancelEvent = True`, virtually avoiding the cascading event phenomenon.

```
Public Sub DefineButtons(bolEnabled As Boolean)
    ...
    mbolCancelEvent = True
    ...
End Sub
```

Synchronizing the frmWorksheets Control Interface

Now that you understand how to avoid that cascading events execute when they unexpectedly fire, the last instruction executed by the `Sub FilllstSheetTabs( )` procedure is to call the `Sub DefineButtons( )` procedure to synchronize the `frmWorksheets` interface. Let's see how this works by analyzing the code (bold instructions define the cascading event technique, setting `mbolCancelEvent=True` and where the cascade event will fire):

```
Public Sub DefineButtons(bolEnabled As Boolean)
    Dim ws As Worksheet
    Dim intIndex As Integer
    Dim bolVisible As Boolean
    Dim bolVeryHidden As Boolean
    Dim bolProtected As Boolean

    mbolCancelEvent = True
    Me.cmdCopyTo.Enabled = bolEnabled
    Me.cmdProtect.Enabled = bolEnabled
    Me.cmdUnhideAll.Enabled = (mintHidden > 0)
    Me.cboVisible.Enabled = bolEnabled

    If bolEnabled Then
        intIndex = Me.lstSheetTabs.ListIndex
        Set ws = Worksheets(Me.lstSheetTabs)

        bolVisible = (ws.Visible = xlSheetVisible)
        bolVeryHidden = (ws.Visible = xlSheetVeryHidden)
        If bolVisible Then
            Me.cboVisible = xlSheetVisible
        ElseIf bolVeryHidden Then
```

```
            Me.cboVisible = xlSheetVeryHidden
        Else
            Me.cboVisible = xlSheetHidden
        End If

        bolProtected = ws.ProtectContents
        Me.cmdProtect.Caption = IIf(bolProtected, "Unprotect Worksheet", "Protect
Worksheet")
    Else
        Me.cmdProtect.Caption = "Protect Worksheet"
        Me.cboVisible = Null
    End If

    Me.cmdMoveLeft.Enabled = bolEnabled And bolVisible And (intIndex > 0)
    Me.cmdMoveRight.Enabled = bolEnabled And bolVisible And (intIndex < Worksheets.Count - 1)
    Me.cmdMoveTo.Enabled = bolEnabled And bolVisible
    Me.cmdDelete.Enabled = bolEnabled And bolVisible
    Me.cmdSort.Enabled = (Sheets.Count > 1)
End Sub
```

Note that `DefineButtons( )` receives the `bolVisible` Boolean argument and uses its value to set the `Enabled` property of some `frmWorksheets` interface controls. After the procedure declares the variables it needs, it synchronizes the availability of the `cmdCopyTo`, `cmdProtect`, `cmdUnhideAll`, and `cboVisible` controls.

```
Public Sub DefineButtons(bolEnabled As Boolean)
    ...
    Me.cmdCopyTo.Enabled = bolEnabled
    Me.cmdProtect.Enabled = bolEnabled
    Me.cmdUnhideAll.Enabled = (mintHidden > 0)
    Me.cboVisible.Enabled = bolEnabled
```

Also note that the `cmdUnhideAll` `CommandButton` `Enabled` property is not synchronized by the `bolEnabled` argument, using the `mintHidden` Integer variable value instead. If there is at least one hidden sheet tab, it will be enabled, and that is why `mintHidden` was declared as a module-level variable. Its value is cleared and set by the `FilllstSheetTabs( )` procedure and used by the `DefineButtons( )` procedure!

The procedure then verifies whether the `bolEnabled` argument is `True`. If it is, it stores the selected item position on the `lstSheetTabs` `ListBox` into the `intIndex` variable and initializes the `ws` as `Worksheet` object variable using the selected `lstSheetTabs` item value, which returns the selected sheet `Index` inside the `Worksheets` collection.

```
If bolEnabled Then
    intIndex = Me.lstSheetTabs.ListIndex
    Set ws = Worksheets(Me.lstSheetTabs)
```

■ **Attention** At first, the selected `lstSheetTabs` `ListIndex` (the position of the selected item on the `ListBox`) and the `lstSheetTab` value (which is the Index of the selected item inside the `Worksheets` collection) seem to be the same, but this is not always true as you play with sheet tabs' position and visibility.

The procedure variables bolVisible, bolVeryHidden, and bolProtected are initialized according to the Visible property of the referenced sheet tab.

```
bolVisible = (ws.Visible = xlSheetVisible)
bolVeryHidden = (ws.Visible = xlSheetVeryHidden)
```

Did you notice that the Boolean variables' state was defined using a logical test, inside parentheses? To verify whether the selected sheet tab is visible, the test compares the sheet tab's Visible property with the Excel xlSheetVisible constant (xlSheetVisible = -1). If they are the same, bolVisible = True. But if the sheet isn't visible, it must be hidden or very hidden. Since any very hidden sheet is also hidden, just the bolVeryHidden variable state is set.

Once thebolVisible and bolVeryHidden variables' values are defined, the code uses them to set the value that cboVisible must show according to the Visible property of the selected sheet on thelstSheetTabs ListBox.

```
If bolVisible Then
    Me.cboVisible = xlSheetVisible
ElseIf bolVeryHidden Then
    Me.cboVisible = xlSheetVeryHidden
Else
    Me.cboVisible = xlSheetHidden
End If
```

▒ **Attention** This is the point where the cboVisible_Change( ) event will cascade, but since mbolCancelEvent = True, the code will immediately return to the next End If instruction of the DefineButtons( ) procedure.

The code also verifies whether the selected sheet is protected, attributing the Worksheet object's ProtectContents property to the bolProtected variable.

```
bolProtected = ws.ProtectContents
```

And bolProtected is used to set the cmdProtect.Caption property, changing what the command button says to the user (the code uses the underscore continuation character to use more than one code line for the same instruction).

```
Me.cmdProtect.Caption = IIf(bolProtected, "Unprotect Worksheet", "Protect Worksheet")
```

If bolEnabled = False (the condition used to disable most frmWorksheets controls), the Else clause of the If bolEnabled Then instruction will be executed. This will happen whenever the lstSheetTabs is filled and no list item is selected, and in this case, the cmdProtect.Caption property must be Protect Worksheet, and cboVisible must be set no Null (no visible value selected).

```
Else
    Me.cmdProtect.Caption = "Protect Worksheet"
    Me.cboVisible = Null
End If
```

Some interface CommandButtons deserve special attention. The two buttons used to change the sheet tab order in Excel sheet tabs, cmdMoveRight and cmdMoveLeft, must have their Enabled property set to True or False according to three different conditions: bolEnabled must be True, the sheet tab must be visible, and the sheet tab must be capable of being moved to the left or right on the Excel sheet tabs bar.

The DefineButtons( ) procedure stored earlier in the intIndex variable the selected lstSheetTabs.ListIndex value(which is a 0-based value): intIndex must be greater than zero so the sheet tab can be left moved and must be smaller than Worksheets.Count - 1 to be right moved.

```
Me.cmdMoveLeft.Enabled = bolEnabled And bolVisible And (intIndex > 0)
Me.cmdMoveRight.Enabled = bolEnabled And bolVisible And (intIndex < Worksheets.Count - 1)
```

Two other CommandButtons, cmdProtect and cmdDelete, also need special attention. To be enabled, two conditions must be met: bolEnabled = True and the sheet tab must be visible (Excel can't move or delete hidden sheet tabs).

```
Me.cmdMoveTo.Enabled = bolEnabled And bolVisible
Me.cmdDelete.Enabled = bolEnabled And bolVisible
```

The cmdSort command button must be enabled whenever at least two sheet tabs are visible.

```
Private Sub lstSheetTabs_Click()
    Sheets(CInt(Me.lstSheetTabs)).Activate
    Call DefineButtons(True)
End Sub
```

That is all frmWorksheets needs to keep its interface synchronized using a single procedure that must be called every time anything changes from user action. Let's see now how to use some Worksheet object methods.

Selecting an Item in the lstSheetTabs ListBox

Whenever you select an item on the lstSheetTabs ListBox, its Change( ) event fires, executing this code:

```
Private Sub lstSheetTabs_Click()
    Sheets(CInt(Me.lstSheetTabs)).Activate
    Call DefineButtons(True)
End Sub
```

The code is quite simple: once you select a sheet tab reference on the lstSheetTabs ListBox, it uses the Sheets collection's Activate method to select the sheet tab in the Excel interface and then calls the DefineButtons(True) procedure (by making the procedure argument bolEnabled = True, some CommandButtons of the UserForm interface will be enabled).

■ **Attention** To help you see the sheet tab activation as you click lstSheetTabs ListBox items, Sheet2 and Sheet3 from the Worksheet Referencing.xlsm workbook have a different back color on their cells. This makes it easy to see the selection in the Excel interface behind the UserForm.

Adding Sheet Tabs

To add a new sheet tab on any Excel workbook, use the Worksheets collection's Add method, which has this syntax:

```
Worksheets.Add(Before, After, Count, Type) as Object
```

In this code:

Before: This is optional; it is an object that specifies the sheet before which the new sheet is added.

After : This is optional; it is an object that specifies the sheet after which the new sheet is added.

Count : This is optional; it is the number of sheets to be added (the default is one).

Type : This is optional; it specifies the sheet type and can be one of the following constants: xlWorksheet (default value), xlChart, xlExcel4MacroSheet, or xlExcel4IntlMacroSheet. It can also insert a sheet based on an existing template, specifying the path to the template.

The Before and After arguments must be Worksheet object pointers. If you use the Before argument, do not use the After argument, and vice versa. If you omit both Before and After, the new worksheet will be inserted *before* the active sheet.

When you click the Add Worksheet CommandButton (cmdAddNew) of frmWorksheets, the cmdAddNew_Click( ) event fires and executes this code:

```
Private Sub cmdAddNew_Click()
    Dim ws As Worksheet
    Dim strMsg As String
    Dim strTitle As String
    Dim intBeforeTab As Integer
    Const conNumbers = 1

    strMsg = "Add a new Sheet tab before sheet:"
    strTitle = "Add a new Sheet tab"
    intBeforeTab = Application.InputBox(strMsg, strTitle, 1, , , , , conNumbers)
    If intBeforeTab > 0 Then
        If intBeforeTab > Sheets.Count Then
            intBeforeTab = Sheets.Count
        End If
        Worksheets.Add Worksheets(intBeforeTab)
       Call FilllstSheetTabs
    End If
End Sub
```

The cmdAddNew_Click( ) event uses the Before argument of the Worksheets.Add method to insert a new sheet tab on the current workbook. After cmdAddNew_Click declares the variables it needs, it uses the Application.InputBox method to show a dialog asking to insert a new sheet tab as the first tab of the Worksheets collection, using 1 as the default value.

```
strMsg = "Add a new Sheet tab before sheet:"
strTitle = "Add a new Sheet tab"
intBeforeTab = Application.InputBox(strMsg, strTitle, 1, , , , , conNumbers)
```

Since the `Application.Inputbox` method allows the value to be restrained, the code uses `Type = 1` on its last argument to enforce numeric values, employing another valuable programming practice. Instead of using a "magic number," the code declares the constant `conNumbers = 1` and uses it on the last argument of the `Application.InputBox` method, making the code far more legible!

The value returned by the `Application.InputBox` method will be stored into the `intBeforeTab` `Integer` variable. If the user clicks the Cancel or Close button of the `InputBox` dialog, `intBeforeTab` will receive zero (default value for numeric variables), so the code verifies whether `intBeforeTab` is greater than zero, which will mean a valid position to insert the new sheet tab.

```
If intBeforeTab > 0 Then
```

Since you have no control of the number the user will type, the code also verifies whether it is greater than the sheets count. If it is, the `intBeforeTab` receives the `Sheets` collection's `Count` property value (note that now it uses the `Sheets` collection instead the `Worksheets` collection, which are the same, with different syntaxes).

```
If intBeforeTab > Sheets.Count Then
    intBeforeTab = Sheets.Count
End If
```

At this point, the code has a valid position to insert the new sheet tab, and it evokes the `Worksheets.Add` method, passing to its `Before` argument a worksheet pointer with an `Index = intBeforeTabs` variable.

```
Worksheets.Add Worksheets(intBeforeTab)
```

■ **Attention** You can also use the `Sheets` collection's `Add` method to add a new sheet tab, using this syntax:

```
Sheets.Add Sheets(intBeforeTab)
```

After the new sheet tab has been added, it calls the `FilllstSheetTabs( )` procedure, which will update the `UserForm` interface, showing on the `lstSheetTabs` `ListBox` the new inserted worksheet, which can also be seen on the Excel sheet tabs bar (Figure 4-12).

```
        Call FilllstSheetTabs
    End If
End Sub
```

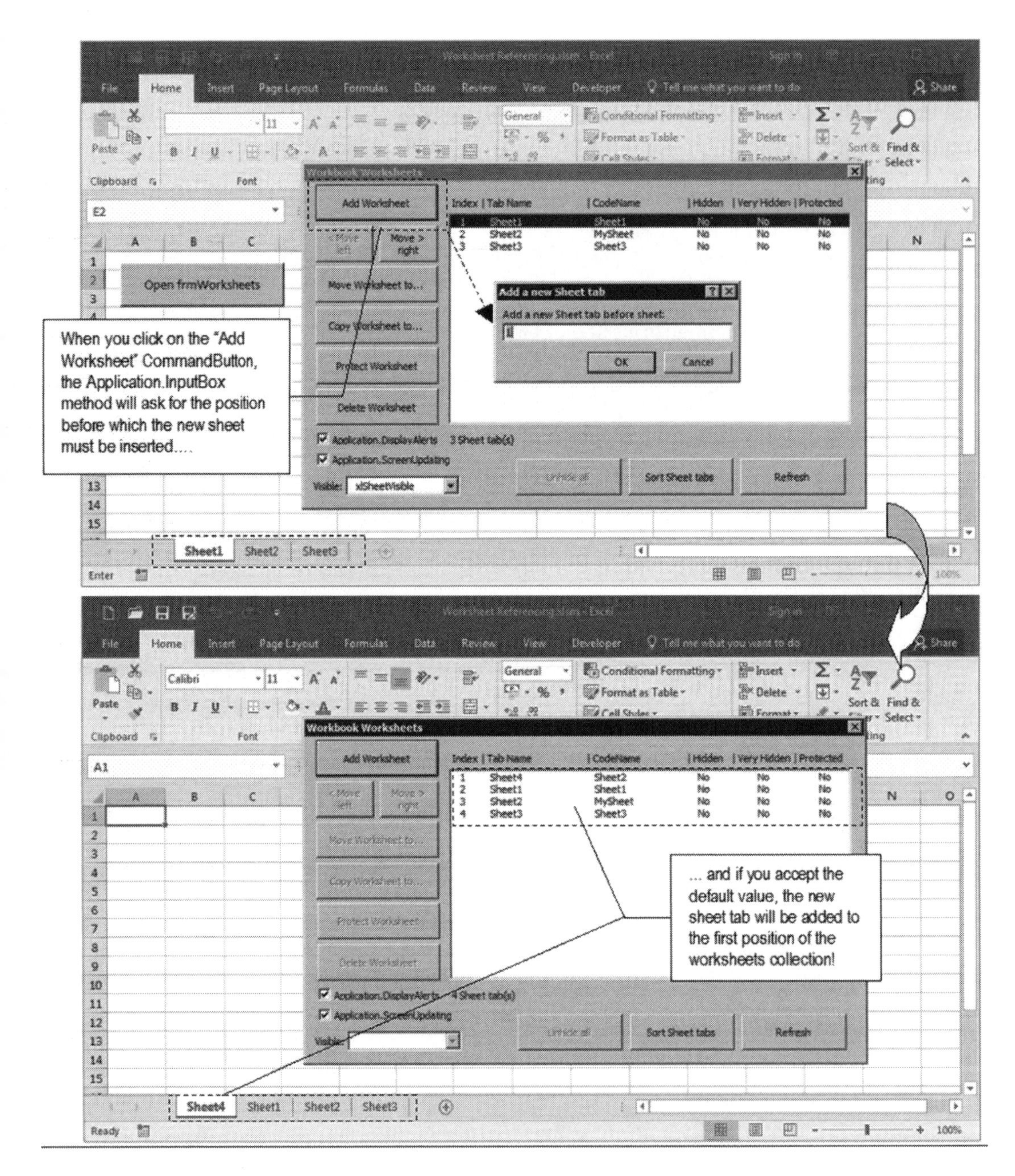

Figure 4-12. Use Add Worksheet CommandButton from frmWorksheets to add a new sheet tab using the Before argument of the Worksheets.Add method. The code will use the Application.InputBox method to validate the user input, forcing it to type a numeric value. The default position for the new sheet tab is 1, which will insert it as the first sheet tab. If the user types a number greater than the sheet tabs count, the new sheet tab will be inserted before the last sheet tab of this workbook

---

■ **Attention** The `FilllstSheetTabs( )` procedure uses `DefineButtons(False)` as its last instruction, which will disable most `UserForm CommandButtons`.

Moving Sheet Tabs

To move any visible sheet tab to a new position inside the Excel sheet tabs bar, use the `Worksheet` object's `Move` method, which has this syntax:

```
Worksheet.Move(Before, After) as Object
```

In this code:

`Before`: This is optional; it is an object that specifies the sheet before which the move will be placed.

`After` : This is optional; it is an object that specifies the sheet after which the move will be placed.

The `Before` and `After` arguments must be `Worksheet` object pointers. If you use the `Before` argument, do not use the `After` argument, and vice versa. The `Move` method returns an object pointer to the moved sheet tab.

The `frmWorksheets` interface has three buttons that allow you to move the selected sheet tab on `lstSheetTabs ListBox`: < Move Left (`cmdMoveLeft`), Move Right > (`cmdMoveRight`), and Move Worksheet to... (`cmdMoveTo`). The first two buttons will be enabled according to the selected sheet tab, while the third will be always enabled to any sheet.

Suppose that you click the first item of the `lstSheetTabs ListBox`. Just the Move Right > and Move Worksheet to... buttons will be enabled. This is the code executed when you click the Move Right > `CommandButton` and the `cmdMoveRight_Click( )` event fires:

```
Private Sub cmdMoveRight_Click()
    Dim ws As Worksheet
    Dim intIndex As Integer

    intIndex = Me.lstSheetTabs.ListIndex
    Set ws = Worksheets(Me.lstSheetTabs)
    If Me.lstSheetTabs < Sheets.Count Then
        ws.Move , Worksheets(Me.lstSheetTabs + 1)
        Call FilllstSheetTabs
        Me.lstSheetTabs.Selected(intIndex + 1) = True
    End If
End Sub
```

That was easy, huh? But there is a trick: the first code instruction gets the selected item position on the `lstSheetTabs ListBox` and stores it on the `intIndex` variable so it can move it and select it again.

```
intIndex = Me.lstSheetTabs.ListIndex
```

Then it gets a reference to the selected sheet tab so it can use the `Move` method.

```
Set ws = Worksheets(Me.lstSheetTabs)
```

Now the code has what is called a good programming practice: although the interface disables the cmdMoveRight button when the last sheet tab is selected on the list box, it makes a double verification if the sheet can be moved right, comparing its position (returned by its Index value, which is also the lstSheetTabs value) with the total number of sheets. If it is lower than Sheet.Counts, it can be moved, turning it into "bulletproof" code.

```
If Me.lstSheetTabs < Sheets.Count Then
```

Since the sheet must be moved to the right, it uses the After argument of the Worksheet.Move method to move it to the new position.

```
ws.Move , Worksheets(Me.lstSheetTabs + 1)
```

And once the sheet tab is moved, the code makes a call to the FilllstSheetTabs( ) procedure to update the UserForm interface.

```
Call FilllstSheetTabs
```

And since FilllstSheetTabs( ) calls DefineButtons(False), disabling most CommandButtons of the UserForm interface, the procedure selects the moved sheet tab on the list, using intIndex+1 to select the right item.

```
Me.lstSheetTabs.Selected(intIndex + 1) = True
```

And when this happens, the cascading event phenomenon happens again, indirectly firing the lstSheetTabs_Click( ) event, which will call again DefineButtons(True), perfectly synchronizing the UserForm interface according to the new position of the moved sheet tab (Figure 4-13).

That is a lot of action! This is instant action; you need to add a VBA interruption point to see when the cascade event fires.

Now look at the code executed when you click the < Move left CommandButton and the cmdMoveLeft_Click( ) event fires:

```
Private Sub cmdMoveLeft_Click()
    Dim ws As Worksheet
    Dim intIndex As Integer

    intIndex = Me.lstSheetTabs.ListIndex
    Set ws = Worksheets(Me.lstSheetTabs)
    If Me.lstSheetTabs > 1 Then
        ws.Move Worksheets(Me.lstSheetTabs - 1)
        Call FilllstSheetTabs
        Me.lstSheetTabs.Selected(intIndex - 1) = True
    End If
End Sub
```

As you can see, the cmdMoveLeft_Click( ) event uses the same technique employed by cmdMoveRigh_Click( ). Although the interface disables the cmdMoveLeft CommandButton whenever the first lstSheetTabs item is selected, the code is turned "bulletproof" by first testing whether the sheet tab can be moved to the left by verifying whether Index > 1 (the lstSheetTabs value). If it is, the sheet can be moved left.

```
If Me.lstSheetTabs > 1 Then
    ws.Move Worksheets(Me.lstSheetTabs - 1)
```

The code also calls the `FilllstSheetTabs( )` procedure and reselects the item on the `lstSheetTabs` `ListBox`, cascade-firing the `lstSheetTabs_Click( )` event, which will again perfectly synchronize the `UserForm` interface).

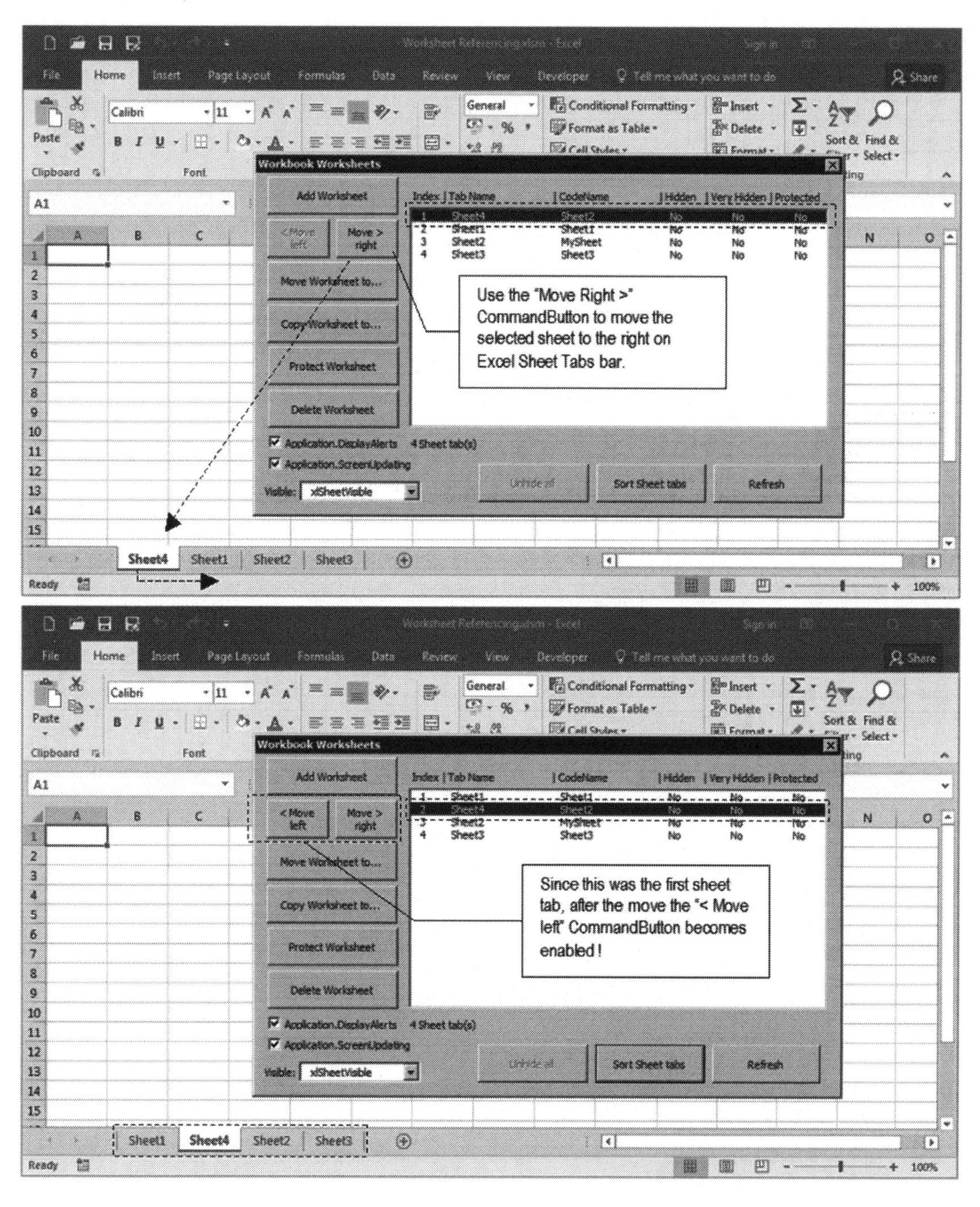

Figure 4-13. *When you click the cmdMoveRight CommandButton, the selected sheet on the lstSheetTabs ListBox is moved right on the Excel sheet tabs bar. The UserForm is updated to reflect the new order of the sheet tabs inside the workbook. Note that all sheet tabs at the right of the moved sheet tab have their Index properties changed*

Now appreciate the code associated to the cmdMoveTo_Click( ) event, which will fire whenever you click the Move Worksheet to... CommandButton of the UserForm interface.

```
Private Sub cmdMoveTo_Click()
    Dim ws As Worksheet
    Dim strMsg As String
    Dim strTitle As String
    Dim intBeforeTab As Integer
    Const conNumbers = 1
    Set ws = ThisWorkbook.Worksheets(Me.lstSheetTabs)
    If ws.Visible Then
        strMsg = "Move " & ws.Name & " tab before sheet:"
        strTitle = "Move " & ws.Name & " tab"
        intBeforeTab = Application.InputBox(strMsg, strTitle, 1, , , , , conNumbers)
        If intBeforeTab > 0 Then
            If intBeforeTab > Sheets.Count Then
                intBeforeTab = Sheets.Count
            End If
            ws.Move Worksheets(intBeforeTab)
            Call FilllstSheetTabs
        End If
    End If
End Sub
```

There is nothing new here. After setting the selected worksheet reference to the ws object variable, the codes uses a "bulletproof" technique to verify whether the sheet is visible because just visible sheets can be moved (although the FilllstSheetTabs( ) procedure disables all move CommandButtons when the sheet isn't visible).

```
Set ws = ThisWorkbook.Worksheets(Me.lstSheetTabs)
If ws.Visible Then
```

The code also uses the Application.InbutBox( ) method with the declared constant conNumbers = 1 to just allow the user to type a numeric value. By default the code offers to move the selected sheet to the Worksheets collection's first position, but you can change it to any number.

```
Const conNumbers = 1
...
   intBeforeTab = Application.InputBox(strMsg, strTitle, 1, , , , , conNumbers)
```

If the number typed is greater than zero, it is a valid sheet position. If it is greater than Sheets.Count, it will be set to Sheets.Count, and the Worksheet object's Move method will be used with its Before argument, always moving the selected worksheet to the left of the desired position.

```
If intBeforeTab > 0 Then
    If intBeforeTab > Sheets.Count Then
        intBeforeTab = Sheets.Count
    End If
    ws.Move Worksheets(intBeforeTab)
```

The `FilllstSheetTabs( )` procedure will be called again to synchronize the `UserForm` interface. Since the position of the move is unpredictable this time, the item will be not selected on the `lstSheetTabs ListBox`.

■ **Attention** Since the `frmWorksheets` was opened in a nonmodal state, you can click Excel sheet tabs and drag sheet tabs to different positions. You can also right-click any sheet tab and add a new sheet or make a copy. To update the `frmWorksheets` interface, click the Refresh button.

Sorting Sheet Tabs

The operation related to sort Excel sheet tabs is made by the `Worksheets` or `Sheets` collection's `Move` method using any sorting algorithm, like *bubble sort*, which is easy to understand, small, and efficient.

To put any items sequence in ascending order, the bubble sort algorithm must take the first item of the sequence and compare it to the next. If it is greater than the next item, they both change orders. The first item continues to compare itself with the next item until it reaches the last item and puts itself on the right position of the sequence. The process begins again with the new first item of the sequence until no change is made on item orders and the sequence is sorted.

This process is very efficient if there are not many items to be sorted. Although Excel 2007 or later versions can have an unlimited number of worksheets, you will seldom find hundreds of them inside a workbook, which makes the bubble sort method very applicable to sorting sheet tabs.

When you click the Sort Sheet Tabs `CommandButton` of `frmWorksheets`, the `cmdSort_Click( )` event fires and executes the bubble sort method, comparing sheet tab names and using the `Sheets` collection's `Move` method to change sheet tabs order, never forgetting that it can't be applied to hidden sheets. Both sheet tabs must be visible before the move operation takes place. Here is the code:

```
Private Sub cmdSort_Click()
    Dim ws As Worksheet
    Dim intI As Integer
    Dim intVisible1 As Integer
    Dim intVisible2 As Integer
    Dim bolChanged As Boolean

    Application.ScreenUpdating = Me.chkScreenUpdating
        Do
            bolChanged = False
            For intI = 1 To Sheets.Count - 1
                If Sheets(intI).Name > Sheets(intI + 1).Name Then
                    intVisible1 = Sheets(intI).Visible
                    intVisible2 = Sheets(intI + 1).Visible
                    Sheets(intI).Visible = True
                    Sheets(intI + 1).Visible = True
                        Sheets(intI + 1).Move Sheets(intI)
                    Sheets(intI).Visible = intVisible1
                    Sheets(intI + 1).Visible = intVisible2
                    bolChanged = True
                End If
            Next
        Loop Until Not bolChanged
```

```
        Sheets(1).Activate
        Call FilllstSheetTabs
    Application.ScreenUpdating = True
End Sub
```

When you move sheet tabs, the Excel screen can flicker; to avoid this behavior, the first procedure instruction after its variable declaration is to set the Application.ScreenUpdating property according to what is selected in the chkScreenUpdating CheckBox of the frmWorksheets UserForm.

```
Application.ScreenUpdating = Me.chkScreenUpdating
```

Note that by default Excel screen updating is active since chkScreenUpdating is checked by default, but you can uncheck it to watch the fast sort process.

The bubble sort algorithm is based on two concentric loops: an external Do...Loop instruction controls the entire sorting process, and an internal For...Next loop takes the first sheet tab name and compares it to all others.

The sorting process begins by making the Boolean variable bolChange = False inside the Do...Loop structure and will run another loop if bolChanged = True at the end of the loop using the Not VBA operator to verify that bolChanged = False.

```
Do
    bolChanged = False
...
Loop Until Not bolChanged
```

The internal For...Next loop runs across each sheet tab of this workbook using the intI Integer variable to generate the Index used to reference sheet tabs on the Sheets collection.

```
For intI = 1 To Sheets.Count - 1
...
Next
```

Inside the For...Next loop, the first sheet tab's Name property is compared to the next. If it is greater, they both must change order on the Excel sheet tabs bar. But since hidden sheet tabs can't be moved, the Visible property of both sheet tabs is first saved in the intVisible1 and intVisible2 variables, so you can put them on the default visible state after the sort.

```
If Sheets(intI).Name > Sheets(intI + 1).Name Then
    intVisible1 = Sheets(intI).Visible
    intVisible2 = Sheets(intI + 1).Visible
```

Both sheet tabs have Visible = True, and they change places using the Sheets collection's Move method with the Before argument to move Sheet(intI+1) before Sheet(intI):

```
Sheets(intI).Visible = True
Sheets(intI + 1).Visible = True
    Sheets(intI + 1).Move Sheets(intI)
```

Once both sheets have changed places, their previous Visible property value is restored, and the bolChanged variable is set to true, signaling to the external Do...Loop structure that it must perform another For...Next internal loop until the sequence is sorted.

```
                Sheets(intI).Visible = intVisible1
                Sheets(intI + 1).Visible = intVisible2
                bolChanged = True
            End If
        Next
Loop Until Not bolChanged
```

When the sort is done, the Do...Loop exits, and the first sorted sheet tab is selected.

```
Sheets(1).Activate
```

Since the sorting process can change sheet tabs orders, the procedure calls again FilllstSheetTabs( ) to update the UserForm interface, and Excel screen updating is turned on again before the procedure ends.

```
        Call FilllstSheetTabs
    Application.ScreenUpdating = True
End Sub
```

■ **Attention** You must always set Application.ScreenUpdating = True on the end of any procedure that is set to false or the Excel interface will seem to be frozen. Don't forget that if this ever happen to you (and it will), you can press Alt+F11 to show the VBA interface and use the Immediate window to set Application.ScreenUpdating = True again.

Try to insert (adding or coping) new sheet tabs in the Worksheet Referencing.xlsm file and see for yourself how the bubble sort algorithm performs. Since Sheet2 and Sheet3 have different back colors, change sheet tab order more than once, and try again the sorting process by unchecking the chkScreenUpdating check box (Figure 4-14).

Copying Sheet Tabs

To set a copy of any sheet tab to a specified position of the Excel sheet tabs bar, use the Worksheet object's Copy method, which has this syntax:

```
Worksheet.Copy(Before, After) as Object
```

In this code:

Before: This is optional; it is an object that specifies the sheet before which the copy will be placed.

After: This is optional; it is an object that specifies the sheet after which the copy will be placed.

The Before and After arguments must be Worksheet object pointers. If you use the Before argument, do not use the After argument, and vice versa.

When you click the Copy Worksheet to... CommandButton of the UserForm interface, the cmdCopyTo_Click( ) event will fire, executing this code:

```
Private Sub cmdCopyTo_Click()
    Dim ws As Worksheet
    Dim strMsg As String
    Dim strTitle As String
    Dim intBeforeTab As Integer
    Const conNumbers = 1

    Set ws = ThisWorkbook.Worksheets(Me.lstSheetTabs)
    If ws.Visible Then
        strMsg = "Copy " & ws.Name & " tab before sheet:"
        strTitle = "Copy " & ws.Name & " Sheet tab"
        intBeforeTab = Application.InputBox(strMsg, strTitle, 1, , , , , conNumbers)
        If intBeforeTab > 0 Then
            If intBeforeTab > Sheets.Count Then
                intBeforeTab = Sheets.Count
            End If
            ws.Copy Worksheets(intBeforeTab)
            Call FilllstSheetTabs
        End If
    End If
End Sub
```

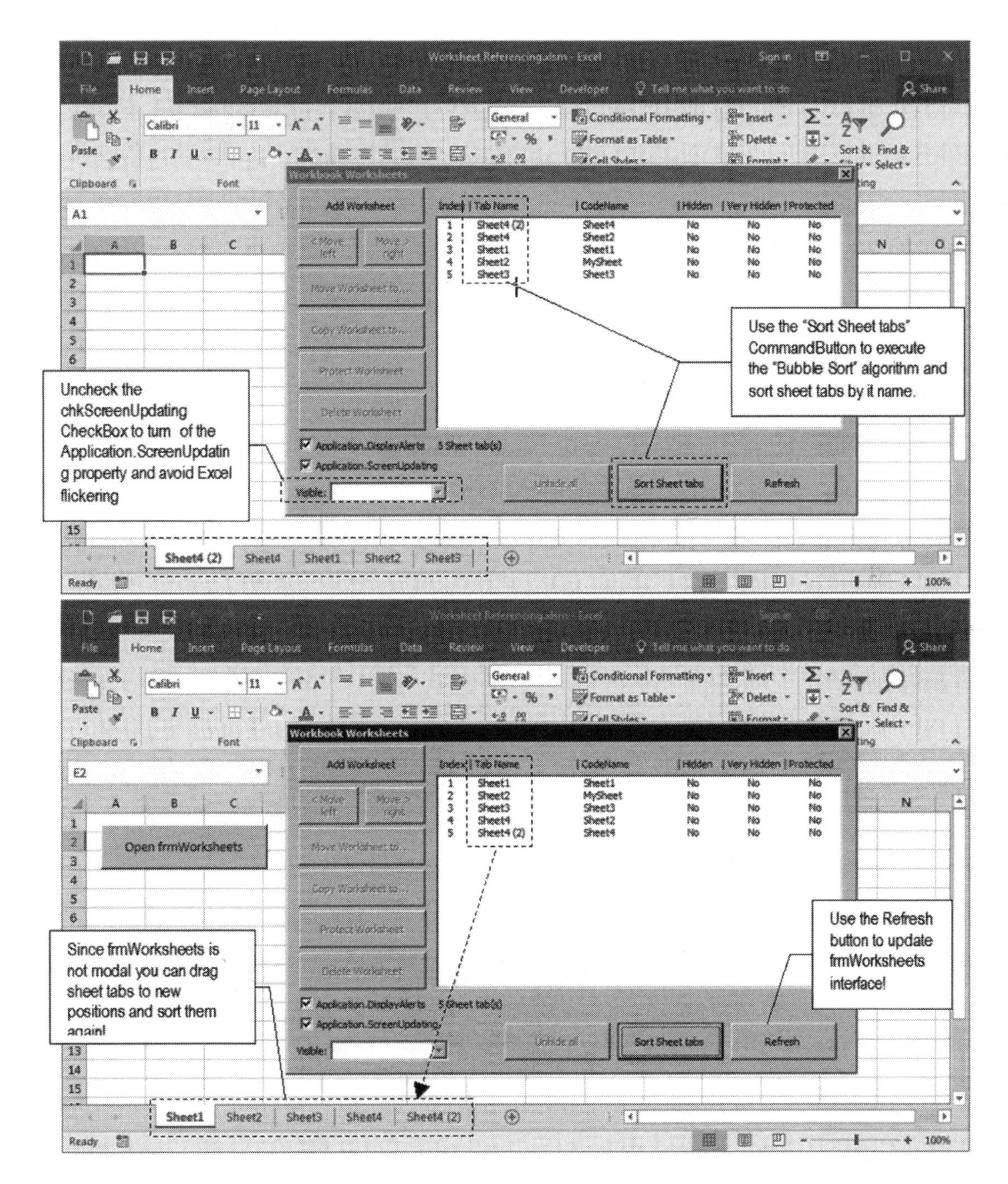

Figure 4-14. *The Sort Sheet tabs command button uses the bubble sort algorithm to sort sheet tabs by name. Use the UserForm Move buttons to change the sheet tab order (or click and drag sheet tabs to new positions on the Excel sheet tabs bar), add or copy new sheet tabs, and then click the Sort Sheet tab button to sort them. Repeat the procedure and uncheck the Application.ScreenUpdating check box (chkScreenUpdating control)*

This is almost identical to the code used by the cmdMoveTo_Click( ) event, except the text message is different and you use the Worksheet object's Copy method with its Before argument (in bold) to copy the selected sheet to the left of the selected position.

The name of the copied worksheet will be composed by the original name concatenated to the (2) suffix (the Sheet1 tab name will copied to Sheet1(2)). Figure 4-14 also shows what happens after the added Sheet4 worksheet was copied to the first (default) position of the Excel sheet tabs bar.

Deleting Sheet Tabs

Use the Sheets collection or the Worksheet object's Delete method to delete the desired sheet tabs. They have these syntaxes:

```
Sheets(Index).Delete
Worksheet.Delete
```

If Application.DisplayAlerts = True (default state), Excel will warn you that the deletion is permanent and cannot be undone, before the deletion process is executed. So, be careful when you delete sheet tabs with VBA code.

When you select any sheet tab in the lstSheetTabs ListBox and click the Delete Worksheet command button, the cmdDelete_Click( ) event will fire, executing this code:

```
Private Sub cmdDelete_Click()
    Dim ws As Worksheet
    Dim strMsg As String
    Dim strTitle As String

    Set ws = Worksheets(Me.lstSheetTabs)
    If ws.CodeName = "Sheet1" Then
        strMsg = "Can't delete Sheet1." & vbCrLf
        strMsg = strMsg & "It has the Command button to reopen frmWorksheets UserForm!"
        MsgBox strMsg, vbCritical, "Can't delete Sheet1"
    Else
        strMsg = "Do you really want to delete " & ws.Name & "?" & vbCrLf
        strMsg = strMsg & "(This operation can be undone!)"
        strTitle = "Delete " & ws.Name & " worksheet?"
        If MsgBox(strMsg, vbCritical + vbYesNo + vbDefaultButton2, strTitle) = vbYes Then
            Application.DisplayAlerts = Me.chkExcelWarnings
                ws.Delete
            Application.DisplayAlerts = True
        End If
        Call FilllstSheetTabs
    End If
End Sub
```

Note that the code begins by setting a reference to the selected sheet tab in the frmWorksheets interface to the ws object variable and then compares its CodeName property to Sheet1 to avoid Sheet1 deletion since it contains the command button that runs the UserForm again.

```
Set ws = Worksheets(Me.lstSheetTabs)
If ws.CodeName = "Sheet1" Then
    strMsg = "Can't delete Sheet1." & vbCrLf
    strMsg = strMsg & "It has the Command button to reopen frmWorksheets UserForm!"
    MsgBox strMsg, vbCritical, "Can't delete Sheet1"
```

■ **Attention** There is no VBA protection to avoid you changing the `Sheet1` object's `CodeName` property in the VBA Properties window and making this code fail, so be careful.

If the selected sheet tab doesn't have `CodeName = "Sheet1"`, the procedure makes a first check of your delete intentions using `MsgBox( )` with the "No" button selected by default.

```
strMsg = "Do you really want to delete " & ws.Name & "?" & vbCrLf
strMsg = strMsg & "(This operation can't be undone!)"
strTitle = "Delete " & ws.Name & " worksheet?"
If MsgBox(strMsg, vbCritical + vbYesNo + vbDefaultButton2, strTitle) = vbYes Then
```

And if you really want to delete the selected sheet tab, the code sets the `Application.DisplayAlerts` property according to the state of the `chkDisplayAlerts CheckBox` and then executes the `Worksheet` object `Delete` method.

```
Application.DisplayAlerts = Me.chkExcelWarnings
    ws.Delete
```

At this moment, if `Application.DisplayAlerts = True` (`chkDisplayAlerts` is checked), Excel will send its own delete confirmation dialog, warning you that the deletion can't be undone (Figure 4-15).

`Application.DisplayAlerts` is turned on again to guarantee that Excel will continue to show the warning messages, and the code finishes updating the `UserForm` interface by calling the `FilllstSheetTabs( )` procedure to reflect the possible deletion of the worksheet.

```
            Application.DisplayAlerts = True
        End If
         Call FilllstSheetTabs
    End If
End Sub
```

■ **Attention** As with the `Application.ScreenUpdating` property, don't ever forget to turn on again the `Application.DisplayAlerts` property at the end of any procedure that needs to disable it or Excel will stop to show alert messages to the user of your application.

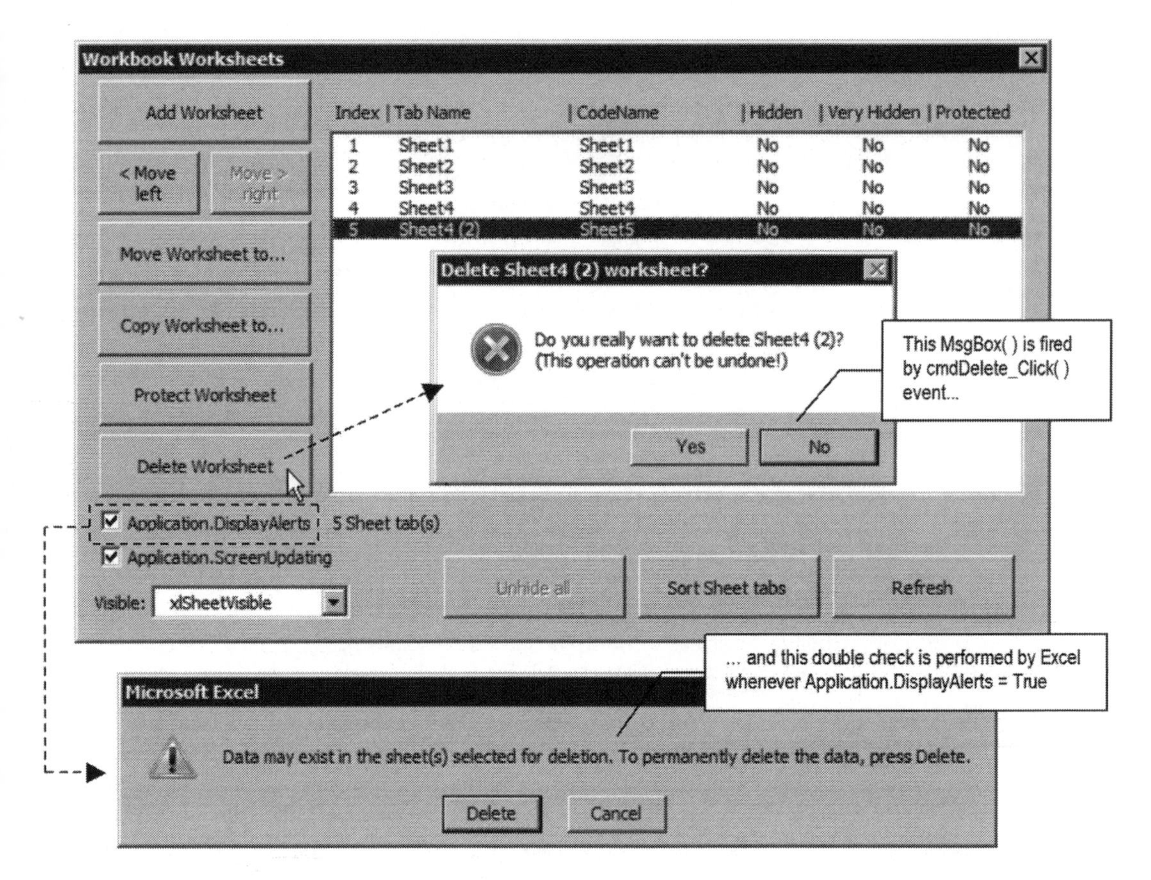

Figure 4-15. *Use the Delete Worksheets CommandButton (cmdDelete) to execute the Worksheet object Delete method. The code has a protection to not delete the Sheet1 object since it contains the button control that reopens the UserForm, using the Sheet tab CodeName property to identify it. If you keep the Application. DisplayAlerts CheckBox checked, Excel will fire a double-check before the sheet tab is permanently deleted from the workbook*

Protecting and Unprotecting Sheet Tabs

When you use the Excel Review tab's Protect command to protect any sheet tab, Excel answers with the Protect Sheet dialog box, where you can set what kind of cells will be selected and which type of operations can be performed on the protected worksheet (Figure 4-16).

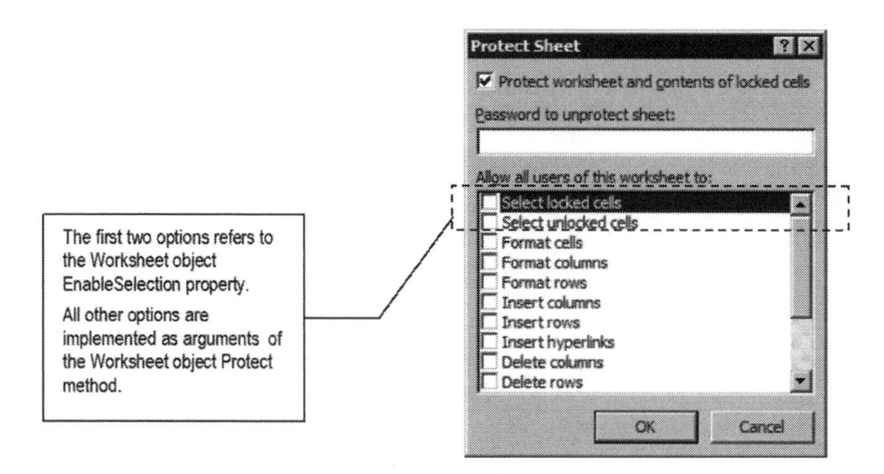

Figure 4-16. *Whenever you execute Excel Protect Worksheet from the Review tab, you receive the Protect Sheet dialog box, which is where you decide what will be selected and what kind of change is allowable on the worksheet*

To deal with Excel worksheet protection using VBA, including all options of the Protect Sheet dialog box, you must use two `Worksheet` object properties (`EnableSelection` and `ProtectContents`) and two methods (`Protect` and `Unprotect`).

The first two options of Excel Protect Sheet dialog box, "Select locked cells" and "Select unlocked cells," relate to the `Worksheet` object's `EnableSelection` property, which can be set using the Excel constants `xlNoSelection` (the default state when both options are unchecked), `xlUnlockedCells` (when just "Select locked cells" is selected) and `xlNoRestrition` (when both options are selected).

Every other option in the Protect Sheet dialog box maps to optional `Boolean` arguments of the `Worksheet` object's `Protect` method, which has this (enormous) syntax:

```
Worksheet.Protect( Password, DrawingObjects, Contents, Scenarios, UserInterfaceOnly,
AllowFormattingCells, AllowFormattingColumns, AllowFormattingRows, AllowInsertingColumns,
AllowInsertingRows, AllowInsertingHyperlinks, AllowDeletingColumns, AllowDeletingRows,
AllowSorting, AllowFiltering, AllowUsingPivotTables)
```

In this code:

`Password`: This is optional; it is the password used to unprotect the worksheet.

`DrawingObjects`: This is optional; it is used to protect shapes inserted on the worksheet (the default is `True`).

`Contents`: This is optional; it protect the contents of all locked cells (the default is `True`).

`Scenarios` : This is optional; it protects scenarios (the default is `True`).

`UserInterfaceOnly` : This is optional; it protects the user interface but not macros (the default is `True`).

`AllowFormattingCells` : This is optional; it allows you to format any cell on a protected worksheet (the default is `False`).

`AllowFormattingColumns` : This is optional; it allows you to format any column on a protected worksheet (the default is `False`).

`AllowFormattingRows` : This is optional; it allows you to format any row on a protected worksheet (the default is `False`).

`AllowInsertingColumns` : This is optional; it allows you to insert columns on the protected worksheet (the default is `False`).

`AllowInsertingRows` : This is optional; it allows you to insert rows on the protected worksheet (the default is `False`).

`AllowInsertingHyperlinks` : This is optional; it allows you to insert hyperlinks on the worksheet (the default is `False`).

`AllowDeletingColumn` : This is optional; it allows you to delete columns on the protected worksheet where every cell in the column to be deleted is unlocked (the default is `False`).

`AllowDeletingRows` : This is optional; it allows you to delete rows on the protected worksheet, where every cell in the row to be deleted is unlocked (the default is `False`).

`AllowSorting` : This is optional; it allows you to sort on the protected worksheet. Every cell in the sort range must be unlocked or unprotected (the default is `False`).

`AllowFiltering` : This is optional; it allows you to set filters on the protected worksheet. Users can change filter criteria but cannot enable or disable an auto filter. Users can set filters on an existing auto filter (the default is `False`).

`AllowUsingPivotTables` : This is optional; it allows you to use pivot table reports on the protected worksheet (the default is `False`).

■ **Attention** Although the `Worksheet.Protect` method has a lot of arguments, they are all optional and have a default value that protects the worksheet against changes, so you do not need to use them unless you want to diminish the protection and allow specific changes.

To verify whether any sheet tab is protected, use the `Worksheet` object's `ProtectContents` property. To unprotect a protected worksheet, use the `Worksheet` object's `Unprotect` method, which has this syntax:

```
Worksheet.Unprotect(Password)
```

In this code:

`Password`: This is optional; it is the password used when the worksheet protection was set.

If the worksheet was protected with a password, you must pass it to the `Worksheet.Unprotect` method in your VBA code or Excel will raise a dialog box asking for it before unprotecting the worksheet.

Let's see these properties and methods in action! When you select any sheet tab in the `lstSheetTabs ListBox` of `frmWorksheets UserForm`, the `lstSheetTabs_Click( )` event will fire and synchronize the `cmdProtect CommandButton Caption` property according to the `ProtectContents` property of the selected sheet tab, alternating the text between Protect Worksheet and Unprotect Worksheet. And when you click the `cmdProtect` command button, the `cmdProtect_Click( )` event fires, executing this code:

```
Private Sub cmdProtect_Click()
    Dim ws As Worksheet
    Dim intIndex As Integer
```

```
    With Me.lstSheetTabs
        intIndex = .ListIndex
        Set ws = Worksheets(CInt(.Value))
        If ws.ProtectContents Then
             ws.Unprotect
        Else
            Load frmPassword
            frmPassword.Worksheet = (CInt(.Value))
            frmPassword.Show vbModal
        End If

        Call FilllstSheetTabs
        .Selected(intIndex) = True
    End With
End Sub
```

The cmdProtect_Click( ) event begins by setting a With lstSheetTabs... End With instruction so lstSheetTabs can be referenced only once in the code. It then stores the position of the selected item on the intIndex Integer variable using the lstSheetTabs.ListIndex property (note that there is a dot before .ListIndex).

```
    With Me.lstSheetTabs
        intIndex = .ListIndex
```

■ **Attention** Whenever you use a With...End With instruction to reference an object in the VBA code, VBA will show the object interface whenever you type the dot (.) inside this instruction so you can easily select the property or method you need.

The code then sets a reference to the desired Worksheet object that must be protected/unprotected to the ws object variable, using the lstSheetTabs.Value property to return its Index inside the Worksheets collection (note the dot before the word Value inside the parentheses).

```
Set ws = Worksheets(CInt(.Value))
```

■ **Attention** Note that any ListBox Value property (default property) returns a String value. Since the Worksheets collection can receive both an Integer value (meaning the Index position of the item) or a String value (meaning the item name) when it receives the lstSheetTabs.Value property, it is really receiving a String numeric value (like "1" instead of 1). Since there is no such sheet tab name, VBA will raise an error. That is why you need to use the VBA CInt(.Value) function: to convert the String value to Integer!

Once the desired Worksheet object is referenced, the code uses it to verify its ProtectContents property. If it is True, the sheet tab is already protected, so the Worksheet object's Unprotect method is called to unprotect the sheet tab.

```
If ws.ProtectContents Then
     ws.Unprotect
```

If the sheet tab is unprotected, the `Else` clause is executed, and `frmPassword` is shown on a modal state to allow you to set the `Worksheet` object protection. The code now stops at this point until `frmPassword` is closed!

```
Else
    Load frmPassword
    frmPassword.Worksheet = (CInt(.Value))
    frmPassword.Show vbModal
End If
```

The code stops it execution here: when frmPassword is exhibited on a modal state

Using the frmPassword UserForm

To deal with `Worksheet` object protection, the `Worksheet Referencing.xlsm` workbook also has the `frmProtection UserForm`, which has the `txtPassword` and `txtConfirmation TextBox` controls, both with the `PasswordChar` property defined to *, so the typed password and confirmation characters can be masked (Figure 4-17).

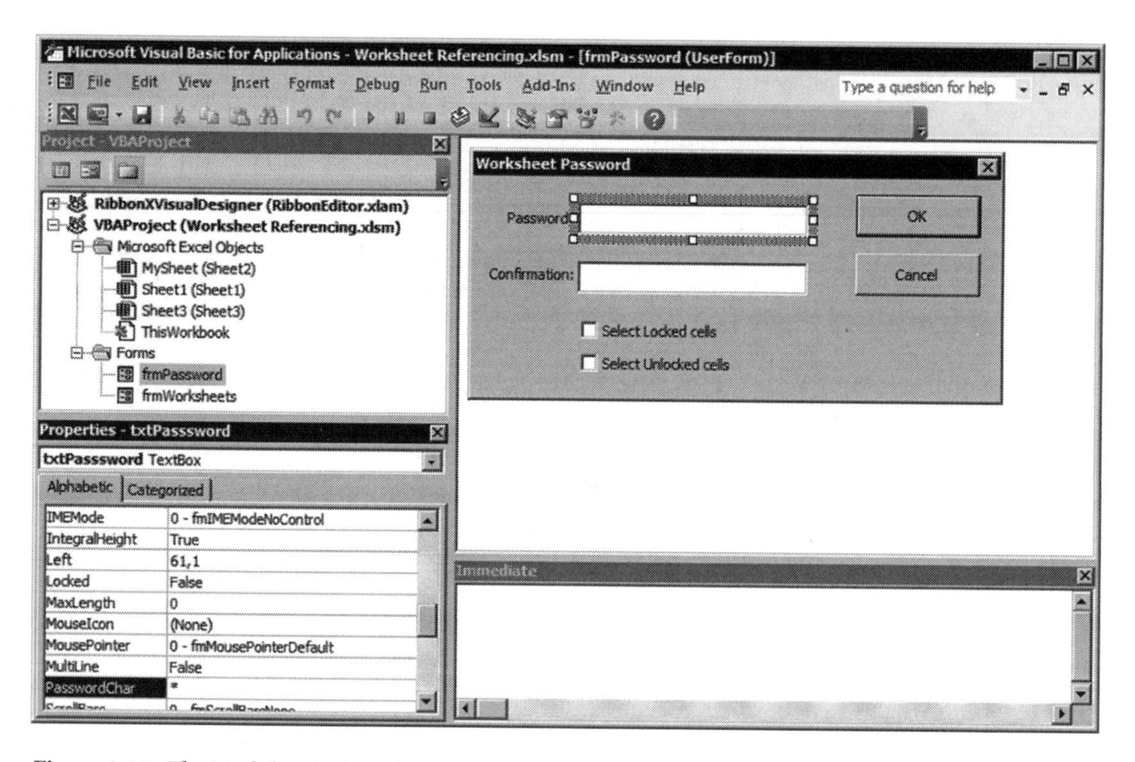

Figure 4-17. *The Worksheet Referencing.xlsm workbook also has the frmPassword, which has the txtPassword and txtConfirmation TextBox controls, both with the PasswordChar property defined to *, to mask the password and confirmation typed characters*

The `frmPassword UserForm` code module implements the `Worksheet` property, which is used to set a reference to the desired `Worksheet` object into the `mWks` module-level object variable, implemented with a `Public property Let Worksheet( )` procedure.

```
Option Explicit

Dim mWks As Worksheet

Public property Let Worksheet(ByVal Index As Integer)
    Set mWks = Worksheets(Index)
End property
```

Note that the frmPassword.Worksheet property receives the Index arguments and uses this Integer value to set a reference to the desired Worksheet object inside the Worksheets collection.

So, the technique to use frmPassword consists of loading it into memory, setting its Worksheet property (which sets a reference to the desired sheet tab), and then showing it on a modal state. Returning to thecmdProtect_Click( ) event procedure, when the selected sheet tab isn't protected, the If ws.ProtectContents instruction executes its Else clause, executing these steps:

1. Load frmPassword with the UserForm Load method.

   ```
   Else
       Load frmPassword
   ```

2. Set the frmPassword.Worksheet property using the Index of the selected sheet tab, which is defined by the lstSheetTabs.Value property. Since this property is a String value, you must first convert it to an Integer using the VBA CInt( ) function.

   ```
   frmPassword.Worksheet = (CInt(.Value))
   ```

3. Show frmPassword on a modal state, passing vbModal to the Modal argument of the UserForm.Show method.

   ```
   frmPassword.Show vbModal
   ```

This last operation will stop the cmdProtect_Click( ) event at this point while frmPassword is opened.

Use the two CheckBox controls to select what kind of protection you want to disallow when you protect a worksheet: chkSelectLockedCells and chkSelectUnlockedCells. Note that they can be both checked or unchecked, but just chkSelectUnLockedCells can be checked while ckdSelecLockedCells is unchecked. chkSelectLockedCells can't be checked alone!

This interface control is made by programming both CheckBox Click( ) events, as follows:

```
Private Sub chkLockedCells_Click()
    If Me.chkLockedCells Then
        Me.chkUnlockedCells = True
    End If
End Sub

Private Sub chkUnlockedCells_Click()
    If Not Me.chkUnlockedCells Then
        Me.chkLockedCells = False
    End If
End Sub
```

The frmPassword UserForm also has two command buttons: cmdCancel (whose Cancel property was set to True) and cmdOK (whose Default property was set to True). When it is opened and you press the Esc key on your keyboard or click the Cancel button, the cmdCancel_Click( ) event will fire, and the UserForm will be closed without setting the worksheet protection.

```
Private Sub cmdCancel_Click()
    Unload Me
End Sub
```

But if you press the Enter key on your keyboard or click the OK button, the cmdOK_Click( ) event will fire, executing this code:

```
Private Sub cmdOK_Click()
    If StrComp(Me.txtPasssword, Me.txtConfirmation, vbBinaryCompare) <> 0 Then
        MsgBox "Confirmation password is not identical!", _
               vbQuestion, _
               "Invalid Password Confirmation"
    Else

        If (Me.chkLockedCells = False And Me.chkUnlockedCells = False) Then
            mWks.EnableSelection = xlNoSelection
        ElseIf (Me.chkLockedCells = True And Me.chkUnlockedCells = True) Then
            mWks.EnableSelection = xlNoRestrictions
        ElseIf Me.chkUnlockedCells Then
            mWks.EnableSelection = xlUnlockedCells
        End If
        mWks.Protect Me.txtPasssword
        Unload Me
    End If
End Sub
```

When you click cmdOK to protect the selected worksheet (defined by the frmPassword.Worksheet property), the code first compares the text typed on the txtPassword and txtConfirmation TextBox controls.

```
If StrComp(Me.txtPasssword, Me.txtConfirmation, vbBinaryCompare) <> 0 Then
```

You may note that it uses the StrComp( ) VBA function instead of the = operator. This is necessary because the Worksheet object password is case sensitive and the = operator is not. The VBA StrComp( ) function has this syntax:

```
StrComp(string1, string2[, compare])
```

In this code:

String1: This is required; it is the first string to compare.

String2: This is required; it is the second string to compare.

Compare: This is optional; it specifies the comparison type, using one of these constants:

vbUseCompareOption = -1: This performs a comparison according to the Option Compare statement.

vbBinaryCompare = 0: This performs a binary comparison, which is case sensitive.

vbTextCompare = 1: This performs a textual comparison.

The StrComp( ) function returns False (0) whenever both strings are equal to the type of comparison performed. It returns 1 when String1 > String2 and returns -1 when String1 < String2.

Attention If you type the instruction `Option Compare Binary` at the beginning of any code module, all comparisons inside the module become case sensitive.

If the `txtPassword` and `txtConfirmation` values differ, the code raises a `MsgBox( )` asking the user to type them again.

```
If StrComp(Me.txtPasssword, Me.txtConfirmation, vbBinaryCompare) <> 0 Then
    MsgBox "Confirmation password is not identical!", _
           vbQuestion, _
           "Invalid Password Confirmation"
```

But if `txtPassword` and `txtConfirmation` match (which both include an empty string), the procedure verifies the state of both `CheckBox` controls to define the `Worksheet.Enabled Selection` property. The first possible state is when both `CheckBox` controls are unchecked, meaning restrained to any cell selection, which is implied by setting the `Worksheet.EnableSelection = xlNoSelection` constant.

```
If (Me.chkLockedCells = False And Me.chkUnlockedCells = False) Then
    mWks.EnableSelection = xlNoSelection
```

The second possible case is when both `CheckBox` controls are checked, meaning no worksheet restrains selection (every cell can be selected), which is implied by setting the `Worksheet.EnableSelection = xlNoRestrictions` constant.

```
ElseIf (Me.chkLockedCells = True And Me.chkUnlockedCells = True) Then
    mWks.EnableSelection = xlNoRestrictions
```

The third and last possible case is when just `chkSelectUnlockedCells` is selected, which is implied by the setting `Worksheet.EnableSelection = xlUnlockedCells`.

```
    mWks.EnableSelection = xlUnlockedCells
End If
```

And once you set the `Worksheet.EnableSelection` property, you can protect the worksheet using the `Worksheet.Protection` method and then type the password and unload the form, ending the procedure.

```
        mWks.Protect Me.txtPasssword
        Unload Me
    End If
End Sub
```

When the form is unloaded, the code returns to the point where it stops, with a great chance of the selected worksheet being now protected (the user cancels `frmPassword`), so it must synchronize the `frmWorksheets` interface calling the `FilllstSheetTabs( )` procedure and reselect the same sheet tab in the `lstSheetTabs ListBox`. Once again, at this point the `lstSheetTabs_Click( )` event will cascade-fire, synchronizing the enabled state of the `UserForm` controls.

```
        Call FilllstSheetTabs
        .Selected(intIndex) = True
    End With
End Sub
```

Figure 4-18 shows how this process happens to protect the `Sheet1` worksheet from the `Worksheet Referencing.xlsm` macro-enabled workbook, implying just unlocked cells to be selected (just cell `E2` of the `Sheet1` worksheet is unlocked).

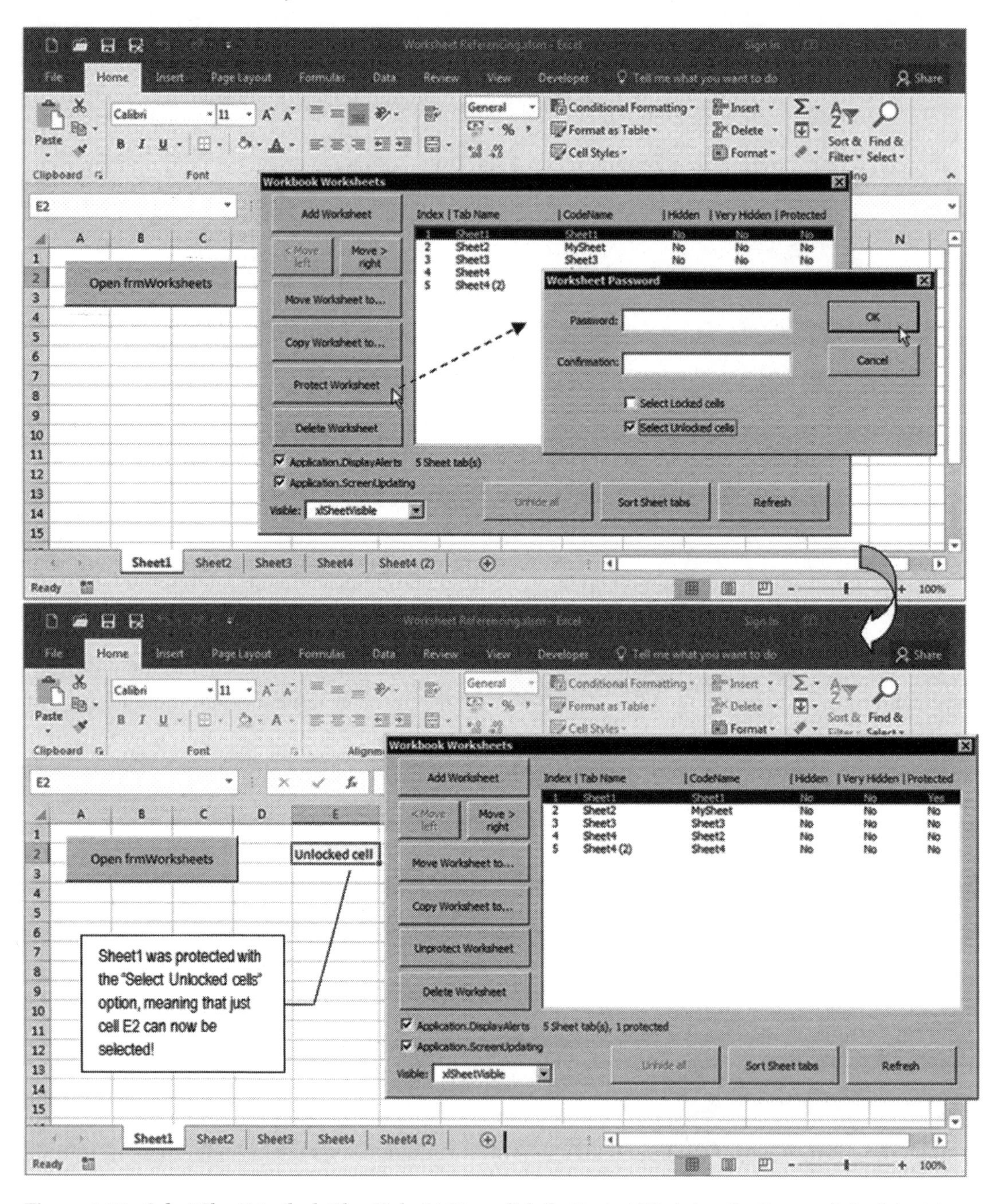

Figure 4-18. *Select Sheet1 in the lstSheetTabs ListBox, click the Protect Worksheet button, and check just the "Select unlocked cells" check box, with or without a password. When you close frmPassword, Sheet1 will be protected and the frmWorksheets interface will be updated to reflect the changes. Note that now just cell E2, the only unlocked cell of Sheet1, can be selected*

Changing a Sheet Tab's Visible Property

To change the visible state of any sheet tab, use the Worksheet.Visible property, which can be set using the Excel constants xlSheetVisible = -1, xlSheetHidden = 0, and xlSheetVeryHidden = 1. These constant values were used to populate the first (integer value) and second (constant name) columns of the cboVisible ComboBox control on the UserForm_Initialize( ) event, as shown earlier in this chapter.

Whenever you select any item in lstSheetTabs ListBox, the cboVisible value is synchronized to the sheet Visible property, and the cboVisible ComboBox will become enabled, meaning that you can change the visible state of the selected sheet tab by choosing another option in the ComboBox control. Whenever you do this action, the cboVisible_Change( ) events fire, executing this code:

```
Private Sub cboVisible_Change()
    Dim intIndex As Integer
    Const conErrHidden = 1004

    On Error GoTo cboVisible_Error

    If mbolCancelEvent Then
        mbolCancelEvent = False
    Else
        With Me.lstSheetTabs
            intIndex = .ListIndex
            Worksheets(CInt(.Value)).Visible = CInt(Me.cboVisible)
            Call FilllstSheetTabs
            .Selected(intIndex) = True
        End With
    End If

cboVisible_End:
    Exit Sub
cboVisible_Error:
    Select Case Err
        Case conErrHidden
            MsgBox "At least one Sheet tab must be visible", vbCritical, "Can't hide " & Me.
lstSheetTabs.Column(1)
        Case Else
            MsgBox "Error " & Err & ": " & Error(Err), vbCritical, "cboVisible_Change event"
    End Select
End Sub
```

There is a catch when you deal with hidden sheet tabs: at least one sheet tab must be visible on the workbook or VBA will raise error = 1004: "Application-defined or object-defined error." That is why cboVisible_Change( ) declares the constant conErrHidden = 1004 to avoid that this "magic number" appears suddenly inside the code.

```
Private Sub cboVisible_Change()
    Dim intIndex As Integer
    Const conErrHidden = 1004
```

The code then sets the error trap to begin at the cboVisible_Error label, using the On Error GoTo instruction.

```
    On Error GoTo cboVisible_Error
```

Once this is made, whenever an error happens on the code, the execution flow will jump to this label so the error can be treated or the procedure ends.

```
cboVisible_Error:
    Select Case Err
        Case conErrHidden
            MsgBox "At least one Sheet tab must be visible", vbCritical, "Can't hide " & Me.
lstSheetTabs.Column(1)
        Case Else
            MsgBox "Error " & Err & ": " & Error(Err), vbCritical, "cboVisible_Change event
error"
    End Select
End Sub
```

Note that the cboVisible_Change( ) event does not treat any errors. It just uses a Select Case statement to verify the error and gives a clear message to the user if it tries to hide the last visible worksheet of the workbook.

```
Select Case Err
    Case conErrHidden
        MsgBox "At least one Sheet tab must be visible", vbCritical, "Can't hide " & Me.
lstSheetTabs.Column(1)
```

But if any other unpredictable error happens, a standard MsgBox( ) will show the error code and the error message and will put "cboVisible_Change event error!" in the message box title.

```
        Case Else
            MsgBox "Error " & Err & ": " & Error(Err), vbCritical, "cboVisible_Change event
error"
    End Select
End Sub
```

After the error message, the code will not be redirected, ending normally.

As commented before, the code first verifies that the cboVisible_Change( ) event was cascade-fired by checking the state of the code module-level variable mbolCanceEvent. If it is False, the event is fired by the user action selecting another value on the ComboBox control, and it uses a With lstSheetTabs... End With instruction to reference the ListBox control just once, making the code more concise.

```
If mbolCancelEvent Then
    mbolCancelEvent = False
Else
    With Me.lstSheetTabs
```

Since there is a selected item on the lstSheetTabs ListBox, the ListIndex property is stored into the intIndex Integer variable so you can select it again, and the Visible property of the selected sheet tab is changed according to the value selected in the cboVisible ComboBox. Note once again that both the ListBox and ComboBox values are numeric strings, meaning that both must be converted to an Integer value with the VBA CInt( ) function, before being used to define the desired sheet inside the Worksheets collection (left side) and to set the desired visible option to the Worksheet object's Visible property (right side).

```
intIndex = .ListIndex
Worksheets(CInt(.Value)).Visible = CInt(Me.cboVisible)
```

Since the visible property of the selected sheet had been changed, the frmWorksheets interface must be synchronized by calling the FilllstSheetTabs( ) procedure, and the sheet is selected again on the lstSheetTabs ListBox, which will cascade-fire again the lstSheetTabs_Click( ) event, synchronizing all UserForm controls.

```
        Call FilllstSheetTabs
        .Selected(intIndex) = True
    End With
End If
```

Once this happen, the code needs an exit door, represented by the Exit Sub statement so that the code inside the error trap will not be inadvertently executed.

```
cboVisible_End:
    Exit Sub
```

■ **Attention** The cboVisible_End label was defined here as good programming practice. It is not used by the procedure to treat the code by any means.

Changing a Sheet Tab's CodeName Property

The lstSheetTabs of the frmWorksheets UserForm also has the CodeName column that shows each Worksheet.CodeName property, which is the value of the Name property defined in the VBA Properties window to the selected Sheet object on VBA Project Explorer tree.

Just the Sheet1 worksheet uses the CodeName property on the cmdDelete_Click( ) event to avoid it being deleted since this sheet tab has the CommandButton open frmWorksheets again whenever you close it (see the "Delete the Sheet Tab" section earlier in this chapter).

But if you change any sheet tab's CodeName in the VBA interface, you will need to refresh the frmWorksheets UserForm either by closing and opening it again or by clicking its Refresh CommandButton to fire the cmdRefresh_Click( ) event, which executes this code:

```
Private Sub cmdRefresh_Click()
    mbolCancelEvent = True
    Me.cboVisible = Null
    Call FilllstSheetTabs
End Sub
```

The code is now simple: it changes the module-level variable to mbolCancelEvent = True, changes cboVisible = Null to avoid a cascade cboVisible_Change( ) event, and then synchronizes the UserForm interface by calling the FilllstSheetTabs( ) procedure.

Figure 4-19 shows that the Sheet2 tab worksheet had its CodeName value changed to MySheet. Every UserForm operation continues to perform well because just the Index property of the sheet tab inside the Worksheets collection is used to reference it in the VBA code.

■ **Attention** Once more, you can avoid changes to the CodeName property of any sheet tab by protecting your VBA code by executing the VBA Tools ➤ Project Properties menu command and setting a password on the Protection tab.

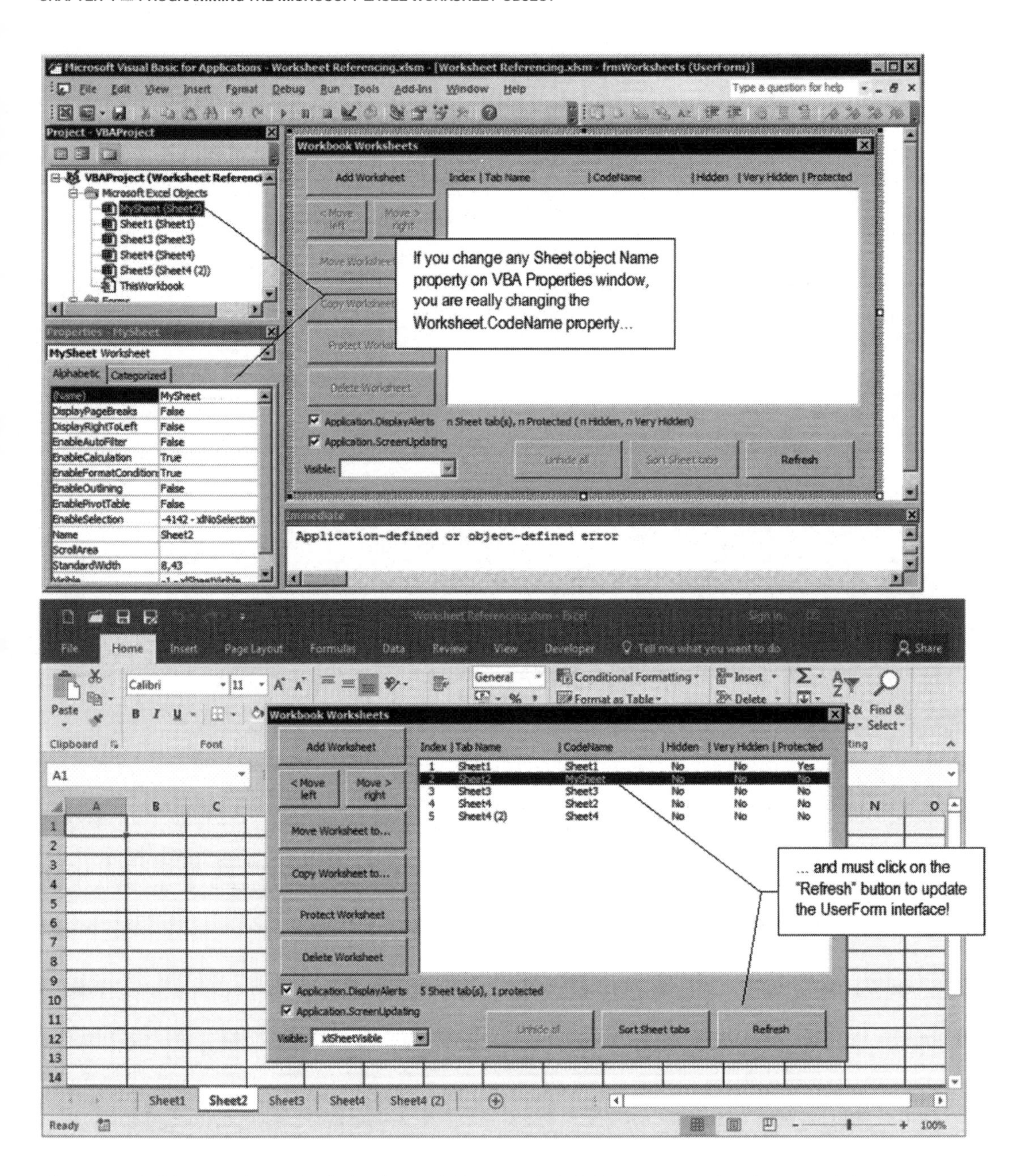

Figure 4-19. *You can change both the sheet tab Name and CodeName properties of any worksheet except Sheet1 CodeName, because it is used to protect the sheet deletion from a frmWorksheets UserForm cmdDelete action. Whenever you do it, click the Refresh button to update the interface*

Chapter Summary

In this chapter, you learned about the following:

- The sequence order of `Worksheet` object events
- The many ways you can make a VBA reference to a `Worksheet` object
- How to program `Class` modules
- How you can use a `Class` module to avoid any sheet tab name from being changed
- How to use the `Worksheet.Name` and `Worksheet.CodeName` properties
- How to create a VBA `UserForm` interface to deal with the `Worksheet` object
- How to add, move, copy, and protect worksheet tabs; change their visible states; and sort and delete worksheet tabs from a workbook using VBA code
- How to synchronize a `UserForm` interface using centralized procedures
- How to deal with a multicolumn `ListBox` control

In the next chapter, you will learn about the last object on the Microsoft Excel object model, the `Range` object, which can represent one or more cells of any Excel worksheet.

CHAPTER 5

Programming the Microsoft Excel Range Object

The Microsoft Excel Range object is where the real action of your worksheet applications takes place. It can represent any number of cells, and in this chapter you will learn about the Range object and how to programmatically interact with it using VBA and its many properties and methods (it has no events), using some of the numerous `Worksheet` object methods to automate your worksheet and give it a professional look and feel. You can obtain all the procedure code in this chapter by downloading the `Chapter05.zip` file from the book's Apress.com product page, located at `www.apress.com/9781484222041`, or from `http://ProgrammingExcelWithVBA.4shared.com`.

The Range Object

To the Microsoft Excel object model, a *range* is any number of worksheet cells, and since it is the basic worksheet unit, it is referenced by different `Collection` objects, returned as an argument from many Excel `Worksheet`, `Workbook`, and `Application` object events, properties, and methods. Table 5-1 shows some Excel `Application` properties and `Workbook` collections that return a Range object.

Table 5-1. *Excel Collection and Object Properties and Methods That Return a Range Object*

Object	Value	Used to
`Application.ActiveCell`	`Property`	Returns a Range object representing the active cell
`Application.Range`	`Property`	Returns a Range object referenced by its address or name
`Application.Selection`	`Property`	Returns the selected object in the active window, which can be a Range object
`Application.ThisCell`	`Property`	Returns a Range object representing the cell from which the user-defined function is being called as a Range object
`Range.Offset`	`Property`	Returns another Range object offset from the current Range object
`Workbook.Names`	`Collection`	Stores workbook named ranges
`Worksheet.Names`	`Collection`	Stores worksheet named ranges
`Worksheet.Cells`	`Property`	Returns a Range object referenced by cell `Row` and `Column` numbers
`Worksheet.Range`	`Property`	Returns a Range object referenced by its address or name

F. Morgado, *Programming Excel with VBA*, DOI 10.1007/978-1-4842-2205-8_5

To programmatically deal with the Range object using VBA, you must first declare a Range object variable, initialize it with the Set keyword, and then use it to deal with the Range object properties and methods, as follows:

```
Dim rg as Range
Set rg = Application.ActiveCell
```

Table 5-2 shows some of the most important properties of the Range object.

▒ **Attention** Search the Internet with the keywords *Range properties* or *Range methods* to find a complete list of Excel Range object properties. Table 5-2 and 5-3 come from the following location on the Microsoft MSDN web site:

http://msdn.microsoft.com/en-us/library/microsoft.office.interop.excel.range_properties(v=office.15).aspx

---

Table 5-2. *Some of the Most Important Microsoft Excel Range Object Properties*

Range Object Property	Value	Used to
Address	String	Returns the range reference
AllowEdit	Boolean	Determines whether the range can be edited on a protected worksheet
Application	Application	Returns an Application object representing the Microsoft Excel windows
Cells	Range	Returns a Range object representing the cells in the specified range
Column	Long Integer	Returns the number of the first column in the first area of the range
Columns	Range	Returns a Range object representing all columns in the range
Count	Long Integer	Returns the number of cells in the range
CountLarge	Decimal	Returns the number of cells in the range for .xlsx workbooks
CurrentArray	Range	If the specified range is part of an array, returns a Range object representing the entire array
CurrentRegion	Range	Returns a Range object representing the current region
End	Range	Returns a Range object representing the cell at the end of the region that contains the source range
EntireColumn	Range	Returns a Range object representing the entire column (or columns) that contains the specified range
EntireRow	Range	Returns a Range object representing the entire row (or rows) that contains the specified range

(*continued*)

Table 5-2. *(continued)*

Range Object Property	Value	Used to
Formula	String	Returns or sets the cell's formula in A1-style
FormulaArray	Boolean	Returns or sets the array formula of a range
FormulaHidden	Boolean	Determines whether the formula will be hidden when the worksheet is protected
HasArray	Boolean	Determines whether the specified cell is part of an array formula
HasFormula	Boolean	Determines whether all cells in the range contain formulas
Hidden	Boolean	Determines whether the rows or columns are hidden
Item	Range	Returns a Range object representing a range at an offset from the specified range
ListHeaderRows	Long Integer	Returns the number of header rows for the specified range
Locked	Boolean	Determines whether the object is locked
MergeArea	Range	Returns a Range object representing the merged range containing the specified cell
MergeCells	Boolean	Determines whether the range or style contains merged cells
Name	String	Returns or sets the name of the referenced range
Offset	Range	Returns a Range object representing a range that's offset from the specified range
Parent	Worksheet	Returns the Worksheet that is the parent of the specified range
Range	Range	Returns a Range object representing a range address
Resize	Range	Resizes the specified range (does not resize a named range)
Row	Long Integer	Returns the number of the first row of the first area in the range
Rows	Range	Returns a Range object representing the rows in the specified range
Value	Variant	Default property; returns or sets the value of the specified range
Value2	Variant	Returns or sets the cell value of the specified range; discards Currency and Data formatting options, returning the range pure value
Worksheet	Worksheet	Returns a Worksheet object representing the worksheet containing the specified range

▒ **Attention** The Range object has a lot of other formatting properties relating to the appearance of the worksheet cells and its contents. Use the MSDN web site to access all the Range object properties.

The Range object has also a lot of methods that you will use to perform actions using VBA code. Table 5-3 lists some of the most important Range object methods.

Table 5-3. *Some Important Microsoft Excel Range Object Methods*

Range Object Method	Action Performed
`Activate`	Activates the upper-left cell of the range
`AdvancedFilter`	Filters or copies data from a list based on a criteria range
`ApplyNames`	Applies names to the cells in the specified range
`AutoFilter`	Filters a list using `AutoFilter`
`AutoFit`	Changes the width of the columns or the height of the rows in the range to achieve the best fit
`AutoFormat`	Automatically formats the specified range, using a predefined format
`AutoOutline`	Automatically creates an outline for the specified range
`BorderAround`	Adds a border to a range and sets the `Color`, `LineStyle`, and `Weight` properties for the new border
`Calculate`	Calculates a specified range of cells on a worksheet
`Clear`	Clears the entire range
`ClearComments`	Clears all cell comments from the specified range
`ClearContents`	Clears the formulas from the range
`ClearFormats`	Clears the formatting of the object
`ClearHyperlinks`	Removes all hyperlinks from the specified range
`ClearNotes`	Clears notes and sound notes from all the cells in the specified range
`ClearOutline`	Clears the outline for the specified range
`Consolidate`	Consolidates data from multiple ranges on multiple worksheets into a single range on a single worksheet
`Copy`	Copies the range to the specified range or to the clipboard
`CreateNames`	Creates names in the specified range, based on text labels in the sheet
`Cut`	Cuts the range values to the clipboard
`Delete`	Deletes the range values
`Find`	Finds specific information in a range and returns a Range object representing the first cell where that information is found
`FindNext`	Continues a search to the next cell that was begun with the `Find` method
`FindPrevious`	Continues a search to the previous cell that was begun with the `Find` method
`ListNames`	Pastes a list of all displayed names onto the worksheet, beginning with the first cell in the range
`Merge`	Creates a merged cell from the specified Range object
`PasteSpecial`	Pastes a range that was copied or cut from the clipboard into the specified range

(*continued*)

Table 5-3. (*continued*)

Range Object Method	Action Performed
Replace	Returns a Boolean indicating characters in cells within the specified range
Select	Selects the specified range
Sort	Sorts a range
SpecialCells	Returns a Range object that represents all the cells that match the specified type and value

Using the Application.Range Property

The easiest way to access any cell value is using the Application object's Range property. Since the Application object is the top-level object in the Microsoft Excel object hierarchy, you do not need to type it when using the Range property, which has this syntax:

```
Range(Cell1, Cell2)
```

In this code:

Cell1: This is required; it is the range address or name between double quotes. It can include the range operator (a colon), the intersection operator (a space), or the union operator (a comma). If it includes dollar signs, they will be ignored.

Cell2: This is optional; use it when both Cell1 and Cell2 are valid Range objects identifying the cell in the upper-left and lower-right corners of the desired range, respectively. Both Cell1 and Cell2 Range objects can contain a single cell, an entire column, or an entire row, or it can be a string that names a single cell.

The default property of the Range object is the Value property, so when you use the Application.Range property and pass it the address of the desired cell, you will receive the cell value. If you pass it a range of cells, you will receive a variant with an array of values that cannot be printed in the VBA Immediate window (which will return the code Error=13, "Type Mismatch").

▒ **Attention** When used without an object qualifier, the Application.Range property is a shortcut for ActiveSheet.Range, returning a range from the active sheet. If the active sheet isn't a worksheet, the property fails.

Also note that when applied to a Range object, the property is relative to the Range object. For example, if the selection is cell C3, then Selection.Range("B1") returns cell D3 because it's relative to the Range object returned by the Selection property. On the other hand, the code ActiveSheet.Range("B1") always returns cell B1.

Figure 5-1 shows what happens when you use the Application.Range property (with or without the Application qualifier) and the Application.Selection property to return cell A1's value, which contains the formula =Today( ), returning the current system date. Also note that the Range property returns the Value property of the range, and when you use the Value2 property, VBA does not return the expected formatted date value but the real number stored into the cell. In this case, this is the integer value that represents the date (days counting from 1-1-1900). You can also use other Range object properties, like HasFormula (which indicates whether the range has a formula), and use the Formula property to return its formula (if any).

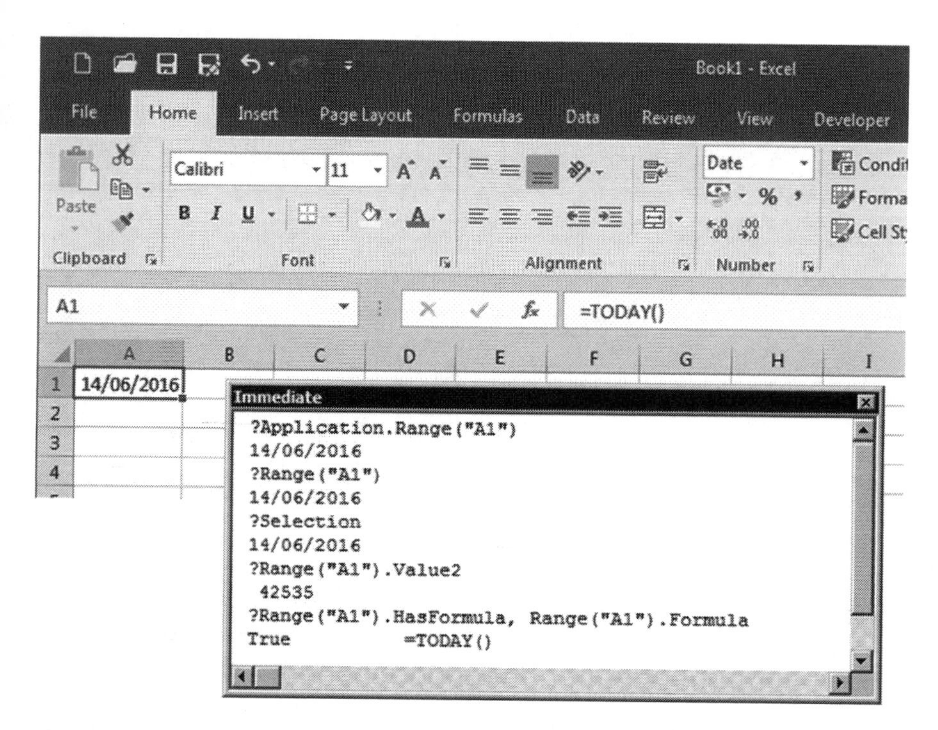

Figure 5-1. *Use the Application.Range property or the Application.Selection property (with or without the Application qualifier) to recover the current cell value. Note that the Application.Range property is a shortcut to the ActiveSheet.Range property, meaning that if the selected sheet is a Chart sheet, the method will fail and VBA will raise an error*

■ **Attention** Use the `Application.Selection` property when you do need to know what is currently selected on the active sheet of Microsoft Excel. To know what is the cell address, use the `Range.Address` property (the `ActiveSheet.Name` property returns the worksheet name of the selected sheet).

Use `Application.Range` whenever you want to know the contents of a specific cell. To know a specific worksheet cell value, precede the cell address with the sheet name (tab name) between single quotes and an exclamation character (with everything inside double quotes), as follows (Figure 5-2):

```
?Range("'Sheet1'!A1")
```

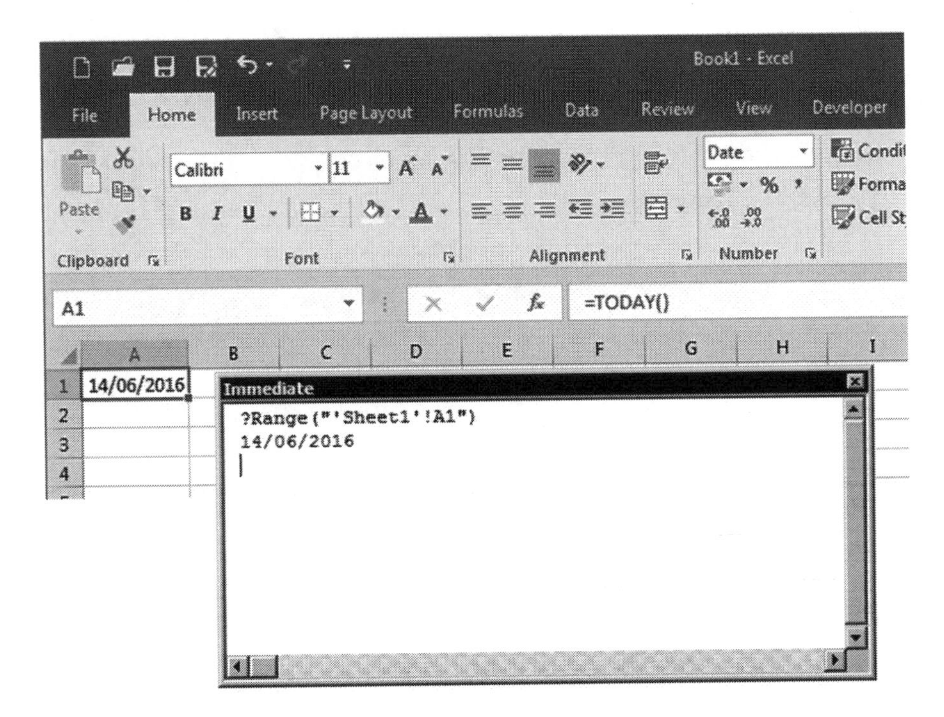

Figure 5-2. *To verify a specific worksheet cell value, type a string that contains the sheet name inside single quotes, followed by an exclamation character and the cell address*

Let's see how to program some Range object methods and properties.

Using Range Object Properties and Methods

You can get the information from the selected cell range with VBA using two different sources: from the `Application.Selection` property or from the `Target` argument returned by some events raised by the `Worksheet`, `Workbook`, or `Application` objects.

They return the same information about what is selected inside Excel, differing only by the fact that the first is based on some user action and the last is gathered from the permanent state of the active sheet.

The `Range Properties.xlsm` macro-enabled workbook, which you can extract from `Chapter05.zip`, has the `frmRange UserForm`, which you can use to learn how to implement some Range object properties and methods using both the `Application.Selection` event and the `Target` argument from some `Application` object events.

When you open this workbook, the `ThisWorkbook` object's `Workbook_Open( )` event fires and loads `frmRange` in a nonmodal state (meaning that you can interact with the sheet tabs while the `UserForm` is opened, as shown in Figure 5-3).

```
Private Sub Workbook_Open()
    frmRange.Show False
End Sub
```

Using the same programming technique described so far this book, the `frmRange UserForm` declares the module-level variable `WithEvents mApp as Application` and initializes it on the `UserForm_Initialize( )` event so it can catch cell range information whenever the user selects any cell range (using the `Application.SheetSelectionChange( )` event) or selects another sheet tab (using the `Application.SheetActivate( )` event) on any opened workbook.

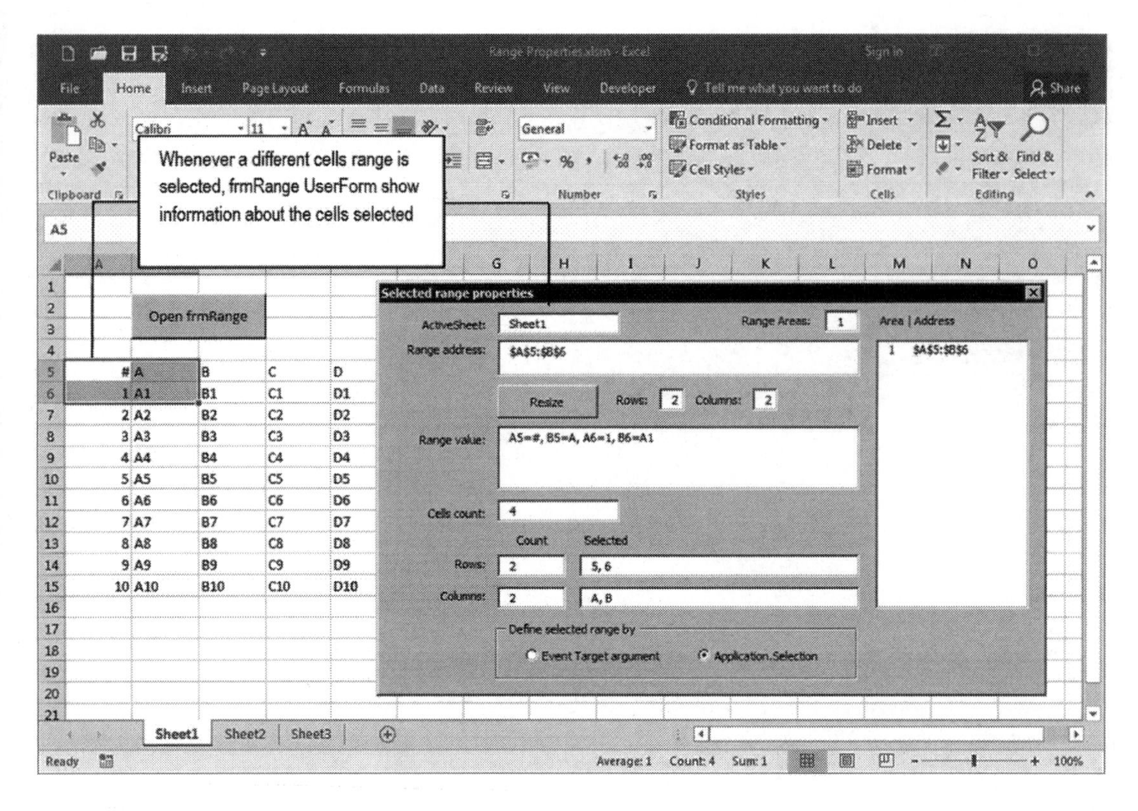

Figure 5-3. *This is the frmRange UserForm from Range Properties.xlsm macro-enabled workbook, where you can learn how to implement some Range object properties and methods*

Note that the frmRange UserForm interface has the range information updated with the aid of the Sub UpdateInterface( ) procedure (called from these three events):

```
Option Explicit

Dim WithEvents mApp As Application

Private Sub UserForm_Initialize()
     Set mApp = Application
     Call UpdateInterface
End Sub
Private Sub mApp_SheetActivate(ByVal Sh As Object)
    Call UpdateInterface(Sh, Application.Selection)
End Sub
Private Sub mApp_SheetSelectionChange(ByVal Sh As Object, ByVal Target As Range)
    Call UpdateInterface(Sh, Target)
End Sub
```

■ **Attention** Note that Sub UpdateInterface( ) receives two optional arguments used just by the mapp_ SheetActivate( ) and mApp_SheetSelectionChange( ) events.

Once frmRange is loaded, whenever you select another cell's range on any workbook sheet tab, a lot of range information is displayed on the frmRange UserForm, like the sheet tab name, range address, range values, selected cells, rows and columns count, rows number, and column letters of the range areas selected.

Figure 5-4 shows what happened to the frmRange UserForm interface when more than one noncontiguous cells range was selected by keeping the Ctrl key pressed while the mouse was dragged over the worksheet cells. Note that the Range Areas text box indicates the number of different ranges selected and that the lstAreas ListBox at its right shows the Area Index and Address information.

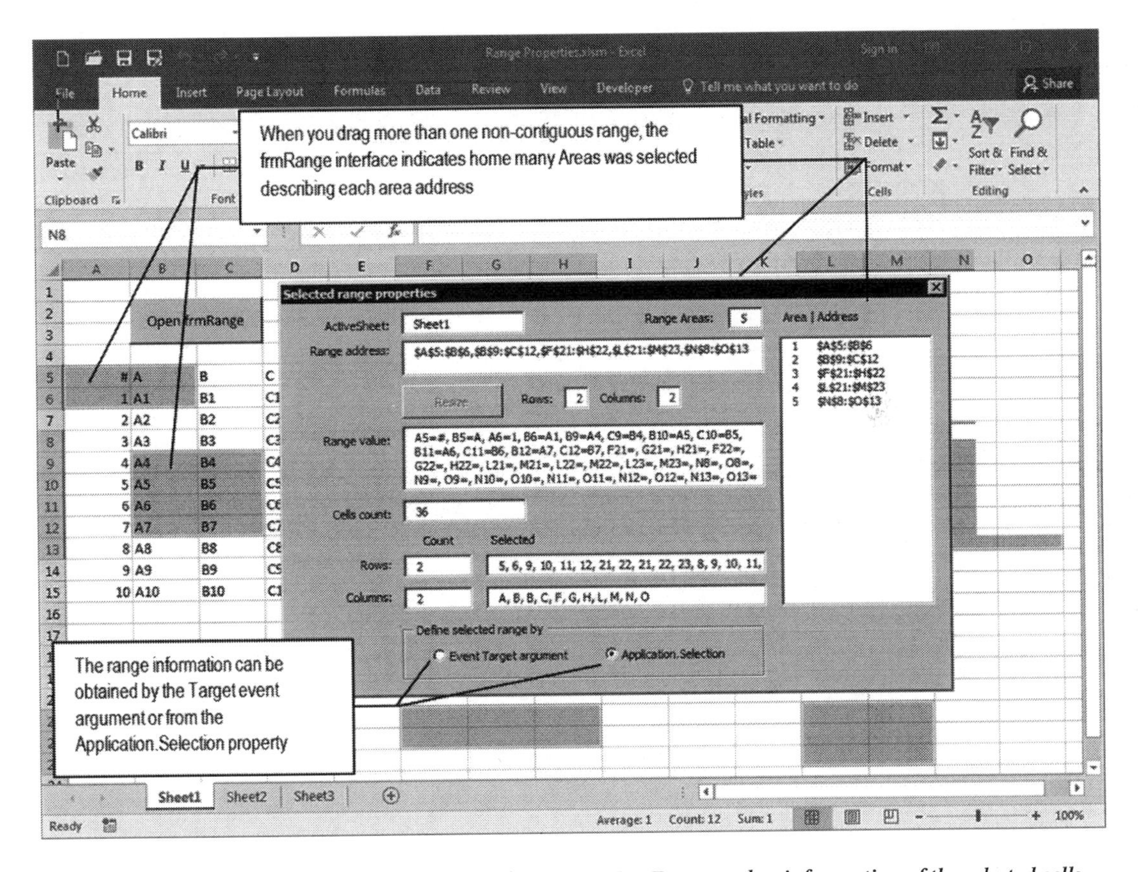

Figure 5-4. *When you select any cell range, the frmRange UserForm catches information of the selected cells, like the sheet tab name, the range address selected, how many range areas were selected, and describes each area index and address; how many cells, rows, and columns were selected; and the rows number and columns letters selected*

Updating the UserForm Interface

The Private Sub UpdateInterface( ) procedure is responsible for synchronizing the information about the cell range selected in the Excel active sheet in the frmRange UserForm. It does this by executing this code:

```
Private Sub UpdateInterface(Optional Sh As Object, Optional Target As Range)
    Dim rg As Range
    Dim varItem As Variant
    Dim strRows As String
```

```
    Dim strColumns As String
    Dim strValues As String
    Dim intI As Integer
    Dim fComma As Boolean

    If Me.optEventTarget Then
        Set rg = Target
        Me.txtActiveSheet = Sh.Name
    Else
        Set rg = Application.Selection
        Me.txtActiveSheet = rg.Worksheet.Name
    End If

    Me.txtRowsSelected = ""
    Me.txtColumnsSelected = ""
    Me.txtValue = ""

    Me.txtAddress = rg.Address
    Me.txtCellsCount = rg.CountLarge
    Me.txtRowsCount = rg.Rows.CountLarge
    Me.txtColumnsCount = rg.Columns.CountLarge

    If rg.Cells.CountLarge > 1000 Then
        MsgBox "Too much cells selected!", vbCritical, "Select less cells"
    Else
        For Each varItem In rg.Rows
            If fComma Then
                strRows = strRows & ", "
            End If
            strRows = strRows & varItem.Row
            fComma = True
        Next

        fComma = False
        For Each varItem In rg.Columns
            If fComma Then
                strColumns = strColumns & ", "
            End If
            strColumns = strColumns & ColumnNumberToLetter(varItem.Column)
            fComma = True
        Next

        fComma = False
        For Each varItem In rg
            If fComma Then
                strValues = strValues & ", "
            End If
            strValues = strValues & varItem.Address(False, False) & "=" & varItem.Value
            fComma = True
        Next
```

```
        Me.txtRowsSelected = strRows
        Me.txtColumnsSelected = strColumns
        Me.txtValue = strValues
    End If

    Me.txtRangeAreas = rg.Areas.Count
    Me.cmdResize.Enabled  = (rg.Areas.Count = 1)
    Me.lstAreas.Clear
    For intI = 1 To Selection.Areas.Count
        Me.lstAreas.AddItem intI
        Me.lstAreas.Column(1, lstAreas.ListCount - 1) = rg.Areas(intI).Address
    Next
End Sub
```

Note that the procedure may receive two optional arguments (`Sh as Object` and `Target as Range`) so it can receive information about the worksheet and cell range affected by the user action whenever the `mApp_SheetActivation( )` and `mapp_SheetSelectionChange( )` events fire, or it can use the `Application.Selection` property, which also returns a `Range` object that reflects the cells selected on the active sheet.

To show that both objects represent the same thing, the procedure begins by declaring the `rg as Range` variable to represent the selected range.

```
Private Sub UpdateInterface(Optional Sh As Object, Optional Target As Range)
    Dim rg As Range
```

If the "Event Target argument" option is selected on the `UserForm` bottom (`optEventTarget OptionButton`), the procedure uses the `Target` argument and the source of the selected range and defines the `txtActiveSheet` text box using the `Name` property of the `Sh` object.

```
If Me.optEventTarget Then
    Set rg = Target
    Me.txtActiveSheet = Sh.Name
```

But if the `Application.Selection` option is selected, the procedure uses the `Application.Selection` property as the source to the selected range, and once the range is defined, it uses its `Worksheet.Name` property to recover the sheet tab name.

```
Else
    Set rg = Application.Selection
    Me.txtActiveSheet = rg.Worksheet.Name
End If
```

Once it has a reference to the selected range, it clears the `txtRowsSelected`, `txtColumnsSelected`, and `txtValues` text boxes.

```
Me.txtRowsSelected = ""
Me.txtColumnsSelected = ""
Me.txtValue = ""
```

And then it recovers the `Range.Address` property, which indicates the cells currently selected.

```
Me.txtAddress = rg.Address
```

The Range.Address property has this syntax:

```
Range.Address((RowAbsolute, ColumnAbsolute, ReferenceStyle, External, RelativeTo)
```

In this code:

RowAbsolute: This is optional; it is a Boolean value indicating whether the row part of the reference must be an absolute reference. The default value is True.

ColumnAbsolute: This is optional; it is a Boolean value indicating whether the column part of the reference must be an absolute reference. The default value is True.

ReferenceStyle: This is optional; it indicates the reference style to be used: xlA1 or xlR1C1. The default value is xlA1.

External: This is optional; it indicates whether the reference must be local or external. Use True to return an external reference or False to a local reference. The default value is False.

RelativeTo: This is optional; it must be a Range object that defines the starting point to relative references when RowAbsolute and ColumnAbsolute are False and when ReferenceStyle=xlR1C1.

So, whenever the procedure uses rg.Address, it returns the range address of the selected cells using absolute references (look to the Range Address text box of Figures 5-3 and 5-4).

To return how many cells are used by the selected range, you can use the Count or CountLarge property of the Range object.

```
Me.txtCellsCount = rg.CountLarge
```

To return how many rows and columns are selected, use the Rows or Columns collection's Count or CountLarge property of the Range object.

```
Me.txtRowsCount = rg.Rows.CountLarge
Me.txtColumnsCount = rg.Columns.CountLarge
```

▒ **Attention** The CountLarge property was introduced in the Excel object model for the .xlsx file of Excel 2007 or newer versions, which has 1,048,576 rows x 1,024 columns = 1,073,741,824 possible cells (compared to the 65,536 rows x 256 columns = 16,777,216 possible cells of Excel 2003 or older versions). Excel 2003 and older versions don't recognize the CountLarge property, while Excel 2007 or newer versions will raise an error if you use the Count property to select more rows, columns, and cells than available to Excel 2003.

Since UpdateInterface( ) returns the selected range values, it has a provision to avoid selecting more than 1,000 cells using the CountLarge property (since the user can select the gray square at the right of column A header—above the row 1 header—to select all worksheet cells).

```
If rg.Cells.CountLarge > 1000 Then
    MsgBox "Too much cells selected!", vbCritical, "Select less cells"
Else
```

Getting the Rows Used by the Selected Range

If fewer than 1,000 cells are selected, every row number selected is retrieved using the Range.Row property and stored in the strRows String variable. The procedure uses a For Each...Next loop to run through all selected rows inside the Range.Rows property (which behaves like a collection).

```
For Each varItem In rg.Rows
```

Inside the For Each... Next loop, the fComma Boolean variable is used to verify the need to add a comma to the strRows String variable, which is needed before the second row number is added. Note that fComma becomes true after the first row is processed.

```
    If fComma Then
       strRows = strRows & ", "
    End If
    strRows = strRows & varItem.Row
    fComma = True
Next
```

Getting the Columns Used by the Selected Range

When the first For Each...Next loop ends, the procedure turns fComma = False and begins a second For Each... Next loop through the Range.Columns property (which also behaves like a collection) to retrieve all the columns used by the selected range.

```
fComma = False
For Each varItem In rg.Columns
    If fComma Then
       strColumns = strColumns & ", "
    End If
```

The Range.Column property returns the column number, and you need to turn this value into a column letter, which is made by the Function ColumnNumberToLetter( ) procedure before storing it into the strColumns string variable:

```
    strColumns = strColumns & ColumnNumberToLetter(varItem.Column)
    fComma = True
Next
```

Changing a Column Number to a Letter

There are a lot of algorithms on the Internet destined to transform the Range.Column Integer value to the associated column letter, but you don't need them. Use the Application.Cells( ) property instead, which returns a Range object for a given row and column number and has this syntax:

```
Application.Cells(RowIndex, ColumnIndex)
```

In this code:

RowIndex: This is required; it is a long integer positive value indicating the row number of the reference.

ColumnIndex: This is required; it is a long integer positive value indicating the column number of the reference.

Since the Cells( ) property returns a range object, use the Range object's Address property to return the cell address for row = 1 and the desired column number, using this syntax:

```
strColumn = Cells(1, rg.Column).Address(False, False)
```

Once the address is returned, just take off the 1 row number at the end of the address and you will get the column letters. Note that the Address property is using False for its RowAbsolute and ColumnAbsolute arguments, forcing the address to be returned as a relative reference.

This is the code used by the Function ColumnNumberToLetter( ) of the basColumnNumberToLetter module:

```
Public Function ColumnNumberToLetter(Optional ColumnNumber As Variant) As String
    Dim strColumn As String

    If IsEmpty(ColumnNumber) Then
        ColumnNumber = Application.Selection.Column
    End If

    strColumn = Application.Cells(1, ColumnNumber).Address(False, False)
    ColumnNumberToLetter = Left(strColumn, Len(strColumn) - 1)
End Function
```

Did you get it? The ColumnNumberToLetter( ) Function declares the optional ColumnNumber as Variant argument. If the argument is missing, it receives the Application.Selection.Column property, which returns the column number of the selected cell in the Excel interface.

```
    If IsEmpty(ColumnNumber) Then
        ColumnNumber = Application.Selection.Column
    End If
```

It then uses the Application.Cells( ) property for the first row and desired column, which returns a Range object, and then uses the Range.Address property to return the relative reference to the desired cell.

```
strColumn = Application.Cells(1, ColumnNumber).Address(False, False)
```

The Application.Cells property returns a range object
... and the Address(False, False) Property of the range object returns a relative reference to the desired cell

The address returned will be relative to row 1, so the next instruction uses the VBA Left( ) and Len( ) functions to get just the column letters, which are the value returned by the function.

```
    ColumnNumberToLetter = Left(strColumn, Len(strColumn) - 1)
End Function
```

Considerations About the Range Rows and Columns Properties

Before we continue, you must be aware of some discrepancies about the Range Rows and Columns properties, which behave like a collection.

- The Count property of both the Rows and Columns properties return the row/column count for the first range, if more than one noncontiguous ranges are selected.
- Both Rows and Columns properties may return repeated values of row and column numbers used by the selected ranges if you select more than one noncontiguous range that uses the same rows and columns.

Look again to Figures 5-3 and 5-4 and see for yourself. Figure 5-3 selects just the cells in the range A5:B6, while Figure 5-4 selects many other cell ranges. But the count value of both the Range.Rows and Columns properties continue to refer to cells A5:B6, while the selected row numbers and column letters have many duplicates.

Getting a Cell's Address and Values for the Selected Range

Use the Range.Address(False, False) and Range.Value properties to get the relative cell addresses and values of the selected range, performing a For Each...Next loop through all cells of the Range object (note that the varItem Variant variable retrieves each cell in the range, which is also a Range object).

```
fComma = False
For Each varItem In rg
    If fComma Then
       strValues = strValues & ", "
    End If
    strValues = strValues & varItem.Address(False, False) & "=" & varItem.Value
    fComma = True
Next
```

And once all row numbers, column letters, and cell addresses and values have been retrieved, update the UserForm interface.

```
    Me.txtRowsSelected = strRows
    Me.txtColumnsSelected = strColumns
    Me.txtValue = strValues
End If
```

Getting Selected Range Areas

The UpdateInterFace( ) procedure ends by using the Range.Areas collection to know how many different noncontiguous ranges have been selected and the addresses of each one.

The Range.Areas.Count property is used to inform the Areas.Count property and to enable the cmdResize button (which can just be used when a single contiguous range is selected).

```
Me.txtRangeAreas = rg.Areas.Count
Me.cmdResize.Enabled  = (rg.Areas.Count = 1)
```

The lstAreas ListBox is used to show each Range.Area Index value and its associated Address property. So, it first clears the lstAreas ListBox with its Clear method and then performs a For...Next loop through all areas of the Range.Areas collection using intI as the Areas counting.

```
Me.lstAreas.Clear
For intI = 1 To Selection.Areas.Count
```

The lstAreas value has two columns. The Area Index (represented by the intI variable) is added to the lstAreas item value using its AddItem method.

```
Me.lstAreas.AddItem intI
```

The Areas.Address property is added to the second column of the same item of the lstAreas ListBox using intI to reference the desired area and using the lstAreas.ListCount-1 property to correctly reference the last added item.

```
        Me.lstAreas.Column(1, lstAreas.ListCount - 1) = rg.Areas(intI).Address
    Next
End Sub
```

Resizing the Selected Range

Use the Range.Resize property to add or delete cells rows and/or columns selected by the range. It has this syntax:

```
Range.Resize(RowSize, ColumnSize)
```

In this code:

> RowSize: This is optional; it is a long integer positive value indicating the number of rows on the new range. If omitted, the number of rows remains the same.
>
> ColumnSize: This is optional; it is a long integer positive value indicating the number of columns on the new range. If omitted, the number of rows remains the same.

When you select just one range and click the Resize CommandButton (cmdResize) of the frmRange UserForm, you will accept the values defined on the resized Rows and Columns text boxes (txtAddRows and txtAddColumns, usually = 1) to resize the selected range, executing this code:

```
Private Sub cmdResize_Click()
    Dim intRows As Integer
    Dim intCols As Integer

    intRows = Selection.Rows.Count + CInt(Me.txtAddRows)
```

```
    intCols = Selection.Columns.Count + CInt(Me.txtAddColumns)
    If intCols <= 0 Then intCols = 1
    Selection.Resize(intRows, intCols).Select
End Sub
```

Note that the procedure declares the intRows and intColumns variables to hold the new range row and column values. It uses the Application.Selection property (without the Application qualifier) to return the selected range and uses the Range.Rows.Count property to return how many rows are selected. It then adds the txtAddRows value to the row count selected, using the VBA CInt( ) function to convert the text box String value to Integer:

```
intRows = Selection.Rows.Count + CInt(Me.txtAddRows)
```

Since the txtAddRows value can be negative, the procedure verifies whether the intRows value is equal to or lower than zero. If it is, it is turned to 1 (meaning that the resized range will have at least one row).

```
If intRows <= 0 Then intRows = 1
```

The same steps are repeated to the intCols variable, and the resize method is executed. Note that when this is done, the new Range object returned by the Selection.Resize property is resized, but it is not selected on the worksheet. So, the Range.Selection property is also executed, so the new defined range is selected.

```
Selection.Resize(intRows, intCols).Select
```

When this is done, the "cascading event" phenomenon happens again, firing the Application_SheetSelectionChange( ) event, captured by the mApp module-level variable, which will call the UpdateInterface( ) procedure and update the UserForm interface (Figure 5-5).

```
Private Sub mApp_SheetSelectionChange(ByVal Sh As Object, ByVal Target As Range)
    Call UpdateInterface(Sh, Target)
End Sub
```

▒ **Attention** If Worksheet_SelectionChange( ) or ThisWorkbook_SheetSelectionChange( ) are programmed, they will also fire at this moment.

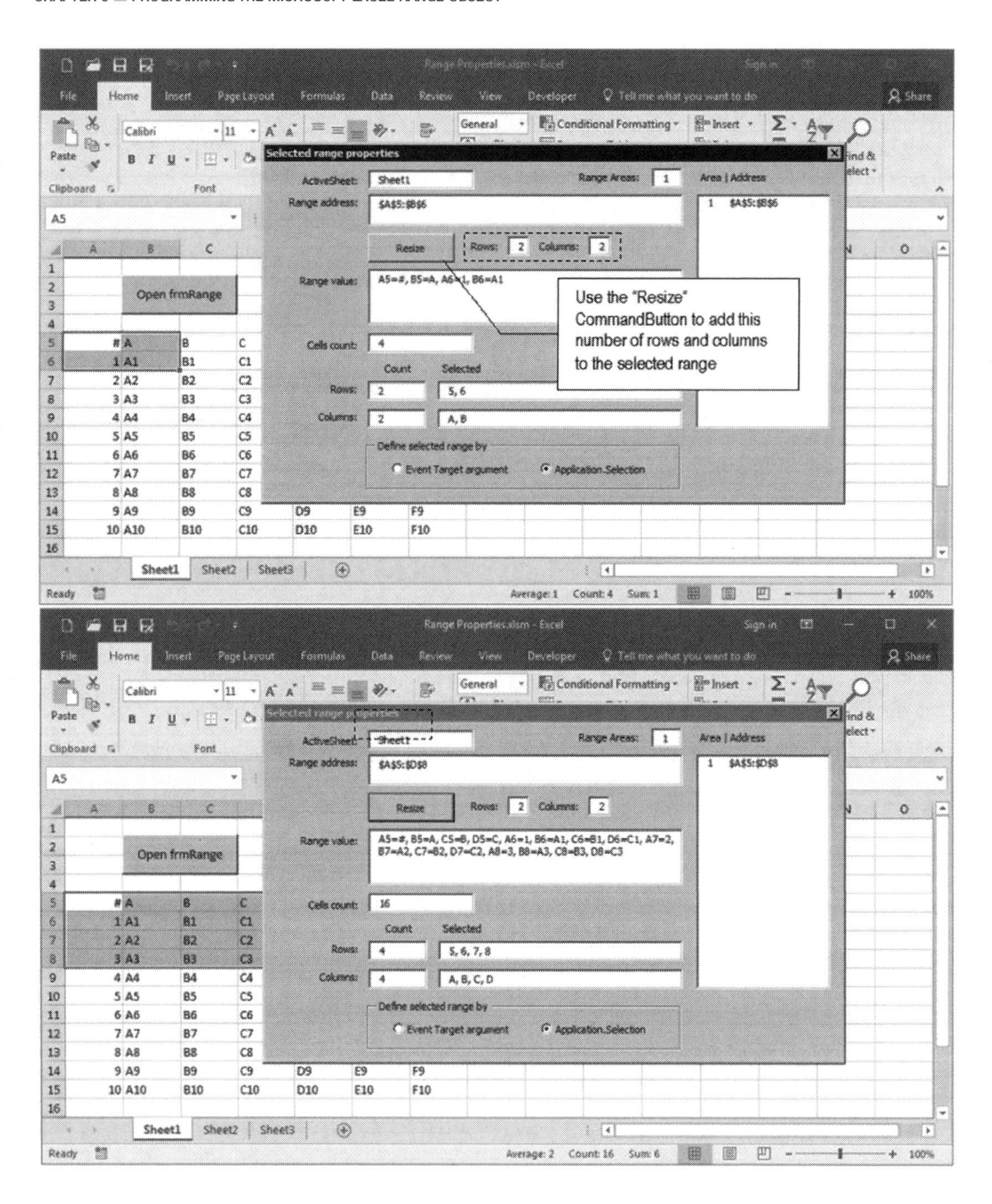

***Figure 5-5.** Use the Range.Resize property to resize the selection (or any other Range variable). Remember that after the resize operation, if you want to see the new range selected on the worksheet, you must use the Range.Select property (you can use negative values on the Rows and Columns text boxes at the right of the button)*

Changing the Range Reference

By default the frmRange UserForm uses the Application.Selection property to return the selected range. This is necessary because when frmRange is opened, it fires the UserForm_Initialize( ) event, which doesn't have any argument and needs to catch the Selection range to update its interface.

```
Private Sub mApp_SheetSelectionChange(ByVal Sh As Object, ByVal Target As Range)
    Call UpdateInterface(Sh, Target)
End Sub
```

When you click the "Event Target argument" (optEventTarget) option, the optEventTarget_Click( ) event fires, executing this code:

```
Private Sub optEventTarget_Click()
    Dim rg As Range

    Set rg = Selection
    Application.EnableEvents = False
        Range("A1").Select
    Application.EnableEvents = True
    rg.Select
End Sub
```

Note that the code declares and uses the rg as Range variable to hold the current selection range, returned by the Application.Selection property.

```
Private Sub optEventTarget_Click()
    Dim rg As Range

    Set rg = Selection
```

Now that the selected range is stored, to force the Application_SheetSelectionChange( ) event to fire you must make a fake selection and then select again what is stored on the rg variable. To avoid that the cascade event fires twice, the code uses the Application.Enabled Event property to disable Excel events, selects cell A1, enables the property again, and then reselects the current range, firing the cascading event.

```
Application.EnableEvents = False
   Range("A1").Select
Application.EnableEvents = True
rg.Select
```

When mApp_SheetSelectionChange( ) fires, it will pass the Sh and Target arguments to the UpdateInterface( ) procedure, which will use them to synchronize the UserForm interface.

```
Private Sub mApp_SheetSelectionChange(ByVal Sh As Object, ByVal Target As Range)
    Call UpdateInterface(Sh, Target)
End Sub
```

Using the Names Collection

Great worksheet applications need named ranges that you set using the Excel interface or VBA. All named ranges are stored on the `Names` collection and exist as different entities for both the `Workbook` and `Worksheet` objects: application scope named ranges must be stored on the `Workbooks.Names` collection, while worksheet scope named ranges must be stored on the `Worksheets.Names` collection. Each collection must have unique named ranges, but a `Workbook` named range can exist with the same name on any `Worksheet.Names` collection, representing different cell ranges.

Each named range must follow this syntax:

- The first character of a name must be a letter, an underscore character (_), or a backslash (\). Remaining characters in the name can be letters, numbers, periods, and underscore characters.
- You cannot use the uppercase and lowercase characters *C*, *c*, *R*, or *r* as a defined name because they are all used as shorthand for selecting a row or column for the currently selected cell when you enter them in the Excel Name or Go To text box.
- Names cannot be the same as a cell reference, such as `A$1` or `R1C1`.
- Spaces are not valid. Use the underscore character (_) or period (.) as word separators instead.
- The name length can contain up to 255 characters.

Using the Excel interface, you create a workbook scope named range by first selecting the cell range and then entering the range name in the Excel Names `ListBox`. To create a worksheet scope named range, precede the name with the sheet tab and an exclamation point (like `Sheet1!Test`, as shown in Figure 5-6).

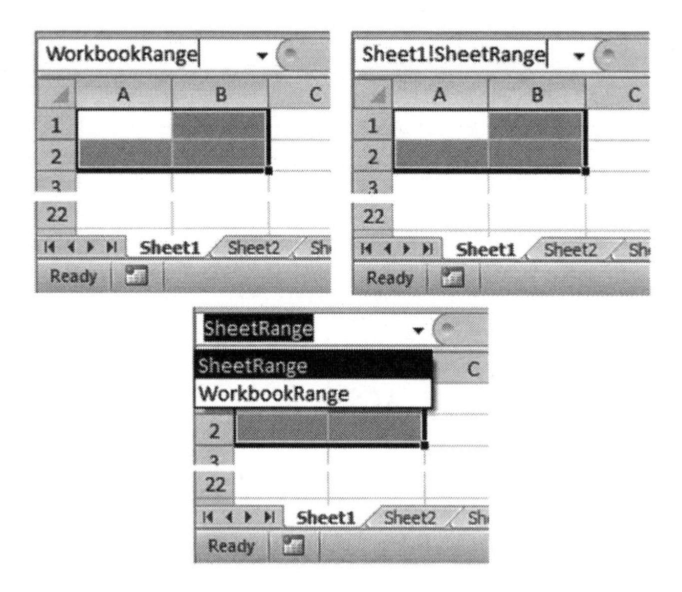

Figure 5-6. *Use the Excel Names ListBox to create workbook or worksheet scope named ranges. To create a worksheet named range, precede the name with the sheet tab and an exclamation point*

Note that Figure 5-6 shows you can create names with different scopes that point to the same cells (both `WorkbookRange` and `SheetRange` point to cells `A1:B2`). Whenever you do this and select these cells range on the sheet tab, Excel will always refer to the worksheet name.

To create a named range with VBA code, use the `Workbook` or `Worksheet` object's `Names.Add` method, which has this syntax:

```
Object.Add(Name, RefersTo, Visible, MacroType, ShortcutKey, Category, NameLocal,
RefersToLocal, CategoryLocal, RefersToR1C1, RefersToR1C1Local)
```

In this code:

`Name`: This is required; it is the range name to be created.

`RefersTo`: This is required; it is the range address that the name refers to, using A1-style notation, if the `RefersToLocal`, `RefersToR1C1`, and `RefersToR1C1Local` parameters are not specified.

`Visible`: This is optional; it is the `Boolean` value that specifies whether the name is visible (hidden names do not appear in the Define Name, Paste Name, or Goto dialog box). The default value is `True`.

`MacroType`: This is optional ; it associates the name with a macro using the following values:

- 1: User-defined function (`Function` procedure).
- 2: Macro (`Sub` procedure).
- 3 or omitted: This is the default value; the name does not refer to a user-defined function or macro.

`ShortcutKey`: This is optional; it specifies the macro shortcut key. It must be a single letter, such as *z* or *Z*. This applies only for command macros.

`Category`: This is optional; it is the category of the macro or function if the `MacroType` argument is defined to 1 or 2. The category is used in the Function Wizard.

`NameLocal`: This is optional; it specifies the localized text to use as the name if the `Name` parameter is not specified.

`RefersToLocal`: This is optional; it describes what the name refers to, in localized text, using A1-style notation, if the `RefersTo`, `RefersToR1C1`, and `RefersToR1C1Local` parameters are not specified.

`CategoryLocal`: This is optional; it specifies the localized text that identifies the category of a custom function if the `Category` parameter is not specified.

`RefersToR1C1`: This is optional; it describes what the name refers to, in English using R1C1-style notation, if the `RefersTo`, `RefersToLocal`, and `RefersToR1C1Local` parameters are not specified.

`RefersToR1C1Local`: This is optional; it describes what the name refers to, in localized text using R1C1-style notation, if the `RefersTo`, `RefersToLocal`, and `RefersToR1C1` parameters are not specified.

The Names.Add method returns a Name object referencing the newly added named range. And although it has a lot of arguments, you need to use just the Name and RefersTo arguments to create or edit an existing name.

You can use the Names collection's Add method to add or modify an existing range name. If the name is new, Excel will create it; if it already exists, Excel will change the RefersTo property to whatever you type in the Names.Add RefersTo argument.

To add (or modify) the WorkbookRange range name indicated in Figure 5-6 (a workbook scope range name) associated to Sheet1 cells A1:B2, use one of these syntaxes in the VBA Immediate window. Here's an example:

```
?ThisWorkbook.Names.Add("WorkbookRange", "=A1:B2")
```

Here's the other example:

```
?Names.Add("WorkbookRange", "=A1:B2")
```

To add (or modify) the SheetRange range name indicated in Figure 5-6 (a worksheet scope range name) associated to cells A1:B2, use this syntax in the VBA Immediate window:

```
?Sheet1.Names.Add("SheetRange", "=A1:B2")
```

Or use this syntax:

```
?Names.Add("Sheet1!SheetRange", "=A1:B2")
```

Did you notice that you can use just the Names collection's Add method to insert both workbook or worksheet range names? To add or modify any worksheet scope name, just precede its name with the sheet name followed by an exclamation point and Excel will add it to the desired sheet Names collection.

■ **Attention** Always precede the cell reference by the equal sign so Excel can interpret it as a cell range.

If the worksheet name contains the space character, you must enclose it in single quotes before using the Names.Add method (like you must do using the Excel Name box). For example, to insert the SheetRange named range into the My Sheet worksheet's Names collection pointing to cells A1:B2, use this syntax:

```
?Names.Add("'My Sheet'!SheetRange", "=A1:B2")
```

Hiding Named Ranges

The same way Excel does with the Worksheet object, the Name object has a Visible property, which is also an argument of the Names.Add collection, that you can set to False to hide the named range from the user view. There is no way to do this operation using the Excel interface.

Once you have created a range name, you can change the Visible property using the VBA Immediate window with this syntax (note that you should not use the ? character in the VBA Immediate window so the Name.Visible property can be changed).

```
Names("WorkbookRange").Visible = False
```

And once you do this, the name does not appear anymore in the VBA Name box or the Name Manager dialog box (Figure 5-7).

To hide or show all named ranges used by your Excel application from user eyes, create a single procedure that receives two arguments: the sheet name and a Boolean argument that indicates what you want to do, such as the Sub HideRangeNames( ) procedure from the basHideRangeNames.bas file that you can also extract from the Chapter02.zip file.

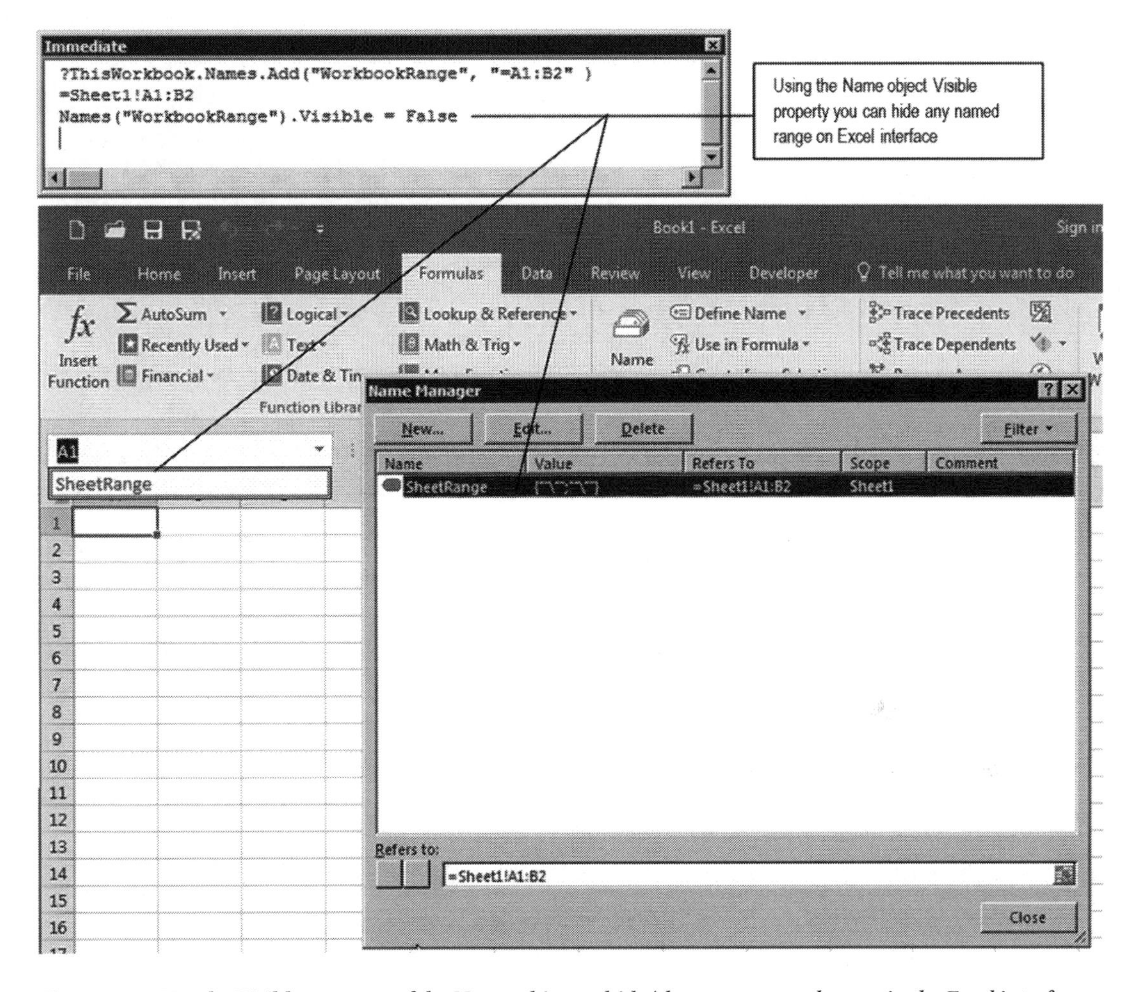

Figure 5-7. Use the Visible property of the Name object to hide/show any named range in the Excel interface, either in its Name box or in the Name Manager dialog box

```
Public Function HideRangeNames(strWorksheet As String, fShow As Boolean)
    Dim nm As Name

    For Each nm In Worksheets(strWorksheet).Names
        nm.Visible = fShow
    Next
End Function
```

To hide all Sheet1 worksheet names, use the VBA Immediate window this way (don't use the ? character):

```
HideRangeNames("Sheet1", False)
```

To show again all Sheet1 range names, call the procedure again using True for the second argument.

```
HideRangeNames("Sheet1", True)
```

Resizing Named Ranges

Another operation that you will need to perform on your Excel applications is to resize an existing range name, inserting or deleting rows and columns on its RefersTo property. And you can do this in two different ways.

- Using the Names.Add method to re-create the existing range name with a new cell range reference
- Using the Range.Resize property to resize a given range and then using the Range object Name property to attribute a name to the resized range

The Range.Resize property has this syntax:

```
Range.Resize(RowSize, ColumnSize)
```

In this code:

RowSize : This is optional; it is the number of rows in the resized range. If this argument is omitted, the number of rows in the range remains the same.

ColumnSize: This is optional; it is the number of columns in the new range. If this argument is omitted, the number of columns in the range remains the same.

Note that the Range.Resize property specifies the total number of rows and columns that the range must have. It will resize the desired range regarding its top-left cell as the first range cell.

The next operation resizes the range returned by the WorkbookRange named range created in Figure 5-6 that begins on cell A1, so it now has five rows and ten columns. Since it returns a Range object, you cannot perform this in the VBA Immediate window.

```
Range("WorkbookRange").Resize(5,10)
```

The last instruction *does not* change the WorkbookRange named range reference. Just the range returned by this operation has changed, resizing it to five rows and ten columns. To see this resizing operation in action, use the Range.Select method, typing this syntax in the VBA Immediate window (note again that it does not use the ? print character, as shown in Figure 5-8):

```
Range("WorkbookRange").Resize(5,10).Select
```

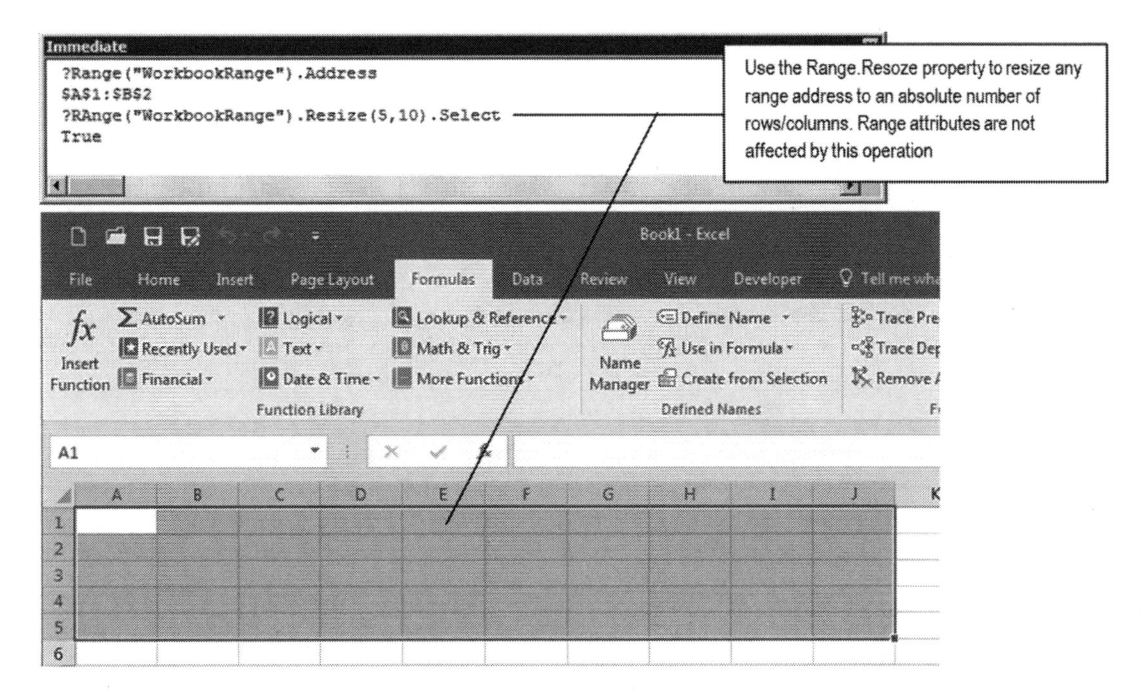

Figure 5-8. *Use the Range.Resize property to resize the address returned by a given named range to an absolute number of rows and columns (keeping the original top-left cell range). To select the resized range, use the Range.Select method*

Most operations you will do to change an existing named range size will be made by adding or deleting one or more rows or columns in the current reference. To do this, you need to specify the `Range.Resize` method's `RowSize` and `ColumnSize` arguments using the range's `Rows.Count` and `Columns. Count` properties as default values that must be added by a positive or negative number of rows/columns.

The next instruction will change the range returned by the `WorkbookRange` named range by adding just one row to its current reference, keeping it with the same column count (note that it uses the `Range.Resize` method's `RowSize` argument, as shown in Figure 5-9):

```
Range("WorkbookRange").Resize(Range("WorkbookRange").Rows.Count + 1).Select
```

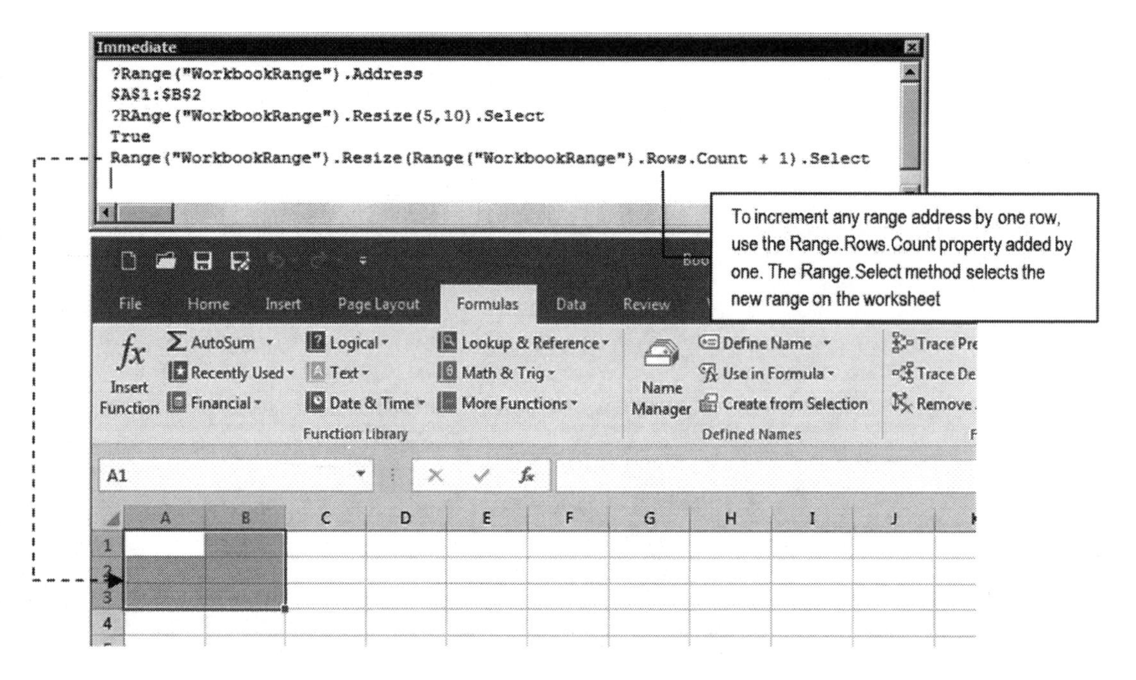

Figure 5-9. *To resize the address returned by a given named range by adding to it a defined number of rows/ columns, use the Range.Rows.Count or Range.Columns.Count property to return the current number of rows/ columns, adding or subtracting an integer to resize them to the desired size. This figure shows how to add one row to the range address returned by the WorkbookRange named range created in Figure 5-6*

To change any named range address by adding to it any number of rows or columns, use the `Range.Resize` operation to resize it and then use the `Range.Name` property to name the new resized range to the desired saved range.

The next instruction resizes the `WorkbookRange` named range by adding to it two rows, keeping it as the same column counting (use the Excel Name box to see the result, as shown in Figure 5-10):

```
Range("WorkbookRange").Resize(Range("WorkbookRange").Rows.Count + 2).Name = "WorkbookRange"
```

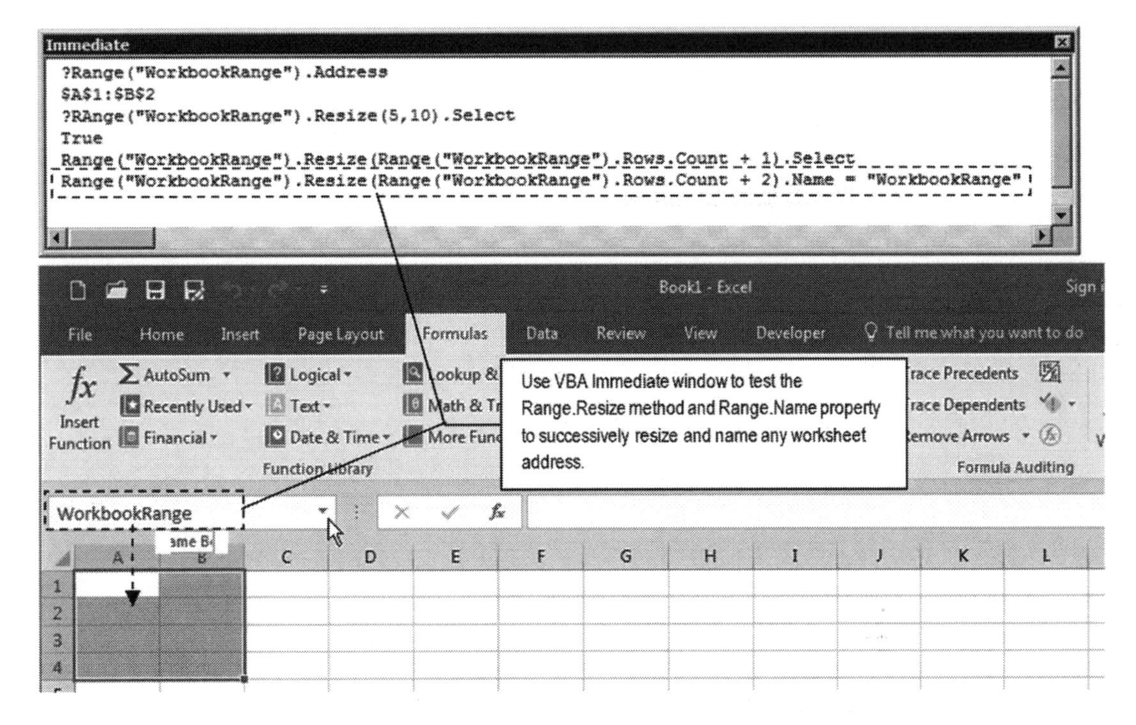

Figure 5-10. *To really resize a saved range, use the Range.Resize method to resize the address and then use the Range.Name property to name the resized range. Once the operation is done, use the Excel Name box to select the range and watch the result. This figure shows how to resize the WorkbookRange (A1.B2) created in Figure 5-6 by adding two more rows (resulting in A1:B4)*

Now that you already have a good understanding of how to do some specific `Name` object operations, let's see all these `Name` object properties and methods in action!

Using Name Object Properties and Methods

Microsoft Excel 2007 or newer versions use the Name Manager command of the "Define names" area of the Formula tab of the ribbon to control the range names you create on any workbook. This command raises the Name Manager dialog box (which is like a `UserForm` object), which is where you can add, edit, or delete range names; filter range names by scope; or change the selected range name address (Figure 5-11).

To add or edit range name details, the Name Manager counts, with the aid of the New/Edit Name dialog box (another `UserForm`), and allows you to select the name scope and add a comment to any range name.

Note the interface behavior. When you click the New/Edit CommandButton of the Name Manager dialog box, it disappears from the screen and loads the New Name dialog box. And if you insert a new name or edit an existing one, when you close the New Name dialog box, the inserted/edited name is selected in the Name Manager dialog box, which shows most of its properties.

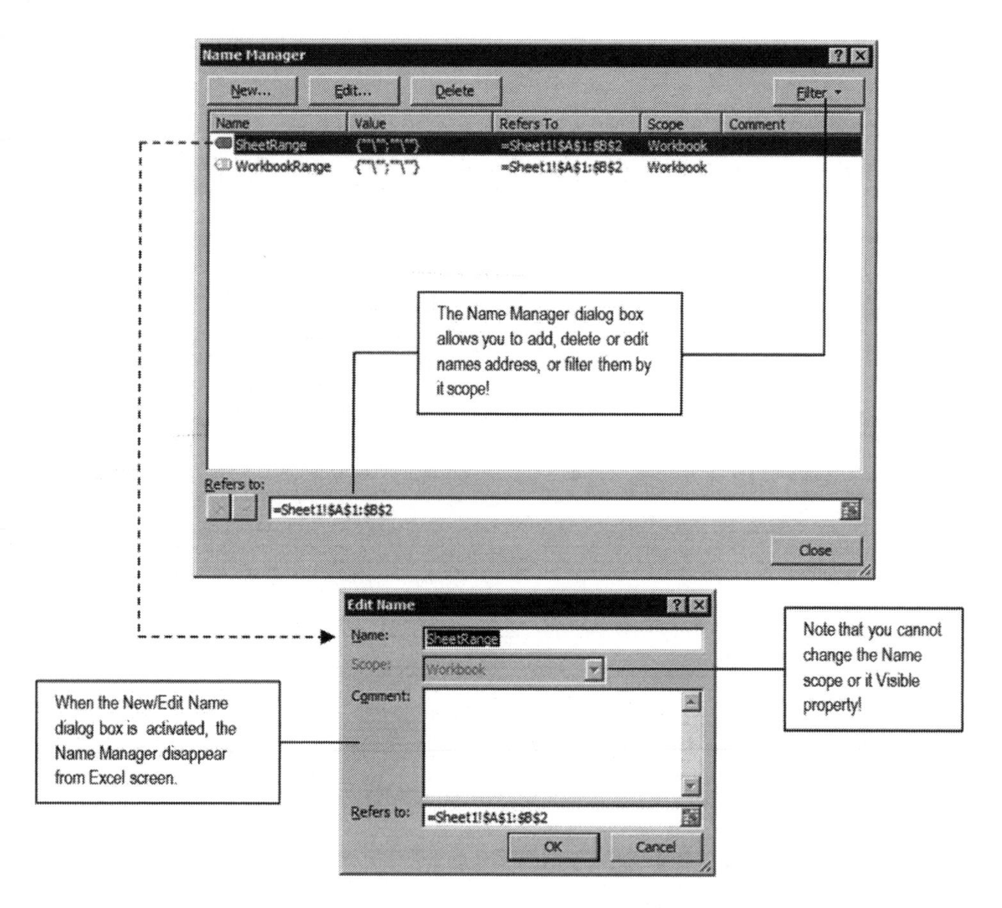

Figure 5-11. *This is the Excel Name Manager dialog box that uses the New Name dialog box to insert/edit range names of any opened workbook. It does not have the ability to hide range names*

■ **Attention** Both the Name Manager and New Name dialog box don't offer any means by which you can hide a range name. And once a range name is created, you also cannot change its scope from workbook to worksheet, and vice versa. To do such operations, you must delete and re-create the range name.

The aim of this section is to try to duplicate the Name Manager behavior with some improvements that can help you to manage the range names of your workbook application, as follows:

- Hide/show one or more range names
- Change the range name scope
- Apply the same comment to one or more range names at once

■ **Attention** Before you begin to read this section, I want reiterate that what you are about to study was not result of chance. It took hard work, study, practice, and experimentation of Excel VBA programming, using a trial-and-error approach that took me many days to finish. When you begin to create your own solutions, remember that to build solid, good, and reliable software, you must access many different knowledge sources and strive to perfection to achieve the desired results.

To see the Excel `Names` collection and `Name` object in action, extract the `Names Collection.xlsm` macro-enabled workbook from the `Chapter05.zip` file, which uses two `UserForms` to improve the Excel Name Manager dialog box: `frmNames` (to mimic the Excel Name Manager) and `frmEditName` (to mimic the New/Edit Name dialog box), as shown in Figure 5-12.

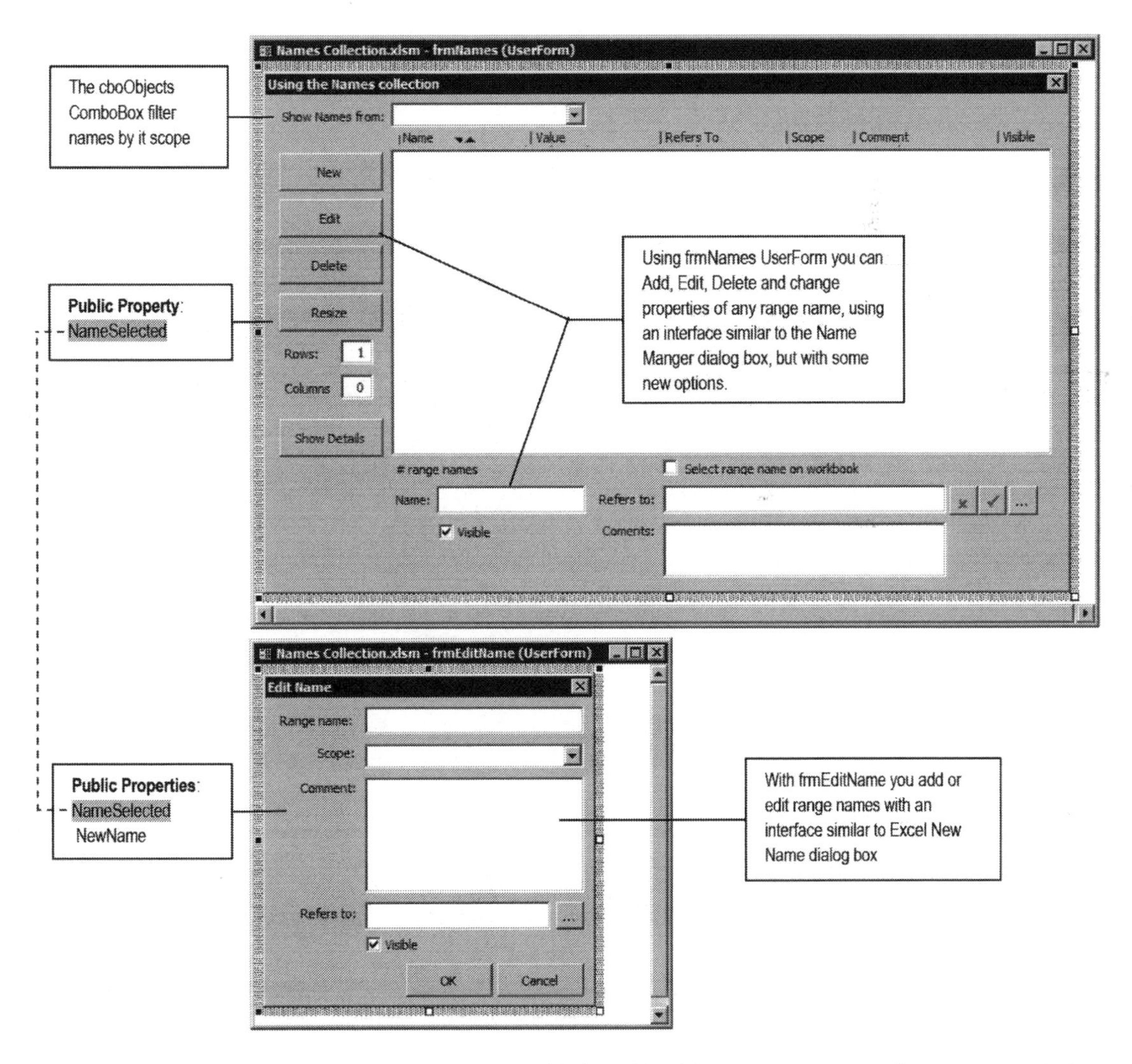

Figure 5-12. *The Names Collection.xlsm macro-enabled workbook has two UserForms: frmNames to mimic the Name Manager dialog box and frmEditName to mimic the New Name dialog box*

Whenever you need to open two successive modal UserForms, the first UserForm must hide itself using the Userform.Hide method before showing the second modal UserForm, which will stop the code at this point until the second modal UserForm can be unloaded, and when this happens, the first UserForm must apply to itself to the UserForm.Show method to show again its interface.

To keep the frmNames and frmEditName UserForm interfaces synchronized, you must use Public Properties declared on both UserForms so one can synchronize the other interface to mimic the way the Excel Name Manager and New/Edit Name dialog boxes behave.

- The frmNames.NameSelected and frmEditName.NamesSelected properties set the connection between the two UserForms regarding the Name object that is being created or edited.
- The frmEditName.NewName property is used to signal the insertion of a new name.

Whenever you open the Names Collection.xlsm macro-enabled workbook, the frmNames UserForm is shown to you, ready to deal with most Excel Name object properties and methods. It synchronizes its interface using two main Sub procedures.

- Sub FilllstNames( ): To fill the lstNames ListBox with current Name object information
- Sub DefineControls( ): To synchronize the Enabled property of most frmNames controls regarding what is currently selected in the lstNames ListBox

Figure 5-13 shows how frmNames should look when you open the Names Collection.xlsm workbook for the first time and no name has been created.

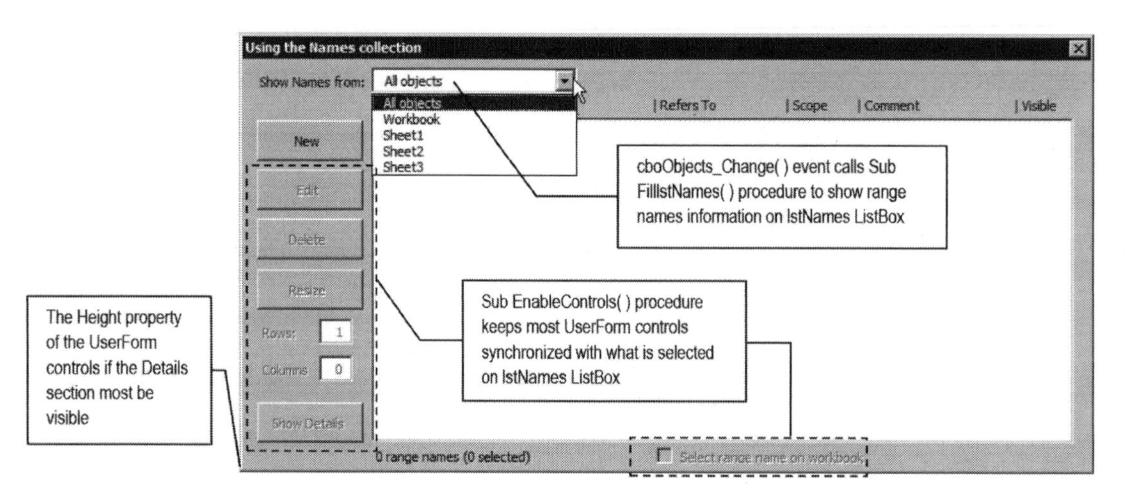

Figure 5-13. *When you open the Names Collection.xlsm macro-enabled workbook, frmNames is shown by the ThisWorkbook.Workbook_Open event*

Once more, frmNames is automatically opened by the ThisWorkbook.Workbook_Open( ) event, which has this code:

```
Private Sub Workbook_Open()
    frmNames.Show
End Sub
```

When the UserForm opens, its UserForm_Initialize( ) event fires, adding two main options to the cboObjects ComboBox ("All objects" and "Workbook") and the name of each sheet tab of the active workbook using a For Each...Next loop to run through all objects of the Application.Worksheets collection.

```
Private Sub UserForm_Initialize()
    Dim varItem As Variant

    mintLastColumn = 1
    Me.cboObjects.AddItem "All objects"
    Me.cboObjects.AddItem "Workbook"
    For Each varItem In Worksheets
        Me.cboObjects.AddItem varItem.Name
    Next
    Me.Height = mconHeight1
    Me.cboObjects.ListIndex = 0
End Sub
```

You should note three important things that happen on the frmNames UserForm_Initialize( ) event.

- It Height property is changed to the mconHeight1 constant, making it become a small vertical dimension and hiding its Details section (compared to frmNames in Figures 5-12 and 5-13).
- The last instruction, which sets cboObject.ListIndex = 0, will cascade-fire the cboObjects_Change( ) event, filling the lstNames ListBox with all range name information by calling the Sub FilllstNames( ) procedure.
- After the cboObjects_Change( ) event is fired, the UserForm interface is synchronized, and most controls become unavailable because of the action's Sub EnableControls( ) procedure.

The cboObjects ComboBox is responsible for defining the scope of the range names that the frmNames must show: The "All objects" option will return all range names from the Application.Names collections (meaning all Workbook and Worksheet scope range names); the "Workbook" option will return just Workbook scope range names, and by selecting any sheet tab name in the list, only the selected Worksheet object range names will be show.

Whenever the user—or the code—changes the cboObjects ComboBox value, the cboObjects_Change( ) event will fire, executing this code:

```
Private Sub cboObjects_Change()
    Call FilllstNames
    Call lstNames_Change
End Sub
```

Recovering Name Object Properties

Whenever another name scope is selected on the cboObjects ComboBox, a new Name object is inserted, or any Name property is changed, the code needs to call Sub FilllstNames( ) to clear and fill lstNames ListBox with the current Name object information. To keep this procedure as short as possible, it uses Function GetNameValue( ) to return the ListBox's Value column using the Excel Name Manager style, which also uses Function EvaluateRange( ) to eventually return a string with the associated error code exhibited by any formula or cell (Figure 5-14).

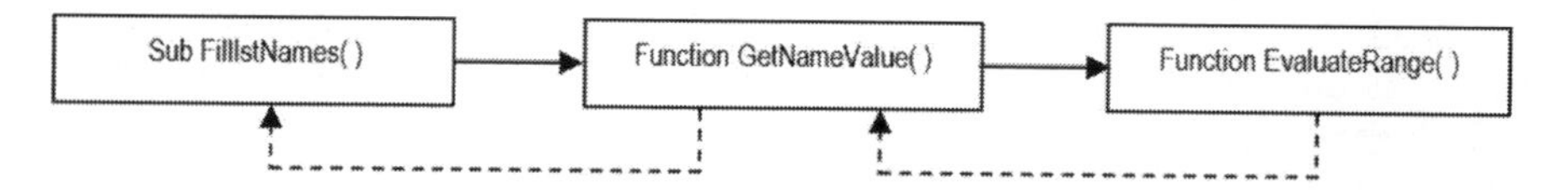

Figure 5-14. *To keep the Sub FilllstNames( ) as short as possible, it was divided by two other specialized Function procedures. GetNameValue( ) must return the name value using the same style of Excel Name Manager dialog box, while EvaluateRange( ) is used to return any formula or cell value as is, including a string with an associated error code (if any)*

This is `Public Sub FilllstNames( )` procedure code, executed to clear and fill the `lstNames ListBox` with Name object properties (since it is a `Public` procedure, it is considered as a `frmNames` method):

```
Public Sub FilllstNames()
    Dim nm As Name
    Dim intIndex As Integer
    Dim intPos As Integer
    Const conWorkbook = 1

    intIndex = Me.cboObjects.ListIndex
    With Me.lstNames
        .Clear
        For Each nm In Names
            intPos = InStr(1, nm.Name, "!")
            If (intIndex < conWorkbook) Or _
               ((intIndex = conWorkbook) And (intPos = 0)) Or _
               ((intIndex > conWorkbook) And (nm.Parent.Name = Me.cboObjects)) Then
                .AddItem nm.Name
                .Column(1, .ListCount - 1) = Mid(nm.Name, intPos + 1)
                .Column(2, .ListCount - 1) = GetNameValue(nm)
                .Column(3, .ListCount - 1) = nm.RefersTo
                .Column(4, .ListCount - 1) = IIf(nm.Parent.Name = ThisWorkbook.Name,
"Workbook", nm.Parent.Name)
                .Column(5, .ListCount - 1) = nm.Comment
                .Column(6, .ListCount - 1) = IIf(nm.Visible, "Yes", "No")
            End If
        Next
    End With
End Sub
```

`Sub FilllstNames( )` declares all the variables it needs plus the `Const conWorkbook = 1` constant to avoid the appearance of a "magic number" inside the code.

```
Private Sub FilllstNames(Optional varListIndex As Variant)
   ...
    Const conWorkbook = 1
```

It then uses the `intIndex Integer` variable to reference the `cboObjects.ListIndex` property just once (the item selected on the `cboObjects ComboBox`), defines a `With Me.lstName...End With` loop to also reference the `lstName` ListBox only once, and clears the `ListBox` by calling the `Clear` method.

```
    intIndex = Me.cboObjects.ListIndex
    With Me.lstNames
        .Clear
```

Once the lstNames ListBox is cleared, it begins a For Each nm in Names...Next loop to run through all names stored in the Application.Names collection, using the nm as Name object variable to easily reference each Name object.

```
Dim nm as Name
...
    For Each nm In Names
```

Since any worksheet name is preceded by the sheet name and an exclamation character, FilllstNames( ) verifies the current Name object scope by searching for an exclamation character (!) on its Name property using VBA InStr( ) function.

```
        intPos = InStr(1, nm.Name, "!")
```

Once this is done, the intPos Integer variable will receive a 0 as an indication of the absence of the ! character (meaning the workbook name scope) or the position of the ! inside the Name property (worksheet named scope).

Then it must make the decision to insert the name in the lstNames ListBox according to three possible scopes selected on the cboObjects ListBox: all objects, workbook names, or any sheet tab names. The name must be inserted if:

1. cboObjects = "All objects", meaning that all names must be inserted. Since this is the first option of the cboObjects ComboBox (ListIndex = 0) and conWorkbook = 1, it makes this test:

   ```
   If (intIndex < conWorkbook) Or _
   ```

2. cboObjects = "Workbook" option, meaning that just workbook scope Name objects should be inserted. Since this is the second option of cboObjects (ListIndex = 1 = conWorkbook), the Name object properties must be inserted if it also does not contain a ! character on its name (intPos = 0):

   ```
   ((intIndex = conWorkbook) And (intPos = 0)) Or _
   ```

3. cboObjects = <SheetTabName> option, meaning that just the selected sheet tab names must be inserted (ListIndex > 1 > conWorkbook) and the selected sheet tab name on cboObjects equals the Worksheet object that the name belongs to, which is given by the Name.Parent.Name property.

   ```
               ((intIndex > conWorkbook) And (nm.Parent.Name = Me.
               cboObjects)) Then
   ```

Look at the VBA properties window and note that the lstNames ListBox was defined with seven columns (ColumnCount = 7) and that its first column is hidden (ColumnWidths = 0 pt;85 pt;...). The first (hidden) column must receive the Name object's Name property as it is: with or without the preceding sheet name for worksheet named scope. The next six columns must show six different Name object properties or information, as their names imply: Name, Value, Refers To, Scope, Comment, and Visible (as indicated in Figure 5-12).

So, the next procedure instruction uses the ListBox AddItem method to add the Name object's Name property to the first lstNames column (hidden):

```
            .AddItem nm.Name
```

Using the ListBox Column property, the procedure begins to add another Name object property to the last item inserted in the list. The second lstNames ListBox column (Column=1) receives the Name.Name property without the sheet name that precedes it, if the name has a worksheet scope. It uses the VBA Mid( ) function to extract the name from the position after the ! character (intPos+1) to the end of the name.

```
.Column(1, .ListCount - 1) = Mid(nm.Name, intPos + 1)
```

■ **Attention** If intPos = 0, meaning that no ! character was found on the name, the Mid(nm.Name, intPos+1) function will return all name characters beginning from the first one.

The second lstNames column must receive the Name object value, or some of it first values if it represents more than one cell, using the same style of the Name Manager dialog box (see Figure 5-11), which lead us to two different situations:

- Name objects that represent a value, like a single cell range or a formula that must be evaluated
- Name objects that represent a range of cells must show values inside braces ({}); each value must be inside double quotes ("") separated by semicolons; row breaks are identified by a backslash (\); the values must reflect what is seen on the worksheet

To keep the procedure smaller, it uses Function GetNameValue( ) to recover the current Name object value.

```
.Column(2, .ListCount - 1) = GetNameValue(nm)
```

Recovering Name Values with GetNameValue( )

The function GetNameValue( ) receives a Name object as an argument and executes this code:

```
Private Function GetNameValue(nm As Name) As String
    Dim rg As Range
    Dim strItem As String
    Dim intI As Integer
    Dim intJ As Integer
    Dim intK As Integer
    Const conMaxItens = 6

    On Error Resume Next

    If Not IsArray(nm.RefersToRange) Then
        GetNameValue = EvaluateRange(nm.RefersTo)
    Else
        Set rg = nm.RefersToRange
        strItem = "{"
        For intI = 1 To rg.Rows.Count
            strItem = strItem & IIf(intI > 1, "\", "")
            For intJ = 1 To rg.Columns.Count
                strItem = strItem & IIf(intJ > 1, ";", "")
                strItem = strItem & Chr(34) & EvaluateRange(rg.Cells(intI, intJ)) & Chr(34)
                intK = intK + 1
```

```
                'Provision to not add more than conMaxItens itens
                If intK >= conMaxItens Then
                    Exit For
                End If
            Next
            If intK >= conMaxItens Then
                Exit For
            End If
        Next
        If intK <= conMaxItens Then
            strItem = strItem & "}"
        Else
            strItem = strItem & "..."
        End If
        GetNameValue = strItem
    End If
End Function
```

After declaring its variables, the procedure disables any VBA raised error using an `On Error Resume Next` instruction and then uses the VBA `IsArray( )` function to verify whether the `Name.RefersToRange` property (which returns a Range object) *does not* refers to multiple cells. If this is true, the returned range refers to a single cell or to a `Name` constant formula (which does not return a `Range` object, raising a VBA error), and `Name.RefersToProperty` (which returns a string) is passed to `Function EvaluateRange( )` to evaluate the reference and see whether it returns any Excel error.

```
    On Error Resume Next

    If Not IsArray(nm.RefersToRange) Then
        GetNameValue = EvaluateRange(nm.RefersTo)
```

Evaluating Excel Values with the Function EvaluateRange( )

Excel cells can represent a wide range of values, including text, numbers, dates, hours, formulas that return any of these values, and...errors!

So, to correctly use the `Application.Evaluate( )` method to evaluate any cell value and show it on the `lstNames ListBox`, you must evaluate the formula or range the name represents and verify whether it returns any Excel error. If this is true, the error must be displayed in the `lstNames` Value column *as is*.

Table 5-4 shows all possible Excel errors, the error code, its VBA constants, and its meaning.

Table 5-4. *Excel Error Constants, Error Types, and Values*

Excel Error	Error Code	Error Constant	Error Type
#DIV/0!	2007	xlErrDiv0	Division by zero
#N/A	2042	xlErrNA	Not available
#NAME?	2029	xlErrName	Name does not exist
#NULL!	2000	xlErrNull	A NULL value
#NUM!	2036	xlErrNum	Number is expected
#REF!	2023	xlErrRef	Range or reference is wrong
#VALUE	2015	xlErrValue	Value is missing

To verify whether a cell address or formula evaluation returns an error, you must compare its value with the Excel CVErr( ) function, which has this syntax:

```
CVErr(Expression)
```

In this code:

> Expression: This is an error code or application constant associated with the error you want to generate.

As Table 5-4 specifies, CVErr( ) receives any Excel error code or error constant as an argument and returns the associated Excel error. For example, to generate the famous #DIV/0! Excel error as a returned value in any VBA procedure, use CVErr( ) this way:

```
CVErr(xlErrDiv0)
```

If you want to know whether cell A1 is returning a #DIV/0! error, use this syntax:

```
If Range("A1") = CVErr(xErrDiv0) then
```

This is the Function EvaluateRange( ) code, which expects to receive a formula or single cell reference and returns its expected value, including error codes as a string:

```
Private Function EvaluateRange(varValue As Variant) As String
    If VarType(varValue) = vbString Then
        varValue = (Evaluate(varValue))
    End If

    If IsError(varValue) Then
        Select Case varValue
            Case CVErr(xlErrDiv0)
                varValue = "#DIV/0!"
            Case CVErr(xlErrNA)
                varValue = "#N/A"
            Case CVErr(xlErrName)
                varValue = "#NAME?"
            Case CVErr(xlErrNull)
                varValue = "#NULL!"
            Case CVErr(xlErrNum)
                varValue = "#NUM!"
            Case CVErr(xlErrRef)
                varValue = "#REF!"
            Case xlErrValue
                varValue = "#VALUE!"
            Case Else
                varValue = "#VALUE!"
        End Select
    End If
    EvaluateRange = varValue
End Function
```

Since EvaluateRange( ) receives the varValue as Variant argument, you can pass it the Name.RefersTo string (which is always a formula that must be evaluated) or the Name.RefersToRange property, which returns the associated Range object (which is a value already evaluated).

Once varValue is received, EvaluateRange( ) uses the VBA VarType( ) function to verify whether the varValue is a text string. If it is true, the Name.RefersTo property string was passed and must be evaluated using the Excel Application.Evaluation method.

```
If VarType(varValue) = vbString Then
    varValue = Evaluate(varValue)
End If
```

At this point, varValue was evaluated to a value or any Excel error, which is tested by the VBA IsError( ) function. If IsError(varValue)=True, a Select Case statement comparing the varValue error with the value returned by the Excel CVErr( ) function for each possible constant error using a Select Case statement.

The first comparison uses the xlErrDiv0 constant, and if the comparison is true, the formula or cell has or returns a #DIV/0! error, and the "#DIV/0!" string is stored into varValue and returned as the function result.

```
If IsError(varValue) Then
        Select Case varValue
            Case CVErr(xlErrDiv0)
                varValue = "#DIV/0!"
                ...
        End Select
    End If
    EvaluateRange = varValue
End Function
```

Getting Back to GetNewName( )...

You must return to the GetNameValue( ) procedure and continue with the Else clause of the If Not IsArray(nm.RefersToRange) Then... instruction to see how a Name object that returns multiple cell values is processed.

```
If Not IsArray(nm.RefersToRange) Then
  GetNameValue = EvaluateRange(nm.RefersTo)
Else  <-------------------------------------  Getting back to here...
```

The Excel Name Manager uses a particular way to show a multiple-cell range name in its interface: its values are shown inside braces ({}), with each value inside double quotes ("") and separated by semicolons. Row breaks are identified by a backslash (\), and the values must reflect what is seen on the worksheet (including Excel errors).

This time you will need to run across all rows and columns of the range name, evaluating each cell value and returning them with the expected format. To select each cell used by the cell range, you must use the Range.Cells property inside two nested For...Next loops: an outer loop to process each range row and an inner loop to process each range column.

But before we dive into the code technique, you must be aware that a range can have an excessive number of cells, so the procedure must put a limit on what can be seen in the `lstNames` Value column. That is why `GetNameValue( )` declares so many `Integer` variables and the `Const conMaxItens = 6` constant: to execute the two For...Next loops and limit the list to no more than six individual cell values.

```
Private Function GetNameValue(nm As Name) As String
    Dim rg As Range
    Dim strItem As String
    Dim intI As Integer
    Dim intJ As Integer
    Dim intK As Integer
    Const conMaxItens = 6
```

The `GetNameValue( )` procedure will use the `strItem` string variable to compound the list of the first range values. Since the list be enclosed by brace characters, an open brace ({) is added to `strItem` before starting to loop through all rows of the range name using the `intI Integer` variable as the row counting. To visually separate each row from the next, `strItem` will receive a backslash (\) character after the first row is entirely processed (which happens when `intI>1`).

```
Else
    Set rg = nm.RefersToRange
    strItem = "{"
    For intI = 1 To rg.Rows.Count
        strItem = strItem & IIf(intI > 1, "\", "")
```

Another For...Next loop is initiated to run through all columns of each range row and return its cells values, using the `intJ` Integer variable as the column counting. Note that after the first item is added to `strItem` (`inJ>1`), the procedure adds a colon to separate each item from the next.

```
For intJ = 1 To rg.Columns.Count
    strItem = strItem & IIf(intJ > 1, ";", "")
```

At this point, the procedure is positioned on the cell range represented by `(intI, intJ)` coordinates, which is perfect for being used by the `Range.Cells` property to return the cell value, which also must be sent to the `EvaluateRange( )` procedure to verify whether it returns any Excel error. Since the Excel Name Manager encloses each cell value in double quotes, the procedure adds a `Chr(34)` (") character before and after the cell value.

```
strItem = strItem & Chr(34) & EvaluateRange(rg.Cells(intI, intJ)) & Chr(34)
```

■ **Attention** Whenever you need to add a single or a double quote as part of a `String` value, use the VBA function `Chr(34) = "` (double quote) or `CHR(39) = '` (single quote) to concatenate it into the string without generating an Excel error, regarding the string close character.

The next steps use the `intK` counter to count how many values have been inserted on the `strItem` variable. When `intK >= conMaxItens`, the procedure uses an `Exit For` instruction to interrupt the inner loop.

```
intK = intK + 1
'Provision to not add more than conMaxItens itens
```

```
    If intK >= conMaxItens Then
        Exit For
    End If
Next
```

This will force the code to exit to the outer loop, which verifies again whether `intK >= conMaxItens`. If it is, all items have already been inserted on the `strItem` variable, and the outer loop must also end with another `Exit For` instruction.

```
    Next
    If intK >= conMaxItens Then
        Exit For
    End If
Next
```

All desired items have been inserted. If they count equal to or less than `conMaxItens`, `strItem` receives a closing brace character (}). If the `strItem` variable does not hold all range values, it receives a reticence.

```
        If intK <= conMaxItens Then
            strItem = strItem & "}"
        Else
            strItem = strItem & "..."
        End If
        GetNameValue = strItem
    End If
End Function
```

Getting Back Again to FilllstNames( )...

We now need to get back again to the `FilllstNames( )` procedure, which at this time has already defined its third `Value` column. The next operations to add `Name` object properties to `lstName` columns for its last-inserted item are easy to understand. Take a look at the next instructions:

```
                .Column(3, .ListCount - 1) = nm.RefersTo
                .Column(4, .ListCount - 1) = IIf(nm.Parent.Name = ThisWorkbook.Name,
"Workbook", nm.Parent.Name)
                .Column(5, .ListCount - 1) = nm.Comment
                .Column(6, .ListCount - 1) = IIf(nm.Visible, "Yes", "No")
            End If
        Next
    End With
End Sub
```

Note that the fifth `lstNames` column's Scope value is added using a `IIF( )` instruction to verify whether the `Name` object's `Parent.Name` property equals the `ThisWorkbook.Name` property, meaning a workbook scope range name. If it isn't, it adds the `Parent.Name` property to this column.

And once all desired `Name` object properties have been recovered by the `For Each nm in Names` loop, the procedure ends and returns to the `cboObject_Change( )` event, which will call the `lstNames_Change( )` event to synchronize the `UserForm` interface.

```
Private Sub cboObjects_Change()
```

```
    Call FilllstNames
    Call lstNames_Change
End Sub
```

■ **Attention** The `lstNames_Change( )` event will be analyzed in the section "Selecting a Name on lstNames ListBox" later in this chapter.

Adding a New Name Object

Now that you have a good understanding about how the `FilllstNames( )` procedure works (calling `GetNameValue( )` to evaluate each Name object's `RefersTo` property, which calls `EvaluateRange( )` to evaluate each cell value), to see `frmNames` in action you need to add some range names by clicking the New `ControlButton`, which will fire the `cmdNew_Click( )` event and execute this code:

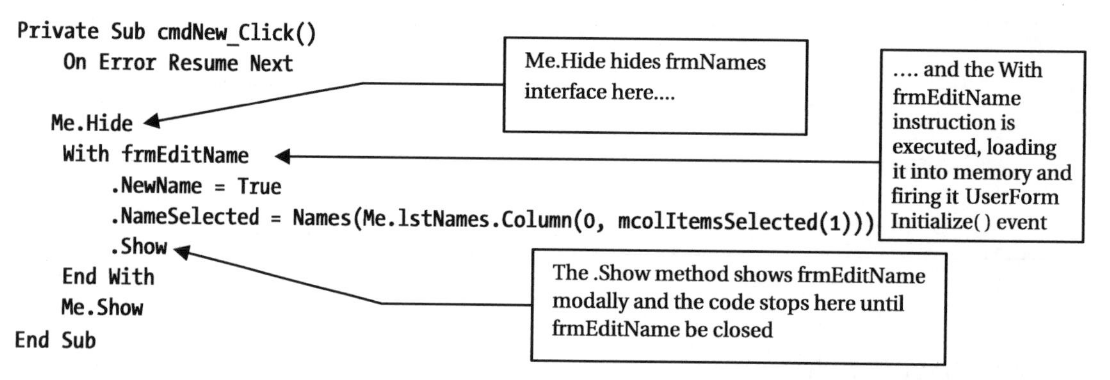

Private Sub cmdNew_Click()
On Error Resume Next
Me.Hide
With frmEditName
.NewName = True
.NameSelected = Names(Me.lstNames.Column(0, mcolItemsSelected(1)))
.Show
End With
Me.Show
End Sub
Me.Hide hides frmNames interface here....
.... and the With frmEditName instruction is executed, loading it into memory and firing it UserForm Initialize( ) event
The .Show method shows frmEditName modally and the code stops here until frmEditName be closed

Note the technique: VBA errors are disabled using an `On Error Resume Next` instruction and then use the `UserForm.Hide` method to hide `frmNames` from the Excel interface and allow `frmEditName` to be loaded. Then it uses the `With frmEditName...End With` loop to reference `frmEditName` just once and sets three properties: `NewName`, `NameSelected`, and `NameFilter`.

When the code uses the `With frmEditName` instruction to reference the `UserForm`, VBA will immediately load it into memory, firing the `UserForm_Initialize( )` event and executing this code:

```
Private Sub UserForm_Initialize()
    Dim varITem As Variant

    Me.cboObjects.AddItem "Workbook"
    For Each varITem In ThisWorkbook.Worksheets
        Me.cboObjects.AddItem varITem.Name
    Next
    Me.cboObjects.ListIndex = 0
    Me.txtRefersTo = "='" & ActiveSheet.Name & "'!" & Selection.Address
End Sub
```

As you can see, `frmEditName` fills the `cboObjects ComboBox` with the word `Workbook` and the name of each workbook sheet tab and defines `Workbook` as the default selection and the `txtReferTo` text box to the address of what is currently selected on the active worksheet. Note that to deal with sheet names that have spaces, it uses a single quote after the = sign, concatenates the `ActiveSheet.Name` property, concatenates another single quote and an exclamation point, and finally concatenates the `Selection.Address` property.

```
Me.txtRefersTo = "='" & ActiveSheet.Name & "'!" & Selection.Address
```

When the `frmEditName Initialize( )` event finishes executing, the code returns to the first instruction of the `With frmEditName` instruction on the `frmNames` code module, setting the `frmEditName.NewName` property to `True` to indicate to the `UserForm` that it must create a new name.

This will fire the `frmEditName Public Property Let NewName( )` procedure to execute, storing the `True` value into the `mbolNewName` module-level variable and setting the `UserForm Caption` property to New Name.

```
On Error Resume Next                Public Property Let NewName(ByVal vNewValue As Boolean)
With frmEditName                        mbolNewName = vNewValue
    .NewName = True                     Me.Caption = "New Name"
                                    End Property
```

When the `Property Let NewName( )` procedure ends, the code returns to the next `With frmEditName` instruction, this time trying to set the `frmEditName.NameSelected` property to the name selected on the `frmNames lstNames ListBox`.

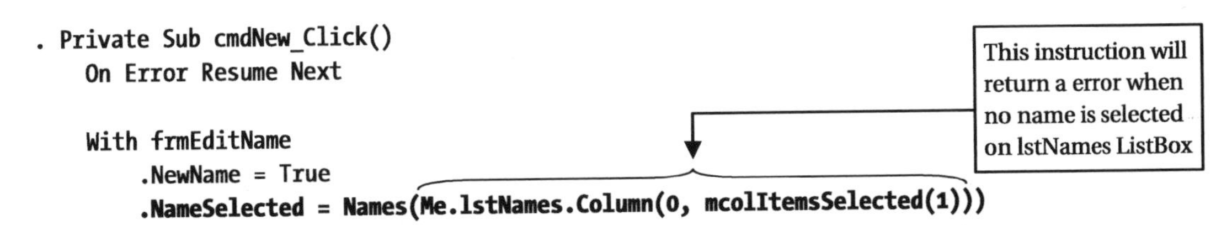

Since no name has been inserted on the `Names Collection.xlsm` workbook and the `lstNames ListBox` has no name selected, the `mcolItemsSelected(1)` instruction on the right will raise an error (Error = 5, "Invalid procedure call or argument"), which will be ignored because of the `On Error Resume Next` instruction executed on the procedure beginning, and the `frmEditName Public Property Let NameSelected( )` procedure will not be executed!

■ **Attention** The `mcolItemsSelected(1)` instruction will be explained in the section "Using Collection Variables" later in this chapter.

When the `frmEditName.Show` method is executed, the `frmEditName UserForm` has its property `ShowModal = True`, the window is showed modally, and the `cmdNew_Click( )` code stops on this instruction until `frmEditName` is closed by the user action (Figure 5-15).

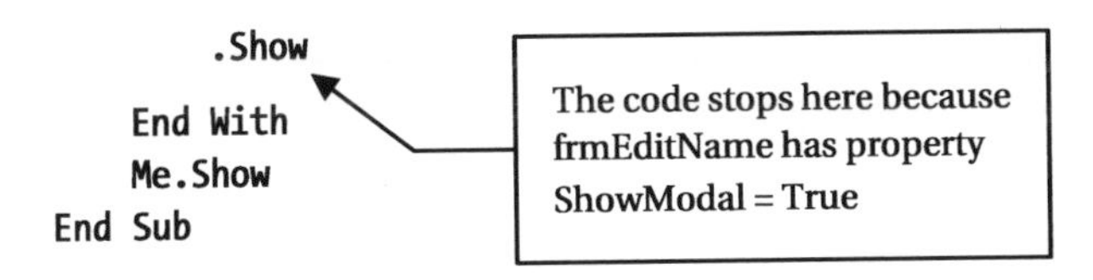

.Show
End With
Me.Show
End Sub
The code stops here because frmEditName has property ShowModal = True

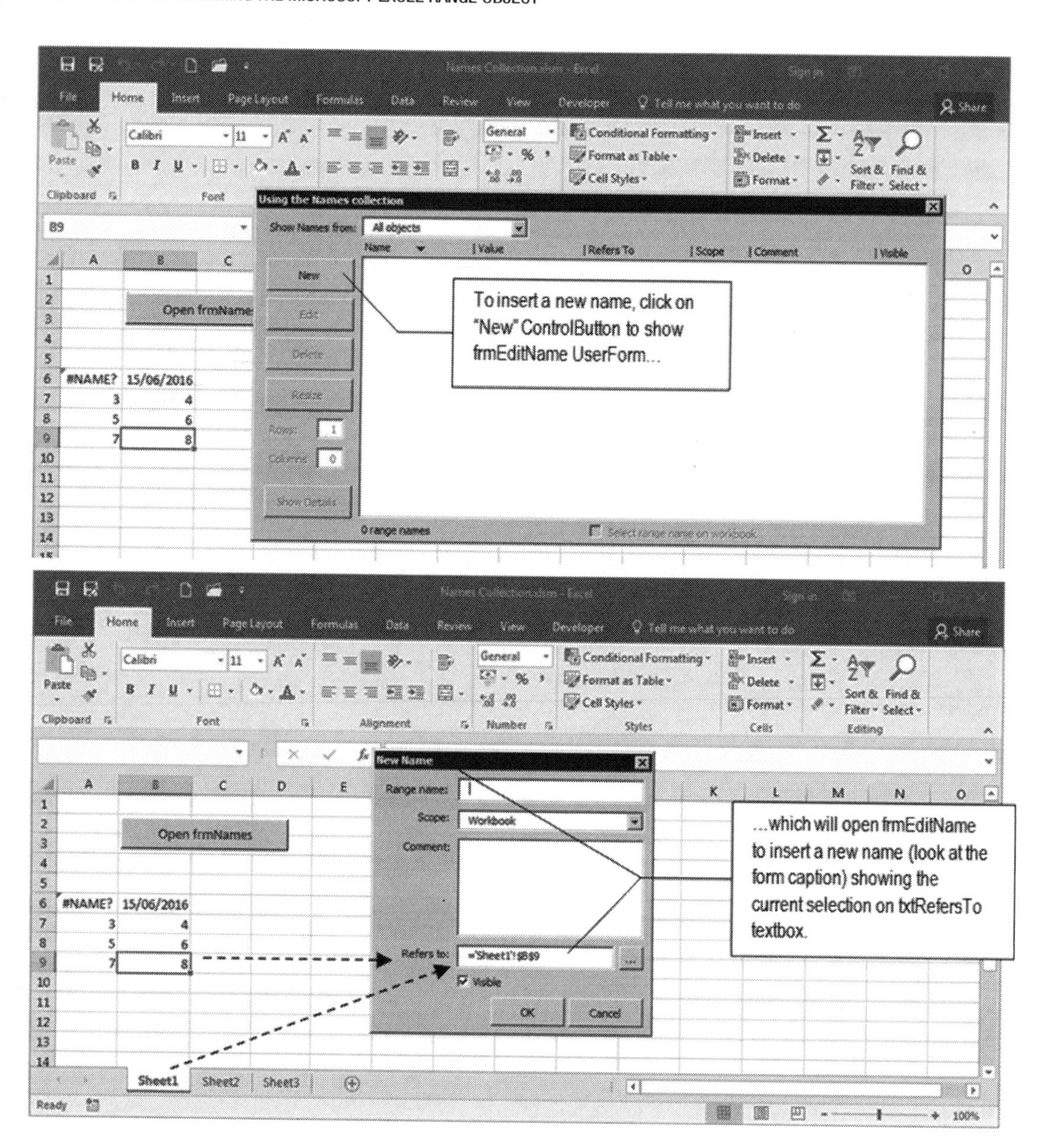

Figure 5-15. *When you click the New CommandButton to create a new name, frmEditName is loaded while frmNames is unloaded from memory*

Let's try to insert a new name that returns a constant with a `#DIV/0!` error so you can try the `frmNames FilllstNames( )` procedure. In the Range Name text box (`txtName`) of `frmEditName`, type `DivOConstant`, and in the "Refers to" text box (`txtRefersTo`), type this formula (Figure 5-16):

```
=2/0
```

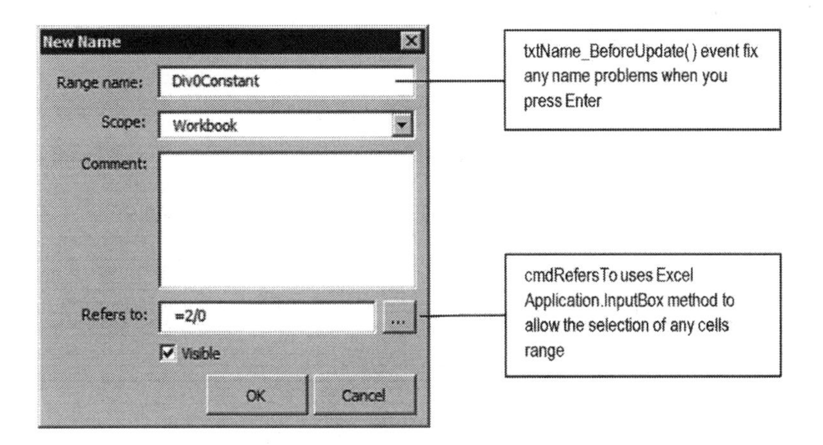

Figure 5-16. *Insert a Name constant that raises an Excel #DIV/0! error to see how the frmNames Sub FilllstNames( ) procedure performs when such a range name is created*

Validating Names

After you type the desired name in the `txtName` text box and press Enter, the `txtName_BeforeUpdate( )` event fires, executing this code:

```
Private Sub txtName_BeforeUpdate(ByVal Cancel As MSForms.ReturnBoolean)
    If Len(Me.txtName) Then
        Me.txtName = FixName(Me.txtName)
    End If
End Sub
```

Note that like any other `Before` event, it passes a `Cancel` argument, meaning that it can be canceled by the VBA code. The event uses the VBA `Len( )` function to verify whether anything was typed in the `txtName` text box, and if this is true, it passes the name typed to the `Function FixName( )` procedure of `basFixName` code module, which takes out any invalid characters from the name before executing.

The function `FixName( )` procedure executes this code:

```
Public Function FixName(ByVal strName As String)
    'Invalid characters inside range names: @#$%&()+~`"':;,.|!?_-/\*[]{}
    Dim strInvalidChars As String
    Dim strChar As String
    Dim intI As Integer

    'Search for invalid characters
    strInvalidChars = "@#$%&()+~`´':;,.|!?-/\*[]{}" & """"
    For intI = 1 To Len(strInvalidChars)
        'Get each invalid character and take it out
        strChar = Mid(strInvalidChars, intI, 1)
        strName = Replace(strName, strChar, "")
    Next

    'Now change spaces to underscores
    strName = Replace(strName, " ", "_")
    FixName = strName
End Function
```

The search technique is quite simple; it stores all invalid characters on a string variable and uses the VBA Replace( ) function to substitute them, which has this syntax:

```
Replace(Expression, Find, Replacement, [Start], [Count], [Compare]) As String
```

In this code:

- Expression: This is required; it is the string expression containing the substring to replace.
- Find: This is required; it is the substring being searched for.
- Replacement: This is required; it is the replacement substring.
- Start: This is optional; it is the position within the expression where the substring search is to begin. If omitted, 1 is assumed.
- Count: This is optional; it is the number of substring substitutions to perform. If omitted, the default value is -1, which means "make all possible substitutions."
- Compare: This is optional; it is the numeric value indicating the kind of comparison to use.
  - Binary: This performs a binary comparison (case sensitive).
  - Text: This performs a textual comparison.

The FixName( ) procedure works this way: it stores all invalid characters to be extracted from the strName argument on the strInvalidChars String variable.

```
'Search for invalid characters
strInvalidChars = "@#$%&()+~`´':;,.|!?-/\*[]{}" & """"
```

A For...Next loop runs through all invalid characters extracting them one by one to the strChar variable and using the VBA Mid( ) function, and it uses the VBA Replace( ) function to search and replace it with an empty string ("").

```
For intI = 1 To Len(strInvalidChars)
    'Get each invalid character and take it out
    strChar = Mid(strInvalidChars, intI, 1)
    strName = Replace(strName, strChar, "")
Next
```

The VBA Replace( ) function is used again to change any space to an underscore character (_) and returns the fixed name.

```
'Now change spaces to underscores
strName = Replace(strName, " ", "_")
FixName = strName
```

Using Names Collection Add Method

Since you are now inserting a new constant name that returns a #DIV/0! Excel error, when you click the OK ControlButton, the cmdOk_Click( ) event fires, executing this code:

```
Private Sub cmdOK_Click()
    Dim nm As Name
    Dim strName As String

    On Error Resume Next

    If Len(Me.txtName) = 0 Then
        MsgBox "Type the range name", vbCritical, "Range name?"
        Exit Sub
    End If

    If Len(Me.txtRefersTo) = 0 Then
        MsgBox "Define range address", vbCritical, "Range address?"
        Exit Sub
    End If

    strName = Me.txtName
    If Me.cboObjects <> "Workbook" Then
        strName = "'" & Me.cboObjects & "'!" & strName
    End If

    If Not Me.NewName Then
        Call FixNameChange
        Me.NameSelected.Delete
    End If

    Set nm = Names.Add(strName, Me.txtRefersTo, Me.chkVisible)
    nm.Comment = Me.txtComment & ""
    Set mName = nm
    Unload Me
End Sub
```

As you can see, the cmdOK_Click( ) event procedure begins using an On Error Resume Next to disable any VBA raised errors and then verifies with the VBA Len( ) function if any text was typed in the txtName or txtRefersTo text box (note that it doesn't care if you type an invalid reference on txtRefersTo).

Once you have typed the name and it references the formula, the procedure stores the desired name into the strName String variable and verifies the name scope. If cboObject = "Workbook", the name is inserted as is, but if you select a sheet name, the sheet name is enclosed by single quotes, suffixed by a ! character and used as a prefix to the name.

```
On Error Resume Next

strName = Me.txtName
If Me.cboObjects <> "Workbook" Then
   strName = "'" & Me.cboObjects & "'!" & strName
End If
```

Next the code uses frmEditName.NewName property to verify whether it is inserting a new Name by verifying it. If this is true, the Name object stored on the frmEditName.NameSelected property will be deleted.

```
If Not Me.NewName Then
    Me.NameSelected.Delete
End If
```

And the name is added to the Application.Names collection using the Names.Add method with strName for the Name.Name property, txtRefersTo to the Name.RefersTo property, and chkVisible refers to the Name.Visible property, storing a reference to it into the nm as Name object procedure variable.

```
Set nm = Names.Add(strName, Me.txtRefersTo, Me.chkVisible)
```

Since the Name.Comment property can't be set by the Names.Add method, it is defined on the nm variable that represents the new added name.

```
nm.Comment = Me.txtComment & ""
```

The local nm object variable is then associated to the mName object variable, and the frmEditName UserForm is unloaded from memory using the VBA Unload method.

```
    Set mName = nm

    Unload Me
End Sub
```

When frmEditName is unloaded, the UserForm_Terminate( ) event fires, executing this code:

```
Private Sub UserForm_Terminate()
    With frmNames
        .FilllstNames
        If Not (Me.NameSelected Is Nothing) Then
            .NameSelected = Me.NameSelected
        End If
    End With
End Sub
```

A With frmNames instruction is used to reference frmNames only once, call it method and change it NameSelected property.

Before frmEditName unloads from memory, it uses a With frmNames...End With instruction to reference frmNames only once and updates the lstNames ListBox to reflect any Name object changes, calling the frmNames.FilllstNames method.

```
With frmNames
    .FilllstNames
```

Remember that the frmEditName cmdOK_Click( ) event stored the nm object variable reference to the new name on its mName module-level variable? This reference is now used when the procedure calls the Property Get NameSelected( ) procedure and compares it to Nothing.

```
If Not (Me.NameSelected Is Nothing) Then
```

```
Public Property Get NameSelected() As Name
    Set NameSelected = mName
End Property
```

Since a new name has been added, (Not (Me.NameSelected) is Nothing) = True, this new name is used to define the frmNames.NameSelected property, which will execute the frmEdifName Property Get NameSelected( ) on the right side of the equation, while executing frmNames Property Let NameSelected( ) on the left side.

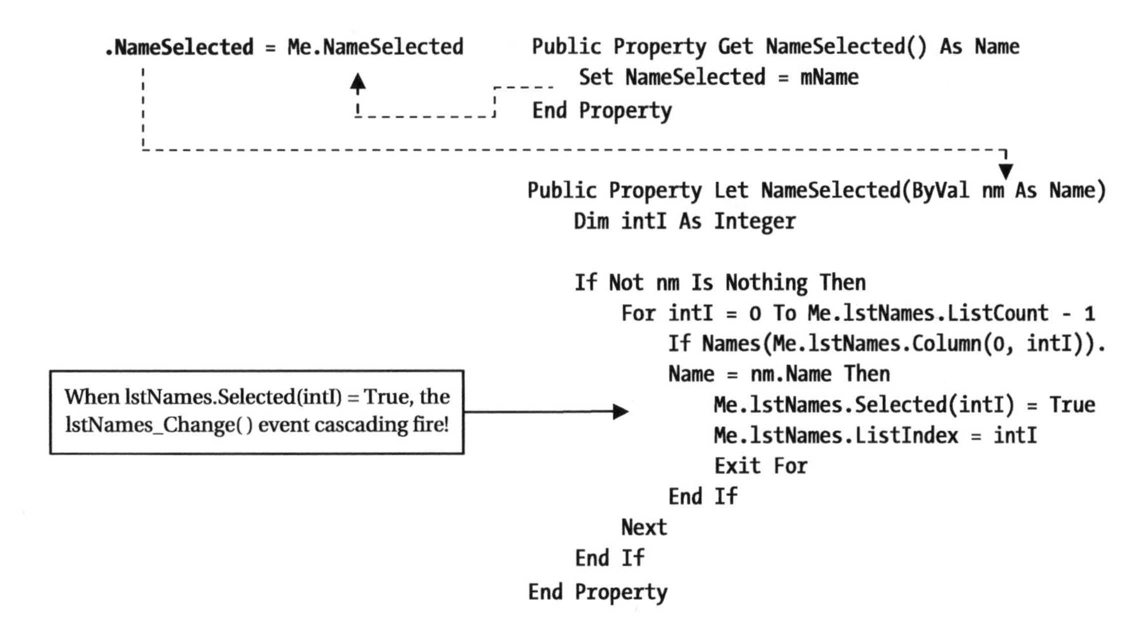

The code for `frmNames Property Let NameSelected( )` is quite simple: it first verifies whether the `nm as Name` argument has some `Name` object reference. If it does, it uses a `For...Next` loop to run through all `lstNames` items, using the `Names` collection to compare each `Name` object's `Name` property with the recently added `nm` object `Name` property (note that it uses `lstNames.Column(0), intI)` to return each `Name.Name` property).

```
For intI = 0 To Me.lstNames.ListCount - 1
    If Names(Me.lstNames.Column(0, intI)).Name = nm.Name Then
```

When it finds a match, it selects the item using the `lstNames ListBox.Selected` property and makes the list scroll to the selected item by setting the `ListIndex` property before exiting the `For...Next` loop and uses the `frmNames.Show` method to show the interface on the screen.

```
Me.lstNames.Selected(intI) = True      <---- lstNames_Change( ) event will cascade fire!
Me.lstNames.ListIndex = intI
Exit For
```

And, when this happens, the `lstNames.Change( )` event cascade-fires because another item was selected in the `lstNames ListBox`, synchronizing the `frmNames` interface to the selected item.

Attention You will learn about the `lstNames_Change( )` event in the next section.

When the `frmEditName UserForm_Terminate( )` event ends, it returns code control to the `cmdOK_Click( )` event, which also returns code control to the `frmNames cmdNew_Click( )` event that executes the `frmNames.Show` method, showing `frmNames` with a synchronized interface.

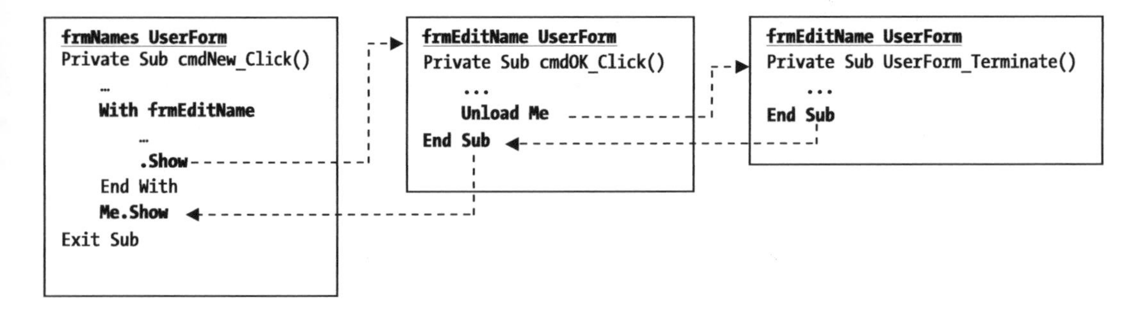

Figure 5-17 shows how `frmNames` appears in Excel after the `DivOConstant` `Name` object was added using `frmEditName`. Note that the `frmNames` `CommandButtons` are now enabled, since the new added name is selected in the `lstNames` `ListBox`

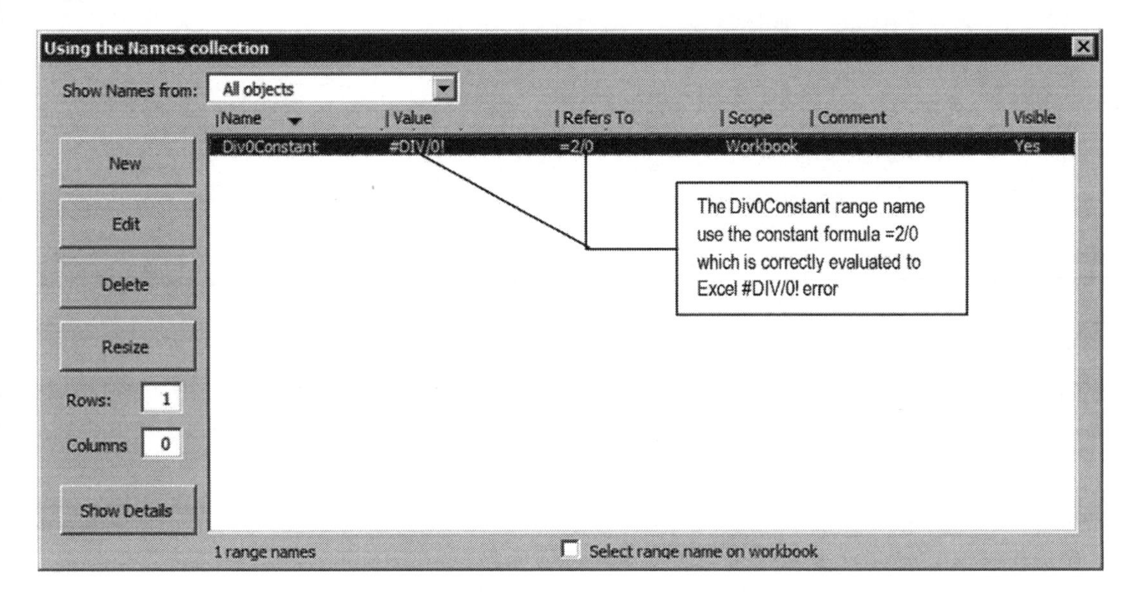

Figure 5-17. *This is frmNames after the DivOConstant Name object was added. Since it produces a division by zero, the lstNames Value column shows a #DIV/0! Excel error*

Inserting a New Name by Selecting a Range Address

To see how `frmNames` performs, you need to add a few more names to the `Names Collection.xlsm` workbook. Since the first `Name` object (`DivOConstant`) was associated to a constant value (the `#DIV/0!` Excel error), let's insert one name that has just a range of valid numbers. Click again the `frmNames` New `ControlButton`, type `MyData` in the `txtName` text box, keep the `Workbook` scope, and click `cmdRefersTo` (the small `ControlButton` at the right of the `txtRefersTo` text box). `frmEditNames` will hide itself and show `Application.Inputbox`, which is where you can click and drag the desired cell addresses to associate the range name. Select cells `A7:B9` from `Sheet1` (Figure 5-18).

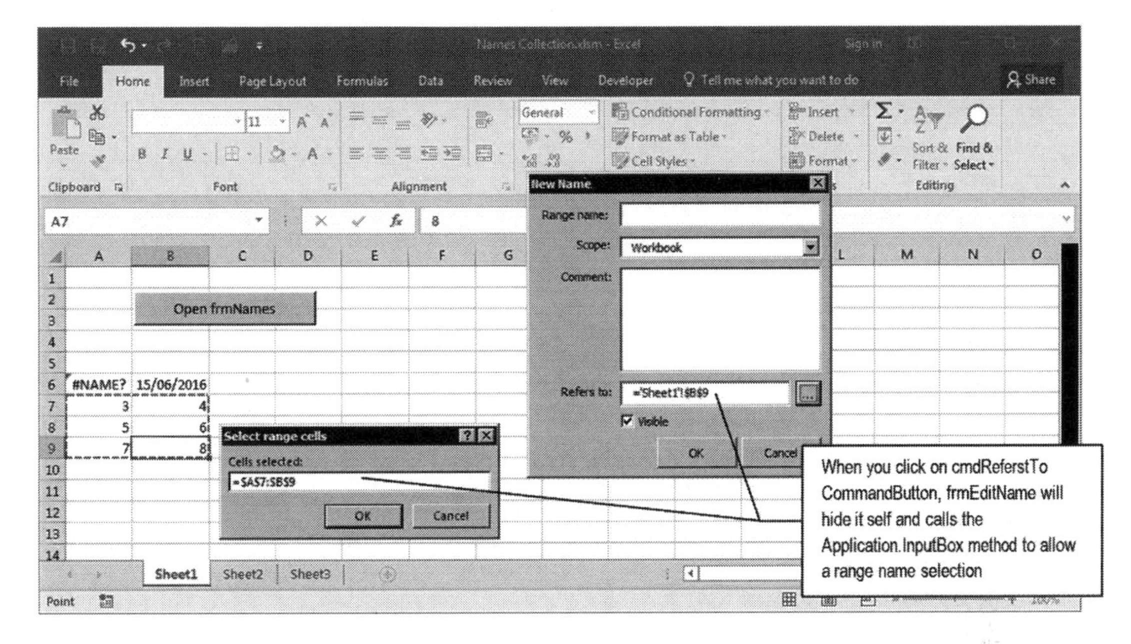

Figure 5-18. *Use frmNames to create a new name associated with just valid numbers (with no error cells), like MyData, with workbook scope, associated to cells A7.B9*

This is the code behind the frmEditName cmdRefersTo_Click( ) event, which allows the selection of the cell's range that will be associated to the range name:

```
Private Sub cmdRefersTo_Click()
    Dim varRange As Variant
    Dim intPos As Integer
    Const conFormula = 0

    Me.Hide
    varRange = Application.InputBox("Cells selected:", "Select range cells",
    Me.txtRefersTo, , , , , conFormula)
    If varRange <> False Then
        varRange = Application.ConvertFormula(varRange, xlR1C1, xlA1)
        'Search for Workbook reference
        intPos = InStr(1, varRange, "]")
        If intPos > 0 Then
            varRange = "'" & Mid(varRange, intPos + 1)
        End If

        'Search for Sheet name
        intPos = InStr(1, varRange, "!")
        If intPos = 0 Then
            varRange = "'" & ActiveSheet.Name & "'!" & Mid(varRange, 2)
        End If

        'Search for "='
        If Left(varRange, 1) <> "=" Then
            varRange = "=" & varRange
        End If
```

```
        Me.txtRefersTo = varRange
    End If
    Me.Show
End Sub
```

When you click the cmdRefersTo ControlButton, the first instruction executed by its Click( ) event is to hide itself by calling the UserForm.Hide method. It then calls the Application.Inputbox method using the txtRefersTo value for the InputBox Default argument (what is currently selected on the active sheet; see the frmEditName UserForm_Initialize( ) event code in section "Adding a New Name" earlier in this chapter). For its last Type argument, the constant conFormula = 0 to avoid the appearance of any magic number on the code.

```
Private Sub cmdRefersTo_Click()
    Dim varRange As Variant
    Dim intPos As Integer
    Const conFormula = 0

    Me.Hide
    varRange = Application.InputBox("Cells selected:", "Select range cells",
    Me.txtRefersTo, , , , , conFormula)
```

If the user clicks the InputBox Cancel button, varRange will receive False, and the procedure will call the UserForm.Show method and end normally. But if the InputBox OK button is selected, the next code instruction uses the Application.ConvertFormula method to change the range selected from R1C1 to A1 style.

```
    If varRange <> False Then
        varRange = Application.ConvertFormula(varRange, xlR1C1, xlA1)
```

■ **Attention** There is no indication in the Application.InputBox method that whenever it uses Type=0 to get a range address by dragging the mouse over any sheet cells, the formula returned will use the R1C1 style. But Excel does this, and you must convert it to A1 style so it appears like most users expect to see it.

If the range you are trying to select belongs to a sheet tab that is different from the active sheet, the Application.InputBox method will also return on the formula the workbook name inside double braces, like this:

```
='[Names Collection.xlsm]Sheet1'!$A$7:$B$9
```

So, you need to search the varRange variable for a closing brace (]) using the VBA InStr( ) function, and if it's found, you take it out from the selected range using the VBA Mid( ) function.

```
'Search for Workbook reference
intPos = InStr(1, varRange, "]")
If intPos > 0 Then
   varRange = "'" & Mid(varRange, intPos + 1)
End If
```

The next instruction will verify whether the returned address is already prefixed by the sheet tab name using again the VBA InStr( ) function to search for a ! character. If it does not exist (which happens when you select any range on the active sheet), it must be inserted in the formula. The new formula is composed by the active sheet name enclosed by single quotes, an exclamation character (!), and the current address without its first = character, which is extracted by the VBA Mid( ) function.

```
        'Search for Sheet name
        intPos = InStr(1, varRange, "!")
        If intPos = 0 Then
            varRange = "'" & ActiveSheet.Name & "'!" & Mid(varRange, 2)
        End If
```

The code then searches for the = character that must be at the very first position of the returned address, and if it is not there, it is added again, and frmEditName txtRefersTo receives the selected range with the appropriate format.

```
'Search for "="'
If Left(varRange, 1) <> "=" Then
   varRange = "=" & varRange
End If

Me.txtRefersTo = varRange
```

The Sub cmdRefersTo_Click( ) event finishes by using the UserForm.Show method to show the frmEditName Userform interface again.

```
    End If
    Me.Show
End Sub
```

When you click the frmEditName cmdOK ControlButton, the MyData range name will be added to the workbook Names collection, and frmNames will rebuild its lstNames ListBox with the newly added name selected in its interface (Figure 5-19).

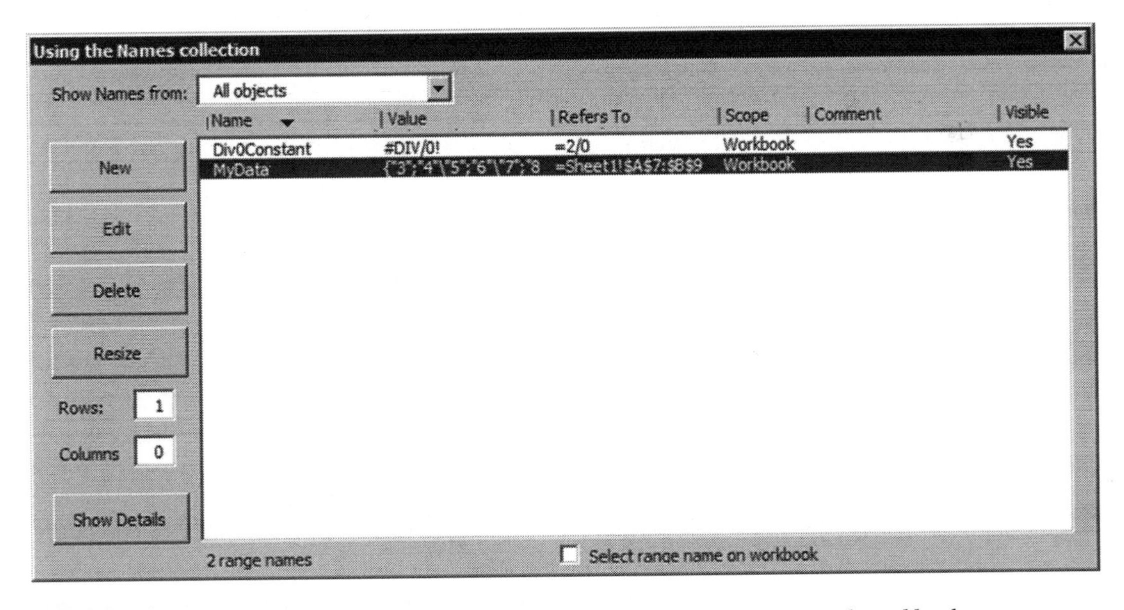

Figure 5-19. *This is frmNames after you have inserted the MyData range name, with workbook scope, associated to cells A7.B9 to return just valid numbers*

To continue with the next sections of this book, please insert some more range names with different scopes. Table 5-5 shows the range names that will be used in the next figures of this chapter.

Table 5-5. *Using frmNames and frmEditName, Insert These Range Names*

Range Name	Scope	Refers To
SumMyDataq	Sheet1	=Sum(MyData)
DataWithError	Sheet1	=Sheet1!$A$6:$B$9
RangeSheet2	Sheet2	=Sheet2!$A$1

Figure 5-20 shows frmNames with all five range names created so far in this book section.

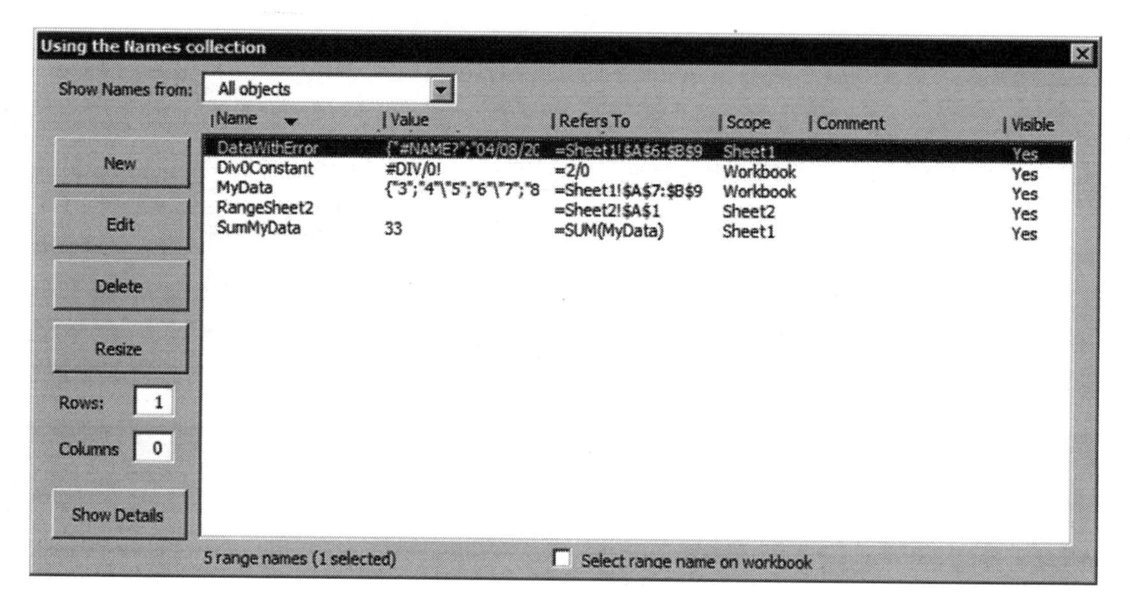

Figure 5-20. *This is the frmNames interface after all the proposed range names of Table 5-5 are inserted in the Names Collection.xslm macro-enabled workbook*

■ **Attention** Note that the SumMyData range name correctly sums all MyData range name values.

Selecting Items in the lstNames ListBox

The lstNames ListBox of the frmNames UserForm was set to allow multiple selections in the lstNames ListBox by setting the MultiSelect property to 2 - frmMultiSelectExtend, meaning that you can click and drag the mouse over the list or use the Ctrl or Shift key to select any combination of items.

Whenever a VBA UserForm ListBox is defined to allow multiselection, it does not return a Value property anymore; the lstNames.Value will now return Null.

Since the VBA UserForm ListBox does not have an ItemsSelected property to indicate which items are selected (like Microsoft Access ListBox does), you need to use a loop to run through all ListBox items verifying whether the Selected property is True, and then take the desired action, as follows:

```
For intI = 0 to Me.lstNames.ListCount - 1
   If Me.lstNames(intI).Selected = True Then
      ' Do something here!
   End If
Next
```

Now you need a way to store all ListBox-selected items in a variable so you can easily process multiple names at once (changing the Visible property, for example), call the FilllstNames( ) procedure to fill lstNames with the new Name object properties, and reselect them after the process is completed. This time you will use the VBA Collection object to hold the items selected.

Using Collection Variables

VBA offers the Collection object as a way to group and manage related objects. It has been widely used to collect object references created with Class modules, and you see it in action every time you use the Excel Workbooks, Worksheets, or Names collections that hold references to different types of Excel objects.

The Collection object offers the Add and Remove methods to manage items, a Count property to indicate how many items it currently holds, and an Item property as an easy way to instantly recover any collected data inside the Collection object.

To be useful, a Collection variable must be declared as a Public or Private variable in the Declaration section of a code module so it can be accessed by all its procedures. To test how a Collection object variable works, you can declare it as a Public variable of the ThisWorkbook code module on any Excel workbook, using the VBA New keyword, by typing the next instruction in its Declaration section (note that the variable name was prefixed with mcol, which is a common way to identify code module collection variables).

```
Option Explicit
Public mcolMyCollection as New Collection
```

■ **Attention** If you did not use the VBA New keyword to declare an object variable (like mcolMyCollection as New Collection), you need to use the VBA Set instruction to instantiate it or add some item with the Add method before trying the Count property.

From this point on, the ThisWorkbook.mcolMyCollection Collection variable can be easily accessed from any part of your code or from the VBA Immediate window, and the mcolMyCollection.Count property will return zero items since no one has already been inserted on it (Figure 5-21)!

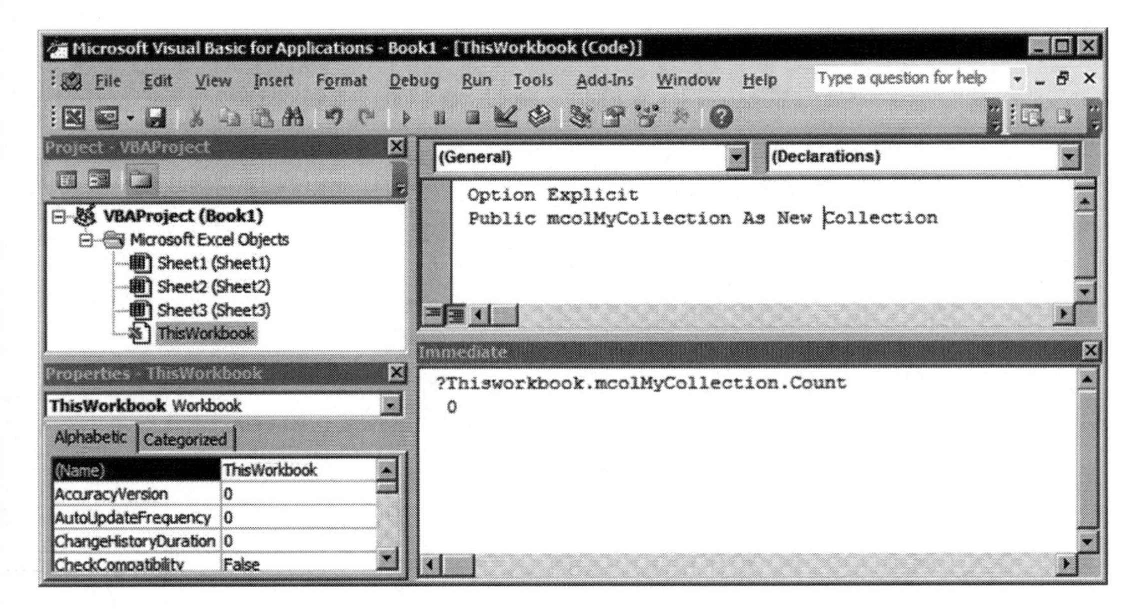

Figure 5-21. *Once a code module has a Public Collection variable declared with the New keyword, you can test it using the VBA Immediate window. Note that when you evoke the mcolMyCollection.Count property, the object variable is automatically instantiated by VBA*

Adding Collection Items

To add a new item to a `Collection` object variable, use the `Add` method, which has this syntax:

```
Object.Add Item, Key, Before, After
```

In this code:

`Object`: This is the name of the object variable declared as `Collection`.

`Item`: This is required; it is an expression that specifies an object reference or any type of value that represents the member to add to the collection.

`Key`: This is optional; it is a string expression that specifies an identification key that can be used, instead of a positional index, to access a member of the collection and return its `Item` property.

`Before, After`: This is optional; it is an expression that specifies an existing member position in the collection where the new member should be placed before or after (you can specify a before position or an after position, but not both). If a numeric expression, `Before` must be a number from 1 to the value of the collection's `Count` property. If a string expression, `Before` or `After` must correspond to the `Key` specified for the desired existing member.

Supposing you want to add the A `Item` to the `mcolMyCollection` variable identified by the `ItemA Key`, type this instruction in the VBA Immediate window:

```
ThisWorkbook.mcolMyCollection.Add "A", "ItemA"
```

Note that A is used as the Item argument of the Collection.Add method (meaning Item value) while ItemA is used as Key argument. Figure 5-22 shows what happened after the mcolMyCollection object variable received item values A, B, and C, identified by the ItemA, ItemB, and ItemC keys, respectively, using the VBA Immediate window. Note that the mcolMyCollection.Count property returns three items.

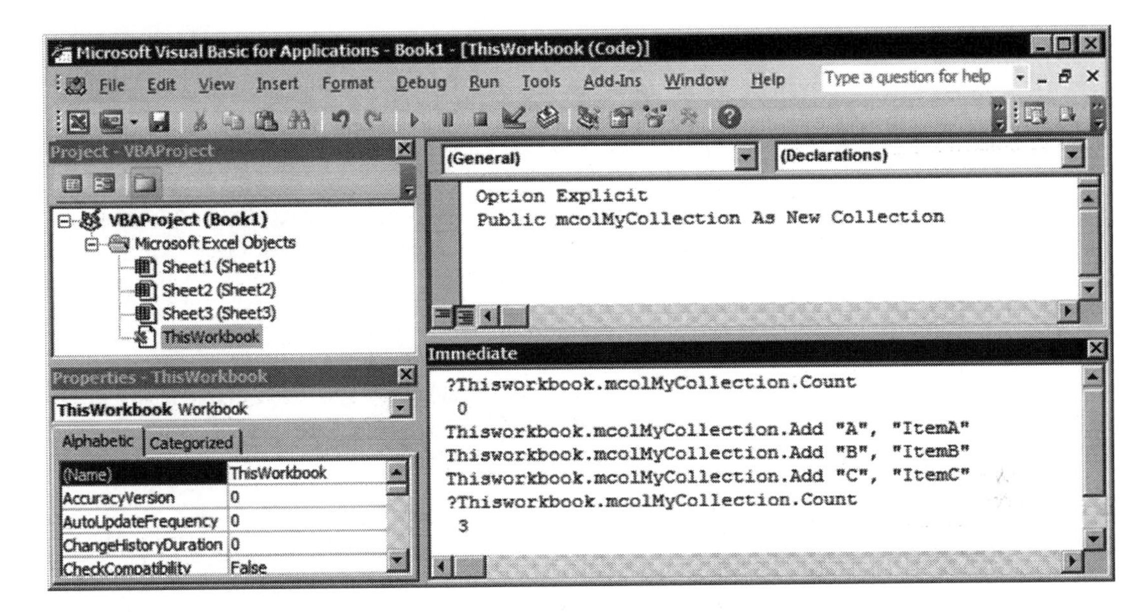

Figure 5-22. Using the VBA Immediate window, you can add, remove, and count items inserted on any Public Collection variable. This figure shows that three items have been added to the ThisWorkbook mcolMyCollection variable, using the Key argument to identify each item

Recovering Collection Items

To recover any collection Item, you can use its 1-based Index position or its Key (if any). That is why you may use for the Key argument of the Collection.Add method unique identifiers: to easily recover any desired item.

To recover the first mcolMyCollection Item using the numerical Index position, type this instruction in the VBA Immediate window:

```
?ThisWorkbook.mcolMyCollection(1)
A
```

To recover the first mcolMyCollection Item using it Key, type this instruction in the VBA Immediate window:

```
?ThisWorkbook.mcolMyCollection("ItemA")
A
```

In both cases, VBA will print in the Immediate window the desired Item value.

Removing Collection Items

To remove any collection, use the Remove method, which has this syntax:

```
object.Remove({Index | Key})
```

In this code:

Object: This is the name of the object variable declared as Collection.

Index, Key: This is the index position (1-based) or the key string associated to the item to delete.

So, to delete the first item of the mcolMyCollection variable, use one of the next two syntaxes with the Index or Key string associated to the item you want to remove in the VBA Immediate window. Here's an example:

```
?ThisWorkbook.mcolMyCollection.Remove(1)
Here's another example:?ThisWorkbook.mcolMyCollection.Remove("ItemA")
```

When you do this, every item remaining in the Collection variable will be re-indexed, but the Key value will remain the same, and that is the best reason to associate a unique Key to each Collection item.

Clearing a Collection

The VBA Collection object *does not have* a Clear method. In fact, you don't need it! To clear any Collection variable of all its items, use the VBA Set and New keywords to instantiate it again. The next instruction will automatically release all items of the mcolMyCollection variable when you type it in the VBA Immediate window (or execute it on any code procedure):

```
Set mcolMyCollection = New Collection
```

It can't be easier than that!

■ **Attention** Some web sites advise you to associate Nothing to the Collection variable as a way to clear its items, as follows:

```
Set mcolMyCollection = Nothing
```

Although this works well to remove all items from the Collection variable, it also destroys the association of the object variable with the Collection object. If the variable was not declared with the New keyword, if you try to use its Count property immediately before this operation, instead of returning zero items, VBA will return an error since the mcolMyCollection variable is still not instantiated.

Using a Collection Variable to Store ListBox Selected Items

The frmNames UserForm has the ability to process multiple Name objects selected in the lstNames ListBox at once to delete them or change the Visible and Comment properties. Since the FilllstNames( ) procedure always shows the current properties of each Name object, you need to hold the selected lstNames items using a Collection object variable, process them, call FilllstNames( ) again to update its new properties, and reselect them again in the ListBox.

That is why frmNames has the Private mcolItemsSelected as Collection object variable declared in its Declaration section.

```
Option Explicit

Dim mcolItemsSelected As Collection
```

Note that the Collection variable was not declared with the New VBA keyword, meaning that it must be instantiated to associate it to a VBA Collection object.

Whenever you select one or more items on the lstNames multiselect ListBox, its Change( ) event fires, executing this code:

```
Private Sub lstNames_Change()
    Dim strNames As String
    Dim intI As Integer
    Dim bolVisible As Boolean
    Dim bolHidden As Boolean

    If mbolCancelEvent Then
        mbolCancelEvent = False
        Exit Sub
    End If

    Set mcolItemsSelected = New Collection
    With Me.lstNames
        For intI = 0 To .ListCount - 1
            If .Selected(intI) Then
                mcolItemsSelected.Add intI
                If .Column(6, intI) = "Yes" Then
                    bolVisible = True
                Else
                    bolHidden = True
                End If
            End If
        Next
    End With

    strNames = intI & " range names"
    strNames = strNames & IIf(mcolItemsSelected.Count > 0, " (" & mcolItemsSelected.Count &
" selected)", "")
    Me.lblNamesCount.Caption = strNames

    If mcolItemsSelected.Count = 1 Then
        Call ShowNameProperties(mcolItemsSelected(1))
    Else
        Call ClearNameProperties
        Me.chkVisible = IIf(bolVisible And bolHidden, Null, bolVisible = True)
    End If

    Call EnableControls((mcolItemsSelected.Count = 1))
End Sub
```

After declaring the variables it needs, the code performs the now famous trick to avoid a cascading event. It verifies whether the module-level variable mbolCancelEvent = True, and if it does it turns it false and exits the event code.

```
If mbolCancelEvent Then
    mbolCancelEvent = False
    Exit Sub
End If
```

The mcolItemsSelected Collection variable is then instantiated or cleared from all its items.

```
Set mcolItemsSelected = New Collection
```

Attention Every time another item is selected in the lstNames ListBox, the Change( ) event fires, destroying and re-creating the mcolItemsSelected Collection variable.

The procedure begins a With lstNames...End With instruction to reference the lstNames ListBox only once and then begins a For...Next loop through all the items.

```
With Me.lstNames
    For intI = 0 To .ListCount - 1
```

If any lstNames item's Selected property is True, the item is selected and added to the mcolItemsSelected Collection variable using the ListIndex value (represented by the intI Integer variable) as the Item value.

```
If .Selected(intI) Then
    mcolItemsSelected.Add intI
```

Since the user can select an undefined number of Name objects on the lstNames ListBox, they can have different Visible properties. The code then tries to recover each Name.Visible property analyzing lstNames seventh column value (Column=6, Visible). If it has a Yes, Name.Visible = True and the bolVisible variable also becomes True. If Name.Visibe = "No", the bolHidden variable becomes True:

```
If .Column(6, intI) = "Yes" Then
    bolVisible = True
Else
    bolHidden = True
End If
```

When all lstNames items have been processed, strNames receives how many Name objects are listed in the ListBox (using intI as the Name counting) and how many are selected (using the mcolItemsSelected.Count property), and the value is associated to the lblNamesCount.Caption property.

```
    strNames = intI & " range names"
    strNames = strNames & IIf(mcolItemsSelected.Count > 0, " (" & mcolItemsSelected.Count &
    " selected)", "")
    Me.lblNamesCount.Caption = strNames
```

There are now two different possibilities: just one item was selected on lstNames or more than one was selected.

If just one item is selected, the selected Name object properties must be shown in the frmNames Detail section. This is made by calling the ShowNameProperties( ) procedure, which receives as an argument the selected Name ListIndex (represented by mcolItemsSelected(1), the first and only Collection variable Item):

```
If mcolItemsSelected.Count = 1 Then
    Call ShowNameProperties(mcolItemsSelected(1))
```

■ **Attention** The ShowNameProperties( ) procedure will be analyzed in the "Showing Name Properties" section later in this chapter.

If more than one item is selected, the frmNames Detail section must have its controls cleared because of the impossibility of showing ambiguous values, by calling the ClearNameProperties( ) procedure.

```
Else
    Call ClearNameProperties
```

■ **Attention** The ClearNameProperties( ) procedure will be analyzed in the "Clearing Name Properties" section later in this chapter.

The frmNames Detail section has the chkVisible check box (see Figure 5-12), which now must reflect the Visible property of all selected Name objects on the lstNames ListBox.

For your information, any check box can have three different states: Checked (=True), Unchecked (= False), and Undetermined (= Null, becoming gray). If both bolVisible and bolHidden variables are True, it means that both visible and hidden names have been selected in the lstNames ListBox. Otherwise, they are all visible or all hidden, and the chkVisible value must be set accordingly, by verifying just whether bolVisible=True.

```
    Me.chkVisible = IIf(bolVisible And bolHidden, Null, bolVisible = True)
End If
```

The lstNames_Change( ) event ends up making a call to the EnableControls( ) procedure, which receives a Boolean argument (True/False) to enable/disable UserForm controls. Note that it receives a Boolean comparison against the mcolItemsSelected.Count property; the frmNames controls will be available if just one item is selected on lstNames ListBox.

```
    Call EnableControls((mcolItemsSelected.Count = 1))
End Sub
```

Now that you already know that lstNames ListBox items selected are held by the mcolSelectedItems Collection variable, let's play for a while with the lstNames selection and its interface.

■ **Attention** Now that you know that lstNames ListBox selected items are held by the mcolItemsSelected Collection variable, return to the section "Adding a New Name Object" and take a look at the cmdNew_Click( ) event procedure and the way it uses mcolItemsSelected(1) to retrieve the selected item and return the Name object that must be set on the frmEditName.NameSelected property.

Showing Name Properties

Most times we use the frmNames UserForm to select and edit one Name object on its lstNames ListBox. Figure 5-23 shows what happens when the first lstNames item in Figure 5-20 is selected and you click the Show Details ControlButton (cmdDetails) to show the frmNames Detail section.

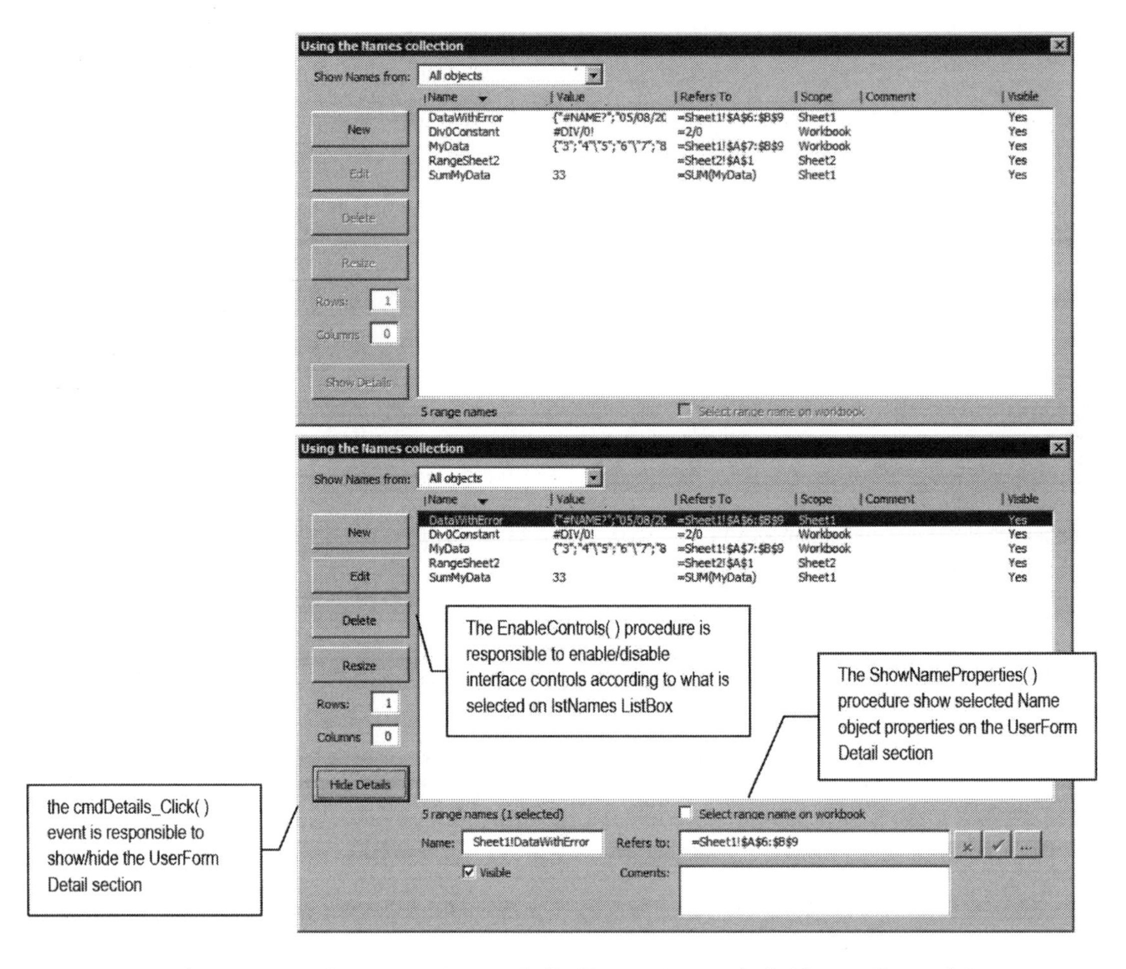

Figure 5-23. *Whenever you select just one item in the* lstNames *ListBox, the* lstNames_Change( ) *event fires, calling the* ShowNameProperties( ) *procedure to show the selected Name properties and the* EnableControls(True) *procedure to enable all the controls*

Showing and Hiding the UserForm Detail Section

When you click, the cmdDetails_Click( ) event fires, executing this code:

```
Const mconHeight1 = 268
Const mconHeight2 = 338

Private Sub cmdDetails_Click()
    Static sbolExended As Boolean
```

```
    Me.Height = IIf(sbolExended, mconHeight1, mconHeight2)
    Me.cmdDetails.Caption = IIf(sbolExended, "Show Details", "Hide Details")
    sbolExended = Not sbolExended
End Sub
```

The cmdDetails_Click( ) event uses the sbolExtended Static Boolean variable to hold the last state of the UserForm Height property (a Static variable does not lose its value between procedure calls).

If sbolExtended = True, the frmNames Detail section is shown and must be hidden, and vice versa. So, the UserForm Height property is changed according to the module-level constants mconHeigh1 and mconHeight2, defined by trial and error by dragging the frmNames bottom margin down and up in the VBA interface and noting the Height property.

```
Me.Height = IIf(sbolExended, mconHeight1, mconHeight2)
```

The same is made to the cmdDetails.Caption property: the sbolExtended value is used to determine whether the frmNames Details section is visible or hidden, alternating its caption from "Show Details" to "Hide Details":

```
Me.cmdDetails.Caption = IIf(sbolExended, "Show Details", "Hide Details")
```

And the sbolExtended variable value alternates its value between True and False each time cmdDetails_Click( ) fires.

```
    sbolExended = Not sbolExended
End Sub
```

Quite simple, huh?

Showing Selected Name Properties

To show the selected Name object properties, the lstNames_Change( ) event calls the ShowNameProperties( ) procedure, which executes this code:

```
Private Sub ShowNameProperties(intIndex)
    Dim rg As Range
    Dim strRefersTo As String
    Dim intPos As Integer

    With Me.lstNames
        Me.txtName = .Column(0, intIndex)
        Me.txtRefersTo = .Column(3, intIndex)
        Me.txtComment = .Column(5, intIndex)
        Me.chkVisible = (.Column(6, intIndex) = "Yes")
    End With

    If Me.chkSelectRangeName Then
        strRefersTo = Mid(txtRefersTo, 2)
        On Error Resume Next
        Set rg = Range(strRefersTo)
        If Err = 0 Then
            intPos = InStr(1, strRefersTo, "!")
```

```
            Worksheets(Left(strRefersTo, intPos - 1)).Activate
            Range(strRefersTo).Select
        Else
            Worksheets(1).Activate
            Range("A1").Select
        End If
    End If
End Sub
```

After declaring its variables, a `With lstNames...End With` loop is used to reference `lstNames` only once, and the `txtName`, `txtRefersTo`, `txtComment`, and `chkVisible` controls of the `frmNames` Detail section receive the associated `Name` object properties, using the appropriate `lstNames ListBox` column.

```
With Me.lstNames
     Me.txtName = .Column(0, intIndex)
     Me.txtRefersTo = .Column(3, intIndex)
     Me.txtComment = .Column(5, intIndex)
     Me.chkVisible = (.Column(6, intIndex) = "Yes")
End With
```

Did you note that `frmNames` has a "Select range name on workbook" check box at the bottom of the `lstNames ListBox` (`chkSelectRangeName`)? If `chkSelectRangeName` is checked, it means that whenever you select any `Name` object on `lstNames`, its cell range must be selected on the worksheet it belongs to. So if `chkSelectRangeName=True`, the procedure uses the VBA `Mid( )` function to remove the = character that precedes the `Name.RefersTo` property and stores the result in the `strRefersTo` variable.

```
If Me.chkSelectRangeName Then
    strRefersTo = Mid(txtRefersTo, 2)
```

An `On Error Resume Next` statement is executed before the code tries to set a reference to the cells range it represents.

```
On Error Resume Next
Set rg = Range(strRefersTo)
```

If the `Name` object is associated to any cell range, no error will be raised, and the sheet name must be extracted from `strRefersTo` so the desired worksheet can be activated before the range is selected. The code stores the position of the ! character that suffixes the sheet name on the `intPos Integer` variable, uses the VBA `Left( )` function to extract the sheet name, and uses it to reference it on the `Worksheets` collection before calling the `Activate` method.

```
If Err = 0 Then
    intPos = InStr(1, strRefersTo, "!")
    Worksheets(Left(strRefersTo, intPos - 1)).Activate
```

And once the worksheet is activated, the cell range is selected.

```
Range(strRefersTo).Select
```

But if the `Name` object is associated to a constant formula, VBA will raise an error (Error: 1004: "method 'Range' of object '_Global' failed"), and you make the decision to select cell `A1` of the first sheet tab whenever this happens. Note that this time I use `Worksheets(1)` to reference the first sheet tab:

```
        Else
            Worksheets(1).Activate
            Range("A1").Select
        End If
    End If
End Sub
```

Figure 5-24 shows the selection of the `DataWithError` Name object range address (`Sheet1!A6:B9`) when `chkSelectRangeName` is checked.

■ **Attention** Whenever you click the `chkSelectRangeName` check box, the `Click ( )` event fires, executing this simple code:

```
Private Sub chkSelectRangeName_Click()
    Call ShowNameProperties(mcolItemsSelected(1))
End Sub
```

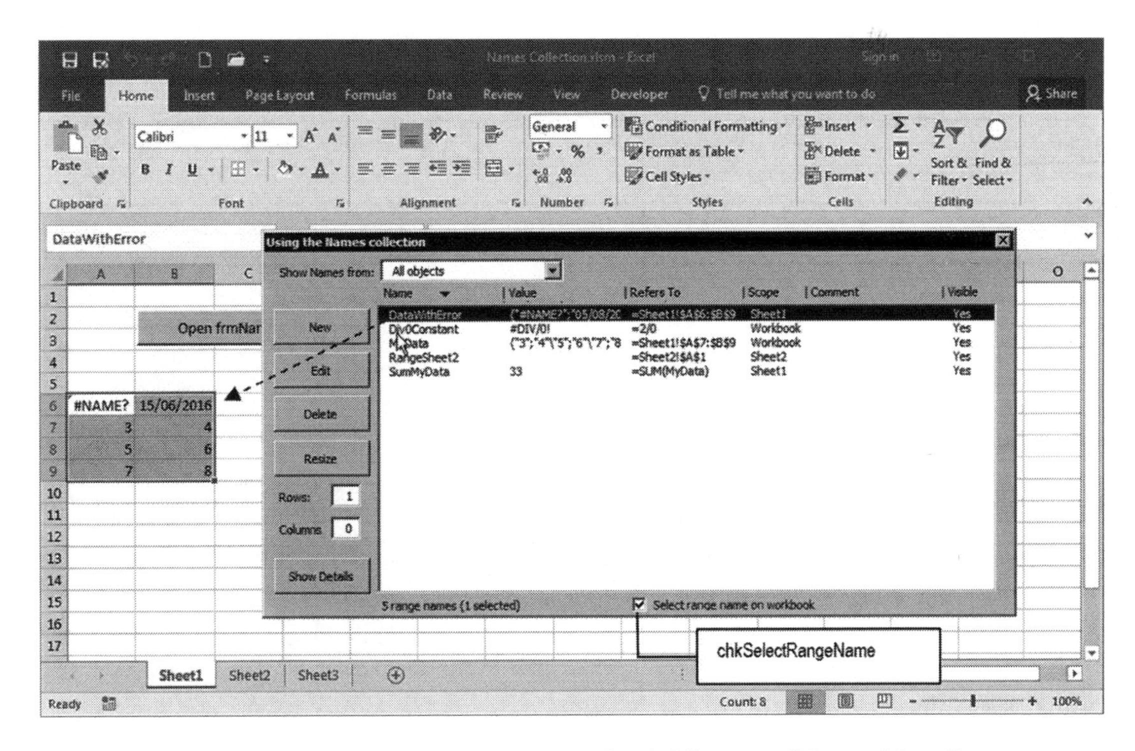

Figure 5-24. *Whenever you check chkSelectRangeName, the Click( ) event will fire, and the cell range associated to the Name object will be selected on the sheet it belongs to. If the Name is associated to a constant formula, cell A1 of the first sheet tab will be selected instead*

Enable/Disable UserForm Controls

The last procedure called by the lstNames_Change( ) event, whenever one or more names are selected on the lstNames ListBox, is the EnableControls( ) procedure.

The EnableControls( ) procedure uses a quite useful, popular, and interesting technique to enable/disable controls on the frmNames UserForm interface. Every VBA control (and I believe that almost any possible control and object) has a Tag property, which is a read/write text string value that you can use to store anything you want, using up to 2,048 characters.

So, each frmNames control that I want to synchronize receives a special value on its Tag property (1 or 2) according to the type of synchronization it is performing.

- Tag = 1 is used for every control that must be enabled when *only one item* is selected on the lstNames ListBox (or disabled when no item is selected).
- Tag = 2 is used on every control that can be used when *more than one item* is selected on the lstNames ListBox (or disabled when any item is selected), as you can see in Figure 5-25.

The EnableControls( ) procedure executes this code:

```
Private Sub EnableControls(bolEnabled  As Boolean)
    Dim intI As Integer

    For intI = 0 To Me.Controls.Count - 1
        Select Case Me.Controls(intI).Tag
            Case "1"
                Me.Controls(intI).Enabled  = bolEnabled
            Case "2"
                Me.Controls(intI).Enabled  = (bolEnabled  Or mcolItemsSelected.Count > 0)
        End Select
    Next
End Sub
```

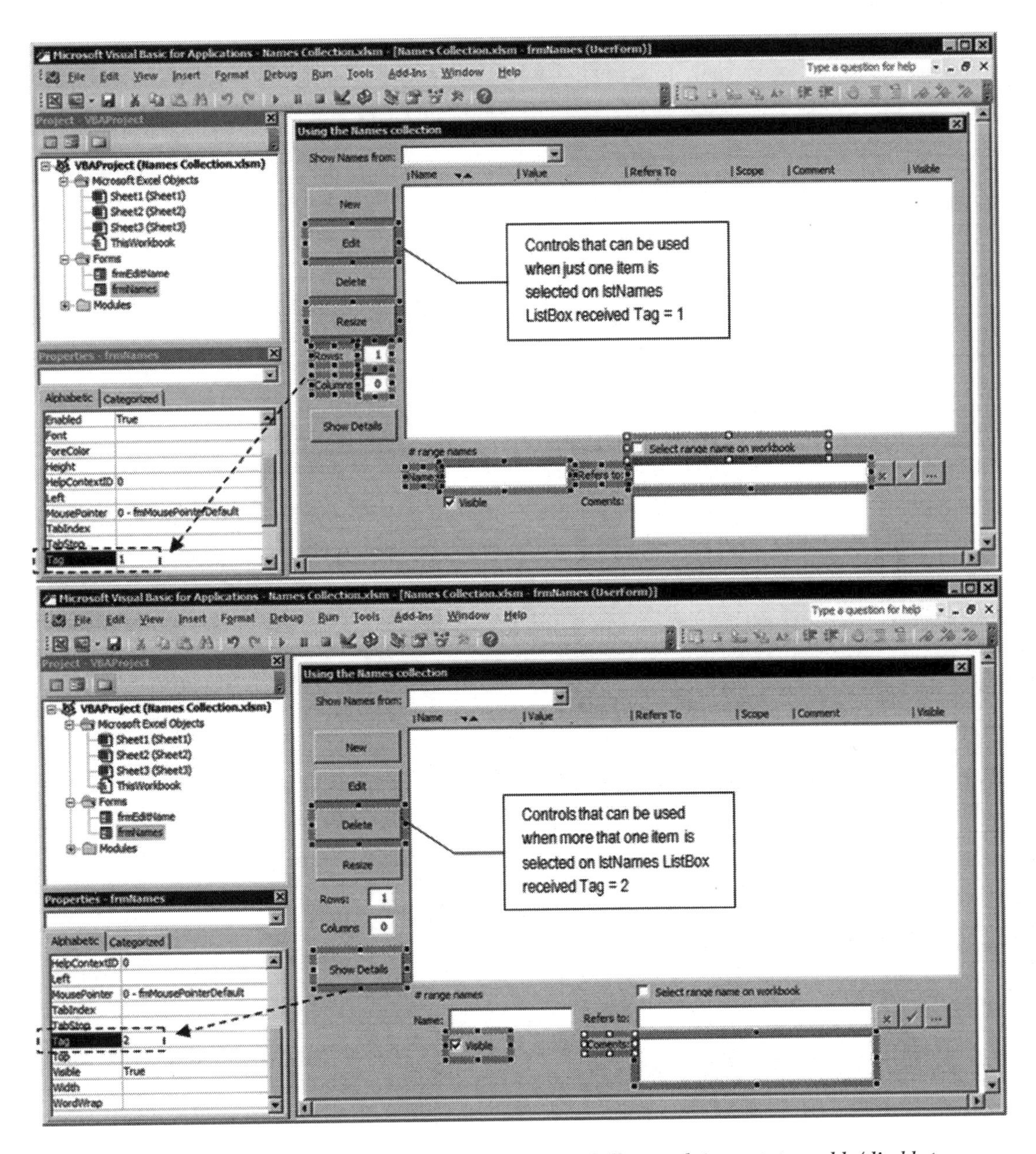

Figure 5-25. *The frmNames UserForm uses the Tag property of all controls it wants to enable/disable to synchronize its interface. Whenever any control Tag = 1, this means that the control can be enabled/disabled by the UserForm EnableControls( ) procedure*

The EnableControls( ) procedure uses the frmNames Controls collection to run through all its controls using a For int=0 to Controls.Count-1...Next loop. At each loop passage, it takes the current control Tag property and verifies its value using a Select Case statement.

```
For intI = 0 To Me.Controls.Count - 1
    Select Case Me.Controls(intI).Tag
```

If the control Tag = "1" (the Tag property is a String value), just one item was selected on the lstNames ListBox, and the control Enabled property receives the bolEnabled argument, enabling or disabling the control.

```
Case "1"
     Me.Controls(intI).Enabled  = bolEnabled
```

But if the control Tag = "2", you have more than one item selected in the lstNames ListBox, so the control must be enabled if bolEnabled = True *or* mcolItemsSelected.Count > 1 (to disable all controls that must be available for just a single selected item).

```
            Case "2"
                Me.Controls(intI).Enabled  = (bolEnabled  Or mcolItemsSelected.Count > 0)
        End Select
    Next
End Sub
```

Attention Note that whenever bolEnabled = False and mcolItemsSelected.Count = 0, all tagged controls will be disabled in the UserForm interface, which will happen whenever you open frmNames or delete a Name object and no item is selected in the lstNames ListBox.

Figure 5-26 shows EnableControls( ) in action selecting just controls with Tag = "2" whenever more than one item is selected in the lstNames ListBox.

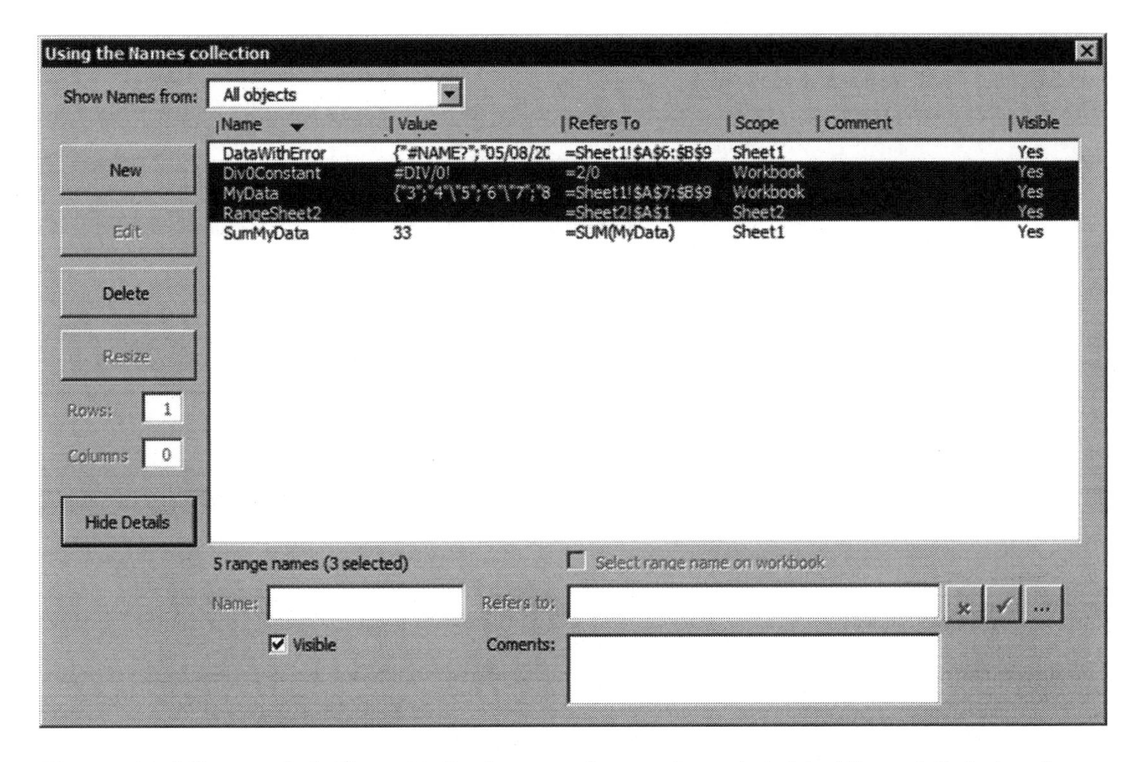

Figure 5-26. Whenever the lstNames ListBox has more than one item selected, just the controls that can be used to operate on one or more Name objects becomes enabled (the ones with property Tag= "2")

Clearing frmNames Detail Section

To clear all the `frmNames` Detail controls whenever more than one item is selected in the `lstNames ListBox`, the `frmNames_Change( )` event calls the `ClearNameProperties( )` procedure, which executes this code:

```
Private Sub ClearNameProperties()
    Me.txtName = ""
    Me.txtRefersTo = ""
    Me.txtComment = ""
    Me.chkVisible = False
End Sub
```

▒ **Attention** Whenever you have cleaned up an undefined number of controls, consider using the same technique described in the previous section: use the controls' `Tag` property to specifically tag them and clear all controls at once using a `For...Next` loop that runs through the `UserForm Controls` collection. Since text box and check box controls are cleared using different values ("" for text boxes and 0 for check boxes), you can add different `Tag` property values to each control type or use the VBA `TypeOf(Controls(intI))` function to verify the control type and act accordingly.

Now that you know how many different procedures and VBA instructions run whenever you select one or more `lstNames ListBox` items, you must think about this unquestioning truth: computers are really fast!

Editing an Existing Name Object

Try to select any `lstName ListBox` item alone (like the last one) and click the Edit (`cmdEdit`) `ControlButton` to edit it in the `frmEditName UserForm` (Figure 5-27).

This happens because the `cmdEdit_Click( )` event fires, executing this code:

```
Private Sub cmdEdit_Click()
    Me.Hide
    If mcolItemsSelected.Count > 0 Then
        With frmEditName
            .NameSelected = Names(Me.lstNames.Column(0, mcolItemsSelected(1)))
            .Show
        End With
    End If
    Me.Show
End Sub
```

The bold instruction indicates that `cmdEdit_Click( )` gets the `Index` of the selected `lstNames` item from the `mcolItemsSelected(1)` item collection and uses this item to get the `Name.Name` property (stored on the hidden `lstNames.Column(0)` column), uses the `Application.Names` collection to return the selected `Name` object reference, and stores it in the `frmEditName.NameSelected` property.

```
With frmEditName
    .NameFilter = Me.cboObjects
    .NameSelected = Names(Me.lstNames.Column(0, mcolItemsSelected(1)))
```

One single code line for so much activity!

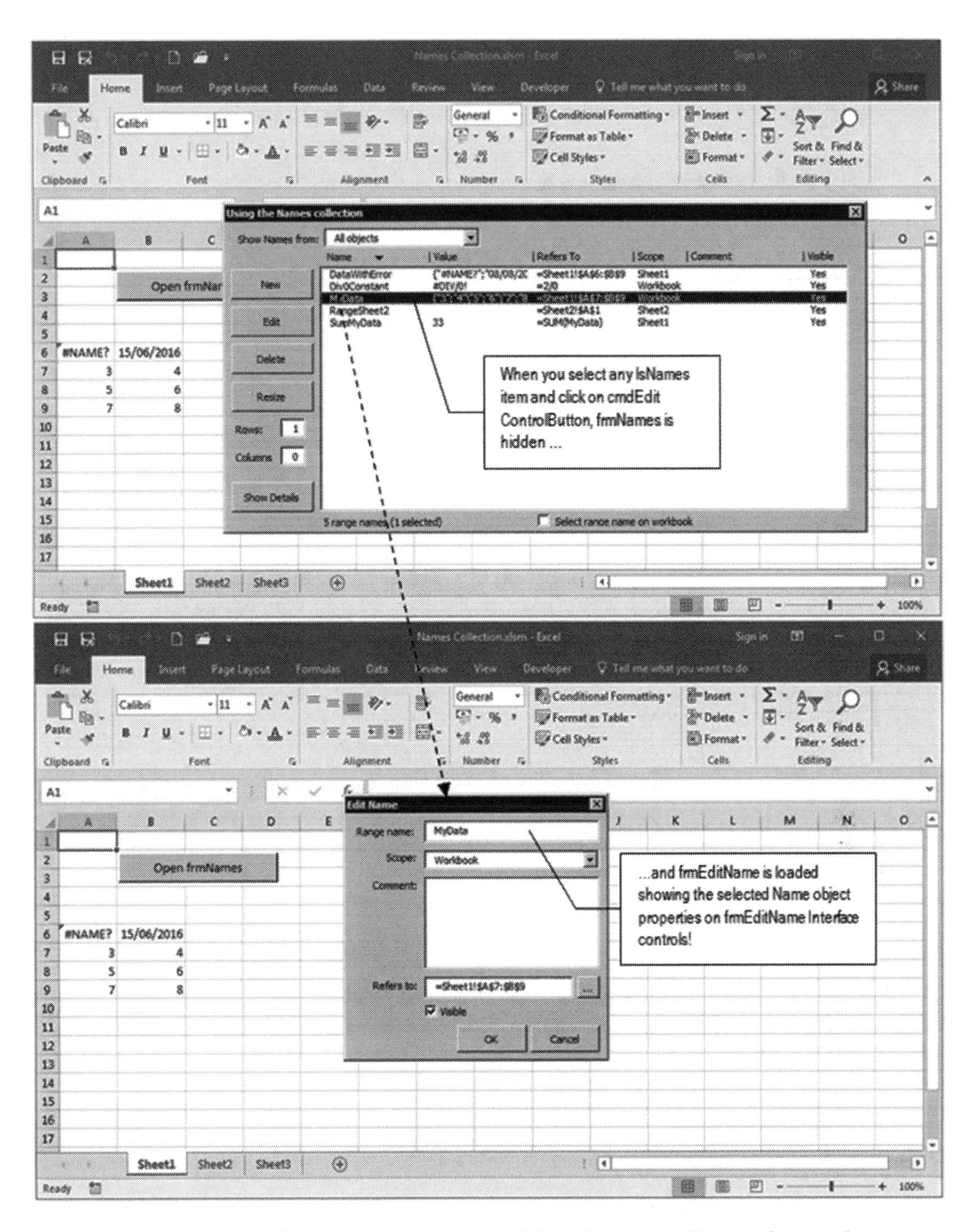

Figure 5-27. *Whenever you click any lstNames item and click the frmNames Edit ControlButton, the cmdEdit_Click( ) event fires and passes the selected Name object reference to the frmEditName.NameSelected property, which gets all it basic properties (Parent, RefersTo, Comment and Visible) and shows it in the frmEditName interface*

When this happens, the frmEditName.NameSelected property's Let( ) event fires and executes this code:

```
Public Property Let NameSelected(ByVal nm As Name)
    Dim intPos As Integer

    Set mName = nm

    If Not Me.NewName Then
        intPos = InStr(1, nm.Name, "!")
        If intPos = 0 Then
            Me.txtName = nm.Name
            Me.cboObjects = "Workbook"
        Else
            Me.txtName = Mid(nm.Name, intPos + 1)
            Me.cboObjects = nm.Parent.Name
        End If

        Me.txtComment = nm.Comment
        Me.txtRefersTo = nm.RefersTo
        Me.chkVisible = nm.Visible
    End If
End Property
```

Note that frmEditName Property Let NameSelected( ) receives the Name object reference on its nm As Name argument and passes this object reference to the frmEditName module-level object variable mName, so it can be used by the cmdOK and cmdCancel Click events.

```
Public Property Let NameSelected(ByVal nm As Name)
    Dim intPos As Integer

    Set mName = nm
```

The code verifies the frmEditName.NewName property, testing Not Me.NewName, which will run the frmEditName.NewName property's Get( ) procedure. Since Not Me.NewName = True, the Name object properties must be retrieved and exhibited on the UserForm interface.

To know the Name object scope, the code searches the Name.Name property for the ! character used to separate the sheet tab name from the Name.Name property using the VBA InStr( ) function and stores the result into the intPos Integer variable.

If there is no such character on the Name.Name property, the Name object has workbook scope, and the cboObject ComboBox is set accordingly.

```
intPos = InStr(1, nm.Name, "!")
If intPos = 0 Then
    Me.txtName = nm.Name
    Me.cboObjects = "Workbook"
```

If the Name object has a worksheet scope and txtName receives just the Name.Name property (without the preceding sheet name), using the VBA Mid( ) function to extract it, and cboObject ComboBox has its value set by the Name.Parent.Name property, which returns the worksheet object sheet tab name.

```
        Else
            Me.txtName = Mid(nm.Name, intPos + 1)
```

```
            Me.cboObjects = nm.Parent.Name
        End If
```

The procedure finishes defining other Name object properties to the appropriate frmEditName controls.

```
        Me.txtComment = nm.Comment
        Me.txtRefersTo = nm.RefersTo
        Me.chkVisible = nm.Visible
    End If
End Property
```

Editing the Name Object

Since the desired Name object reference is set to the frmEditName module-level mName variable, you can change any of its properties, including its scope, which the Excel Edit Name dialog box doesn't allow.

The Microsoft Excel New/Edit Name dialog box does not allow you to change any Name object scope but allows you to change the Name object's Name property if the selected name is not used by any other Name object's RefersTo property.

Let's see a special case. The MyData range name is used by the SumMyData range name on its RefersTo property constant formula. You want to use Excel Edit Name dialog box to change the MyData name to anything else, like the MyNewData name, but Excel will not allow you to do that because this name is used by other Name.RefersTo property. Neither allows you to change the MyData scope, from Workbook to Sheet1 (Figure 5-28).

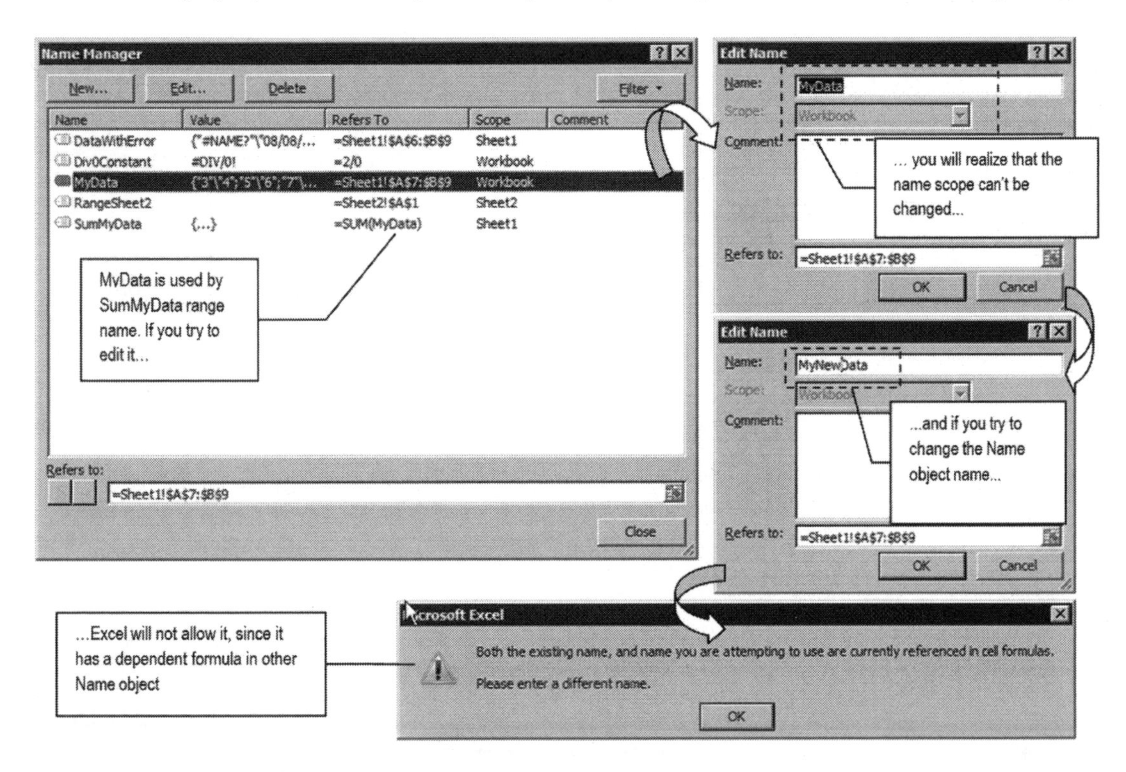

Figure 5-28. The Excel Name Manager and Edit Name dialog box don't allow you to change any Name object's Name property when it is used by any other Name object (which happens on the MyData range name used by the SumMyData range name constant formula). In addition, they don't allow you to change the name scope (Edit Name dialog box's Scope ComboBox is disabled)

My vision of frmNames and frmEditName is to allow such interesting and desirable changes both on any Name object's Name property and on its scope, fixing any possible formula conflict that may arise.

Supposing that you want to change the MyData Name object to MyNewData and its scope from Workbook to Sheet1 using frmNames, double-click the name to show it on frmEditName, type the new name in the txtName text box, select Sheet1 in the cboObjects ComboBox, and press the frmEditName OK button, which will run the cmdOK_Click( ) event (already explored in "Adding a New Name Object" earlier on this chapter), executing this partial code (where the first validating instructions were removed):

```
Private Sub cmdOK_Click()
    Dim nm As Name
    Dim strName As String
   ...
    If Not Me.NewName Then
        Call FixNameChange
        Me.NameSelected.Delete
    End If

    Set nm = Names.Add(strName, Me.txtRefersTo, Me.chkVisible)
    nm.Comment = Me.txtComment & ""
    Set mName = nm

    Unload Me
End Sub
```

You now have opened another great "parenthesis" on the frmEditName cmdOK_Click( ) event to explain how Sub FixNameChange( ) works to change the proposed name change on any Excel formula of the entire workbook.

Searching and Replacing Formula Content

Since you are editing an existing Name object, frmEditName.NewName = False. The code first calls the Sub FixNameChange( ) procedure to replace the old name with the new name inside any formula used on the workbook.

```
    If Not Me.NewName Then
        Call FixNameChange
```

The Sub FixNameChange( ) procedure executes this code:

```
Private Sub FixNameChange()
    Dim nm As Name
    Dim ws As Worksheet
    Dim rg As Range
    Dim rgInitial As Range
    Dim strName As String
    Dim strRefersTo As String

    strName = mName.Name
    strName = Mid(strName, InStr(1, strName, "!") + 1)
    If strName <> Me.txtName Then
        'Change Name references
```

```
        For Each nm In Names
            strRefersTo = nm.RefersTo
            If InStr(1, strRefersTo, strName) Then
                strRefersTo = Replace(strRefersTo, strName, Me.txtName)
                nm.RefersTo = strRefersTo
            End If
        Next

        'Change cells formulas references
        For Each ws In ActiveWorkbook.Worksheets
            Set rg = ws.Cells.Find(strName, , xlFormulas, xlPart)
            If Not rg Is Nothing Then
                Set rgInitial = rg
                Do
                    rg.Formula = Replace(rg.Formula, strName, Me.txtName)
                    Set rg = ws.Cells.FindNext(rg)
                    If rg Is Nothing Then Exit Do
                Loop While (rg.Address <> rgInitial.Address)
                Set rgInitial = Nothing
            End If
        Next
    End If
End Sub
```

Whenever you change the `Name.Name` property, you must make this change in two different places in any Excel workbook: inside any `Name.RefersTo` property and inside any worksheet cell formula.

Since any worksheet scope `Name` object has its `Name` property prefixed with the sheet name, the procedure first searches for the sheet name existence and removes it, storing just the name on the `strName` variable.

```
strName = mName.Name
strName = Mid(strName, InStr(1, strName, "!") + 1)
```

Then `FixNameChange( )` must first verify whether the `Name` object you want to save suffers any change on its `Name` property, comparing the module-level object variable `mName.Name` property to `frmEditName` `txtName` text box value.

```
If strName <> Me.txtName Then
```

If both name strings don't match, you are proposing a `Name` property change, and the procedure will first use a `For Each....Next` loop to run through the `Names` collection searching for the current `Name` property in any `Name.RefersTo` property. Note that `strRefersTo` holds the `Name.RefersTo` property, and the VBA `InStr( )` function is used to find the current `Name` property stored in the `strName` variable.

```
For Each nm In Names
    strRefersTo = nm.RefersTo
    If InStr(1, strRefersTo, strName) Then
```

Since `InStr( )` returns the initial position of `strRefersTo` inside `strName`, whenever `InStr( ) > 0` (`strRefersTo` found inside `strName`), the code uses the VBA `Replace( )` function to make the desired change and stores the replaced property again in the `Name.RefersTo` property.

```
strRefersTo = Replace(strRefersTo, strName, Me.txtName)
nm.RefersTo = strRefersTo
```

The VBA Replace( ) function is quite useful because it can quickly search and replace any substring inside a desired string, returning another string with the desired replacement. It has this syntax:

```
Replace (Expression, Substring, Replacement, [Start, [Count, [Compare]]] )
```

In this code:

Expression: This is required; it is the string expression containing the substring to replace.

Substring: This is required; it is the substring being searched for.

Replacement: This is required; it is the replacement string.

Start: This is optional; it is the position in Expression where the substring search is to begin. If omitted, 1 is assumed.

Count: This is optional; it is the number of substring substitutions to perform. If omitted, the default value is -1, meaning "all possible substitutions."

Compare: This is optional; it is the type of comparison to use when evaluating substrings.

Vbbinarycompare: This makes a binary comparison (case sensitive).

Vbtextcompare: This makes a textual comparison (case insensitive).

And once the For...Next loop ends, all possible formulas inside any workbook Name object will be correctly replaced by the new proposed Name property.

And once this is made, it is time to search all workbook cells formulas and make the same substitution, and this process is made by the Range.Find method.

Using the Range.Find Method

Microsoft Excel has the Find and Replace dialog box, which is used to search and replace items inside a single worksheet or on the entire workbook. This dialog box uses the Excel Range object's Find method, which has this syntax:

```
Expression.Find(What, After, LookIn, LookAt, SearchOrder, SearchDirection, MatchCase,
MatchByte, SearchFormat)
```

In this code:

Expression: This is an object variable that represents a Range object.

What: This is required; it is the data to search for.

After: This is optional; it is a single cell after which you want the search to begin. It corresponds to the position of the active cell when a search is done from the user interface. The search will begin after this cell, which will not be searched until the method wraps back around to it. If you don't specify this argument, the search starts after the cell in the upper-left corner of the range (the default is the current selection).

LookIn: This is optional; it is where the information will be searched (values, formulas, or comments) and can be set to xlValue (default), xlFormula, or xlComments.

LookAt: This is optional; it is the scope to be searched in LookIn, all or part, and can be set to the following XlLookAt constants: xlWhole (default) or xlPart.

SearchOrder: This is optional; it is the search order, by row or column. It can be set to one of the following XlSearchOrder constants: xlByRows (default) or xlByColumns.

SearchDirection: This is optional; it is the search direction to be made and can be set to the following XlSearchDirection constants: xlNext (default) or xlPrevious.

MatchCase: This is optional; it allows the search to be case sensitive. The default value is False.

MatchByte: This is optional; it is used only if you've selected or installed double-byte language support. Set it toTrue to have double-byte characters match only double-byte characters. Set it to False (default) to have double-byte characters match their single-byte equivalents.

SearchFormat: This is optional; it is for making search use specific character formatting options.

■ **Attention** All Range.Find arguments are used by the Excel Find and Replace dialog box, and every change you make to one of them will be shown in the Excel Find and Replace interface.

The Range.Find method returns Nothing if any cell is found or if a Range object representing a *single cell* was found. You can continue to search more cells with the same settings using the Range.FindNext or Range.FindPrevious method.

Note that when the search reaches the end of the specified search range, it wraps around again to the beginning of the range. So, to stop a search when this wraparound occurs, save the address of the first found cell and then test each successive found-cell address against this saved address.

The Range.Find is method is quite interesting because it searches any range size. Since any Worksheet object has a Cells property, which returns a Range object with all worksheet cells, any time you use Cells. Find you are in fact searching all active sheet cells!

So, the procedure needs to search every Worksheet object inside the workbook using a For...Each loop to run through the Worksheets collection and then use the Worksheet object's Cells property to return a range object and apply its the Range.Find method, as follows:

```
'Change cells formulas references
For Each ws In ActiveWorkbook.Worksheets
    Set rg = ws.Cells.Find(strName, , xlFormulas, xlPart)
```

This last instruction will search all cells of a single Worksheet object (represented by the ws variable), *looking in* the cell formula (xlFormulas) for any part (xlPart) that has what is stored in the strName variable (which holds the current Name.Name property).

If no cell is found having strName inside any part of its formula, the rg as Range variable will hold Nothing, and the procedure will end the If...End If instruction, selecting another Worksheet object in the Worksheets collection.

```
For Each ws In ActiveWorkbook.Worksheets
    Set rg = ws.Cells.Find(strName, , xlFormulas, xlPart)
    If Not rg Is Nothing Then
        ...
    End If
Next
```

But if any cell is found having strName inside any formula part, the rg variable will have a reference to it, and in this case, you must store the reference of this first cell found in the rgInitial variable.

```
If Not rg Is Nothing Then
   Set rgInitial = rg
```

And a Do...Loop is initiated to change this first cell formula using the VBA Replace( ) function to change any current Name.Name property (strName) by the new proposed name (Me.txtName).

```
Do
  rg.Formula = Replace(rg.Formula, strName, Me.txtName)
```

Now is the interesting part. You must continue the search through all other worksheet cells using the Range.FindNext method, which has this syntax:

```
Expression .FindNext(After)
```

In this code:

Expression: This is an object variable that represents a Range object.

After: This is optional; it is a single cell after which you want to search (the position of the active cell when a search is done from the user interface). The search will begin after this cell, meaning that it will not be searched until the method wraps back around to it. If this argument is not specified, the search starts after the cell in the upper-left corner of the range.

■ **Attention** The Range.FindPrevious method obeys the same rules of Range.FindNext when searching backward.

So, you must continue to search the current worksheet *after* the last cell found using the Range.FindNext method, passing to its After argument a reference to the last found cell.

```
Set rg = ws.Cells.FindNext(rg)
```

And when the Range.FindNext must be executed, one of these conditions may occur: no other cell is found (rg = Nothing) or the first found cell searches again. And you need to make a double comparison.

Whenever you need to verify whether an object variable is Nothing (doesn't point to any object), you must make this test alone. Let me put this in other words: you cannot test on a single instruction if (rg = Nothing) or (rg.Address = rgInitial.Address) because if rg = Nothing, rg.Address returns a VBA error. rg points to no Range object, so it cannot have an Address property! So, the next row verifies whether the Range.FindNext method finds another cell. If it is equal to Nothing, you can stop searching on this Worksheet object and the Exit Do instruction will be executed.

```
Set rg = ws.Cells.FindNext(rg)
If rg Is Nothing Then Exit Do
```

Otherwise, another cell is found, and the Do...Loop must continue changing its Formula property. Note that this time the Address property is used to compare the initial range object with the new found one as a condition to end the Do...Loop.

```
Do
  rg.Formula = Replace(rg.Formula, strName, Me.txtName)
  Set rg = ws.Cells.FindNext(rg)
  If rg Is Nothing Then Exit Do
Loop While (rg.Address <> rgInitial.Address)
```

▒ **Attention** The Range.Value property (the default property) returns the range value, which can be either a single value or an array of values, so you cannot compare range values directly. The same applies to the Range.Name property, which just exists for any individual named cell. So, since Range.Find and Range.FindNext return a single cell reference, you can surely use the Range.Address property to compare the initial and found ranges.

If the last found cell is the same cell processed (rg.Address = rgInitial.Adress), it can't be processed again, because if you are changing a name just by adding a suffix (like changing the MyData name to MyData2), when the search returns to the first cell, it will find again the MyData prefix inside the now changed MyData2 formula and will substitute it again and again, until you end up with a range name of MyData222222... (with 2 repetitions enough to fill up the maximum 256 characters formula limit), and Excel will raise an error.

So, when the code returns to the first changed cell, the Do...Loop ends, the rgInitial variable is set to Nothing, and another worksheet object is processed again, until all worksheet cell formulas are correctly processed for the proposed name change.

```
        Loop While (rg.Address <> rgInitial.Address)
        Set rgInitial = Nothing
    End If
Next
```

▒ **Attention** Go back to Sub FixNameChange( ) and note how easy it is to use the Range.Find method to search and replace any workbook cell value using VBA, the same way the Excel Find and Replace dialog box does!

Now that you know how FixNameChange( ) works, so let's return to the frmEditName cmdOK_Click( ) event and continue the VBA process of editing an existing Name object.

Getting Back to the frmEditName cmdOK_Click Event

After all the workbook formulas have been adequate processed, the cmdOK_Click( ) event must delete the currently edited Name object, because if you change the Name.Name property or the Name.RefersTo property by changing just its scope (from Workbook to any sheet name, and vice versa), you end up with two names (two different Name objects, having different Name properties with the same Name property but different scopes).

```
Private Sub cmdOK_Click()
    Dim nm As Name
    Dim strName As String
   ...
    If Not Me.NewName Then
        Call FixNameChange
        Me.NameSelected.Delete
    End If
```

And once the Name object is deleted, it is re-created with the new desired Name properties, the same way you created a new Name.

```
    Set nm = Names.Add(strName, Me.txtRefersTo, Me.chkVisible)
    nm.Comment = Me.txtComment & ""
```

After the Name object is re-created, the frmEditName mn object module-level variable (which returns the frmEditName.NameSelected property value) is associated to the new added name, and the UserForm is unloaded from memory.

```
    Set mName = nm

    Unload Me
End Sub
```

■ **Attention** According to the Microsoft Excel documentation, you do not need to delete a name to change its properties. Just use again the Names collection's Add method to re-create it. But you must delete it before using the Add method whenever you want to change its Name property or its scope.

Canceling Name Editing

If you decide not to make changes to the edited Name object, you can press the keyboard Esc key or click the frmEditName Cancel ControlButton (cmdCancel, which has its Cancel property set to True), and this code will be executed:

```
Private Sub cmdCancel_Click()
    Unload Me
End Sub
```

Synchronizing the frmEditName and frmNames Interfaces

Whenever you click the frmEditName cmdOK of cmdCancel control buttons, the UserForm is unloaded from memory and its UserForm_Terminate( ) event fires, executing this code:

```
Private Sub UserForm_Terminate()
     With frmNames
        .cboObjects = Me.NameFilter
        If Not (Me.NameSelected Is Nothing) Then
            .NameSelected = Me.NameSelected
        End If
        .Show
    End With
End Sub
```

Note that frmEditName.NameFilter (what was selected on the frmNames cboObjects ComboBox) is used to define again the frmNames.cboObjects value (which will cascade-fire the cboObjects_Change( ) event).

```
     With frmNames
        .cboObjects = Me.NameFilter
```

Next, the code verifies frmEditName.NameSelected has any name associated with it (which will not happen whenever frmNames has no item selected). If this is true, the frmNames.NameSelected Property Let( ) procedure receives the frmEditName.NameSelected Property Get( ) procedure, selecting the appropriate item on in frmNames lstNames ListBox, which is shown by frmNames whenever the Show method finally executes.

```
        If Not (Me.NameSelected Is Nothing) Then
            .NameSelected = Me.NameSelected
        End If
        .Show
    End With
End Sub
```

■ **Attention** To see a Name.Name property change in action, you must create one cell formula that references the name (like =Sum(MyData)) and then copy and paste this cell to other multiple cells of any sheet tab and then perform the name change. You will see that VBA will change all Name objects and cell formula references to the new desired name, keeping everything working as it should on the workbook. Try it!

Resizing an Existing Name Object

One common operation on most Excel applications is to resize a given Name object whose RefersTo property is associated with a contiguous worksheet cell addresses by adding or deleting one or more rows or columns to/from it. This operation is performed by the Range.Resize property, which has this syntax:

```
Expression .Resize(RowSize, ColumnSize)
```

In this code:

> Expression: This is required; it is an expression that returns a Range object to be resized.
>
> RowSize: This is optional; it is the total number of rows in the new range. If omitted, the number of rows in the range remains the same.
>
> ColumnSize: This is optional; it is the number of columns in the new range. If omitted, the number of columns in the range remains the same.

So, whenever you want to resize a given range name, you must use the Range Rows and/or Columns collection's Count property to retrieve the current number of rows/columns and then add/subtract to these values to/from the desired final range dimensions.

The next syntax resizes a hypothetical Name object Range1 address, by adding to it one more row:

```
Dim rg as Range
Set rg = Range("Range1").Resize(Range("Range1").Rows.Count + 1)
```

Note that since the second resize argument is missing, the Range1 Name object will keep its current column number. This operation can be shortened by first setting the desired Name object to a Range object variable, as follows:

```
Dim rg as Range
Set rg = Range("Range1")  ' Range1 is the Name.Name property
Set rg = rg.Resize(rg.Rows.Count +1)
```

As you can see, the final operation ends with an rg object variable with the desired dimensions, but no change had been made to the Name object's RefersTo property: it remains the same!

To really resize a Name object by adding/subtracting it to/from one or more rows/columns, you must use the Range.Name property to name the new range with the Name.Name property. The next instructions really change the Range1 Name object, adding one more row to its RefersTo property:

```
Dim rg as Range
Set rg = Range("Range1")  ' Range1 is the Name.Name property
rg.Resize(rg.Rows.Count +1).Name = "Range1"
```

You saw that? You must perform a double operation: rg.Resize returns a new Range object with the desired dimensions, which has a Name property, which is set to the desired Name object, effectively resizing it!

Now let's return to frmNames. Whenever you select a valid Name object in the lstNames ListBox (one that is associated to any cell address, not a constant formula) and click the Resize ControlButton (cmdResize), the cmdResize_Click( ) event fires, executing this code:

```
Private Sub cmdResize_Click()
    Dim nm As Name
    Dim rg As Range
    Dim intIndex As Integer

    On Error Resume Next

    intIndex = mcolItemsSelected(1)
    Set nm = Names(Me.lstNames.Column(0, intIndex))
    Set rg = Range(nm.RefersTo)
```

```
    If Err = 0 Then
        rg.Resize(rg.Rows.Count + Me.txtAddRows, rg.Columns.Count + Me.txtAddColumns).Name =
        nm.Name
        Call FilllstNames
        Call SelectItems
        Call chkSelectRangeName_Click
    End If
End Sub
```

Since cmdResize will be enabled only when just one item is selected in the lstNames ListBox, after declaring the variable it needs, the code begins by disabling VBA errors, stores the selected lstName.Index value from the mcolItemsSelected(1) collection on the intIndex variable, and uses this value to set a reference to the selected Name object:

```
On Error Resume Next

intIndex = mcolItemsSelected(1)
Set nm = Names(Me.lstNames.Column(0, intIndex))
```

Once the selected Name object is retrieved, the code tries to set a Range object reference to it, using the Name.Name property.

```
Set rg = Range(nm.RefersTo)
```

Since the selected Name object can be associated to a constant formula, if you select such types of Name, VBA will raise an error (Error = 1004, "method 'Range' of object '_Global' failed"). So, the next instruction verifies whether there is any error setting the Range object reference, and if it is true, the procedure ends doing nothing.

```
    If Err = 0 Then
       ...
    End If
End Sub
```

But if the Range object reference is adequately set, the Name is associated to a valid range address: the Range.Resize property is evoked, adding to the Range.Address property the default values of the txtAddRows (1) and txtAddcolumns (0) text boxes (the ones below cmdResize in the frmNames interface).

This will resize the Range object, which is then named using the Name.Name property, effectively resizing the Name object address.

```
rg.Resize(rg.Rows.Count + Me.txtAddRows, rg.Columns.Count + Me.txtAddColumns).Name = nm.Name
```

And once this operation is done, the frmNames interface must be synchronized by calling FilllstNames( ) (to update lstNames ListBox), SelectedItems( ) (to reselect the resized name on lstNames), and chkSelectRangeName_Click( ) events (to select the new dimensions of the Name object on the worksheet it belongs to if chkSelectRangeName is checked).

```
Call FilllstNames
Call SelectItems
Call chkSelectRangeName_Click
```

This is simple code used by the Sub SelectItems( ) procedure to reselect the same lstNames items after any change is made to one or more Name object properties.

```
Private Sub SelectItems()
    Dim varItem As Variant

    'Reselect names on ListBox
    Me.lstNames.ListIndex = mcolItemsSelected(1)
    For Each varItem In mcolItemsSelected
        mbolCancelEvent = True
        Me.lstNames.Selected(varItem) = True
    Next
End Sub
```

As you can see, the code begins by selecting the first Collection item in the ListBox, changes the ListIndex property, and then performs a For Each...Next loop through all the mcolItemsSelected collection items, setting mbolCancelEvent = True to avoid cascading the lstNames_Change( ) event and then reselecting the desired items on the lstNames ListBox.

■ **Attention** If the chkSelectRangeName check box is checked before you click the frmNames Resize ControlButton, you will see the new name cells selected on the worksheet it belongs to. The chkSelectRangeName_Click( ) event calls the Sub ShowNameProperties( ) procedure, analyzed in the section "Show Selected Name Properties" earlier in this chapter.

Note that you can use negative values on the txtAddRows and txtAddColumns text boxes to shrink any range name.

Figure 5-29 shows how the MyData Name object has the RefersTo property resized by clicking the cmdResize button, accepting the frmNames default resizing values (adding to it one row).

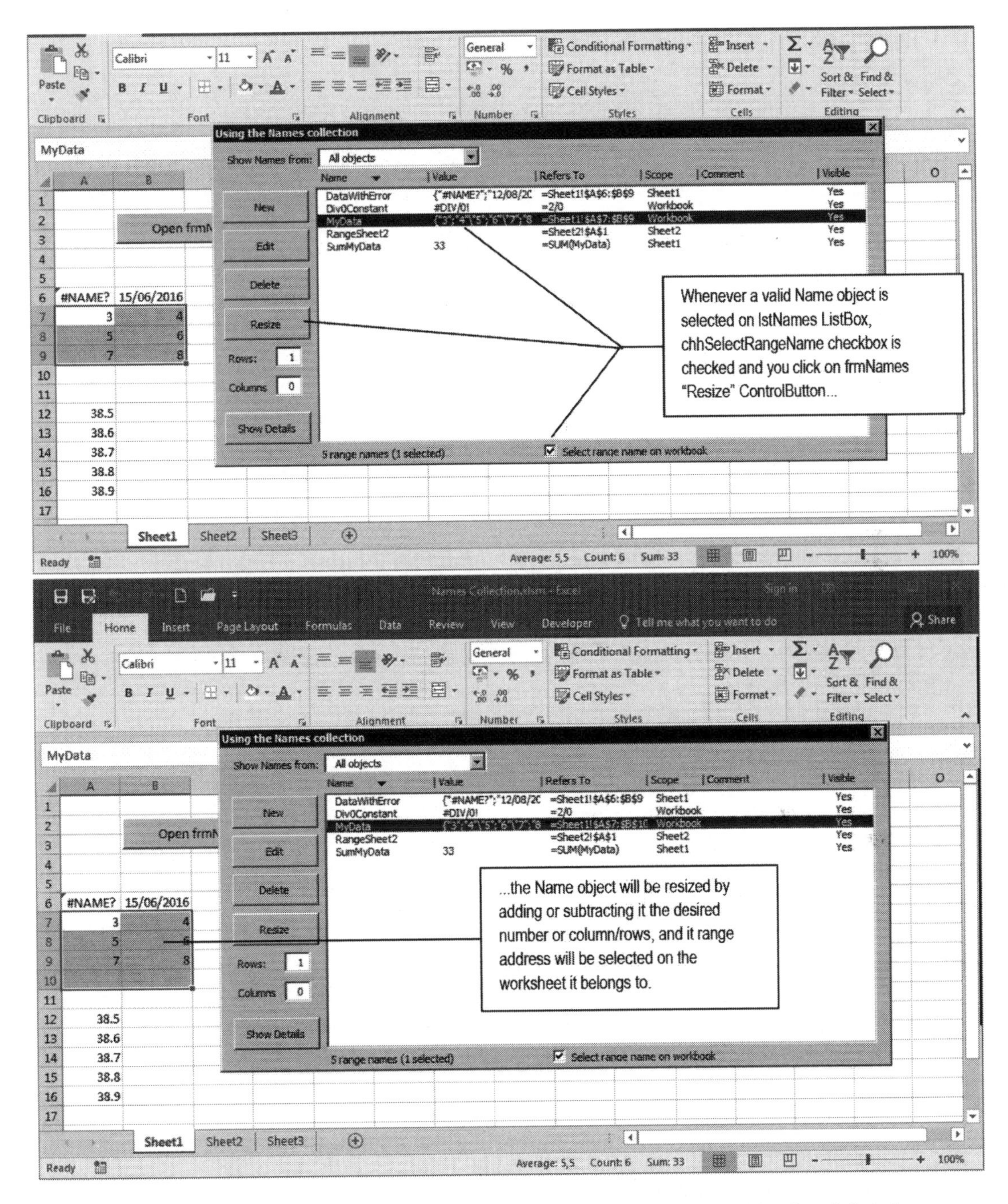

Figure 5-29. *When you select a valid Name object (one associated to any range address) and click the frmNames Resize ControlButton, the cmdResize_Click( ) event fires and adds the specified number of rows and columns to the Name.RefersTo property. This time the MyData Name was added by one row, which can be seen on the worksheet since the "Select range name on workbook" option is checked*

Performing Multiple Name Properties Changes

The frmNames UserForm allows you to delete one or more range names or change the Visible and Comment properties at once: a desired Name object operation not performed by the Excel Name Manager dialog box (which allows multiple deletions).

Since all selected Name objects are stored in the mcolItemsSelected collection, whenever desired you can run through all collection items and perform the same operations on all of them. Let's see this in action!

Changing the Name.Visible Property

Although every Name object has a Visible property, Excel doesn't allow the user to change it, unless you use VBA code to do it. When any name has Visible = False, it doesn't appears anymore in the Excel Name box or Excel Name Manager, which is quite desirable for most Excel applications to hide development details from users' eyes.

To change any Name object's Visible property using the frmNames interface, follow these steps:

1. Select the desired Name objects on the lstNames ListBox.
2. Click the frmNames Show Details ControlButton.
3. Click the Visible check box.

Whenever you click the chkVisible check box, the chkVisible_AfterUpdate( ) event fires, executing this simple code:

```
Private Sub chkVisible_AfterUpdate()
    Dim varItem As Variant

    For Each varItem In mcolItemsSelected
        Names(CStr(Me.lstNames.Column(0, varItem))).Visible = Me.chkVisible
    Next
    Call FilllstNames
    Call SelectItems
End Sub
```

Quite simple, huh? A For Each...Next loop is performed through the mcolItemsSelected collection, and each selected Name object has the Visible property changed to the chkVisible state, effectively making it hidden/visible each time the chkVisible check box has its value changed.

Note that Sub FilllstNames( ) is called to update the frmNames interface with the new Name object properties, and Sub SelectItems( ) is called to select again the same lstNames items.

▒ **Attention** The chkVisible_AfterUpdate( ) event does not cascade-fire whenever the chkVisible value is changed, which happens whenever one or more items is selected in the lstName ListBox.

Changing the Name.Comment Property

This is also quite simple. Whenever you want to add the same Comment property to more than one Name object, just select the desired items in the lstNames ListBox, click the frmNames Show Details ControlButton, type the desired comment in txtComment, and press Enter.

The `txtComment_ AfterUpdate ( )` event will fire, executing this code:

```
Private Sub txtComment_AfterUpdate()
    Dim varItem As Variant
    Dim strMsg As String
    Dim strTitle As String

    If mcolItemsSelected.Count > 1 Then
        strMsg = "Apply the same comment to all " & mcolItemsSelected.Count & " selected
        names?"
        strTitle = "Comment all selected names?"
        If MsgBox(strMsg, vbYesNo + vbDefaultButton2 + vbCritical, strTitle) = vbNo Then
            Exit Sub
        End If
    End If

    Call FilllstNames
    Call SelectItems
End Sub
```

Note that this time a warning is raised every time there is more than one item selected on the `lstName` `ListBox`.

```
    If mcolItemsSelected.Count > 1 Then
        strMsg = "Apply the same comment to all " & mcolItemsSelected.Count & " selected
names?"
        strTitle = "Comment all selected names?"
        If MsgBox(strMsg, vbYesNo + vbDefaultButton2 + vbCritical, strTitle) = vbNo Then
            Exit Sub
        End If
    End If
```

If the `MsgBox( )` Yes button is selected, all selected `Name` objects will receive the same comment, executing a `For Each...Next` loop through all the `mcolItemsSelected` collection items and changing the `Comment` property—the same way you changed one or more `Name.Visible` properties.

```
    For Each varItem In mcolItemsSelected
        Names(CStr(Me.lstNames.Column(0, varItem))).Comment = Me.txtComment & ""
    Next
```

Figure 5-30 shows how easy it is to change one or more `Name` objects' `Visible` property, hiding it from being seen in the Excel Name box or Name Manager dialog box, as well as associating the same comment to all of them.

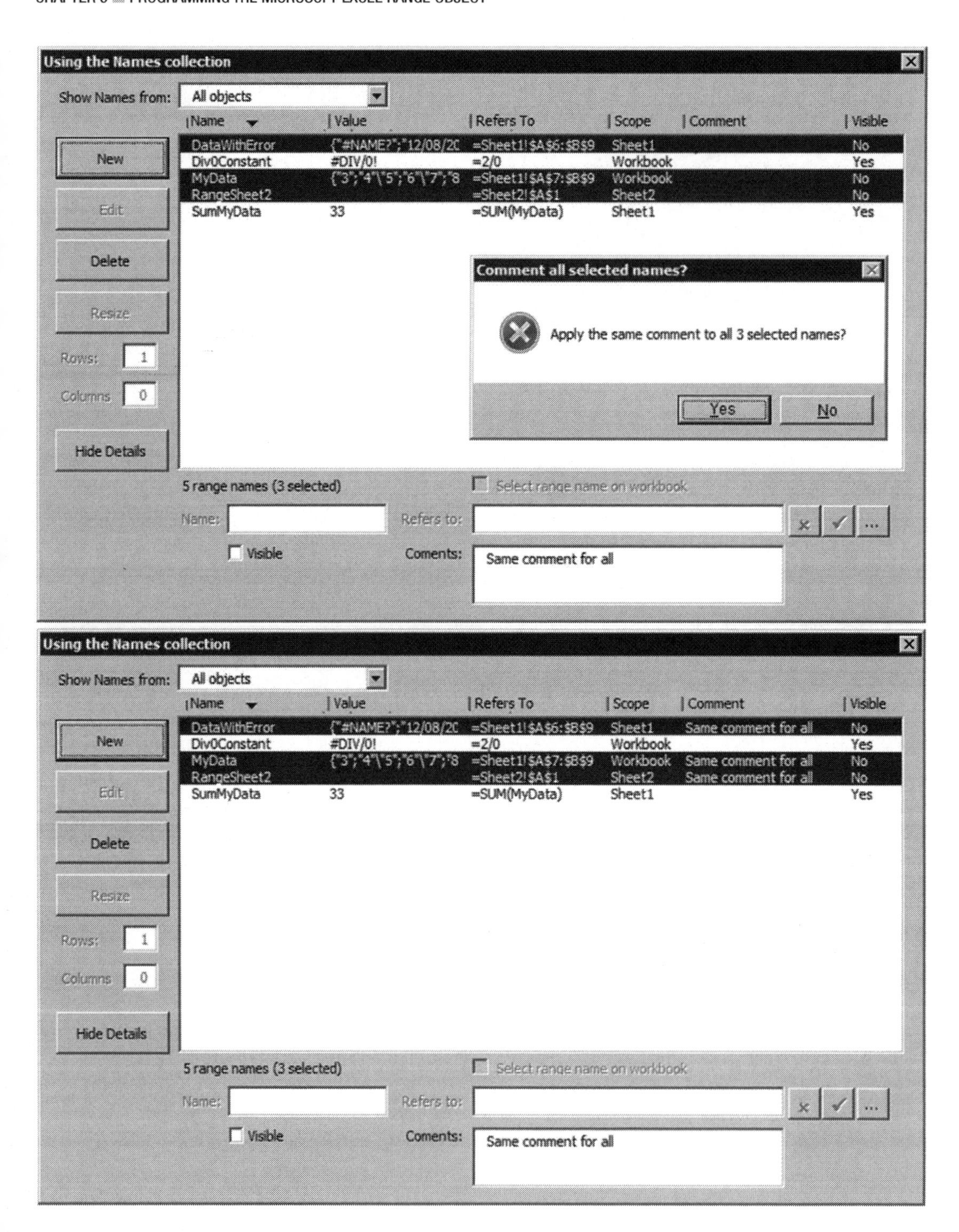

Figure 5-30. *If you select one or more Name objects in the lstNames ListBox, you can change its Visible and Comment properties at once. A hidden range name can't be seen in the Excel Name box or Name Manager dialog box*

Deleting Name Objects

To delete one or more name objects, select the desired names and click the frmNames Delete button, which will fire the cmdDelete_Click( ) event, executing this simple code:

```
Private Sub cmdDelete_Click()
    Dim varItem As Variant
    Dim strMsg As String
    Dim strTitle As String

    If mcolItemsSelected.Count = 1 Then
        strMsg = "Confirm deletion of selected name?"
        strTitle = "Delete Name?"
    Else
        strMsg = "Confirm deletion of all " & mcolItemsSelected.Count & " names selected?"
        strTitle = "Delete selected names?"
    End If

    If MsgBox(strMsg, vbYesNo + vbDefaultButton2 + vbCritical, strTitle) = vbYes Then
        For Each varItem In mcolItemsSelected
            Names(CStr(Me.lstNames.Column(0, varItem))).Delete
        Next

        Set mcolItemsSelected = New Collection
        Call FilllstNames
        Call ClearNameProperties
        Call EnableControls(False)
    End If
End Sub
```

Since one or more Name object is about to be deleted—and this operation can be undone, unless you close the workbook without saving it and reopen it again—the code asks for a confirmation before proceeding.

```
If mcolItemsSelected.Count = 1 Then
    strMsg = "Confirm deletion of selected name?"
    strTitle = "Delete Name?"
Else
    strMsg = "Confirm deletion of all " & mcolItemsSelected.Count & " names selected?"
    strTitle = "Delete selected names?"
End If

If MsgBox(strMsg, vbYesNo + vbDefaultButton2 + vbCritical, strTitle) = vbYes Then
```

If the MsgBox( ) Yes button is selected, a For Each...Next loop runs through all the mcolItemsSeleted collection items and performs the Names collection's Delete method, removing all selected Name objects from the workbook.

```
For Each varItem In mcolItemsSelected
    Names(CStr(Me.lstNames.Column(0, varItem))).Delete
Next
```

And then it clears the mcolItemsSelected collection by attributing to it a New Collection (with zero items selected).

```
Set mcolItemsSelected = New Collection
```

Changing the Name.RefersTo Property

The last VBA technique you will see performed by the frmNames UserForm mimics the way the Excel Name Manager dialog box allows the addition of the Name.RefersTo property using a text box control and some command buttons.

Whenever you select any Name object on frmNames and click the Details ControlButton, you will see its properties in the UserForm Details section. Note that the txtRefersTo text box is enabled, and on the right side you can see three CommandButtons mimicking how the Excel Name Manager does it (Figure 5-31).

Note in Figure 5-31 that cmdRefersTo is the only enabled CommandButton. If you click it, the cmdRefersTo_Click( ) event will fire and execute this code:

```
Private Sub cmdRefersTo_Click()
    Dim varRange As Variant
    Dim intPos As Integer
    Const conFormula = 0

    varRange = Application.InputBox("Cells selected:", "Select range cells",
    Me.txtRefersTo, , , , , conFormula)
    If varRange <> False Then
        varRange = Application.ConvertFormula(varRange, xlR1C1, xlA1)
        'Search for Workbook reference
        intPos = InStr(1, varRange, "]")
        If intPos > 0 Then
            varRange = "'" & Mid(varRange, intPos + 1)
        End If

        'Search for Sheet name
        intPos = InStr(1, varRange, "!")
        If intPos = 0 Then
            varRange = "'" & ActiveSheet.Name & "'!" & Mid(varRange, 2)
        End If

        'Search for "='
        If Left(varRange, 1) <> "=" Then
            varRange = "=" & varRange
        End If

        Me.txtRefersTo = varRange
        Call EnableEditing(True)
    End If
End Sub
```

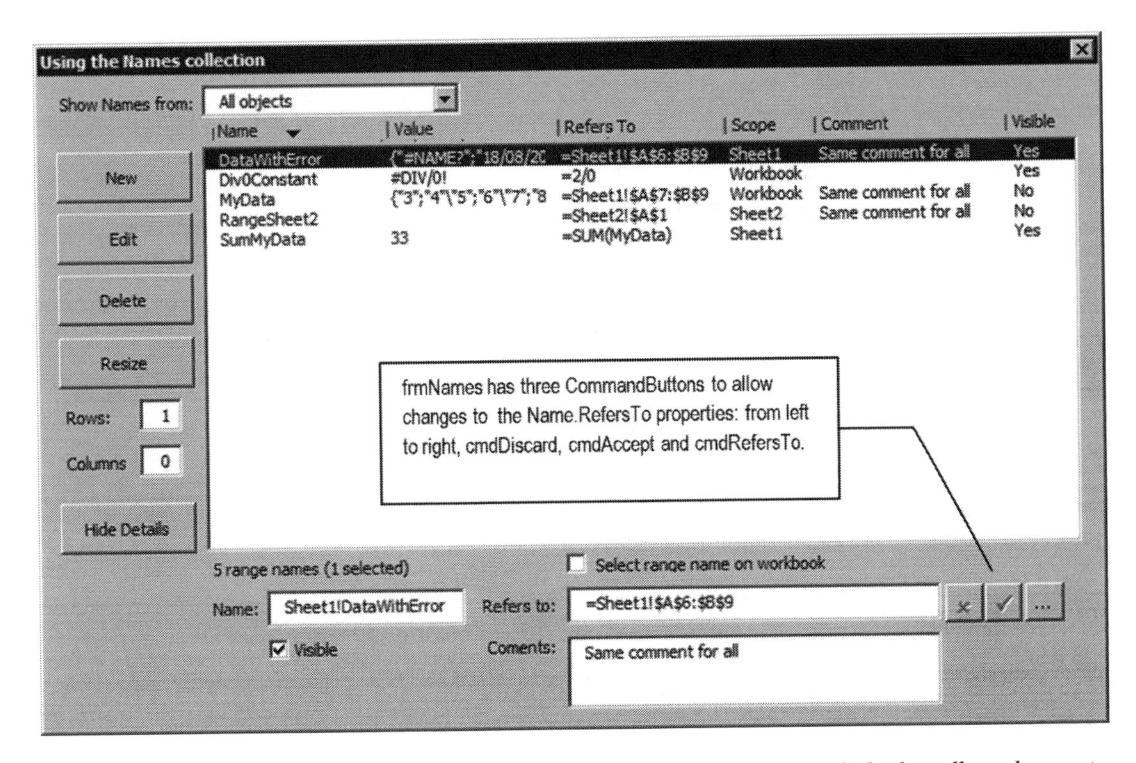

Figure 5-31. *The frmNames UserForm mimics the way the Excel Name Manager dialog box allows changes to the Name.RefersTo property, using a TextBox control (txtRefersTo) along with three CommandButtons (from left to right, cmdDiscard, cmdAccept, and cmdRefersTo)*

If you look at the section "Inserting a New Name by Selecting a Range Address" earlier in this chapter, you will see that this code is quite similar to that used on the `frmEditName` `cmdRefersTo CommandButton` (it uses the `Application.Inputbox` method to allow the selection of worksheet cells), except with its last instruction (bold in the last listing), which is used to enable/disable the `frmName UserForm` controls whenever you try to change any `Name.RefersTo` property.

The `Sub EnableEditing( )` procedure executes this code:

```
Private Sub EnableEditing(bolEnabled  As Boolean)
    Dim intI As Integer

    For intI = 0 To Me.Controls.Count - 1
        If Me.Controls(intI).Name <> "txtRefersTo" Then
            If Len(Me.Controls(intI).Tag) > 0 Then
                Me.Controls(intI).Enabled  = Not bolEnabled
            End If
        End If
    Next

    Me.cmdAccept.Enabled  = bolEnabled
    Me.cmdDiscard.Enabled  = bolEnabled
End Sub
```

Note that it uses a For...Next loop to run through the frmNames.Controls collection testing if the control Name property is different from txtRefersTo:

```
For intI = 0 To Me.Controls.Count - 1
    If Me.Controls(intI).Name <> "txtRefersTo" Then
```

It verifies whether the control Tag property has something inside using the VBA Len( ) function. If this is true, the control is enabled/disabled, according to the procedure bolEnabled argument value.

```
If Len(Me.Controls(intI).Tag) > 0 Then
    Me.Controls(intI).Enabled  = Not bolEnabled
End If
```

And when the loop finishes, it changes the Enabled property of both cmdAccept and cmdDiscard ControlButtons.

```
    Me.cmdAccept.Enabled  = bolEnabled
    Me.cmdDiscard.Enabled  = bolEnabled
End Sub
```

This operation has the effect of disabling all controls except txtRefersTo and lstNames, whenever bolEnabled = True, and enabling them otherwise.

But the frmNames cmdRefersTo ControlButton is not the only way to change the Name.RefersTo property. You can also type any value to the txtRefersTo property to change the value. And whenever anything is typed in the txtRefersTo property, three different key events fire:

- KeyDown( ), which fires when any printable key is pressed
- KeyUp( ), which fires when any printable key is released
- KeyPress( ), which fires when any keyboard key is pressed, including Control, Alt, and Tab, and distinguishes the keys from the numeric keypad on the left of any keyboard

In other words, the KeyDown and KeyUp events report the exact physical state of the keyboard: pressed or released, respectively. The KeyPress event does not report if the keyboard state for the key is up or down; it simply supplies the character that any key represents.

■ **Attention** Although the txtRefersTo_KeyPress( ) event will fire whenever you press the Shift, Ctrl, or Alt keys, the text box content will not be changed, so you don't use this event to catch text box changes.

Both the KeyDown( ) and KeyUp( ) events receive two "by value" arguments: KeyCode and Shift.

1. KeyCode means the printable ASCII code key, which can be verified in the VBA Immediate window using the VBA Chr( ) function when the code is in Break mode, this way:

   ```
   ?Chr(KeyCode)
   ```

2. Shift means the pressing state of Shift, Ctrl, and Alt keys, where Shift = 1, Ctrl = 2, and Alt = 4, and by combining the values, you know which keys were pressed along with any other keyboard key. (For example, if the KeyDown( ) event argument Shift=3, it means that the Shift+Ctrl keys were pressed.)

■ **Attention** Yes, there is a VBA `Asc( )` function that does the opposite. Given a key string like A, it returns the associated ASCII code.

The Backspace key is part of the ANSI character set, but the Delete key isn't. If you delete a character in a control by using the Backspace key, you cause a `KeyPress` event; if you use the Delete key, you don't.

The `KeyPress( )` event receives just one argument, `KeyAscii`, which returns the associated ASCII code of any keyboard pressed (including keys Esc, Tab, Shift, Ctrl, Alt, Enter, and so on).

Whenever you type anything in the `frmNames` `txtRefersTo` text box, the `KeyDown( )` event fires, executing this code:

```
Private Sub txtRefersTo_KeyDown(ByVal KeyCode As MSForms.ReturnInteger, ByVal Shift As
Integer)
    If Not mbolEditRefersTo Then
        mbolEditRefersTo = True
        Call EnableEditing(True)
    End If
End Sub
```

As you can see, the code uses the `mbolEditRefersTo` module-level variable to verify whether `txtRefers to` is in editing mode. If `mbolReferTo = False`, then `Not mbolRefersTo = True`, meaning that it was the first key pressed to change the `txtRefersTo` value. So, `mbolReferTo` becomes `True`, and the procedure makes a call to `Sub EnableEditing( )` just for the first key pressed.

Figure 5-32 shows what happens to the DataWithError `Name` object when the `RefersTo` property is changed from `Sheet1!$B$6:$B$9` (see Figure 5-31) to `Sheet1!$B$6:$B$91` (note that all `UserForm` controls were disabled except the `txtRefersTo` and `lstNames ListBoxes`)!

Whenever you make any change to the `txtRefersTo` text box, you have three possibilities:

- Discard the change by clicking `cmdDiscard` CommandButton
- Accept the change by clicking cmdAccept CommandButton
- Click the New command button or in the `lstNames ListBox`

When you click `cmdDiscard`, the `Click( )` event fires, executing this code:

```
Private Sub cmdDiscard_Click()
    mbolEditRefersTo = False
    Me.txtRefersTo = Me.lstNames.Column(3, Me.lstNames.ListIndex)
    Call EnableEditing(False)
End Sub
```

Quite simple, huh? It sets `mbolEditRefersTo = False`, as a clear indication that `txtRefersTo` is not in edit mode anymore, updates the `txtRefersTo` value to the current `Name.RefersTo` property (which is stored in `lstNames.Column(3)`), and calls `Sub EnableEditing(False)`, which will revert the `frmNames` control's state.

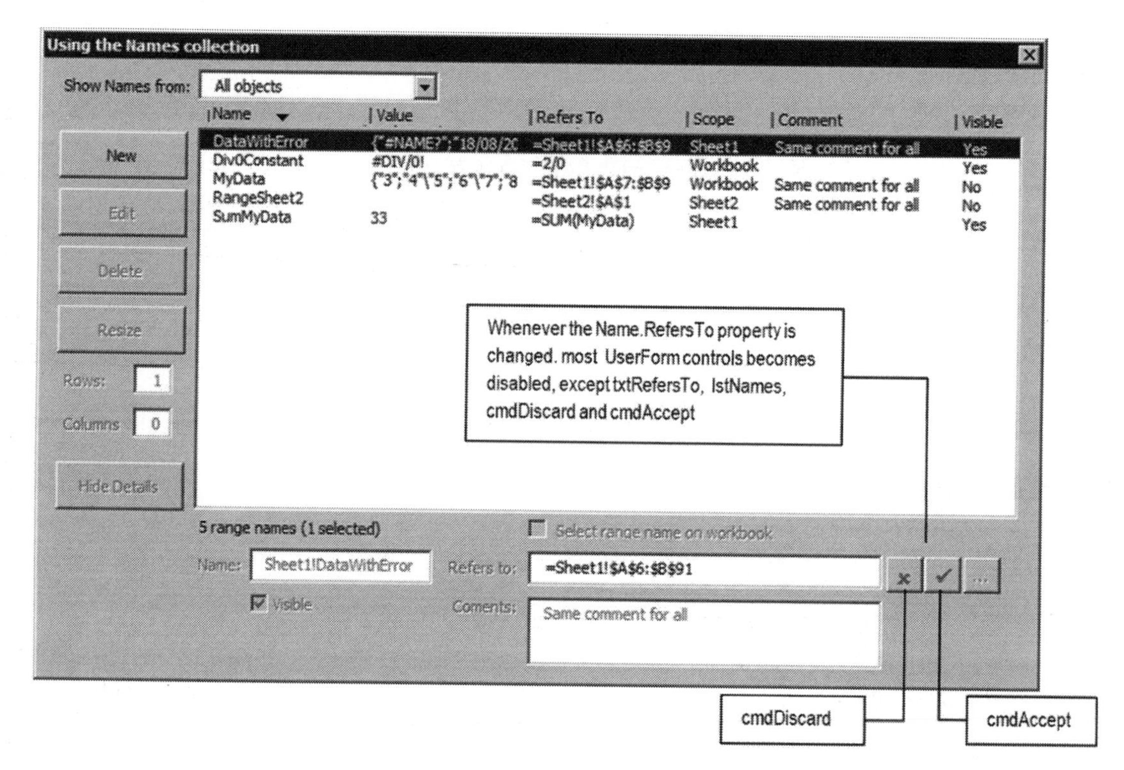

Figure 5-32. *Whenever you click cmdRefersTo and select another cell range or type anything inside txtRefersTo to change the selected Name.RefersTo property, Sub Enabled Editing( ) is executed, disabling most UserForm controls, except the txtRefersTo and lstNames ListBoxes, while the cmdDiscard and cmdAccept command buttons become enabled!*

When you click `cmdAccept`, you want to make changes to the `Name.txtRefersTo` property, so the `cmdAccept_Click( )` event must execute this code:

```
Private Sub cmdAccept_Click()
    Dim nm As Name

    mbolEditRefersTo = False
    Set nm = Names(Me.lstNames.Column(0, mcolItemsSelected(1)))
    nm.RefersTo = Me.txtRefersTo
    Call EnableEditing(False)
    Call FilllstNames
    Call SelectItems
End Sub
```

This time, the `Name.RefersTo` property is updated according to the new value typed in the `txtRefersTo` text box, `Sub EnableEditing(False)` is called to return `frmNames` controls to their default state, `Sub FilllstNames( )` is called to update the `lstNames ListBox`, and `Sub SelectItems( )` is called to reselect the edited `Name` object on `lstNames`. Easy, huh?

The last special case happens when the user tries to abandon the `txtRefers to` change by clicking the `frmNames Edit CommandButton` or `lstNames ListBox`. If the user clicks the Edit button, you must do nothing, because `frmNames` will be unloaded, while `frmEditName` is opened.

But if the user clicks the lstNames ListBox, it is trying to abandon the txtRefersTo new value, so the code uses the lstNames_Enter( ) event to raise a MsgBox( ) with the same message used by the Excel Name Manager, executing this code:

```
Private Sub lstNames_Enter()
    Dim strMsg As String
    Dim strTitle As String

    If mbolEditRefersTo Then
        strMsg = "Do you want to save the changes you made to the name reference?"
        strTitle = "Change name reference?"
        Select Case MsgBox(strMsg, vbYesNo + vbDefaultButton2 + vbCritical, strTitle)
            Case vbYes
                Call cmdAccept_Click
            Case Else
                Call cmdDiscard_Click
        End Select
    End If
End Sub
```

This time, the MsgBox( ) offers two options: accept or discard the change (discard is the default option, by clicking No). Whichever decision the user makes, cmdAccept or cmdDiscard Click( ) events are called to accept or discard the Name.RefersTo editing! Figure 5-33 shows the message received whenever an attempt is made to abandon the txtRefersTo editing.

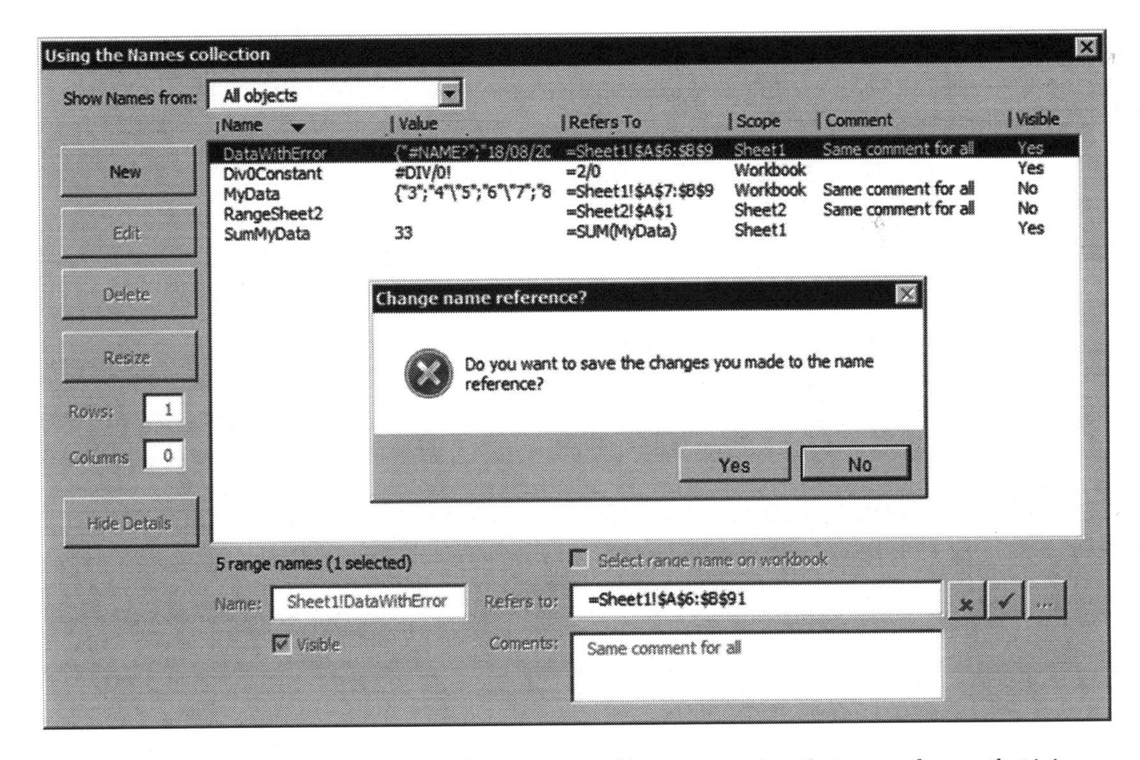

Figure 5-33. This message is issued by the lstNames_Enter( ) event procedure that warns the user that it is about to abandon the changes made to the selected Name.RefersTo property

---

Attention If you are wondering why I used the `lstNames_Enter( )` event procedure instead of any `txtRefersTo` event to detect any `txtRefersTo` change, it is because if I used any `txtRefersTo` event, the user could not click `cmdDiscard` or `cmdAccept` without firing the event. By transferring the event to the only control it can click (`ListNames`), I use exactly the same programming technique employed by the Excel Name Manager. The user can change the `txtRefersTo` content, click `cmdAccept` or `cmdDiscard`, or be warned that Excel doesn't make such decisions after attempt to change the `Name.RefersTo` property.

Chapter Summary

In this chapter, you learned about the following:

- How to use two VBA `UserForm` objects (`frmNames` and `frmEdit`) to mimic the way Excel Name Manager behaves
- How to use Excel `Name` object properties and methods to create and edit any `Name` object
- How to use an Excel `Names` collection to select a given `Name` object
- That Excel `Names` collection allows the insertion of different scope `Name` objects (workbook or sheet scope)
- That you can use the `Application.InputBox` method to select a valid range address to any `Name` object
- That to change a `Name` address by adding or removing rows/columns, you must successively use the `Range.Resize` and `Range.Name` properties
- That when `Name.Visible` property is set to `False`, you can hide the `Name` object on Excel Name box or Excel Name Manager dialog box
- That to change the `Name.Visible` property you must use VBA code
- How to use VBA `Collection` object variable to hold programming data
- How to use the `Range.Find` method to search for worksheet information
- How to use the Excel `Replace( )` function to easily find and replace data on any string variable
- Many different VBA programming techniques to master VBA programming and worksheet applications development

In the next chapter, you will learn about special `Range` object methods and properties that can be used to enhance the data management of your worksheet applications.

CHAPTER 6

Special Range Object Properties and Methods

Although the Range object is the last item in the Microsoft Excel object model hierarchy (shown earlier in Figure 2-1 in Chapter 2), it is the most important one because it represents the source of all information stored in any Excel file, including each cell value and any group of cell values.

The Range object fires no event but holds the most useful properties and methods to deal with the workbook data set, and its programmable interface is implemented by many different commands on the Microsoft Excel ribbon.

In this chapter, you will learn how to program some of the Range object properties and methods using VBA, so you can implement certain Excel tasks into your solutions, creating really great worksheet applications. Among the tasks you'll learn are range selection; cut, copy, and paste operations; sort and filter operations; and cell selection operations. With this information, you will be able to navigate (*walk*) your worksheet data and manipulate it in any way you need.

I could use many different types of worksheet data, such as business sales data, human resources data, scholar grade data, quality control data, scientific research data, and any other type of data that a worksheet application is based on to allow the user make data selections and perform useful calculations, eventually generating the associated chart analysis.

To accomplish this task, this chapter will use as an example a very large worksheet data set: the United States Department of Agriculture (USDA) Agriculture Research Service (ARS) food table using Standard Reference (SR) file's 27th version (SR27). This offers about 8,800 food items (rows) using about 180 nutrient (columns) to describe their nutritional value. This is a data set large enough to mimic most real situations that worksheet applications must face. You can find all the files in this chapter in the `Chapter06.zip` file that you can download from the book's Apress.com product page, located at `www.apress.com/9781484222041`, or from `http://ProgrammingExcelWithVBA.4shared.com`.

Defining a Range with VBA

To master Microsoft Excel VBA programming, you must know how to easily determine the size and address of any contiguous cell range inside an Excel workbook. Figure 6-1 shows the `sr27_NutrientsPer100g.xls` Excel 2003 workbook file, containing the 27th version of the ARS-USDA food tables with all the nutrient information available for a 100g portion of food.

F. Morgado, *Programming Excel with VBA*, DOI 10.1007/978-1-4842-2205-8_6

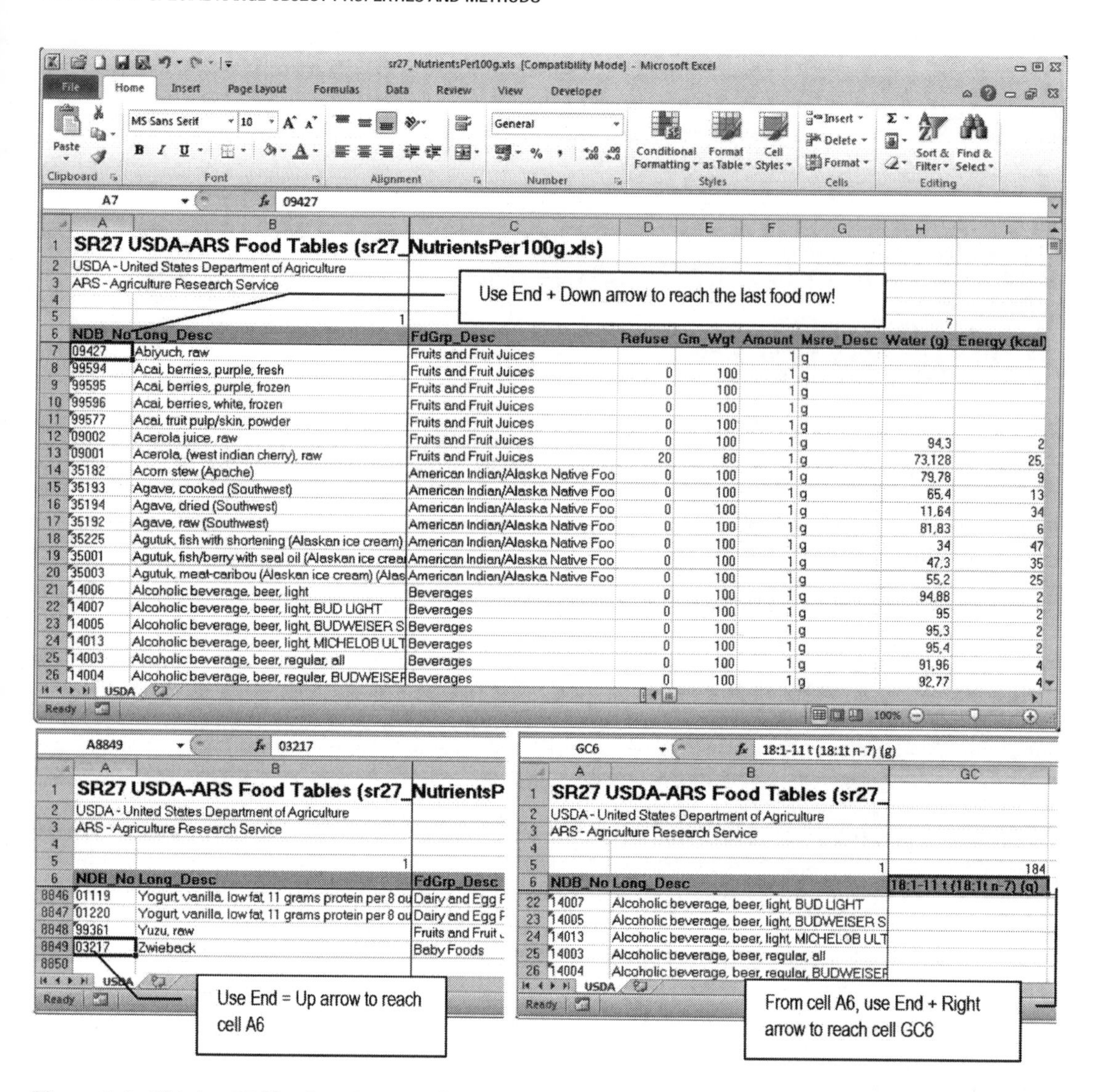

Figure 6-1. *This is sr27_NutrientsPer100g.xls, containing the ARS-USDA food table, that you can extract from the file Chapter06.zip. It has an enormous nutritional food table comprising 8,842 food items and up to 185 columns of nutrient information*

■ **Attention** The `sr27_NutrientsPer100g.xls` Excel 2003 workbook that you can extract from `Chapter06.zip` was generated using the Microsoft Access `USDA Food List Creator.mdb` application, which is capable of opening and exploring any `SRxx.mdb` (Access 2003) or `SRxx.accdb` (Access 2007 or later) file and generating a complete USDA-ARS food table using all nutrient values available, offering a complete set of range names to allow exploration by an Excel application. This workbook file was manipulated by removing all its range names and had its default food item order changed so you can practice this chapter's exercises. Chapter 9 shows how to use this Microsoft Access application to generate other workbook files for newer versions of the SR nutrient database.

When you open the `sr27_NutrientsPer100g.xls` file for the first time, you will notice that it has a large food table on a single USDA worksheet tab, ordered by the `Long_Desc` column (food name Long Description). It has cell A7 selected by default, and to know the size of the food table, you can use Excel's End key capabilities.

1. From cell A7, press the End key followed by the Down Arrow key (or press Ctrl+Down Arrow) to reach cell A8849 (the last filled cell in column A, meaning that the worksheet has 8,849 - 7 = 8,842 food items).

2. From cell A8849, press the End key followed by the Up Arrow key (or press Ctrl+Up Arrow) to reach cell A6 (the first food table row, which counts columns beginning on column B - the `lng_Desc` column, which has each food item name).

3. From cell A6, press End followed by the Right Arrow key (or press Ctrl+Right Arrow) to reach cell GC6 (the last filled cell in row 6, meaning that the worksheet has 184 columns).

▒ **Attention** When you reach cell GC6, you can use the VBA Immediate window to find which column number you are on by typing in this instruction:

```
?Range(ActiveCell.Address).Column 184
```

The Excel End capabilities work like this:

- If the selected cell is empty, when you press the End key followed by any arrow key (or press Ctrl+Arrow key), Excel will move to the next nonempty cell using the keyboard arrow direction. If a nonempty cell can't be found, Excel will select the cell next to the boundary of the worksheet.
- If the selected cell is filled, when you press the End key followed by any arrow key (or press Ctrl+any arrow key), Excel will move to the last filled cell using the keyboard arrow direction. If there is no filled cell in that direction, Excel will select the next empty cell.

Using the Range.End Property

You can mimic this previous navigation exercise using the VBA Immediate window and the Range object End property to move through any worksheet cell range. The `Range.End` property has this syntax:

```
Expression.End(Direction)
```

In this code:

`Expression`: This is required; it is a variable that represents a Range object.

`Direction`: This is required; it is a Microsoft Excel constant that can be set to one of the following:

`xlDown = -4121`, move down
`xlToLeft = -4159`, move to left
`xlToRight = -4161`, move to right
`xlUp = -4162`, move up

To practice with the Range.End property, select again cell A7 of the USDA worksheet from the sr27_NutrientsPer100g.xls Excel 2003 workbook and type this instruction in the VBA Immediate window:

```
?Range(Selection.Address).End(xlDown)
03217
```

The cell that will receive the End property (A7) was returned by Selection.Address, which is an Application property that returns the selected cell. Also note that the Range.End property returns the value of the cell selected by the End property (03217 is the NDB_No value—Nutrient Database Number—of the cell on the last range row), keeping the selected cell on the active sheet. If you want to select the last cell row beginning from cell A7, use the Range.Select method, as follows:

```
?Range(Selection.Address).End(xlDown).Select
True
```

This time, the last cell in the down arrow direction (A8849) will be correctly selected.

You do not need to first select cell A6 to reach the last column used by the USDA worksheet food table. Using VBA, you can begin straight from it using the following syntax in the VBA Immediate window (note that this time I did not use VBA ? print character):

```
Range("A6").End(xlToRight).Select
```

Using the Range.CurrentRegion Property

To get the entire range address used by any contiguous cells (up to the point that Excel finds a blank row or column), use the Range.CurrentRegion property. Using the USDA worksheet, if you select any filled column A cell beginning on cell A6 and use it to identify the range, you can select the entire range by using the Range.Select property, typing this instruction in the VBA Immediate window:

```
Range("A6").CurrentRegion.Select
```

All the cells that the USDA worksheet associated to the entire food table will be automatically selected. To find out the address returned by the Range.CurrentRegion property without selecting it, use the Range.Address property in this way in the VBA Immediate window:

```
?Range("A6").CurrentRegion.Address
$A$5:$GC$8849
```

■ **Attention** Note that Excel selected from cell A5 to cell GC8849, including row 5, which counts the columns of the nutritional food table beginning on column B, the food item Long_Desc column. This column count is important to search the USDA food table using the Excel VLOOKUP( ) function.

Moving Through a Range with VBA

One of the most important operations you need to make when you create a worksheet application solution is to walk along the worksheet data to make some operation. There are two different properties that you can use to run through any worksheet data set: Range.Cells and Range.Offset.

Using the Cells Property

The Cells property is a collection of cells and returns a Range object representing a single cell using a convenient bidimensional matrix. To reference a specific cell by its relative row and column, use this syntax:

```
Object.Cells(RowNumber, ColumnNumber)
```

In this code:

Object: This is optional; it is an object variable representing a Worksheet or Range object.

RowNumber: This is required; it is the cell row number;

ColumnNumber: This is required; it is the cell column number.

When you use the Cells property alone, it returns a collection of all cells from the ActiveSheet object (the sheet that has the focus). If you prefix it with any Worksheet object variable, it will return all cells of the referenced sheet tab; and if you prefix it with any Range object, it will represent all cells associated to that range.

As you can see, the Cells property is for relative navigation to all cells of any Worksheet or Range object. Supposing that you want to set a reference to all cells of the USDA worksheet nutrient table (disregarding rows 1:4 cells), you can use the next instruction, which will return it to the rg object variable:

```
Set rg = Range("A6").CurrentRegion.Cells
```

And you can confirm that by using the Range.Address property, applied to the Cells collection of the last instruction, using the VBA Immediate window in this way:

```
?Range("A6").CurrentRegion.Cells.Address
$A$5:$GC$8849
```

Note that when you use a given Range.Cells property (like Range("A6").Cells), the first row and column of the Cells collection relates to the left cell of the range. To realize this, use the next syntax in the VBA Immediate window to print the first cell address returned by the CurrentRegion property:

```
?Range("A6").CurrentRegion.Cells(1,1).Address
$A$5
```

▒ **Attention** Note that the Range object was set to cell A6, while the CurrentRegion property returned cell A5 as it top-left cell. In fact, if you use any column A cell inside the nutritional table, the CurrentRegion property will always return the same address: $A$5:$GC$8849.

As you saw in the previous chapter, you can reference all individual cells of any range using two nested For...Next instructions: the outer loop runs through all range rows, while the inner loop runs through all range columns (vice versa). To limit the loops, you use the Range.Rows.Count and Range.Columns.Count properties (Count is a property of the Range.Rows and Range.Columns collections).

The next code fragment can perform a loop through all cells of the USDA worksheet nutrient table, returned by the Range.CurrentRegion property:

```
Dim rg as Range
Dim rgCell as Range
Dim intI as Integer
Dim intJ as integer

Set rg = Range("A20").CurrentRegion  '<--This will return address $A$5:$GC$8849
For intI = 1 to rg.Rows.Count
    For intJ = 1 to rg.Columns.Count
       Set rgCell = rg.Cells(intI, inJ)
       'Do something here...
    Next J
Next I
```

This time I used cell A20 inside the USDA worksheet column A nutrient table to get the Range.CurrentRegion collection reference and used the rg.Cells(intI, inJ) collection inside the inner For...Next loop to return a reference for each cell to the rgCell variable.

Using the Range.Offset Property

Another way to navigate through worksheet cells is use the Range.Offset property, which can jump to any point of the worksheet relative to any desired cell. It has this syntax:

```
Expression.Offset(RowOffset, ColumnOffset)
```

In this code:

> Expression: This is required; it is an expression that returns a Range object.
>
> RowOffset: This is optional; it is the number of rows (positive, negative, zero) by which the range is to be offset. Positive values are offset downward, and negative values are offset upward. The default value is 0.
>
> ColumnOffset: This is optional; it is the number of columns (positive, negative, or zero) by which the range is to be offset. Positive values are offset to the right, and negative values are offset to the left. The default value is 0.

As with the Cells property, the Range.Offset property doesn't change the ActiveCell address relative to the cell you want to offset. To select the offset cell, you will need to use the Range.Select property.

The next instructions use Range.Offset to run through all cells of the USDA worksheet nutrient table, relative to the top-left cell returned by the Range.CurrentRegion property:

```
Dim rg as Range
Dim rgInitial as Range
Dim rgCell as Range
Dim intI as Integer
Dim intJ as integer

Set rg = Range("A100").CurrentRegion '<--This will return address $A$5:$GC$8849
Set rgInitial = rg.Cells(1,1)
For intI = 0 to rg.Rows.Count -1
    For intJ = 0 to rg.Columns.Count -1
       Set rgCell = rgInitial.Offset(intI, intJ)
       'Do something here...
    Next J
Next I
```

Did you note the differences? After you get all desired cells using the `Range.CurrentRegion` property, the top-left range cell was set to the `rgInitial` variable. You must also use an `intI` loop from `0 to rg.Rows.Count - 1` (and an `intJ` loop from `0` to `rg.Columns.Count-1`) to run through all `Range.CurrentRegion` cells, since `rgInitial.Offset(0,0)` returns the top-left range cell address (cell `A5`). The inner loop will select each cell of the first range row, while the outer loop will change the range row, virtually selecting every range cell.

Creating the USDA Range Name

Any worksheet application created to explore the `sr27_NutrientsPer100g.xls` file (or any other version) must base the search for any food item—and its nutrient values—on the Excel `VLOOKUP( )` function to search the USDA worksheet using a defined `USDA` range name. This name comprises all `USDA` sheet tab nutrient cells from range `B7:GC8849` (all nutrient cell values beginning with the nutrient column name, `Long_Desc`, and discarding the `Ndb_No` first column, which is column `A`, and the nutrient row names, which is row `6`).

For example, to get the water amount of the food item in cell `B13` of Figure 2-1, which is "Acerola, (west indian cherry), raw," you can use the next `VLOOKUP( )` formula:

```
=VLOOKUP(B13; B7:GC8849; 7, FALSE)
```

`B13` (1st argument) is the item you are vertically searching on in the first column of the `B7:GC8849` range (2nd argument), returning its 7th column (3rd argument) for an exact match (4th argument = `False`). The water amount will be returned if the `B13` value can be found on the searched range.

■ **Attention** Search Excel Help for more information about the `VLOOKUP( )` function.

Considering the `sr27_NutrientsPer100g.xls USDA` sheet tab and using the knowledge you have gathered so far, you already know that you can get all `USDA` worksheet cells beginning on cell `A6` (or any other filled cell beginning on cell `A5`) using the `Range.CurrentRegion.Address` property.

```
?Range("A6").CurrentRegion.Address
$A$5:$GC$8849
```

Since this last instruction gets column `A` and rows `5` and `6` (which you don't want on the final range), you can easily offset this range by one column to the right and two rows below using the `Range.Offset` property, as follows:

```
?Range("A6").CurrentRegion.Offset(2,1).Address
$B$7:$GD$8851
```

Note that this offset operation now returns one extra empty column and two extra empty rows (column `GD` and rows `8850` and `8851`), which can be easily removed using another `Range.Resize` operation (Figure 6-2).

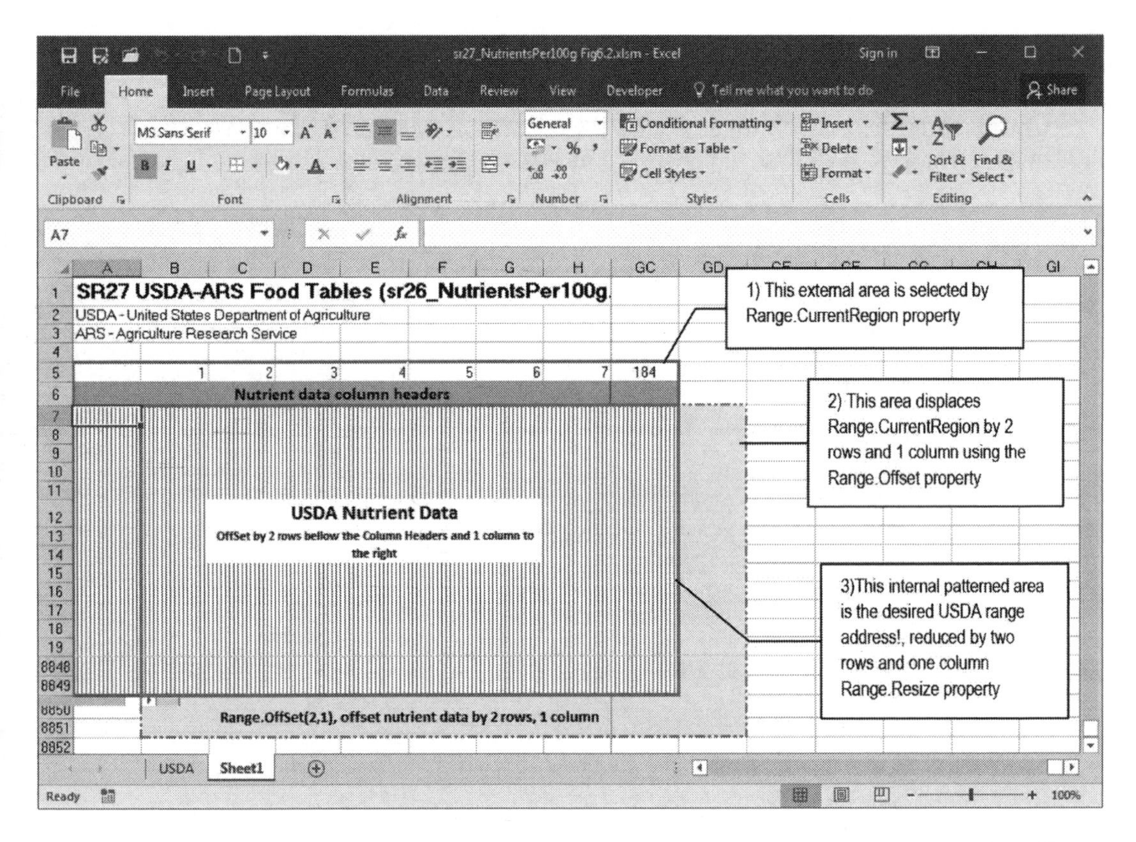

Figure 6-2. *This figure shows what happens when you use the Range.CurrentRegion property to select all USDA worksheet nutrient data and then use the Range.Offset property to displace it by two rows and one column*

So, you can easily create the USDA range address and resize it to the desired row number using a two-step approach: first create the range name and then resize it to the correct row number.

1. The first instruction can create a USDA range name using the VBA Immediate window.

   ```
   Range("A6").CurrentRegion.Offset(2,1).Name="USDA"
   ```

2. The second instruction removes the two extra rows created by the `Range.Offset` property, using the `Range.Resize` property, and re-creates the range. It uses `Range("USDA").Rows.Count -2` and `Range("USDA").Columns.Count-1` for the `Resize` property's `Row` and `Column` arguments.

   ```
   Range("USDA").Resize(Range("USDA").Rows.Count-2, Range("USDA").
   Columns.Count-1).Name = "USDA"
   ```

You can easily create the USDA range name to represent the entire SR27 food table, associating it to cell address `B7:GC8849` using a VBA procedure code, as you can see in the next `Function CreateUSDARangeName( )` procedure code:

```
Public Function CreateUSDARangeName()
    Dim rg As Range
```

```
    Set rg = Sheets("USDA").Range("A6").CurrentRegion
    Set rg = rg.Offset(2, 1)
    Set rg = rg.Resize(rg.Rows.Count - 2, rg.Columns.Count - 1)
    rg.Name = "USDA"
End Function
```

Sorting Range Names

Any range can be easily sorted using the Excel Sort command found in the Edit area of the Home tab of the ribbon. First select the range to be sorted and then apply one of the Sort & Filter options: Sort A to Z, Sort Z to A, or Custom Sort. All these operations can be done by the `Range.Sort` method, which has this (enormous) syntax:

```
Expression .Sort(Key1, Order1, Key2, Type, Order2, Key3, Order3, Header, OrderCustom,
MatchCase, Orientation, SortMethod, DataOption1,DataOption2, DataOption3)
```

In this code:

`Expression`: This is required; it is a variable that represents a `Range` object.

`Key1`: This is optional; it is the first sort field, either as a range name (`String`) or a `Range` object, which determines the values to be sorted.

`Order1`: This is optional; it is an Excel `XlSortOrder` constant that determines the sort order for the values specified in `Key1`.

`Key2`: This is optional; it specifies the second sort field. It cannot be used when sorting a pivot table.

`Type`: This is optional; it specifies which elements are to be sorted when sorting a pivot table.

`Order2`: This is optional; it is an Excel `XlSortOrder` constant that determines the sort order for the values specified in `Key2`.

`Key3`: This is optional; it specifies the third sort field. It cannot be used when sorting a pivot table.

`Order3`: This is optional; it is an Excel `XlSortOrder` constant that determines the sort order for the values specified in `Key3`.

`Header`: This is optional; it is an Excel `XlYesNoGuess` constant, which specifies whether the first row contains header information. `xlNo` is the default value; specify `xlGuess` if you want Excel to attempt to determine the header presence.

`OrderCustom`: This is optional; it determines specific sort orders for known values (like "Sunday, Monday Tuesday..." to sort weekdays).

`MatchCase`: This is optional; it is set to `True` to perform a case-sensitive sort, and it is set to `False` to perform a non-case-sensitive sort. It cannot be used with pivot tables.

`Orientation`: This is optional; it is an Excel `XlSortOrientation` constant that specifies whether the sort should be in ascending or descending order.

SortMethod: This is optional; it is an Excel XlSortMethod that specifies the sort method.

DataOption1: This is optional; it is an Excel XlSortDataOption constant that specifies how to sort text in the range specified in Key1; it does not apply to pivot table sorting.

DataOption2: This is optional; it is an Excel XlSortDataOption constant that specifies how to sort text in the range specified in Key2; it does not apply to pivot table sorting.

DataOption3: This is optional; it is an Excel XlSortDataOption constant that specifies how to sort text in the range specified in Key3; it does not apply to pivot table sorting.

Although the Range.Sort method seems to be quite complex, it is not! You just need to specify the range address to be sorted and use the Key, Key2, and Key3 arguments to identify the columns used in the sort process (supposing that all three key columns must be sorted ascending).

No matter the range you select to be sorted, to specify any sort key column, you must use a range object variable that returns any cell of the desired column or a range that specifies the entire column—as long as the column belongs to the sort key. To indicate that you want to sort the range by column C value, you can use either Range("C1") or Range("C:C") in the Range.Sort Key1 argument.

The next instruction allows you to sort the USDA range name (or the range represented by cells B7:GB8693) by its FdGrp_Desc column value—column C, the food group description—using the VBA Immediate window. (Note that although the C1 cell does not belong to the USDA range name, it can be used to indicate the desired sort column.)

```
Range("USDA").Sort Range("C1")
```

To sort the USDA range name first by its FdGrp_Desc column (column C) and then by its Long_desc column (food name or column B), both in ascending order, use both the Key1 and Key2 arguments, as follows:

```
Range("USDA").Sort Range("C1"), , Range("B1")
```

Figure 6-3 shows the state of the USDA sheet tab after the last sort command, becoming sorted first by the FdGrp_Desc column (column C) and then by the Long_Desc column (column B)—each food group has its food items sorted alphabetically in ascending order!

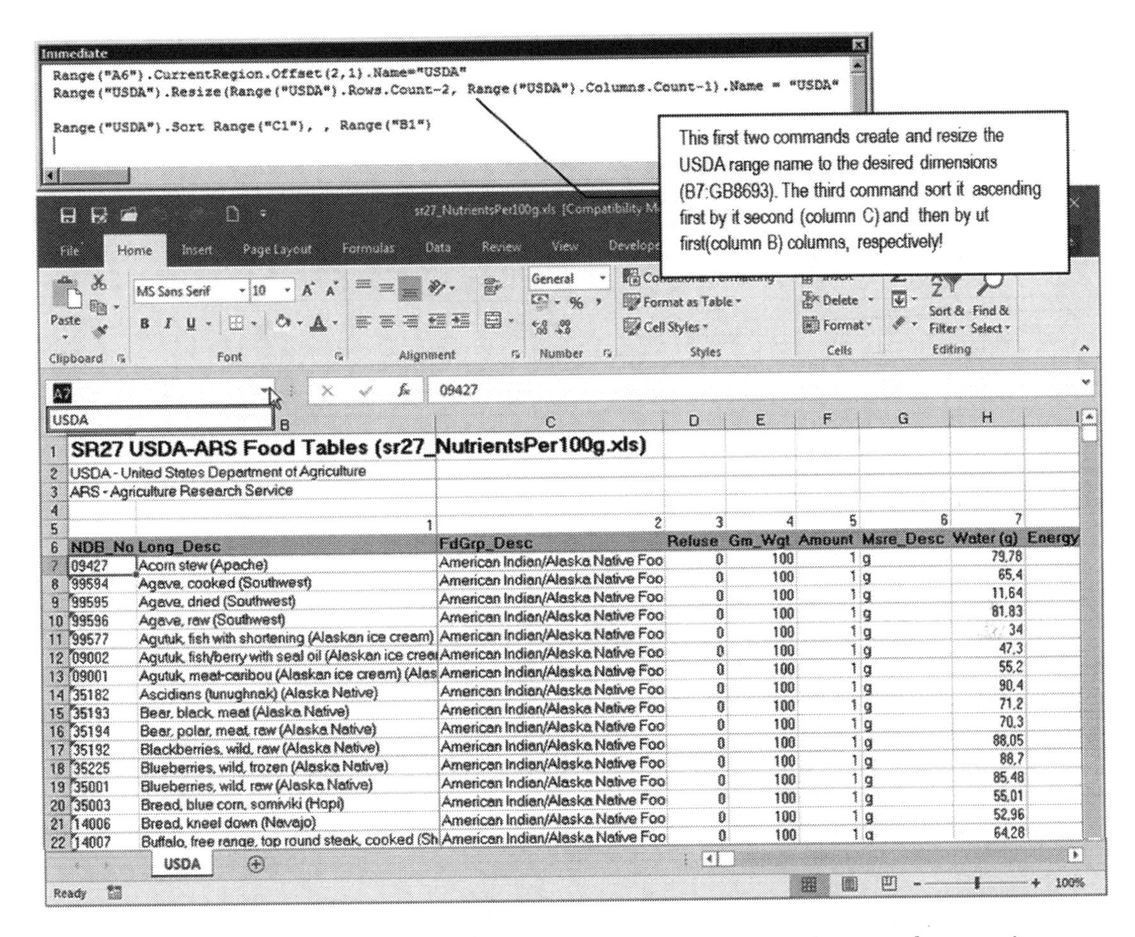

Figure 6-3. *You can test the Range.Sort method using the VBA Immediate window to apply any sorting to a specified range. After you have created the USDA range name (B7:GB8693), you can easily sort it by its Fd_Grp column first and then by its Lng_Desc column using this instruction: Range.Sort Range("C1"), Range("B1")*

Using Cascading Data Validation List Cells

You can create great worksheet applications to deal with large data sets by using the Excel cascading data validation list strategy, which allows the user to select the desired data category in the first data validation list cell (the master list) and the associated data on another data validation list cell (the dependent list). Consider the Figure 6-4 example, which shows a list of food categories and range names that identify food items associated to each food category.

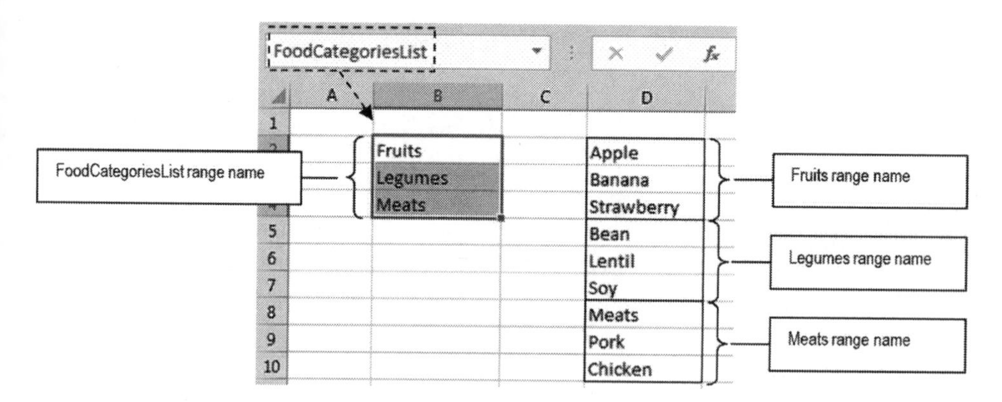

Figure 6-4. *The food category names pertain to a FoodCategoriesList range name, and each food category is associated with its own set of cells*

By using the Excel data validation list option, you can create two cascading data validation list cells on cells `B12` and `D12`, using the `Allow = List` option.

- The master list (cell `B12`) has it Source option set to the formula `=FoodCategoriesList`, which points to the `FoodCategoriesList` range name (a range name that has a list of valid range names).
- The dependent list (cell `D12`) has its Source option set to the formula `=Indirec($B$12)`.

This way, whenever you select an item (valid range name) on the master list cell, the dependent list cell is automatically filled with all cells of the range name selected in the master list (Figure 6-5).

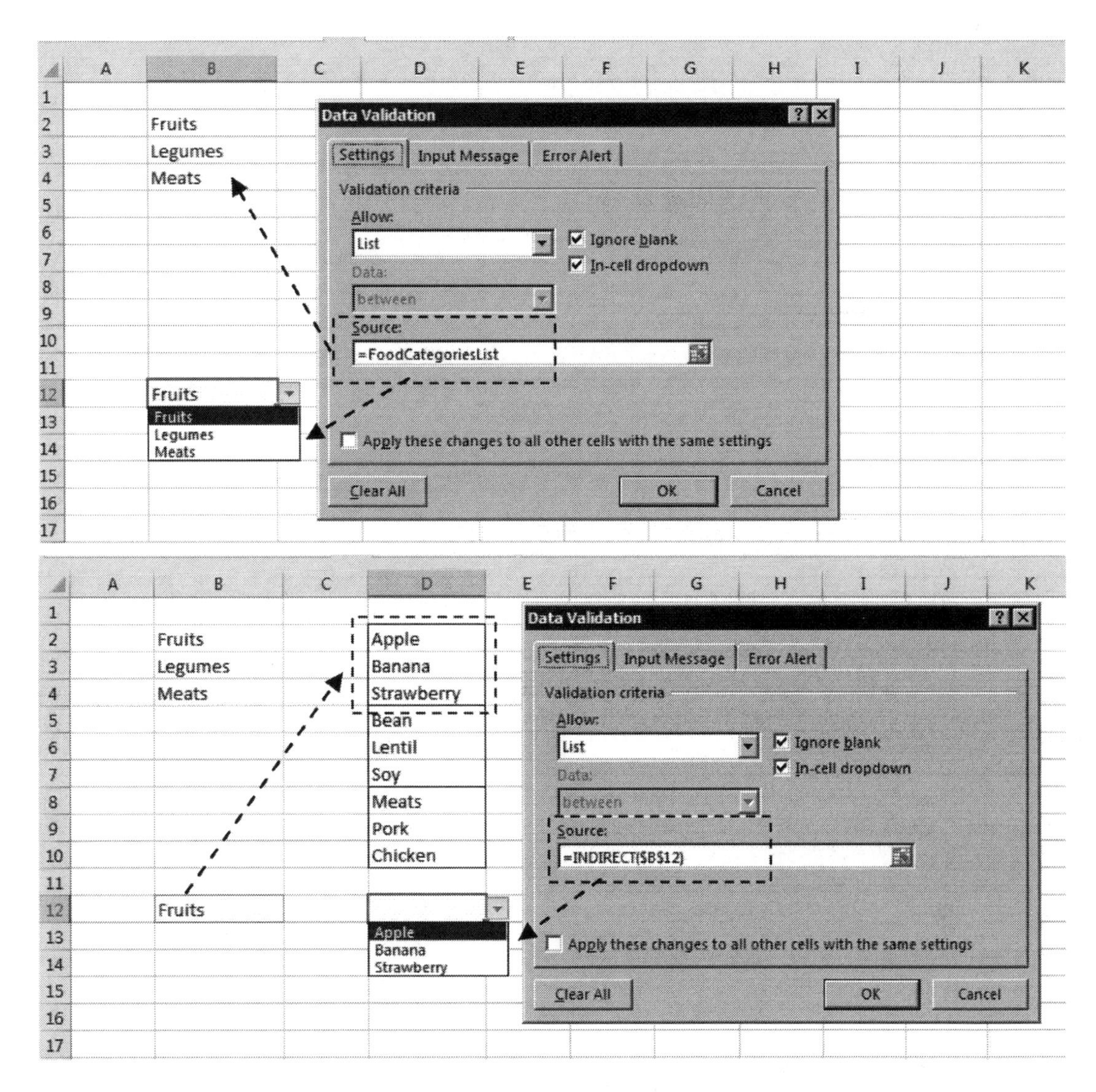

Figure 6-5. *Using cascading data validation list cells, you can fill the master list with a range name that points to a list of valid range names and use the Excel Indirect( ) function to fill the dependent data validation list. Whenever a range name is selected on the master list, the dependent list is automatically filled with the cell values associated to the selected range name*

The USDA Food Composer.xlsm Worksheet Application

Figure 6-6 shows a worksheet application that uses such approach. This is the `USDA Food Composer.xlsm` macro-enabled workbook, which allows you to compose any recipe using up to 18 different food items.

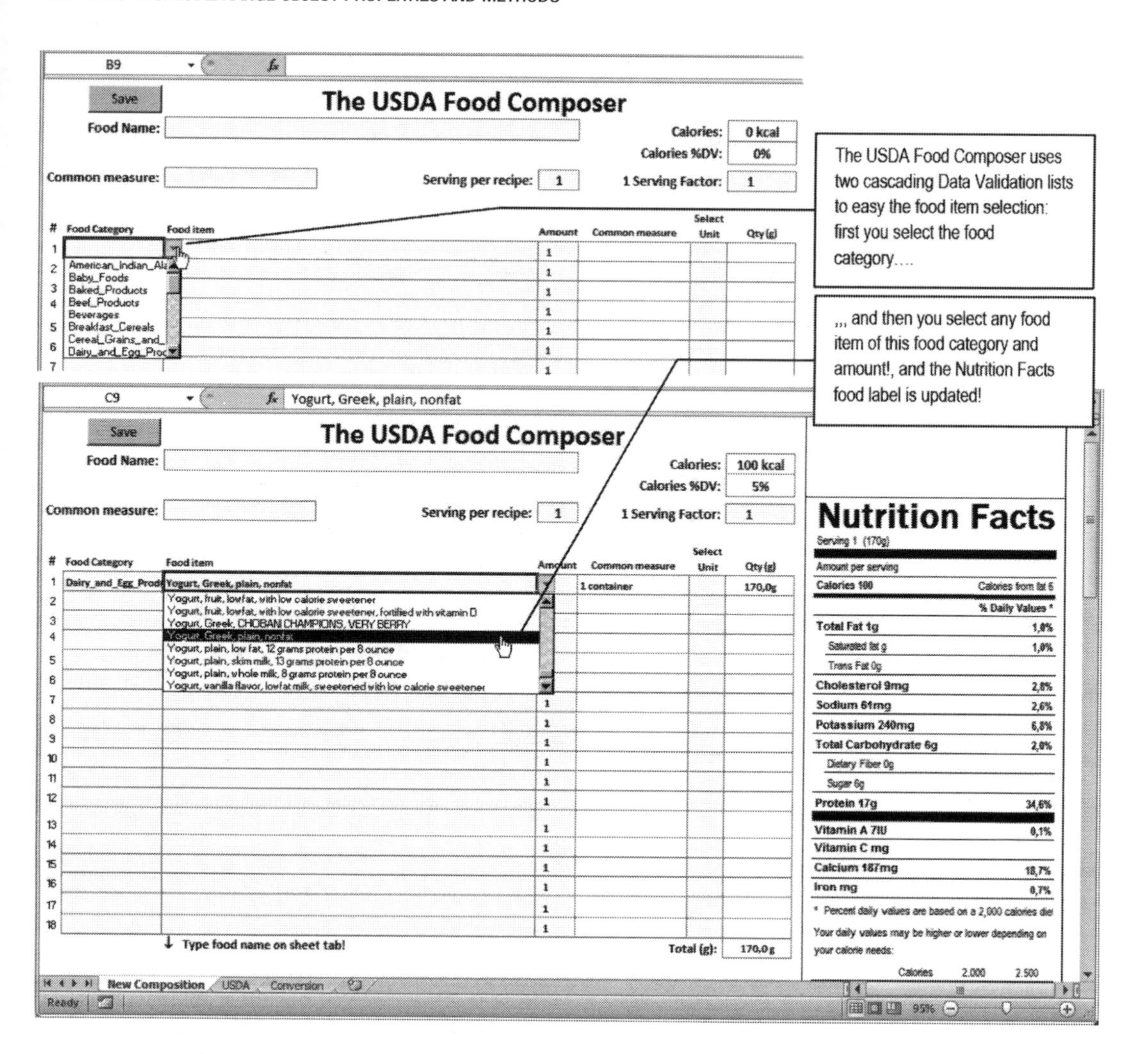

Figure 6-6. *This is the USDA Food Composer.xlsm macro-enabled workbook, which allows you to compose any recipe using up to 18 different food items, determining some recipe data, like the portion number and type, and reducing it to a single portion. The application creates the Nutrition Facts food label for any recipe amount, along with a detailed nutrient analysis using all possible nutrients granted by the ARS-USDA SRxx version used by the associated USDA worksheet*

There are places to define the recipe name, specify the common measure, specify the servings per recipe, and reduce it to a single portion using the one-serving factor (a fraction value that reduces the servings per recipe to a single recipe). The application composes the recipe's Nutrition Facts food label and gives the best nutrient analysis available for a single recipe portion recovering nutrient data from the USDA worksheet (found in the `Chapter06.zip` file).

Note in Figure 6-6 that you first select the food category (`Dairy_and_Egg_Products`) and then select the food item (yogurt, Greek, plain, nonfat) and then the worksheet application does the job.

This is possible because the USDA worksheet used by this application has these range names:

- One range name to represent all food items of each food category (like `Dairy_and_Egg_Products`).

- One range name that contains all food category range names; it is the `FoodCategoriesList` range name, which is used to populate the data validation `ListBox` of each food category row on the worksheet.
- A USDA range name to represent the entire ARS-USDA food table, beginning on column B (`Long_Desc`, or food item description), using all its nutrient columns and discarding it headers—the way you did in the previous section. This USDA range name is where `VLOOKUP( )` will search for each food item's nutritional information.

▒ **Attention** A big set of Excel `VLOOKUP( )` functions was inserted on hidden columns `V9:XFD26` for each food item nutrient value amount per 100g. Each formula uses a direct rule to recalculate the value based on the food item weight used by the recipe. A `Sum( )` formula sums all food items' nutrient columns, returning values to the Nutrition Facts food label and the Nutrient Composition area below the application input cells. The worksheet application hides these calculated cells, all unused columns and rows, and worksheet gridlines and headers to give it a more professional touch. It also protects the worksheet (no password) allowing just unprotected, input cells to be selected. You are invited to explore how the application was created and works.

After the recipe is composed, you can click the Save button to run standard macro-recording VBA code and store it inside the `My_Recipes` food category—which does not belong to the original ARS-USDA food table but is part of the USDA range name, added as the last food category range name on the table bottom.

Once a recipe is saved, you can select the `My_Recipes` food category and use it as any other food item—along with all its nutrient information—to compose another recipe or meal.

By keeping the `New Composition` sheet tab empty and making copies of it (right-click it, select Move or Copy, check Create a Copy, and press Enter), the user can reuse the worksheet using multiple sheet tabs to store independent recipes in a single Excel file. Such an approach was used to create the `USDA Food Composer (with Recipes).xlsm` worksheet application (which you can also extract from the `Chapter06.zip` file to see how it composes recipes and uses those recipes to compose meals).

Creating USDA Worksheet Range Names

If you are wondering whether all these USDA worksheet range names were created by the `USDA Food List creator.accbd` Microsoft Access application, you are absolutely right. This Microsoft Access application first created the food table, and then it exported the table to Excel and used the techniques described in this book to create all these range names. So, anyone can create such smart nutritional applications.

This is what you need to do with `sr27_NutrientsPer100g.xls` to leave it ready to be used on such nutritional applications:

1. Delete all range names already defined in the `sr27_NutrientsPer100g.xls USDA` worksheet (if any).
2. Sort the USDA worksheet with the `Range.Sort` method to show the food items sorted first by `FdGrp_Desc` and then by `Long_Desc` (the way you did in Figure 6-3).
3. Create all food category range names using the `Range.Offset` property to run through the `FdGrp_Desc` column, identifying each range boundary.
4. Create the `My_Recipes` food category at the bottom of the USDA food table.
5. Create the `FoodCategoriesList` range name, which contains an alphabetical list of all food category range names.

6. Create the USDA range name.

7. Save the worksheet, ready to be used by other worksheet applications.

You can see how to apply all these actions using VBA procedure code by extracting the sr27_Nutrients Per100g.xlsm macro-enabled workbook from the Chapter06.zip file and selecting the basCreateRange Names code module, where you will find Function CreateRangeNames( ), which has this code:

```
Public Function CreateRangeNames()
    Dim ws As Worksheet
    Dim nm As Name
    Dim rg As Range
    Dim strCategory As String
    Dim rgFirstCell As Range
    Dim rgLastCell As Range
    Dim rgFoodCategoriesList As Range

    'Delete all range names
    For Each nm In Names
        nm.Delete
    Next

    'Create USDA range name
    Set ws = Sheets("USDA")
    Set rg = ws.Range("A6").CurrentRegion
    Set rg = rg.Offset(2, 1)
    Set rg = rg.Resize(rg.Rows.Count - 1, rg.Columns.Count - 1)
    rg.Name = "USDA"

    'Set initial cell of rgFoodCategoriesList
    Set rgFoodCategoriesList = ws.Cells(7, Range("USDA").Columns.Count + 3)

    'Sort USDA
    Range("USDA").Sort ws.Range("C1"), , ws.Range("B1")

    'Create food categories range names
    Set rg = ws.Range("C7")
    Do
        'Define the food category name
        strCategory = FixName(rg)
        Set rgFirstCell = rg
        Do
            Set rg = rg.Offset(1, 0)
        Loop While rg = rgFirstCell

        'Get the last item of the food category
        Set rgLastCell = ws.Cells(rg.Row - 1, rg.Column)
        Range(ws.Cells(rgFirstCell.Row, "B"), ws.Cells(rgLastCell.Row, "B")).Name = strCategory

        'Save new food category on rgFoodCategoriesList
        rgFoodCategoriesList = strCategory
        Set rgFoodCategoriesList = rgFoodCategoriesList.Offset(1, 0)
    Loop Until rg = ""
```

```
    'Create My_Recipes range name
    rg.Offset(0, -1).Name = "My_Recipes"
    rgFoodCategoriesList = "My_Recipes"

    'Define, sort and create range FoodCategoriesList
    Set rgFoodCategoriesList = ws.Range(ws.Cells(7, rgFoodCategoriesList.Column),
    rgFoodCategoriesList.Address)
    rgFoodCategoriesList.Sort rgFoodCategoriesList
    rgFoodCategoriesList.Name = "FoodCategoriesList"

    'Save the workbook
    ThisWorkbook.Save
End Function
```

The code begins by deleting all names already created in the workbook (if any), using a `For Each...Next` loop through the `Application.Names` collection.

```
'Delete all range names
For Each nm In Names
    nm.Delete
Next
```

It then sets a reference to the USDA worksheet so all cells can be referenced to this sheet tab without needing to activate it, and the USDA range name is created using the same techniques described earlier in this chapter.

```
'Create USDA range name
Set ws = Sheets("USDA")
Set rg = ws.Range("A6").CurrentRegion
Set rg = rg.Offset(2, 1)
Set rg = rg.Resize(rg.Rows.Count - 1, rg.Columns.Count - 1)
rg.Name = "USDA"
```

▒ **Attention** The USDA range name created will keep a blank row at the bottom as a provision to insert the `My_Recipes` range name.

Once the USDA range name is created, it sets the position where the `FoodCategoriesList` range name (which will keep a list of all individual food category range names) will be placed: at row 7 and two columns to the right of the last USDA range name column.

```
'Set initial cell of rgFoodCategoriesList
Set rgFoodCategoriesList = ws.Cells(7, Range("USDA").Columns.Count + 3)
```

▒ **Attention** The expression `Range("USDA").Columns.Count` will return how many columns the USDA range name has. Since it starts at column `B`, you must add it 3 to set the position two columns to the right of the last USDA range name column.

The USDA range name is then sorted to put it in the desired sort order: first by FdGrp_desc (column C) and then by Long_Desc (column B).

```
'Sort USDA
Range("USDA").Sort ws.Range("C1"), , ws.Range("B1")
```

Once the USDA range name is sorted, you are ready to begin creating all food category range names. To do this, you must begin on cell C7 (first food category), store the food category range name, and loop down through column C until the food category name changes. When this happens, you will have the first and last rows of the food category, and you will set the food category range name on these rows of column B, where the food item names reside.

The food category range name will be added to the current position of the FoodCategoriesList range name, and the next food category will then be equally processed, until the food category name becomes an empty string ("", two successive double quotes), as an indication that the code reached the last USDA food table row on column C, indicating that the job was done.

This entire process is made using two nested Do...Loop instructions. The outer loop controls the entire process and finishes when the food category becomes an empty string. The inner loop uses the Range.Offset method to run down through column C.

The entire process begins by setting the initial position on cell C7, attributed to the rg object variable.

```
'Create food categories range names
Set rg = ws.Range("C7")
```

The outer Do...Loop begins getting the current food category name to be created. Note that it uses the FixName( ) procedure to remove invalid characters from the range name (spaces are changed to underscores) and to store it into the strCategory string variable.

```
Do
  'Define the food category name
  strCategory = FixName(rg)
```

■ **Attention** The FixName( ) Function procedure resides in the basFixName code module (available in the CHAPTER05.zip file) and was commented on in Chapter 5.

The search for the food category boundaries begins by storing the rg object variable into the rgFirstCell object variable, and an inner Do...Loop instruction runs down through column C using rg,Offset(1,0) until a different food category name is found. Note that it does this by comparing two range object variables, rg and rgFirstCell.

```
Set rgFirstCell = rg
Do
    Set rg = rg.Offset(1, 0)
Loop While rg = rgFirstCell
```

■ **Attention** Don't forget that when the Range object points to a single cell, it returns the cell value, and when it points to more than one cell, it returns an array of values.

When a different food category is found, the inner loop ends, and you have the boundaries of the last processed food category, which is stored in the rgLastCell object variable. Note that it uses the Cells property along with the rg.Row-1 property to correctly reference the last food category row.

```
'Get the last item of the food category
Set rgLastCell = ws.Cells(rg.Row - 1, rg.Column)
```

The entire range is then created using the Range(cell1, cell2) arguments. To define the first range cell (cell1), it uses the Cells property with rgFirstCell.Row and column B. To define the last range cell, it uses the Cells property with rgLastCell.Row and column B, which will return all food item names of this food category. The defined range is then named using the Range.Name property.

```
Range(ws.Cells(rgFirstCell.Row, "B"), ws.Cells(rgLastCell.Row, "B")).Name = strCategory
```

And once the new food category range name is created, its name is stored in the cell value associated to the rgFoodCategoriesList object variable.

```
'Save new food category on rgFoodCategoriesList
rgFoodCategoriesList = strCategory
```

The rgFoodCategoriesList is offset one row down, and the outer loop verifies whether the new food category selected that is responsible for ending the inner Do...Loop instruction is equal to an empty string (""). If *it is not*, another food category is found, and the loop runs again to create it. If it is, the entire USDA food table has been processed, and all food category range names have been created.

```
        Set rgFoodCategoriesList = rgFoodCategoriesList.Offset(1, 0)
    Loop Until rg = ""
t
```

When this happens, the code is on the last USDA range row. This is an empty row, reserved with the My_Recipes range name, which is then created and added to the cell pointed at by the rgFoodCategoriesList object variable.

```
'Create My_Recipes range name
rg.Offset(0, -1).Name = "My_Recipes"
rgFoodCategoriesList = "My_Recipes"
```

After the last food category has been created and added to the rgFoodCategoriesList cell, the rgFoodCategoriesList is redefined to include all food category range names, beginning on row 7 and ending in the row associated to the My_Recipes range name. Note that it uses Range(cell1, cell2) arguments to define the entire range. The cell1 address is defined by the Cells( ) property, using row 7 and rgFoodCategoriesList.Column to define the first cell address, while cell2 is defined by rgFoodCategoriesList.Address.

```
'Define, sort and create range FoodCategoriesList
Set rgFoodCategoriesList = ws.Range(ws.Cells(7, rgFoodCategoriesList.Column),
rgFoodCategoriesList.Address)
```

Once the rgFoodCategoriesList range is defined, it is sorted to put the My_Recipes food category in the appropriate order. Note that the Range.Sort Key1 argument is defined to the rgFoodCategoriesList range itself, because it has just one column.

```
rgFoodCategoriesList.Sort rgFoodCategoriesList
```

The `FoodCategoriesList` range name is created using the `Range.Name` property, and the workbook is saved, ending the `CreateRangeNames( )` procedure.

```
    rgFoodCategoriesList.Name = "FoodCategoriesList"

    'Save the workbook
    ThisWorkbook.Save
End Function
```

To see the `CreateRangeNames( )` procedure in action, type this command in the VBA Immediate window:

```
?CreateRangeNames()
```

In a blink of an eye, the USDA food table will be sorted, and all desired range names will be created! Your computer is quite fast, huh?

To confirm that the range names was created correctly, use the Excel Name box to select any range name, or use the Excel Name Manager dialog box, which can be found on the Excel Formulas tab (Figure 6-7).

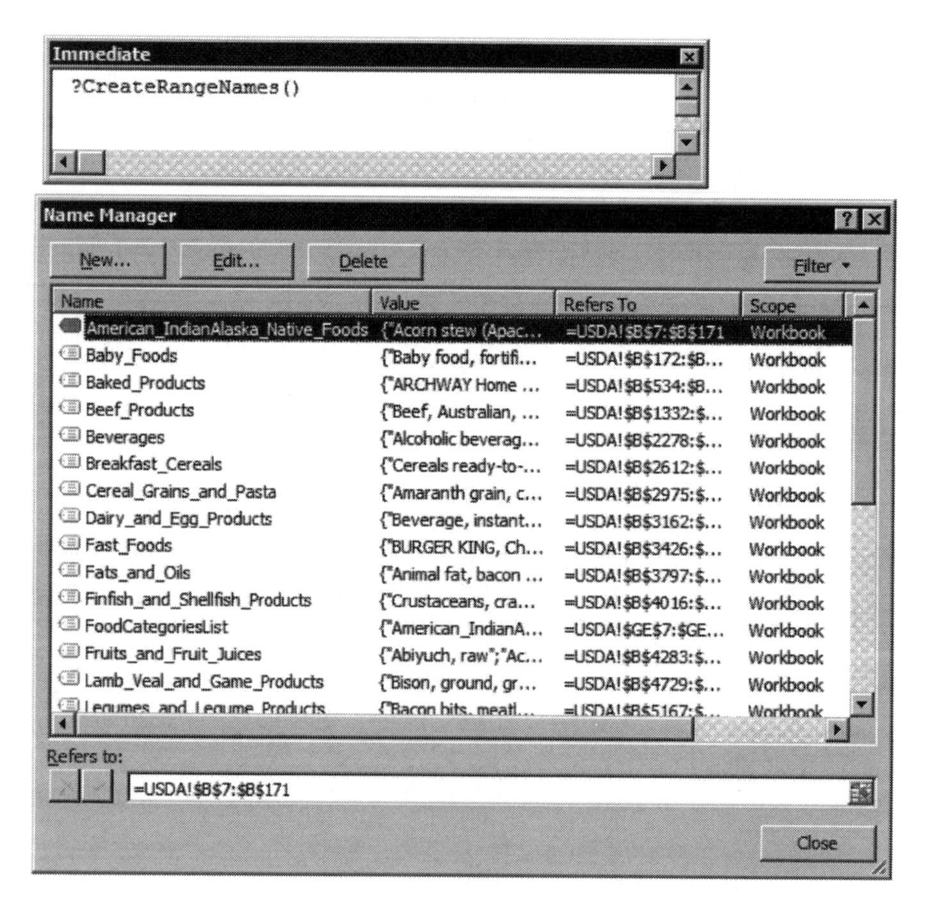

Figure 6-7. *Use the VBA Immediate window to run the CreateRangeNames() Function procedure and then use the Excel Name box or the Name Manager dialog box to see all food categories names it created*

■ **Attention** The USDA worksheet range names associated to each food category will remain the same as long as the USDA range name remains sorted by the food category column (`FdGrp_Desc`).

Finding the Last Worksheet Used Cell

When you use large tables like the USDA-ARS food table using an Excel worksheet, there are times when you will need to programmatically find the last-used worksheet cell, which will always be relative to the user point of view. It can be the last-used cell on a specific row or column or the last-used cell on the last-used row or last-used column of the worksheet. Figure 6-8 details these points of view on the `Find Last Cell.xlsm` Excel macro-enabled workbook that you can find inside the `Chapter06.zip` file.

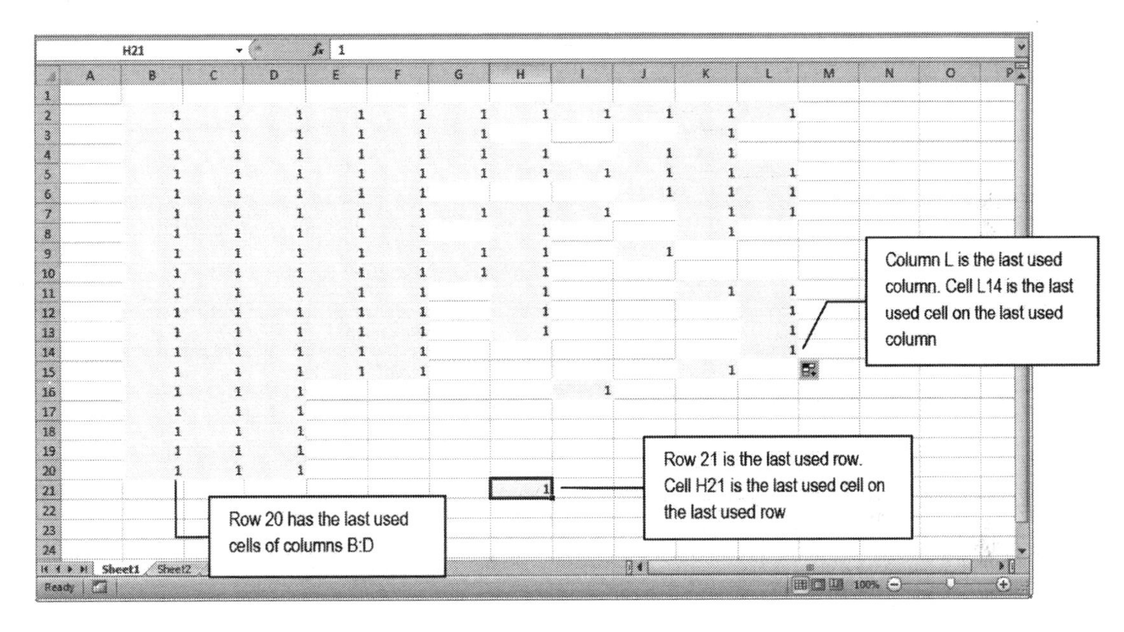

***Figure 6-8.** The last-used cell relates to a specific point of view. It can be the last cell on a specific row or column or the last cell on the last-used row or last-used column*

Finding the Last Worksheet Row/Column with Range.End

To find the last-used cell on a specific row or column on any worksheet, many Excel programming books and Internet sites recommend this strategy:

1. Select the last cell on the last worksheet row or column.
2. Once the last possible cell is selected on the desired row (or column), use the `Range.End(xlUp)` method (or `Range.End(xlToLeft)`) to select the last-used cell on a specific column (or row).

The `Find Last Cell.xlsm` Excel macro-enabled workbook uses this strategy on functions `LastRow( )` and `LastColumn( )` from the `basLastCell` code module.

```
Public Function LastRow(rg As Range) As Long
    Dim lngRowsCount As Long
```

```
    lngRowsCount = ActiveSheet.Rows.Count
    LastRow = ActiveSheet.Cells(lngRowsCount, rg.Column).End(xlUp).Row
End Function

Public Function LastColumn(rg As Range) As Long
    Dim lngColumnsCount As Long

    lngColumnsCount = ActiveSheet.Columns.Count
    LastColumn = ActiveSheet.Cells(rg.Row, lngColumnsCount).End(xlToLeft).Column
End Function
```

Note that both `LastRow( )` and `LastColumn( )` expect to receive a Range object indicating the column (or row) you want to inspect. You can use both functions to inspect the last-used row or column of the `Find Last Cell.xlsm` workbook using the VBA Immediate window, with instructions like these (Figure 6-9):

```
?LastRow(Range("B1"))
20
?LastColumn(Range("A5"))
```

■ **Attention** `Function LastColumn( )` of `basLastCell` uses similar code to return the number of the last-used column on an specific row.

```
Immediate
?LastRow(Range("B1"))
 20
?LastColumn(Range("A5"))
 12
```

Figure 6-9. *Using the VBA Immediate window, you can find the last-used cell of any row or column, using a Range object as an argument to the functions*

Finding the Last Worksheet Row/Column with Worksheet.UsedRange

If you want to find the last-used cell on the last-used row or column of any worksheet, use `Worksheet.UsedRange`. This property returns a Range object according to the following:

- If the worksheet is empty, `UsedRange` returns `$A$1` as its default address.
- If the worksheet has just one used cell, `UsedRange` returns the absolute reference to the cell address.
- Otherwise, `UsedRange` returns the biggest rectangle necessary to encompass all used cells on the worksheet, where the bottom-right cell represents the last-used column and row addresses.

You can verify the range address returned by the `Worksheet.UsedRange` property for `Sheet1` of the `Find Last Cell.xlsm` workbook using the VBA Immediate window, using the `ActiveSheet` object and the `Range.Address` property, as follows:

```
?Activesheet.UsedRange.Address
$B$2:$L$21
```

The `Worksheet.UsedRange` property will return the last row and column with data, indicating on the bottom-right range corner the last-used cell and last-used column (Figure 6-10).

■ **Attention** Note that `UsedRange` also returns the address of the first-used row and column.

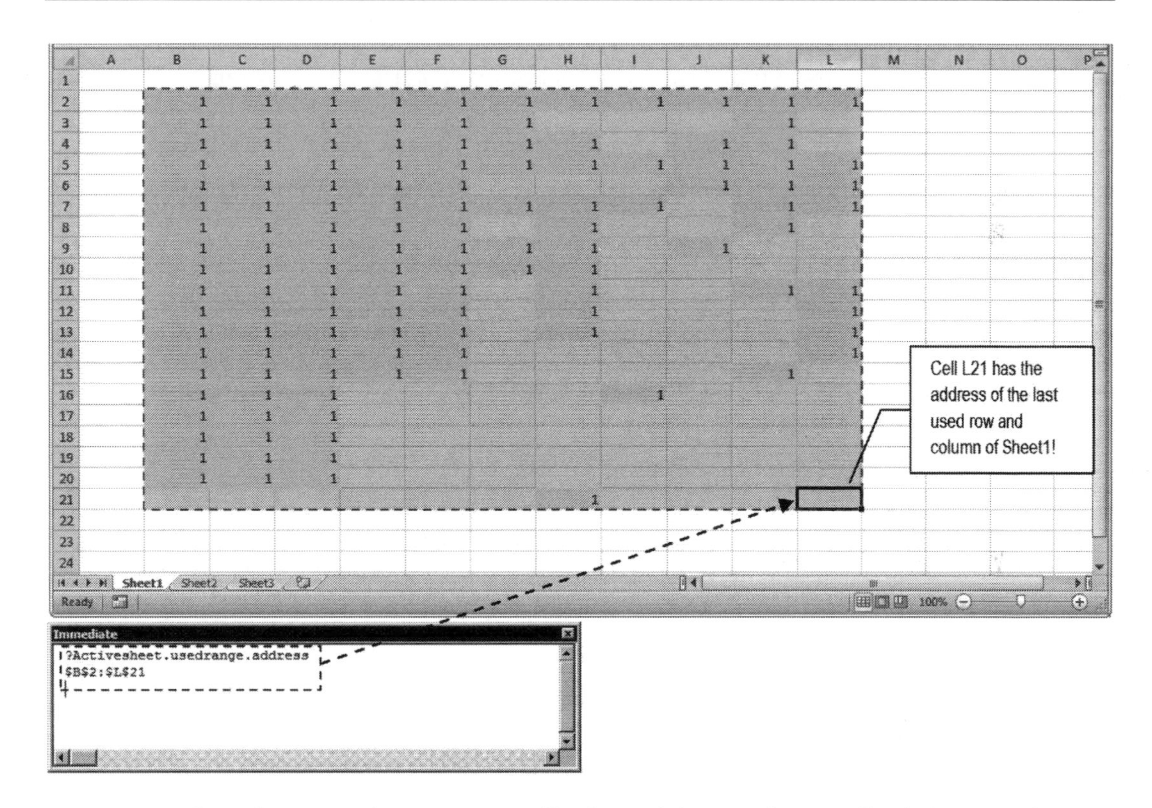

Figure 6-10. *This is the rectangular range returned by the Worksheet.UsedRange cells. The bottom-right corner coordinates (cell L21) indicate the last-used cell on the last-used column and last-used row*

Functions `LastUsedRow( )` and `LastUsedColumn( )` from the `basLastCell` code module of the `Find Last Cell.xlsm` workbook use this property to return the number of the last-used worksheet row or column, respectively:

```
Public Function LastUsedRow() As Long
    LastUsedRow = ActiveSheet.UsedRange.Rows.Count
End Function
```

```
Public Function LastUsedColumn() As Long
    LastUsedColumn = ActiveSheet.UsedRange.Columns.Count
End Function
```

Note that both functions use the Count property of the Rows or Columns collection to return the desired last row or column addresses. And with the knowledge you gathered so far in this chapter, you may note that you can use the UsedRange property with the Cells( ) and End( ) properties to find the last-used cell on the last-used row by typing in the VBA Immediate window an instruction like this:

```
?Activesheet.UsedRange.Cells(Activesheet.UsedRange.Rows.Count, Activesheet. UsedRange.
Columns.Count).End(xlToLeft).Address
$H$21
```

Did you note that? The UsedRange.Cells( ) property received as arguments UsedRange.Rows.Count and UsedRange.Columns.Count to select the last cell returned by the UsedRange property, which is located on the last-used worksheet row. The End(xlToLeft) property applies to the UsedRange.Cells( ) property and selects the last-used cell on the last-used worksheet row. The Address property returns the address of this specific cell.

▒ **Attention** Using the End(xlUp) property, you easily select the last-used cell on the last-used worksheet column.

The functions LastUsedRowCell( ) and LastUsedColumnCell( ) from the basLastCell code module of the Find Last Cell.xlsm workbook use this strategy to return the address of the last-used cell on the last-used worksheet row or column, respectively.

```
Public Function LastUsedRowCell() As String
    Dim lngRow As Long
    Dim lngColumn As Long

    lngRow = ActiveSheet.UsedRange.Rows.Count
    lngColumn = ActiveSheet.UsedRange.Columns.Count
    LastUsedRowCell = ActiveSheet.UsedRange.Cells(lngRow, lngColumn).End(xlToLeft).Address
End Function

Public Function LastUsedColumnCell() As String
    Dim lngRow As Long
    Dim lngColumn As Long

    lngRow = ActiveSheet.UsedRange.Rows.Count
    lngColumn = ActiveSheet.UsedRange.Columns.Count
    LastUsedColumnCell = ActiveSheet.UsedRange.Cells(lngRow, lngColumn).End(xlUp).Address
End Function
```

Warning: Range.End Method and Hidden Rows

These strategies work well on most worksheets, but they seem to misbehave when you hide worksheet rows or columns that have used cells. When this happens, the Range.End method fails and the functions LastRow( ), LastColumn( ), LastUsedRow( ), LastUsedColumn, LastUsedRowCell( ), and LastUserColumnCell( ) will begin to systematically fail to return the desired information. Instead, they will return the last used *and visible* cell.

Try this experiment:

1. On Sheet1 of the Find Last Cell.xlsm workbook, hide rows 10:21 and columns H:L. (Click the row 10 header and drag the mouse to row 21; right-click click any selected row header and choose Hide on the context menu to hide the selected rows. Do the same to hide columns H:L.)

2. In the VBA Immediate window, type again ?LastRow(Range("B1")) and ?LastColumn (Range("A5")).

Now the VBA Immediate window must print row 9 and column 7 as the last-used cells (Figure 6-11), which is not correct, since you are not asking for *the last-used and visible cell!* Interesting, huh?

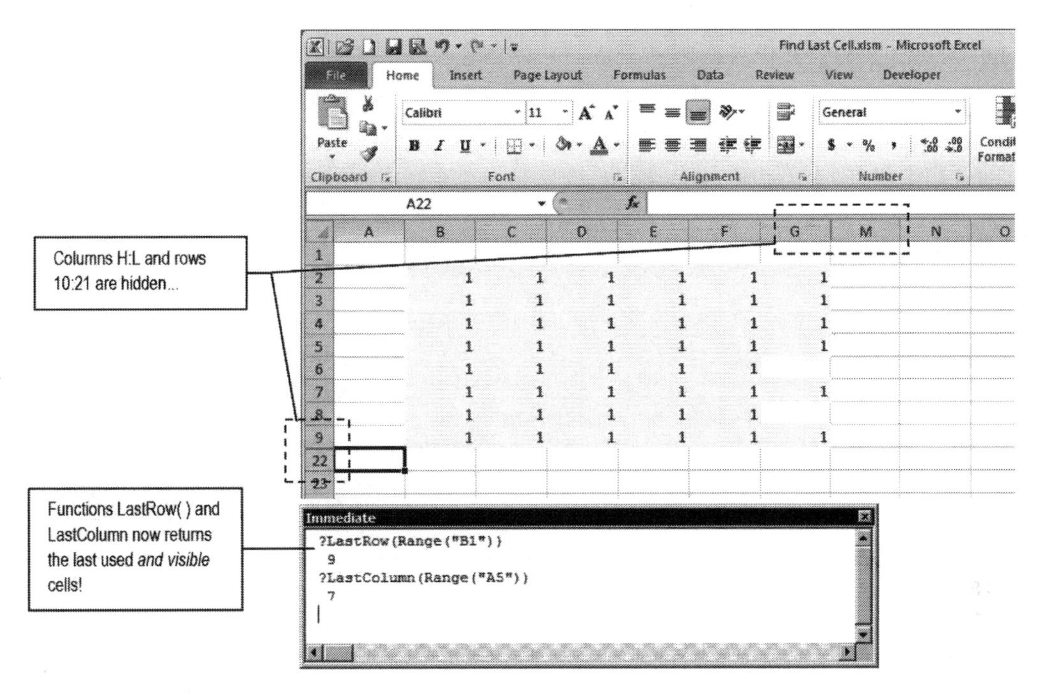

Figure 6-11. *When you hide worksheet rows or columns that contain used cells, the code used by functions LastRow( ) and LastColumn( ) fails to return the last-used row (or column) on the worksheet. They both return the last-used and visible cell*

The only way you can resolve this Excel conflict in your VBA code is to first turn visible all hidden rows/columns, then execute the desired function, and hide again all rows/columns that must be hidden.

The functions LastUsedRowCellHidden( ) and astUsedColumnCellHidden( ) from basLastCell do this job, expecting to receive as an argument the address of the first row/column that must be hidden on the worksheet.

```
Public Function LastUsedRowCellHidden(FirstHiddenRow As Long) As String
    Dim lngRow As Long
    Dim lngColumn As Long

    Application.ScreenUpdating = False
        Range(Cells(FirstHiddenRow, 1), Cells(ActiveSheet.Rows.Count, 1)).EntireRow.Hidden = 
False
```

```
            lngRow = ActiveSheet.UsedRange.Rows.Count
            lngColumn = ActiveSheet.UsedRange.Columns.Count
            LastUsedRowCellHidden = ActiveSheet.UsedRange.Cells(lngRow, lngColumn).
End(xlToLeft).Address
        Range(Cells(FirstHiddenRow, 1), Cells(ActiveSheet.Rows.Count, 1)).EntireRow.Hidden =
True
    Application.ScreenUpdating = True
End Function

Public Function LastUsedColumnCellHidden(FirstHiddenColumn As Long) As String
    Dim lngRow As Long
    Dim lngColumn As Long

    Application.ScreenUpdating = False
        Range(Cells(1,FirstHiddenColumn), Cells(ActiveSheet.Rows.Count, 1)).EntireColumn.
        Hidden = False
            lngRow = ActiveSheet.UsedRange.Rows.Count
            lngColumn = ActiveSheet.UsedRange.Columns.Count
            LastUsedColumnCellHidden = ActiveSheet.UsedRange.Cells(lngRow, lngColumn).
            End(xlUp).Address
        Range(Cells(1, FirstHiddenColumn), Cells(ActiveSheet.Rows.Count, 1)).EntireColumn.
        Hidden = True
    Application.ScreenUpdating = True
End Function
```

Note that both functions use the Range object to select the entire rows/columns that must become visible before applying the End( ) property to the UsedRange property. They will also hide all worksheet rows/columns beginning with the row/column that they receive as an argument. Since the entire process happens with ScreenUpdating = False, this will work as if nothing happens to the worksheet, and the last-used cell will always be returned.

Finding Range Information

Once you have created on the USDA sheet tab the individual food category range names, the USDA range name, and the FoodCategoriesList range names, you can use two cascading Excel data validation lists to easily select the food category on the first data validation list (the list is filled using the FoodCategoriesList range name), then choose any food item of the selected food category on the second, or *cascade*, data validation list (which is filled with the appropriate food items using the Excel INDIRECT( ) function), and finally return any nutritional food item information you want, as long as it can exist inside the USDA range name.

You can see such a food selection system by opening the sr27_NutrientsPer100g_RangeFind.xlsm macro-enabled workbook (which can be extracted from the Chapter06.zip file) and using the Select Food Item sheet tab input cells (the ones formatted with a light yellow background and a blue border). Select any food category on the first available column and any food item of this food category on the second available column. As you select food items, the worksheet recovers four nutrient values for each one: Energy (kcal), Protein (g), Total Lipid (g), and Carbohydrate by difference (g) (Figure 6-12).

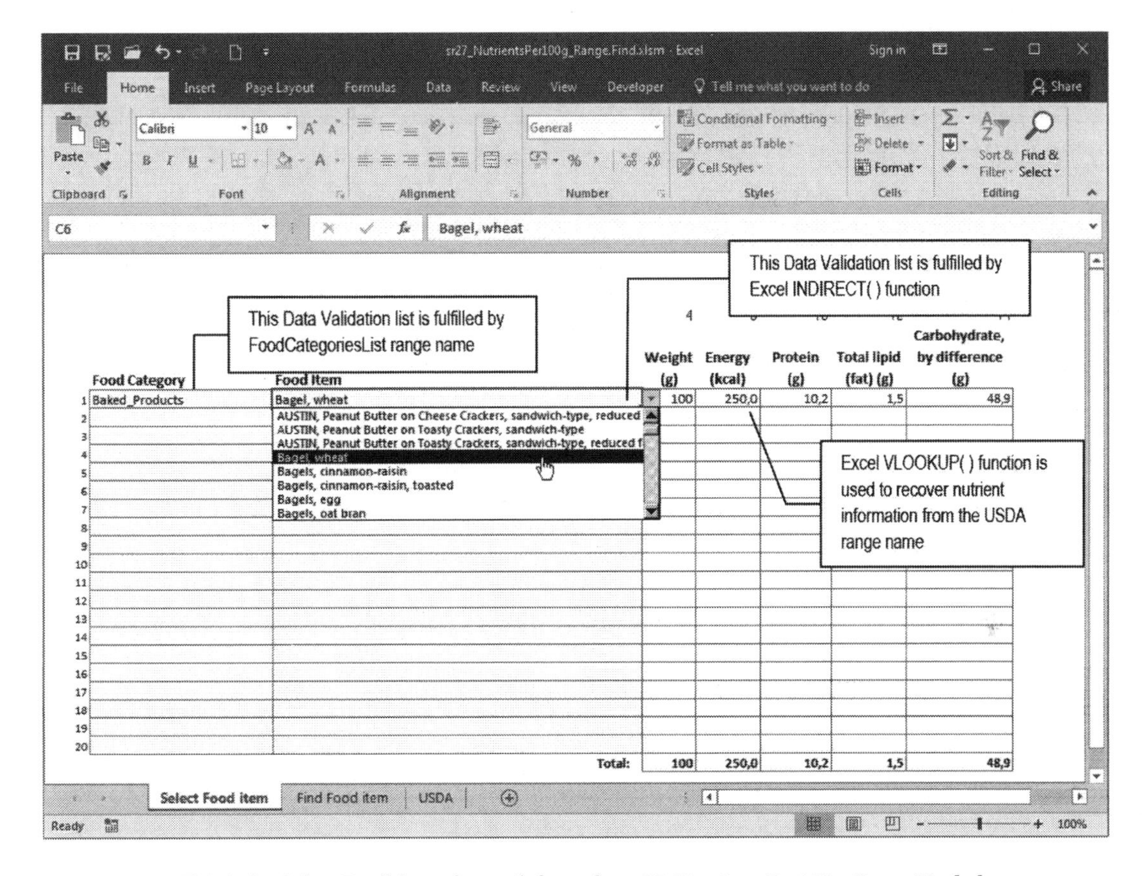

Figure 6-12. *This is the Select Food Item sheet tab from the sr27_NutrientsPer100g_RangeFind.xlsm macro-enabled workbook, where you can use the power of Excel range names and data validation lists to easily search and select individual food items stored in the USDA worksheet and return all available nutrient information*

Cool, isn't it? But if you try to find specific food items or you need to compose specific recipes or whatever, you will realize that it is not too easy to find them inside the USDA food table. Which food category does a food belong to? How it is named inside the table?

Here is the perfect moment to use VBA power associated to Excel range find methods to produce a `UserForm` and give you such answers.

The Range.Find Method

You already used the `Range.Find` method in Chapter 5; it has this syntax:

```
Expression.Find(What, After, LookIn, LookAt, SearchOrder, SearchDirection, MatchCase,
MatchByte, SearchFormat)
```

In this code:

> `Expression`: This is the range to be searched.
>
> `What`: This is the search string.

LookIn: This is where the information will be searched (xlValue , which is the default; xlFormula; or xlComments).

LookAt: This is the scope to be searched in LookIn, all or part of each cell value (xlPart, which is the default, or xlWhole).

By keeping all other arguments on their default values, the Range.Find method will make a substring search inside all range cell values. If any cell value matches the search string, Range.Find will return a Range object that represents the first found cell. Otherwise, it returns Nothing.

▒ **Attention** All values used by the last Range.Find method executed by VBA will be reflected in the Excel Find dialog box options.

To keep searching the range by the same search string, use Range.FindNext, which has this syntax:

```
Expression .FindNext(After)
```

In this code:

Expression: This is a variable that represents a Range object.

After: This is a Range object that represents a single cell after which you want to search. This corresponds to the position of the active cell when a search is done from the user interface. The specified cell is not searched until the method wraps back around to it again. If this argument is not specified, the search starts after the cell in the upper-left corner of the range.

The Range.FindNext method expects to receive a range object on the After argument to keep searching the range using the same initial arguments. When the search reaches the end of the specified search range, it wraps around again to the beginning of the range and will find the first occurrence.

Let's try a simple search on the USDA range name to find the first occurrence of the word *apple* in all its cells. Since the Range.Find method returns a Range object and that word exists inside the USDA range name, you can type this command in the VBA Immediate window to find the address of the first found cell:

```
?Range("USDA").Find("apple", , ,xlPart).Address
$B$173
```

Cell B173 is the first food item with *apple* in its name. So, to keep searching the USDA range name for cells that have the word *apple*, use the Range.FindNext method with the B173 address on its After argument, as follows:

```
?Range("USDA").FindNext(Range("B173")).Address
$B$174
```

Now the next cell is B174. Using the Immediate window, you need to manually keep changing the cell address of the Range,FindNext After argument until all matches are found and the search returns to cell B173, which will be quite a tedious task.

Using a VBA code procedure, you can search any range with the Range.Find and Range.FindNext methods using three Range object variables.

- One Range object variable (rg) to represent the searched range.
- A second Range object variable (rgFound) to represent the range found by the Range.Find and Range.FindNext methods. It will also be used by the Range.FindNext After argument.
- A third Range object variable (rgInitial) to keep a reference of the first found range, so you can stop the search when it is find again.

The code will look like this:

```
Set rg = Range("USDA")
Set rgFound = rg.Find("apple",,,xlPart)
If Not  rgFound Is Nothing Then
    Set rgInitial = rgFound
    Do
       '... Do something here with the found cell!
        Set rgFound = rg.FindNext(rgFound)
    Loop Until rgFound.Address = rgInitial.Address
End If
```

Did you see that? After executing the Range.Find method to search for the desired string (*apple*) inside the USDA range name, if the code found it (Not rgFound Is Nothing), a reference to the first found cell is stored in the rgInitial variable (Set rgInitial = rgFound), and a Do...Loop will keep searching the range with the Range.FindNext method, using rgFound on the After argument, until it goes back to the first found cell (Loop Until rgFound.Address = rgInitial.Address).

Note that you must compare both ranges using the Range.Address property because the default Range property returns the range value. If the searched range eventually has duplicated values, it may falsely stop the loop on a wrong cell (a common VBA bug you may insert in your code).

Let's see this in action! The Find Food Item sheet tab of the sr27_NutrientsPer100g_RangeFind.xlsm macro-enabled workbook has the Find Food Item command button to assist you in selecting the food item you want using the Range.Find method. The worksheet is protected, meaning that just its input cells (Food Category or Food Item) can be selected, and any food item in the data validation list must be selected before clicking the Button control—or you will receive a warning message (Figure 6-13).

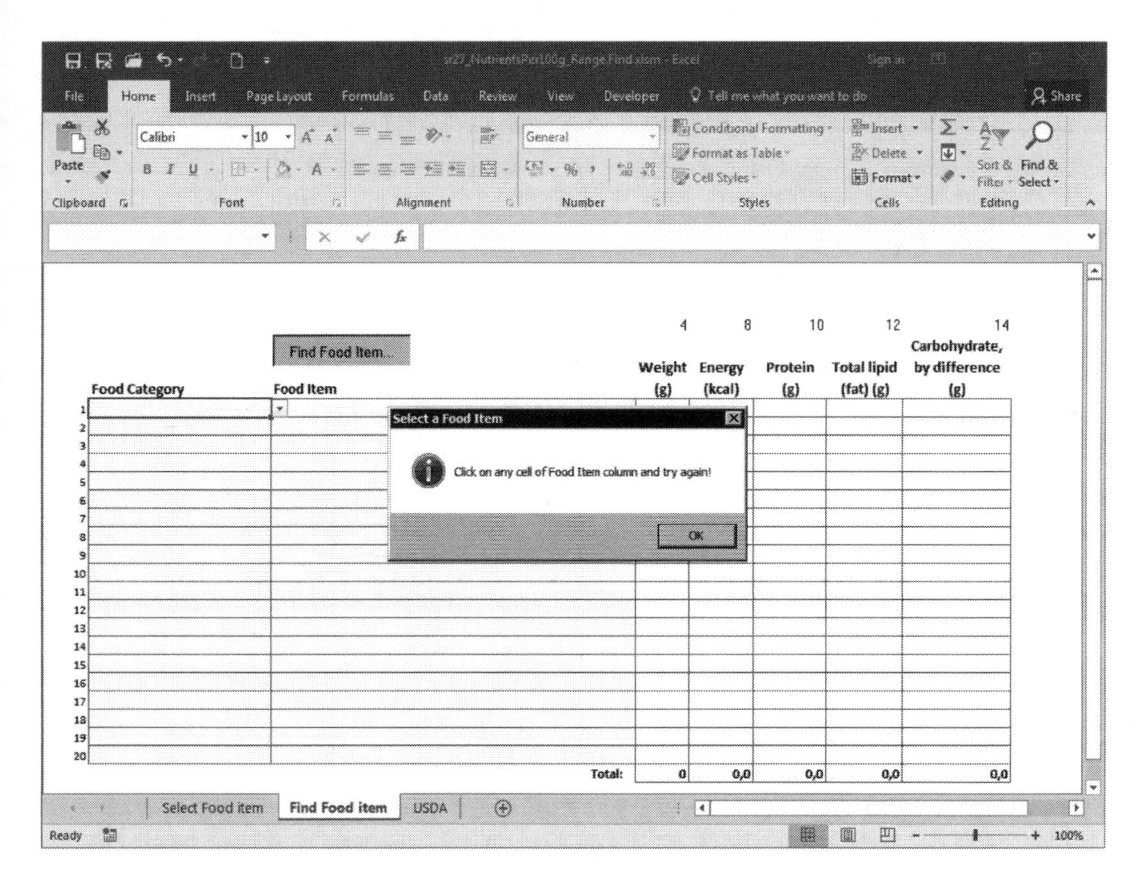

***Figure 6-13.** This is the Find Food Item sheet tab of the sr27_NutrientsPer100g_RangeFind.xlsm macro-enabled workbook. It has the Find Food Item Button control that raises a warning message if you do not select any food item input cell*

This is not really needed but is an excellent moment to show how you can impose to the user of your application the selection of a given cell before they click a Button control to execute an action. When you click the Find Food Item Button control, Excel runs the Function SelectFoodItem() procedure of the Sheet3 worksheet (Find Food Item tab), and this code is executed:

```
Public Sub SelectFoodItem()
    Const conC = 3

    If Selection.Column = conC Then
        frmRangeFind.Show vbModal
    Else
        MsgBox "Click on any cell of Food Item column and try again!", vbInformation,
        "Select a Food Item"
    End If
End Sub
```

As you can see, it declares the conC =3 constant and uses the Application.Selection.Column property (which returns a column number) to guarantee that the user is in the right column before it loads the frmRangeFind UserForm (see Figure 6-14).

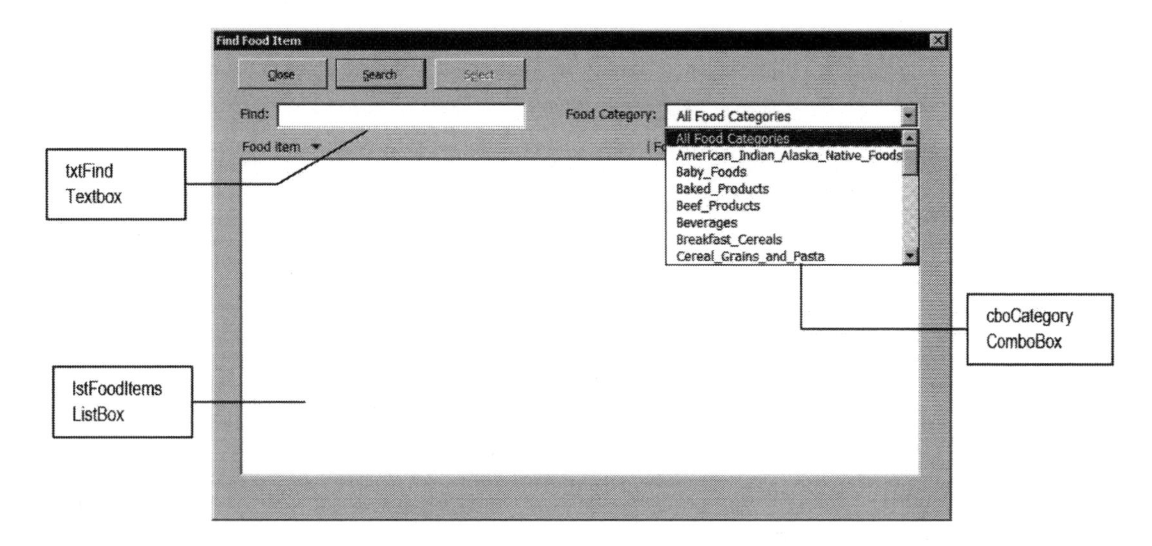

Figure 6-14. *This is frmRangeFind UserForm interface. It allows you to search food items with the Range.Find method using the entire USDA food table or just one of its food categories*

Preparing frmRangeFind

When `frmRangeFind` is loaded, it fires the `UserForm_Initialize( )` event, which has this code:

```
Private Sub UserForm_Initialize()
    Dim rg As Range
    Const conNameDoesntExist = 1004

    On Error GoTo Initialize_Error

    mintLastColumn = 1
    Me.cboCategory.AddItem "All Food Categories"
    Me.cboCategory = "All Food Categories"

    'Verify if range names already exist
    For Each rg In Range("FoodCategoriesList")
        Me.cboCategory.AddItem rg
    Next

Initialize_End:
    Exit Sub
Initialize_Error:
    Select Case Err
        Case conNameDoesntExist
            Call CreateRangeNames
            Resume
        Case Else
            MsgBox "Error " & Err & " in UserForm_Initialize"
    End Select
    Resume Initialize_End
End Sub
```

It sets an error trap, initializes the `mintLastColumn` module-level variable (you will see it later in the section "Sorting ListBox Items"), adds All Categories as the first item of the `cboCategories ListBox` using the `AddItem` method, and defines it as it default value.

```
msngLastX = Me.lblAZ.Left
mintLastColumn = 1
Me.cboCategory.AddItem "All Food Categories"
Me.cboCategory = "All Food Categories"
```

Next it tries to initiate a `For Each...Next` loop through all cells of the `FoodCategoriesList` range name. If this range name does not exist, VBA will raise error 1004 ("Application-defined or object-defined error"), which will be caught by the error trap.

Inside the error trap, a `Select Case` instruction verifies the error code. If it is equal to `conNameDoesntExist = 1004`, it will call the `CreateRangeNames( )` procedure to create it and use a `Resume` statement to try to execute the loop again.

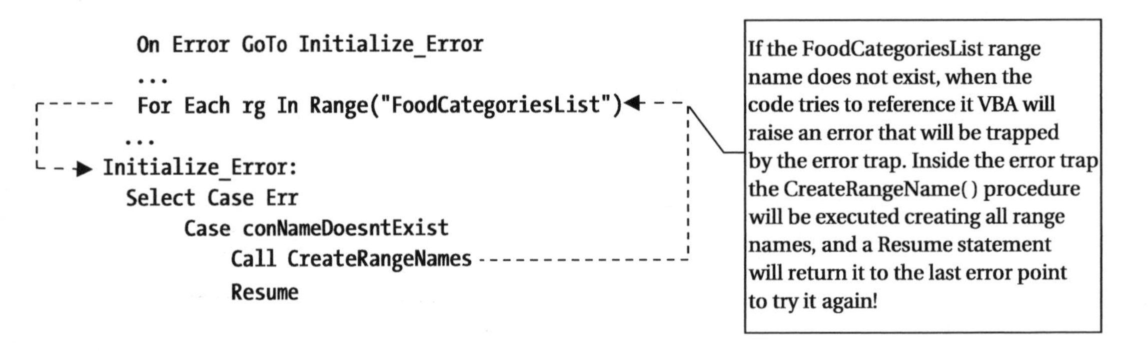

■ **Attention** By default, the `sr27_NutrientsPer100g_RangeFind.xlsm` macro-enabled workbook has no range names. They are created the first time you open the `frmRangeFind UserForm`, and I bet that you didn't notice it (unless by the worksheet saving process).

Searching with frmRangeFind

Let's try a simple search. Type **apple** in the Find text box and press Enter to find every food item (exactly 173) on the USDA worksheet that has this word on the `Long_Dsc` column, as well as the food category each belongs to. Since the USDA worksheet is sorted by food category first and then by the food item, the results will appear using this sort order (Figure 6-15).

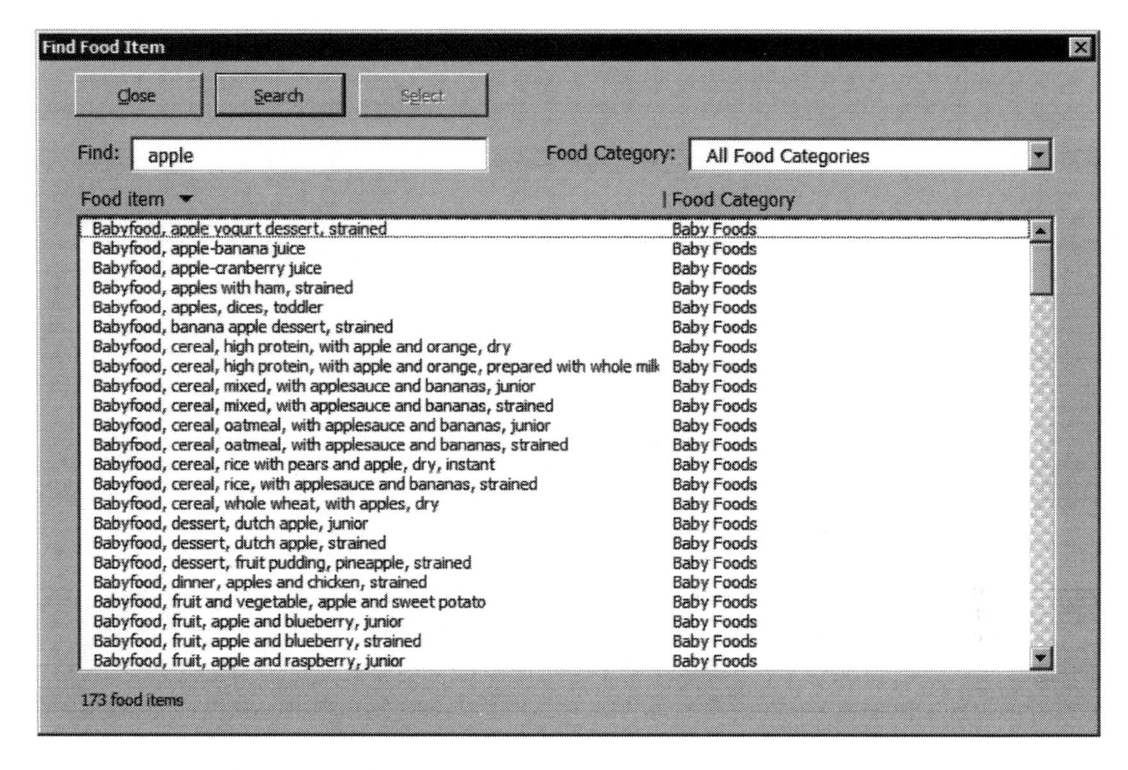

Figure 6-15. *Using frmRangeFind UserForm, you can make a substring search on the USDA range name, or any of its food categories, for every food item that has a given word on it food name*

As you may be wondering, this search happens because the `cmdSearch CommandButton` has the `Default` property set to `True`, and when you press Enter, the `cmdSearch_Click( )` event fires, executing this code:

```
Private Sub cmdSearch_Click()
    Dim rg As Range
    Dim rgInitial As Range
    Dim rgFound As Range

    Me.lstFoodItems.Clear

    If Me.cboCategory.ListIndex = 0 Then
        Set rg = Range("USDA")
    Else
        Set rg = Range(Me.cboCategory)
    End If

    Set rgFound = rg.Find(Me.txtFind, , , xlPart)
    If rgFound Is Nothing Then
        MsgBox "Can't find " & Me.txtFind
    Else
        Set rgInitial = rgFound
        Do
```

```
            Me.lstFoodItems.AddItem rgFound
            Me.lstFoodItems.Column(1, lstFoodItems.ListCount - 1) = rgFound.Offset(0, 1)
            Set rgFound = rg.FindNext(rgFound)
        Loop Until rgFound.Address = rgInitial.Address
    End If
    Me.lblFoodItems.Caption = Me.lstFoodItems.ListCount & " food items"
    Me.lblFoodItems.Visible = True
End Sub
```

Quite simple, huh? The code declares the three Range object variables it will need to perform the search, clears the lstFoodItems ListBox, and uses the rg object variable to set the search scope, which is the USDA range name or selected food category on the cboCategory ComboBox.

```
Me.lstFoodItems.Clear


If Me.cboCategory.ListIndex = 0 Then
    Set rg = Range("USDA")
Else
    Set rg = Range(Me.cboCategory)
End If
```

The search begins by attributing to the rgFound variable the rg.Find result, using what you type in the txtFind text box as the search string. Note that it uses the LookAt = xlPart argument to make a substring search.

```
Set rgFound = rg.Find(Me.txtFind, , , xlPart)
```

If rgFound has a valid range (a range name representing a single cell address), this range is set to the rgInitial object variable. A Do...Loop begins and adds the first cell food item name using the ListBox AddItem method and adds the food item category using the ListBox Column property to fill the first and second columns of the lstFoodItems ListBox. Note that it uses rgFound.Offset(0,1) to displace the food item name by one column to the right and returns its associated food category.

```
If rgFound Is Nothing Then
    MsgBox "Can't find " & Me.txtFind
Else
    Set rgInitial = rgFound
    Do
        Me.lstFoodItems.AddItem rgFound
        Me.lstFoodItems.Column(1, lstFoodItems.ListCount - 1) = rgFound.Offset(0, 1)
```

▪ **Attention** When you use the Range.Offset method on a Range object variable, the original range address stays the same. Offset displaces the range to the desired row/column and returns another range address.

The search for another food item continues inside the Do...Loop structure using the Range.FindNext method, passing to it the After argument of the last found cell represented by the rgFound object variable. The loop will end when it reaches again the first found cell, represented by the rgInitial object variable. Once more, to guarantee that the first range found was found again, the code must compare both Range.Address properties.

```
        Set rgFound = rg.FindNext(rgFound)
    Loop Until rgFound.Address = rgInitial.Address
End If
```

This is all you need to fill the lstFoodItems list box with all food item names and categories that have the searched string inside the food name identification. When the loop ends, the lblFoodItems label below the ListBox has its Caption property updated to reflect how many food items have been found, using the ListBox ListCount property, and the label becomes visible.

```
    Me.lblFoodItems.Caption = Me.lstFoodItems.ListCount & " food items"
    Me.lblFoodItems.Visible = True
End Sub
```

Returning the Selected Food Item in frmRangeFind

The reason you limit the use of frmRangeFind to any food item input cell of the Find Food Item sheet tab is to allow frmRangeFind to automatically fill both the food category and the food item on the worksheet after you select the desired food in the lstFoodItem ListBox.

Note that the Select CommandButton (cmdSelect) is disabled by default, becoming enabled after you select any food item in the lstFoodItems ListBox. To return the desired item to the Food Item cell currently selected on the Find Food Item sheet tab, you must first select it in the list box and click the cmdSelect CommandButton, or just double-click it, firing the lstFoodItems_DblClick( ) event, which will make a call to the cmdSelect_Click( ) procedure.

```
Private Sub lstFoodItems_DblClick(ByVal Cancel As MSForms.ReturnBoolean)
    Call cmdSelect_Click
End Sub

Private Sub cmdSelect_Click()
    Application.Selection = Me.lstFoodItems
    Application.Selection.Offset(0, -1) = Me.lstFoodItems.Column(1, Me.lstFoodItems.
    ListIndex)
    Unload Me
End Sub
```

To return the selected food item to the current selected worksheet cell, use the Application.Select property, which has the address of the selected cell(s). Note again that this time the code uses Application.Selection.Offset(0, -1) to displace the cell currently selected by one column to the last and returns the food category (Figure 6-16).

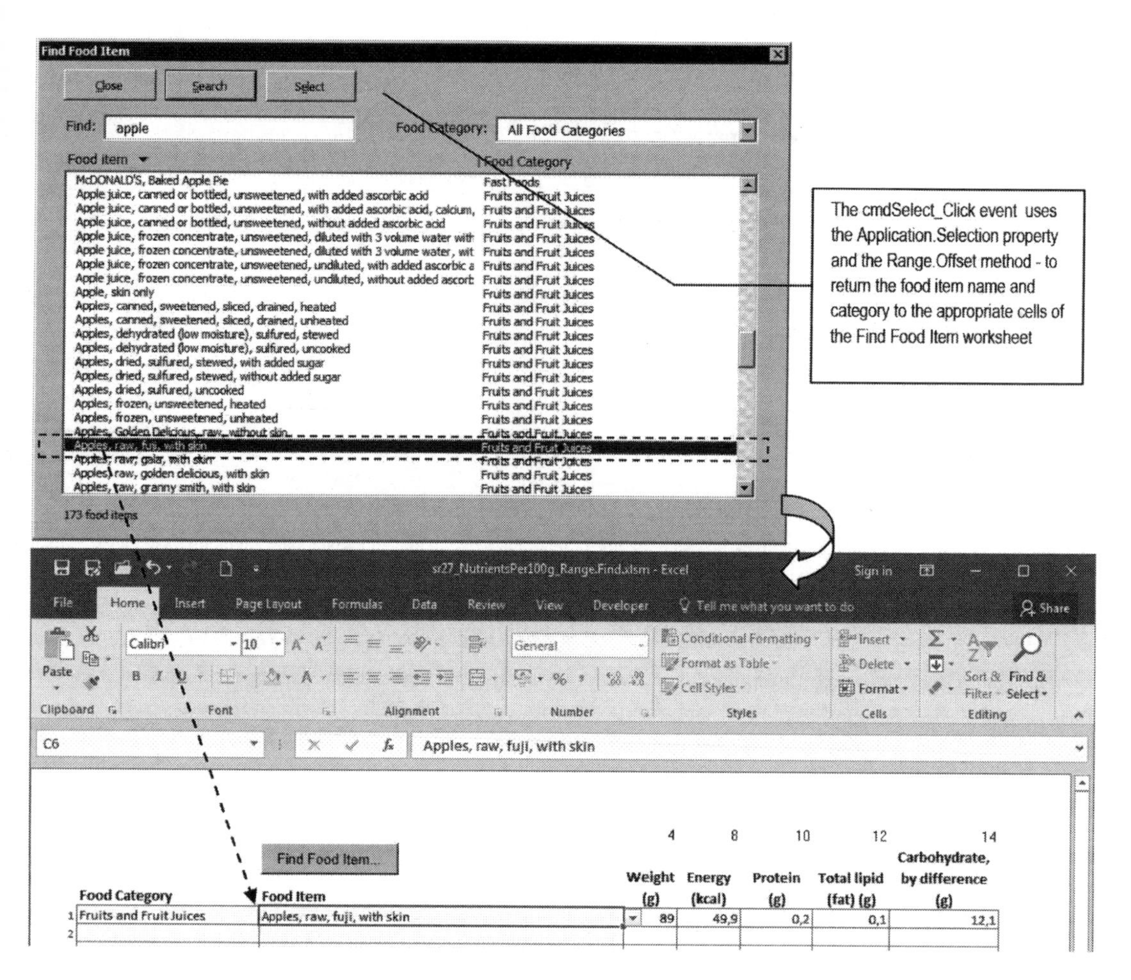

Figure 6-16. *After you find the desired food item using the frmRangeFind* `UserForm`*, double-click the item to return the food category and food name to the appropriate cells of the Find Food Item worksheet. The worksheet formulas will do its job and return the food item nutritional information from the USDA worksheet*

Sorting ListBox Items

When you make a search with the `Range.Find` method and use the results to fill a `ListBox`, they will be presented using the originally searched range order, which may not be the most convenient way to exhibit the results.

It will be nice if the `ListBox` control could offer a kind of sort method so you could easily change the presentation order. Instead, you must count with VBA and some program skills to surpass this challenge. Let's face it!

No matter how the data to be sorted is presented, you must always take into account that you first need to put it inside a VBA array variable, structured in rows and columns, because this is the most effective way to manipulate it using code. Besides that, you need to implement an efficient sort method so the sort operation can run as fast as possible.

Among the many sorting algorithms available, one of the most efficient in large data sets is the bubble sort method (already covered in Chapter 4), which works by taking the first array item, comparing it with every other array item, and swapping each pair found in the wrong order. Once this item is correctly positioned on the array order, it takes the second array item and repeats the same process, until all array items are correctly positioned, as an indication that the array is sorted. This algorithm has this name because of the way the smaller elements move to the top of the array like a bubble floating through the air.

To implement the bubble sort algorithm, you need two `For...Next` loops. The outer loop is responsible for putting the item in it correct position, while the inner loop takes care of comparing it with every other array item. Supposing that you have a one-dimensional array variable of *n* items represented by the `varArray` variable, this simple code mimics the bubble sort algorithm (note that it uses `UBound(varArray) - 1` to limit each loop because it uses `intI+1` and `intJ+1` to compare each item with the next):

```
For intJ = LBound(varArray) To UBound(varArray) - 1
    For intI = LBound(varArray) To UBound(varArray) - 1
        If varArray(intI) > varArray(intI + 1) Then
                strTemp = varArray(intI)
                varArray(intI) = varArray(intI + 1)
                varArray(intI + 1) = strTemp
        End If
    Next intI
Next intJ
```

Now that you have the basic idea, let's see how it was implemented in the user interface of `frmRangeFind` `UserForm`. Make a search to any food item that contains a desired word (like *peas*) and try to click the label headers at the top of the `lstFoodItems` `ListBox`. If you click the "Food item" header, the `ListBox` will become ascending sorted by this column. Click it again, and it will become descending sorted. A small triangle at the right side of the header indicates the sort order. Repeat the process by clicking the Food Category header to sort the list by its second column, in ascending or descending order. All these operations use the bubble sort algorithm to sort the `ListBox`.

Note that when you click the sorted column, it alternates the sort in ascending/descending order, and whenever you change the sorted column, the new column is always sorted ascending first. Fast and cool, isn't it (Figure 6-17)?

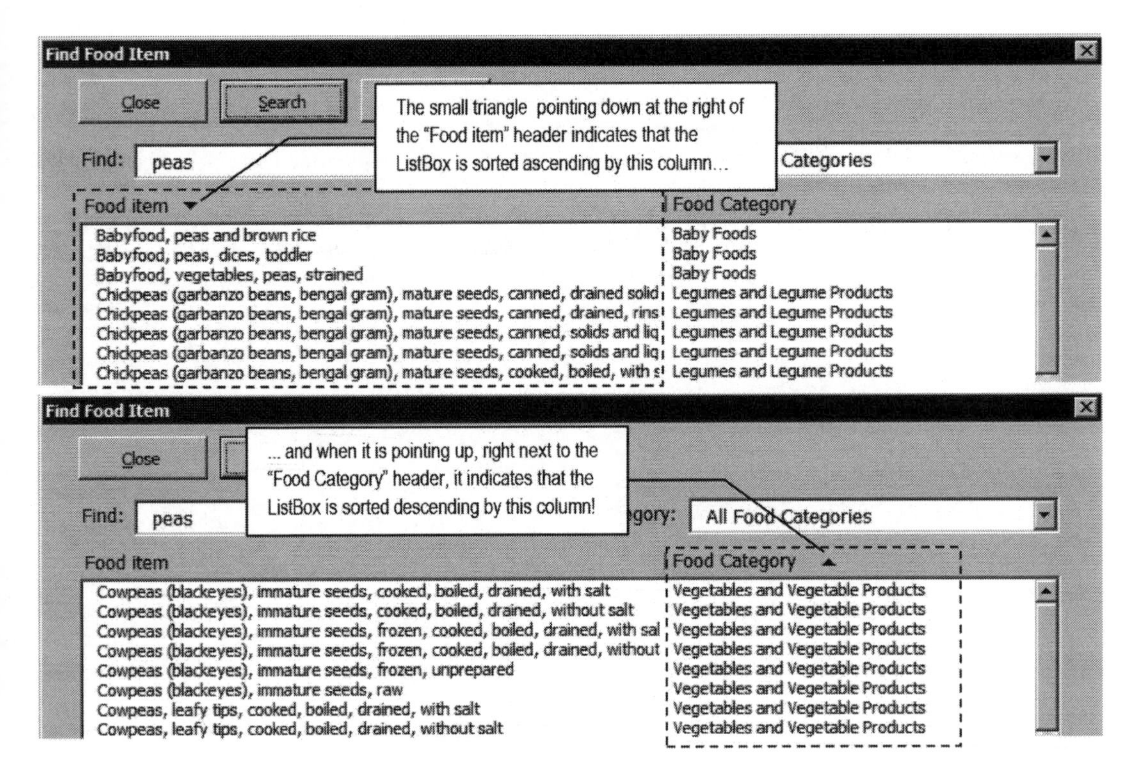

Figure 6-17. *The frmRangeFind UserForm allows you to change its items' sort order by clicking the label headers right above it. As you click any column header, the selected column becomes sorted ascending or descending, while a small triangle positioned at the right of the header indicates the sort order*

Let's first understand this interface mechanism. Everything begins with the use of different `Label` controls positioned on the right place and a specific Z-order in the `UserForm` interface (the Z-order is the stacking order that controls overlap each other by in the `UserForm`).

If you look to the `frmRangeFind UserForm` in design mode, you will realize that the `ListBox` header is composed of five different **Label** controls, stacked in this Z-order (from back to top, as shown in Figure 6-18).

- `lbl0` and `lbl1`: These **Label** controls name the Food Item and Food Category columns.
- `lblAZ`: This label control stacks over `lblFoodCategory`. It was formatted using Webdings, 11pt font, and is responsible for showing the small black triangle that points up (`Caption=5`) or down (`Caption=6`).
- `lbllstFoodItems`: This `Label` control must be over every column label name and `lblAZ`. Since its width extends through all `lstFoodItems ListBox` widths, you don't have to click all the controls behind it.
- `lblCol1`: This `Label` control is responsible for showing the vertical bar character (|) that is used to indicate the division between the first and second columns of the `lstFoodItems ListBox` and is on top of `lbllstFoodItems`.

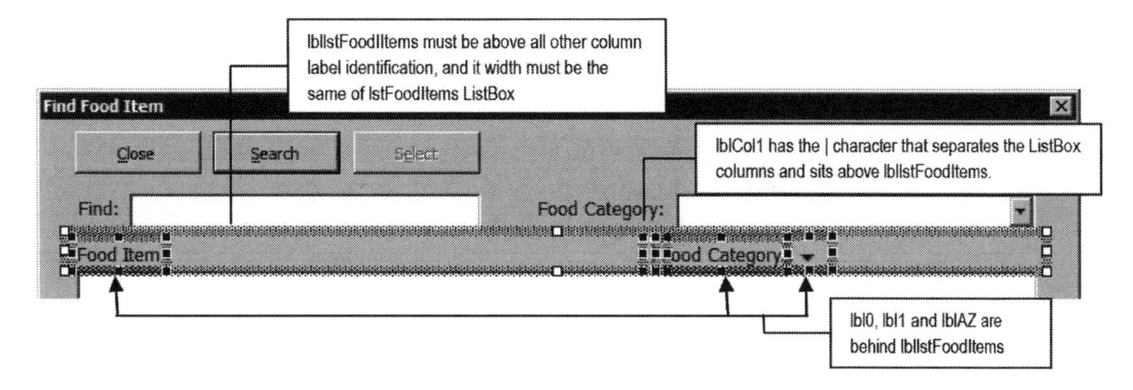

Figure 6-18. *To implement the lstfFoodItems sorting, the header is composed of five different Label controls, stacked in this order: lbl0, lbl1, lblAZ, lbllstFoodItems, and lblCol1*

Since `lbllstFoodItems` is on top of all other column label identifiers, you can just click it or in any other control that is right above it in the `UserForm` Z-order. Whenever you click any `Label` control, three successive mouse events fire, in this order: `MouseDown( )`, `MouseUp( )`, and `Click( )`.

Among this three events, just `MouseDown( )` and `MouseUp( )` pass arguments to indicate the `X` and `Y` coordinates where the click occurred, relative to the top-left corner of the control, which has `(X,Y)` coordinates equal to `(0,0)`. So, to know which column must be sorted, you must choose the `lbllstFoodItems_MouseUp( )` event, which is the one that fires immediately before the `lbllstFoodItems_Click( )` event, and execute this code:

```
Private Sub lbllstFoodItems_MouseUp(ByVal Button As Integer, ByVal Shift As Integer, ByVal X
As Single, _
                                                                ByVal Y As Single)
    Dim intLeft As Integer
    Dim intColumn As Integer
    Static sbolDesc As Boolean
    Const conTriangleUp = 5
    Const conTriangleDown = 6

    If Me.lstFoodItems.ListCount > 1 Then
        Application.Cursor = xlWait
        Select Case X + Me.lbllstFoodItems.Left
            Case Is < Me.lbl1.Left
                intColumn = 0
            Case Else
                intColumn = 1
        End Select
        intLeft = Me("lbl" & intColumn).Left + Me("lbl" & intColumn).Width

        If mintLastColumn = intColumn Then
            sbolDesc = Not sbolDesc
            Call SortListBox(Me.lstFoodItems, intColumn, IIf(sbolDesc, Desc, Asc))
            Me.lblAZ.Caption = IIf(sbolDesc, conTriangleUp, conTriangleDown)
        Else
            mintLastColumn = intColumn
```

```
            sbolDesc = False
            Call SortListBox(Me.lstFoodItems, intColumn, Asc)
            Me.lblAZ.Caption = conTriangleDown
        End If
        Me.lblAZ.Left = intLeft
        Application.Cursor = xlDefault
    End If
End Sub
```

This code works based on some assumptions:

- The module-level variable mintLastColumn holds the last ListBox-sorted column. Since by default the ListBox appears sorted by its second column (column 1), this variable is defined to 1 on the UserForm_Initialize event.
- If you click a column that is currently sorted ascending, it must be sorted descending, and vice versa.
- If you change the ListBox-sorted column, the new selected column must be sorted ascending.

This procedure declares three variables.

- intLeft is used to position lblAZ horizontally over the ListBox column, graphically showing the small black triangle that indicates the sorted column and sort direction.
- intColumn indicates what column must be sorted, according to the point the user clicked on the lbllstFoodItems Label control.
- sbolDesc is a static Boolean variable that holds the last sorting order (False = ascending).

The ListBox will be sorted if it has at least two items, which is controlled using the ListCount property. If the list must be sorted, the procedure uses the Application.Cursor property to change the mouse pointer to the traditional hourglass cursor before sorting it and turns it again to the default mouse pointer when the process finishes.

```
    If Me.lstFoodItems.ListCount > 0 Then
        Application.Cursor = xlWait
        ...
        Application.Cursor = xlDefault
    End If
End Sub
```

To know where the user clicked lbllstFoodIterms, relative to the left border of the UserForm, the procedure takes into account the X argument (horizontal position inside the Label control) plus the lbllstFoodItem.Left property.

```
Select Case X + Me.lbllstFoodItems.Left
```

Now it is a matter of comparing the point where the user clicked the UserForm with the lbl1.Left property. If the user clicks left of lbl1 (the Food Category label), the ListBox must be sorted by its first column (Food Item name, intColumn=0); otherwise, it must be sorted by its Food Category name (intColumn=1).

```
Case Is < Me.lbl1.Left
      intColumn = 0
   Case Else
      intColumn = 1
End Select
```

Knowing the column to be sorted, the code uses the `intLeft` variable to define the new horizontal position of `lblAZ` (the small black triangle). Note that it uses the indirect syntax to compose this position adding the `Label` control (`lbl0` or `lbl1`) `Left` and `Width` properties, which will put the triangle character right next the appropriate column label.

```
intLeft = Me("lbl" & intColumn).Left + Me("lbl" & intColumn).Width
```

The procedure then verifies if the new clicked column is the same sorted column, comparing the module-level variable `mintLastColumn` with `intColumn`. If they are the same, it means that the column sort order must be changed, which is made by alternating the `sbolDesc` static `Boolean` variable value.

```
If mintLastColumn = intColumn Then
    sbolDesc = Not sbolDesc
```

The `ListBox` is then sorted using the `SortListBox()` procedure of the `basSortListBox` code module, which receives three arguments: a reference to the `ListBox` control to be sorted, the column to be sorted, and the sorting order. Note that it uses a VBA `IIF( )` function to verify the `sbolDesc` value and pass the correct sort procedure argument.

```
Call SortListBox(Me.lstFoodItems, intColumn, IIf(sbolDesc, Desc, Asc))
```

Once the list is sorted, the character used by the small black triangle of `lblAZ` alternates between pointing down or up as a visual clue to the column sorting order and is correctly positioned to the right of the appropriate column name by changing the `Left` property.

```
    Me.lblAZ.Caption = IIf(sbolDesc, conTriangleUp, conTriangleDown)
Else
   ...
End If
Me.lblAZ.Left = intLeft
```

Note that if the user clicks another column, the `mintLastColumn` module-level variable will be updated to reflect the new sorted column, and the new column will always be sorted ascending.

```
Else
    mintLastColumn = intColumn
    sbolDesc = False
    Call SortListBox(Me.lstFoodItems, intColumn, Asc)
    Me.lblAZ.Caption = conTriangleDown
```

Using the Bubble Sort Algorithm

The `basSortListBcx` code module (which is also available in the `Chaper06.zip` file) implements the bubble sort algorithm to any `ListBox` column, with this syntax:

```
SortListBox(lst As MSForms.ListBox, Optional intColumn As Integer, Optional intOrder As
Order = Asc)
```

In this code:

> `lst`: This is required; it is a reference to the `ListBox` control to be sorted.
>
> `intColumn`: This is optional; it is the `ListBox` column number (0-based) to be sorted.
>
> `intOrder`: This is optional; it is the sort order. Use `Asc` for ascending (default) and `Desc` for descending order.

The code module declares the `Order` enumerator to adequately define the sorting order.

```
Public Enum Order
    Asc = 1
    Desc = 2
End Enum
```

The `SortListBox()` procedure is quite simple and fast, executing this code:

```
Public Sub SortListBox(lst As MSForms.ListBox, _
                       Optional intColumn As Integer, _
                       Optional intOrder As Order = Asc)
    Dim varArray() As Variant
    Dim intI As Integer
    Dim intJ As Integer
    Dim intX As Integer
    Dim strTemp As String
    Dim fSort As Boolean

    With lst
        varArray = .List
        For intJ = LBound(varArray) To UBound(varArray) - 1
            For intI = LBound(varArray) To UBound(varArray) - 1
                If intOrder = Asc Then
                    If IsNumeric(varArray(intI, intColumn)) And IsNumeric(varArray(intI + 1,
                    intColumn)) Then
                        fSort = Val(varArray(intI, intColumn)) > Val(varArray(intI + 1,
                        intColumn))
                    Else
                        fSort = varArray(intI, intColumn) > varArray(intI + 1, intColumn)
                    End If
                Else
                    If IsNumeric(varArray(intI, intColumn)) And IsNumeric(varArray(intI + 1,
                    intColumn)) Then
                        fSort = Val(varArray(intI, intColumn)) < Val(varArray(intI + 1,
                        intColumn))
                    Else
                        fSort = varArray(intI, intColumn) < varArray(intI + 1, intColumn)
                    End If
                End If

                If fSort Then
                    For intX = 0 To (.ColumnCount - 1)
```

```
                        strTemp = varArray(intI, intX)
                        varArray(intI, intX) = varArray(intI + 1, intX)
                        varArray(intI + 1, intX) = strTemp
                    Next intX
                End If
            Next intI
        Next intJ
        .List = varArray
    End With
End Sub
```

The procedure speed is fundamentally based on attributing the entire ListBox content to the varArray variable, declared as a Variant array, using the ListBox List property, which will allow the code to run faster.

```
Dim varArray() As Variant
...
varArray = .List
```

■ **Attention** Although the lstFoodItems ListBox is defined to use the two columns, the List property returned a ten-column array. You can see this by defining a VBA breakpoint immediately before the varArray variable is initialized and using this instruction in the VBA Immediate window (remember that arrays are 0-based):

```
?UBound(varArray,2)

9
```

Now that all ListBox content is inside varArray, you must use the VBA LBound( ) and UBound( ) functions to create two successive For...Next loops. The outer loop takes each ListBox item and compares it to the next; the inner loop compares the item with every other list item to put it in the desired sort order (ascending or descending).

```
For intJ = LBound(varArray) To UBound(varArray) - 1
    For intI = LBound(varArray) To UBound(varArray) - 1
  ...
    Next intI
Next intJ
```

Inside the inner loop, the comparison is made according to the procedure's intOrder argument, using intI and intI+1 values to reference two successive array items and the intColumn argument to make the comparison using the desired ListBox column. The procedure uses the VBA IsNumeric( ) function to verify whether both values are text or numbers. If they are numbers, the code uses the VBA Val( ) function to compare the numeric values instead of the numeric text strings. The fSort variable will become true whenever two successive values must change its position.

```
If intOrder = Asc Then
   If IsNumeric(varArray(intI, intColumn)) And IsNumeric(varArray(intI + 1, intColumn)) Then
```

```
        fSort = Val(varArray(intI, intColumn)) > Val(varArray(intI + 1, intColumn))
    Else
        fSort = varArray(intI, intColumn) > varArray(intI + 1, intColumn)
    End If
Else
    If IsNumeric(varArray(intI, intColumn)) And IsNumeric(varArray(intI + 1, intColumn)) Then
        fSort = Val(varArray(intI, intColumn)) < Val(varArray(intI + 1, intColumn))
    Else
        fSort = varArray(intI, intColumn) < varArray(intI + 1, intColumn)
    End If
End If
```

Whenever any item must change its position in the list (`fSort=True`), the code must execute another `For...Next` loop to run through all item columns, changing the position of the entire array row, no matter how many columns it has (note that it uses the `ListBox ColumnCount` property to limit the loop).

```
If fSort Then
    For intX = 0 To (.ColumnCount - 1)
        strTemp = varArray(intI, intX)
        varArray(intI, intX) = varArray(intI + 1, intX)
        varArray(intI + 1, intX) = strTemp
    Next intX
End If
```

And once this is made, the code is ready for the next comparison, until all items are in the right position. When this happens, the procedure takes the now-sorted `varArray` variable and attributes it to the `ListBox List` property. VBA and Windows will instantly update the `ListBox` interface to the desired sort order!

```
                    Next intX
                End If
            Next intI
        Next intJ
        .List = varArray
    End With
End Sub
```

Quite fast, huh?

Changing ListBox Column Widths

There is one more technique that deserves to be mentioned on `frmRangeFind`: it allows you to change the `ListBox` column widths by dragging the | character used to indicate where one column ends and another column begins (Figure 6-19).

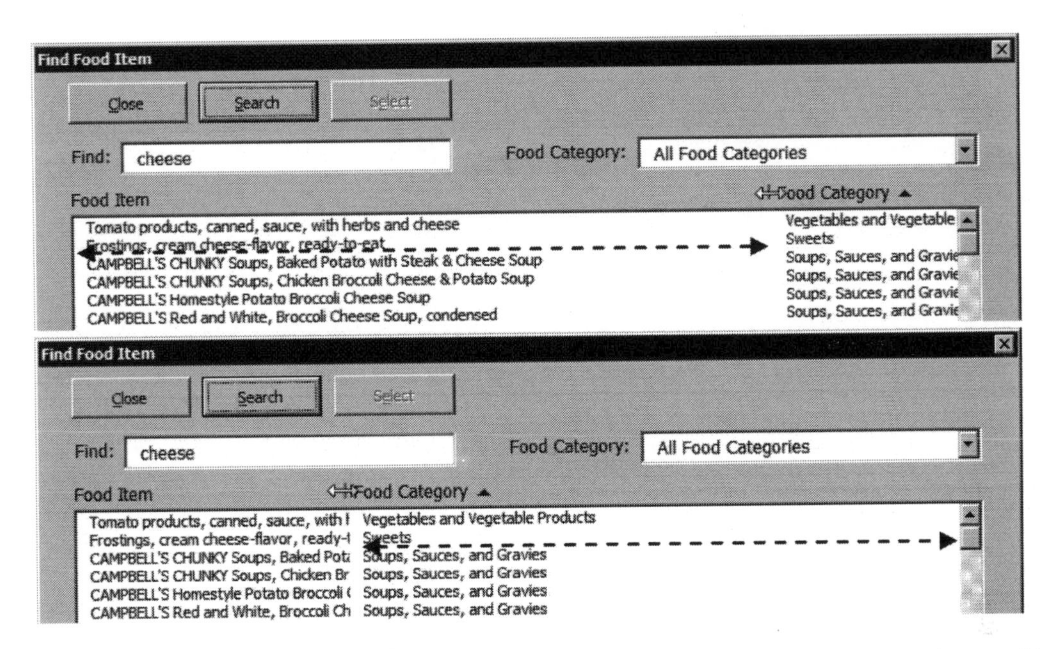

__Figure 6-19.__ Drag the pipe character of the lstFoodItems ListBox to any side to change the column widths

Did you notice that the mouse icon changes to an east-west arrow when you point it over the | character and that the code limits how long you can drag the column separator to the left or right position?

This happens because the `lblCol1 Label` control has the `MousePointer` property defined to `9 fmMousePointerSizeWE`, while the code implements three `lblCol1` mouse events: `MouseDown( )`, `MouseMove( )`, and `MouseUp( )`.

The code uses the `mbolPressed Boolean` module-level variable to signal whether the mouse is pressed, setting it to `True` on the `lblCol1_MouseDown( )` event (which fires when a mouse button is pressed) and setting it to `False` on the `lblCol1_MoudeUp( )` event (which fires when the mouse button pressed is released).

```
Private Sub lblCol1_MouseDown(ByVal Button As Integer, ByVal Shift As Integer, ByVal X As
Single, ByVal Y As Single)
    mbolPressed = True
End Sub

Private Sub lblCol1_MouseUp(ByVal Button As Integer, ByVal Shift As Integer, ByVal X As
Single, ByVal Y As Single)
    mbolPressed = False
End Sub
```

Now look at the `lblCol1_MouseMove( )` event procedure, which fires when the mouse is pressed and dragged over the `lblCol1 Label` control.

```
Private Sub lblCol1_MouseMove(ByVal Button As Integer, ByVal Shift As Integer, ByVal X As
Single, ByVal Y As Single)
    Dim intLeft As Integer
    Const conMinLeft = 100
    Const conMaxLeft = 400
```

```
    If mbolPressed Then
        intLeft = Me.lblCol1.Left + X
        If intLeft < conMinLeft Then
            intLeft = conMinLeft
        End If
        If intLeft > conMaxLeft Then
            intLeft = conMaxLeft
        End If
        Me.lblCol1.Left = intLeft
        Me.lbl1.Left = intLeft + Me.lblCol1.Width
        If mintLastColumn = 1 Then
            Me.lblAZ.Left = Me.lbl1.Left + Me.lbl1.Width
        End If
        Me.lstFoodItems.ColumnWidths = intLeft - Me.lbllstFoodItems.Left
    End If
End Sub
```

The mouse move event passes the X and Y arguments that indicate the (X,Y) coordinates of the mouse pointer regarding the size of Label control, where the point (0,0) means the control's top-left corner. Note, however, that when the mouse is down and you drag it farther than the control left border, the X argument will become negative. Conversely, when you drag it farther than the right border, the X argument will become greater than the control's Width property.

So if the mbolPressed module-level variable is True, the X argument that indicates the horizontal mouse position regarding the Label control is added to the control Left property and stored into the intLeft variable. Note that intLeft will hold a value that relates to the horizontal position of the UserForm.

```
If mbolPressed Then
    intLeft = Me.lblCol1.Left + X
```

The procedure then tests whether the intLeft position is smaller than the conMinLeft constant or greater than the conMaxLeft constant. If it is, intLeft is redefined to the minimum or maximum constant, literally restricting the horizontal drag movement of the pipe (|) character:

```
If intLeft < conMinLeft Then
    intLeft = conMinLeft
End If
If intLeft > conMaxLeft Then
    intLeft = conMaxLeft
End If
```

And once you have the new intLeft position, two or three controls of the ListBox header must be repositioned: lblCol1 (the | character), lbl1 (the Food Category label), and eventually lblAZ, if the last column sorted was column 1, Food Category:

```
Me.lblCol1.Left = intLeft
Me.lbl1.Left = intLeft + Me.lblCol1.Width
If mintLastColumn = 1 Then
    Me.lblAZ.Left = Me.lbl1.Left + Me.lbl1.Width
End If
```

And once the header is repositioned, the ListBox control will have its ColumnWidths property redefined for the first column to the intLeft value minus the ListBox Left property, putting its second column beginning exactly below lblAZ (the | character).

```
        Me.lstFoodItems.ColumnWidths = intLeft - Me.lbllstFoodItems.Left
    End If
End Sub
```

The Range.AutoFilter Method

The Range.Find method is quite fast to make simple substring searches using just one search criteria. But in such large and complex data tables such as the ARS-USDA, there are moments where it can fail to help you find some specific food items because you simply don't know where (Food Category) and/or how (Food Item name) the desired information is written in the food table.

When this moment comes, you will need to make more complex searches using two criteria strings and Boolean operators (AND/OR) to compare and try to find what you want. And here is where the Range.AutoFilter method comes in!

The Range.AutoFilter method is how VBA implements the Excel Sort & Filter ➤ Filter command, which you can find in the Editing area of the Home tab. But before you dive in to how to programmatically interact with this method, let's look at how it works in the Excel interface so you can correctly implement it using the USDA worksheet data as an example.

Supposing that you have any USDA worksheet nutrient data cell selected (any cell below row 5) and apply the Excel Sort & Filter ➤ Filter command, you will notice that Excel will use its Range.CurrentRegion property to set the data table to be filtered and will put drop-down ListBox filters where it considers the first data table row to be (Figure 6-20).

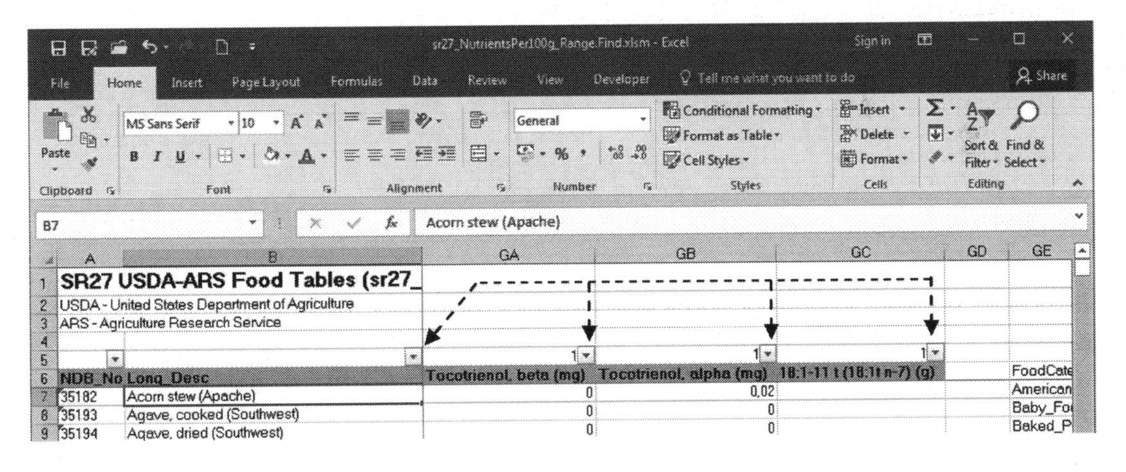

Figure 6-20. When you apply the Sort & Filer ➤ Filter command, Excel uses the Range.CurrentRegion property to set the data table to be filtered and uses what it considers as the first table row to add drop-down ListBox filters

Although Excel can be considered very smart, it does not always make the perfect choice: since the USDA worksheet is produced by the USDA Food List Creator.mdb application, which reserves row 5 to count the nutrient columns (to help create the VLOOKUP( ) formulas to return the desired nutrient value of any food item), Excel considers this row as the first table row and puts the drop-down ListBox filters on it, including the nutrient name row (row 6) as part of the table to be filtered.

To guarantee that Excel will filter the USDA worksheet correctly, you must execute the Excel Sort & Filter command again to remove the ListBox filters, select the correct data table (click cell B6, press Ctrl+Shift+Down arrow followed by Ctrl+Shift+Right arrow), and reapply the Excel Sort & Filter ➤ Filter command (Figure 6-21).

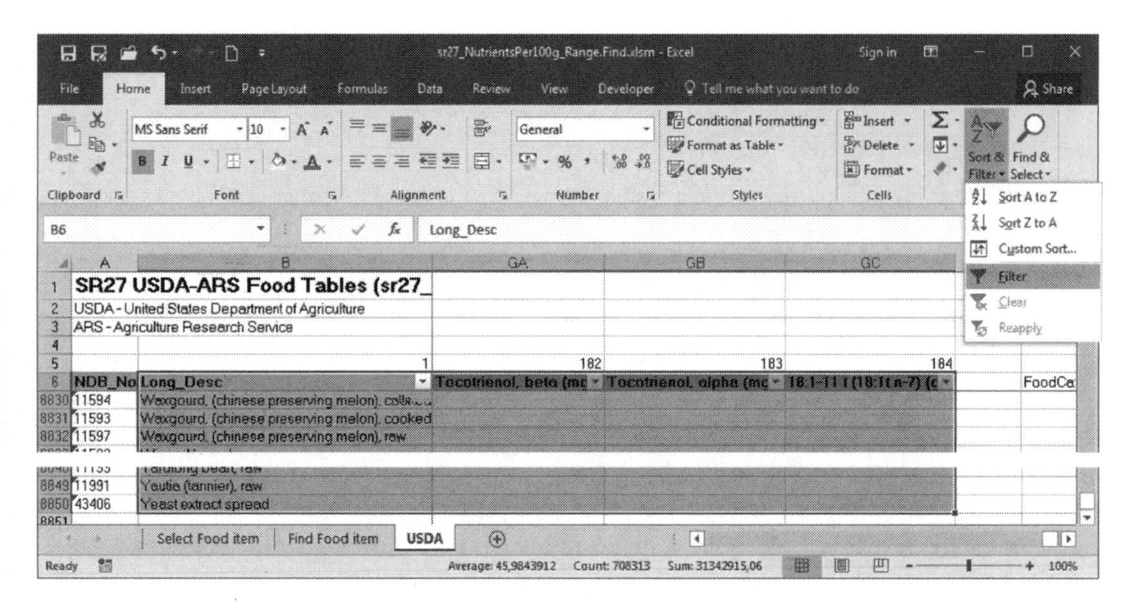

Figure 6-21. *To guarantee that Excel puts its filter drop-down ListBoxes on the correct row of the USDA data table, you first need to select cell B6, press Ctrl+Shift+Down arrow followed by Ctrl+Shift+Right arrow, and apply the Excel Sort & Filter ➤ Filter command*

Therefore, before you use the Range.AutoFilter method, you need to correctly select the USDA range name, including row 6, which is its row headers and not part of this range name. And as you might know by now, you will need to use the Range Offset and Resize methods.

Let's use the VBA Immediate window to see how to do this selection using VBA code. Begin by typing this instruction to print the current USDA range name address:

```
?Range("USDA").Address
$B$7:$GC$8850
```

Now displace the USDA range name one row up using the Range.Offset method and print the new range address (note that since the new range address was offset by one row up, it also has one less row at the bottom).

```
?Range("USDA").Offset(-1).Address
$B$6:$GC$8849
```

Finally, use the Range.Resize method to resize the range, using the Rows.Count property to add an extra row at the bottom of the displaced range.

```
?Range("USDA").Offset(-1).Resize(Range("USDA").Rows.Count+1).Address
$B$6:$GC$8850
```

■ **Attention** If you want to select the range on the USDA worksheet instead to print it in the VBA Immediate window, use the `Range.Select` method instead of the `Range.Address` property.

That's it! Using `Range.Offset` and `Range.Resize` in sequence, you can use VBA code to reference the desired cells on the USDA worksheet before applying the Excel Sort & Filter ➤ Filter command to put its `ListBox` filters on row 6. You are now ready to filter the USDA nutrient data table for specific food items.

Suppose that you need to find a food item called *black beans*. You want to filter the USDA worksheet showing all food items whose name contains both *black* and *bean*, appearing in any sequence inside the food item name. To do this, you must click the `Long_Desc` column drop-down `ListBox` filter to expand it, point the mouse pointer to the Text Filter option, and click the Contains option.

Excel will show the Custom AutoFilter dialog box where you can select two different criteria, using AND/OR to combine them (black AND bean), and will perform the desired search and filter over the selected cell range (Figure 6-22).

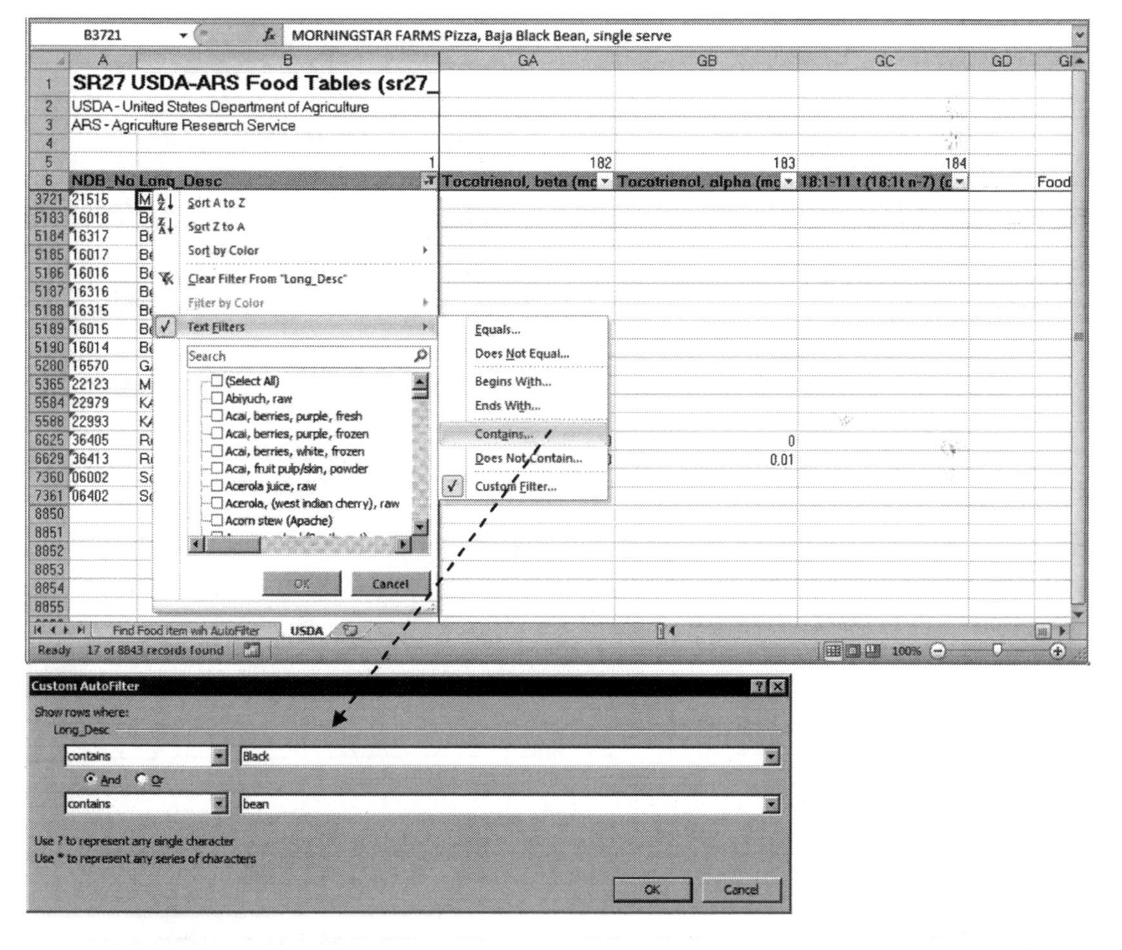

Figure 6-22. *Once you have selected the desired cells and applied them with the Sort & Filter ➤ Filter command, use the Long_Desc drop down ListBox filter, select the Text Filters ➤ Contains option to expose the Custom AutoFilter dialog box, and select each food item whose name has both black and bean, in any order inside the food name*

After defining what you want to search, click OK in the Custom AutoFilter dialog box so Excel can perform the search and hide all table rows that do not match the filter criteria. Note that the Long_Desc column now spots a small filter icon, as a visual clue that this column has a criteria filter, while the worksheet row headers for the rows that match the desired criteria are now blue (Figure 6-23).

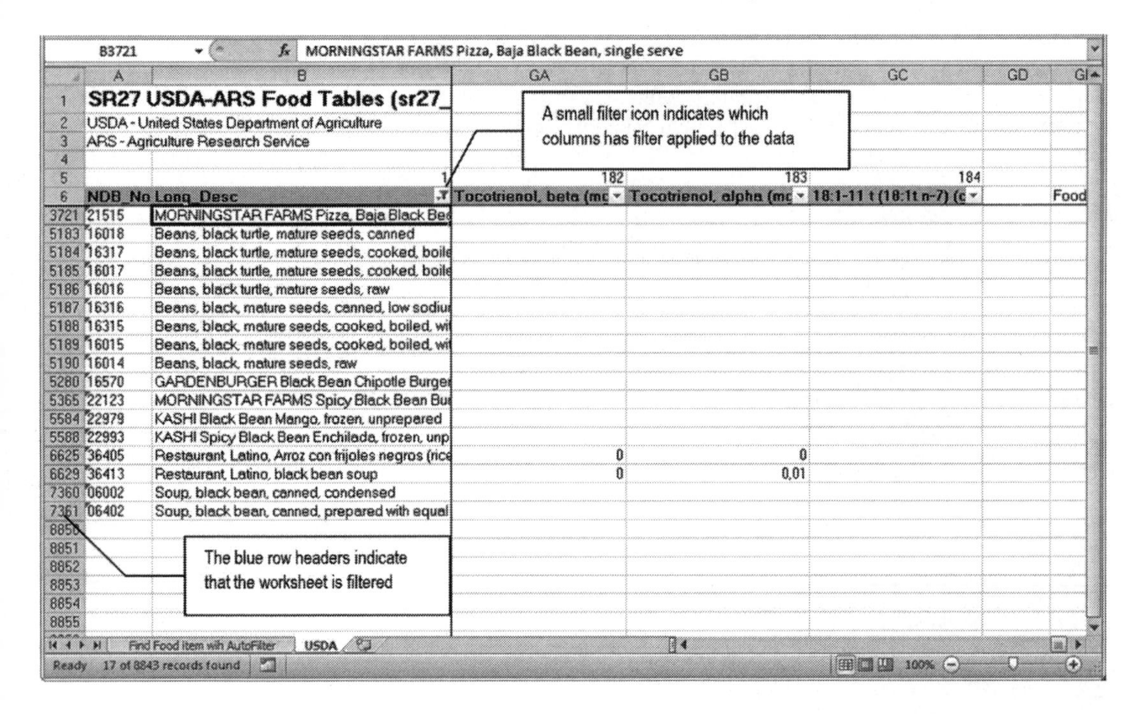

Figure 6-23. *This is the USDA worksheet after you have applied the Excel Filter command to show just rows that have both words black and bean in any order inside the Long_Desc column (food item name)*

If you want to use another filter criteria for the rows already selected, to show food items of just one food category, apply another filter to the already filtered list using the FdGrp_Desc column drop-down ListBox filter to select the desired category (like Legumes and Legume Products). It works like filtering a filtered list!

■ **Attention** To remove the current filter, just apply the Excel Sort & Filter ➤ Filter command again. Go ahead and do this right now.

The VBA implementation for the Excel Custom AutoFilter dialog box shown in Figure 6-22 is the Range.AutoFilter method, which has this syntax:

```
Expression.AutoFilter(Field, Criteria1, Operator, Criteria2, VisibleDropDown)
```

In this code:

> Expression: This is required; it is an expression that returns a Range object.
>
> Field: This is optional; it is the integer offset of the field on which you want to base the filter (from the left of the list; the leftmost field is field 1).

Criteria1: This is optional; it is the string criteria to be filtered. Use = to find blank fields; use <> to find nonblank fields. If omitted, the criteria means All. If Operator is xlTop10Items, Criteria1 specifies the number of items (for example, 10). Use an asterisk (*) as a wildcard to perform a substring search.

Operator: This is optional: it is an XlAutoFilterOperator specifying the type of filter:

xlAnd = 1 : Logical AND of Criteria1 and Criteria2

xlBottom10Items = 4: *n* lowest-valued items displayed (number of items specified in Criteria1)

xlBottom10Percent = 6: % Lowest-valued items displayed (percentage specified in Criteria1)

xlFilterCellColor = 8: Color of the cell

xlFilterDynamic = 11: Dynamic filter

xlFilterFontColor = 9: Color of the font

xlFilterIcon = 10 : Filter icon

xlFilterValues = 7: Filter values

xlOr = 2: Logical OR of Criteria1 or Criteria2

xlTop10Items = 3: Highest-valued items displayed (number of items specified in Criteria1)

xlTop10Percent = 5: Highest-valued items displayed (percentage specified in Criteria1)

Criteria2: This is optional; it is the second criteria (a string) and must be used with Criteria1 and Operator to construct compound criteria.

VisibleDropDown: This is optional; it uses True (default) to display the AutoFilter drop-down arrow for the filtered field and it uses False to hide the AutoFilter drop-down arrow for the filtered field.

To verify whether any sheet tab is currently filtered by the Excel AutoFilter command and eventually remove the Excel AutoFilter drop-down ListBox arrows, use the Worksheet object AutoFilterMode property, which has this syntax:

```
expression.AutoFilterMode [= False]
```

In this code:

expression: This is required; it refers to any object variable that represents a Worksheet object.

The Worksheet.AutoFilterMode property returns True if the drop-down arrows are currently displayed. You can set this property to False to remove the arrows but cannot set it to True to apply the AutoFilter.

Having removed the Excel Filter option from the USDA worksheet and supposing that the rg object variable refers to the desired cell range of the USDA worksheet, to perform the same search using the Range.AutoFilter method, you can use an instruction like this, where 1 refers to the first range column (long_Desc, or food item name), *black* and *bean* are Criteria1 and Criteria2, and xlAnd is the operator that joins both criteria:

```
rg.AutoFilter 1, "*black*", xlAnd, "*bean*"
```

■ **Attention** Note that to perform a substring search inside the `Long_Desc` column, you must use the * as the substring operator. An * before and another after any the search string means that the Find method must search and return all cells that have this word.

Knowing that you must first define the desired range name to be filtered and then apply the desired `Range.AutoFilter` criteria, you can type the next instruction in the VBA Immediate window to define and filter the USDA worksheet the same way you did in Figure 6-22:

```
?Range("USDA").Offset(-1).Resize(Range("USDA").Rows.Count+1).AutoFilter(1, "*black*", xlAnd,
"*bean*")
```

■ **Attention** To use the interrogation character to print the `Range.AutoFilter` returned result in the VBA Immediate window, you must enclose all its arguments in parentheses. To discard the result and just apply the `Range.AutoFilter` criteria, don't use the interrogation character or put the `Range.Find` method arguments inside parentheses.

Selecting Filtered Cells with the Range.SpecialCells Property

When you press the F5 function key on an unprotect sheet tab and Excel shows the Go To dialog box, you can click the Special button to display the Excel Go To Special dialog box and select many different types of cell contents, such as cells with comments, constants, formulas, blanks, and so on (Figure 6-24).

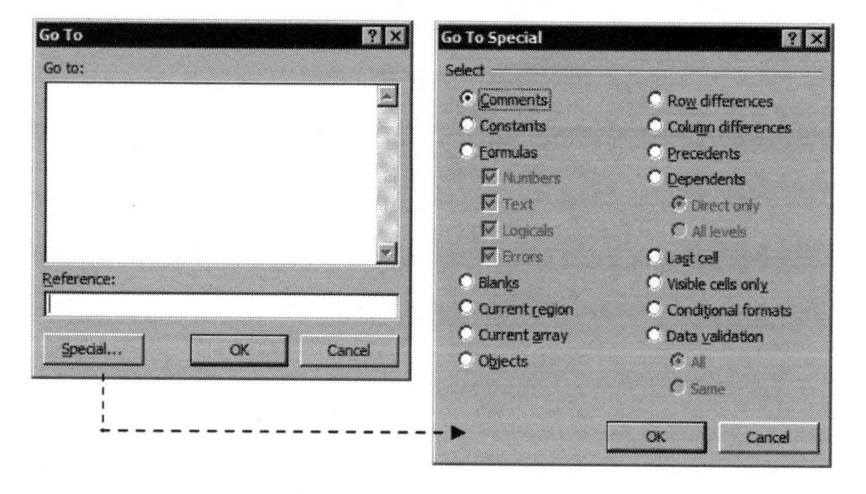

Figure 6-24. *If you click the Special button of the Excel Go To dialog box, Excel will show the Go To Special dialog box, where you can select all worksheet cells that have a special type of content, like cells with comments, with constant values, and with formulas*

When you click OK, Excel offers the Go To Special operation to select the cells with the desired content using the Range.SpecialCells method, which has this syntax:

```
Expression.SpecialCells(Type, Value)
```

In this code:

Expression: This is required; it is a variable that represents a Range object.

Type: This is required; it is an XlCellType constant indicating the cells to include.

xlCellTypeAllFormatConditions = 4172, cells of any format

xlCellTypeAllValidation = -4174, cells having validation criteria

xlCellTypeBlanks = 4, empty cells

xlCellTypeComments = -4144, cells containing notes

xlCellTypeConstants = 2, cells containing constants

xlCellTypeFormulas = -4123, cells containing formulas

xlCellTypeLastCell = 11, the last cell in the used range

xlCellTypeSameFormatConditions = -4173, cells having the same format

xlCellTypeSameValidation = -4175, cells having the same validation criteria

xlCellTypeVisible = 12, all visible cells

Value: This is optional; if Type is either xlCellTypeConstants or xlCellTypeFormulas, this argument is used to determine which types of cells to include in the result using the XlSpecialCellsValue constants, which can be added together to return more than one type. The default is to select all constants or formulas, no matter what the type.

```
xlErrors = 16
xlLogical= 4
xlNumbers = 1
xlTextValues =
```

Note that most Go To Special options can be set by selecting one of the Range.SpecialCells Type arguments (which are mutually exclusive), while the Value argument can be used when Type = xlCellTypeFormulas and is not mutually exclusive. This is exactly how the Go To Special dialog box implements them.

So, to select all cells with formulas on the active sheet using the VBA Immediate window, you must type this instruction:

```
Cells.SpecialCells(xlCellTypeFormulas).Select
```

Note in the previous instruction that it uses the Cells collection (when not used, it refers to a given Range object, meaning the ActiveSheet object, or all active sheet cells) and its SpecialCells property to produce a range object that has all the desired cells and then uses the Range.Select method to select all the cells on the active sheet that have formulas. Figure 6-25 shows what happens when you unprotect the Find Food Item sheet tab of the sr27_NutrientsPer100g_Range.Find.xlsm macro-enabled workbook and use this instruction.

Now that you know that Range.SpecialCells exists and that the Range.AutoFilter method hides all sheet rows that don't match the desired criteria and shows the desired ones, to programmatically get the range addresses returned by the Range.AutoFilter method, you must use Range.SpecialCells to set the Type argument to the xlCellTypeVisible constant value.

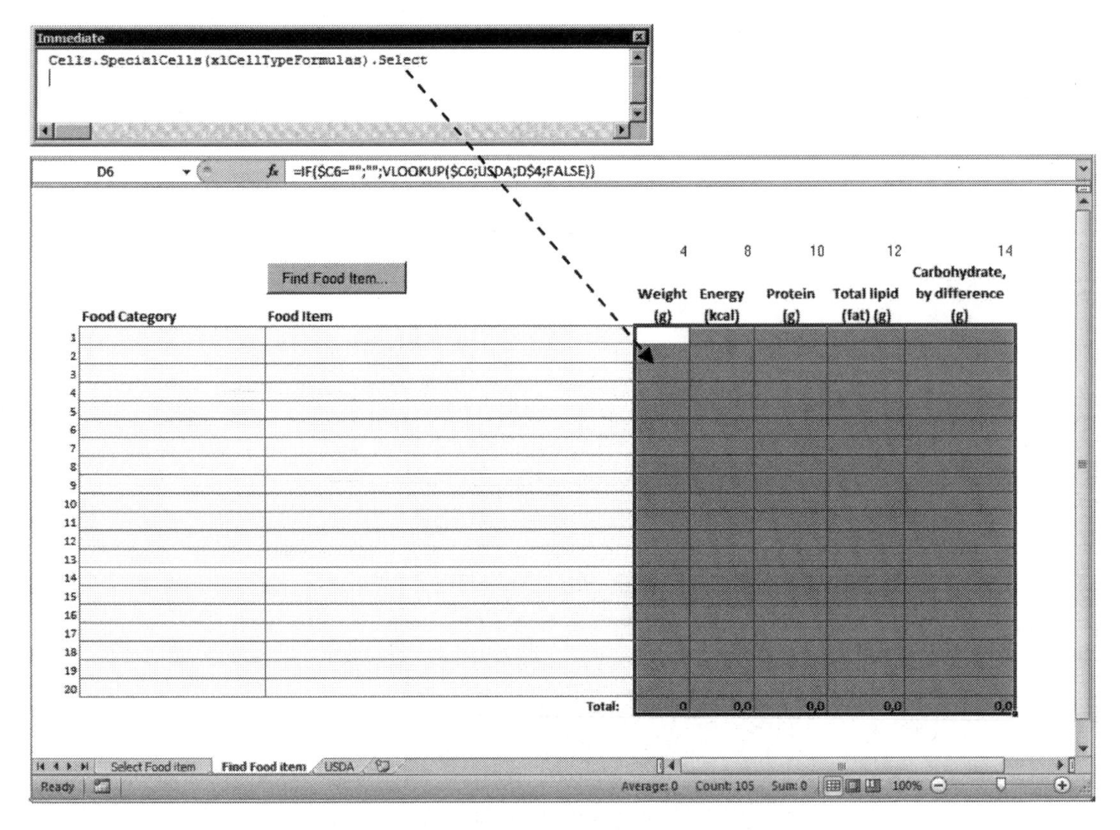

Figure 6-25. You can use the Cells collection and the Range.SpecialCells property to define a range that contains all cells in the active sheet that have specific content. By using the Range.Select method, you can select them in the worksheet, exactly the same way the Excel Go To Special dialog box does

The next two instructions will use the VBA Immediate window to first filter the USDA range name showing all food items that have both the words *black* and *bean* (in any order) and then use the same filtered range to show all range addresses returned by the Range.SpecialCells(xlCellTypeVisible) property:

```
Range("USDA").Offset(-1).Resize(Range("USDA").Rows.Count+1).AutoFilter(1, "*black*", xlAnd,
"*bean*")
?Range("USDA").Offset(-1).Resize(Range("USDA").Rows.Count+1).Specialcells(xlCellTypeVisible
).Address
$B$6:$GC$6,$B$3721:$GC$3721,$B$5183:$GC$5190,$B$5280:$GC$5280,$B$5365:$GC$5365,$B$5584:$GC$5
584,$B$5588:$GC$5588,$B$6626:$GC$6626,$B$6630:$GC$6630,$B$7361:$GC$7362,$B$8851:$GC$8851
```

To know how many different range addresses have been returned, use the Range.Areas.Count property applied to the Range.SpecialCells method, as follows:

```
?Range("USDA").Offset(-1).Resize(Range("USDA").Rows.Count+1).Specialcells(xlCellTypeVisible).
Areas.Count
 11
```

As you can see, Figure 6-22 is composed of 11 different, noncontiguous range addresses, where the first range address (Range.Area(1)) will always be the row with the drop-down ListBox headers ($B$6:$GC$6) and will contain more than one row when the first row below it (the first table data row) matches the Range.AutoFilter criteria.

So, to run through all the cells returned by the `Range.AutoFilter` method applied to the USDA sheet tab using VBA code, you need to use the `Range.Speciallcells` property and use four range object variables.

- `rgUSDA`, to represent the entire USDA food table (including its row headers) where the `Range.AutoFilter` will be applied
- `rgFilter`, to represent all `rgUSDA.SpecialCells(xlCellTypeVisible)` cells, meaning the ones that match the desired filter criteria and are visible on the sheet tab
- `rgArea`, to represent each range address returned by the `rgFilter.Areas` collection
- `rg`, to represent any cell of the `rgArea` object variable

Let's see this in action!

Using the frmRangeFilter

Open the `sr27_NutrientsPer100g_Range.AutoFilter.xlsm` macro-enabled workbook (which you can extract from the `Chapter06.zip` file) and click the Find Food Item button on the `Find Food Item with AutoFilter` sheet tab to show the `frmRangeFilter UserForm`, which uses the `Range.AutoFilter` method to search the entire USDA sheet tab for any food item name using two different words joined by the AND/OR operator (Figure 6-26).

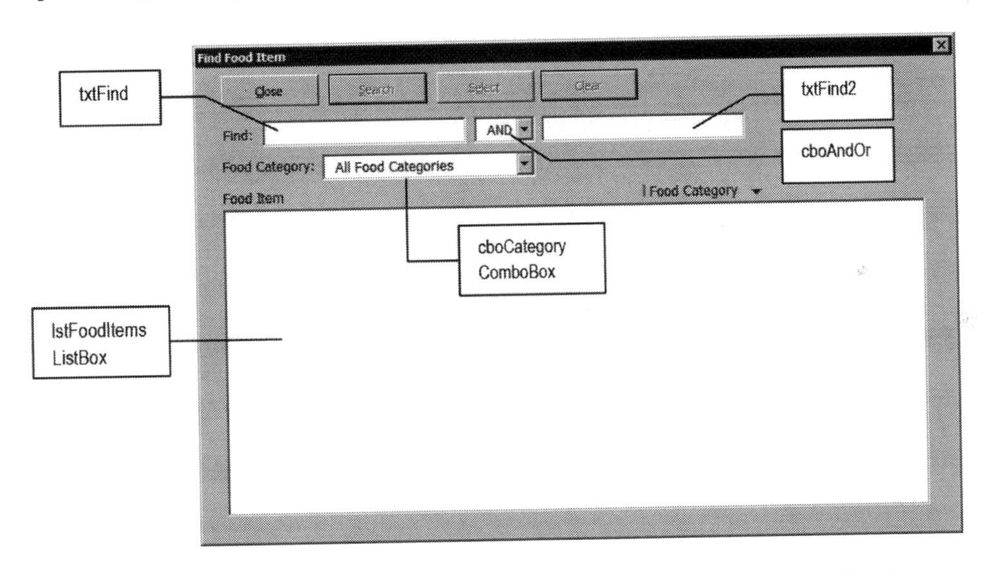

Figure 6-26. *This is frmRangeFilter from the sr27_NutrientsPer100g_Range.AutoFilter.xlsm macro-enabled workbook. Note that it has two text boxes (txtFind and txtFind2) to allow you to search for any food item using two different words joined by the AND/OR operator*

Note in Figure 6-26 that the food category names were not inserted in `cboCategory` using their respective range names (where invalid characters, like spaces, were changed by underscores) but as they appear in the USDA worksheet `FdGrp_Desc` column.

This is necessary because if you need to use `Range.AutoFilter` to search inside a given food category, you must use it exactly as it appears in the `FdGrp_Desc` column, which is quite different from the food category range name. Also note that by default `txtFind2` is disabled and that the `cboAndOr ComboBox` has just two options: AND and OR.

Look at how the frmRangeFilter UserForm_Initialize( ) event was programmed to correctly insert these values in both the cboAndOr and cboCategory ComboBoxes:

```
Private Sub UserForm_Initialize()
    Dim ws As Worksheet
    Dim rg As Range
    Dim rgUSDA As Range
    Dim rgFound As Range
    Dim strFind As String
    Const conNameDoesntExist = 1004

    On Error GoTo Initialize_Error

    mintLastColumn = 1
    Me.cboAndOr.AddItem "AND"
    Me.cboAndOr.AddItem "OR"
    Me.cboAndOr = "AND"

    Me.cboCategory.AddItem "All Food Categories"
    Me.cboCategory = "All Food Categories"

    'Verify if range names already exist
    Set rgUSDA = Range("USDA")
    Set rgUSDA = Range(rgUSDA.Cells(1, 2), rgUSDA.Cells(rgUSDA.Rows.Count, 2))
    For Each rg In Range("FoodCategoriesList")
        If InStr(1, rg, "_") Then
            strFind = Mid(rg, 1, InStr(1, rg, "_") - 1) & "*"
            Set rgFound = rgUSDA.Find(strFind, , , xlWhole)
            If Not rgFound Is Nothing Then
                Me.cboCategory.AddItem rgFound
            End If
        Else
            Me.cboCategory.AddItem rg
        End If
    Next

Initialize_End:
    Exit Sub
Initialize_Error:
    Select Case Err
        Case conNameDoesntExist
            Call CreateRangeNames
            Resume
        Case Else
            MsgBox "Error " & Err & " in UserForm_Initialize"
    End Select
    Resume Initialize_End
End Sub
```

After setting the error trap, the procedure uses the ListBox AddItem method to add the AND and OR options to the cboAndOr ComboBox and to add All Categories as the first option of cboCategories.

But since there is no range name in the sr27_NutrientsPer100g_Range.AutoFilter.xlsm macro-enabled workbook, the first time it tries to set a reference to the USDA range name, which does not exist, VBA will raise error = 1004, and the code will be directed to the error trap, running the CreateRangeNames( ) procedure to create them.

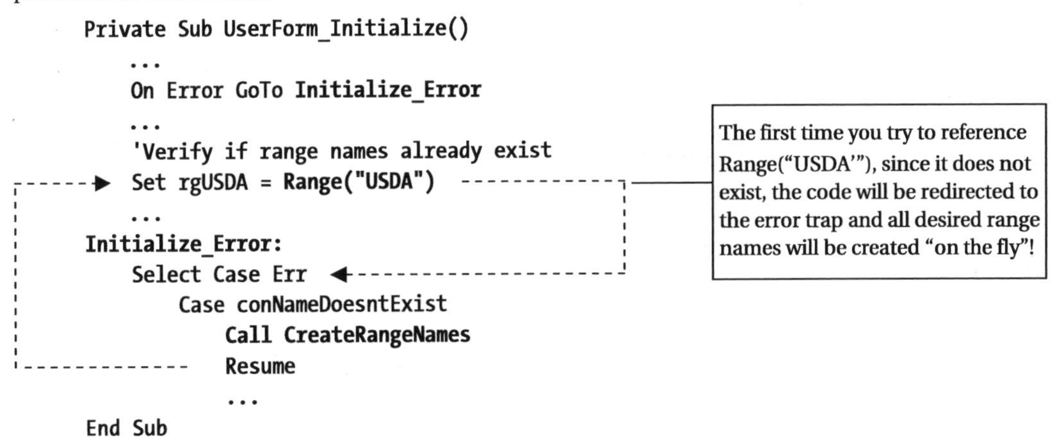

Private Sub UserForm_Initialize()
...
On Error GoTo Initialize_Error
...
'Verify if range names already exist
Set rgUSDA = Range("USDA")
...
Initialize_Error:
Select Case Err
Case conNameDoesntExist
Call CreateRangeNames
Resume
...
End Sub
The first time you try to reference Range("USDA'"), since it does not exist, the code will be redirected to the error trap and all desired range names will be created "on the fly"!

Having created all the necessary range names, a reference is set to the USDA range name, and the Range(Cell1, Cell2) property is used to create a single range that takes all FdGrp_Column (column C) of the USDA range name. Note that for the Cell1 argument it uses rgUSDA.Cells(1,2), which returns cell C7, while for the Cell2 property it uses rgUSDA.Cells(rgUSDA.Rows.Count,2), which will return cell C8850.

```
Set rgUSDA = Range("USDA")
Set rgUSDA = Range(rgUSDA.Cells(1, 2), rgUSDA.Cells(rgUSDA.Rows.Count, 2))
```

To fill the cboCategory ComboBox with the appropriate food categories names, a For Each...Next loop begins to run through all FoodCategoriesList range names.

```
For Each rg In Range("FoodCategoriesList")
```

Not all food category names needs to be corrected to be transformed on a range name, so the code tries to find an underscore inside the food category range name using the VBA InStr( ) function. If it does, the first word of the range name before its first underscore is extracted to the strFind variable. Note that the code concatenates a * wildcard after the search word to search for any cell that begins with that word.

```
If InStr(1, rg, "_") Then

    strFind = Mid(rg, 1, InStr(1, rg, "_") - 1) & "*"
```

▒ **Attention** This works because there are no two food categories beginning with the same name prefix.

The Range.Find method is then used to find the first reference to the food category that has this word using the LookAt = xlWhole argument, to guarantee that only items that begin with that word will be found.

```
Set rgFound = rgUSDA.Find(strFind, , , xlWhole)
```

If Range.Find succeeds on the search (and it will), rgFound will have the desired food category, which will be added to the cboCategory ComboBox.

```
If Not rgFound Is Nothing Then
    Me.cboCategory.AddItem rgFound
End If
```

For the cases that the food category is composed of just one word (like Beverages or Sweets), its range name is directly inserted on the list.

```
For Each rg In Range("FoodCategoriesList")
    If InStr(1, rg, "_") Then
    ...
    Else
        Me.cboCategory.AddItem rg
    End If
Next
```

Once frmRangeFilter is loaded, try to find all food items that have *black* and *bean* using the two Find text boxes and click the Search (cmdSearch) CommandButton. All 17 food items that match the desired criteria will be instantly recovered and shown in the lstFoodItems ListBox (Figure 6-27).

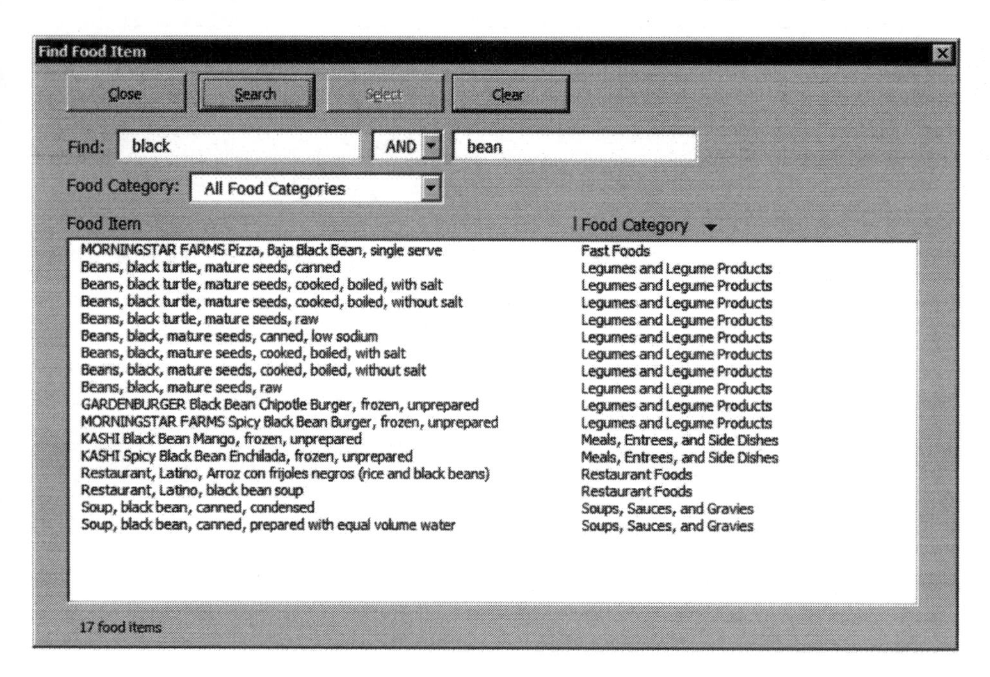

Figure 6-27. *frmRangeFind allows you to use two different criteria to find food items inside the USDA range name, like the 17 that have both black and bean (note that you use the AND operator to include the food item names on the ListBox)*

Use a cboCategory ComboBox to try to select any of the food categories on the list, like Legumes and Legume Products. Just food items of the selected food category will be shown (Figure 6-28).

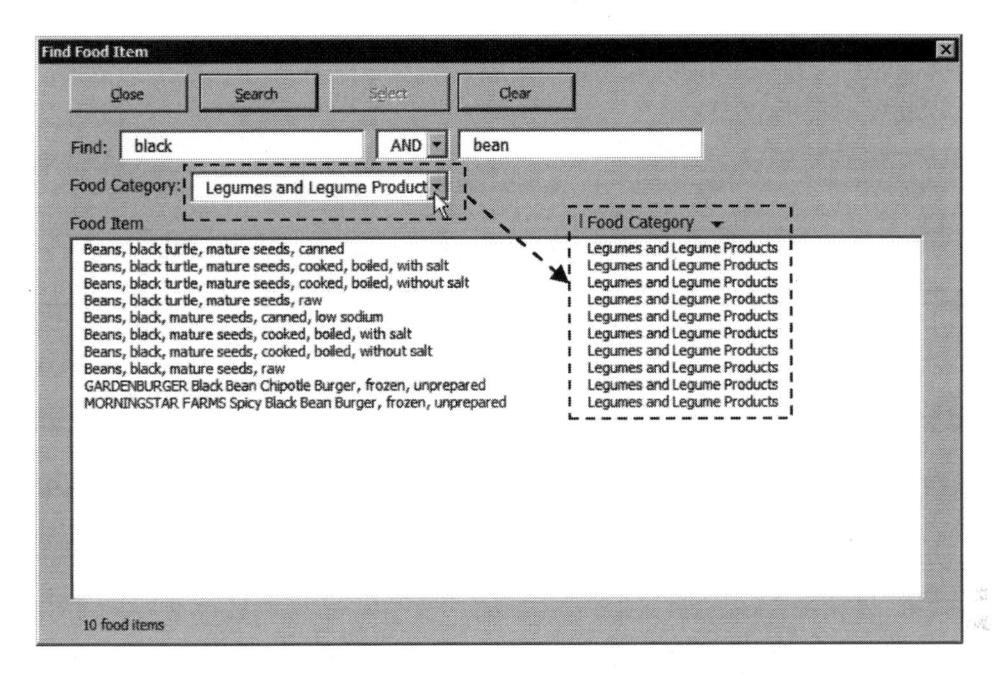

Figure 6-28. *If you select any of the listed food categories in the cboCategory ComboBox, just food items of this category will be shown in the ListBox*

When you type something in `txtFind` (the left text box) and press Enter, the `txtFind_AfterUpdate( )` event fires, executing this code:

```
Private Sub txtFind_AfterUpdate()
    Dim bolEnabled As Boolean

    If Me.txtFind = "" Then
        Me.txtFind2 = ""
    End If

    bolEnabled = (Me.txtFind <> "")
    Me.txtFind2.Enabled = bolEnabled
    Me.cmdSearch.Enabled = bolEnabled
    Me.cmdClear.Enabled = (bolEnabled Or Me.lstFoodItems.ListCount > 0)
    mbolFiltered = False
End Sub
```

If `txtFind` becomes cleared (`""`), `txtFind2` must also be cleared. And `txtFind2` (the right TextBox), `cmdSearch`, and `cmdClear` will become enabled when `txtFind` has something typed in it. Note that `cmdClear` will become enabled if `lstFoodItems` shows any food item in its list and that the module-level variable `mbolFiltered` changes to `False`.

The Clear (`cmdClear`) `CommandButton` allows you to clear the interface and begin another search, executing this code:

```
Private Sub cmdClear_Click()
    mbolFiltered = False
```

```
    Me.lstFoodItems.Clear
    Sheets("USDA").AutoFilterMode = False
    Me.txtFind.SetFocus
    Me.txtFind = ""
    Me.txtFind2 = ""
    Me.cboAndOr = "AND"
    Me.cboCategory.ListIndex = 0
    Me.cmdSearch.Enabled = False
    Me.cmdSelect.Enabled = False
    Me.cmdClear.Enabled = False
End Sub
```

Note that it turns the module-level variable `mbolFiltered` to `False`, clears the `lstFoodItems` `ListBox`, and removes any `AutoFilter` imposed on the USDA worksheet; it clears and disables all controls of the `UserForm` interface. Try it!

Now while you keep `txtFind` and `txtFind2` empty, try to select any food category and click the Search (`cmdSearch`) `CommandButton`. You will notice that all food items of the selected food category will be instantly returned in the `lstFoodItems` `ListBox`. Select another category, and the `ListBox` will be automatically updated (Figure 6-29).

■ **Attention** Note that if you select again All Food Categories, the `UserForm` interface is cleared, and no food item is returned.

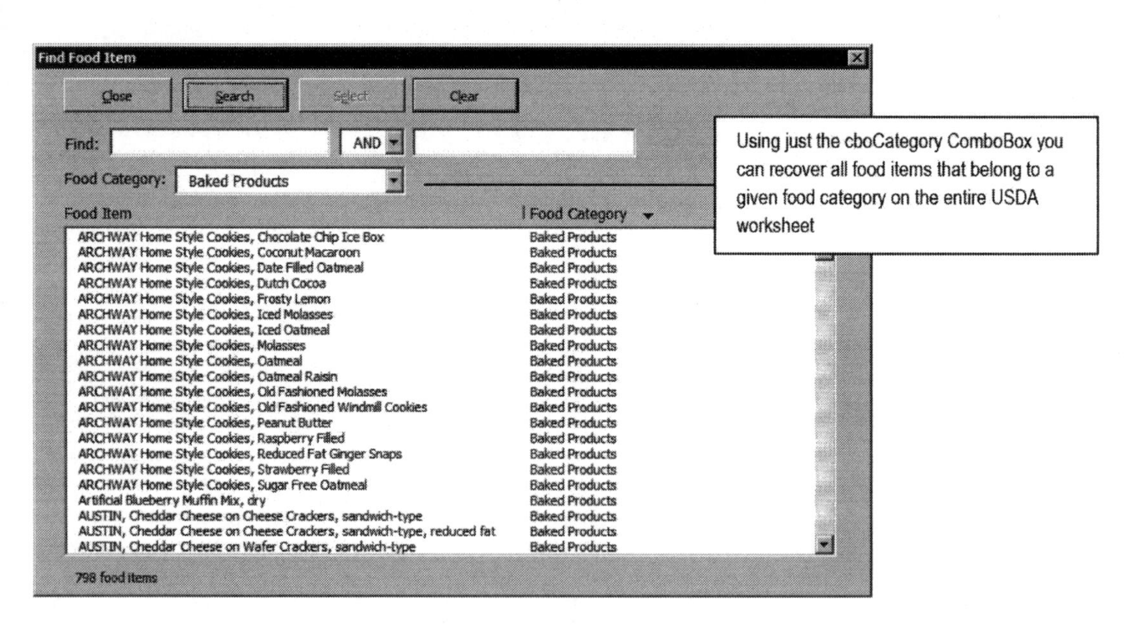

Figure 6-29. *You can also use the frmRangeFilter UserForm to select all food items of a given food category. Click the Clear (cmdClear) CommandButton to clear the interface, select any food category on the cboCategory ComboBox, and the click Search (cmdSearch) CommandButton. All food items of the selected food category will be returned. Keep selecting food in other food categories to update the list. The All Food Categories option clears the UserForm interface*

When you select any food category in the cboCategory ComboBox, the cboCategory_Change( ) event fires, executing this code:

```
Private Sub cboCategory_Change()
    Dim bolEnabled As Boolean

    If mbolFiltered Then
        If Me.cboCategory.ListIndex > 0 Then
            Call cmdSearch_Click
        Else
            Call cmdClear_Click
        End If
    Else
        bolEnabled = (Me.cboCategory.ListIndex > 0)
        Me.cmdSearch.Enabled = bolEnabled
        Me.cmdClear.Enabled = bolEnabled
    End If
End Sub
```

Perhaps you now understand why you need the mbolFiltered module variable. If it is True *and* cboCategory has a specific food category selected, it calls cmdSearch_Click, automatically filtering again the USDA worksheet for all food items that belong to the selected food category.

```
If mbolFiltered Then
    If Me.cboCategory.ListIndex > 0 Then
        Call cmdSearch_Click
```

But if mbolFiltered = False, frmRangeFilter is still not filtered in the USDA worksheet, and its interface must be synchronized, enabling the cmdSearch and cmdClear CommandButtons.

```
Else
    bolEnabled = (Me.cboCategory.ListIndex > 0)
    Me.cmdSearch.Enabled = bolEnabled
    Me.cmdClear.Enabled = bolEnabled
```

Now that you know that you can show food items either by part of a name and/or by food category, you know the search and filter process happens when the cmdSearch_Click( ) event fires, executing this code:

```
Private Sub cmdSearch_Click()
    Dim rgUSDA As Range
    Dim rgFilter As Range
    Dim rgArea As Range
    Dim rg As Range
    Dim intI As Integer
    Const conFoodName = 1
    Const conFoodCategory = 2

    Sheets("USDA").AutoFilterMode = False
    Me.lstFoodItems.Clear

    Set rgUSDA = Range("USDA").Offset(-1).Resize(Range("USDA").Rows.Count + 1)
    If Me.txtFind2 = "" Then
        rgUSDA.AutoFilter conFoodName, "*" & Me.txtFind & "*"
```

```
    ElseIf Me.txtFind <> "" Then
        rgUSDA.AutoFilter conFoodName, "*" & Me.txtFind & "*", IIf(Me.cboAndOr = "AND",
        xlAnd, xlOr), "*" & Me.txtFind2 & "*"
    End If

    If Me.cboCategory.ListIndex > 0 Then
        rgUSDA.SpecialCells(xlCellTypeVisible).AutoFilter conFoodCategory, Me.cboCategory
    End If

    Set rgFilter = rgUSDA.SpecialCells(xlCellTypeVisible)

    For Each rgArea In rgFilter.Areas
        For intI = 1 To rgArea.Rows.Count
            Set rg = rgArea.Cells(intI, 1)
            If rg <> "" And rg.Row > rgFilter.Row Then
                Me.lstFoodItems.AddItem rg
                Me.lstFoodItems.Column(1, Me.lstFoodItems.ListCount - 1) = rg.Offset(0, 1)
            End If
        Next
    Next

    Me.lblFoodItems.Caption = Me.lstFoodItems.ListCount & " food items"
    Me.lblFoodItems.Visible = True
    If Me.lstFoodItems.ListCount = 0 Then
        MsgBox "Food item not found"
    End If

    mbolFiltered = True
End Sub
```

After declaring all variable and constants needed, the first instructions prepare the stage. If there is any filter applied to the USDA worksheet, it is removed, and the `lstFoodItems` `ListBox` is cleared.

```
Sheets("USDA").AutoFilterMode = False
Me.lstFoodItems.Clear
```

The `rgUSDA` object variable is then defined to contain the nutrient headers and all nutritional information of the entire USDA range name, as you did before using the VBA Immediate window.

```
Set rgUSDA = Range("USDA").Offset(-1).Resize(Range("USDA").Rows.Count + 1)
```

And once the `rgUSDA` variable is defined, it is time to impose the `Range.AutoFillter` method, according to what was typed in the `txtFind` and `txtFind2` text boxes. If just `txtFind` was filled, `txtFind2` is empty, and the `AutoFilter` method must use just the `txtFind` `TextBox` on the `Criteria1` argument. Note that what you type in the `txtFind` `TextBox` is enclosed between asterisks (*). This is the wildcard character that will allow a substring search (like the Text ➤ Contains option offered by the drop-down `ListBox` filter in the Excel interface).

```
If Me.txtFind2 = "" Then
    rgUSDA.AutoFilter conFoodName, "*" & Me.txtFind & "*"
```

But if both `txtFind` and `txtFind2` `TextBoxes` have been filled, the `Criteria1`, `Operator`, and `Criteria2` arguments of the `Range.AutoFilter` method must be used to compose the desired filter criteria. Once again, note that both criteria are enclosed in asterisks and that it uses the VBA IIF( ) function to test the cboAndOr ComboBox value and to decide whether the Operator argument must be defined to xlAnd or xlOr.

```
ElseIf Me.txtFind <> "" Then
    rgUSDA.AutoFilter conFoodName, "*" & Me.txtFind & "*", IIf(Me.cboAndOr = "AND",
    xlAnd, xlOr), "*" & Me.txtFind2 & "*"
End If
```

At this moment, if anything has been typed in the txtFind and txtFind2 text boxes, the rgUSDA range can be filtered to show all food items that match the desired search criteria, and the code verifies whether cboCategory has a specific food category selected (its first list option is All Categories, which received ListIndex = 0).

If this is true, if rgUSDA is already filtered, it will be filtered again to show among the selected food items just the ones that belong to the selected food category. If it has not been filtered yet, it will be filtered to show all food items of the selected food category. In both cases, all rgUSDA visible cells must be submitted to the Range.AutoFilter method, using the SpecialCells(xlCellTypeVisible) property to select all visible cells of the range and using the selected food category on the Criteria1 argument.

```
If Me.cboCategory.ListIndex > 0 Then
    rgUSDA.SpecialCells(xlCellTypeVisible).AutoFilter conFoodCategory, Me.cboCategory
End If
```

■ **Attention** As a good programming practice, the code declares and uses the conFoodName = 1 and conFoodCategory = 2 constants to indicate the first or second rgUSDA range column as the Range.AutoFilter Field argument, avoiding the appearance of “magic numbers” in the code.

This is enough to filter the USDA worksheet using all possibilities allowed by the frmRangeFilter interface, and the code needs to set a reference to the rows filtered by the Range.AutoFilter methods. They belong to all visible rows of the filtered rgUSDA object variable and can be defined using the Range.SpecialCells(xlCellTypeVisible) property.

```
Set rgFilter = rgUSDA.SpecialCells(xlCellTypeVisible)
```

At this point, rgFilter has all visible cells of the filtered rgUSDA object variable, and the code needs to run through all visible and independent range addresses, getting the food item name to insert on the lstFoodItems ListBox. It does this using two successive For...Next loops. An outer For Each...Next loop gets each independent range address using the rgFilter.Areas collection and sets it to the rgArea object variable, while the inner loop uses a For intI = 1 To rgArea.Rows.Count...Next loop to run through all rows of each independent address returned.

```
For Each rgArea In rgFilter.Areas
    For intI = 1 To rgArea.Rows.Count
    ..
    Next
Next
```

To fill the lstFoodItems ListBox with the food item names and its food category, each cell on the first column of the current rgArea address is set to the rg object variable using the rgArea.Cells(intI, 1) property. The rg value is then tested. If it is not empty *and* if its row number is greater than the first rgFilter row (the drop-down ListBoxes row), the food item is added to the first column in the lstFoodITem ListBox, and its food category is added to the second ListBox column using rg.Offset(0,1).

```
Set rg = rgArea.Cells(intI, 1)
If rg <> "" And rg.Row > rgFilter.Row Then
    Me.lstFoodItems.AddItem rg
    Me.lstFoodItems.Column(1, Me.lstFoodItems.ListCount - 1) = rg.Offset(0, 1)
End If
```

■ **Attention** This last test to see whether the `rg` is not empty (`rg <> ""`) is necessary because when the `USDA` range name was created, it received a blank row on the bottom as a provision to the `My_Recipes` food category, which was not used yet. This blank row confuses the `Range.AutoFilter` method and is always returned as the last `rgFilter.Area` address.

To know how many food items have been returned, you just use `lstFoodItems.ListCount`. If it is first used to change the `lblFoodITems Label` control's `Caption` property and if `lstFoodItems.ListCount = 0`, no food item matches the desired criteria, and a `MsgBox( )` function will warn the user. To signal to other events that the `USDA` worksheet is now filtered, the `mbolFiltered` module-level variable is set to `True`.

```
    Me.lblFoodItems.Caption = Me.lstFoodItems.ListCount & " food items"
    Me.lblFoodItems.Visible = True
    If Me.lstFoodItems.ListCount = 0 Then
        MsgBox "Food item not found"
    End If

    mbolFiltered = True
End Sub
```

And this is all you have to know about `frmRangeFilter UserForm` VBA code.

Did you notice that with the `frmRangeFilter UserForm`, using the `Range.AutoFilter` method is faster than `frmRangeFind` using the `Range.Find` method? Oh, it is! And it is more powerful too because it allows a two-word substring search and the possibility of filtering the food items again by a single food category.

Cool, huh?

Finding Food Items with the Range.Sort Method

Sometimes when you have a large data table, you need to find the *n* greatest (or lowest) values of a given data column. Speaking in terms of the USDA-ARS food table, say you need to find the first *n* food item that has the greatest (or lowest) values of a given nutrient, such as protein, Vitamin E, a given amino acid, and so on.

In such moments, you need to first sort the USDA food table by the desired column using the appropriate sort method (ascending for lowest values or descending for greatest values) and then look for the first *n* food items that appear on the table.

To perform such a search on a given food category, you must first filter the USDA food table to show just the desired food category food items and then apply the appropriate sort method on the filtered items to find the ones with the greatest or lowest amount of a given nutrient.

Let's do these operations using Excel's Sort & Filter commands on the `sr27_NutrientsPer100g.xlsm` workbook.

Suppose that you want to find the first *n* food items on the USDA food table stored inside the `USDA` sheet tab that have the greatest amount of protein (column `K`). You can easily do this by following these steps:

1. Select the `USDA` range name in the Excel Name box.

■ **Attention** You may need to execute `Function CreateRangeName( )` in the VBA Immediate window to create the USDA and food category range names on the `sr27_NutirentsPer100g.xlsm` workbook.

2. Apply the Excel Sort & Filter ➤ Custom Sort command to show the Excel Sort dialog box.
3. In the Excel Sort dialog box, delete all the existing sorting levels (if any), click the Add Level button, select Protein in the Sort by ListBox, select Largest to Smallest in the Order column, and press Enter to sort the USDA range name.

After you have sorted the USDA range name by the Protein column using descending order (from largest to smallest) and scroll it to the right to show the Protein column. You will see that the food item that has the greatest amount of protein is "Soy protein isolate, potassium type, crude protein basis" (88.3g of protein for each 100g), followed by "Gelatins, dry powder, unsweetened" and some "Egg dried" products (Figure 6-30).

■ **Attention** Use the Smallest to Largest option of the Excel Sort dialog box so you can find the first *n* food items with the least amount of any desired nutrient.

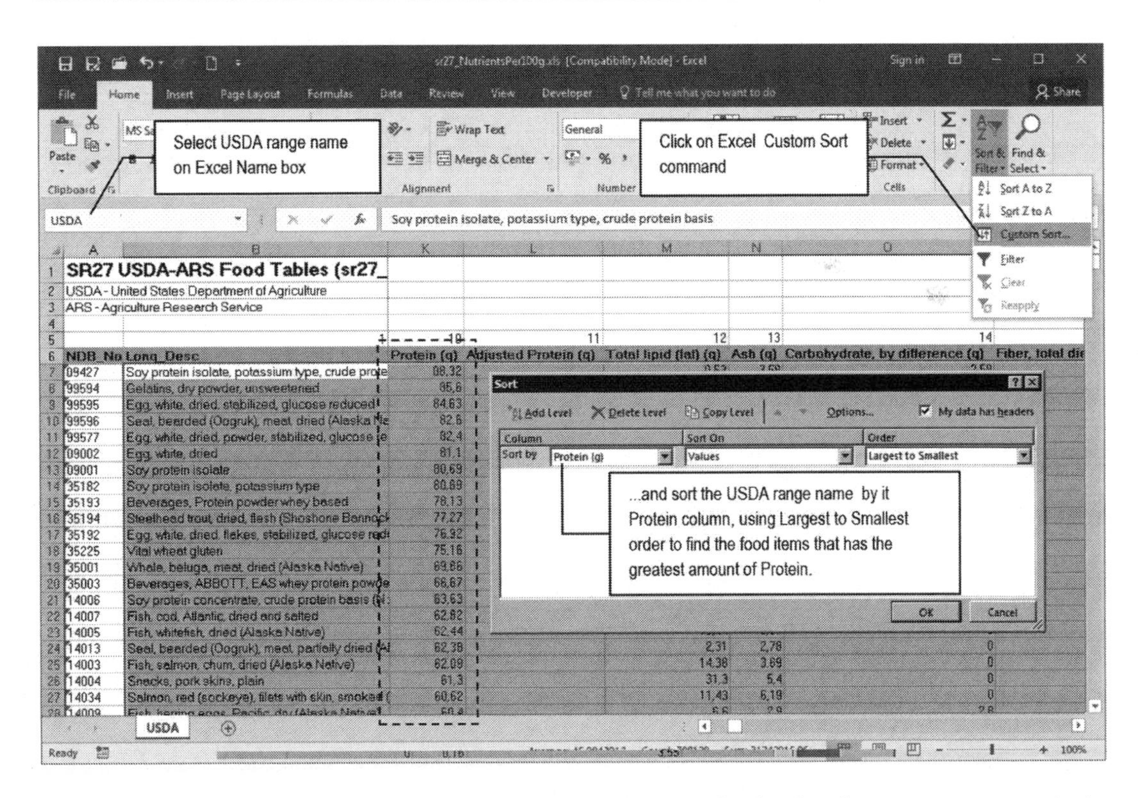

Figure 6-30. In the sr27_NutrientsPer100g.xlsm macro-enabled workbook, select the USDA range name in the Excel Name box, apply the Sort & Filter ➤ Custom Sort command, and use the Excel Sort dialog box to sort the USDA range name by its Protein column (column K), using Order = Largest to Smallest, to find the first food items with the greatest amount of protein

---

■ **Attention** When you sorted the USDA range name by its Protein column, the NDB_No column (column A) of the USDA worksheet will be desynchronized from the rest of the food table. Don't bother with this! You can easily resynchronize it by sorting again the USDA range name by Food Category and Food Item.

Now suppose you want to find, in the Fruit and Fruit Juices food category, the food items that have the greatest amount of protein per 100g of food. This time you first need to filter the USDA worksheet by the desired food category and then apply the Excel Custom Sort command to sort the filtered food items by the desired nutrient column. Follow these steps:

1. On the USDA sheet tab, click cell B6 and press Ctrl+Shift+Right Arrow and Ctrl+Shift+Page Down to select the entire USDA food table, including its nutrient identification row (row 6).
2. Click the Excel Sort & Filter ➤ Filter command to put Excel ListBox arrows on each nutrient name of row 5 of the USDA sheet tab.
3. Click the FdGrp_Desc column filter arrow, uncheck the Select All option, and check the desired food item category: Fruit_and_Fruit_Juices. Click OK to have Excel apply the filter and show just these food category food items (the Excel status bar must show 446 of 8,843 records found).
4. Click Excel's Sort & Filter ➤ Custom Sort command to show the Excel Sort dialog box, select the Protein column, set Order = Greatest to Smallest, and click OK to sort the filtered items.

When you have finished these operations, you must see just the Fruit and Fruit Juices food category's food items sorted by the USDA sheet tab's Protein column in descending order. The fruit that has the greatest amount of protein is "Apricots, dehydrated (low-moisture), sulfured, uncooked" (4.9g of Protein per 100g), followed by other dried fruits. The first raw fruit with the greatest amount of protein is "Guavas, common, raw" (1,989g of protein per 100g) (Figure 6-31).

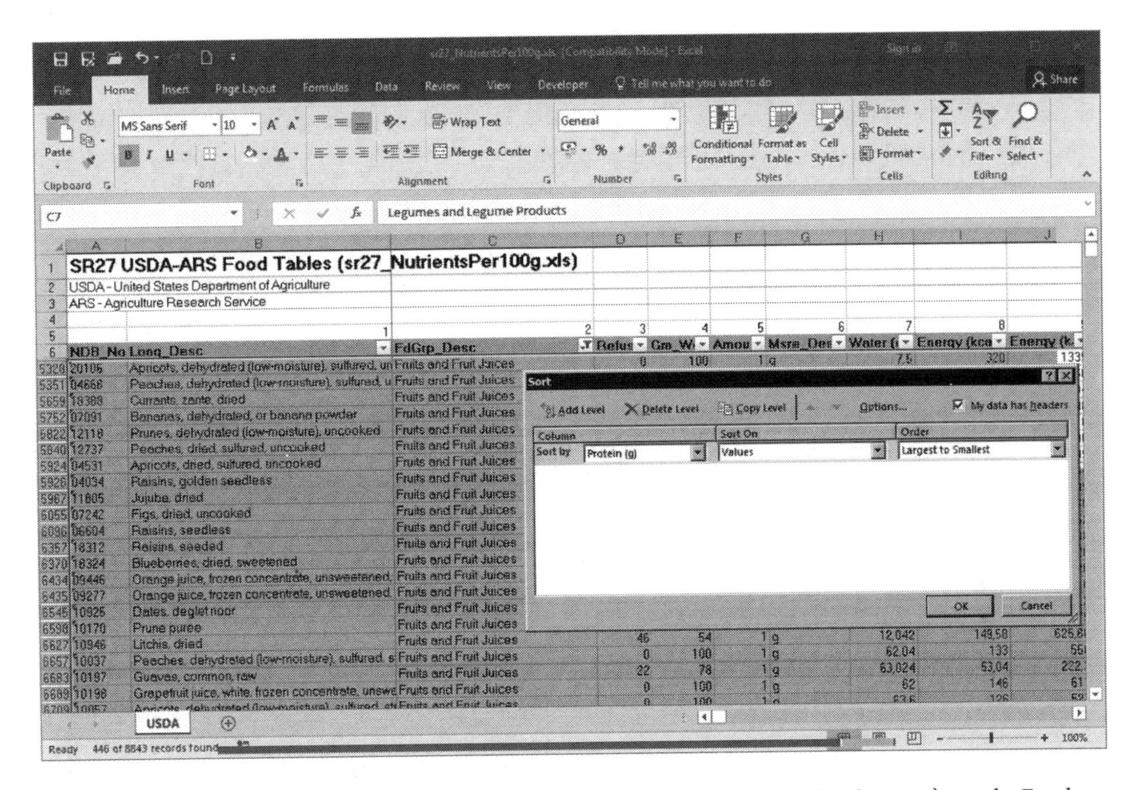

Figure 6-31. *Select the entire USDA range name (including row 5, the nutrient headers row), use the Excel Filter command to show just the food items from the food category Fruit and Fruit Juices, and then apply the Excel Sort command to show the filtered food items with the greatest amount of protein*

Now suppose that you want to find the food items that have the greatest amount of two different nutrients, like sucrose and glucose, in this order: the ones that have the greatest amount of sucrose and, then within those items, the ones that also have the greatest amount of glucose.

You cannot always sort the USDA food table by two different nutrient columns using just Excel Sort dialog box because Excel will use its own criteria of double sorting. It sorts the table by the first nutrient column and then for each amount of this first nutrient sorts by the second nutrient column.

To find the first *n* food items in the USDA range name that have the greatest amount of sucrose and glucose, you will need to use the Excel Sort dialog box to sort the entire USDA range name by columns R and S in descending order. Figure 6-32 shows that the first two food items with the greatest amount of sucrose have no glucose at all! The food items you are interested are *the ones that have the greatest amount of both nutrients*—those with both nutrients greater than zero. These do not appear on the top of the list sorted by Excel.

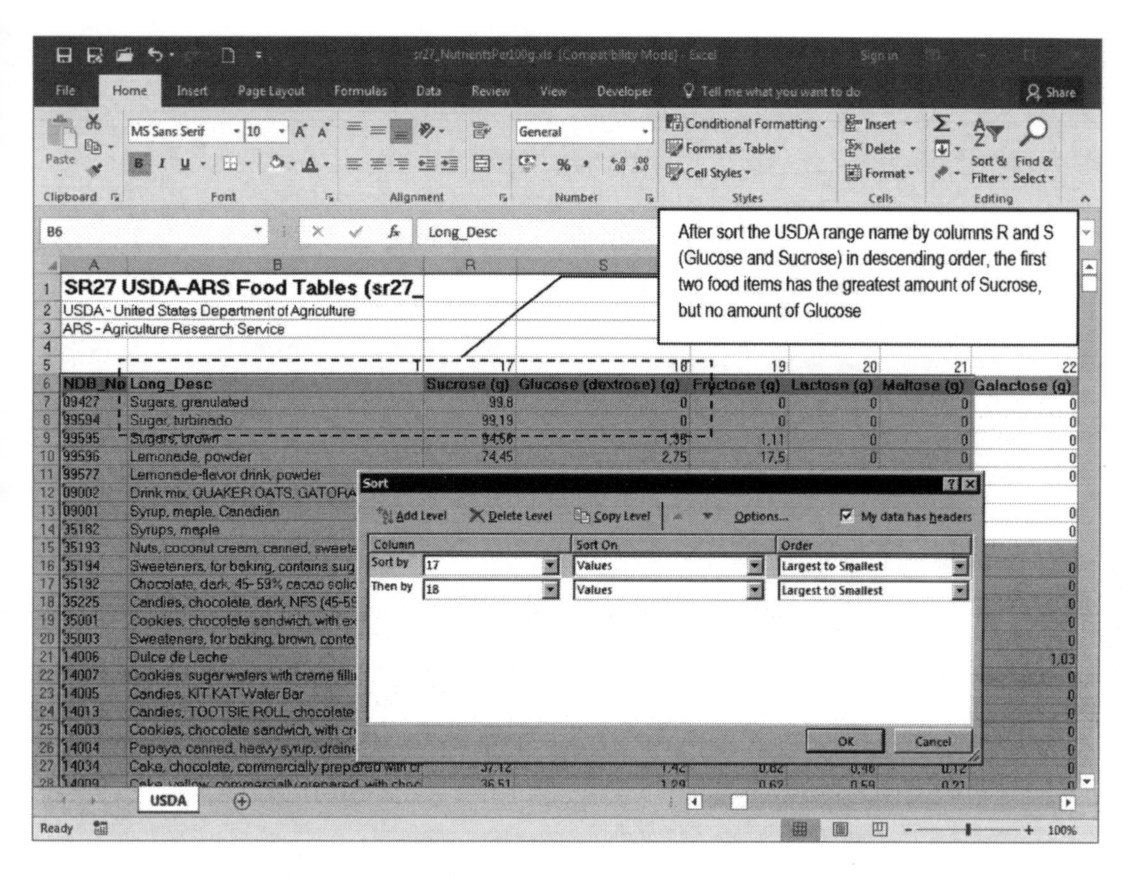

Figure 6-32. *Sometimes the Excel Sort dialog box does not return the desired results. If you try to sort the USDA range name by columns R and S in descending order to find food items with the greatest amount of sucrose and glucose, you realize that the first two food items have no amount of glucose, although they appear on the top of the sorted table*

There is no such simple method to find food items that have the greatest (or lowest) amount of two different columns using this kind of double sorting in a table like the USDA worksheet. After thinking about this problem for a while, it comes to me that to find the desired food items you can base the sort process on many different strategies, like the next four methods (there are probably many others...):

- Sort the USDA range name by the Nutrient 1 and Nutrient 2 columns and use the first *n* food items (a standard double sorting method).
- Create a new nutrient column that sums the Nutrient 1 + Nutrient 2 amounts at the right of the last USDA range column, sort the USDA range name by this calculated column, and use just the first *n* food items that have both Nutrient 1 and Nutrient 2 amounts greater than zero.
- Sort the USDA range name by Nutrient 1 and Nutrient 2 in ascending order, but use just the first *n* food items that have Nutrient 2 amounts greater than zero.
- Invert the sorting process: sort the USDA range name by Nutrient 2 and Nutrient 1 (inverse sort order), and use just the first *n* food items that have Nutrient 1 amounts greater than zero.

Sorting a Range by a Calculated Column

Let's suppose you want to insert a calculated column with a formula that sums the calcium (column Y) and iron (column Z) amounts of the first food item and then copy this formula to all other food items of the USDA worksheet. You will need to follow these steps:

1. Find the first empty cell to the right of the first USDA range row (row 7): the one that must receive this formula (cell GD7).
2. Create on this cell the formula that adds the cell address of both nutrient amounts (=Y7+Z7).
3. Copy and paste the cell GD7 formula to all other food item rows in the same column of the USDA range (cells GD8:GD8949).
4. Expand the USDA range address to include this new column (GD) and sort the entire range by this calculated column, using descending or ascending order for maximum or minimum amounts of both nutrients, respectively.

This can be easily hand-made, but what about with VBA code? You will need to use the knowledge you received so far about the Excel `Range.Cells` and `Range.Resize` properties and the `Range.Copy` method.

The Range.Copy and Range.PasteSpecial Methods

The Excel object model exposes the `Range.Copy` and `Range.PasteSpecial` methods to easily allow you to execute copy and paste operations on any worksheet. They have the next syntax:

```
Expression.Copy(Destination)
```

In this code:

`Expression`: This is a variable that represents a Range object.

`Destination`: This is optional; it defines the new range to which the range defined by `Expression` will be copied. If omitted, Excel will copy the range to the clipboard.

```
Expression.PasteSpecial(Paste, Operation, SkipBlanks, Transpose)
```

In this code:

`Expression`: This is required; it is a variable that represents a Range object.

`Paste`: This is optional; it is a constant of `XlPasteType` type that specifies the part of the range to be pasted and can be one of these constants:

`xlPasteAll`, to paste everything

`xlPasteAllUsingSourceTheme`, to paste everything using the source theme

`xlPasteAllMergingConditionalFormats`, to paste just conditional formats

`xlPasteAllExceptBorders`, to paste everything except border styles

`xlPasteFormats`, to paste just number formats

`xlPasteFormulas`, to paste just formulas

`xlPasteComments`, to paste just comments

xlPasteValues, to paste just values

xlPasteColumnWidths, to paste just the column width of the source cell to the destination cells

xlPasteValidation, to paste just the data validation option of the source cell on the destination cells

xlPasteFormulasAndNumberFormats, to paste formulas and number formats

xlPasteValuesAndNumberFormats, to paste values and number formats

Operation: This is optional; it is a constant of XlPasteSpecialOperation that indicates the kind of paste operation to be performed.

SkipBlanks: This is optional; use True to tell Excel to not paste blank cells into the destination range. The default value is False (blank cells will be pasted).

Transpose: This is optional; use True to indicate that Excel must transpose rows and columns when the range is pasted. The default value is False.

As you can see, you can use just the Range.Copy method to copy and paste the desired information with a single line of code. Use the Range.PasteSpecial method when you want to choose what to paste on the destination range, like just the values of the copied cells.

Using the VBA Immediate Window to Sort by a Calculated Column

Let's try to do each of the last four steps using the VBA Immediate window.

1. To find the first empty column to the right of the last USDA range column, use the Worksheet object's Cells property this way (note that it uses the Range.Columns.Count +1 property to find the right column and the Range.Address property to print the cell address in the VBA Immediate window).

```
?range("USDA").Cells(1, range("USDA").Columns.Count+1).Address
$GD$7
```

2. Use the Range.Formula property to insert the desired formula in the right cell; to do this, *do not* use the VBA print character (?) in the Immediate window.

```
range("USDA").Cells(1, range("USDA").Columns.Count+1).Formula = "=Y7+Z7"
```

3. Use the VBA Range.Copy method to copy the cell GD7 formula and paste the formula into all the other desired cells (note that the destination range was resized using the Range.Resize property).

```
range("GD7").Copy Range("GD7").Resize(Range("USDA").Rows.Count)
```

4. Once all cells have received the appropriate formula, it is time to sort the USDA range by the new calculated column. So, you need to resize it to include this column and sort it descending, as follows:

```
Range("USDA").Resize(, Range("USDA").Columns.Count + 1).Sort
Range("GD7"), xlDescending
```

Figure 6-33 shows how the USDA worksheet should look after you make these three operations using the VBA Immediate window.

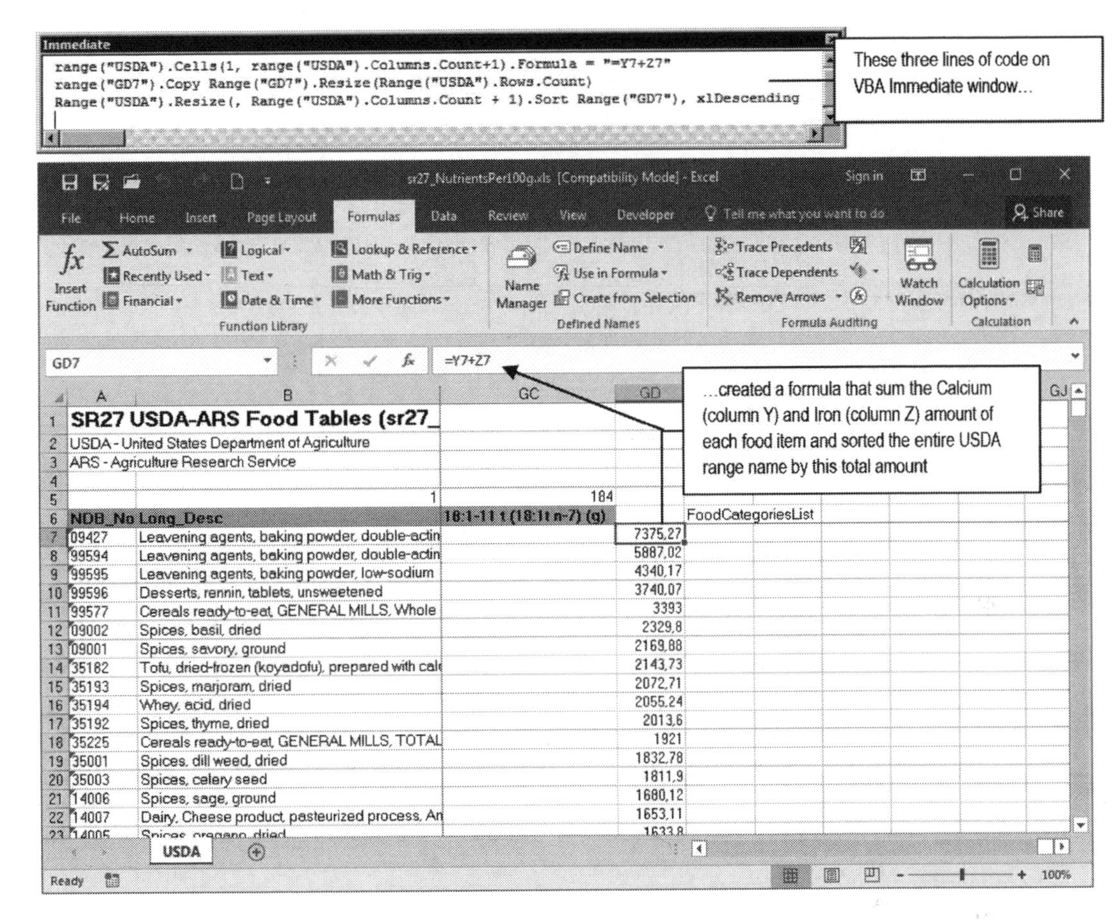

Figure 6-33. Using the VBA Immediate window, the Excel Range.Cells, Range.Formula, and Range.Resize properties, as well as the Range.Copy method, you can create a new calculated column and sort the entire USDA range name using just three lines of code

Sorting by a Calculated Column with VBA

But there is a catch on this programmable operation because of another Excel bug! If you try to follow this exact operation sequence using VBA code, Excel will copy all the formulas but will not update their references when you use the `Range.Sort` method.

The file `sr27_NutrientsPer100g_Sort Sum of Nutrients.xlsm` macro-enabled workbook that you can extract from the `Chapter06.zip` file has the `frmSortBySum` UserForm, which allows you to select two different nutrients, create a sum of nutrient formulas, and sort the entire USDA food table by this new column (Figure 6-34).

Figure 6-34. This is the frmSortBySum UserForm from the sr27_NutrientsPer100g_Sort Sum of Nutrients.xlsm macro-enabled workbook that allows you to select two different nutrients with its cboNutrient1 and cboNutrient2 ComboBoxes to create a new sum of nutrients formula column and sort the USDA food table by it

The `frmSortBySum` uses the `UserForm_Initialize( )` event to run through all USDA worksheet row 7 columns (nutrient column names) and fill both the `cboNutrient1` and `cboNutrient2` ComboBoxes with all available nutrient names, executing this code:

```
Private Sub UserForm_Initialize()
    Dim rg As Range
    Dim rgUSDA As Range
    Dim intI As Integer
    Const conFirstNutrient = 7

    Set rgUSDA = Range("USDA").Resize(1, Range("USDA").Columns.Count).Offset(-1)
    For intI = conFirstNutrient To rgUSDA.Columns.Count
        Set rg = rgUSDA.Cells(1, intI)
        Me.cboNutrient1.AddItem rgUSDA.Parent.Name & "!" & rg.Offset(1).Address(False,
        False)
        Me.cboNutrient1.Column(1, Me.cboNutrient1.ListCount - 1) = rg
        Me.cboNutrient2.AddItem rgUSDA.Parent.Name & "!" & rg.Offset(1).Address(False,
        False)
        Me.cboNutrient2.Column(1, Me.cboNutrient2.ListCount - 1) = rg
    Next
End Sub
```

This code begins by resizing the entire USDA range name for just one row and all its columns using the `Range.Resize` method and then sets a reference to the USDA table headers (nutrient names) using the `Range.Offset(-1)` property.

```
Set rgUSDA = Range("USDA").Resize(1, Range("USDA").Columns.Count).Offset(-1)
```

And once the range reference was correctly set to the `rgUSDA` object variable, it performs a `For intI... Next` loop to run through all nutrient column names, beginning on column 7, which is associated to the `conFirstNutrient` constant (the 8th worksheet column: column H; nutrient = "Water (g)"), to avoid a *magic number* inside the code. The selected nutrient column is attributed to the `rg` object variable using the `rgUSDA.Cells( )` property.

```
For intI = conFirstNutrient To rgUSDA.Columns.Count
    Set rg = rgUSDA.Cells(1, intI)
```

Both the cboNutrient1 and cboNutrient2 ComboBoxes have two columns: the first column (hidden) must contain a reference formula to the first nutrient value, which is stored in the USDA range name row 7; the second column (visible) must receive the nutrient name.

To add the first column formula with the address of the first nutrient value (row 7; the first USDA range row), the procedure uses the ComboBox.AddItem method, employing the rgUSDA.Parent.Name property to identify the worksheet name, followed by an exclamation mark and the range address—which is returned as a relative reference using the rg.Offset method with both Address arguments (row and column reference type) set to False.

```
Me.cboNutrient1.AddItem rgUSDA.Parent.Name & "!" & rg.Offset(1).Address(False, False)
```

Each nutrient name is associated to the first USDA range name row, using the ComboBox.Column property and the rg object variable value, which contain the nutrient column name.

```
Me.cboNutrient1.Column(1, Me.cboNutrient1.ListCount - 1) = rg
```

This process is repeated to fill the cboNutrient2 ComboBox with the same information, until all nutrient column names are processed inside the For...Next loop.

```
        Me.cboNutrient2.AddItem rgUSDA.Parent.Name & "!" & rg.Offset(1).Address(False,
        False)
        Me.cboNutrient2.Column(1, Me.cboNutrient2.ListCount - 1) = rg
    Next
End Sub
```

The cmdSort CommandButton executes the code needed to create the sum of nutrient formula on the new calculated column, creating it on the first nutrient row and then copying it to all other nutrients. It then sorts the USDA range name by this new calculated column, running this code:

```
Private Sub cmdSort_Click()
    Dim rg As Range
    Dim rgPaste As Range
    Dim rgUSDA As Range

    If IsNull(Me.cboNutrient1) Or IsNull(Me.cboNutrient2) Then
        MsgBox "Please, select Nutrient 1 and Nutrient 2 columns!", vbInformation, "Select
        both nutrients!"
        Exit Sub
    End If

    'Create formula to sum both nutrients
    Set rgUSDA = Range("USDA")
    Set rg = rgUSDA.Cells(1, rgUSDA.Columns.Count + 1)
    Set rgPaste = rg.Resize(rgUSDA.Rows.Count)
    rg.Formula = "=" & Me.cboNutrient1 & "+" & Me.cboNutrient2
    rg.Copy rgPaste
```

```
    'Expand rgUSDA columns to contain the new column formula and sort it by the formula
    results
    Set rgUSDA = rgUSDA.Resize(, rgUSDA.Columns.Count + 1)
    rgUSDA.Sort rg, xlDescending
End Sub
```

Note that it first looks if both cboNutrient1 and cboNutrient2 have some nutrient selected.

```
If IsNull(Me.cboNutrient1) Or IsNull(Me.cboNutrient2) Then
    MsgBox "Please, select Nutrient 1 and Nutrient 2 columns!", vbInformation, "Select both
    nutrients!"
    Exit Sub
End If
```

It then sets a reference to the calculated column using the rg object variable. It uses the Range.Columns.Count + 1 property to refer to the right column and sets another reference to all USDA range rows using the Range.Rows.Count property.

```
Set rgUSDA = Range("USDA")
Set rg = rgUSDA.Cells(1, rgUSDA.Columns.Count + 1)
Set rgPaste = rg.Resize(rgUSDA.Rows.Count)
```

The formula that adds the selected nutrient is then created on the first nutrient row using the rg.Formula property and then copied to all other nutrient rows using the Range.Copy method with the rgPaste object variable as the destination.

```
rg.Formula = "=" & Me.cboNutrient1 & "+" & Me.cboNutrient2
rg.Copy rgPaste
```

To finish the process, the rgUSDA object variable has its column count increased by 1 using the Range.Resize method, and then it is sorted by this new column value (a formula with the sum of selected nutrients) in descending order.

```
    'Expand rgUSDA columns to contain the new column formula and sort it by the formula
    results
    Set rgUSDA = rgUSDA.Resize(, rgUSDA.Columns.Count + 1)
    rgUSDA.Sort rg, xlDescending
End Sub
```

There is nothing wrong with that code strategy, but when you use the frmSortBySum UserForm from the sr27_NutrientsPer100g_Sort Sum of Nutrients.xlsm macro-enabled workbook to execute this process, it seems to work right—but it doesn't! Figure 6-35 shows what happens to the calculated column formulas after you select two nutrients (protein and fiber) and try to sort the USDA food table. The sort happens, but the formula references were not updated like they should be.

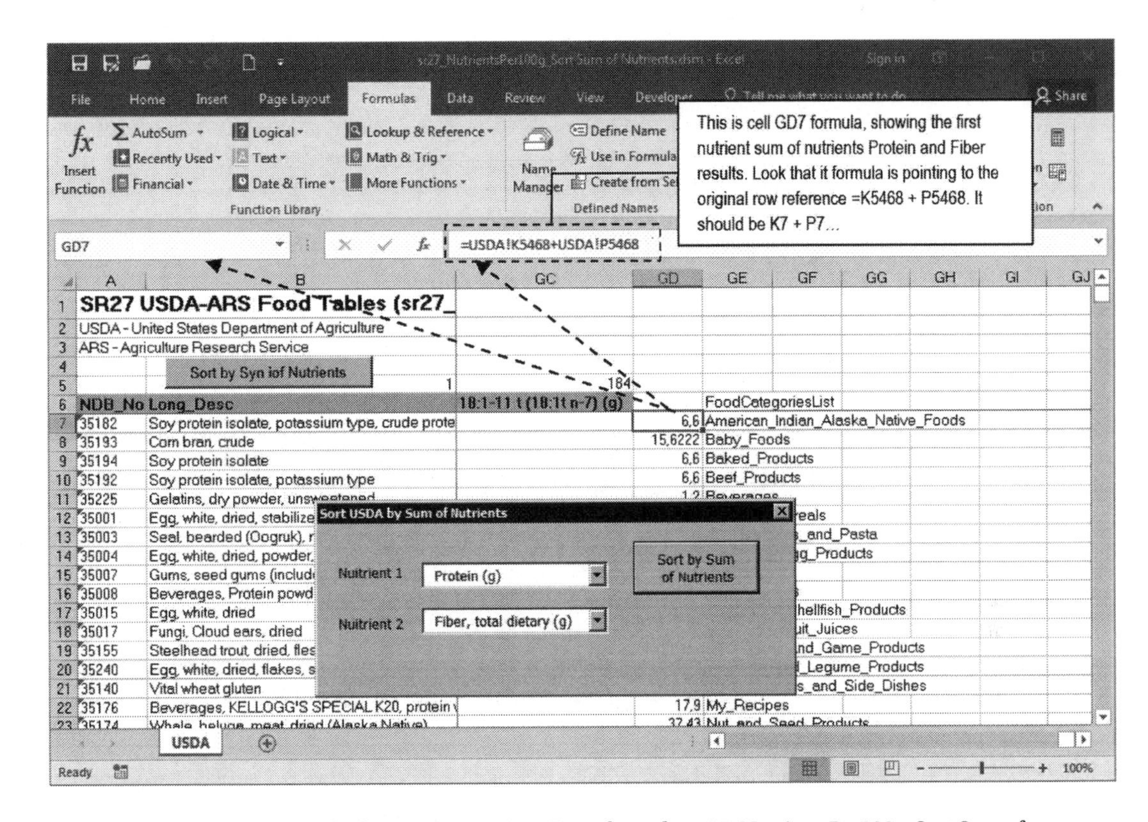

***Figure 6-35.** When you use the frmSortBySum UserForm from the sr27_NutrientsPer100g_Sort Sum of Nutrients.xlsm macro-enabled workbook to sort the entire USDA range name, the sort process is done, but all the calculated column formula references are not updated to reflect the sum of nutrients for each food item row*

■ **Attention** Note that `frmSortBySum UserForm` is a no-modal window, allowing you to click any worksheet cell while it is opened (the `ShowModal` property was set to `False`). Also note that when you close it, it fires the `Terminate( )` event, re-sorting the USDA worksheet by its default order.

Also note that if you click the `cmdSort` command button again, the values will be changed.

To fix this bad Excel VBA behavior, you must do an extra process before applying the `Range.Sort` method to the sum of the nutrient formula column: you must copy all the sum of nutrient formulas and use the `Range.PasteSpecial` method *to paste its values before sorting the range*!

The file `sr27_NutrientsPer100g_Sort Sum of Nutrients_PasteSpecial.xlsm` Excel macro-enabled workbook has this fix on its `cmdSort_Click( )` event. Observe the bold rows:

```
Private Sub cmdSort_Click()
    Dim rg As Range
    Dim rgPaste As Range
    Dim rgUSDA As Range

    If IsNull(Me.cboNutrient1) Or IsNull(Me.cboNutrient2) Then
```

```
        MsgBox "Please, select Nutrient 1 and Nutrient 2 columns!", vbInformation, "Select
        both nutrients!"
        Exit Sub
    End If

    'Create formula to sum both nutrients
    Set rgUSDA = Range("USDA")
    Set rg = rgUSDA.Cells(1, rgUSDA.Columns.Count + 1)
    Set rgPaste = rg.Resize(rgUSDA.Rows.Count)
    rg.Formula = "=" & Me.cboNutrient1 & "+" & Me.cboNutrient2
    rg.Copy rgPaste
    rgPaste.Copy
    rgPaste.PasteSpecial xlPasteValues

    'Expand rgUSDA columns to contain the new column formula
    Set rgUSDA = Range("USDA").Resize(, rgUSDA.Columns.Count + 1)
    rgUSDA.Sort rg, xlDescending
    'rgPaste.Clear
End Sub
```

This time, before implementing the search process, all calculated sums of nutrient formulas are copied and pasted using the Range.PasteSpecial method and the xlPasteValues constant, changing all calculated formulas by their real values. The result can be seen in Figure 6-36, which shows a very different sorting result.

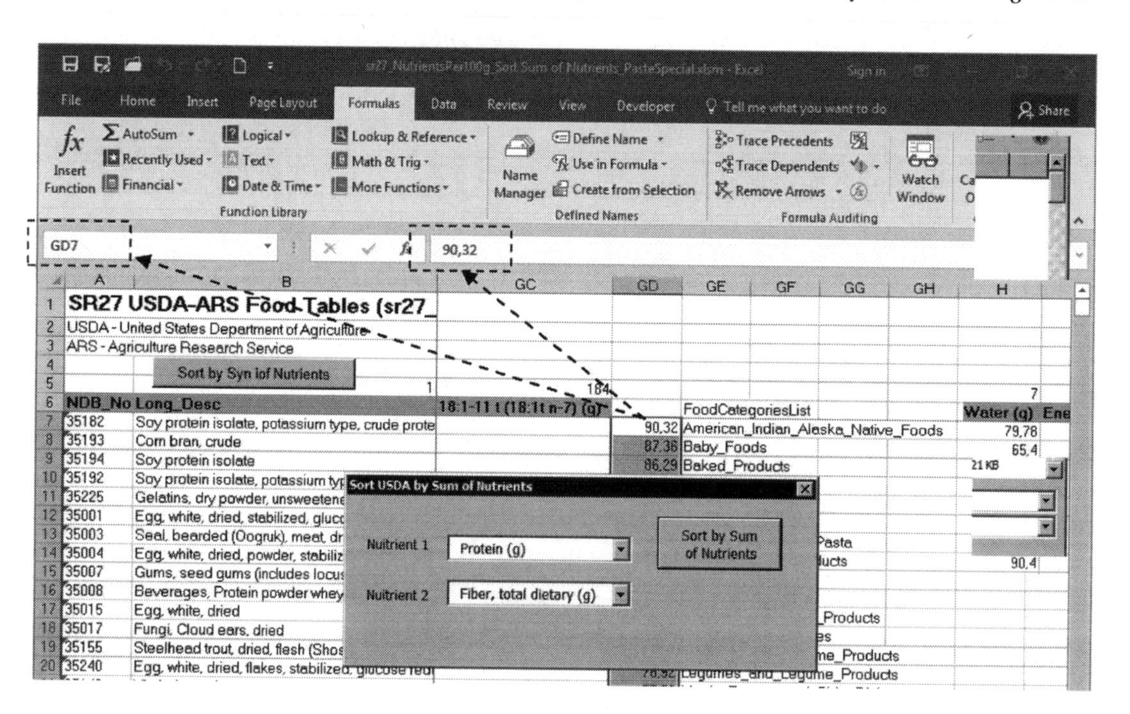

***Figure 6-36.** To correctly sort any range by a calculated formula, you must first use the Range.PasteSpecial method with the xlPasteValues constant to change the formula references by its returned results and then sort the range. You can try this new approach using the sr27_NutrientsPer100g_Sort Sum of Nutrients_PasteSpecial.xlsm Excel macro-enabled workbook*

Using frmRangeSort

Now that you know you can sort the USDA food table by one or two different nutrient values, you are ready to understand how to implement a good all-in-one `UserForm` solution to sort food items by different methods using VBA code.

The file `sr27_NutrientsPer100g_Range.Sort.xlsm` Excel macro-enabled workbook offers the `frmRangeSort` `UserForm`, which allows you to sort food items using two different nutrient values via four different sorting methods: Sort by Nutrient1 and Nutrient2 columns, sum of both nutrient column amounts, sort by Nutrient1 and find top *n* items of Nutrient2, or sort by Nutrient2 and find top *n* items of Nutrient1 (Figure 6-37).

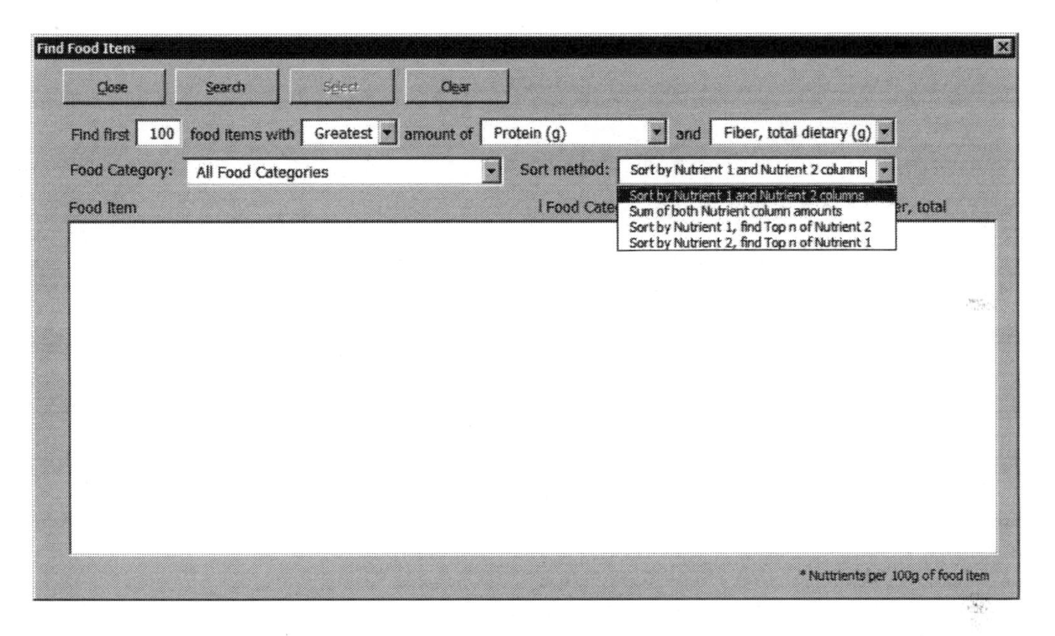

Figure 6-37. *This is frmRangeSort* `UserForm` *from the sr27_NutrientsPer100g_Range.Sort.xlsm Excel macro-enabled workbook, which allows you to select two different nutrients, choose among four different sorting methods, and recover the first n food items using a ListBox control*

The `frmRangeSort` offers all the tricks already covered in this chapter with the `frmRangeFind` and `frmRangeFilter` `UserForms`, so I will just comment on how it sorts and fills the `lstFoodItems` `ListBox` with the food items it finds with one of its four different sort processes

The `frmRangeSort` `UserForm` requires that you select just one or two different nutrients to sort the `USDA` range name. If you select just one nutrient, you can just use its first sorting method: sort by Nutrient 1 and then by Nutrient 2 columns. If you select two different nutrients, you can sort by any method.

After you have selected the desired nutrients and sort method, click the Search button (`cmdSearch`) to sort the `USDA` range name and fill the `ListBox` with the first *n* food items found (100 is the default value). Change the sort method, and the list will be filled again (Figure 6-38).

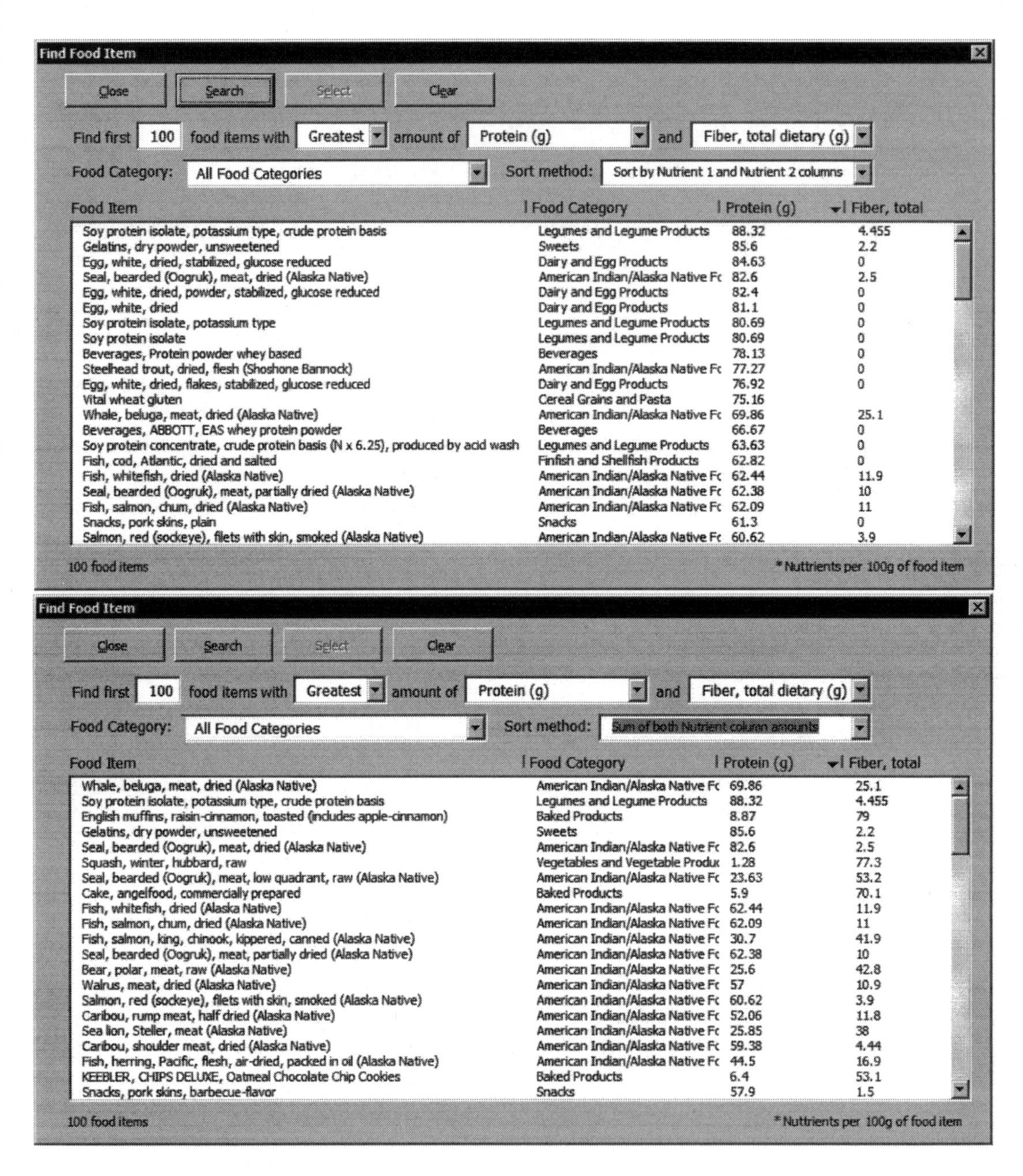

***Figure 6-38.** When you select two different nutrients, choose the sort method and click Search to fill the UserForm ListBox with the first n food items found. Select another sort method and look at how the list is changed, regarding the selected method*

This happens because the first time you select a nutrient and click the cmdSearch CommandButton, it defines the module-level variable mbolSorted = True, and when you select another sort method, the cboSortMethod_Change( ) event fires, running this code:

```
Private Sub cboSortMethod_Change()
    If mbolSortedThen
        Call cmdSearch_Click
    End If
End Sub
```

And this technique is reused whenever you select another food category, firing the cboCategory_Change( ) event.

```
Private Sub cboCategory_Change()
    If mbolSorted Then
        Call cmdSearch_Click
    End If
End Sub
```

Note in Figure 6-38 how the food item list changes as you select another sort method. The sort process is made by the cmdSearch_Click( ) event, which has a long list that employs many programming techniques described so far in this book. It has four main parts: declare the variable and clean up the interface, define the range to be sorted, sort the range, and fill the lstFoodItems ListBox. Let's look at each one of them, beginning with the first part.

```
Private Sub cmdSearch_Click()
    Dim rgUSDA As Range
    Dim rg As Range
    Dim rgPaste As Range
    Dim intArea As Integer
    Dim intSortOrder As Integer
    Dim intI As Integer
    Dim intItems As Integer
    Dim intMaxItems As Integer
    Dim bolInsert As Boolean
    Const conTriangleDown = 6
    Const conFoodCategory = 2
    Const conSortNutrients = 0
    Const conNutrientSum = 1
    Const conSortByNutrient1 = 2
    Const conSortByNutrient2 = 3

    If Me.cboNutrient1 = Me.cboNutrient2 Then
        MsgBox "Select two different nutrients!", vbInformation, "Nutrients are equal"
        Exit Sub
    End If

    Me.lstFoodItems.Clear
    Me.lblAZ.Caption = conTriangleDown
    Me.lblAZ.Left = Me.lbl2.Left + Me.lbl2.Width
    Sheets("USDA").AutoFilterMode = False
```

```
'Define the range to be searched
Set rgUSDA = Range("USDA")
```

Note that this first part requires that the user select two different nutrients and that the rgUSDA object variable be defined to the entire USDA range name.

The second code part disables screen updating and defines the range to be sorted. It first verifies whether the first sort method was selected and requires that a sum of nutrient formula must be created. If this is true, the rgUSDA object variable is increased by one column.

```
Application.ScreenUpdating = False
'Verify it rgUSDA must be sorted by nutrients sum!
If Not IsNull(Me.cboNutrient2) And _
   Me.cboSortMethod.ListIndex = conNutrientSum Then
        'Create formula to sum both nutrients
        Set rg = rgUSDA.Cells(1, rgUSDA.Columns.Count + 1)
        Set rgPaste = rg.Resize(rgUSDA.Rows.Count)
        rg.Formula = "=" & Me.cboNutrient1 & "+" & Me.cboNutrient2
        rg.Copy rgPaste
        rgPaste.Copy
        rgPaste.PasteSpecial xlPasteValues
        'Expand rgUSDA columns to contain the new column formula
        Set rgUSDA = Range("USDA").Resize(, rgUSDA.Columns.Count + 1)
End If
```

Still in this second code part, the procedure verifies whether the user selected a specific food category. If this is true, the rgUSDA object variable must be filtered to show just food items of the selected food category. It is now time to use the Range.AutoFilter method, filtering the rgUSDA object variable by its second column (represented by the conFoodCategory constant) and by the food category selected in the cboCategory ComboBox.

```
'Verify if user selected a single category
If Me.cboCategory.ListIndex > 0 Then
    'Filter rgUSDA by selected category and redefine it
    Set rgUSDA = rgUSDA.Offset(-1).Resize(rgUSDA.Rows.Count + 1)
    rgUSDA.AutoFilter conFoodCategory, Me.cboCategory

    'SpecialCells.Area(1) has the nutrient names, with just one row, unless the user selected
    'the first food category.
    If rgUSDA.SpecialCells(xlCellTypeVisible).Areas(1).Rows.Count > 1 Then
        Set rgUSDA = rgUSDA.SpecialCells(xlCellTypeVisible).Areas(1)
        Set rgUSDA = rgUSDA.Offset(1).Resize(rgUSDA.Rows.Count - 1)
    Else
        'User did not select first food category!
        Set rgUSDA = rgUSDA.SpecialCells(xlCellTypeVisible).Areas(2)
    End If
End If
```

Note in the previous code how the rgUSDA object variable is resized to show just the food items returned by the Range.AutoFilter method, using the SpecialCells(xlCellTypeVisible) property and the Range.Areas collection, as stated in the section "Selecting Filtered Cells with the Range.SpecialCells Property" earlier in this chapter.

And once the rgUSDA object variable points to all desired cells, the procedure is ready to execute its third code part: sorting the rgUSDA object variable. Note that the code first defines the sort order and then verifies whether just one nutrient was selected, redefining the lstFoodItems.ColumnCount property to just three columns and sorting the rgUSDA accordingly.

```
'rgUSDA now contains all desired food items and columns
'Define sort order
intSortOrder = IIf(Me.cboSortOrder = "Greatest", xlDescending, xlAscending)
'Verify if just one nutrient was selected and sort rgUSDA accordingly
If IsNull(Me.cboNutrient2) Then
    Me.lstFoodItems.ColumnCount = 3
    rgUSDA.Sort Range(Me.cboNutrient1), intSortOrder
```

If two different nutrients were selected, the code needs to redefine the lstFoodItems.ColumnCount property to four columns and use one of the four available sort methods. It first verifies whether the second sort method, "Sum of both nutrient column amount," was selected. If it is, rgUSDA is sorted by the calculated column, represented by the rg object variable, as you did earlier in the section "Sorting by a Calculated Column with VBA." Note that once the sort is made, the rgPaste object variable that represents all calculated cells is cleared.

```
Else
    'Two nutrients was selected
    Me.lstFoodItems.ColumnCount = 4
    If Me.cboSortMethod.ListIndex = conNutrientSum Then
        'Sort by sum of both nutrients
        rgUSDA.Sort rg, intSortOrder
        rgPaste.Clear
```

If the user selected the first or second sort method, rgUSDA must be sorted first by Nutrient 1 and then by Nutrient 2 (both methods differ in the way they choose the food items that will be inserted in the lstFoodItems ListBox).

```
Else
    If (Me.cboSortMethod.ListIndex = conSortNutrients) Or _
       (Me.cboSortMethod.ListIndex = conSortByNutrient1) Then
       rgUSDA.Sort Range(Me.cboNutrient1), intSortOrder, _
                   Range(Me.cboNutrient2), , intSortOrder
```

The fourth sort method is when the food items must be sorted first by Nutrient 2 and then by Nutrient 1.

```
            Else
                rgUSDA.Sort Range(Me.cboNutrient2), intSortOrder, _
                            Range(Me.cboNutrient1), , intSortOrder
                Me.lblAZ.Left = Me.lbl3.Left + Me.lbl3.Width
            End If
        End If
    End If
```

And once the rgUSDA object variable is sorted by the selected method, it is time to fill the lstFoodItems ListBox. To do this, the procedure first verifies whether rgUSDA (that may be filtered) has at least the number of items defined by the txtFoodItemsNumber text box and uses the smaller value:

```
intI = 1
intMaxItems = IIf(rgUSDA.Rows.Count > Me.txtFoodItemsNumber, Me.txtFoodItemsNumber, rgUSDA.
Rows.Count)
```

The first food item available in rgUSDA is attributed to the rg object variable using the rgUSDA.Cells collection and the intI Integer variable, and a Do...Loop structure is executed to insert all possible food items, until the maximum number of items is recovered or a blank item is achieved (rg = "").

```
Set rg = rgUSDA.Cells(intI, 1)
Do While intItems < intMaxItems And rg <> ""
```

Each food item must be verified against the selected sort method before being inserted in the ListBox. To help with this selection process, the code uses the bolInsert Boolean variable. The item must be inserted if:

1. The first sort method was selected (cboSortMethod = conSorNutrients) *or* the fourth method was selected (cboSortMethod = conSortByNutrient2) *and* Nutrient 1 value is greater than zero.

```
'Verify if food item must be inserted on the list
bolInsert = (Me.cboSortMethod.ListIndex = conSortNutrients) Or _
            (Me.cboSortMethod.ListIndex = conSortByNutrient2 And _
            rg.Offset(, Range(Me.cboNutrient1).Column - 2) > 0)
```

2. The first condition was met (bolInsert = True) *or*

   a. The second sort method was selected (cboSortMethod = conNutrientSum) *and* both Nutrient 1 and Nutrient 2 values are greater than 0; or

   b. The third sort method was selected (cboSortMethod = conSortByNutrient1), and Nutrient 2 value is greater than 0.

```
If Not IsNull(Me.cboNutrient2) Then
     bolInsert = bolInsert Or _
                 (Me.cboSortMethod.ListIndex = conNutrientSum And _
                 rg.Offset(, Range(Me.cboNutrient1).Column - 2) > 0
                 And _
                 rg.Offset(, Range(Me.cboNutrient2).Column - 2) > 0)
                 Or _
                 (Me.cboSortMethod.ListIndex = conSortByNutrient1
                 And _
                 rg.Offset(, Range(Me.cboNutrient2).Column - 2) > 0)
End If
```

3. A food category was selected in the cboFoodCategory ComboBox and the item pertain to it:

```
If Me.cboCategory.ListIndex > 0 Then
    bolInsert = bolInsert And rg.Offset(, 1) = Me.cboCategory
End If
```

The bolInsert variable will be true if the item must be inserted. So, do it when bolInsert = True:

```
If bolInsert Then
    With Me.lstFoodItems
        .AddItem rg
        .Column(1, .ListCount - 1) = rg.Offset(, 1)
        .Column(2, .ListCount - 1) = rg.Offset(, Range(Me.cboNutrient1).Column - 2)
        If Not IsNull(Me.cboNutrient2) Then
            .Column(3, .ListCount - 1) = rg.Offset(, Range(Me.cboNutrient2).Column - 2)
        End If
    End With
    intItems = intItems + 1
End If
```

And once the item was inserted, increment the intI counter and get the next food item using the rgUSDA.Cells collection before trying the next loop passage.

```
    intI = intI + 1
    Set rg = rgUSDA.Cells(intI, 1)
Loop
```

The procedure finishes indicating how many items were found, enabling interface command buttons, returning the USDA range name to its original sort order, and reactivating the screen updating.

```
Me.lblFoodItems.Caption = Me.lstFoodItems.ListCount & " food items"
Me.lblFoodItems.Visible = True
Me.cmdSelect.Enabled = False
Range("USDA").Sort Range("USDA!C1"), , Range("USDA!B1")
mbolSorted = True
Application.ScreenUpdating = True
```

Go ahead and play for a while with the frmRangeSort UserForm interface. Note how it works smoothly and how to do its job quickly. Oh, sure: don't forget to try to change the ListBox column widths and sort order by clicking and dragging the column names! Also try to double-click any food item or select it and click the Select ControlButton to see how it is promptly stored in the selected cell of the Find Food Item by Nutrient sheet tab.

Using frmSearchFoodItems.xlsm

Throughout this chapter you have used the sr27_NutrientsPer100g.xls Excel workbook inserted as the USDA sheet tab to search for food items and their nutrient content per 100g of food. But on most nutritional applications you will seldom use nutrients per 100g of food. Instead, you need to select them by one of their possible common measures, available from two different USDA worksheets: sr27_NutrientsPerFirstCommonMeasure.xls (which lists food items by most popular common measure and amount) and sr27_FoodItemsCommonMeasures.xls (which lists all possible common measures for many, but not all, USDA food items). Both are produced by the Microsoft Access USDA Food List Creator.mdb application.

■ **Attention** The `NutrientsPerFirstCommonMeasure.xls` and `sr27_FoodItemsCommonMeasures` Excel 2003 workbooks were generated using the Microsoft Access `USDA Food List Creator.mdb` application.

The `frmSearchFoodItems.xlsm` Excel macro-enabled workbook, which you can extract from the `Chapter06.zip` file, encompasses both methods studied in this chapter to search for a specific nutrient: by its name or nutrient content using food item amounts regarding its available common measures. Figure 6-39 shows that it has three sheet tabs: `Find Food Item`, `USDA`, and `USDACommonMeasures`. It also has a new `frmSearchFoodItems UserForm`, where you can note these changes:

- It has a `Frame` control with two `Option` buttons (`opptFindContain` and `optFindAmount`) that allow the selection of the search method: by name or nutrient content.
- It has a "Qty g" column in the `lstFoodItems ListBox`, indicating the amount of the most popular food item common measure.
    - It has the `lstCommonMeasures ListBox` on its bottom, showing all possible common measures for the selected food item.

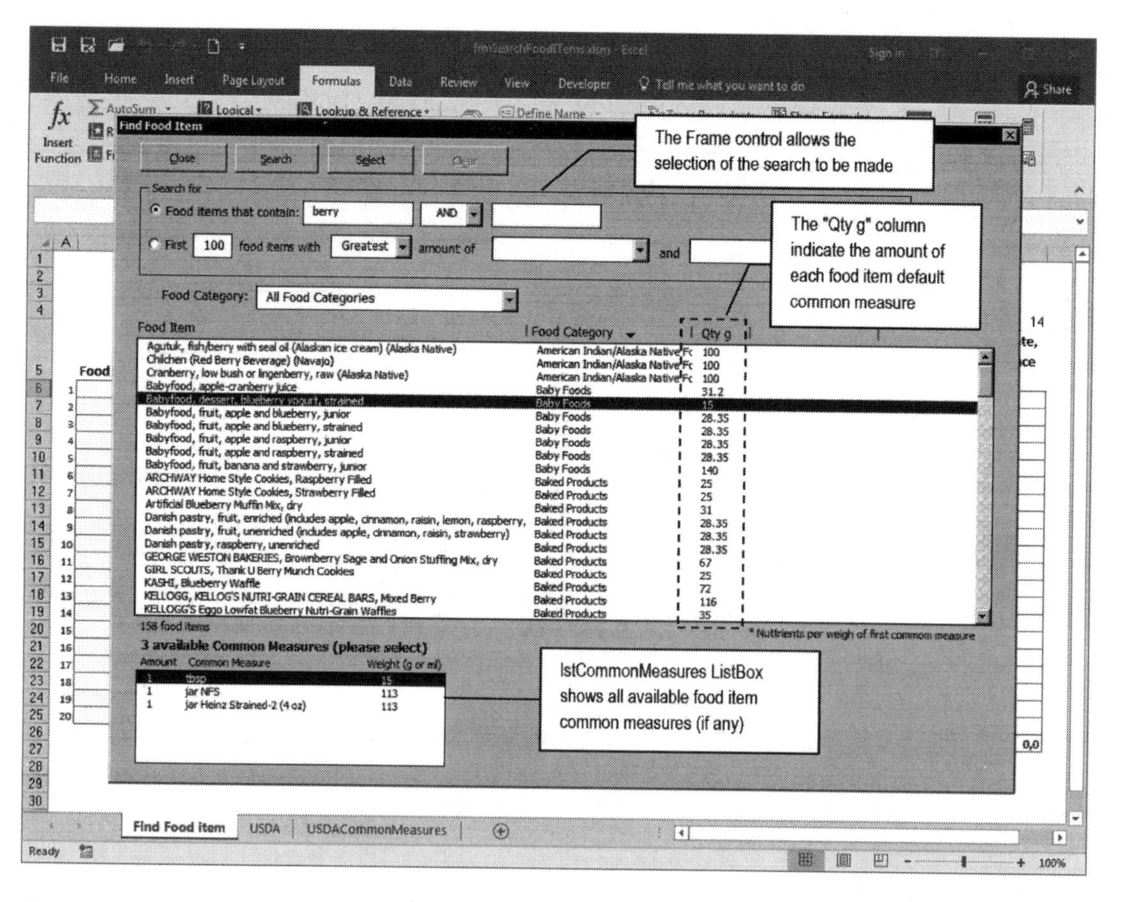

Figure 6-39. This is the fmrSearchFoodItems.xlsm Excel macro-enabled workbook with its three sheet tabs, used to find food items by name or nutrient content, allowing the user to select a food item amount by all its available common measures

■ **Attention** The USDACommonMeasure sheet tab uses about 15,200 rows to store all USDA available food item common measures, ordered by the food item name.

After you select the desired Option button search method (optFindContain or optFindAmount), type the food item name (or select the desired nutrient names), and click the UserForm Search button, the cmdSearch_Click( ) event will fire, executing this code:

```
Private Sub cmdSearch_Click()
    Application.Calculation = xlCalculationManual
    If Me.optFindContain = True Then
        Call FindByAutoFilter
    Else
        Call FindBySorting
    End If
    Me.lstCommonMeasures.Clear
    Application.Calculation = xlCalculationAutomatic
End Sub
```

■ **Attention** Since the Excel AutoFilter method forces a full workbook recalculation, you must always set Application.Calculation = xlCalculationManual before using this method. This will avoid the cascade execution of any Function procedure set to Volatile in your VBA project.

Showing Selected Food Item Common Measures

Both the FindByAutoFilter( ) and FindBySorting( ) procedures execute the same code described earlier in this chapter in the sections "Using frmRangeFilter" and "Using frmRangeSort" to find the desired food item.

Having a list of possible food items to choose from, whenever you select the desired food item on the lstFoodItems ListBox, the lstFooditems_Click( ) event will fire, calling the ShowCommonMeasures( ) procedure, which will search inside the USDACommonMeasures worksheet for all possible food item common measures, exhibiting them (if any) in the lstCommonMeasures ListBox using this code:

```
Private Sub lstFoodItems_Click()
    If Not IsNull(Me.lstFoodItems) Then
        Call ShowCommonMeasures(Me.lstFoodItems)
    End If
    Me.cmdSelect.Enabled = Not IsNull(Me.lstFoodItems)
End Sub
Public Sub ShowCommonMeasures(strItem as String)
    Dim ws As Worksheet
    Dim rg As Range
    Dim rgFilter As Range
    Dim lngRows As Long
    Dim intI As Integer

    'Clear ListBox
    Me.lstCommonMeasures.Clear
```

```
    'clean any auto filter in action
    Set ws = Sheets("USDACommonMeasures")
    ws.AutoFilterMode = False

    'Find last used row of USDACommonMeasures sheet tab
    lngRows = ws.UsedRange.Rows.Count

    'Filter common measures for selected food item (including headers)
    Set rg = ws.Range("$A$5:$E$" & lngRows)
    rg.AutoFilter 1, strItem

    'Loop through all filtered range names
    For Each rgFilter In rg.SpecialCells(xlCellTypeVisible).Areas
        'Discard row 5 (column names)
        If rgFilter.Row > 5 Then
            For intI = 0 To rgFilter.Rows.Count - 1
                With Me.lstCommonMeasures
                    .AddItem rgFilter.Cells(intI + 1, 2)
                    .Column(1, intI) = rgFilter.Cells(intI + 1, 3)
                    .Column(2, intI) = rgFilter.Cells(intI + 1, 4)
                End With
            Next
        End If
    Next

    If Me.lstCommonMeasures.ListCount > 0 Then
        Me.lstCommonMeasures.ListIndex = 0
    End If
    Me.lblCommomMeasures.Caption = Me.lstCommonMeasures.ListCount & " available Common
    Measures (please select)"
    'Remove USDACommonMeasures autofilter
    ws.AutoFilterMode = False
End Sub
```

As you can see, ShowCommonMeasures( ) uses the Range.AutoFilter method to select all possible food item common measures. To do this, it first clears the lstCommonMeasures ListBox, sets an object variable reference to the USDACommonMeasure worksheet, and removes any AutoFilter that may have been applied.

```
'Clear ListBox
Me.lstCommonMeasures.Clear

'clean any auto filter in action
Set ws = Sheets("USDACommonMeasures")
ws.AutoFilterMode = False
```

It then finds the last USDACommonMeasure worksheet used row using the UsedRange.Rows.Count property, sets the range address to be filtered, and uses the selected food item lstFoodItems ListBox as the Range.Autofilter Criteria1 argument to filter the entire worksheet data to just the selected food item common measures (if any).

```
'Find last used row of USDACommonMeasures sheet tab
lngRows = ws.UsedRange.Rows.Count
```

```
'Filter common measures for selected food item (including headers)
Set rg = ws.Range("$A$5:$E$" & lngRows)
rg.AutoFilter 1, Me.lstFoodItems
```

To fill the lstCommonMeasures ListBox with all possible food item common measures, the procedure executes a For Each...Next loop to select each range address returned by the Areas collection of Range.AutoFilter.SpecialCells. Note that it discards the first area (row 5) and uses an internal For...Next loop to run through all rows of each returned Area range. The rgFilter object variable captured on each passage through the For Each... Next outer loop has a Cells( ) collection of all filtered worksheet cells, which is used to fill the lstCommonMeasures ListBox using the AddItem method and Column property.

```
'Loop through all filtered range names
For Each rgFilter In rg.SpecialCells(xlCellTypeVisible).Areas
    'Discard row 5 (column names)
    If rgFilter.Row > 5 Then
        For intI = 0 To rgFilter.Rows.Count - 1
            With Me.lstCommonMeasures
                .AddItem rgFilter.Cells(intI + 1, 2)
                .Column(1, intI) = rgFilter.Cells(intI + 1, 3)
                .Column(2, intI) = rgFilter.Cells(intI + 1, 4)
            End With
        Next
```

When all food item common measures have been recovered, the procedure verifies whether there is at least one common measure inserted in the lstCommonMeasures ListBox, selects the first item in the list, and fills the lblCommonMeasures.Caption property with the number of possible food item common measures.

```
If Me.lstCommonMeasures.ListCount > 0 Then
    Me.lstCommonMeasures.ListIndex = 0
End If
Me.lblCommomMeasures.Caption = Me.lstCommonMeasures.ListCount & " available Common Measures
(please select)"

'Remove USDACommonMeasures autofilter
ws.AutoFilterMode = False
```

Returning the Selected Food Item

The frmSearchFoodItems UserForm uses a simple strategy to return the selected food item on its lstFoodItems ListBox to the calling worksheet.

- It has a Public Property CallingSheet that stores a reference to the calling worksheet as an Object variable, declared this way on frmSelectFoodItem UserForm.

```
Option Explicit
...
Dim mWks As Object

Public Property Let CallingSheet(ByVal wks As Worksheet)
    Set mWks = wks
End Property
```

- If the mwks as Object variable has a calling Worksheet object reference, it fills an array variable with the selected food item name, food category, selected common measure quantity, and name, and it returns this array to a supposed SelectedFoodItem public property on the calling worksheet, which is bound to a Variant module-level variable, declared on the worksheet code module. The Find Food Item sheet tab has such a variable procedure declaration.

```
Option Explicit
Dim SelectedFoodITem As Variant
```

It works this way: when you click the Find Food Item button of the Find Food Item sheet tab, the Sub SelectFoodItem( ) procedure is called, executing this code:

```
Public Sub SelectFoodItem()
    Dim frm As New frmSearchFoodItems
    Const conC = 3

    If Selection.Column = conC Then
        With frm
            .CallingSheet = Me
            .Show vbModal
        End With
        If IsArray(mavarFoodItem) Then
            With Application.Selection
                .Value = mavarFoodItem(0)
                .Offset(0, -1) = mavarFoodItem(1)
                .Offset(0, 1) = mavarFoodItem(2)
                .Offset(0, 2) = mavarFoodItem(3)
            End With
        End If

    Else
        MsgBox "Click on any cell of Food Item column and try again!", _
                vbInformation, _
                "Select a Food Item"
    End If
End Sub
```

The SelectedFoodItem( ) procedure declares the frm as New frmSearchFoodItem object variable, passes to it the CallingSheet( ) property a self-reference (using the VBA keyword Me), and then shows the frmSearchFoodItems UserForm modally.

```
Public Sub SelectFoodItem()
    Dim frm As New frmSearchFoodItems
    ...
        With frm
            .CallingSheet = Me
            .Show vbModal
```

Selecting a Food Item on the UserForm

This last instruction (`.Show vbModal`) will stop the code execution until the `frmSearchFoodItems` `UserForm` instance is closed. If you select a food item on the `lstFoodItems` `ListBox` and click the Select command button, the `cmdSelect_Click( )` event of the `UserForm` will fire, executing this code:

```
Private Sub cmdSelect_Click()
    Dim avarFoodItem(3) As Variant
    Const conErrObjectDoesntSupportMethod = 438

    On Error GoTo cmdSelect_Error

    If Not mWks Is Nothing Then
        'Selected food item name
        avarFoodItem(0) = Me.lstFoodItems
        'Selected Food category
        avarFoodItem(1) = Me.lstFoodItems.Column(1, Me.lstFoodItems.ListIndex)
        If Not IsNull(Me.lstCommonMeasures) Then
            'Qty and Common measure
            avarFoodItem(2) = Me.lstCommonMeasures.Column(2, Me.lstCommonMeasures.ListIndex)
            avarFoodItem(3) = Me.lstCommonMeasures.Column(1, Me.lstCommonMeasures.ListIndex)
        Else
            avarFoodItem(2) = Me.lstFoodItems.Column(2, Me.lstFoodItems.ListIndex)
            avarFoodItem(0) = "g"
        End If
        mWks.SelectedFoodITem = avarFoodItem
    End If

cmdSelect_End:
    Unload Me
    Exit Sub
cmdSelect_Error:
    Select Case Err
        Case conErrObjectDoesntSupportMethod
            MsgBox "The calling worksheet '" & mWks.Name & "' must have a Public Property
            SelectedFoodItem( ) procedure!", _
                    vbCritical, _
                    "SelectedFoodItem property not found on active sheet"
        Case Else
            MsgBox "Error " & Err & ":" & Error(Err), vbCritical, "Error on Select Food
Item"
    End Select
    Resume cmdSelect_End
End Sub
```

Note that the code declares the `avarFoodItems(3) as Variant` array variable and the `conErrObjectDoesntSupportMethod = 438` constant error (to avoid the appearance of a *magic number* in the code), activates the error trap, and then verifies whether the `mWks` module-level object variable has some object reference on it, comparing it to the keyword `Nothing`.

```
Private Sub cmdSelect_Click()
    Dim avarFoodItem(3) As Variant
    Const conErrObjectDoesntSupportMethod = 438

    On Error GoTo cmdSelect_Error

    If Not mWks Is Nothing Then
```

If there is such an object reference on the mWks object variable, the mavarFoodItems array is filled with the selected food item data, containing the food item name, food category, common measure quantity, and name, in this index order. Note that since some food items don't have any common measures stored in the USDACommonMeasures sheet tab, when this happens, nothing will be selected in the lstCommonMeasure ListBox, and it must get this information from the lstFoodItems ListBox.

```
If Not mWks Is Nothing Then
    'Selected food item name
    avarFoodItem(0) = Me.lstFoodItems
    'Selected Food category
    avarFoodItem(1) = Me.lstFoodItems.Column(1, Me.lstFoodItems.ListIndex)
    If Not IsNull(Me.lstCommonMeasures) Then
        'Qty and Common measure
        avarFoodItem(2) = Me.lstCommonMeasures.Column(2, Me.lstCommonMeasures.ListIndex)
        avarFoodItem(3) = Me.lstCommonMeasures.Column(1, Me.lstCommonMeasures.ListIndex)
    Else
        avarFoodItem(2) = Me.lstFoodItems.Column(2, Me.lstFoodItems.ListIndex)
        avarFoodItem(0) = "g"
    End If
```

Then it tries to return this array to the supposed worksheet's SelectedFoodItem property.

```
mWks.SelectedFoodITem = avarFoodItem
```

This is the critical moment where the code may fail, if the calling worksheet doesn't have the supposed property. If this happens, VBA error = 438 ("Object doesn't support this property or method") will be raised, and the code will jump to the error trap and will issue a warning message to the user:

```
cmdSelect_Error:
    Select Case Err
        Case conErrObjectDoesntSupportMethod
            MsgBox "The calling worksheet '" & mWks.Name & "' must have a Public Property
            SelectedFoodItem( ) procedure!", _
                   vbCritical, _
                   "SelectedFoodItem property not found on active sheet"
        Case Else
            MsgBox "Error " & Err & ":" & Error(Err), vbCritical, "Error on Select Food Item"
    End Select
    Resume cmdSelect_End
End Sub
```

If the calling worksheet's `SelectedFoodItem` property will receive the `avarFoodItem` array variable, the `UserForm Terminate( )` event will fire, and the code will return to the calling procedure.

Processing the Selected Food Item on the Worksheet

Now the code returns to the `Find Food Item` sheet tab's calling procedure, which will verify whether the `mavarFoodItem` variable has an array reference. If this is true, the array content is processed, associating its expected content to the appropriate worksheet cells. Note that this is done using the `Application.Selection` property to set a reference to the worksheet cell selected before calling the `UserForm`, using the `Range.Offset` method to return the appropriate results to the desired worksheet cells.

```
With frm
    .CallingSheet = Me
    .Show vbModal
End With
If IsArray(mavarFoodItem) Then
    With Application.Selection
        .Value = SelectedFoodITem (0)
        .Offset(0, -1) = SelectedFoodITem (1)
        .Offset(0, 1) = SelectedFoodITem (2)
        .Offset(0, 2) = SelectedFoodITem (3)
    End With
End If
```

Researching for a Selected Food Item

Once a food item is selected and returned to the desired cell of the `Find Food Item` worksheet sheet tab, chances are that you need to change its default common measure. You can force the automatic research for the selected food item using the `Application.Selection` property to get its name and automatically search for it on the `frmSearchFoodItems_Initialize( )` event.

Figure 6-40 shows that there is a food item selected on cell `C6` of the `Find Food Item` worksheet ("Babyfood, dessert, blueberry yogurt, strained"), and if you click the Find Food Item button, `frmSearchFoodItems` will automatically search for it, showing all its available common measures (if any).

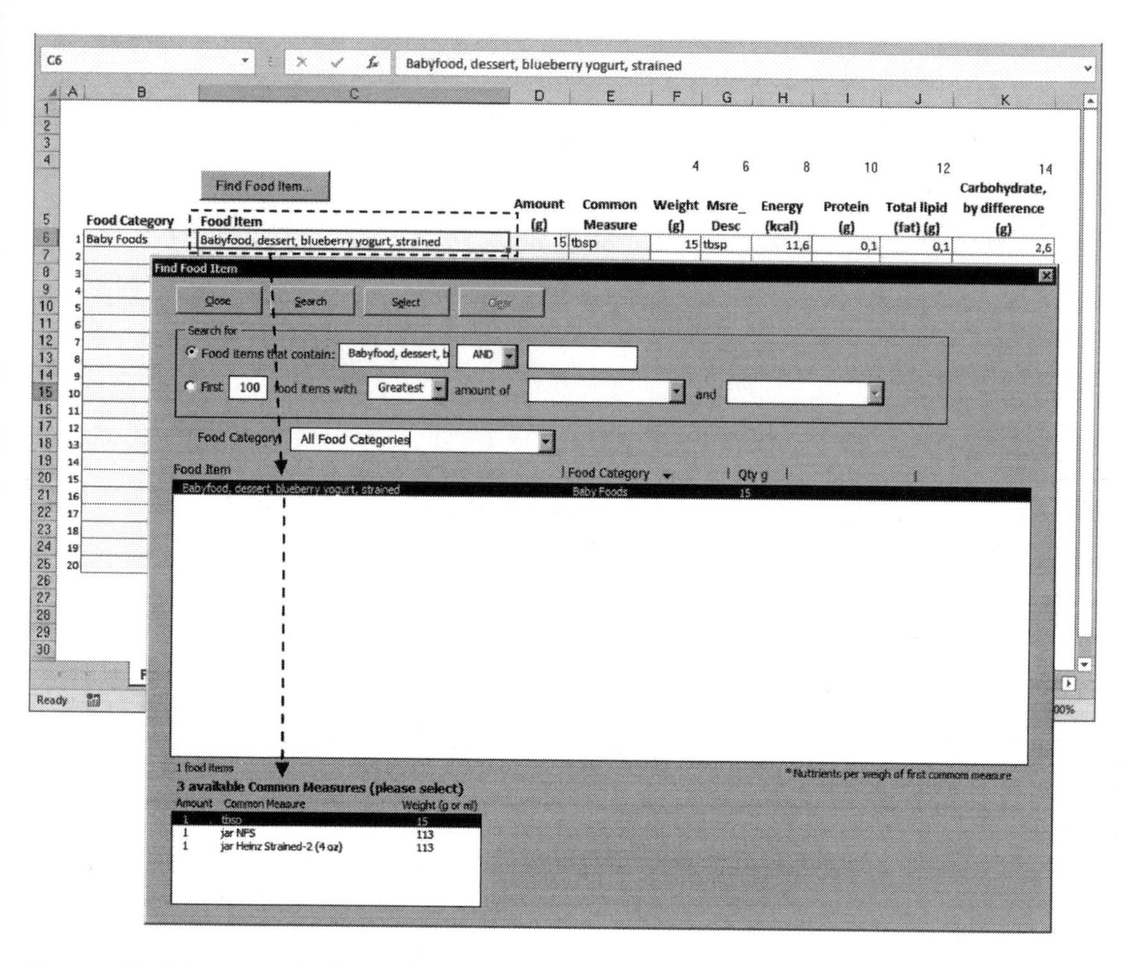

Figure 6-40. If there is any food item selected on the calling worksheet, the frmSearchFoodItems UserForm will automatically research it and show all its available common measures

This is done on the `frmSearchFoodItems_Initialize( )` event that has these last lines of code:

```
Private Sub UserForm_Initialize()
    ...
    'Look if any food item had been selected on Dietary Planner or New Recipe interface
    strFoodITem = Application.Selection.Cells(1, 1)
    If Len(strFoodITem) > 0 Then
        Me.txtFind = strFoodITem
        Call cmdSearch_Click
        Me.lstFoodItems.ListIndex = 0
        ShowCommonMeasures (strFoodITem)
    End If

Initialize_End:
    Exit Sub
```

As you can see, it uses `Application.Selection.Cells(1,1)` to return the first selected cell of any merged range to the `strFoodItem String` variable (this is necessary because if the selected input cell is merged, VBA can return an empty string). If there is anything stored in the selected cell, this value is set to the `txtFind TextBox` (which will cascade-fire the `txtFind_Change( )` event), and a call to `cmdSearch_Click( )` is made. Since this is a precise food item name, just one food item will be returned to `lstFoodItems`, which will be selected (`lstFoodItems.ListIndex = 0`) and passed as an argument to the `ShowCommonMeasures( )` procedure, where all its possible common measures will be recovered (if any).

Attention All `Application` object properties and methods don't need to be prefixed by the `Application` object. Instead of using `Application.Selection`, you can use just `Selection` on the code, and it will work as expected.

When the `lstFoodItems.ListIndex = 0` property is changed to select the desired food item, this will cascade-fire the `lstFoodItems_Click( )` event. For an unknown reason, when this is done on the `UserForm_Initialize( )` event, the `lstFoodItems` value will return an empty string (""), although it has the food item name as the default control value. That is why the code made a second call to the `ShowCommonMeasures( )` procedure, passing as an argument the `Selection` value: to guarantee that all selected food item common measures will be returned.

And that is all you need to know about the `frmSearchFoodItem` interface!

Chapter Summary

In this chapter, you learned about the following:

- How you can execute most Excel Range properties and methods from the VBA Immediate window
- How to define a range object variable with VBA to refer to a large table, such as the USDA food table of nutrients
- How to use the `Range.End` property to navigate through a range of cells
- How to automatically select a range address using the `Range.CurrentRegion` property
- How to programmatically move through a range of cells using the `Range.Cells` collection or the `Range.Offset` property
- How to programmatically create the USDA range name to encompass all USDA food items and their nutrients
- How to sort a range name with the `Range.Sort` method
- How to create all food category range names on the USDA sheet tab using VBA code
- How you can use the `Range.Find` and `Range.FindNext` methods to find information on any range name or range address
- How to implement a `UserForm` interface to search information inside a big worksheet

- How to use the bubble sort algorithm to sort data inside a `ListBox` control
- How to implement with VBA a `ListBox` control with variable column widths using VBA
- How you can find items on a worksheet using the `Range.AutoFilter` method and the `Worksheet.AutoFilterMode` property
- How to use the `Range.SpecialCells` property
- How to use the `Range.Areas` collection
- How to find food items using two different nutrient values and four different sort methods
- How to create a unified food item search `UserForm` that can search food items by their name or nutrient content
- How to pass a worksheet reference to a `UserForm` property and use it to return `UserForm` information that will be used by the worksheet application
- And a lot of other programming techniques that will turn you into an Excel expert programmer

In the next chapter, you will learn how to apply the VBA knowledge gained so far to create a database management system to store worksheet data inside unused worksheet cells, enhancing the usefulness of your worksheet applications.

CHAPTER 7

Using Excel as a Database Repository

Microsoft Excel was conceived to produce tables of calculated data, and because of its ability to easily produce calculated worksheets to manage business information, many people also use it as a way to store precious business data.

This data storage happens by either making a workbook file copy, managing the data using some file name strategy, making copies of a predefined sheet tab inside the workbook, or creating sets of similar data inside the same Excel workbook file.

Using all the VBA knowledge you've built up so far and a good code strategy, you can implement a data-saving/retrieving system to manage many thousands of sheets of data inside each sheet tab.

This chapter is intended to teach you how to implement such a worksheet database storage system inside any Excel sheet tab; it covers the strategy, the technique, and the action. You can obtain all files and procedure code in this chapter by downloading the `Chapter07.zip` file from the book's Apress.com product page, located at `www.apress.com/9781484222041`, or from `http://ProgrammingExcelWithVBA.4shared.com`.

The Worksheet Database Storage System

Most worksheet applications have a kind of *one-to-many* data record relationship in their implementation. The *one-side* record identifies the aim of the worksheet data, while the *many-side* records store associated data used by the one-side worksheet record.

As an example, we can cite the many "invoice" worksheet templates available on Office.com, which can be easily download by selecting File ➤ New ➤ Invoices in the Microsoft Excel menu. All these invoice templates have in common the fact that they represent a one-to-many record relationship (each *one* invoice record has many record-related invoice items), needing two different worksheet places to store the data.

- A main place used to store invoice identification data (the *one side* of the *one*-to-many data record relationship), such as invoice number, date, customer ID, due date, and billing data information, that represents the purpose of the sheet tab
- A secondary place, used to store the invoice item data using as many rows and columns as needed to store item details (the *many side* of the one-to-many data record relationship) such as item ID, item description, quantity, unit price, and so on

F. Morgado, *Programming Excel with VBA*, DOI 10.1007/978-1-4842-2205-8_7

To create invoices whose data is stored inside Excel worksheet solutions, you must do one of the following:

- *Use an Excel workbook as a template to create multiple Excel invoice files*: Using this strategy, you will need to create a new workbook based on the invoice template for every new invoice needed, and you will end up with a lot of different Excel files to store each invoice's data.
- *Use an Excel sheet tab as a template to create multiple invoice sheet tabs*: By following this strategy, you will need to keep a clean invoice sheet tab as a template inside the workbook and copy it to another sheet tab for every new invoice needed. This way, you end up with just one Excel workbook file storing as many sheets as can fit into the available memory of your computer.

Think about it: instead of creating multiple Excel files or sheet tabs to store the same kind of data, can you create VBA code to copy all invoice data to unused sheet tab rows/columns and save, load, or delete them whenever you need, turning each sheet tab into a data repository system?

Sure you can!

Using this approach and knowing that any Excel `.xls*` file created with version 2007 or newer has up to 1,048,576 available rows and 1,024 columns, you can probably store tens of thousands of worksheet invoice data records on a single sheet tab!

■ **Attention** This is an Excel programming book, and as such] it will show you how to use Excel as a simulated database repository. Although you can use it this way, an Excel worksheet is *not* the recommended way to do such things. It will work but will always be a fragile and incorrect data storage system. You should use a more robust storage system that uses a database file structure, such as Microsoft Access, to do such tasks.

The BMI Companion Chart

To implement the worksheet database storage system, this chapter will use as its first example `BMI Companion Chart.xlsx` and `BMI Companion Chart with Data.xlsx` (which can be extracted from `Chapter07.zip`). These are worksheet applications that have a single sheet tab (`BMI Chart`) created to allow any adult person (whose height does not change with time) to control their weight for up to 20 successive weeks. It offers input cells to insert the person's name and height (the one-side record) and another input cell sequence that allows type pairs of dates/measured weights to control the person's weight tendency by calculating the body mass index (BMI) through consecutive weeks or months with the aid of a beautiful Excel chart (Figure 7-1).

The `BMI Companion Chart.xlsx` worksheet application formats its input cells with a light yellow background and a blue border. Cells that have a white background and a blue border return calculated data. Grid lines and headers are hidden, and the worksheet is protected (with no password) allowing just unlocked cells (the input cells with a yellow background) to be selected. All unused rows and columns are also hidden, leaving a big gray area surrounding its interface if the user tries to navigate the worksheet using Excel scroll bars.

Note how the one-to-many record relationship principles apply to relate the worksheet data (*one* person record has *many* dates/weights measures).

- The *one-side* record is used to store just three input cells: the person name and height (in feet and inches; it is assumed that adults do not change height in a reasonable amount of time).
- The *many-side* records are the ones used to store the pairs of date/weight (in pounds) data associated to each person.

■ **Attention** The main difference between a one-to-many record relationship implemented in a database and one implemented in a worksheet is that the first can have an undefined number or records in its many-side records, while the second—Excel applications—*must* limit the many-side record count.

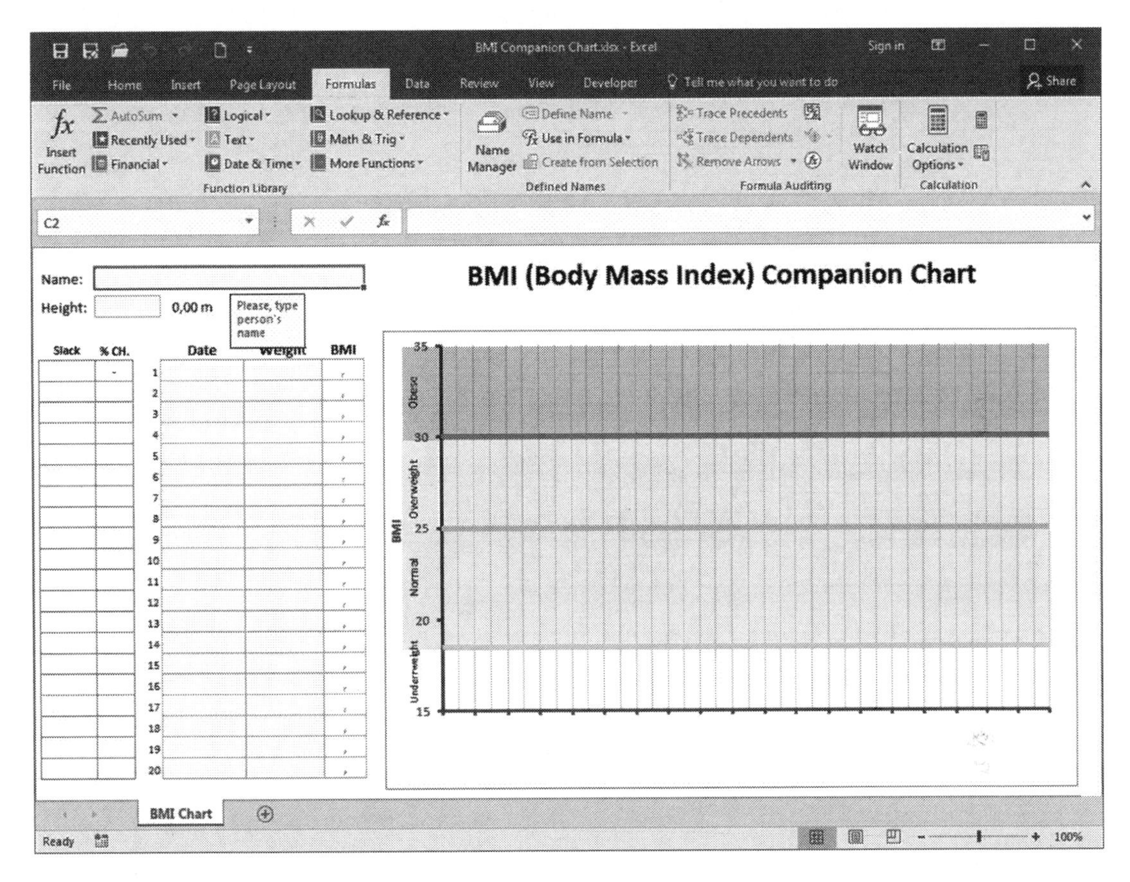

Figure 7-1. This is the BMI Companion Chart.xlsx Excel application. It allows anyone to type some personal data and then control their weight using some simple calculations and an Excel chart

If you fill it with some personal data (name, height, dates, and weights), the person's BMI will be calculated for each date and plotted on the chart to the right. The worksheet will also calculate the person's weight slack in the Slack column (which means how many pounds one must lose/gain to enter/exit the safe, green BMI chart area to achieve a normal weight), and percentage of weight change in the %CH column (which is the percentage of weight loss/gain over two consecutive weight measures). A color scheme using green to indicate safe, blue to indicate loss, and red to indicate gain/danger is used to conditionally format the Slack, %CH, and BMI calculated data (Figure 7-2).

■ **Attention** The `BMI Companion Chart.xlsx` Excel application also uses data validation to give warning messages and avoid the insertion of consecutive dates that are less than seven days apart.

To produce another BMI chart for the same or another person, you must do the following:

1. Right-click the `BMI Chart` tab, select Move or Copy, and create a new copy of the sheet tab.

2. Use another workbook file.

You can take a better approach to this dilemma by storing just the input cell data in the worksheet unused cells and use VBA to load and save it on the worksheet.

Supposing that you want to store Figure 7-2 data on the `BMI Chart` sheet tab, you can devise a single strategy to do it using all the hidden rows below the visible worksheet area. Figure 7-3 shows a diagram that explains how this can be done on the `BMI Chart` sheet tab to store the worksheet data.

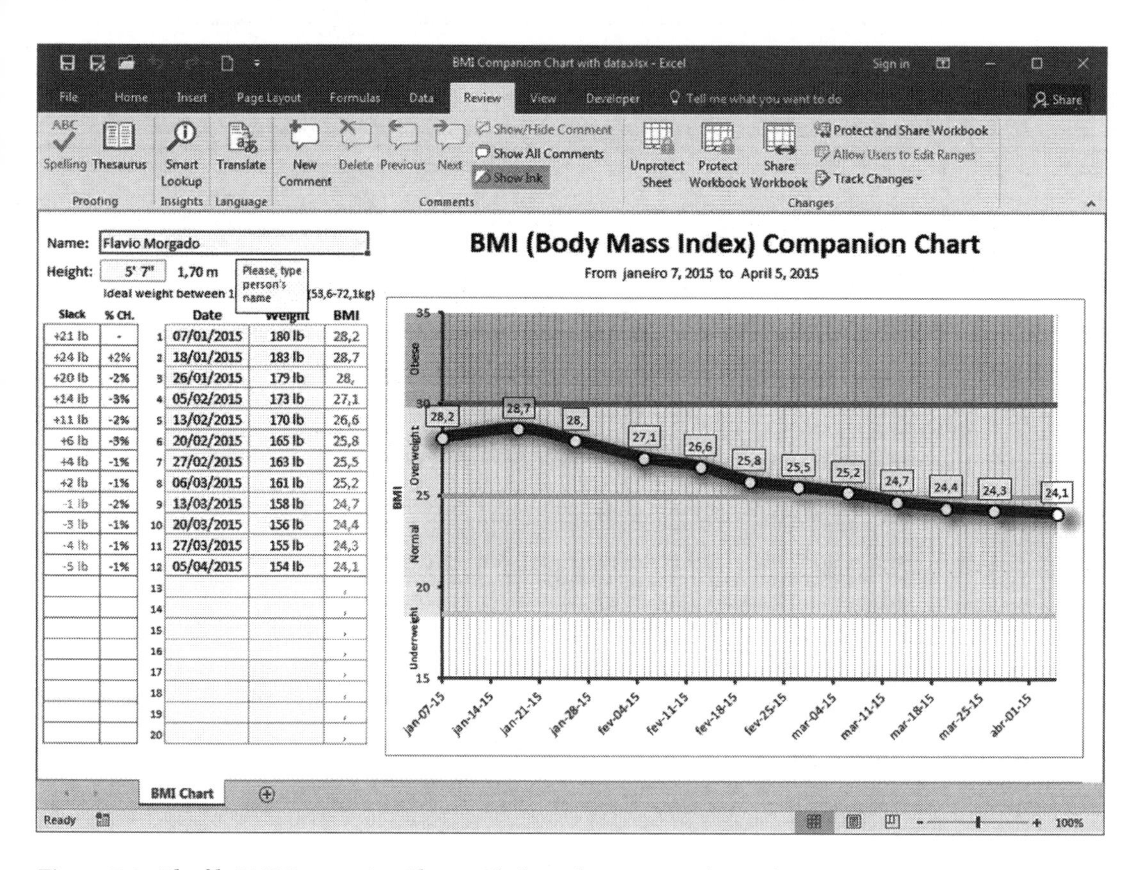

Figure 7-2. The file BMI Companion Chart with data.xlsx uses a traditional one-to-many database relationship to store the data. The one-side record uses three cells to store the person's name and height (in foot and inches), while the many-side records store pairs of the date and weight measured. An Excel chart plots the weight gain/loss using the BMI calculated for each date

This strategy is based on the following:

1. Reserve a single place to identify the name associated to each worksheet data record. This place will be represented by a range name (`mconSavedRecords`), and this range name will be used to fill a data validation list, which can indicate/select each worksheet record saved.

2. Reserve another place to store the *one side* of the *one*-to-many data relationship. This place will use just one row and as many columns as needed to store data in the one-side record cells. Regarding the `BMI Chart` sheet tab, this place will use just three cells, spread by three different worksheet columns, to store the person's name and height in feet and inches.

3. Reserve one more place to store the *many side* of the one-to-many data relationship. This place will use as many rows and columns as needed to store the worksheet detail records. Regarding the `BMI Chart` worksheet data structure, this place will use 2 columns and 20 rows for each worksheet details record (one column for the dates and other for the weights).

Looking at Figure 7-3, you can note that it takes at least 21 worksheet rows to store each `BMI Chart` worksheet data (20 rows for each pair of date-weight data and 1 extra row to separate two successive charts' data). Considering that each Excel 2007 or later version `.xlsx` workbook presents worksheets that have up to 1,048,576 rows, a single `BMI Chart` worksheet tab can store about 1,048.576/21 = 49,932 BMI chart application records using just two of its columns! That is a lot of records to be stored on a single sheet tab, huh?

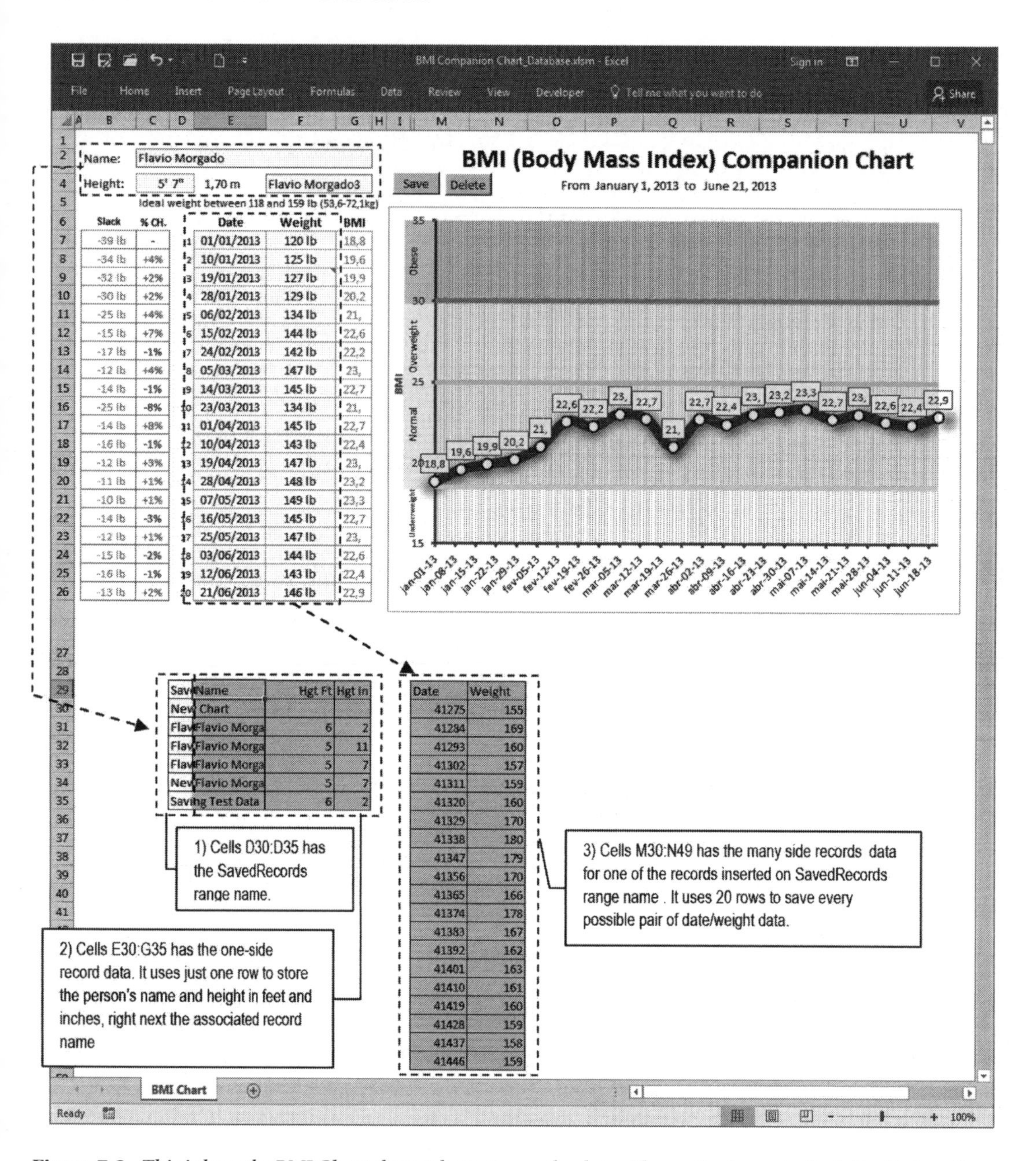

Figure 7-3. *This is how the BMI Chart sheet tab must save the data. The vertical area to the left will store the file record names and the one-side record data, using one row for each saved record. The vertical area to the right will store the many-side worksheet records using 2 columns and 20 rows*

The BMI Companion Chart_Database.xlsm Excel Application

Since an image is worth a thousand words, I want to introduce you to the `BMI Companion Chart_Database.xlsm` Excel macro-enabled workbook. This is an Excel application that you can extract from the `Chapter07.zip` file that uses the strategy briefly discussed in the previous section to store `BMI Chart` worksheet data in its hidden rows. Figure 7-4 shows how it looks when you open it for the first time and click its `CurrentRecord` range name data validation list to show the saved data.

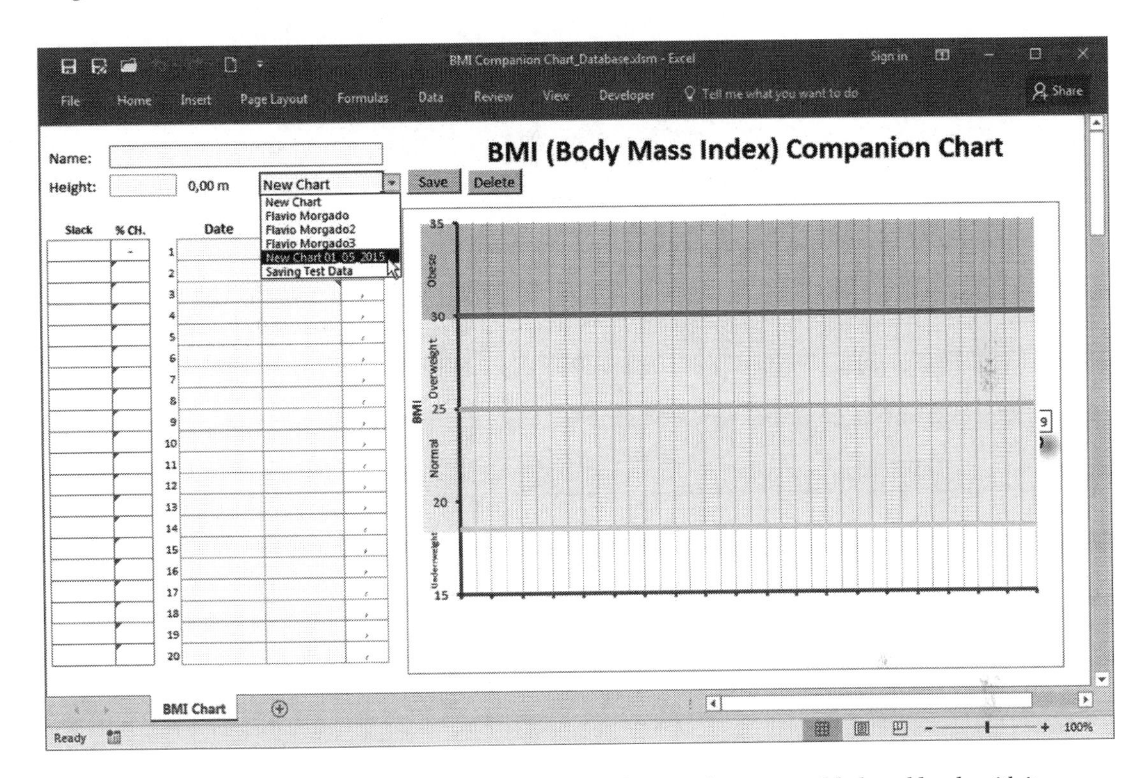

Figure 7-4. *This is the BMI Companion Chart_Database.xlsm Excel macro-enabled workbook, with its CurrentRecord range name data validation list showing all chart data already saved in the worksheet's hidden rows*

If you select any item in the `CurrentRecord` data validation list, the data will be loaded into the worksheet, which will calculate and plot the person's BMI evolution (Figure 7-5).

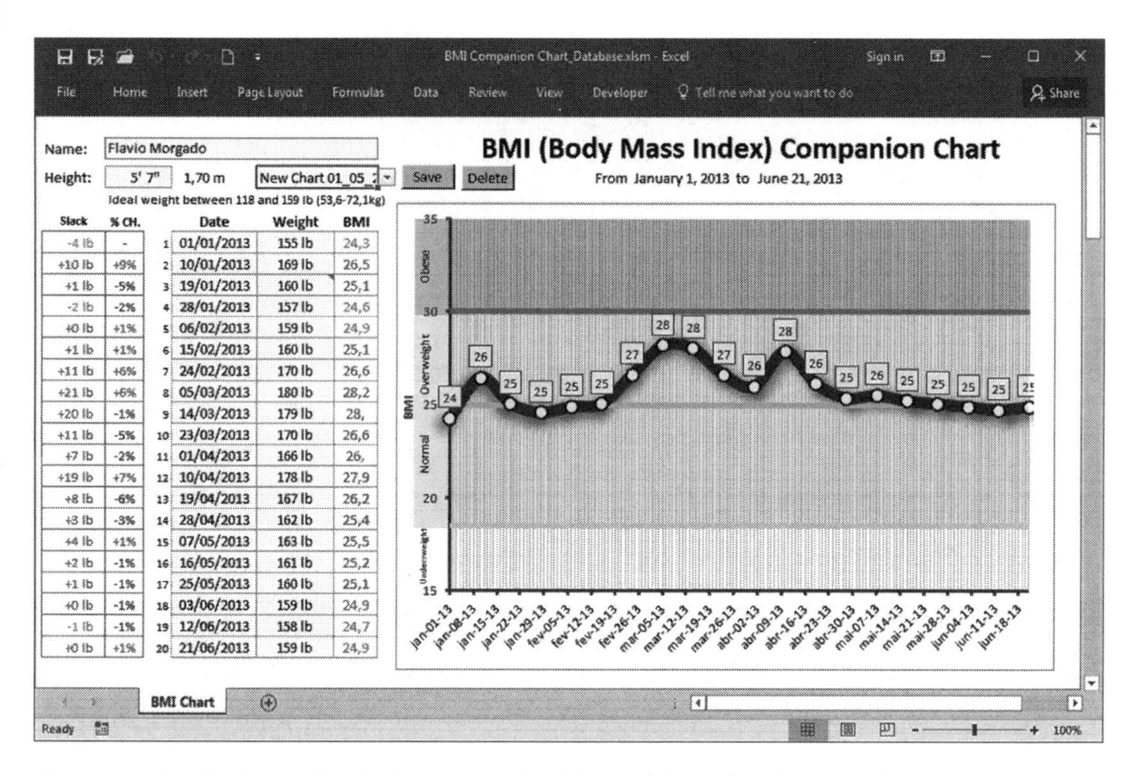

Figure 7-5. *By selecting any item in the CurrentRecord data validation list, the data will be loaded in the appropriate worksheet cells. The worksheet will calculate and plot the person's BMI evolution*

You may continually select another chart data in the `CurrentRecord` data validation list and watch how fast the data is loaded and the chart is changed. To clean up and begin another `BMI Chart` worksheet, select New Chart in the data validation list. And if you want, you can also change any chart data and click Save to save the chart with the same or another name. Or you can select Delete and delete all the selected chart data from the workbook file.

Let's see how this works by first understanding the worksheet mechanism; follow the next steps:

1. Unprotect the `BMI Chart` sheet tab (Review tab ➤ Unprotect).
2. Show the row/column headers (View tab ➤ check Headings).
3. Unhide all `BMI Chart` hidden rows by doing the following:
    - Click row `27` header and drag it down until you see the row `1048576` indicator.
    - Right-click row `27` header and choose Unhide in the context menu.

Figure 7-6 shows the `mconSavedRecords` range name, beginning on cell `D30`, where all chart data record names are stored, the associated *one-side* record area (containing the person's name and height in feet and inches), and the first block of the many-side records that store date/weight pairs of data. It also shows how the `CurrentRecord` range name data validation list is filled with data from the `mconSavedRecords` range name.

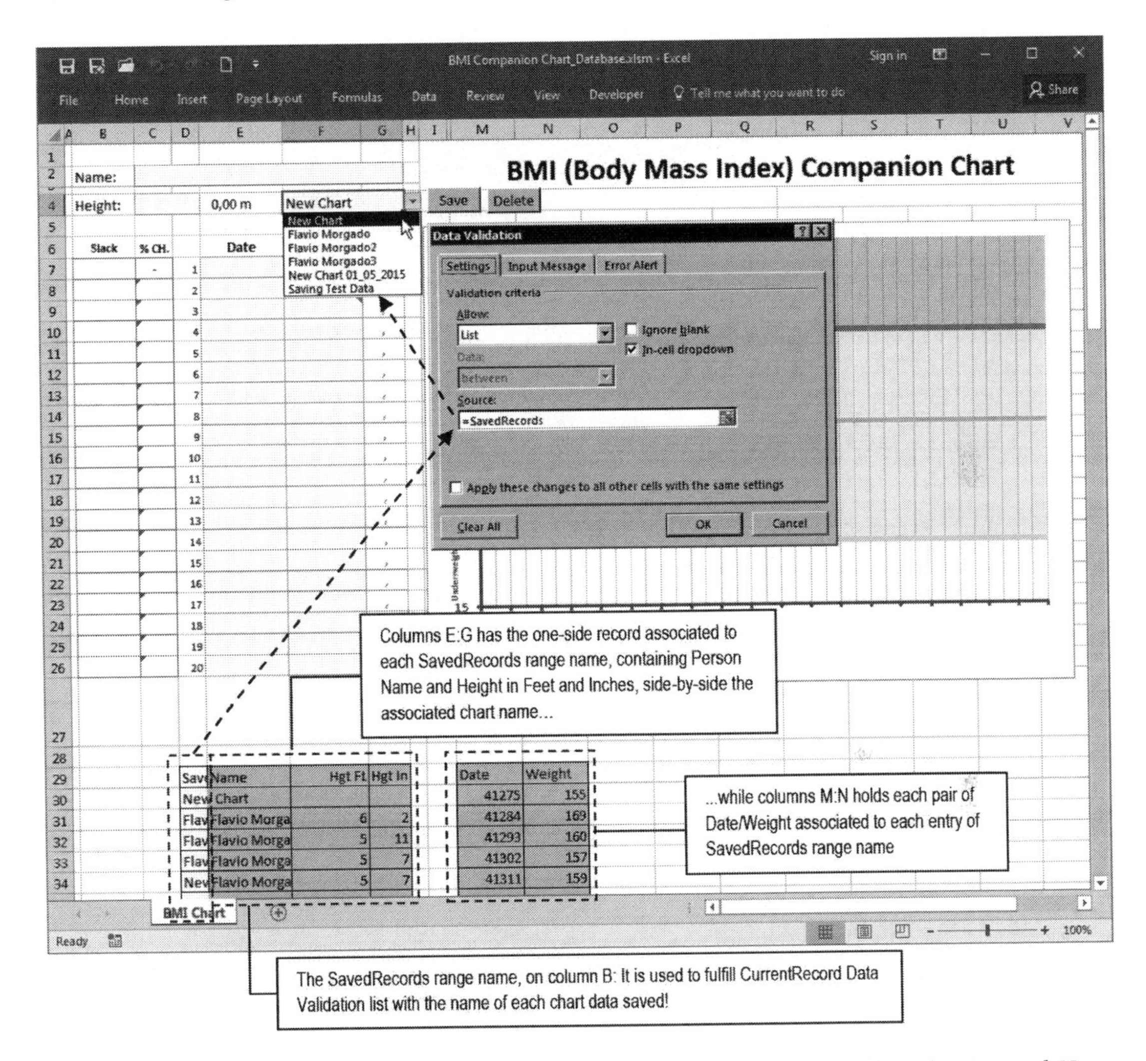

Figure 7-6. *When all BMI Chart worksheets are shown, you will see where the BMI Chart data is stored. Note that the CurrentRecord range name data validation list uses the mconSavedRecords range name to fill its list of saved chart data*

The saving process of any `BMI Chart` worksheet data works this way every time you click the worksheet Save button:

1. The user is asked to give a name to the chart data. If this is a new BMI chart, it will offer the default name `New Chart mm/dd/yyyy`, where `mm/dd/yyyy` is the system date. Otherwise, it will offer the current record name (chart selected in the `CurrentRecord` data validation list).
2. The accepted name will be searched inside the `mconSavedRecords` range name. If it already exists, its row will be selected. Otherwise, the new name will be added to the `mconSavedRecords` range name (so it appears in the `CurrentRecord` range name data validation list).
3. The person name and height in feet and inches are then stored to the right and on the same row of the saved name (columns `E:G`), creating the one-side record of the one-to-many record relationship.
4. A new range name will be created to represent the first entry of all 20 possible pairs of date/weight data (the many side of the one-to-many record relationship) on the next available row of column `M`. This range name will be named as `rec_` followed by the chosen chart name stored inside the `mconSavedRecords` range name. (For example, a chart accepted to be saved as New Chart 5/5/2015 will have an associated range named `rec_New_Chart_5_5_2015`).
5. All 20 pairs of date/weight data are then retrieved for the associated range name on column `M`.

It worth noting that all `BMI Chart` range names have their `Visible` property set to `False`, so they *do not* appear in the Excel Name box or Name Manager dialog box. To help you see what cell each range name is associated to, I also imported a copy of the `frmNames` and `frmEditNames UserForms`, created in Chapter 5.

To use `frmNames` and change any range name's `Visible` property, follow these steps:

1. Press Alt+F11 to open the Visual Basic environment.
2. Double-click the Forms option in the VBA Project Explorer tree to expand it.
3. Double-click the `frmNames` entry to show the `frmNames UserForm` of the VBA editor.
4. Press F5 to run `frmNames` and see all the available range names inside the `BMI Chart` worksheet (Figure 7-7).

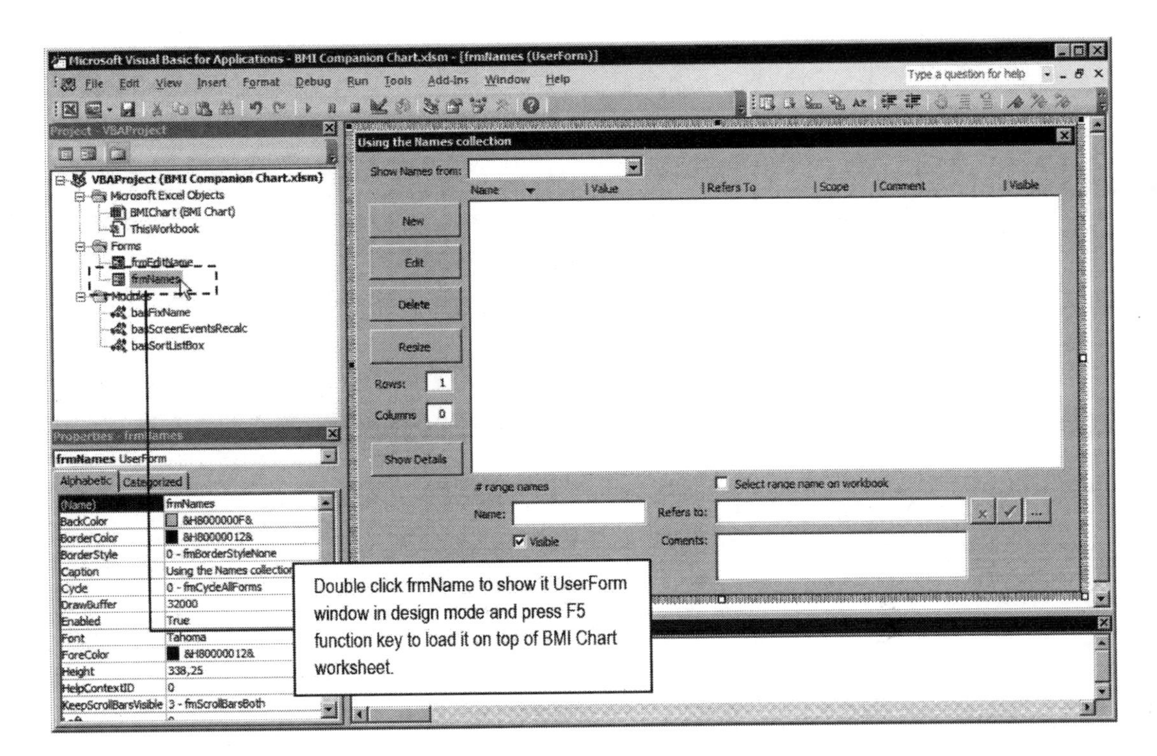

Figure 7-7. Press F11 to open the VBA environment, and double-click the frmNames in the Forms option of the Project Explorer tree to show the UserForm in design mode

5. Double-click any range name (or select it in the list and click the Show Details button) and check the Visible check box to make it visible in the Excel interface (Figure 7-8).

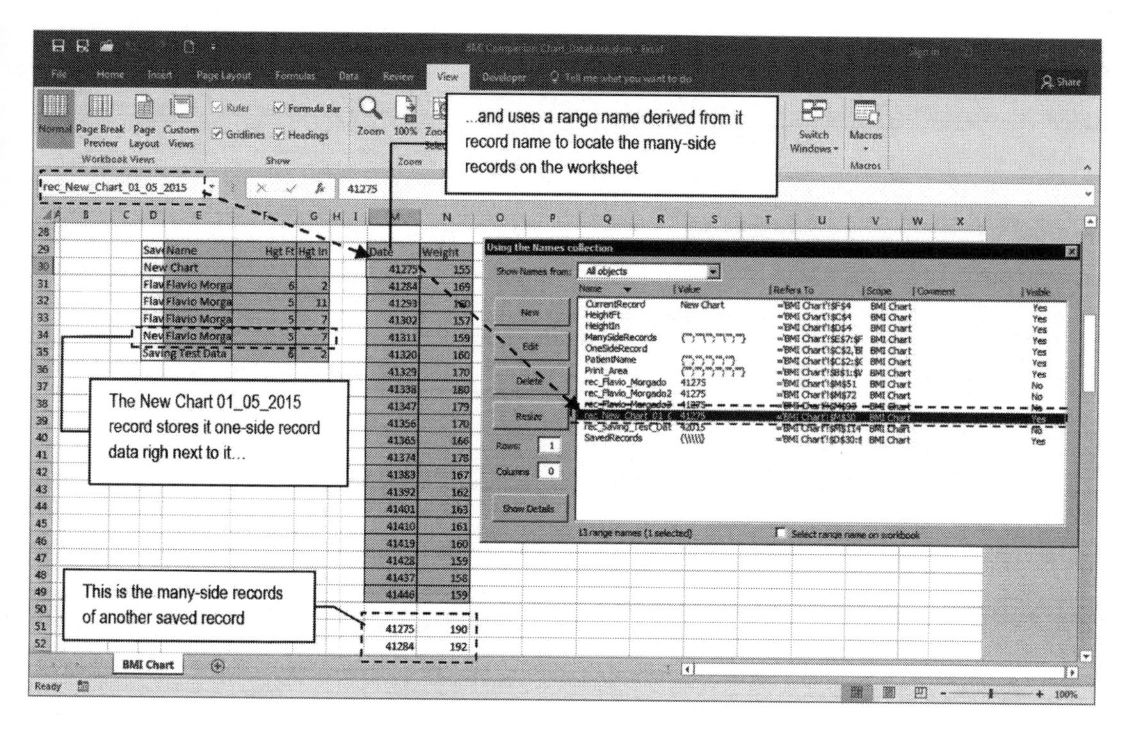

Figure 7-8. *Select the desired range name in the frmNames list box and change its Visible property so you can see it in the Excel Name box*

Figure 7-8 shows that the chart data saved as `New Chart 01_05_2015` in cell `D34` for the `SavedRecords` range name stores its one-side record to its right and uses the range name `rec_New_rec_01_05_2015` to indicate where in the worksheet its many-side records begin.

Every other worksheet data saved on the `BMI Chart` sheet tab obeys this principle.

1. Store a name entry inside the `SavedRecords` range name to represent the `BMI Chart` data.
2. Store the one-side record to its right, using as many columns as needed to do it.
3. Store the may-side records on another worksheet place associated to a range name that is unequivocally derived from the name cited in step 1.

Now is where the fun begins; let's understand how this works using VBA code!

Parameterization of BMI Chart Data

The `BMI Chart` worksheet has its `CodeName` property set to `BMIChart` in the VBA Properties window, so it can be easily accessed in VBA code. It declares four module variables (`mwb`, `mws`, and `mstrLastRecord`) and one private enumerator (`Operation`) to help control worksheet data changes and avoid the appearance of magic numbers on the code.

```
Option Explicit
t
Private WithEvents mwb As Workbook
```

```
Private mws As Worksheet
Public Dirty As Boolean  'Indicate if record dat had been changed
Private mstrLastRecord As String 'Retain the name of current record
Private Enum Operation
    LoadRecord = 1
    SaveRecord = 2
End Enum
```

The Private WithEvents mwb as Workbook object variable is used to capture the Workbook object events, while Private mws as Worksheet is used to set a reference to the ActiveSheet object. The mwb object variable is used to allow you to save a changed record before the workbook is closed. They are instantiated on the BMIChart Worksheet object's SelectionChange( ) or Change( ) event.

```
Private Sub Worksheet_SelectionChange(ByVal Target As Range)
    Set mwb = ThisWorkbook
    Set mws = ActiveSheet
    ...
End Sub
Private Sub Worksheet_Change(ByVal Target As Range)
    Set mwb = ThisWorkbook
    Set mws = ActiveSheet
    ...
End Sub
```

The Public Dirty as Boolean variable acts as a worksheet property to signal whenever the record data is changed. As a public variable, it can be accessed using the Me.Dirty syntax inside the BMIChart code module and *must* be accessed using the BMIChart.Dirty syntax on every other VBA code module.

■ **Attention** Any Worksheet object's Public variables appear in a VBA code object list as a property, while any Worksheet object Public procedures appear as a method.

The Private mstrLastRecord as String variable is used to retain the name of the last loaded record, and since it is declared as Private, it can just be accessed inside the BMIChart code module.

It also declares 15 code module constants in its worksheet module's Declaration section to define special database engine values (parameters) needed to operate the save, load, and delete worksheet records operations (note that they all have the mcondb prefix: m = module scope; con = constant prefix; db = database engine value).

```
'This constants refers to local range names
Const mcondbDataValidationList = "CurrentRecord"         'Data Validation list range
Const mcondbSavedRecords = "SavedRecords"  'Saved records range name
Const mcondbRecordName = "Chart"            'Record name
Const mcondbOneSide = "OneSideRecord"       'One-side record range
Const mcondbOneSideColumsCount = 3          'One-side record columns needed
Const mcondbManySide1 = "ManySideRecords"   'Many-side1 records range
Const mcondbManySide2 = ""                  'Many-side2 records range
Const mcondbManySide3 = ""                  'Many-side3 records range
Const mcondbManySide4 = ""                  'Many-side4 records range
Const mcondbManySidePrefix = "rec_"         'Many-side range name prefix
Const mcondbManySideColumnsCount = 2        'Many-side record columns needed
Const mcondbManySideRowsCount = 21          'Many-side record rows needed (+ 1 blank row)
```

```
Const mcondbRecordsFirstRow = 30              'Row where database begins
Const mcondbManySideFirstColumn = "$M$"       'Many-side record first column
Const mcondbRangeOffset = 3                   'One-side record column offset to
mconSavedRecords
```

■ **Attention** Note that the `BMIChart` code module declares the constants `mcondbManySide1`, `mcondbManySide2`, `mcondbManySide3`, and `mcondbManySide4` to allow you to define up to four different one-to-many record relationships in the worksheet application.

These code module constants are associated to special values used by the database code engine, allowing them to be easily adapted by any other Excel worksheet application that needs to implement the same database storage system. All range names pointed to by these constants must be created as local worksheet range names. Table 7-1 describes each one of these module constants.

Table 7-1. *Constants Declared in the BMIChart Code Module Needed to Parameterize the Database Code Engine*

Constant Name	Purpose
`mcondbDataValidationList`	Range name where the record's data validation list resides
`mcondbSavedRecords`	Range name where the record's name is stored
`mcondbRecordName`	Name associated to each worksheet record (for user interaction)
`mcondbOneSide`	Range name for the worksheet one-side record cells
`mcondbOneSideColumnsCount`	Number of columns needed to save the one-side record data
`mcondbManySide1` to `mcondbManySide4`	Range names for up to four different worksheet many-side records relantionhips ranges
`mcondbManySidePrefix`	String prefix used to identify the range name associated to many-side record data
`mcondbManySideColumnsCount`	Indicates how many worksheet columns are needed to store the many-side records
`mcondbManySideRowsCount`	Indicates how many worksheet rows are needed to store the many-side records
`mcondbRecordsFirstRow`	Sets the worksheet first row where the database records start
`mcondbManySideFirstColumn`	Sets the worksheet column where the many-side records' storage starts
`mcondbRangeOffset`	Set how many columns must be offset between range `mcondbSavedRecords` and the first many-side record column

---

■ **Attention** This chapter will treat the `CurrentRecord` range name associated to the data validation list cell as the `mcondbDataValidationList` constant, in an attempt to acquaint you with the database constant that represents this record selection cell.

Changing BMI Chart Data

The Public Dirty as Boolean module variable works as a BMI Chart worksheet public property (BMIChart.Dirty) and is used to track any changes made by the user in the worksheet record data using the BMIChart object's Worksheet_Change( ) event, which has this code:

```
Private Sub Worksheet_Change(ByVal Target As Range)
    Set mwb = ThisWorkbook
    Select Case Target.Address
        Case Is = mws.Range(mcondbDataValidationList).Address
            'User is trying to load a new Record
            TryToLoadSelectedRecord
        Case Else
            'Sheet data has changed
            Me.Dirty = True
            If mws.Range(mcondbDataValidationList) = "New " & mcondbRecordName Then
                Application.EnableEvents = False
                    mws.Range(mcondbDataValidationList) = ""
                Application.EnableEvents = True
            End If
     End Select
End Sub
```

The Worksheet_Change( ) event receives the Target as Range argument indicating which cell has been changed by the user action, sets a reference to the ThisWorkbook object, and uses a Select Case statement to verify where the change happened. If the Target argument points to the data validation list cell address associated to the mcondbDataValidationList constant range name, it means that a record was selected to be loaded, and the code will call the TryToLoadSelectedRecord( ) procedure.

```
Select Case Target.Address
    Case Is = mws.Range(mcondbDataValidationList).Address
        'User is trying to load a new Record
        TryToLoadSelectedRecord
```

■ **Attention** The TryToLoadSelectedRecord( ) procedure will be analyzed in the section "Discarding Changes by Selecting Another mcondbDataValidationList Item" later in this chapter.

Otherwise, the change happens on any other possible worksheet input cell, leaving the current record in a *dirty* state, meaning that the BMIChart.Dirty property becomes True:

```
Case Else
    'Sheet data has changed
    Me.Dirty = True
```

The code verifies whether the current mcondbDataValidationList value was set to a "New " & mcondbRecordName value (which will be updated to "New Chart"), indicating that a new record has been selected by the user action. If this is true, Application.EnableEvents is set to False (to avoid cascade-firing the Worksheet_Change( ) event again), and the mcondbDataValidationList cell receives an empty space ("").

```
        If mws.Range(mcondbDataValidationList) = "New " & mcondbRecordName Then
            Application.EnableEvents = False
                mws.Range(mcondbDataValidationList) = ""
            Application.EnableEvents = True
        End If
```

That is why every time you select New Chart in the `mcondbDataValidationList` data validation list cell to receive a new record and change any input cell value, the data validation list cell value is cleared, giving a visual clue that the worksheet data is new, has been changed, or has not been saved yet (Figure 7-9).

■ **Attention** The code sets `BMIChart.Dirty = True` whenever the worksheet record data is changed. And on a new, unsaved record, the `mcondbDataValidationList` cell will become empty.

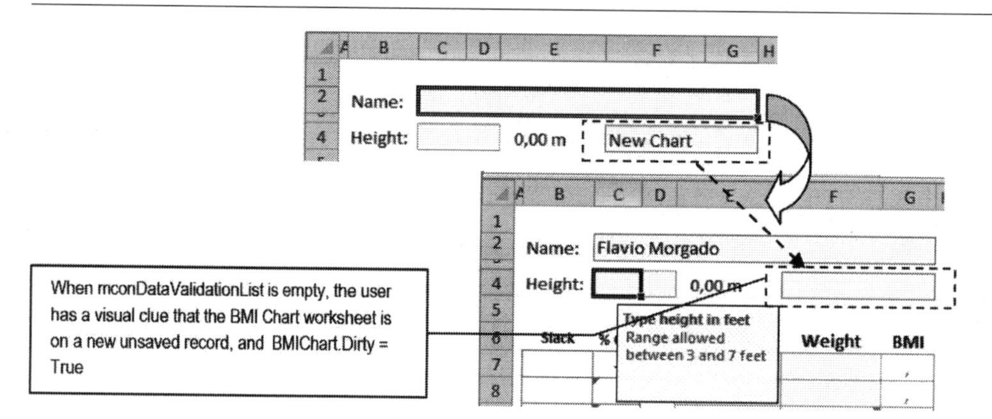

Figure 7-9. *Whenever you select the New Chart option to clear the BMI Chart worksheet and change any of its input cells (like the person's Name), the Worksheet_Change( ) event fires, set the Dirty module-level variable to True, and clears the data validation list value, as a visual indication that a new worksheet data has been inserted by the user*

This is important because now the workbook record is in a *dirty* state, meaning that any possible sheet tab data has been changed. From the Excel application's perspective, the workbook file data has also been changed and not saved yet. If you try to close the workbook at this moment, the `Workbook_BeforeClose( )` event will fire, and the `Private WithEvents mwb as Workbook` object module-level variable will capture this event, executing this code on the `BMIChart` code module:

```
Private Sub mwb_BeforeClose(Cancel As Boolean)
    Dim strMsg As String
    Dim strTitle As String
    Dim strRecord As String
    Dim bolSaved As Boolean

    If Me.Dirty Then
        strRecord = mws.Range(mcondbDataValidationList)
        If strRecord = "" Then strRecord = "New " & mcondbRecordName
        strTitle = "Save " & strRecord & " data?"
        strMsg = strRecord & " data had been changed." & vbCrLf
```

```
        strMsg = strMsg & "Save " & strRecord & " data before close the workbook?"
        Select Case MsgBox(strMsg, vbYesNoCancel + vbQuestion, strTitle)
            Case vbYes
                bolSaved = Save(strRecord)
                Cancel = Not bolSaved
            Case vbCancel
                Cancel = True
        End Select
    End If
End Sub
```

As you can see, whenever you try to close the workbook, the `mwb_BeforeClose( )` event will check whether the `BMIChart.Dirty` property is `True`, meaning that the current record has been changed. If it has, the code will issue a VBA `MsgBox( )` function asking you to save the record. By clicking Yes to the `MsgBox( )` function, the event will call the `Function Save(strRecord)` procedure before closing the workbook (Figure 7-10).

```
        Select Case MsgBox(strMsg, vbYesNoCancel + vbQuestion, strTitle)
            Case vbYes
                bolSaved = Save(strRecord)
                Cancel = Not bolSaved
```

■ **Attention** As you will see later, `Function Save( )` returns `True` if the record is saved and `False` if the saving process is aborted, which in this case means that the workbook close operation must also be aborted. That is why the code sets `Cancel = not boSaved`.

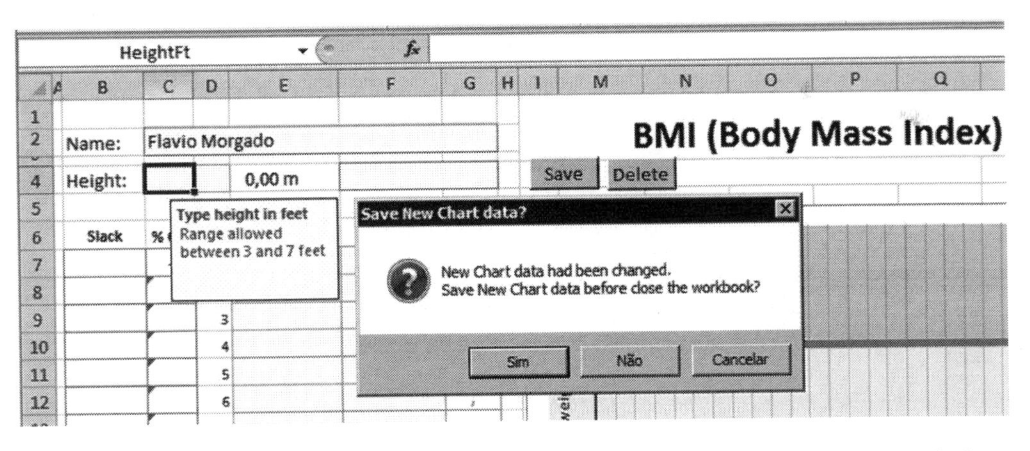

Figure 7-10. *If you try to close the BMI Companion Chart.xlsm Excel macro-enabled workbook, the Workbook_BeforeClose( ) event will fire. If the BMIChart.Dirty variable is True, you will receive a MsgBox( ) function asking to save the chart data before the workbook is closed*

By clicking No, the workbook will be closed without changing the `BMI Chart` state. By clicking Cancel, the `Cancel` argument of the `Workbook_BeforeClose( )` event will be set to `True`, canceling the event and keeping the workbook open.

Saving the Last Selected Record

Every time an item is selected in the mcondbDataValidationList cell, a new record is about to be loaded by the worksheet. Since the current worksheet data can be changed, you need to keep the last record name loaded so the code can give appropriate warnings to the application user.

The mstrLastRecord as String module-level variable is one that retains the name of the last-selected record. Its value changes when any worksheet input cell is selected (or when a record is saved, loaded, or deleted), by programming the BMIChart object's Worksheet_SelectionChange( ) event.

```
Private Sub Worksheet_SelectionChange(ByVal Target As Range)
    Set mwb = ThisWorkbook
    If mws.Range(mcondbDataValidationList) = "" Then
        mstrLastRecord = "New " & mcondbRecordName
    Else
        mstrLastRecord = mws.Range(mcondbDataValidationList)
    End If
End Sub
```

Saving BMI Chart Data

Suppose you are in an empty BMI Chart worksheet record, with the mcondbDataValidationList cell showing New Chart. Go ahead and type some data in it: type your name, type your weight in feet and inches, and then type some hypothetic pairs of date/weight values (without forget that successive dates must be at least seven days apart because of the validation rule used on date cells). Note that after you change any input data cell in the BMI Chart worksheet, the mcondbDataValidationList cell becomes empty, meaning to the database engine this new worksheet record has been changed and not saved yet.

When you are ready, click the Save control button to save the worksheet data, and the VBA code will answer with an InputBox( ) message, asking you to confirm the record name to be saved (every new record will be saved by default as New Chart mm_dd_yyyy n, where mm/dd/yyyy is the system date and n is a default file name counter). Figure 7-11 shows that the proposed default name was changed to "Saving test data."

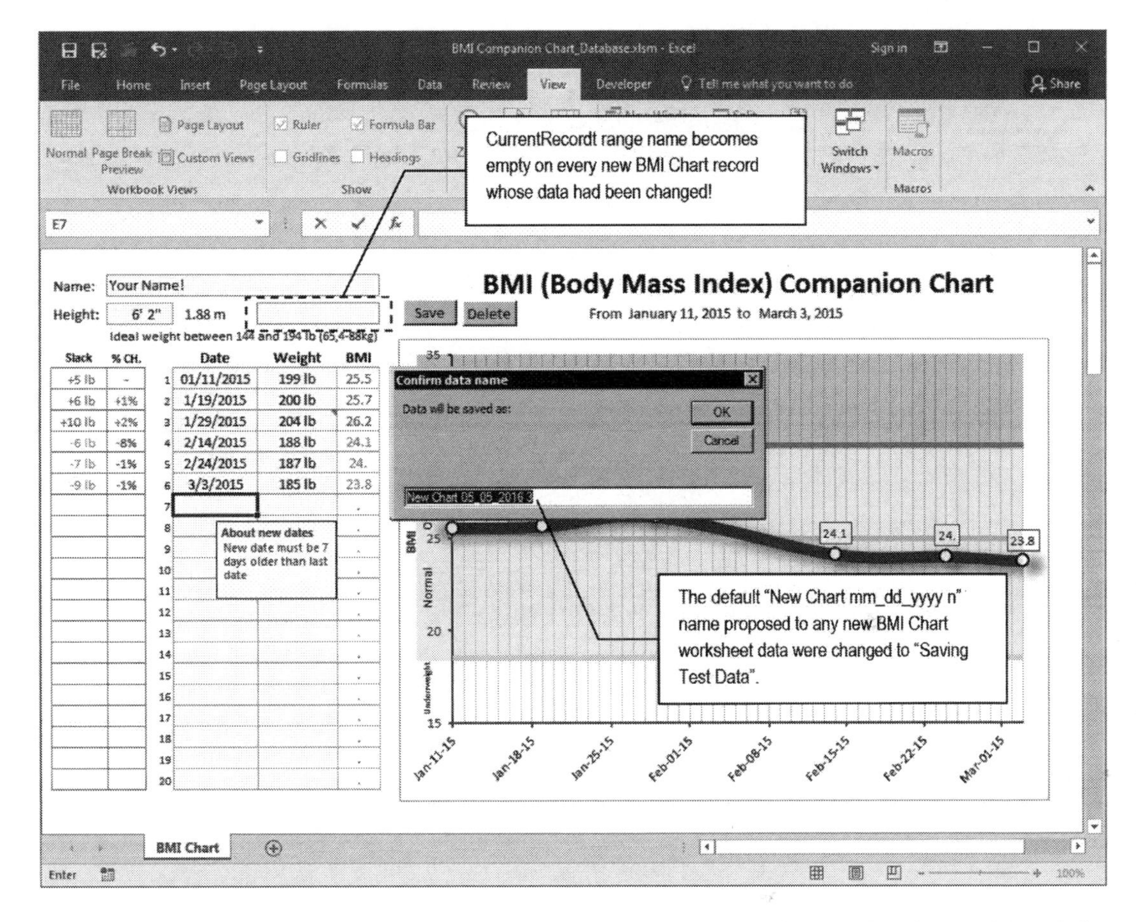

Figure 7-11. Type your name, weight in feet and inches, and some pairs of date/weight values and then click the Save button to save the worksheet data

By clicking the InputBox( ) OK button, the worksheet data will be saved as a new database record and the VBA code will show a saving confirmation message while the record name appears in mcondbDataValidationList cell (Figure 7-12).

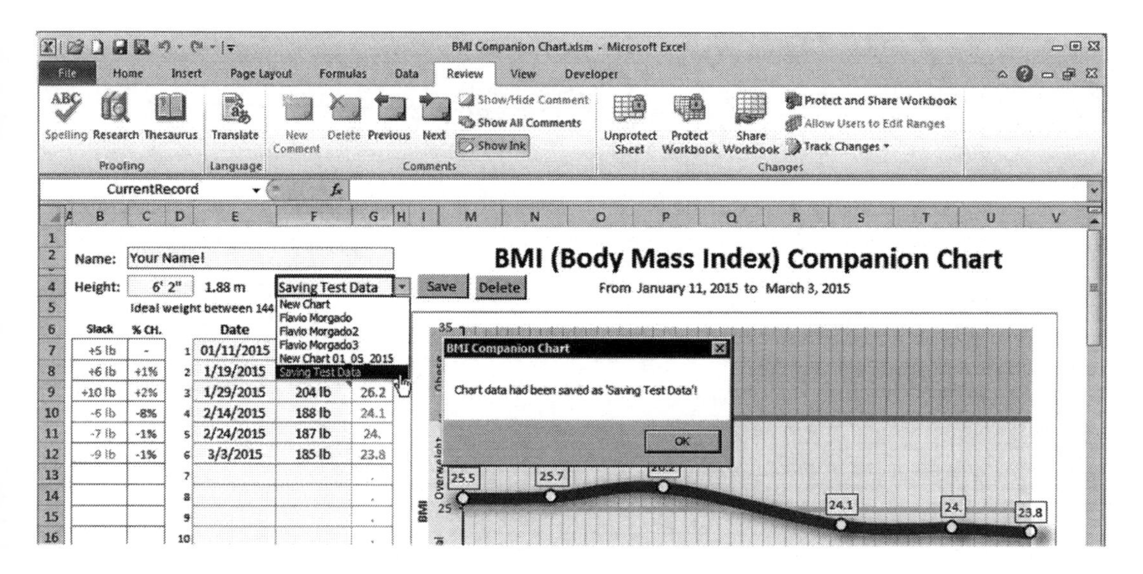

Figure 7-12. After the BMI Chart worksheet data has been saved, you will receive a MsgBox( ) confirmation message, and the saved name will automatically appear in the CurrentRecord range name

The database engine "save record" process is divided into three main procedures.

- `Public Function Save( )`: This is responsible for asking for the record name, and if it is granted, it calls `SaveData( )` to begin the saving record process and gives the saving warning message.
- `Private Function SaveData( )`: This verifies whether the record is new, and if it is, it verifies whether there is still room to save a new worksheet record; if it has enough room, it stores the record name in the `mcondbSavedRange` range and calls the `LoadSaveData( )` procedure to effectively save the record data in the worksheet.
- `Private Sub LoadSaveData( )`: This receives the record name and the operation to perform (load or save), using a generic algorithm to load/save all cell values contained in the `mcondbOneSide` and/or `mcondbManySide` range names.

The `BMI Chart` worksheet's Save control button is associated to the `BMIChart.Save( )` procedure, which executes the saving process using the next detailed steps:

1. Do nothing on a new empty record.
2. Get the record name to be saved on any dirty record by calling `Function GetRecordName( )`.
3. Call the `Function SaveData( )` procedure to effectively save the worksheet data.

■ **Attention** The `BMIChart.Save` method was not named as the `BMIChart.SaveAs( )` procedure because the `Worksheet` object already exposes a default `SaveAs` method that allows any worksheet to be saved in another workbook.

```
Public Function Save(Optional strLastRecord As String) As Boolean
    Dim rg As Range
    Dim strRecord As String
    Dim bolNewRecord As Boolean
    Dim bolRecordSaved As Boolean

    'Verify if Record data is still empty
    strRecord = mws.Range(mcondbDataValidationList)
    If strRecord = "New " & mcondbRecordName Then
        Exit Function
    End If

    If strLastRecord = "" Then
        strLastRecord = strRecord
    End If
    strRecord = GetRecordName(strLastRecord, bolNewRecord)

    If Len(strRecord) Then
        'Disable application events to allow cell change by macro code
        SetScreenEventsRecalc (False)
            mws.Unprotect
                bolRecordSaved = SaveData(strRecord, bolNewRecord)
            mws.Protect
            If bolRecordSaved Then
                'Define current Record as saved Record
                mws.Range(mcondbDataValidationList) = strRecord
                mws.Range(mcondbDataValidationList).Select

                'Save the worbook
                ThisWorkbook.Save
                mstrLastRecord = strRecord
                Me.Dirty = False
                Save = True
                MsgBox mcondbRecordName & " data had been saved as '" & strRecord & "'!", ,
                "BMI Companion Chart"
            Else
                MsgBox "There is no more room to save data on this worksheet!", vbCritical,
                "Can't save data"
            End If
        SetScreenEventsRecalc (True)
    End If
End Function
```

The Public Save( ) procedure begins checking whether the worksheet is on a new empty record and whether the new record data has not been changed. If this is true, nothing must be saved, and the procedure exits graciously using the Exit Function instruction.

```
Public Function Save (Optional strLastRecord As String) As Boolean
    ...
    'Verify if Record data is still empty
    strRecord = mws.Range(mcondbDataValidationList)
```

```
    If strRecord = "New " & mcondbRecordName Then
        Exit Function
    End If
```

But if the record is not new or any change has been made to a new record, it first verifies whether strLastRecord is empty. If it is, it stores the current record name in the strLastRecord string variable and calls GetRecordName( ), passing as arguments strLastRecord and bolNewRecord (which are False at this moment).

```
    If strLastRecord = "" Then
        strLastRecord = strRecord
    End If
    strRecord = GetRecordName(strLastRecord, bolNewRecord)
```

The strRecord variable now has the record name to be saved or an empty string, indicating that the save process was aborted, and it uses the VBA function Len(strRecord) to verify the length of the strRecord variable. If GetRecordName( ) returns an empty string, Len(strRecord) = 0, and the code ends doing nothing.

```
Public Function Save(Optional strLastRecord As String) As Boolean
    ...
    strRecord = GetRecordName(strLastRecord, bolNewRecord)
    If Len(strRecord) Then
        ...
   End If
End Function
```

Let's see how GetRecordName( ) asks for the record name to be saved!

Getting the Record Name

The function GetRecordName( ) receives two arguments passed by reference: strRecord (with current record name, if any), and bolNewRecord, a Boolean argument that must signal to the Save( ) procedure if this is a new record. This procedure must do the following:

1. Propose a default record name if strRecord is an empty string (a new record).
2. Verify whether the current record name or the new proposed name already exists in the mcondbSavedRecords range name.
3. If the name already exists, ask to update it, allowing another name change.
4. Return the desired name or an empty string—as an indication that no name was selected and the record was not saved.

```
Private Function GetRecordName(strRecord As String, bolNewRecord As Boolean) As String
    Dim rg As Range
    Dim strNewRecord As String
    Static sintDefaultName As Integer

    If strRecord = "" Then
        sintDefaultName = sintDefaultName + 1
        strRecord = "New " & mcondbRecordName & " " & Replace(Date, "/", "_")
        If sintDefaultName > 1 Then
```

```
            strRecord = strRecord & " " & sintDefaultName
        End If
        strRecord = InputBox("Data will be saved as:", "Confirm data name", strRecord)
    End If

    If Len(strRecord) Then
        'Verify if strRecord already exist on mcondbSavedRecords
        Set rg = mws.Range(mcondbSavedRecords).Find(strRecord)

        If rg Is Nothing Then
            bolNewRecord = True
        Else
            'Confirm proposed record name
            strNewRecord = InputBox(mcondbRecordName & " '" & strRecord & "' already exist.
            Do you want to overwrite it?", _
                                    "Overwrite " & strRecord & " data?", strRecord)
            If strRecord <> strNewRecord Then
                'Proposed record name changed. Verify if new name alteady exist
                Set rg = mws.Range(mcondbSavedRecords).Find(strNewRecord)
                If rg Is Nothing Then
                    bolNewRecord = True
                Else
                    'New name already exist. Confirm overwrite
                    If MsgBox("The name you typed, '" & strNewRecord & "', already exist.
                    Overwrite it?", _
                              vbYesNo + vbDefaultButton2 + vbQuestion, _
                              "Overwrite '" & strNewRecord & "'?") = vbNo Then
                        strNewRecord = ""
                    End If
                End If
                strRecord = strNewRecord
            End If
        End If
    End If

    GetRecordName = strRecord
End Function
```

The procedure declares three variables: the `rg as Range` object variable, the `strNewRecord` string, and the `Static sintDefaultName as Integer` variable that is responsible for adding a counter to the default proposed name (like New Record 1, New Record 2, ..., for example).

If the worksheet data comes from a new record, `strRecord` will be an empty string, and the code must create a default record name and offer it to the user using an `InputBox( )` VBA function. The name proposed will use this format: `New <mcondbRecordName> mm_dd_yyyy n`, where `mcondbRecordName` is the record name (`Chart`), `mm_dd_yyyy` is the system date, and *n* is the application section counter for the default name. Note that the slashes returned by the system date are changed to underscores using the VBA `Replace( )` function, and if `sintDefaultName > 1`, it will be concatenated to the end of the proposed default name (the first default name on 1/5/2015 will be `New Chart 1_5_2015`, the second default name on the same date will be `New Chart 1_5_2015 2`, and so on).

```
    If strRecord = "" Then
        sintDefaultName = sintDefaultName + 1
        strRecord = "New " & mcondbRecordName & " " & Replace(Date, "/", "_")
        If sintDefaultName > 1 Then
            strRecord = strRecord & " " & sintDefaultName
        End If
        strRecord = InputBox("Data will be saved as:", "Confirm data name", strRecord)
    End If
```

When the user receives the InputBox( ) message with the proposed record name for a new record (or the name has been saved before) and cancels the operation, strRecord will hold an empty string, and GetRecordName( ) will end and return this empty string value to the Save( ) procedure, which will also end doing nothing.

```
Private Function GetRecordName(strRecord As String, bolNewRecord As Boolean) As String
    ...
    If strRecord = "" Then
        ...
        strRecord = InputBox("Data will be saved as:", "Confirm data name", strRecord)
    End If
    If Len(strRecord) Then--------
    ...
    End If

    GetRecordName = strRecord  <--
End Function
```

If the InputBox( ) Cancel button is selected, strChart = "", GetChartName( ) will return this empty value and SaveChart( ) will end doing nothing!

By accepting the proposed name, the code will search inside the range name associated with the mcondbSaveRecords constant (the default is the mcondbSavedRecords range name) using the Range.Find method. If the name is not found, bolNewRecord receives True to indicate to Save( ) that a new record must be inserted on the database, while strRecord is associated to the Function GetRecordName( ) return value.

```
    If Len(strRecord) Then
        'Verify if strRecordName name already exist
        Set rg = ws.Range(mcondbSavedRecords).Find(strRecord)

        If rg Is Nothing Then
            bolNewRecord = True
        Else
          ...
        End If
    End If

    GetRecordName = strRecord
End Function
```

But when the proposed name already exists on the database (it was found in the mcondbSavedRecords range name), the user will receive a warning message in another InputBox( ) function, while asking to overwrite the record also gives a second chance to change it. This final proposed record name will be saved in the strNewRecord string variable.

```
        Else
            'Confirm proposed record name
            strNewRecord = InputBox(mcondbRecordName & " '" & strRecord & "' already exist.
Do you want to overwrite it?", _
                                                "Overwrite " & strRecord & "
data?", strRecord)
```

If the user changes the name proposed by this second InputBox( ), strRecord and strNewRecord will diverge, and the new name must be searched again inside the mcondbSavedRecords range name.

```
            If strRecord <> strNewRecord Then
                'Proposed record name changed. Verify if new name already exist
                Set rg = ws.Range(mcondbSavedRecords).Find(strNewRecord)
```

If the Range.Find method does not find the proposed name inside the mcondbSavedRecords range, the bolNewRecod argument will receive True, signaling that this is a new record.

```
                If rg Is Nothing Then
                    bolNewRecord = True
```

But if the name is found, a MsgBox( ) function will give a final warning that the selected record name will be overwritten.

```
                Else
                    'New name already exist. Confirm overwrite
                    If MsgBox("The name you typed, '" & strNewRecord & "', already exist.
Overwrite it?", _
                                        vbYesNo + vbDefaultButton2 + vbQuestion, _
                                        "Overwrite '" & strNewRecord & "'?") = vbNo Then
```

Note that this Msgbox( ) function offers Yes and No buttons, using No as default button. If the user decides to *not overwrite* the existing name (MsgBox( ) = vbNo), strNewRecord will receive an empty string.

```
                    If MsgBox("The name you typed, ...) = vbNo Then
                        strNewRecord = ""
                    End If
```

No matter if the user accepts or rejects overwriting the name, the strNewRecord value will be associated to strRecord, which now contains either a valid record name or an empty string that is used as the Function GetRecordName( ) return value.

```
                strRecord = strNewRecord
            End If
        End If
    End If

    GetRecordName = strRecord
End Sub
```

Attention If the user cancels the InputBox( ) overwrite warning, strNewRecord ="" and will diverge from strRecord. The Range.Find method will not find this empty name on mcondbSavedRecords, strRecord = strNewRecord = "", and the Save( ) procedure will end returning False.

Disabling/Enabling Screen Updates, Events Firing, and Worksheet Recalculation

Let's suppose that a name has been typed or accepted to save the current record in the worksheet database. When Function GetRecordName( ) returns the name that must be used to save the BMI Chart worksheet data, the saving process begins by calling the SetScreenEventsRecalc(False) procedure to disable screen updating, events firing, and worksheet recalculation.

To allow the VBA code to make changes on the worksheet data without cascade-firing worksheet events (like Worksheet_Change( ) and Worksheet_SelectionChange( )), avoid screen flickering, and worksheet recalculation, you must always disable the Application object's ScreenUpdating, EnableEvents, and Calculation properties. Since you must do these operations from different procedures, these tasks are delegated to a single, centralized procedure stored in the basScreenEventRecalc module called Sub SetScreenEventsRecalc( ), which receives the bolEnabled argument to turn on/off these properties.

```
Public Function Save (Optional strLastRecord As String) As Boolean
    ...
    If Len(strRecord) Then
        'Disable application events to allow cell change by macro code
        SetScreenEventsRecalc (False)
       ...
    End If
End Function

Public Sub SetScreenEventsRecalc(bolEnabled As Boolean)
    With Application
        .ScreenUpdating = bolEnable
        .EnableEvents = bolEnable
        .Calculation = IIf(bolEnable, xlCalculationAutomatic, xlManual)
    End With
End Sub
```

Saving the Record Name with SaveData( ) Procedure

Since the BMI Chart worksheet is protected (without a password), the Save( ) code needs to unprotect it before saving the worksheet data, delegating the first part of the saving process to the SaveData( ) procedure, which receives the record name and an indication of whether this is a new worksheet record.

```
Public Function Save (Optional strLastRecord As String) As Boolean
    ...
            mws.Unprotect
                bolRecordSaved = SaveData(strRecord, bolNewRecord)
```

The SaveData( ) procedure is fully commented to allow you to follow its code easily, executing this code to save the BMI Chart worksheet data:

```
Private Function SaveData(strRecord As String, bolNewRecord As Boolean) As Boolean
    Dim rg As Range
    Dim strRangeName As String
    Dim strAddress As String
    Dim lngRow As Long
    Dim bolWorksheetIsFull As Boolean

    Set rg = mws.Range(mcondbSavedRecords)
    If bolNewRecord Then
        'Define sheet row where next Record data will be stored
        lngRow = NextEntryRow(bolWorksheetIsFull)

        'Verify if sheet can receive more records
        If bolWorksheetIsFull Then
            'No more room to save data
            Exit Function
        End If
        'Insert a new row at bottom of mcondbSavedRecords range name and update rg object
        rg.Resize(rg.Rows.Count + 1).Name = "'" & mws.Name & "'!" & mcondbSavedRecords
        Set rg = mws.Range(mcondbSavedRecords)

        'Position on new cell of mcondbSavedRecords range and save new Record name
        rg.Cells(rg.Rows.Count, 1) = strRecord

        If Len(mcondbManySide) Then
            'Define many-side Record name as 'mcondbManySidePrefix<strRecord>' and create
            tbe range name
            strRangeName = mcondbManySidePrefix & FixName(strRecord)
            strAddress = "='" & mws.Name & "'!" & mcondbManySideFirstColumn & lngRow
            mws.Names.Add strRangeName, strAddress, False
        End If
    End If

    Call LoadSaveData(strRecord, SaveRecord)

    'Sort mcondbSavedRecords range keeping "New <mcondbRecordName>"  on the top of the list
    Set rg = mws.Range(Cells(rg.Row + 1, rg.Column), _
                Cells(rg.Row + rg.Rows.Count, rg.Column + mcondbRangeOffset +
                mcondbOneSideColumsCount))
    rg.EntireRow.Hidden = False
        rg.Sort rg.Cells(, 1)
    rg.EntireRow.Hidden = True

    mws.Range("A1").Select
    SaveData = True
End Function
```

Verify Whether the Worksheet Is Full with NewEntryRow( )

After declaring its variables, SaveData( ) sets a reference to the mcondbSavedRecords range name and verifies whether this is a new record. If it is, you must verify whether the worksheet still has room to store another database record, calling the NextEntryRow( ) procedure.

```
    Set rg = mws.Range(mcondbSavedRecords)
    If bolNewRecord Then
        'Define sheet row where next Record data will be stored
        lngRow = NextEntryRow(bolWorksheetIsFull)
```

The NextEntryRow( ) procedures does two different things.

- Sets the next row where the new record must be inserted on the worksheet database
- Verifies whether there is still room to save the record data

```
Private Function NextEntryRow(bolWorksheetIsFull As Boolean) As Long
    Dim lngRow As Long

    If Len(mcondbManySide) Then
        'Use many-side records to find next entry row
        lngRow = mcondbRecordsFirstRow + (mws.Range(mcondbSavedRecords).Rows.Count - 1) *
        mcondbManySideRowsCount
        If lngRow < mws.UsedRange.Rows.Count Then
            lngRow = mcondbRecordsFirstRow + (mws.Range(mcondbSavedRecords).Rows.Count *
            mcondbManySideRowsCount)
        End If
        bolWorksheetIsFull = (lngRow > (mws.Rows.Count - mcondbManySideRowsCount))
    Else
        'Just one-side record to find next entry row
        lngRow = mcondbRecordsFirstRow + mws.Range(mcondbSavedRecords).Rows.Count
        bolWorksheetIsFull = (lngRow > (mws.Rows.Count - mws.Range(mcondbSavedRecords).Rows.
        Count))
    End If

    NextEntryRow = lngRow
End Function
```

Note that NextEntryRow( ) first verifies the worksheet record has a many side of record data, and if it does, it adds to mcondbRecordsFirstRow (row 30, where the database storage begins) the number of worksheet records already saved on the database, using the ws.Range("mcondbSavedRecords").Rows.Count -1 property (since mcondbSavedRecords' first entry will always be New Chart). It then multiplies this value by mcondbManySideRowsCount (21 rows, meaning 20 rows for all date/weight pairs of data and one extra row to separate two successive worksheet records):

```
    If Len(mcondbManySide) Then
        'Use many-side records to find next entry row
        lngRow = mcondbRecordsFirstRow + (mws.Range(mcondbSavedRecords).Rows.Count-1) * _
                                     mcondbManySideRowsCount
```

There is a critical situation here that happens when BMI Chart has just two sets of saved data and the first possible many-side record data set (the data saved in rows 30:49; see Figure 7-8) is deleted. When this happens, lngRow will wrongly point to the next saved set of data points. That is why the procedure compares lngRow (the next possible saving row) with the last-used worksheet row (UsedRange.Rows.Count) and then recalculates the next row to be saved.

```
If lngRow < mws.UsedRange.Rows.Count Then
    lngRow = mcondbRecordsFirstRow + (mws.Range(mcondbSavedRecords).Rows.Count * _
                            mcondbManySideRowsCount)
End If
```

Once it determines the row where the next record must be saved, it uses it to verify whether the worksheet has still enough empty rows to store the worksheet data. This is made by comparing whether lngRow is smaller than the total worksheet row count (ws.Rows.Count) minus the size of the many-side records (mcondbManySideRowsCount).

```
bolWorksheetIsFull = (lngRow > (mws.Rows.Count - mcondbManySideRowsCount))
```

▒ **Attention** Since bolWorksheetIsFull was passed by reference, it may now contain True as an indication that the worksheet has no more room to store the new record.

But if the worksheet has no many-side records, mcondbManySide will be an empty string, and it must use just the total number of one-side records already stored to determine the next record row and whether the worksheet can store it.

```
If Len(mcondbManySide) Then
    ...
Else
    'Just one-side record to find next entry row
    lngRow = mcondbRecordsFirstRow + mws.Range(mcondbSavedRecords).Rows.Count
    bolWorksheetIsFull = (lngRow > (mws.Rows.Count - mws.Range(mcondbSavedRecords).Rows.
    Count))
End If
```

Aborting the SaveData( ) Procedure

When NextEntreyRow( ) ends, the code control returns to the SaveData( ) procedure, with lngRow receiving the row where the record must be saved and bolWorksheetIsFull indicating whether there is still room to store it. If bolWorksheetIsFull = True, the worksheet is full, and the Function SaveData( ) procedure ends, returning False.

```
Private Function SaveData(strRecord As String, bolNewRecord As Boolean) As Boolean
    ...
    lngRow = NextEntryRow(bolWorksheetIsFull)

    'Verify if sheet can receive more records
    If bolWorksheetIsFull Then
        'No more room to save data
        Exit Function
    End If
```

Code control will return to the Save( ) procedure with bolRecordSaved = False. The worksheet will be protected (with no password), warning the user by a VBA MsgBox( ) function that there is no more room on the worksheet to save application data. The procedure ends calling SetScreenEventsRecalc(True) to restore screen updating, events firing, and automatic recalculation.

```
Public Function Save (Optional strLastRecord As String) As Boolean
    ...
    If Len(strRecord) Then
        ...
        SetScreenEventsRecalc (False)
            ws.Unprotect
                bolRecordSaved = SaveData(strRecord, bolNewRecord)
            ws.Protect
            If bolRecordSaved Then
                ,,,
            Else
                MsgBox "There is no more room to save data on this worksheet!", vbCritical,
                "Can't save data"
            End If
        SetScreenEventsRecalc (True)
    End If
End Function
```

Inserting a New Record on the mcondbSavedRecords Range Name

Having detected enough space to save the record, bolWorksheetFull = False, and SaveData( ) will add a new row at the bottom of the mcondbSavedRecords range name, using the Range.Resize property. Note that the range is resized and renamed to a local worksheet name and that the rg object variable that represents it needs to be updated to reflect its new size.

```
Private Function SaveData(strRecord As String, bolNewRecord As Boolean) As Boolean
    ...
    Set rg = Range(mcondbSavedRecords)
     ...
        If bolWorksheetIsFull Then
           ...
        End If

        'Insert a new row at bottom of mcondbSavedRecords range name and update rg object
        rg.Resize(rg.Rows.Count + 1).Name = "'" & ws.Name & "'!" & mcondbSavedRecords
        Set rg = ws.Range(mcondbSavedRecords)
```

This new row added at the bottom of the mcondbSavedRecords range name receives the new record name.

```
'Position on new cell of mcondbSavedRecords range and save new Record name
rg.Cells(rg.Rows.Count, 1) = strRecord
```

To complete the operation of a new record insertion, the SaveData( ) procedure verifies whether there is a many-side record, and if there is, it passes the record name to Function FixName( ) to derive a unique range name for the range where the many-side records must be saved on the worksheet.

```
If Len(mcondbManySide) Then
    'Define Record name as ' mcondbManySidePrefix<strRecord>' and create it range name
    strRangeName = mcondbManySidePrefix & FixName(strRecord)
```

Then it sets the address for the new record as a worksheet scope range name—considering the worksheet name, first many-side records column (`mcondbManySideFirstColumn`), and next available row (`lngRow`)—and passes this address to the `ws.Names.Add` method. (Note that the third argument of the `Names.Add` method—the `Visible` property of the `Name` object—is defined as `False` to avoid that this range name appears in the Excel Name box or Name Manager dialog box.)

```
        strAddress = "='" & ws.Name & "'!" & mcondbManySideFirstColumn & lngRow
        ws.Names.Add strRangeName, strAddress, False
    End If
End If
```

Once the record name is inserted in the `mcondbSavedRecords` range name, `SaveData( )` calls the `LoadSaveData( )` procedure, passing the record name (`strRecord`) and the operation to be performed (`SaveRecord`), to effectively save the worksheet record data.

```
Call LoadSaveData(strRecord, SaveRecord)
```

Attention Note that `LoadSaveData( )` declares its second argument as the `Perform as Operation` enumerator, which can be set to `LoadRecord =1` or `SaveRecord=2`. When the `SaveData( )` procedure calls `LoadSaveData( )`, it passes `SaveRecord` to the `Perform` argument, indicating that the record must be saved.

Saving Record Data with LoadSaveData( )

Since load and save records are *mirrored* operations executed on the same worksheet cells (changing source and destination places according to the operation being performed), `LoadSaveData( )` uses a generic algorithm that takes the range name (or names) stored on `mcondbOneSide` and/or `mcondbManySide1` to `mcondbManySide4` constants and walks through its cells, saving all application input cell values in the worksheet database or loading database data into these same input cells.

```
Private Sub LoadSaveData(strRecord As String, Perform As Operation)
    Dim rg As Range
    Dim rgCells As Range
    Dim rgArea As Range
    Dim rgAreaColumn As Range
    Dim strRangeName As String
    Dim strRelation As String
    Dim intOffSet As Integer
    Dim intRelation As Integer
    Dim intRow As Integer
    Dim intCol As Integer
    Dim intAreaCol As Integer
    Dim intMaxRows As Integer

    Set rg = mws.Range(mcondbSavedRecords).Find(strRecord, , , xlWhole)
```

```
    'Load/Save one side worksheet records (one cell at a time)
    If Len(mcondbOneSide) Then
        Set rgCells = mws.Range(mcondbOneSide)
        For Each rgArea In rgCells.Areas
            For intRow = 1 To rgArea.Rows.Count
                For intCol = 1 To rgArea.Columns.Count
                    If Perform = SaveRecord Then
                        rg.Offset(0, mcondbRangeOffset + intOffSet) = rgArea.Cells(intRow,
                        intCol)
                    Else
                        rgArea.Cells(intRow, intCol) = rg.Offset(0, mcondbRangeOffset +
                        intOffSet)
                    End If
                    intOffSet = intOffSet + 1
                    If rgArea.Cells(intRow, intCol).MergeCells Then
                       intRow = intRow + rgArea.Cells(intRow, intCol).MergeArea.Rows.Count - 1
                        intCol = intCol + rgArea.Cells(intRow, intCol).MergeArea.Columns.
                        Count - 1
                    End If
                Next
            Next
        Next
    End If

    'Load/Save many side worksheet records
    strRangeName = mcondbManySidePrefix & FixName(strRecord)
    intRow = 0
    'Process each many-side records range Relation
    For intRelation = 1 To 4
        strRelation = Choose(intRelation, mcondbManySide1, mcondbManySide2, mcondbManySide3,
        mcondbManySide4)
        If Len(strRelation) Then
            intCol = 0
            intMaxRows = 0
            Set rgCells = mws.Range(strRelation)
            For Each rgArea In rgCells.Areas
                For intAreaCol = 0 To rgArea.Columns.Count - 1
                    Set rg = mws.Range(strRangeName).Offset(intRow, intCol)
                    Set rg = rg.Resize(rgArea.Rows.Count, 1)
                    Set rgAreaColumn = mws.Range(mws.Cells(rgArea.Row, rgArea.Column +
                    intAreaCol), _
                    mws.Cells(rgArea.Row + rgArea.Rows.Count - 1, rgArea.Column +
                    intAreaCol))
                    If Perform = SaveRecord Then
                        rg.Value = rgAreaColumn.Value
                    Else
                        rgAreaColumn.Value = rg.Value
                    End If

                    If rgArea.Cells(1, intAreaCol + 1).MergeCells Then
                        intAreaCol = intAreaCol + rgArea.Cells(1, intAreaCol + 1).MergeArea.
                        Columns.Count - 1
```

```
                    End If
                    intCol = intCol + 1
                Next

                If intMaxRows < rgArea.Rows.Count Then
                    intMaxRows = rgArea.Rows.Count
                End If
            Next
            intRow = intRow + intMaxRows + 1
        End If
    Next
End Sub
```

Processing the One-Side Record

The strRecord argument with the name of the desired database record is searched inside the mcondbSavedRecords range name using the Range.Find method being associated to the rg object variable. It then verifies whether there is a one-side record range name to operate on.

```
Private Sub LoadSaveData(strRecord As String, Perform As Operation)
    ...
    Set rg = ws.Range(mcondbSavedRecords).Find(strRecord, , , xlWhole)

    'Load/Save one-side worksheet records (one cell at a time)
    If Len(mcondbOneSide) Then
```

The one-side record can be distributed on the worksheet with different layouts, using independent or merged cells to store each one-side record data. In the BMI Chart worksheet, the mcondbOneSide constant points to the OneSideRecord range name, which contains the person's name (using merged cells) and person's height in feet and inches, while the ManySideRecords range name points to all pairs of dates/weights that produce the BMI chart (Figure 7-13).

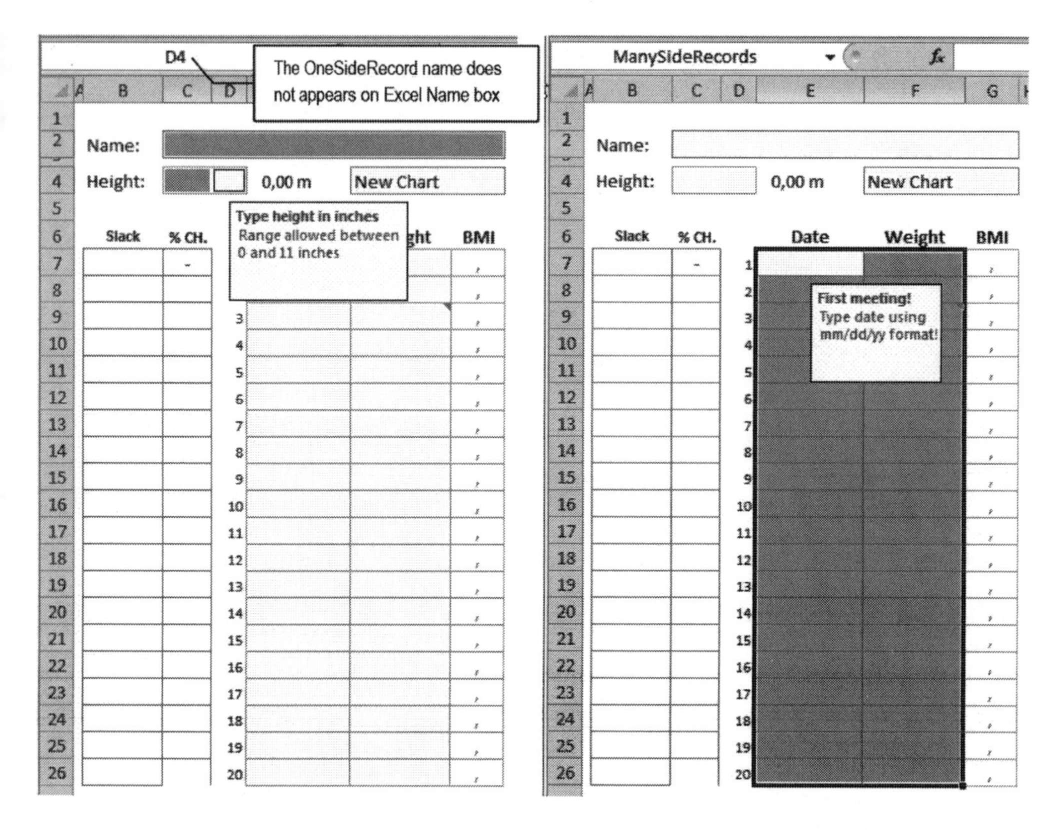

Figure 7-13. These are the OneSideRecord and ManySideRecords range names of the BMI Chart worksheet, used to represent all one-side and many-side record cells saved by the database engine. Note that the OneSideRecord range name does not appear in the Excel Name box because it points to no contiguous worksheet cells

That is why the code sets a reference to the `mcondbOneSide` range in the `rgCells` object variable and then uses an outer `For Each... Next` loop to run through the `rgCells.Areas` collection, operating on one area at a time.

```
Set rgCells = ws.Range(mcondbOneSide)
For Each rgArea In rgCells.Areas
```

Since each one-side record Range.Area can have multiple independent or merged cells, the code uses two `For...Next` nested loops to run through these cells. The outer loop runs through each `Range.Area` row, while the inner loop runs through each `Range.Area` column.

```
For intRow = 1 To rgArea.Rows.Count
    For intCol = 1 To rgArea.Columns.Count
```

At each loop passage, the code verifies the operation to perform. If it's a `SaveRecord` operation, the application input cell (represented by `rgArea.Cells(intRow, intCol)`) must be stored in the database. Since the one-side record is saved on the right of the `mcondbSavedRecords` range name record entry (represented by `rg` object variable), the record saving is done by using the `Range.Offset` property to displace `rg` to the right by the `mcondbRangeOffset + intOffSet` Integer variable (`intOffSet` is incremented at each loop passage).

```
If Perform = SaveRecord Then
    rg.Offset(0, mcondbRangeOffset + intOffSet) = rgArea.Cells(intRow, intCol)
Else
    rgArea.Cells(intRow, intCol) = rg.Offset(0, mcondbRangeOffset + intOffSet)
End If
intOffSet = intOffSet + 1
```

■ **Attention** Note in the previous code of `case Perform = LoadRecord`, the invert operation is done: the application input cell (represented by `rgArea.Cells(intRow, intCol)`) receives the database stored value (represented by the displaced `rg.Offset(0, mcondbRangeOffset + intOffSet)` cell).

Since `rgArea.Rows.Count` takes into account all individual merged columns cells (as `rgArea.Columns.Count` does with merged rows cells), the code needs to verify whether the last operated on cell is merged using the `Range.MergeCells` property. If it is, `intRow` and `intCol` must be incremented by the number or rows/columns merged using `MergeArea.Rows.Count -1` (or `MergeArea.Columns.Count - 1`).

```
                If rgArea.Cells(intRow, intCol).MergeCells Then
                    intRow = intRow + rgArea.Cells(intRow, intCol).MergeArea.Rows.Count - 1
                    intCol = intCol + rgArea.Cells(intRow, intCol).MergeArea.Columns.Count
                    - 1
                End If
            Next
        Next
    Next
End If
```

Processing the Many-Side Records

Once the one-side record is saved (or loaded), it is time to operate all possible four many-side records ranges, treated as "relations" using this very principle. Cells in the same many-side record range will be saved side by side; cells of different many-side ranges will be saved *at the bottom of the previous relation ranges*, with a blank row between them.

This is done by first storing in `strRangeName` the worksheet scope range name associated to all many-side record cells and resetting the `intRow` counter.

```
    'Load/Save many side worksheet records
    strRangeName = mcondbManySidePrefix & FixName(strRecord)
    intRow = 0
```

A `For...Next` loop using the `intRelation` variable as a counter is then used to process each of the four possible many-side relationship range names. At each loop passage, one of the four possible many-side constants is selected using the VBA `Choose( )` function, and its content is verified by the VBA `Len( )` function, being processed if it is not an empty string.

```
For intRelation = 1 To 4
    strRelation = Choose(intRelation, mcondbManySide1, mcondbManySide2, mcondbManySide3,
    mcondbManySide4)
```

```
        If Len(strRelation) Then
            ...
        End If
    Next
End Sub
```

When the selected many-side constant has a range name inside it, it must be processed. Note that at the first loop passage, intRow = 0, while intCol and intMaxRows are reset for each many-side constant processed. Since strRelation now holds the range name associated to the selected many-side constant, it uses the rgCells variable to set a reference to its cells.

```
If Len(strRelation) Then
    intCol = 0
    intMaxRows = 0
    Set rgCells = mws.Range(strRelation)
```

As with the one-side record, each many-side records constant is processed by an outer For Each... Next loop that runs through its Range.Areas collection, one Range.Area at a time (using the rgArea as Range variable), and an inner For...Next loop to process each selected Range.Area column, one column at a time (referenced by the intAreaCol as Integer variable).

```
For Each rgArea In rgCells.Areas
    For intAreaCol = 0 To rgArea.Columns.Count - 1
```

The generic code created to manipulate each possible many-side record cell takes into account that each independent Range.Area must have an identical row count and multiple contiguous columns that may be eventually composed by column merged cells *but ever has row merged cells.*

Attention When producing worksheet applications, avoid merge cells rows on the many-side record cells. Change row heights instead when this effect is necessary. The reader is invited to change this generic code to deal with such complex many-side record cells. Designs.

On the first inner loop passage, intRows = 0 and intCols = 0, so the code takes the worksheet place where the many-side records (the strRangeName) are (or must be) stored as database records and uses Range.Offset(0,0) to select the top-left cell, attributing it to the rg object variable.

```
Set rg = mws.Range(strRangeName).Offset(intRow, intCol)
```

The first Range.Area column has all its cell rows selected using Range.Resize and rgArea.Rows.Count.

```
Set rg = rg.Resize(rgArea.Rows.Count, 1)
```

The same worksheet application input cells are selected, attributing them to the rgAreaColumn object variable (note that this is made using Range(Cell1, Cell2) arguments, with intAreaCol = 0 for the first Range.Area column).

```
Set rgAreaColumn = mws.Range(mws.Cells(rgArea.Row, rgArea.Column + intAreaCol), _
                mws.Cells(rgArea.Row + rgArea.Rows.Count - 1, rgArea.Column + intAreaCol))
```

The Perform as Operation argument contains the operation to be performed: the Load or Save record. Since the code is now saving a record, the rg object variable must receive the rgAreaColumn value to effectively transfer worksheet application input cell values to the database storage system (an inverse operation is performed to load a saved record).

```
If Perform = SaveRecord Then
    rg.Value = rgAreaColumn.Value
Else
    rgAreaColumn.Value = rg.Value
End If
```

Attention This is the only way you can place merged cells values inside independent cells. If rgAreaColumn is composed of merged column cells, rg.Value will receive just the merged cell value, which is associated to the top-left merged cell. You cannot use the Range.Copy and Range.PasteSpecial methods to perform such operations when the source range is merged.

Once the first Range.Area column is processed, the code checks whether its top-left cell is merged (considering that all other cells of the same column must be equal) using the rgArea.Cells(1,1). MergeCells property. If they are, the intAreaCol counter used as a counter for the inner For..Next loop is incremented by the number of merged cells used in this column, and the intCol counter is incremented, ready to process the next Range.Area column.

```
    If rgArea.Cells(1, intAreaCol + 1).MergeCells Then
        intAreaCol = intAreaCol + rgArea.Cells(1, intAreaCol + 1).MergeArea.Columns.Count - 1
    End If
    intCol = intCol + 1
Next
```

When all Range.Area columns are processed, the code checks whether the intMaxRows variable is lower than the last Range.Area rows count. If it is, its value intMaxRow is updated to reflect the greatest row count for the many-side records constant being processed, and intRows is updated accordingly to reflect the greatest number of rows processed so far, to correctly displace the database cells on the next inner loop passage (if any).

```
                If intMaxRows < rgArea.Rows.Count Then
                    intMaxRows = rgArea.Rows.Count
                End If
            Next
            intRow = intRow + intMaxRows + 1
        End If
    Next
End Sub
```

Sorting mcondbSavedRecords After a New Record Insertion

When LoadSaveData( ) ends, program control returns to the SaveData( ) procedure, and the mcondbSavedRecords needs to be sorted, keeping the New <mcondbRecordName> option on top of the record list (New Chart for the BMI Chart worksheet).

```
Private Function SaveData(strRecord As String, bolNewRecord As Boolean) As Boolean
    ...
    Set rg = ws.Range(mcondbSavedRecords)
    ...
    Call LoadSaveData(strRecord, SaveRecord)

    If bolNewRecord Then
```

Since the rg object variable holds a reference to the entire mcondbSavedRecords range name, including the first New Record item, it must be rebuilt. The new range must begin on the second range row (Cells(rg.Row + 1, rgColumn)) and finish on the last range row (Cells(rg.Row + rg.Rows.Count, ...)), selecting all one-side records data columns (Cells(..., rg.Column to rg.Colum + mcondbRangeOffset + mcondbOneSideColumnsCount)).

```
'Sort mcondbSavedRecords range keeping "New <mcondbRecordName>"  on the top of the list
Set rg = ws.Range(Cells(rg.Row + 1, rg.Column), _
             Cells(rg.Row + rg.Rows.Count, rg.Column + mcondbRangeOffset + _
                 mcondbOneSideColumsCount))
```

And once the range is correctly set, its rows are unhidden because Excel may fail to sort ranges on hidden rows. The Range.Sort method is executed, using the first range column (returned by rg.Cells(, 1)) for the Key1 sort argument, and all range rows are hidden again.

```
rg.EntireRow.Hidden = False
    rg.Sort rg.Cells(, 1)
rg.EntireRow.Hidden = True
```

Since the code made a lot of selections on hidden worksheet rows, Excel loses its screen reference, so it must be taken again to cell A1 to guarantee that it will not displace the worksheet vertically when the save process ends with SaveData( ) returning True.

```
    Range("A1").Select
    SaveData = True
End Function
```

When SaveData( ) successfully ends, the program control returns to the Save( ) procedure, where now bolRecordSaved = True. The database engine must select the saved record name on the mcondbDataValidationList data validation list cell while also selecting this cell in the user interface.

```
Public Function Save(Optional strLastRecord As String) As Boolean
      ...
            bolRecordSaved = SaveData(strRecord, bolNewRecord)
        mws.Protect
        If bolRecordSaved Then
            'Define current Record as saved Record
            mws.Range(mcondbDataValidationList) = strRecord
            mws.Range(mcondbDataValidationList).Select
```

And since a record has been saved on the database, it calls the ThisWorkbook.Save method to also save the workbook file.

```
'Save the workbook
ThisWorkbook.Save
```

And once the workbook has been saved, the mstrLastRecord must point to the saved record, the BMIChart.Dirty property must be set to False, Save( ) must return True, and a VBA MsgBox( ) function must warn the user that the record was successfully saved in the database.

```
            mstrLastRecord = strRecord
            Me.Dirty = False
            Save = True
            MsgBox mcondbRecordName & " data had been saved as '" & strRecord & "'!", ,
            "BMI Companion Chart"
            ...
        End If
    SetScreenEventsRecalc (True)
  End If
End Function
```

The saving process operation has finally ended!

Discarding BMI Chart Changes

Let's now suppose you have changed some chart data on a new worksheet record and want to discard it without saving. You can do this by following these steps:

1. Selecting another saved record on the `data validation list` (including the New Chart option)
2. Clicking the Delete button in the `BMI Chart` worksheet

Discarding Record Changes and Loading a Saved Record

Whenever you select any saved record in the `mcondbDataValidationList` cell, the `BMIChart` object `Worksheet_Change( )` event will fire, receiving on its `Target` argument the new data validation list cell value and asking the user to save the current record before loading a new one (Figure 7-14).

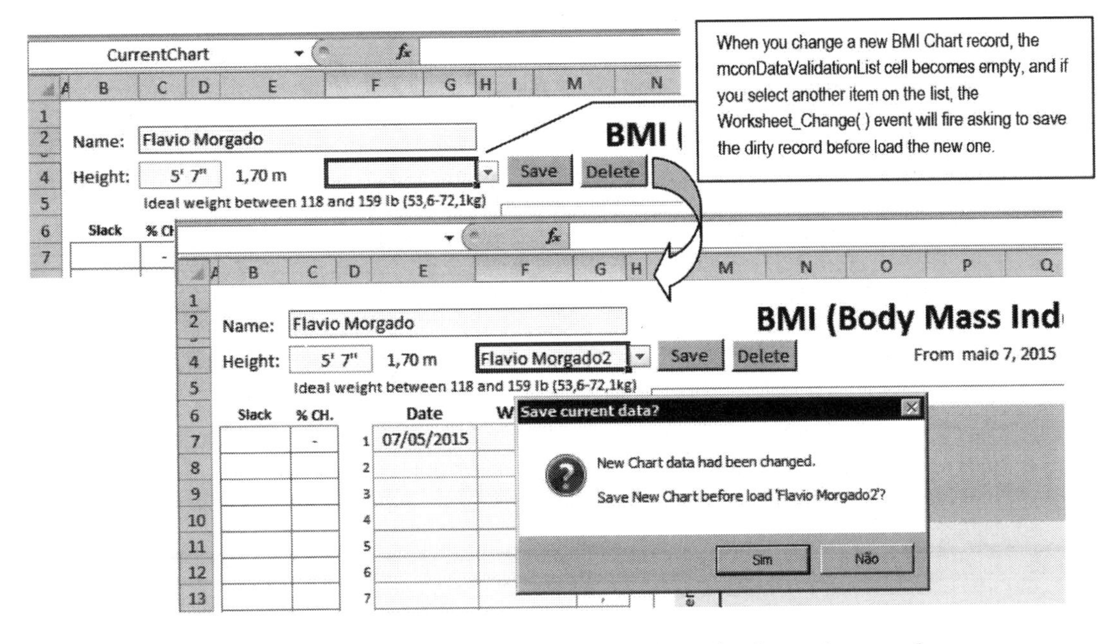

Figure 7-14. Whenever you change any BMI Chart data and select to load any other record on mcondbDataValidationList before saving it, the Worksheet_Change( ) event will fire, asking you to save the data before loading the selected item

The `Worksheet_Change( )` event calls `Sub TryToLoadSelectedRecord( )`, executing this code:

```
Private Sub Worksheet_Change(ByVal Target As
Range)
    ...
            TryToLoadSelectedRecord --------┐
           ...                              ¦
--► End Sub                                 ¦
                                            ¦
    Private Sub TryToLoadSelectedRecord()◄-----┘
        ...
        strNewRecord = mws.Range(mconDataValidationList)

        If Me.Dirty Then
            strMsg = mstrLastRecord & " data had been changed." & vbCrLf & vbCrLf
            strMsg = strMsg & "Save " & mstrLastRecord & " before load '" &
    strNewRecord & "'?"
            If MsgBox(strMsg, vbQuestion + vbYesNo, "Save current data?") = vbYes
    Then
                If Not Save(mstrLastRecord) Then
                    Application.EnableEvents = False
                        mws.Range(mconDataValidationList) = mstrLastRecord
                    Application.EnableEvents = True
◄---------------    Exit Sub
                End If
            End If
            Me.Dirty = False
        End If

        'Load selected Record data
        Call Load(strNewRecord)
----End Sub
```

Note that the item selected in `mcondbDataValidationList` is stored in the `strNewRecord` variable and checks whether the `BMIChart.Dirty` property is true (as an indication that the current worksheet data was changed) and sends a VBA `MsgBox( )` function to asking the user to save the current record before loading a new one.

```
Private Sub TryToLoadSelectedRecord()
    ...
    strNewRecord = ws.ge(mcondbDataValidationList)

    'Verify if current Record had been changed
    If Me.Dirty Then
        'Save current Record before change it?
        strMsg = mstrLastRecord & " data had been changed." & vbCrLf & vbCrLf
        strMsg = strMsg & "Save " & mstrLastRecord & " before load '" & strNewRecord & "'?"
        If MsgBox(strMsg, vbQuestion + vbYesNo, "Save current data?") = vbYes Then
```

By clicking No, the code will just set BMIChart.Dirty = False, and the TryToLoadSelectedRecord( ) record will discard the changes by calling Load(strNewRecord) to load the desired record. By clicking Yes to save the record, it will pass the current record name (mstrLastRecord) as an argument to Function Save( ). If the current record was correctly saved, Save( ) = True, and the selected record will be loaded, following the same code flux.429123_1_En

---

■ **Attention** The Load( ) procedure is detailed in the section "Loading BMI Chart Data" later in this chapter.

Note that if Save( ) is canceled and the current record is not saved, TryToLoadSelectedRecord( ) will disable events firing, and the mcondbDataValidationList cell will receive the mstrLastRecord name and will use an Exit Sub *instruction to exit without loading the record selected in the data validation list cell.*

```
If Not Save(mstrLastRecord) Then
                'Record data not saved!
                Application.EnableEvents = False
                    ws.Range(mconDataValidationList) = mstrLastRecord
                Application.EnableEvents = True
                Exit Sub
            End If
```

Discarding Record Changes with the Delete Control Button

The second way you can discard any BMIChart worksheet change made to its data is by clicking the Delete Button control, which calls the Function DeleteRecord( ) procedure and executes this code:

```
Public Function DeleteRecord() As Boolean
    Dim strRecord As String
    Dim strMsg As String
    Dim strTitle As String
    Dim intCancelDelete As Integer
    Dim intCancelSave As Integer
    Dim bolNewRecord As Boolean

    strRecord = mws.Range(mcondbDataValidationList)
    If strRecord = "" Or strRecord = "New " & mcondbRecordName Then
        If Me.Dirty Then
            bolNewRecord = True
            strMsg = "New " & mcondbRecordName & " data has not been saved yet." & vbCrLf
            strMsg = strMsg & "Do you want to delete it?"
            strTitle = "Delete unsaved record?"
        Else
            Exit Function
        End If
    Else
```

```
        strMsg = "Do you want to delete " & strRecord & " record?"
        strTitle = "Delete record?"
    End If

    If MsgBox(strMsg, vbYesNo + vbDefaultButton2 + vbQuestion, strTitle) = vbYes Then
        'Disable screen updating, events and recalc
        SetScreenEventsRecalc (False)
            Call Clear
            If Not bolNewRecord Then
                Call DeleteRecordData(strRecord)
            End If
            DeleteRecord = True
            Me.Dirty = False
            mstrLastRecord = "New " & mcondbRecordName
        'Enabled screen updating, events and recalc
        SetScreenEventsRecalc (True)

        mws.Range(mcondbDataValidationList) = mstrLastRecord

        'Save workbook after deletion
        ThisWorkbook.Save
    End If
End Function
```

This time, the procedure verifies whether the `BMIChart` worksheet is at a new record by comparing `mcondbDataValidationList` with an empty string or New Chart value. If it is showing a new unchanged record, `Dirty = False`, and the code exits graciously. If `Dirty = True` is set to `bolNewRecord = True` and define the warning messages to be shown by the `MsgBox( )` function (Figure 7-15).

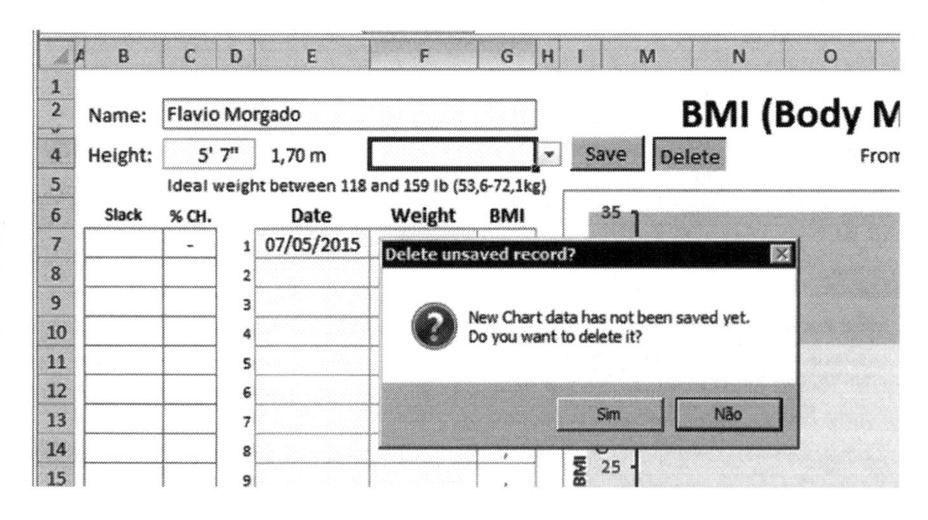

Figure 7-15. *If you click the Delete button on a new unchanged record, nothing will happen. But if the new record has been changed, you will receive a warning message asking you to save before deleting it*

```
Public Function DeleteRecord() As Boolean
    ...
    strRecord = mws.Range(mcondbDataValidationList)
    If strRecord = "" Or strRecord = "New " & mcondbRecordName Then
        If Me.Dirty Then
            bolNewRecord = True
            strMsg = "New " & mcondbRecordName & " data has not been saved yet." & vbCrLf
            strMsg = strMsg & "Do you want to delete it?"
            strTitle = "Delete unsaved record?"
        Else
            Exit Sub
        End If
    Else
       ...
    End If

    If MsgBox(strMsg, vbYesNo + vbDefaultButton2 + vbQuestion, strTitle) = vbYes Then
```

When choosing No, nothing will happen to the worksheet record, but when choosing Yes, the procedure will first call the Sub SetScreenEventsRecalc(False) procedure to disable screen updates, events firing, and worksheet recalculation, and then it will call the Sub Clear( ) procedure to clear all worksheet input cells.

```
    If MsgBox(strMsg, vbYesNo + vbDefaultButton2 + vbQuestion, strTitle) = vbYes Then
        'Disable screenupdating, events and recalc
        SetScreenEventsRecalc (False)
            Call Clear
            ...
    End If
End Function
```

Since you are considering that this is a new unsaved BMI Chart record, bolNewRecord = True, and there is no need to call the Sub DeleteData( ) procedure. After clearing the worksheet data, Function DeleteRecord( ) must return True. Dirty and mstrLastRecord are updated, Sub SetScreenEventsRecalc(True) is called again (to enabled screen updates, events firing, and worksheet recalculation), the data validation list cell is updated (cascade-firing Sub Load( )), and the workbook file is saved.

```
            If Not bolNewRecord Then
                Call DeleteRecordData(strRecord)
            End If
            DeleteRecord = True
            Me.Dirty = False
            mstrLastRecord = "New " & mconRecordName
        'Enabled screen updating, events and recalc
        SetScreenEventsRecalc (True)

        mws.Range(mconDataValidationList) = mstrLastRecord  <-- When the Data Validation
                                                                list cell value is updated,
                                                                it cascade fire Sub Load()

        'Save workbook after deletion
        ThisWorkbook.Save
    End If
End Function
```

Clearing Chart Data

The Clear( ) procedure uses a simple strategy to clear all worksheet cell input data: it first verifies whether there is a one-side record range, uses the rgCells object variable to set a reference to these cells, and attributes to rgCells an empty string to clear all its cells at once. This process is repeated using a For...Next loop to run through all possible four many-side records constants, clearing the many-side record cells.

```
Private Sub Clear()
    Dim rgCells As Range
    Dim strRange As String
    Dim intI As Integer

    'Clear one side worksheet records
    If Len(mcondbOneSide) Then
        Set rgCells = mws.Range(mcondbOneSide)
        rgCells = ""
    End If

    'Clear many side worksheet records
    For intI = 1 To 4
        strRange = Choose(intI, mcondbManySide1, mcondbManySide2, mcondbManySide3,
        mcondbManySide4)
        If Len(strRange) Then
            Set rgCells = mws.Range(strRange)
            rgCells = ""
        End If
    Next
End Sub
```

■ **Attention** Since Sub Clear( ) makes a lot of cell changes, you must first disable events firing to avoid cascading the Worksheet_Change( ) event.

Loading BMI Chart Data

To load any saved BMI Chart worksheet data, you need to select the name in the mcondbDataValidationList data validation list. As explained in section "Discarding Record Changes and Loading a Saved Record" earlier in this chapter, this selection will fire the Worksheet_Change( ) event, which will make a call to the TryToLoadSelectedRecord( ) procedure, which will end up calling the Load( ) procedure to effectively load the desired record. The Load( ) procedure executes this code:

```
Private Sub Load(strRecord As String)
    'Disable screen updating, events and recalc
    SetScreenEventsRecalc (False)
        Select Case strRecord
            Case "", "New " & mcondbRecordName
                'User selected a "New Record"
                Call Clear
                mws.Range(mcondbDataValidationList) = "New " & mcondbRecordName
            Case Else
```

```
                Call LoadSaveData(strRecord, LoadRecord)
                mws.Range(mcondbDataValidationList).Select
                mstrLastRecord = strRecord
        End Select
        Me.Dirty = False
    'Enable screen updating, events and recalc
    SetScreenEventsRecalc (True)
End Sub
```

Note that the entire loading process happens between two calls to the SetScreenEventsRecalc( ) procedure. The first call is to disable screen updating, events firing, and worksheet recalculation, and the second is to enable screen updating, events firing, and worksheet recalculation. After calling SetScreenEventsRecalc(False), it verifies whether the current record is a new one, and if it is, the worksheet is cleared, making a call to the Clear( ) procedure.

```
    SetScreenEventsRecalc (False)
        Select Case strRecord
            Case "", "New " & mcondbRecordName
                'User selected a "New Record"
                Call Clear
                mws.Range(mcondbDataValidationList) = "New " & mcondbRecordName
```

By selecting any other item, a call is made to the LoadSaveData( ) procedure, passing the record name to its strRecord argument and the LoadRecord enumerator to its Perform argument, effectively loading the desired worksheet record.

```
            Case Else
                Call LoadSaveData(strRecord, LoadRecord)
```

■ **Attention** The LoadSaveData( ) procedure was analyzed in the section "Saving Record Data with LoadSaveData( )" earlier on this chapter.

Once all chart data is recovered, it sets the worksheet focus to the mcondbDataValidationList cell, sets mstrLastRecord to the loaded record name, sets Dirty = False, and reenables screen updates, events firing, and worksheet recalculation.

```
                mws.Range("mcondbDataValidationList").Select
                mstrLastRecord = strRecord
        End Select
        Me.Dirty = False
    'Enabled screen updating, events and recalc
    SetScreenEventsRecalc (True)
End Sub
```

Deleting BMI Chart Data

The last operation made in the BMI Chart worksheet is the deletion of a saved record, which requires that the record is selected in the mcondbDataValidationList cell to be loaded into the worksheet data input cells before clicking the Delete Button control to run Function DeleteRecord( ).

```
Public Function DeleteRecord() As Boolean
    '''
    If MsgBox(strMsg, vbYesNo + vbDefaultButton2 + vbQuestion, strTitle) = vbYes Then
        'Disable screen updating, events and recalc
        SetScreenEventsRecalc (False)
            Call Clear
            If Not bolNewRecord Then
                Call DeleteRecordData(strRecord)
            End If
            ...
End Function
```

■ **Attention** The function DeleteRecord( ) was analyzed in the section "Discarding Record Changes with the Delete Control Button" earlier in this chapter.

To effectively remove all worksheet record data from the database, Sub DeleteRecordData( ) needs to follow these steps:

1. Unprotect the worksheet.
2. Find the record name inside the mcondbSavedRecords range name.
3. Delete the record name and all the one-side record columns.
4. Rebuild and resize the mcondbSavedRecords range name.
5. Delete the many-side records stored in the worksheet.
6. Remove the many-side record range name used to point to where these records are stored.

```
Private Sub DeleteRecordData(strRecord As String)
    Dim rg As Range
    Dim rgRecord As Range
    Dim strRecordRange As String
    Dim lngLastRow As Long
    Dim lngSafeRow As Long
    Dim intColumns As Integer

    mws.Unprotect
        Set rg = mws.Range(mcondbSavedRecords)
        'Get the last row used by Database parameters
        lngSafeRow = mcondbRecordsFirstRow
        lngLastRow = rg.Row + rg.Rows.Count - 1
        'Set the last safe sheet row to delete entire row
        If lngSafeRow < lngLastRow Then
            lngSafeRow = lngLastRow
        End If

        'Delete the One-side record from mcondbSavedRecords range
        Set rgRecord = rg.Find(strRecord)
        intColumns = mcondbRangeOffset + mcondbOneSideColumsCount
        rgRecord.Resize(1, intColumns).ClearContents
```

```
        If rgRecord.Row <> lngLastRow Then
            'Reposition other record entries by copy and paste
            mws.Range(Cells(rgRecord.Row + 1, rgRecord.Column), Cells(lngLastRow,
            rgRecord.Column + intColumns - 1)).Copy
            rgRecord.PasteSpecial xlPasteValues
        End If

        'Clear last mcondbSavedRecords record row
        Range(Cells(lngLastRow, rgRecord.Column), Cells(lngLastRow, rgRecord.Column +
        intColumns - 1)).ClearContents
        'Resize mcondbSavedRecords range name without deleted Record
        rg.Resize(rg.Rows.Count - 1).Name = "'" & mws.Name & "'!" & mcondbSavedRecords

        'Delete the Many-side records and it range name
        strRecordRange = mcondbManySidePrefix & FixName(strRecord)
        Set rg = mws.Range(strRecordRange)
        'Verify if record data amd mcondbSavedRecords range use the same rows
        If rg.Row <= lngSafeRow Then
            'This saved records data rows must just be cleaned
            rg.Resize(mcondbManySideRowsCount, mcondbManySideColumnsCount).ClearContents
        Else
            'It is safe to delete entire saved records data rows
            rg.Resize(mcondbManySideRowsCount).EntireRow.Delete
            'Provision to keep rows hidden
            mws.Range(Cells(mcondbRecordsFirstRow, 1), Cells(mws.Rows.Count, 1)).EntireRow.
            Hidden = True
        End If
        'Delete the many-records Range name
        mws.Names(strRecordRange).Delete
        'Scroll to row 1
        ActiveWindow.ScrollRow = 1
    mws.Protect
End Sub
```

After unprotecting the worksheet, the procedure sets an object variable reference (rg) to the mcondbSavedRecords range name and does the following:

```
    mws.Unprotect
        Set rg = ws.Range(mcondbSavedRecords)
```

Now it must determine the safe row: the greatest row number that allows an entire row deletion without causing collateral effects on the database. This is made by getting the first database row (mcondbRecordsFirstRow) and the last row used by the mcondbSavedRecords range name (rg.Row _ rg.Rows.Count - 1), using the greater of these two values as the safe row number (they will be the same when the database is empty).

```
        'Get the last row used by Database parameters
        lngSafeRow = mcondbRecordsFirstRow
        lngLastRow = rg.Row + rg.Rows.Count - 1
        'Set the last safe sheet row to delete entire row
        If lngSafeRow < lngLastRow Then
            lngSafeRow = lngLastRow
        End If
```

The record is then searched in mcondbSavedRecords using the Range.Find method, storing its cell reference on the rgRecord object variable.

```
'Delete the One-side record from mcondbSavedRecords range
Set rgRecord = rg.Find(strRecord)
```

The total number of columns used by the one-side record is stored in the intColumns Integer variable, and this value is used to resize the rgRecord variable so it can encompass both the record name and all its one-side record cells in the database before using Range.ClearContents to remove the data.

```
intColumns = mcondbRangeOffset + mcondbOneSideColumnsCount
rgRecord.Resize(1, intColumns).ClearContents
```

Once the one-side record is deleted from the mcondbSavedRecords range, it is time to resize this range using two different possibilities regarding the deleted record position.

1. If it was at the last mcondbSavedRecords row, the range will be resized by removing the last row.
2. If it was in the middle of the mcondbSavedRecords range, the range must be first rebuilt and then resized.

Since the rgRecord object variable has a reference to the deleted record cell and lngLastRow has the last mcondbSavedRecords used row, the procedure compares these row numbers. If they are different, it means that the record deletion happened in the middle of the mcondbSavedRecords range, which now has a blank row in between, which must be removed before it can be resized (Figure 7-16).

The code needs to select all records below the deleted one, using the Range.Copy and Range.PasteSpecial methods to reposition them one row above.

```
If rgRecord.Row <> lngLastRow Then
    'Reposition other record entries by copy and paste
    mws.Range(Cells(rgRecord.Row + 1, rgRecord.Column), Cells(lngLastRow, rgRecord.Column +
    intColumns - 1)).Copy
    rgRecord.PasteSpecial xlPasteValues
End If
```

Note the previous code uses the Range( ) method with the Worksheet.Cells collection to determine the top-left and bottom-right cells that define the range to be copied. The top-left cell is the next record (Cells(rgRecord.Row + 1, rgRecord.Column)), while the bottom-right cell is at the last mcondbSavedRecords row, using all one-side column cells (Cells(lngLastRow, rgRecord.Column + intColumns - 1)).

And once all records are repositioned, the last mcondbSavedRecords row is now duplicated and must be removed.

```
'Clear last mcondbSavedRecords record row
Range(Cells(lngLastRow, rgRecord.Column), Cells(lngLastRow, rgRecord.Column +
intColumns - 1)).ClearContents
```

The mcondbSavedRecords range must now be resized to update the record deletion, which is made by first using the Range.Resize method followed by the Name property to rename it.

```
'Resize mcondbSavedRecords range name without deleted Record
rg.Resize(rg.Rows.Count - 1).Name = "'" & mws.Name & "'!" & mcondbSavedRecords
```

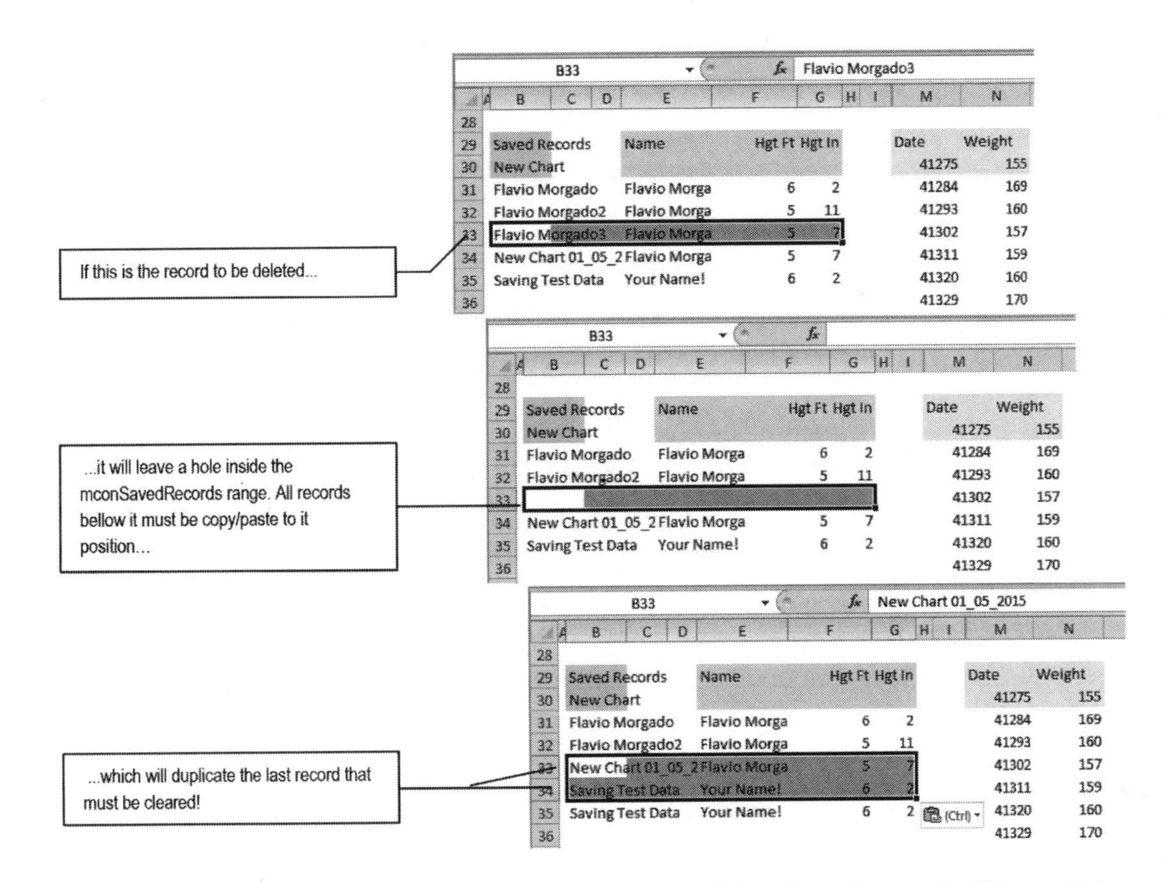

Figure 7-16. *When the deleted record is on the middle of the mcondbSavedRecords range, it will leave a hole inside the range after the data is cleaned. The code needs to copy/paste all other records below to its position and then delete the last remaining chart name*

Once the `mcondbSavedRecords` range has been processed, it is time to delete the many-side worksheet records using two different deletion processes (see Figures 7-6 and 7-8).

- One process happens when the many-side records are saved side by side with the `mcondbSavedRecords` rows (a condition that will always happen to the first database saved record). In this case, these many-side records rows can't be deleted; they must be just cleaned.
- Another process happens otherwise, when the may-side records are saved on independent worksheet rows. In this case, these rows can be entirely deleted from the worksheet without any collateral damage.

To take care of these two possibilities, the procedure sets an object variable reference (`rg`) to the associated many-side record range name and then compares the row number with the last `mcondbSavedRecords` row (`lngLastRow`).

```
'Delete the Many-side records and it range name
strRecordRange = mcondbManySidePrefix & FixName(strRecord)
Set rg = ws.Range(strRecordRange)
'Verify if record data amd mcondbSavedRecords range use the same rows
If rg.Row <= lngSafeRow Then
```

Whenever rg.Row <= lngLastRow, the first many-side records row uses the same mcondbSavedRecords range rows, indicating that these records rows can't be deleted; their content must be cleaned. This is done by first resizing the rg object variable to encompass all many-side records columns and rows and then using the Range.ClearContents method.

```
'This saved records data rows must just be cleaned
rg.Resize(mcondbManySideRowsCount, mcondbManySideColumnsCount).ClearContents
```

Otherwise, the many-side records are on a safe worksheet area, and its rows can be entirely deleted from the worksheet, using the Range.EntireRow property to select them all before applying the Range.Delete method to exclude them from the worksheet.

```
Else
    'It is safe to delete entire saved records data rows
    rg.Resize(mcondbManySideRowsCount).EntireRow.Delete
```

A strange situation now arises because of another Excel bug: whenever you try to delete entire *hidden* rows, Excel will unhide some of them, and I can explain why. So, the code needs to hide them all again, beginning on the first hidden row (mcondbRecordsFirstRow) to the end of the worksheet.

```
'Provision to keep rows hidden
mws.Range(Cells(mcondbRecordsFirstRow, 1), Cells(mws.Rows.Count, 1)).EntireRow.Hidden = True
```

■ **Attention** Try for yourself! Put a comment in the last instruction and try to delete any saved record. You will realize that many hidden rows will become inadvertently visible after the database record deletion.

The deletion process ends by deleting the local range name used to indicate where the many-side records were stored.

```
'Delete the many-records Range name
ws.Names(strRecordRange).Delete
```

And since this deletion process may select entire hidden rows, the code must scroll the worksheet to row 1 using the ActiveWindow.ScrowRow method and reactivate the worksheet protection, ending the Sub DeleteRecordData( ) code procedure.

```
        'Scroll to row 1
        ActiveWindow.ScrollRow = 1
    mws.Protect
End Sub
```

Once the DeleteRecordData( ) procedure ends, it returns the code to the DeleteRecord( ) procedure, which must return True as an indication that the record was removed from the database worksheet.

```
Public Function DeleteRecord() As Boolean
                ...
            If Not bolNewRecord Then
                Call DeleteRecordData(strRecord)
            End If
```

```
        DeleteRecord = True
        ...
End Function
```

Associating Database Procedures to Worksheet Button Controls

Any Button control inserted on a worksheet must be associated to a given Public Sub procedure inserted on the worksheet code module. This works fine for a single worksheet, but if you make a sheet copy, all existing Button controls will not update their code. They will continue to point to the original worksheet code module. When you click the Button control expecting that it executes the code inside the active sheet, it will continue to execute procedures of the original sheet, probably referring to wrong cell ranges.

The Save and Delete Button controls located on the BMI Chart worksheet at the right of the record data validation list use a simple approach to guarantee that if you make copies of this sheet tab, they will always execute the code contained on the active sheet code module. Basically, they are associated to the generic procedures Sub SaveRecord( ) and Sub DeleteRecord( ) stored in the basButtonControls code module, which has this code:

```
Public Sub SaveRecord()
    Dim obj As Object

    Set obj = ActiveSheet
    obj.Save
End Sub

Public Sub DeleteRecord()
    Dim obj As Object

    Set obj = ActiveSheet
    obj.DeleteRecord
End Sub
```

Note that they have been created as Public Sub procedures to appear in the Excel Assign Macro dialog box, and they use a quite simple object declaration technique called *late bound*. This means the procedure declares a Dim obj As Object variable that has no type, initializes it to the ActiveSheet object (meaning a Worksheet object), and then calls an object method (Save or DeleteRecord) that must exist inside the object that the variable represents.

Since the variable represents a generic object, when you compile it, Excel will not generate an error. Any possible error will be raised just when the procedure tries to *late-bound execute* the desired variable object method.

■ **Attention** Right-click any Button control and choose Assign Macro in the context menu to show the Excel Assign Macro dialog box and verify (or associate) the procedure attached to it. Just Public Sub procedures declared in independent code modules will appear on the list (Figure 7-17).

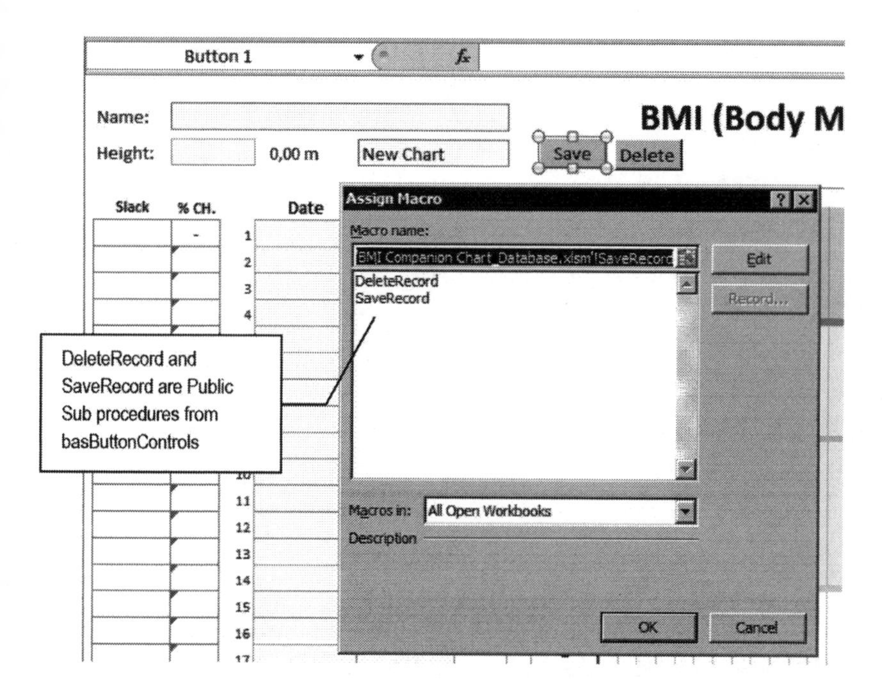

Figure 7-17. The Assign Macro Excel command (located in the Control section of the Developers tab) is used to set or verify the procedure associated to each worksheet's Button controls. By using generic procedures that call active sheet codes, you can make copies of the sheet tab that always execute the active sheet code module procedures

Making Copies of the BMI Chart Worksheet

Knowing that all the `BMI Chart` worksheet code module procedures point to local worksheet range names and that the `Button` controls points to generic procedures that execute code on the `ActiveSheet` object, you can make copies of the `BMI Chart` worksheet, producing another sheet tab with its own database storage system.

1. Right-click the `BMI Chart` sheet tab and choose the Move or Copy menu command.
2. In the Excel Move or Copy dialog box, select the "(Move to the end)" list option, select the Create a Copy check box, and click OK.

Excel will create the `BMI Chart (2)` sheet tab, with all procedure code (and records) of the original copy, but it is now capable of managing its own set of worksheet records (Figure 7-18).

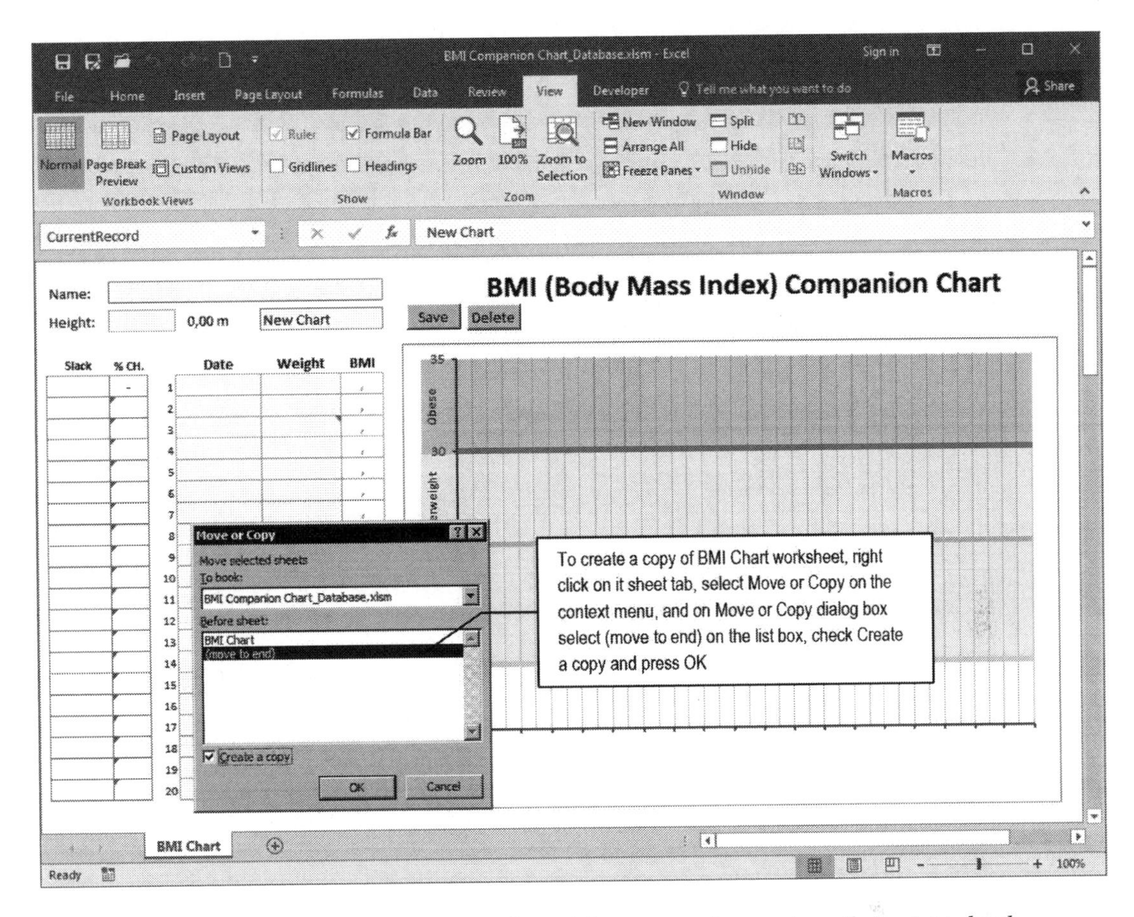

Figure 7-18. *Since the BMI Chart worksheet code module uses generic procedures that point to local worksheet range names and its Button controls use a generic code, you can make copies of this sheet tab to create another worksheet with its own, independent database storage system*

The USDA Food Composer_Database.xlsm Excel Application

In the file `Chapter07.zip` you will find the `USDA Food Composer_Database.xlsm` Excel macro-enabled workbook, another worksheet application produced to store any recipe information (food category, food item, amount by common measure or selected unit) and calculate its nutritional content for a single-serving portion. The USDA Food Composer has this name because it uses the USDA Agricultural Research Services (USDA-ARS) nutrient composition table to retrieve each food item's nutrient profile, using up to 184 different nutrients (Chapter 9 explains how to create updated versions of the hidden USDA worksheet used by this worksheet application).

This USDA Food Composer uses each food item amount to count the total recipe amount (in grams) for the number of servings that the recipe provides, calculate its calories, and offer the one-serving factor cell to reduce the recipe servings to a single serving (for example, a six-serving recipe can use the formula `=1/6` on its one-serving factor cell to recalculate the nutritional value of a single recipe serving).

This worksheet application employs the same interface formatting techniques used by the BMI Companion Chart worksheet application.

- Its input cells are formatted with a light yellow background and a blue border.
- Cells with calculated data have a white background and a blue border.
- Grid lines and headers are hidden.
- The worksheet is protected (with no password), allowing just its input cells (cells with a yellow background) to be selected.
- All unused rows and columns are also hidden, leaving a big gray area surrounding its interface if you try to navigate the worksheet using Excel scroll bars.
- It has three hidden worksheets: `USDA` (with nutritional data), `CommonMeasures` (with food item common measure information), and `Conversion` (to make unit conversions to grams).

It also uses the same database code engine employed by the `BMI Companion Chart_Database.xlsm` application to save each recipe data record.

- It has a `My Recipes` sheet tab, whose `CodeName` property was changed to `MyRecipes` to easily access its properties and methods in VBA.
- The `My Recipes` sheet tab code module has the same set of constants used by the `BMI Chart` worksheet, whose values were updated to reflect its one-side and many-side records.
- It uses the same database strategy: a data validation list cell to allow the selection of previously saved records.
- It has the New, Save, and Delete `Button` controls to easily manage worksheet records.

The `USDA Food Composer_Database.xlsm` workbook philosophy uses the hidden `USDA` worksheet tab (which has the latest USDA-ARS standard reference file for nutrient information) to build and nutritionally analyze any recipe composed by up to 18 different ingredients. Among those, any previously saved recipe record can be selected as a regular food item of the My Recipes food category. It produces the recipe Nutrition Facts food label for one single serving, offering the best-detailed nutrient information available in its Nutrient Composition area.

It was used to compose and save every recipe proposed by two of the most prominent NHLBI diet plans (National Health, Lung and Blood Institute, available at `https://www.nhlbi.nih.gov/health/health-topics/topics/dash`):

- Dietary Approach to Stop Hypertension (DASH)
- Therapeutic Life Changes (TLC)

Many other vegetarian recipes are offered on the EatingWell web site (`www.eatingwell.com`) for the first seven days of its 1,800-calorie EatingWell 28-day vegetarian meal plan. At the right of the USDA Food Composer worksheet title you can note that this application file indicates that it has 123 recipe records saved in its database (Figure 7-19).

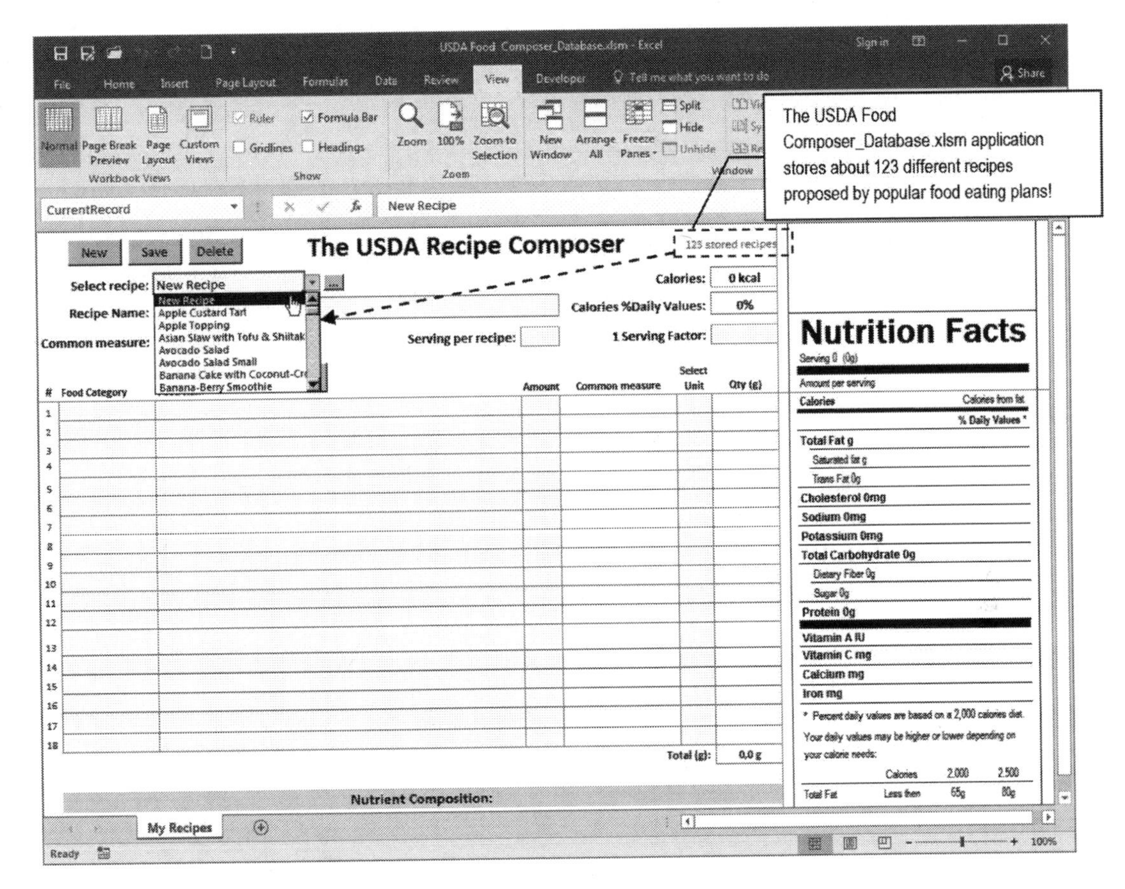

Figure 7-19. *This is the My Recipes sheet tab of the USDA Food Composer_Database.xlsm Excel application. It was used to compose and store all recipes from DASH, TLC, and part of the 1,800-calorie EatingWell 28-day vegetarian meal plan*

To create any recipe, click the New `Button` control or select New Recipe in its data validation list, type the recipe name, common measure, servings per recipe, and 1-serving factor. (This factor is a decimal number; it's a fraction used to reduce the number of servings to a single serving. It is usually typed as =1/Servings.) To compose the recipe ingredients, select the food category, choose the desired food item inside the food category (or click the Find Food Item button), type the amount (for its first common measure), or select a unit (gram, oz, cup, and so on). The amount selected will be always converted to grams (in the Qty column), and the application will search the USDA database to return all food item nutritional information proportional to the amount in grams selected in the recipe.

If you select any saved recipes, they will be automatically loaded from the worksheet database using the same technique and code explained in the section "The BMI Companion Chart.xlsm Excel Application" earlier in this chapter.

When any recipe record is selected in the data validation list cell pointed to by the `mcondbDataValidationList` constant, the one-side record is retrieved to show the recipe name, common measure, servings per recipe, and one-serving factor (a value used to reduce the recipe nutritional information to a single serving), while the many-side records are retrieved to show all food categories, food items, amount, and select unit used by each food item to compose the selected recipe.

Recipe nutritional information is recalculated by Excel, which will show its Nutrition Facts food label and the best nutrient composition available on the bottom of the worksheet, using the latest USDA Standard Reference (SR) file available at the time it was conceived. Figure 7-20 shows the "Banana Cake with Coconut-Cream Frosting" recipe, from the EatingWell web site, with all its food items and Nutrition Facts food label for a single serving (of about 144g = 144/28 = 5,1 oz).

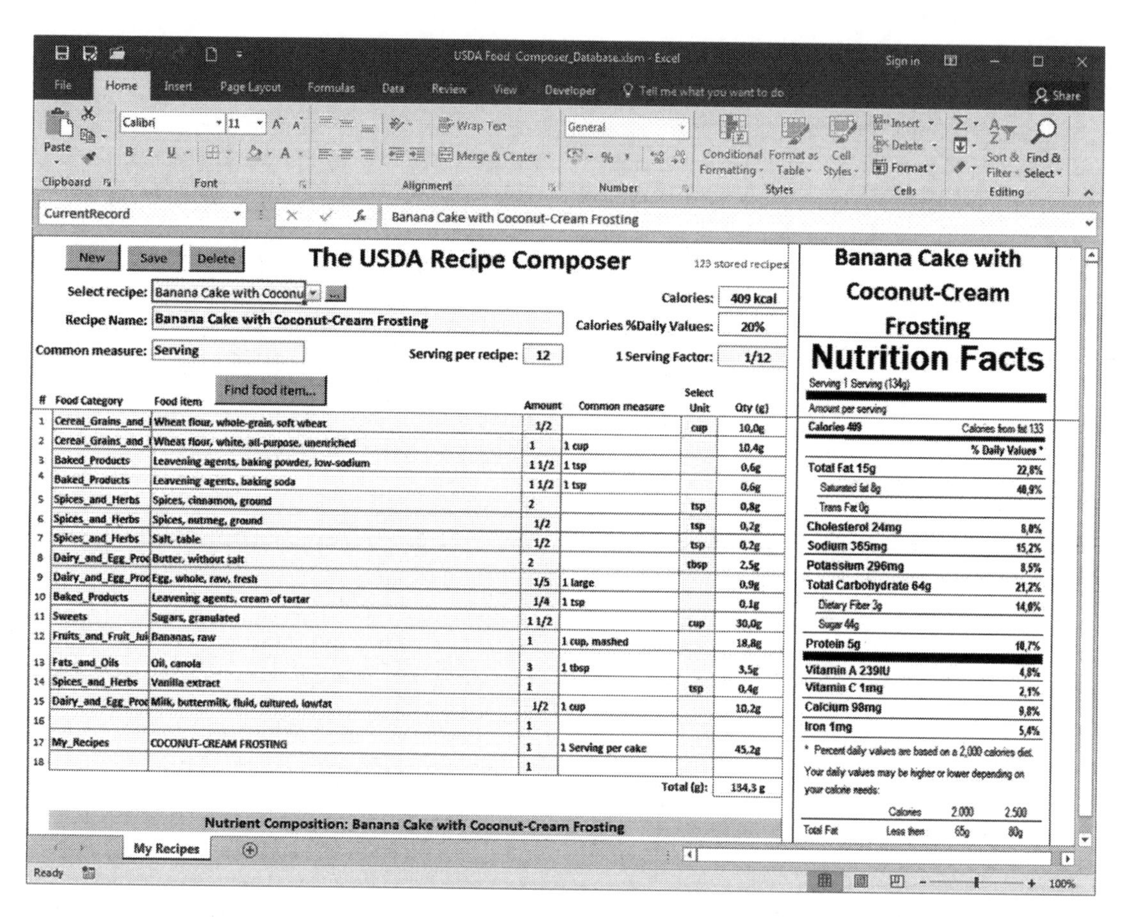

Figure 7-20. This is the USDA Food Composer_Database.xlsm application showing the "Banana Cake with Coconut-Cream Frosting" recipe, along with its Nutrition Facts food label for a single serving of 144g (≅5 oz)

And if you show all the `My Recipes` sheet tab hidden rows, you will see that they use the same technique to store recipe records. The `SaveRecords` range name holds each recipe name, while other sheet columns hold the one-side and many-side record data (Figure 7-21).

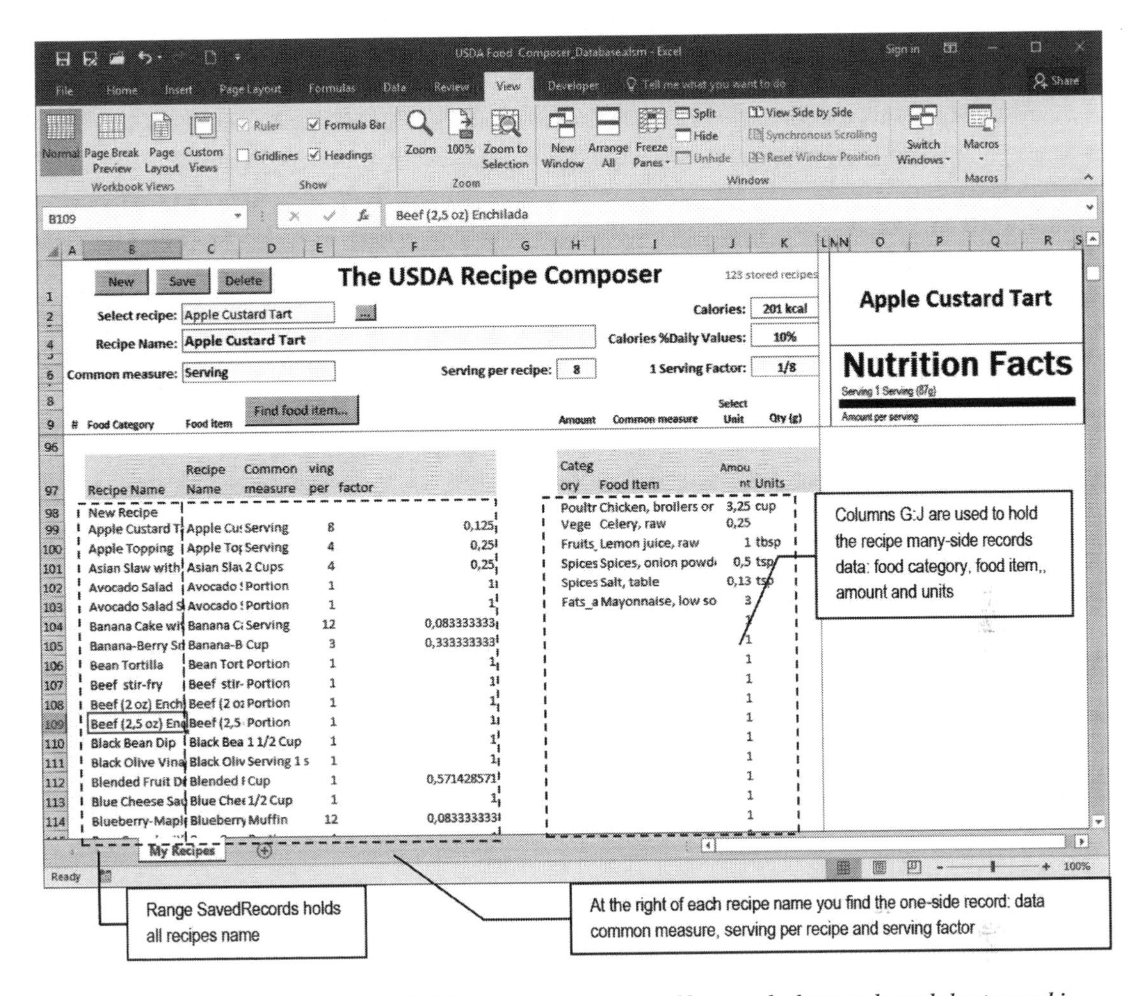

Figure 7-21. *By showing My Recipes hidden rows, you can reveal how and where each worksheet record is saved on the sheet rows. The mcondbSavedRecords range holds the recipe name. Contiguous columns on the same row of each recipe name hold the one-side record, identifying the recipe common measure, serving per recipe, and one-serving factor data. Other sheet columns holds the food category, food item, amount, and select unit used to compose each recipe*

Changing Database Constant Values

To effectively use the same code employed by the `BMI Chart` worksheet to manipulate recipe records stored on unused rows of the `My Recipes` worksheet, you just need to change the `MyRecipes` module constant values to fit them to the `My recipes` database needs. This is the `Declaration` section of the `MyRecipes` code module after such changes have been made:

```
Option Explicit

Private WithEvents mwb As Workbook
Public Dirty As Boolean  'Indicate if record dat had been changed
Private mstrLastRecord As String 'Retain the name of current record
Private Enum Operation
```

```
    LoadRecord = 1
    SaveRecord = 2
End Enum

'This variable receive frmSearchFoodItems selected item
Public SelectedFoodITem As Variant

'This constants refers to local range names
mcondbDataValidationList = "CurrentRecord"       'Data Validation list range
Const mcondbSavedRecords = "SavedRecords      'Saved records range name
Const mcondbRecordName = "Recipe               'Record name
Const mcondbOneSide = "OneSideRecord             'One-side record range
Const mcondbOneSideColumnsCount = 4            'One-side record columns needed
Const mcondbManySide = "ManySideRecords          'Many-side record range
Const mcondbManySidePrefix = "rec_"              'Many-side range name prefix
Const mcondbManySideColumnsCount = 4           'Many-side record columns needed
Const mcondbManySideRowsCount = 19             'Many-side record rows needed (+ 1 blank row)
Const mcondbRecordsFirstRow = 98               'Row where database begins
Const mcondbManySideFirstColumn = "$H$         'Many-side record first column
Const mcondbRangeOffset = 1                      'One-side record column offset to
mcondbSavedRecords
```

The constant values in bold mean that the MyRecipes worksheet database records begin on row 98 (mcondbRecordsFirstRow=98), use four columns to save the one-side worksheet record (mcondbOneSideColumnsCount=4), save the many-side worksheet records beginning on column H (mcondbManySideFirstColumn="H"), and use up to four worksheet columns (mcondbManySideColumnsCount=4) and 19 worksheet rows (mcondbManySideRowsCount=19) to save each recipe's many-side records.

Saving Recipe Data

Every time a recipe record is saved in the worksheet database, its nutritional value (associated to the My Recipes OneSideRecord range name) must also be saved in the hidden USDA worksheet My_Recipes food category so it can be reused as an independent food item in other recipes.

To avoid disturbing the database engine code, the My Recipes worksheet code module has two new distinct procedures: SaveRecipe( ) and SaveInMyRecipes( ).

- Privave Sub SaveRecipe( ): This is responsible for calling Function Save( ) to effectively save the recipe record on the worksheet database, and if Save( ) returns true, it calls SaveInMyRecipes( ).
- Private Sub SaveInMyRecipes( ): This is responsible for saving the recipe nutritional information in the USDA My_Recipes food category.

This is the SaveRecipe( ) code:

```
Private Sub SaveRecipe()
    Dim strRecord As String

    strRecord = Range(mcondbDataValidationList)
    If Save() Then
        'Update USDA My_Recipes range name with recipe nutritional data
```

```
        Call SaveInMyRecipes(strRecord)
    End If
End Sub
```

■ **Attention** The My Recipes worksheet Save button is associated to the `MyRecipes!SaveRecipe( )` procedure.

The `SaveInMyRecipes( )` procedure is also called from the `mwb_BeforeClose( )` event and from `Sub TryToLoadSelectedRecord( )` whenever `Save( )` returns `True`.

To keep things simple, the `My Recipes` worksheet uses the same range names to associate its one-side record cells (which contain recipe information) and many-side records cells (which contain recipe food item details, as shown in Figure 7-22).

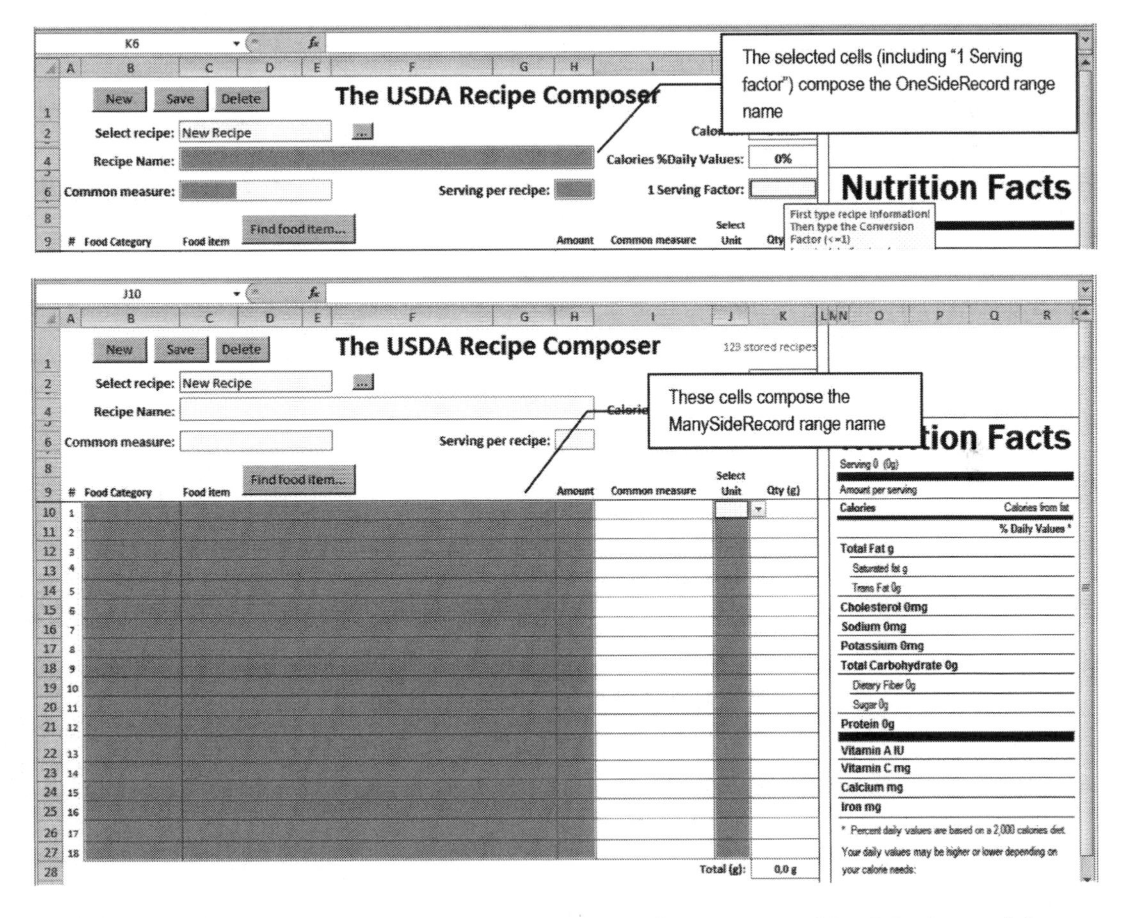

Figure 7-22. These are the OneSideRecord and ManySideRecord range names of the My Recipes worksheet, used by the database engine to save each recipe record. Since both range names select no contiguous cells, they do not appear in the Excel Name box

■ **Attention** Use `frmNames` to inspect the `OneSideRecord` and `ManySideRecords` hidden range names.

Saving Recipe Nutritional Information in My_Recipes Range Name

Every time a recipe record is saved, it must also save its nutritional information in the USDA worksheet `My_Recipes` range name so it can also be used as any other food item to compose new recipes (by first selecting the `My_Recipes` food category).This is done by calling the `Sub SaveOnMyRecipes( )` procedure every time a recipe is successfully saved, executing this code:

```
Public Sub SaveInMyRecipes(strRecord As String)
    Dim ws As Worksheet
    Dim rg As Range
    Dim rgRecipe As Range
    Dim rgUSDA As Range
    Dim strAddress As String
    Dim bolNewRecord As Boolean

    'Update recipe information on USDA worksheet
    'Find recipe name on USDA worksheet
    Set ws = Worksheets("USDA")
    Set rg = ws.Range("My_Recipes")
    Set rgRecipe = rg.Find(strRecord)

    If rgRecipe Is Nothing Then
        rg.Resize(rg.Rows.Count + 1).Name = "My_Recipes"
        'update rg object variable to contain My_Recipes new row
        Set rg = ws.Range("My_Recipes")
        'Position on new cell of My_Recipes range
        Set rgRecipe = rg.Cells(rg.Rows.Count, 1)

        'Resize USDA range name to encompass this new recipe
        Set rgUSDA = ws.Range("USDA")
        rgUSDA.Resize(rgUSDA.Rows.Count + 1, rgUSDA.Columns.Count).Name = "USDA"
        bolNewRecord = True
    End If

    'Copy current recipe nutritional data to Clipboard
    ActiveSheet.Range("NewRecipe").Copy
    'Paste nutritional data for current recipe
    rgRecipe.PasteSpecial xlPasteValues

    If bolNewRecord Then
        'A New Recipe was inserted on USDA My_Recipes range name. Sort it!
        rg.Sort rg
    End If
End Sub
```

After variable declarations, object range variables are initiated to represent the USDA worksheet and its `My_Recipes` range name, and a search is made on `My_Recipes` by the `Range.Find` method for the recipe represented by the `strRecord` argument.

```
    Set ws = Worksheets("USDA")
    Set rg = ws.Range("My_Recipes")
    Set rgRecipe = rg.Find(strRecord)
```

If the recipe is not found inside the My_Recipes range name, rgRecipe = Nothing, and the recipe must be inserted. To do it, first add the new row to the bottom of the My_Recipes range name (represented by the rg object variable) using the Range.Resize method.

```
    If rgRecipe Is Nothing Then
        rg.Resize(rg.Rows.Count + 1).Name = "My_Recipes"
        'update rg object variable to contain My_Recipes new row
        Set rg = ws.Range("My_Recipes")
```

Now attribute to rgRecipe this newly added blank row.

```
        'Position on new cell of My_Recipes range
        Set rgRecipe = rg.Cells(rg.Rows.Count, 1)
```

Since the USDA range name also contains the My_Recipes range name, increment its size by adding to it one row, and use bolNewRecord to signalize that this is a new recipe entry.

```
        'Resize USDA range name to encompass this new recipe
        Set rgUSDA = ws.Range("USDA")
        rgUSDA.Resize(rgUSDA.Rows.Count + 1, rgUSDA.Columns.Count).Name = "USDA"
        bolNewRecord = True
    End If
```

Now rgRecipe has either a new empty row or the desired recipe found by the by Range.Find method. Use the Range.Copy and Range.PasteSpecial methods to copy/paste the recipe nutritional information between the My Recipes and USDA worksheets.

```
    'Copy current recipe nutritional data to Clipboard
    ActiveSheet.Range("NewRecipe").Copy
    'Paste nutritional data for current recipe
    rgRecipe.PasteSpecial xlPasteValues
```

If the recipe is new, sort the My_Recipe range name with the Range.Sort method, using the rg column as the sort column, to place the new recipe in ascending order.

```
    If bolNewRecord Then
        'A New Recipe was inserted on USDA My_Recipes range name. Sort it!
        rg.Sort rg
    End If
End Sub
```

This is everything you need to know about how a recipe is saved!

Deleting a Recipe Data

The same way SaveRecipe( ) needs to call SaveInMyRecipes( ) to save the recipe nutritional information in the USDA My_Recipes food category, to delete a recipe record, this information must be also deleted from USDA My_Recipes.

To not disturb the database engine code, the My Recipes code module offers Sub DeleteRecipe( ), which calls Function DeleteRecord( ) to effective delete the recipe record and, if it returns True, takes care of also deleting the recipe entry in the USDA worksheet's My Recipes range name.

```
Private Sub DeleteRecipe()
    Dim rg As Range
    Dim rgRecipe As Range
    Dim strRecord As String

    strRecord = Range(mcondbDataValidationList)

    If DeleteRecord() Then
        'Delete recipe from USDA My_Recipes range name
        Set rg = Worksheets("USDA").Range("My_Recipes")
        Set rgRecipe = rg.Find(strRecord, , , xlWhole)
        If Not rgRecipe Is Nothing Then
            rgRecipe.EntireRow.Delete
        End If
    End If
End Sub
```

I think that this code deserves further explanation.

Things That Are Worth Being Mentioned

The My Recipes sheet tab of the USDA Food Composer_Database.xlsm Excel application makes use of two other UserForms and one Excel function that are worth mentioning. Let's see them in more detail.

Finding Food Items to Compose Recipes

The first one is frmSearchFooditems, developed in Chapter 6. You can use it to search for food items while composing any new recipe or to change/verify other common measures of any selected food item. Just select the desired Food Item input cell and click the Find Food Item control button to show the UserForm interface (Figure 7-23).

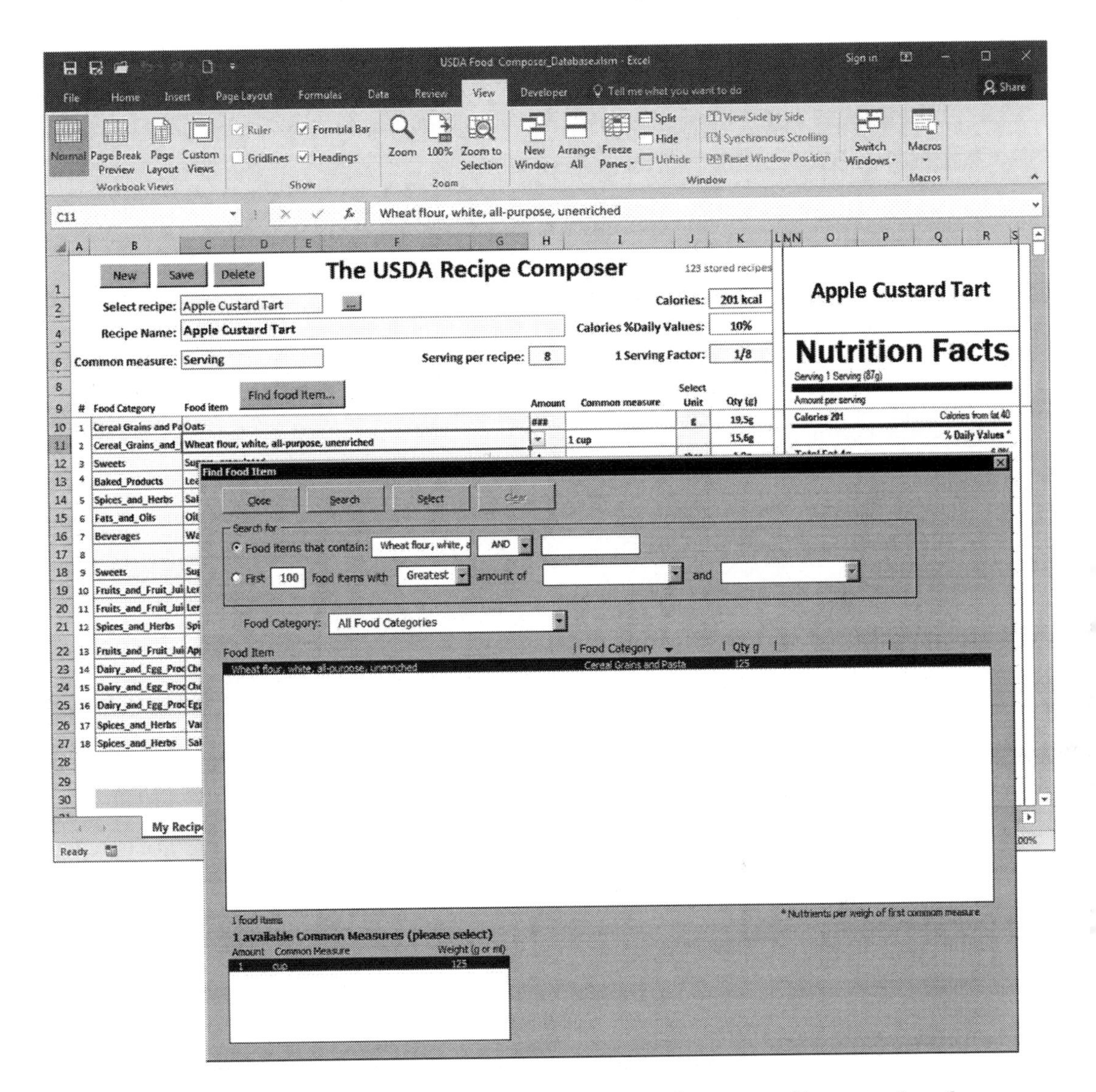

Figure 7-23. *Select any Food Item input cell and click the Find Food Item control button to show the frmSearchFoodItem interface. If the selected cell has any food item selected, this item will be searched and shown in the UserForm, like the oats used in the selected recipe*

If you have not selected any Food Item input cell, a warning message will appear indicating that you should do so. This is made by Sub FindFoodItem( ), which works together with the SelectedFoodItem As Variant module-level variable.

```
Option Explicit
...
'This variable receive frmSearchFoodItems selected item
Public SelectedFoodITem As Variant
...
Private Sub FindFoodItem()
    Dim frm As New frmSearchFoodItems
    Dim rg As Range

    Set rg = Application.Intersect(Selection, Range("RecipeFoodItems"))
    If rg Is Nothing Then
        MsgBox "Click on any cell of Food Item column and try again!", _
                vbInformation, _
                "Select a Food Item"
    Else
        With frm
            .CallingSheet = Me
            .Show vbModal
        End With
        If IsArray(mws.SelectedFoodITem) Then
            With Application.Selection
                .Value = mws.SelectedFoodITem(0)
                .Offset(0, -1) = mws.SelectedFoodITem(1)
                .Offset(0, 1) = mws.SelectedFoodITem(2)
                .Offset(0, 3) = "g"
            End With
            mws.SelectedFoodITem = Empty
        End If
    End If
End Sub
```

To verify whether the selected cell (represented by the Application.Selection property) is inside the range RecipeFoodItems, the procedure uses the Application.Intersect method, which has this syntax:

```
Expression.Intersect(Arg1, Arg2, Arg3, ..., Arg30)
```

In this code:

Expression: This is required; it is a variable that represents an Application object.

Arg1: This is required; it is a Microsoft Excel range object.

Arg2: This is required; it is a Microsoft Excel range object.

Arg3...Arg30: These are optional; they are Microsoft Excel range objects.

The Application.Intersect method requires that at least two ranges (Arg1 and Arg2) be specified and returns a range object that represents the rectangular intersection of two or more of the specified ranges.

So, Sub FindFoodItem( ) uses Intersect(Selection, Range("RecipeFoodItems) to initialize the rg object variable. If Selection is not a cell inside RecipeFoodItems, rg = nothing, and the MsgBox( ) is displayed to the user, asking the user to select any food item cell.

```
    Set rg = Application.Intersect(Selection, Range("RecipeFoodItems"))
    If rg Is Nothing Then
        MsgBox "Click on any cell of Food Item column and try again!", _
                vbInformation, _
                "Select a Food Item"
```

But if any food item cell is selected, a new instance of frmSearchFoodItems is instantiated, passing to the frmSearchFoodItems.CallingSheet property a reference to the current worksheet (represented by the Me keyword), and the UserForm is shown in modal mode, stopping the code execution until it is closed.

```
Private Sub FindFoodItem()
    Dim frm As New frmSearchFoodItems
   ...
    Else
        With frm
            .CallingSheet = Me
            .Show vbModal
        End With
```

As explained in Chapter 6, if the user tries to search a given food item and clicks the UserForm Select CommandButton, the frmSearchFoodItems cmdSelect Click( ) event will fire and try to fill a worksheet property called SelectedFoodItem with a one-dimensional array containing just four rows to indicate the selected food item name, food category, amount in grams, and common measure (if any).

So, immediately after the UserForm is closed, the procedure will verify whether the SelectedFoodItem property is an array using the VBA IsArray( ) function. If it is, it will use the Application.Selection property with the Range.Offset method to correctly return the desired food item values.

```
        If IsArray(mws.SelectedFoodITem) Then
            With Application.Selection
                .Value = mws.SelectedFoodITem(0)
                .Offset(0, -1) = mws.SelectedFoodITem(1)
                .Offset(0, 1) = mws.SelectedFoodITem(2)
                .Offset(0, 3) = "g"
            End With
```

And once this operation is done, the MyRecipes.SelectedFoodItem property is set to Empty, which is the default state of any variant variable, destroying its array of food item characteristics.

```
            mws.SelectedFoodITem = Empty
        End If
```

Finding Recipes Already Composed

The second thing that is worth mentioning is frmSearchRecipes, which can be shown by clicking the small control button with reticence, right next to the "Select recipe" data validation list (mcondbDataValidationList range name). When you click it, the UserForm is loaded, showing all 123 stored recipe names. To select any recipe, scroll the list box or type any text to filter the list and double-click the recipe name to load it in the My Recipes sheet tab (Figure 7-24).

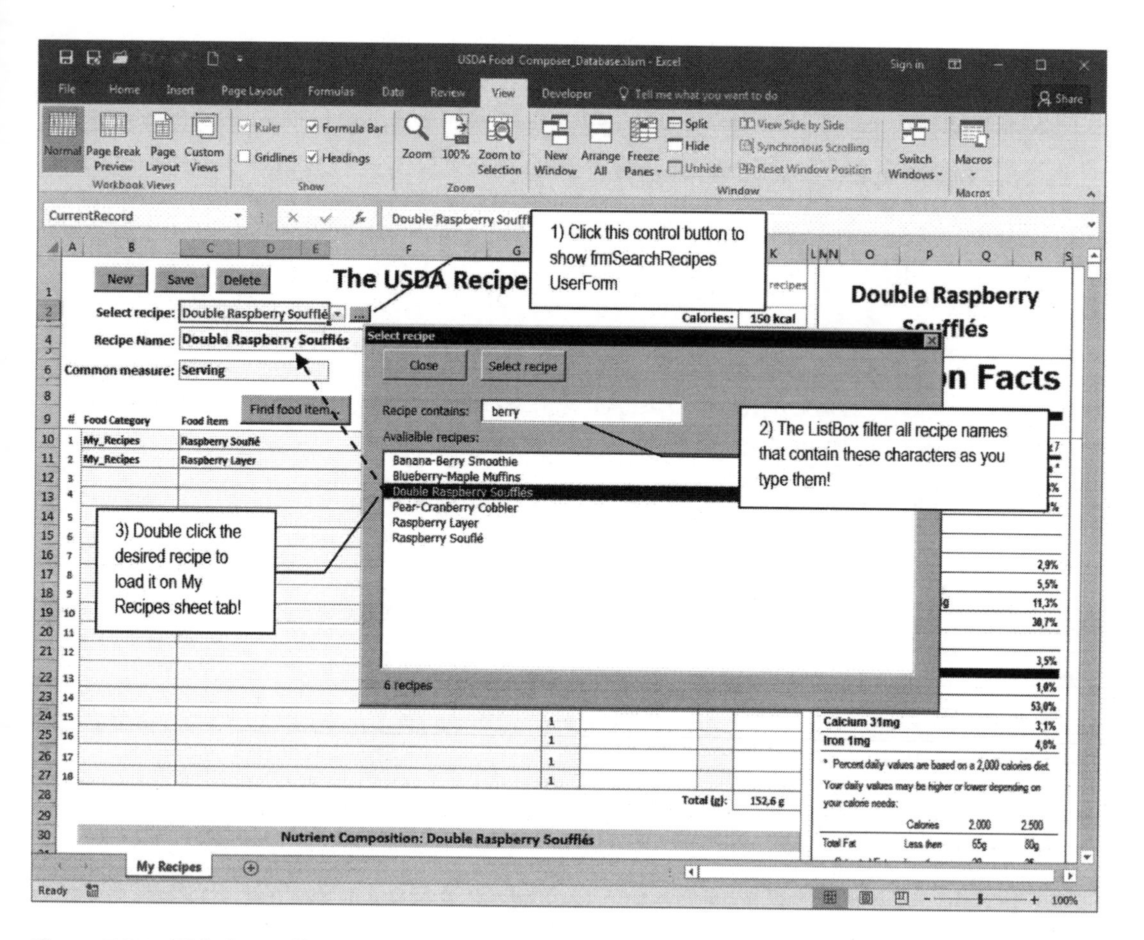

Figure 7-24. *Click the small control button next to the "Select recipe" data validation list to open the UserForm frmSearchRecipes, where you can easily find any recipe by typing part of it name*

By selecting a recipe in the `lstRecipes` `ListBox` or double-clicking a recipe name in the `frmSearchRecipes` `UserForm`, the recipe name is attributed to the `My Recipes` `mcondbDataValidationList` range name, which will fire the `MyRecipes` `Worksheet_Change( )` event and load the recipe, executing what can be considered a kind of *automation* on the worksheet application. It is the same thinking when you click the New `Button` control to begin a new recipe.

This time I will leave it to you, as an exercise, to study and understand how `frmSearchRecipes` works.

Counting Saved Recipes

Although it can be done many different ways, to count how many recipes are currently stored in the worksheet database and show the number in the merged cells `J1:L1` of the `My Recipes` worksheet, the `My Recipes` worksheet uses the simplest approach: a formula with the Excel function `=CountA(My_Recipes)` that counts how many nonempty cells exist inside the `My_Recipes` range name (Figure 7-25).

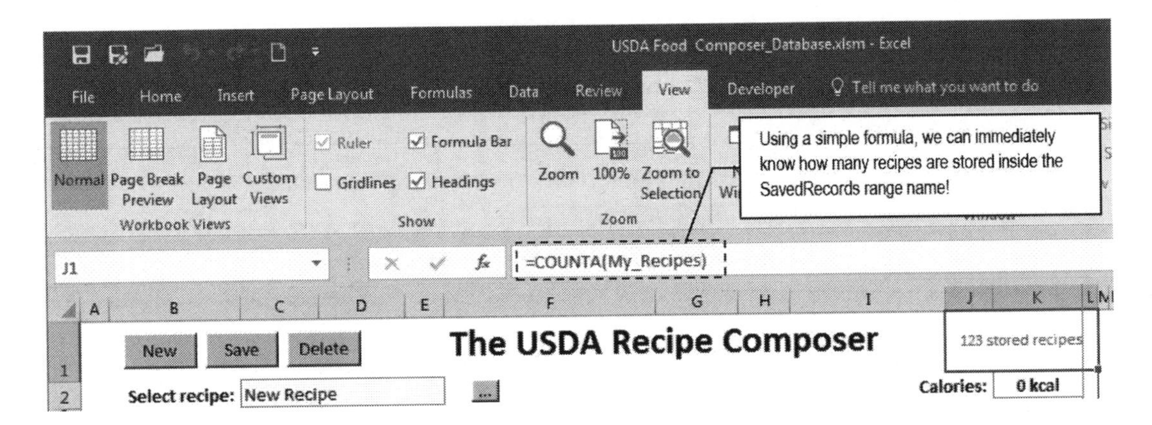

Figure 7-25. *To count and show how many recipes are stored inside the database, the formula =CountA("My_Recipes") counts how many nonempty cells exist in the My_Recipes range name*

Chapter Summary

In this chapter, you learned about the following:

- That almost any worksheet solution uses a default one-to-many record relationship so commonly used on database design tables
- That using the correct strategy, with a data validation list (`mcondbDataValidationList`) and a unique range name (`mcondbSavedRecords`), you can use unused worksheet rows and columns to store the worksheet data, making it more productive
- How to use the `Worksheet` object's `Change` and `SelectionChange` events to take care of a database application system
- How to use the `Application.EnabledEvents` property to avoid that the worksheet code cascade-fires events
- How to avoid screen flickering using the `Application.ScreenUpdates` property
- How to save records using a one-side and many-side record relationship using different places of the worksheet
- How to create a set of procedure codes, based on the declared module `Constant` values, to indicate where the database record is stored
- How to save, load, and delete records stored in a worksheet database system
- How to determine which is the last-used worksheet row, where a new record must be saved
- How to rebuild a range name (`mcondbSavedRecords`) after inserting or removing any of its items

- How to paste values inside a merged range (like `RecipeFoodItems`), attributing a range value to another (see the `LoadSaveData( )` procedure for details)
- How to use the Excel `Application.Intersect` method to verify whether a cell is inside a range name
- How to use a single centralized procedure (`LoadSaveData( )` to load and save worksheet records to/from the worksheet
- That you can use the Excel `CountA( )` function to count the number of nonempty cells inside any range name

In the next chapter, you will learn how to transform the worksheet database code into a worksheet database engine `Class` module, including the creation of a wizard to help users implement it in any worksheet application.

CHAPTER 8

Creating and Setting a Worksheet Database Class

In the previous chapter you studied how to use Excel as a database repository, using a programmable approach based on a data validation list filled with a New Record item followed by all saved records and a simple strategy to save any one-to-many record database relationship on unused worksheet rows. The database engine was based on some Excel object events (`Worksheet.Change`, `Worksheet.SelectionChange`, and `Workbook.BeforeClose`) and a punch of generic procedures that use module-level constant values to load, save, and delete worksheet application records.

The proposed code can be easily copy and adapted from one worksheet layout to another by changing some constant values, letting the database engine deal with different one-to-many worksheet database record relationships.

The drawback of this approach is code duplication: any improvement made on any worksheet application database engine code must be made on every other worksheet application to propagate it.

In this chapter, you will learn how to create and use a `Class` module to produce a robust, generic database engine to easily implement the database storage system in any Excel worksheet. You can obtain all files and procedure code in this chapter by downloading the `Chapter 07.zip` and `Chapter08.zip` files from the book's Apress.com product page, located at `www.apress.com/9781484222041`, or from `http://ProgrammingExcelWithVBA.4shared.com`.

Creating a Database Class

The power of `Class` modules comes from the fact that they work like any other object. You just need to declare a variable of the module's type and use its programmable interface (properties, methods, and events) to make it do something useful, keeping all the code complexity encapsulated inside the `Class` module.

Using the `BMI Companion Chart_Database.xlsm` macro-enabled worksheet that you can find inside `Chapter07.zip` as an example, you can duplicate all the code procedures used to save, load, and delete the `BMI Chart` worksheet records inside a `Class` module. The code will expose `Class.Save`, `Class.Load`, and `Class.Delete` methods to do the same jobs without knowing how they're done.

To convert all `BMIChart` code module procedures to a database engine's `Class` module, you must do the following:

1. Create a new `Class` module and change its `Name` property to give it a precise identity.
2. Cut all `BMIChart` code module procedures and paste them inside the `Class` module.

F. Morgado, *Programming Excel with VBA*, DOI 10.1007/978-1-4842-2205-8_8

3. Change the `Worksheet_Change( )` and `Worksheet_SelectionChange` event declarations to `mws_Change( )` and `mws_SelectionCange( )`.
4. Change each constant associated to the database engine to a local class module variable that has the same name (but a different prefix).
5. Make a search-and-replace operation on the code to change each constant name to the associated variable name.

These are the basic steps necessary to implement the database `Class` code, but to make it work on any kind of one-to-many worksheet record relationship, you must also do the following:

1. Store the database class variable values on unused worksheet rows, using range names to identify each database property.
2. Use the `Class_Initialize( )` event to set a reference between the `mws` object variable and the `ActiveSheet` object, and load each database property range name value to the associated class module variable created in step 6.

Let's do all these steps so you can understand the database `Class` creation process.

Steps 1 and 2: Create the Database Class Module

To change all `BMIChart` worksheet code procedure to a `Class` module, follow these steps:

1. Press Alt+F11 to show the Visual Basic IDE.
2. Double-click the `BMIChart` object in the Project Explorer tree to show the code module.
3. Place the text cursor behind the `Option Explicit` statement and press Ctrl+Shift+End to select all code module declarations and procedures.
4. Press Ctrl+X to cut the code from the `BMIChart` code module.
5. Create a new `Class` module using the Visual Basic Insert ➤ Class module menu command and press Ctrl+V to paste it.
6. Change the `Class1 Name` property to `clsDatabase` to name it.

All declarations and procedures used by the `BMIChart` code module will be transferred to the `clsDatabase Class` module (Figure 8-1).

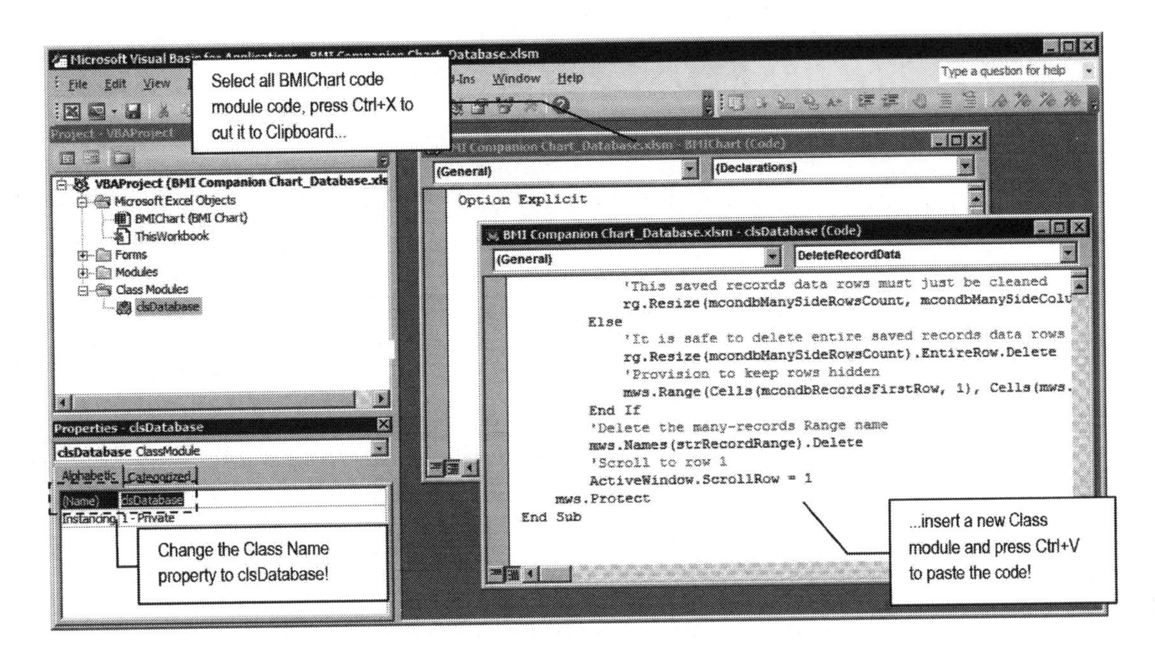

Figure 8-1. *Cut all code below the Option Explicit statement from the BMIChart object code module, create a new Class module, paste the code inside it, and change the Class Name property to clsDatabase*

Step 3: Create an Object Variable to Capture Worksheet Events

Once all the `BMIChart` object declarations and procedures are transferred to the `clsDatabase` class module, it is time to create the module-level object variable that can capture the active sheet events.

> While the `clsDatabase` class module is selected, locate the `Worksheet_SelectionChange( )` and `Worksheet_Change( )` procedures and change their names to the `mws_SelectionChange( )` and `mws_Change( )` events (Figure 8-2).

■ **Attention** You just need to select the `Worksheet` name part of each event procedure and change it to `mws`. After you make that change, the top-left code module `ComboBox` must show the `mws` name selected, while the top-right `ComboBox` must show the `SelectionChange( )` event name.

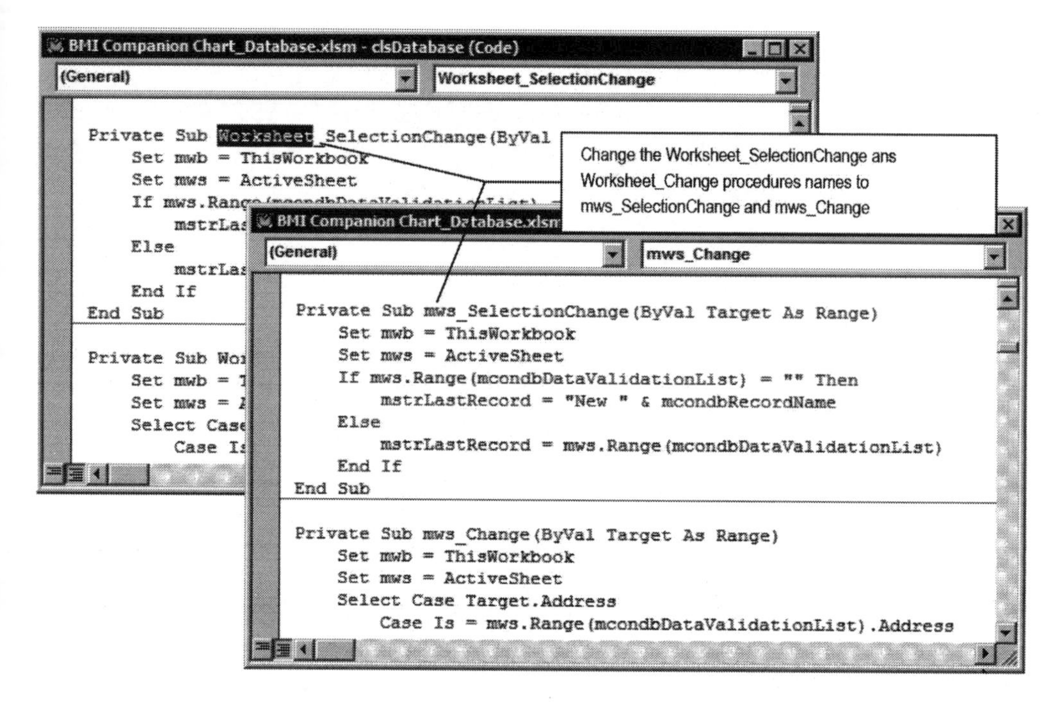

Figure 8-2. *Select the Worksheet_SelectionChange( ) and Worksheet_Change( ) event procedures and change their names to mws_SelectionChange and mws_Change( )*

Steps 4 and 5: Change Constant Names to Variable Declarations

To parameterize the `clsDatabase` database properties, you now need to declare one module-level variable to represent each database `Constant` declaration, changing the con prefix part of its name to the appropriate prefix for the variable data type.

Constants associated to a string value must be declared `As String`; the ones associated to numeric values must be declared `As Integer`. For example, if the `Constant` value stores a string, the `mcondb` prefix of the declared variable must be changed to `mstrdb`. If it stores an integer value, the `mcondb` prefix must be changed to `mintdb`. Table 8-1 states how each variable name must be declared.

Table 8-1. *Constant Names That Must Be Declared as Variable Names*

Constant Name	Variable Name Must Be Declared As	Variable Type
`Const mcondbDataValidationList`	`Dim mstrdbDataValidationList`	`String`
`Const mcondbSavedRecords`	`Dim mstrdbSavedRecords`	`String`
`Const mcondbRecordName`	`Dim mstrdbRecordName`	`String`
`Const mcondbOneSide`	`Dim mstrdbOneSide`	`String`
`Const mcondbOneSideColumsCount`	`Dim mintOneSideColumsCount`	`Integer`
`Const mcondbManySide1`	`Dim mstrdbManySide1`	`String`
`Const mcondbManySide2`	`Dim mstrdbManySide2`	`String`

(continued)

Table 8-1. *(continued)*

Constant Name	Variable Name Must Be Declared As	Variable Type
`Const mcondbManySide3`	`Dim mstrdbManySide3`	`String`
`Const mcondbManySide4`	`Dim mstrdbManySide4`	`String`
`Const mcondbManySidePrefix`	`Dim mstrdbManySidePrefix`	`String`
`Const mcondbManySideColumnsCount`	`Dim mintdbManySideColumnsCount`	`Integer`
`Const mcondbManySideRowsCount`	`Dim mintdbManySideRowsCount`	`Integer`
`Const mcondbRecordsFirstRow`	`Dim mintdbRecordsFirstRow`	`Integer`
`Const mcondbManySideFirstColum`	`Dim mstrdbManySideFirstColum`	`String`
`Const mcondbRangeOffset`	`Dim mintdbRangeOffset`	`Integer`

After all variables have been declared, execute a search-and-replace operation to change all `Constant` occurrences inside the `clsDatabase` code module to the respective variable names.

1. Since `mcondbDataValidationList` stores a string that represents a range name or cell address, declare the `mstrdbDataValidationList as String` variable above its constant declaration.

   ```
   Dim mstrdbDataValidationList as String
   ```

2. Double-click the `mcondbDataValidationList` constant name to select it and press Ctrl+H to show the Visual Basic Replace dialog box with the selection stored in the Find What option.
3. In the "Replace with" option, type `mstrdbDataValidationList`.
4. Use Search = Current Module, Direction = Down and check Find Whole Word Only.
5. Click the Replace All button to change all occurrences of this `Constant` name inside the code. Visual Basic will make 15 substitutions (Figure 8-3).

Repeat the last four operations to all other `Constant` declarations noting that `mcondbOneSideColumnsCount`, `mcondbManySideColumnsCount`, `mcondbManySideRowsCount`, `mcondbRecordsFirstRow`, and `mcondbRangeOffset` must be associated with a variable of the same name, prefixed by `mintdb` and declared `As Integer`.

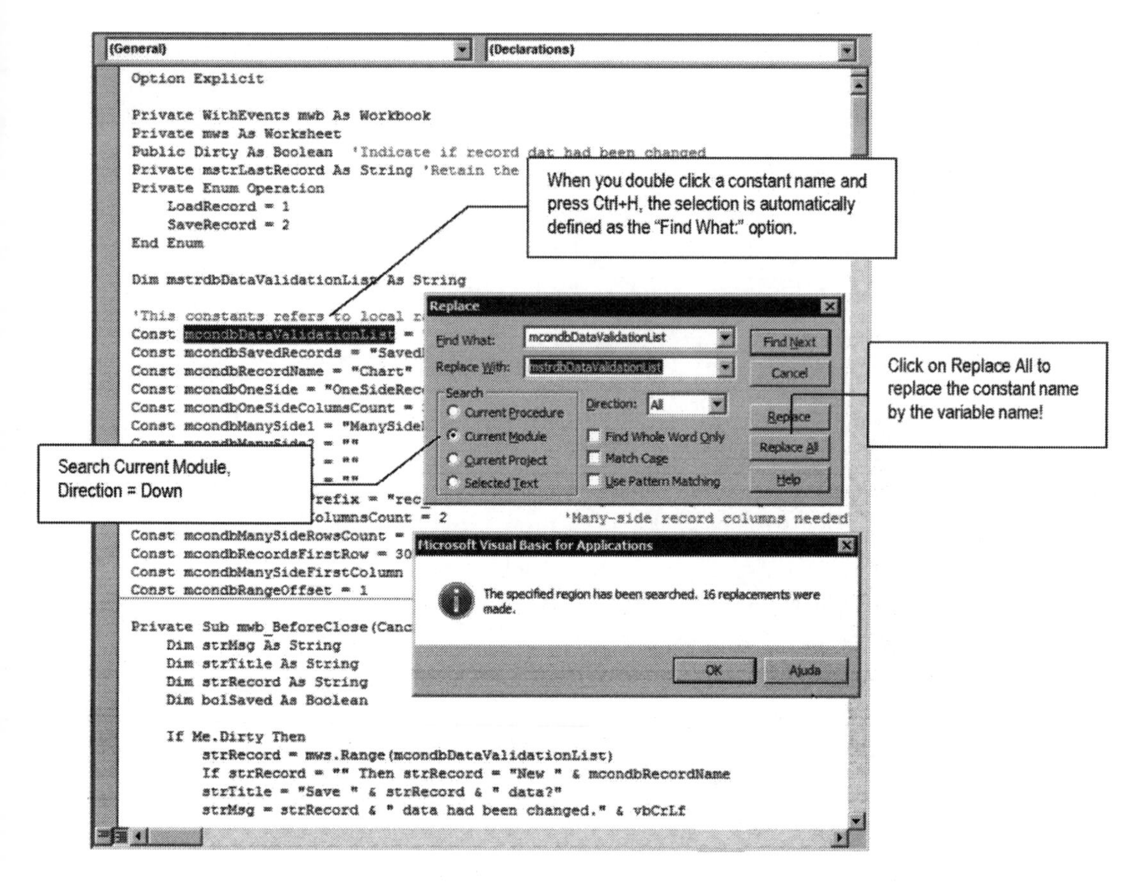

Figure 8-3. For each constant name associated to a database engine property, declare a module-level variable with the same name and different prefix and use Visual Basic's Replace form (Ctrl+H) to change all constant name occurrences by the associated variable name

---

■ **Attention** Although this is a simple operation, you must be careful to avoid making inappropriate `Constant` names changes.

After all search-and-replace operations have been performed to replace constant names with class module variable names, you must comment each `Constant` declaration by prefixing each line with a single quote. Then use the Visual Basic Debug ➤ Compile VBAProject command to check whether all constants have been correctly replaced. The `clsDatabase` code module declaration section will look like Figure 8-4.

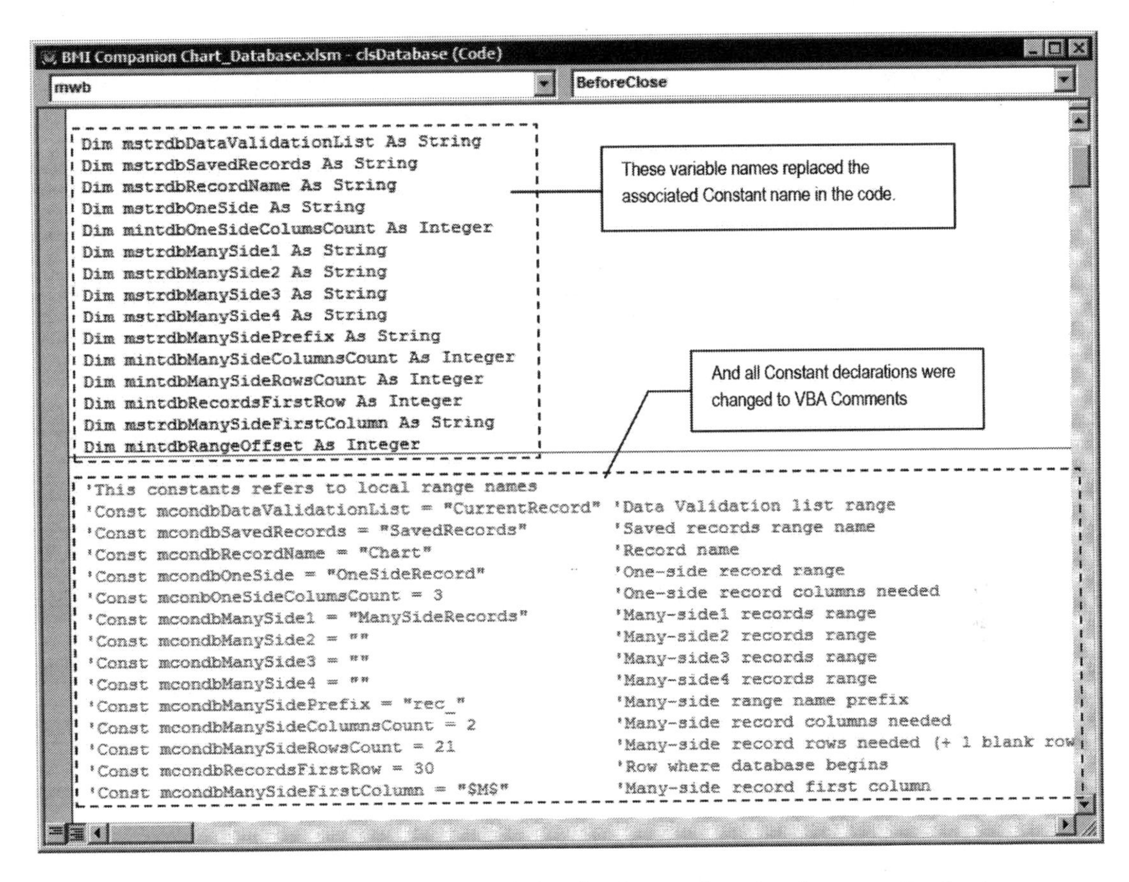

Figure 8-4. *After all variable names have been declared and a search-and-replace operation has been performed to change all constant names to the associated variable names, the clsDatabase declaration section will look like this*

■ **Attention** It is good programming practice to group variables declared with the same type: `Strings` first, then `Integers`.

Step 6: Save Database Properties as Range Names

Since the `clsDatabase` class does not have any database property value (it now has just variable declarations), it must read those values from the worksheet where the records reside. You need to save each variable name (removing the four-letter prefix) and value on worksheet cells and associate worksheet scope range names to these values.

These range names must be placed in safe, unused cells that will be not disturbed by any database operation. A good place are the first `BMI Chart` worksheet columns (`A` and `B`) beginning at the first record row (row `30`, pointed at by the old `mcondbRecordsFirstRow = 30` constant). All range names must be created as local worksheet names, allowing other sheet tabs of the same workbook to have their own database storage system. You need to follow these steps:

1. Unprotect the `BMI Chart` worksheet and show its hidden rows.

2. Use cell `A30:B44` to create the database property range names.
   - Column A receives the database property name (type each variable name without its first four-letter prefix. For example, `mstrdbDataValidationList` must be typed as `dbDataValidationList`).
   - Column B receives the database property value (same constant value).
3. Select cells `A30:B44` (where the range names have been inserted) and execute the command Create from Selection, located in the Defined Names area in the Formula tab of the Excel ribbon to show the Create Range Names from Selection dialog box,
4. Keep just the "Left column" option selected and click OK (Figure 8-5).

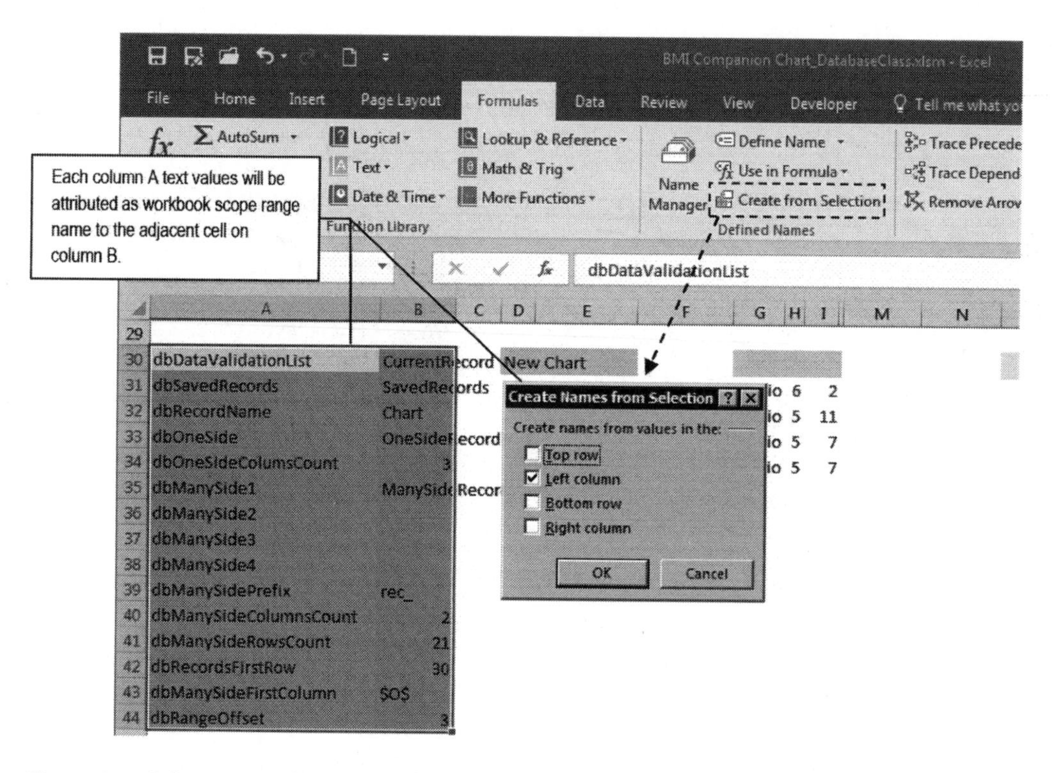

Figure 8-5. *Select a two-column range (where the first column contains the desired range names, and the second column contains the range name values) and use the Create from Selection command in the Defined Names area in the Formula tab of the ribbon to automatically create workbook range names. You will need to use frmNames to change the name scope to the BMI Chart worksheet*

■ **Attention** The Create Range names from Selection dialog box can create just workbook scope range names. All database property range names must be created as local worksheet names to allow other sheet tabs to implement their own database storage system. Since Excel doesn't allow local range name creation with this method or allow changing the name scope using the Names Manager dialog box, use the `frmNames`, located in the VBA project tree, to change each name scope from the `Workbook` worksheet to the `BMI Chart` worksheet.

Figure 8-6 shows how the `BMI Chart` worksheet should look after you have created all range names. Note that cell `B29` is associated to the `dbDataValidationList` cell range name (which is a local name whose scope was changed using the `frmNames` interface.

■ **Attention** Note in Figure 8-6 that the `dbManySideFirstColumn` range name value was defined as `$O$`, as an indication that the many-side record storage for the `BMI Chart` worksheet begins on column O.

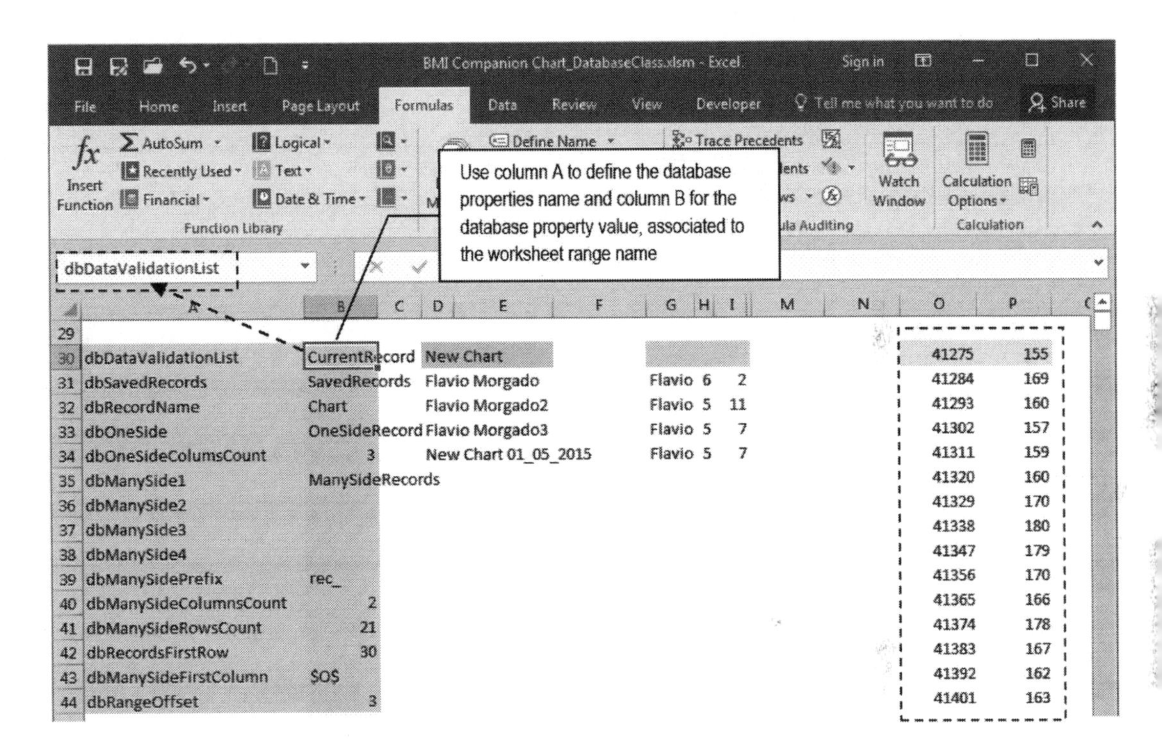

Figure 8-6. *This is how the BMI Chart worksheet should look after you have created the database property range name using cell range A30:B44. Note that each cell on column A is used just to indicate the database property name located at its right, on column B*

Step 7: Use the Class_Initialize( ) Event to Read Database Properties

The last step needed to implement the `clsDatabase` class is to make it associate the active sheet database property range name values to each of its module-level variables, which is a simple approach.

1. In the `clsDatabase Sub Class_Initialize( )` event, initialize the `mws as Worksheet` module-level variable to the `ActiveSheet`.

2. Define each module-level variable to the appropriate local range name, using the `mws.Range` property.

This is how the Class_Initialize( ) event should look to load all database property values:

```
Private Sub Class_Initialize()
    Set mws = ActiveSheet
    mstrdbDataValidationList = mws.Range("dbDataValidationList")
    mstrdbSavedRecords = mws.Range("dbSavedRecords")
    mstrdbRecordName = mws.Range("dbRecordName")
    mstrdbOneSide = mws.Range("dbOneSide")
    mintdbOneSideColumsCount = mws.Range("dbOneSideColumsCount")
    mstrdbManySide1 = mws.Range("dbManySide1")
    mstrdbManySide2 = mws.Range("dbManySide2")
    mstrdbManySide3 = mws.Range("dbManySide3")
    mstrdbManySide4 = mws.Range("dbManySide4")
    mstrdbManySidePrefix = mws.Range("dbManySidePrefix")
    mintdbManySideColumnsCount = mws.Range("dbManySideColumnsCount")
    mintdbManySideRowsCount = mws.Range("dbManySideRowsCount ")
    mintdbRecordsFirstRow = mws.Range("dbRecordsFirstRow")
    mstrdbManySideFirstColumn = mws.Range("dbManySideFirstColumn")
    mintdbRangeOffset = mws.Range("dbRangeOffset")
End Sub
```

Referencing the clsDatabase Class

Now that the clsDatabase class is capable of reading all worksheet database properties, you need to initialize it on the BMI Chart worksheet by following these steps:

1. Declare a Private mdb as clsDatabase object variable in the BMI Chart code module declaration section.

   ```
   Option Explicit

   Dim mdb as clsDatabase
   ```

2. Initialize the mdb object variable using the Worksheet_Activate( ) event and make it a Public procedure so you can call it in the Workbook.Open( ) event, initializing the class every time the workbook is opened.

   ```
   Public Sub Worksheet_Activate()
       If mdb Is Nothing Then
           Set mdb = New clsDatabase
       End If
   End Sub
   ```

3. In the ThisWorkbook code module, create the Workbook_Open( ) event that calls the Public BMIChart.Worksheet_Activate event (note that the code uses the worksheet's CodeName property and that the BMI Chart worksheet is activated).

   ```
   Private Sub Workbook_Open()
       Call BMIChart.Worksheet_Activate
       Sheets("BMI Chart").Select
   End Sub
   ```

▒ Attention The active sheet `Worksheet_Activate( )` event does not fire when the workbook opens. That is why you must explicitly call it from the `Workbook_Open( )` event.

---

4. Create `Public Sub Save( )` and `DeleteRecords( )` procedures on the `BMI Chart` worksheet code module (needed by the worksheet's `Button` controls), making them call the `clsDatabase.Save` and `clsDatabase.DeleterRecord` methods.

```
Public Function Save()
    mdb.Save
End Function

Public Function DeleteRecord()
    mdb.DeleteRecord
End Function
```

To initialize `mdb` as the `clsDatabase` object variable and initiate the database system services, you must do any of these:

- Close and reopen the workbook to force the `ThisWorkbook Workbook_Open( )` event to fire.
- Use the VBA Immediate window to call the `ThisWorkbook.Workbook_Open` or `BMIChart.Worksheet_Activate` event.
- Insert another sheet tab, select the new sheet tab, and select again the `BMI Chart` sheet tab to fire the `Worksheet_Activate` event.

▒ Attention Although it is not necessary, you can also initialize the `clsDatabase` class using the `Worksheet_SelectionChange( )` and `Worksheet_Change( )` events. This will guarantee that any time the user selects another input cell or changes any cell value, the database engine will begin to work.

Figure 8-7 shows a view of the `BMI Chart` code module implementing the `clsdatabase` code to manage its records. It uses just small code snippets to implement the complex task of managing its database records. All the code complexity is now encapsulated inside the `clsDatabase` class. Oh, of course, you can make as many copies as you like of the `BMI Chart` worksheet. Each copy will use its own instance of the `clsDattabase` class module. No duplication code is necessary.

Welcome to the beauty of VBA object class module programming!

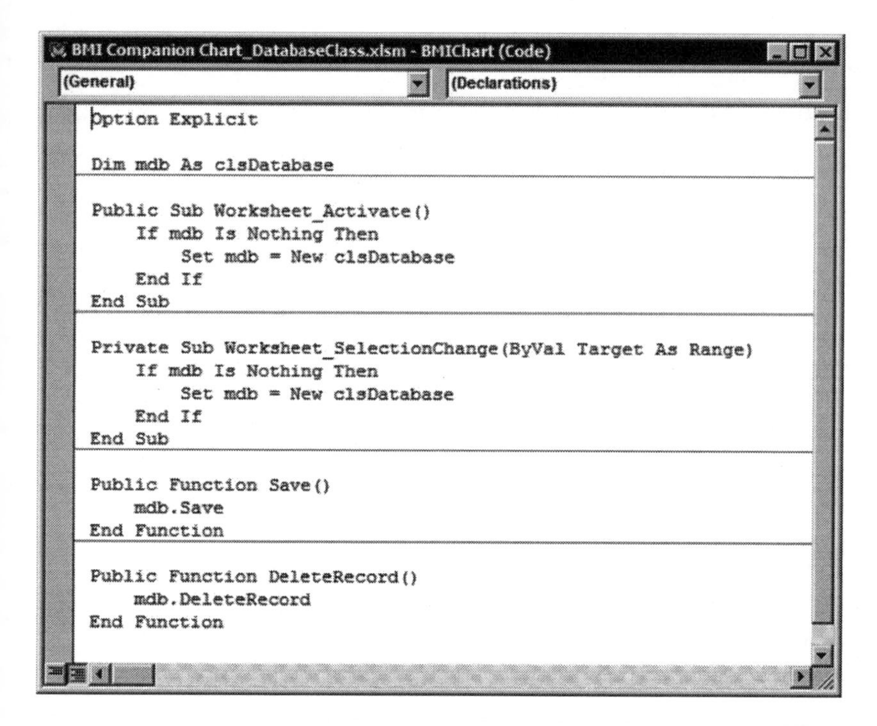

Figure 8-7. *This is the BMI Chart worksheet code module, showing all the code it needs to implement a database storage system using the clsDatabase class module*

■ **Attention** The `BMI Companion Chart_DatabaseClass.xlsm` and `USDA Food Composer_DatabaseClass.xlsm` macro-enabled workbooks that you can extract from the `Chapter08.zip` file implement the `clsDatabase` class created in this section.

Improving the clsDatabase Class Interface

Although the `clsDatabase` class produced in the previous section is quite interesting in terms of encapsulation and one-to-many record relationship capacity, it lacks a lot in customization and functionality:

- It exposes just two methods: Save and `DeleteRecord`.
- It doesn't expose any database properties.
- It doesn't fire events.
- It is quite confusing to set up; you need to understand how to create and define its database property range names.

This is comprehensible because it was not planned to be a database object. You just take a bunch of worksheet code and encapsulate it inside a class module with no planning in advance. The workforce—the database engine code—is already there, but it needs to be improved by defining an object interface that exposes a useful set of properties, methods, and events.

Improving the Object Model

To succeed in producing a robust, reusable worksheet database object, it is necessary to first define the following:

- The object purpose—what the object does
- The object name—how to call the object by name on any code module
- The object programmable interface—the properties, methods, and events that allow anyone to manipulate the object code using VBA
- A simplified interface to define the database properties

After thinking for a while, I decided to build a database object that must do (*object purpose*) worksheet database record manipulation (save, load, delete, and move to records) and call `SheetDBEngine` (*object name*), which exposes a set of properties, methods, and events (*object interface*) and resembles another popular database engine, the Microsoft Access `Forms` object.

Figure 8-8 shows the `SheetDBEngine` interface, while Table 8-2 gives a detailed explanation of each object member (in alphabetical order), including its purpose (properties), what it does (methods), and when it occur (events).

■ **Attention** If you are wondering if I am a genius to anticipate such an object structure, I can surely tell you that I am not. In fact, the `SheetDBEngine` interface exposed in Figure 8-8 and commented on in Table 8-2 was created using a step-by-step approach based on trial and error that lasted for weeks until the code stabilized to a point that could be considered sufficiently trusted to write about it.

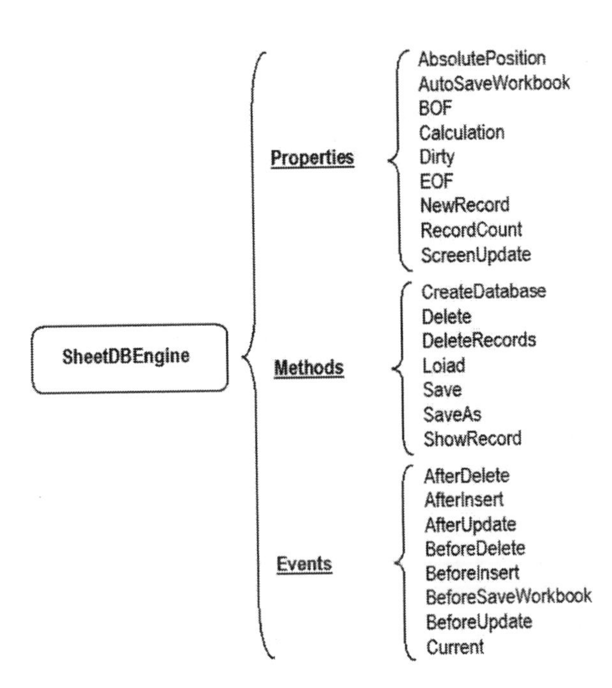

Figure 8-8. *This is the SheetDBEngine programmable interface, showing its properties, methods, and events*

Table 8-2. *SheetDBEngine Interface Members and Utility*

Type	Name	Utility
Property	AbsolutePosition	Long integer, read-only. Indicates the record order in the database.
Property	AutoSaveWorkbook	Boolean. If True, saves the workbook at every save or delete operation. The default is False.
Property	BOF	Boolean, read-only. True at the first record or at a new record.
Property	Calculation	Integer. Sets the Application.Calculation method to xlCalculationManual, xlCalculationAutomatic, lCalculationSemiAutomatic.
Property	Dirty	Boolean, read-only. True when the current record is changed and not saved yet.
Property	EOF	Boolean, read-only. True when at the last record or at a new record.
Property	NewRecord	Boolean, read-only. True at a new record (BOF and EOF properties are True).
Property	RecordCount	Long integer, read-only. Returns how many records the database has.
Property	ScreenUpdate	Boolean. Enables/disables Excel screen updates.
Method	CopyRecord	Copies existing record data to range object variables.
Method	CreateDatabase	Defines the database structure in the worksheet.
Method	Delete	Deletes the current record. Ask for confirmation before deletion.
Method	DeleteRecord	Deletes one record without asking for confirmation.
Method	Load	Loads a desired record. If a record is not specified, shows a new record.
Method	PasteRecord	Pastes record data from range object variables into the database structure.
Method	Save	Saves the current record without asking for confirmation.
Method	SaveAs	Saves the current record. Ask for confirmation before save.
Method	ShowRecord	Moves to first, next, previous, last, or new record.
Method	Sort	Sorts database record names.
Events	AfterDelete	Occurs before a record has been deleted from the database.
Events	AfterInsert	Occurs before a record has been inserted in the database.
Events	AfterUpdate	Occurs after a record has been saved.
Events	BeforeDelete	Occurs before a record can be deleted. Can be canceled.
Events	BeforeInsert	Occurs before a record can be inserted. Can be canceled.
Events	BeforeSaveWorkbook	Occurs before the workbook is saved or after a save or delete operation.
Events	BeforeUpdate	Occurs before a record can be updated.
Events	Current	Occurs after a new or existing record is shown.

■ **Attention** The `SheetDBEngine.xlsm` macro-enabled workbook that you can extract from `Chapter08.zip` contains the `SheetDBEngine` class module, an object covered in this chapter.

Implementing SheetDBEngine Properties

There are `Application` properties, database properties, and record properties. The `Application` properties relate to Excel behavior and the `Application` object: screen updating and calculation. Database properties relate to the records as a whole: how many there are and which is the first and last records regarding the `SavedRecords` range name. Record properties relate to each individual record, indicating the record position inside the `SavedRecords` range name and whether the record data has been changed. Figure 8-9 depicts these record position properties.

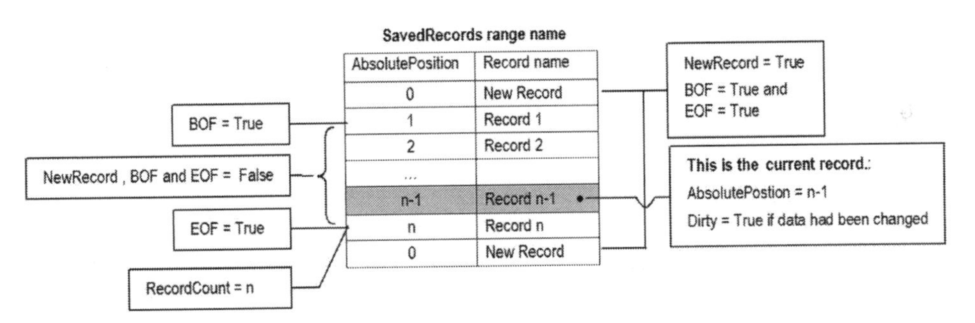

Figure 8-9. *This image describes the Database and Record properties, indicating how they are set regarding the record position inside the SavedRecords range name*

As you can see from Figure 8-9, the `BOF` property (Begin Of File) is positioned on the first record, and the `EOF` property (End Of File) is positioned on the last record. The `SavedRecords` range name is sorted ascending, so both properties may point to different records as they are added to the database. `AbsolutePosition` relates to the record order inside the database, while `RecordCount` relates to the total number of records. Using both properties, you can create an expression such as "Record 3 of 15."

The current record is the one currently exhibited by the worksheet application. Whenever the current record data is changed, its `Dirty` property becomes `True` (`Dirty = True`).

The new record has no defined position. It can be at the beginning or the end of the records. When the current record points to a new record, `NewRecord = True`, and both `BOF = EOF = True` (by default).

The `SheetDBEngine.xlsm` macro-enabled workbook has the `SheetDBEngine` class module, a modified version of the `clsDatabase` class, that implements the database properties in Table 8-2. The next code is part of the `SheetDBEngine` class module declaration section, which uses `Public` and `Private` variables to implement these properties (note that they are organized by data type):

```
'SheetDBEngine Public Properties
'===================================================
Dim mlngCalculation As Long
Dim mlngAbsolutePosition As Long

Public AutoSaveWorkbook As Boolean
```

```
Dim mbolScreenUpdating As Boolean
Dim mbolNewRecord As Boolean
Dim mbolDirty As Boolean
Dim mbolBOF As Boolean
Dim mbolEOF As Boolean
```

■ **Attention** Some people argue about when to use `Dim` or `Private` to declare a private module variable. The answer is that it doesn't matter! When Visual Basic evolved from version 3 to version 4, the `Private` keyword appeared as an alternative to indicate a module-level variable scope. You can use both declaration instructions in module-level variables but can't use `Private` inside a procedure code. Any declaration that does not have an explicit `Public` declaration is considered `Private` by default.

Read/Write Properties

A read/write property must either be a variable name declared with the `Public` keyword or be associated to a pair of `Public Property Let` and `Public Property Get` procedures.

Use a `Public` variable declaration when the property value can be changed by the user without consequences to the object code or the application environment, like the `AutoSaveWorkbook` property.

```
Public AutoSaveWorkbook As Boolean
```

Use a pair of `Property Let( )` and `Property Get( )` procedures when the property value can impact the object code or the application environment, such as `Calculation` (which changes the way Excel calculates by changing the `Application.Calculation` property) and `ScreenUpdating` (which enables or disables the `Application.ScreenUpdating` property). The `Property Let( )` procedure is responsible for making the property value change.

```
Public Property Let Calculation(CalculateMethod As XlCalculation)
    mlngCalculation = CalculateMethod
    Application.Calculation = CalculateMethod
End Property

Public Property Get Calculation() As XlCalculation
    Calculation = mlngCalculation
End Property

Public Property Let ScreenUpdating(Enabled As Boolean)
    mbolScreenUpdating = Enabled
    Application.ScreenUpdating = Enabled
End Property

Public Property Get ScreenUpdating() As Boolean
    ScreenUpdating = mbolScreenUpdating
End Property
```

Read-Only Properties

A read-only property is the one whose value can be read but cannot be changed. It is generally used to indicate object states that change according to the object code and are always implemented by declaring a module-level variable and just a `Public Property Get( )` procedure. The property value is changed by private code that directly interacts with the module-level variable or uses an associated `Private Let` property or `Function/Sub` procedure to do it.

The `AbsolutePosition` property uses the `Private SetAbsolutePosition( )` procedure to change the `mlngAbsolutePostion Long` variable. It receives as an argument a `Range` object related to the record cell.

```
Private Sub SetAbsolutePosition(rg As Range)
    mlngAbsolutePosition = (rg.Row - mintdbRecordsFirstRow)
End Sub

Public Property Get AbsolutePosition() As Long
    AbsolutePosition = mlngAbsolutePosition
End Property
```

■ **Attention** VBA generates an error if the pair of `Property Let( )` and `Property Get( )` procedures receive and return different data types, respectively. The `AbsolutePosition` property uses different procedure types to allow different data types to be used. To set the property value, a `rg as Range` argument is used, and to get the property value, it uses a `Long` integer.

Other properties change the private variable value that holds the property value inside the object code, whenever necessary. This is the case of `BOF`, `Dirty`, `EOF`, `NewRecord`, and `RecordCount` (which manipulates the `mbolBOF`, `mbolDirty`, `mbolEOF`, and `mbolNewRecord Boolean` private variables).

```
Public Property Get BOF() As Boolean
    BOF = mbolBOF
End Property

Public Property Get Dirty() As Boolean
    Dirty = mbolDirty
End Property

Public Property Get EOF() As Boolean
    EOF = mbolEOF
End Property

Public Property Get NewRecord() As Boolean
    NewRecord = mbolNewRecord
End Property

Public Property Get RecordCount() As Long
    RecordCount = Range(mstrdbSavedRecords).Rows.Count - 1
End Property
```

The AbsolutePosition Property

This property value reflects the record order inside the SavedRecords range name (sorted ascending), used to fill the data validation list. For new records, AbsolutePostion = 0 by default. The property value must be changed by the SheetDBEngine when

- The class is initialized to reflect the record currently exhibited by the worksheet application, by the Class_Initialize( ) event
- A new record is exhibited, by Sub Load( )
- A record is loaded or saved, by Private Sub SaveData( ) or Private Sub LoadSaveData( )

In the Class_Initialize( ) event, it is necessary to verify what is currently selected in the data validation list and then use the Range.Find method to search it inside the SavedRecords range name. The entire operation is conducted using local variable values.

```
Private Sub Class_Initialize()
    Dim rg As Range
    ...
        Set rg = mws.Range(mstrdbSavedRecords).Find(mws.Range(mstrdbDataValidationList))
        If Not rg Is Nothing Then
            Call SetAbsolutePosition(rg)
        End If
    End If
End Sub
```

Inside Sub Load( ), the NewRecord, AbsolutePosition, BOF, EOF, and Dirty properties have their values updated for a new record (Dirty is the only property always updated to false). Note that this time the procedure interacts directly with the module-level variable that represents the property.

```
Public Sub Load(Optional strRecord As String)
    'Disable screen updating, events and recalc
    Call Echo(False)
        Select Case strRecord
            Case "", "New " & mstrdbRecordName
                ...
                mbolNewRecord = True
                'Set record position
                mbolBOF = True
                mbolEOF = True
                mlngAbsolutePosition = 0
               ...
        End Select
        mbolDirty = False
       ...
End Sub
```

In the Sub LoadSaveData( ) procedure, after the record had been inserted in the SaveRecords range name and the database engine is about to save the one-side and/or many-side record cells, AbsolutePosition, BOF, and EOF are also updated as the first procedure steps. Note that BOF and EOF have their associated variables directly manipulated by the code, which verifies whether the record is at the first or last database position.

```
Private Sub LoadSaveData(strRecord As String, Perform As Operation)
    ...
    Set rg = mws.Range(mstrdbSavedRecords).Find(strRecord, , , xlWhole)
    'Set record position
    Call SetAbsolutePosition(rg)
    mbolBOF = (rg.Row = mws.Range(mstrdbSavedRecords).Row + 1)
    mbolEOF = (rg.Row = mws.Range(mstrdbSavedRecords).Row + mws.Range(mstrdbSavedRecords).
    Rows.Count - 1)
```

In the Sub SaveData( ) procedure, the AbsolutePosition property is updated as soon as the SavedRecords range name is sorted, repositioning the saved record.

```
Private Function SaveData(strRecord As String, Optional bolNewRecord As Boolean) As Boolean
        ...
        rgData.Sort rg.Cells(, 1)
        Set rg = Range(mstrdbSavedRecords).Find(strRecord, , , xlWhole)
        Call SetAbsolutePosition(rg)
```

The BOF and EOF Properties

Both BOF and EOF properties are commonly used by the database engine as pointers to set records boundaries. Their use is most preeminent to determine the beginning or end of file after a search is made or when a step-by-step forward walk is done through the records to the last record (EOF = True) or a backward walk is done to the first record (BOF = True). As the NewRecord property does, when both BOF and EOF are True or AbsolutePosition = 0, a new record is shown by the database.

As shown before, it value is changed when

- A new record is shown by Sub Load( )
- An existing record data is loaded or saved by Sub LoadSaveData( )

The Dirty Property

This property indicates whether the current record has been changed. It is used to issue a warning message asking to save the record before loading a new one. Its value must be changed whenever any worksheet record input cell is changed, which is controlled by the Worksheet_Change( ) event with the aid of the Dim WithEvents mws as Worksheet object variable.

```
Private Sub mWs_Change(ByVal Target As Range)
    Select Case Target.Address
        ...
        Case Else
            'Sheet data has changed
            mbolDirty = True
            ...
     End Select
End Sub
```

As shown before, the mbolDirty variable is set to False in the Sub Load( ) procedure, after the desired record is shown.

The RecordCount Property

This property just does a record count inside the SaveRecords range name. Since the first range cell is always reserved for the new record, it subtracts one from the Range.Rows.Count property. It is always recalculated on the Property Get( ) event.

```
Public Property Get RecordCount() As Long
    RecordCount = Range(mstrdbSavedRecords).Rows.Count - 1
End Property
```

Implementing SheetDBEngine Events

The events proposed by Table 8-2 were not defined randomly. They reflect the same events fired by Microsoft Access Forms when a record is inserted, selected, saved, or deleted in a user interface. They had a defined occurrence moment and order to fire, as explained by Figure 8-10.

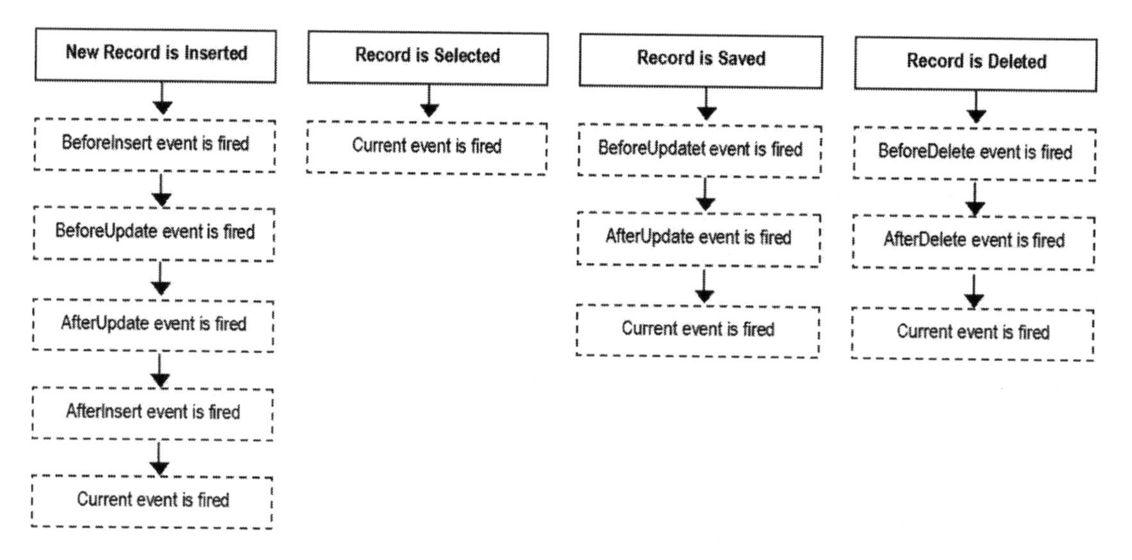

Figure 8-10. *This is the Microsoft Access Forms event order when a record is inserted, selected, saved, or deleted*

■ **Attention** Note that every Before... event must have a Cancel argument to allow the user to cancel it.

To those of you who are wondering what to do with so many events, here are some ideas:

- Use the Before... events to ask for a user confirmation before the record is inserted, saved, or deleted. If necessary, use the event procedure to execute all necessary operations before the event happens.
- Use the Current event to update the user interface according to the type of record that is presented to the user: a new or an existing record.

- Use the `After...` events to give a `MsgBox( )` confirmation that the operation succeeds or to make another operation on your application, after a record is inserted, changed, or deleted.

Attention When a new record is about to be inserted, both the `AfterInsert( )` and `AfterUpdate( )` event will fire in sequence. To avoid boring the user with two successive confirmations, use the `NewRecord` property to avoid the `BeforeUpdate( )` event confirmation.

To implement the `SheetDBEngine` events in Table 8-1, each event name was declared using a VBA `Event` instruction on the class declaration section. Note that some events also declare an argument that is passed by the `SheetDBEngine` object to the calling procedure (like the `Cancel as Integer` argument declared on all `Before...` events or the `Record as String` argument declared on all `After...` events).

```
'SheetDBEngine Events
'======================================================
Event Current()
Event BeforeInsert(Cancel As Integer)
Event AfterInsert(Record As String)
Event BeforeUpdate(Cancel As Integer)
Event AfterUpdate(Record As String)
Event BeforeDelete(Cancel As Integer)
Event AfterDelete(Record As String)
Event BeforeSaveWorkbook(Cancel As Integer)
```

Raising Events When a Record Is Saved

Now let's see what happens in terms of the `SheetDBEngine` class module. Let's begin with a new record that is about to be saved (properties `NewRecord = Dirty = True`) by the `Function Save( )` procedure, which executes this code:

```
Public Function Save(strRecord As String, Optional bolNewRecord As Boolean) As Boolean
    Dim intCancelInsert As Integer
    Dim intCancelUpdate As Integer
    Dim intCancelSave As Integer
    Dim intCancelSaveWorkbook As Integer
    Dim bolRecordSaved As Boolean

    'Raise events BeforeInsert and BeforeUpdate (allow cancel operation)
    If bolNewRecord Then
        RaiseEvent BeforeInsert(intCancelInsert)
        If intCancelInsert Then
            Exit Function
        End If
    End If

    RaiseEvent BeforeUpdate(intCancelUpdate)
    If intCancelUpdate Then
        Exit Function
    End If
```

```
    'Disable application events to allow cell change by macro code
    Call Echo(False)
        mws.Unprotect
            bolRecordSaved = SaveData(strRecord, bolNewRecord)
        If mbolSheetProtected Then
            mws.Protect
        End If

        If bolRecordSaved Then
            'Update record properties
            mbolNewRecord = False
            mbolDirty = False

            'Define current Record as saved Record
            mws.Range(mstrdbDataValidationList) = strRecord
            mws.Range(mstrdbDataValidationList).Select

            'Raise events AfterUpdate. AfterInsert, Current and BeforeSaveWorkbook
            RaiseEvent AfterUpdate(strRecord)
            If bolNewRecord Then
                RaiseEvent AfterInsert(strRecord)
            End If
            RaiseEvent Current
            'Save the worbook after save the record?
            If Me.AutoSaveWorkbook Then
                RaiseEvent BeforeSaveWorkbook(intCancelSaveWorkbook)
                If Not intCancelSaveWorkbook Then
                    ThisWorkbook.Save
                End If
            End If

            Save = True
        Else
            MsgBox "There is no more room to save data on this worksheet!", vbCritical,
            "Can't save data"
        End If
    Call Echo(True)
End Function
```

You may note that the procedure declares independent variables to hold the Cancel argument of each event cited in Figure 8-9, following best programming practices, which say to not reuse variables inside procedure code.

```
Public Function Save(strRecord As String, Optional bolNewRecord As Boolean)
    Dim intCancelInsert As Integer
    Dim intCancelUpdate As Integer
    Dim intCancelSave As Integer
    Dim intCancelSaveWorkbook As Integer
    Dim bolRecordSaved As Boolean
```

The proposed `After...` events fire successively using the VBA `RaiseEvent` statement, passing to the `Cancel` argument the appropriate variable, which is immediately tested to see whether it was canceled by the user action. This is what is happens to a new record that is about to be inserted on the database.

```
Public Function Save(strRecord As String, Optional bolNewRecord As Boolean)
    Dim intCancelInsert As Integer
    ...
    'Raise events BeforeInsert and BeforeUpdate (allow cancel operation)
    If bolNewRecord Then
        RaiseEvent BeforeInsert(intCancelInsert)
        If intCancelInsert Then
            Exit Function
        End If
    End If
```

This is quite simple, isn't it? The procedure abruptly ends whenever the event is canceled, and the new, changed record is not saved, remaining in its current state. But if the event was ignored or not canceled, the `BeforeUpdate( )` event is raised, and the same principle is followed: test the `intCancelUpdate` variable to see whether the event was canceled by the user action.

```
RaiseEvent BeforeUpdate(intCancelUpdate)
If intCancelUpdate Then
    Exit Function
End If
```

If both events were ignored or not canceled, the record will be saved by the `SaveData( )` procedure. To allow it to quietly save in the worksheet database, the code needs to disable Excel events firing, which is made by the `Sub Echo( )` procedure (which replaced `Sub SetScreenEventsRecalc ( )` used by the `clsDatabase` class).

```
Call Echo(False)
    mws.Unprotect
        bolRecordSaved = SaveData(strRecord, bolNewRecord)
    If mbolSheetProtected Then
        mws.Protect
    End If
```

Immediately after a trial was made to save the record, it checks the `bolRecordSaved` variable value if the saving operation succeeded. If this is true, the current record is not anymore a new record or is dirty, so both properties `NewRecord` and `Dirty` are updated, interacting with the module-level variables that represent them.

```
If bolRecordSaved Then
    'Update record properties
    mbolNewRecord = False
    mbolDirty = False
```

The data validation list cell value is updated to the saved record and selected, and the `AfterUpdate( )`, `AfterInsert( )` (for a new record) and `Current( )` events are raised in succession.

```
    'Raise events AfterUpdate. AfterInsert, Current and BeforeSaveWorkbook
    RaiseEvent AfterUpdate(strRecord)
    If bolNewRecord Then
        RaiseEvent AfterInsert(strRecord)
    End If
    RaiseEvent Current
```

■ **Attention** It is important to take care of the order in which database properties are changed and events are raised, because when these events are programmed by the user, it may base its decisions on the record property values.

When all events have been raised, it is time to verify the property's AutoSaveWorkbook value, which has as a default value False. If it is True, it signals that the user wants the workbook to be automatically saved after every record operation. So, it is time to raise the BeforeSaveWorkbook( ) event, which can be canceled by the user. If the event is ignored or when intCancelSaveWorkbook = False, the workbook will be saved.

```
If Me.AutoSaveWorkbook Then
    RaiseEvent BeforeSaveWorkbook(intCancelSaveWorkbook)
    If Not intCancelSaveWorkbook Then
        ThisWorkbook.Save
    End If
End If
```

The Sub Echo( ) Procedure

The Sub Echo( ) procedure used by the SheetDBEngine class module substituted the SetScreenEventsRecalc( ) procedure because both the Calculation and ScreenUpdating properties may be set by the user to a desired state.

```
Private Sub Echo(fEnable As Boolean)
    With Application
        .ScreenUpdating = (fEnable And Me.ScreenUpdating)
        .EnableEvents = fEnable
        .Calculation = IIf(fEnable, Me.Calculation, xlManual)
    End With
End Sub
```

Now, any call to Echo(False) will always disable the Application object's ScreenUpdating and EnableEvents properties and change Calculation to xlManual. But if SheetDBEngine.ScreenUpdating = False, any call to Echo(True) *will not* activate the Application.ScreenUpdating.

```
.ScreenUpdating = (fEnable And Me.ScreenUpdating)
```

This strategy allows the user to keep ScreenUpdating disabled in some desired circumstances. The same is true for the Calculation property, which will alternate between xlManual when the database engine is working and the user setting for Excel calculations when a call to Echo(False) is made.

Attention Did you notice that `Function Save( )` doesn't use the VBA `MsgBox( )` function to ask for the record name like it did in the `clsDatabase` class? This change was made to produce a `Save( )` method that silently saved a record by receiving up to two arguments: the required record name to be saved and an optional indication if it is a new record. The `Public Function Sub SaveAs( )` continue to asks for the record name, as it should, and after a name is granted, it calls `Save( )` to do the task. Take a look at it:

```
Public Function SaveAs(Optional strLastRecord As String) As Boolean
    Dim strRecord As String
    Dim bolNewRecord As Boolean

    'Verify if Record data is still empty
    strRecord = mws.Range(mstrdbDataValidationList)
    If strRecord = "New " & mstrdbRecordName Then
        Exit Function
    End If

    If strLastRecord = "" Then
        strLastRecord = strRecord
    End If
    strRecord = GetRecordName(strLastRecord, bolNewRecord)

    If Len(strRecord) Then
        SaveAs = Save(strRecord, bolNewRecord)
   End If
End Function
```

Raising Events When a Record Is Deleted

Now you'll see what happens to a record that is about to be deleted from the database by `Function DeleteRecord( )`, which executes this code:

```
Public Function DeleteRecord(strRecord As String, Optional NewRecord As Boolean) As Boolean
    Dim intCancelDelete As Integer
    Dim intCancelSaveWorkbook As Integer

    'Raise event BeforeDelete
    RaiseEvent BeforeDelete(intCancelDelete)
    If intCancelDelete Then
        Exit Function
    End If

    'Disable screen updating, events and recalc
    Call Echo(False)
        Call Clear
        If Not NewRecord Then
            Call DeleteRecordData(strRecord)
        End If
```

```
        'Update record properties
        mbolNewRecord = True
        mbolDirty = False

        'Define current Record as New Record
        mstrLastRecord = "New " & mstrdbRecordName
        mws.Range(mstrdbDataValidationList) = mstrLastRecord

        'Raise events AfterDelete, Current and BeforeSaveWorkbook
        RaiseEvent AfterDelete(strRecord)
        RaiseEvent Current
        'Save workbook after deletion?
        If Me.AutoSaveWorkbook Then
            RaiseEvent BeforeSaveWorkbook(intCancelSaveWorkbook)
            If Not intCancelSaveWorkbook Then
                ThisWorkbook.Save
            End If
        End If
    'Enabled screen updating, events and recalc1
    Call Echo(True)
End Function
```

It uses the same technique explained before to allow the user to cancel the record deletion: declaring the `intCancelDeletion` variable and raising the `BeforeDelete( )` event, passing it by reference. If the variable becomes `True`, the user has canceled the event, and the procedure ends abruptly.

```
Dim intCancelDelete As Integer
Dim intCancelSaveWorkbook As Integer

'Raise event BeforeDelete
RaiseEvent BeforeDelete(intCancelDelete)
If intCancelDelete Then
    Exit Function
End If
```

If the event was ignored or not canceled, the code disables Excel reactions with `Echo(False)`, clears all input cells, and if it is not a new record, calls `Sub DeleteRecordData( )` to remove the record from the database.

```
    'Disable screen updating, events and recalc
    Call Echo(False)
        Call Clear
        If Not NewRecord Then
            Call DeleteRecordData(strRecord)
```

And immediately after the record deletion, when the user already has a cleared, new record interface, the `NewRecord` and `Dirty` properties are updated, using direct manipulation of the variables.

```
'Update record properties
mbolNewRecord = True
mbolDirty = False
```

To update the worksheet application interface, the data validation list cell receives "New Record."

```
 'Define current Record as New Record
mstrLastRecord = "New " & mstrdbRecordName
 mws.Range(mstrdbDataValidationList) = mstrLastRecord
```

The `AfterDelete( )` and `Current( )` events are raised.

```
'Raise events AfterDelete, Current and BeforeSaveWorkbook
RaiseEvent AfterDelete(strRecord)
RaiseEvent Current
```

And once again, it verifies the `AutoSaveWorkbook` property value. If it is true, it raises the `BeforeSaveWorkbook( )` event. If the event is ignored or not canceled, the workbook is saved after the record deletion.

```
'Save workbook after deletion?
If Me.AutoSaveWorkbook Then
    RaiseEvent BeforeSaveWorkbook(intCancelSaveWorkbook)
    If Not intCancelSaveWorkbook Then
        ThisWorkbook.Save
    End If
End If
```

■ **Attention** Did you notice again that `Function DeleteRecord( )` doesn't issue a warning message before the record is deleted, as it did on the `clsDatabase` class? This change was made to produce the `DeleteRecord( )` method that needs to receive a record name that must be silently deleted. The `Public Function Delete( )` continues to give such warning, and if the user confirms the deletion, it will call `DeleteRecord( )` to do the task. This is the function's `Delete( )` code:

```
Public Sub Delete()
    Dim strRecord As String
    Dim strMsg As String
    Dim strTitle As String
    Dim bolNewRecord As Boolean

    strRecord = mws.Range(mstrdbDataValidationList)
    If strRecord = "" Or strRecord = "New " & mstrdbRecordName Then
        If Dirty Then
            bolNewRecord = True
            strMsg = "New " & mstrdbRecordName & " data has not been saved yet." & vbCrLf
            strMsg = strMsg & "Do you want to delete it?"
            strTitle = "Delete unsaved record?"
        Else
            Exit Sub
        End If
    Else
```

```
        strMsg = "Do you want to delete " & strRecord & " record?"
        strTitle = "Delete record?"
    End If

    If MsgBox(strMsg, vbYesNo + vbDefaultButton2 + vbQuestion, strTitle) = vbYes Then
        Call DeleteRecord(strRecord, bolNewRecord)
    End If
End Sub
```

Raising an Event When a Record Is Loaded

Whenever a record is selected in the data validation list cell, it becomes the current record, so the Current( ) event must be fired. This is done on Sub Load( ).

```
Public Sub Load(Optional strRecord As String)
    ...
    'Raise Current event
    RaiseEvent Current
End Sub
```

Implementing SheetDBEngine Methods

Methods are Public Sub or Function procedures implemented on the SheetDBEngine class module. They constitute the database engine core procedures used to load (Public Sub Load( )), save (Public Function SaveAs( ) and Save( )), delete (Public Function Delete( ) and DeleteRecord( )), and show records (Public Sub ShowRecord( )), as well as copy and paste record data (CopyRecord( ) and PasteRecord( )). All these procedures were described in Figure 8-8 and Table 8-2, and most of them (except ShowRecord( ), CopyRecord( ), PasteRecord( ), and Sort( )) were commented on in the previous sections, while the code was carefully analyzed in Chapter 7.

The ShowRecord Method

The Public Sub ShowRecord( ) procedure allows you to move to the first, last, previous, next, or new record in the database storage system, according to the argument it receives. It does this by declaring as an argument the Record as RecordPosition variable, which represents the Public Enum Record Position enumerator declared at the beginning of the SheetDBEngine class.

```
Public Enum RecordPosition
    FirstRec = 1
    PreviousRec = -1
    NextRec = 2
    LastRec = 3
    NewRec = 0
End Enum
...
Public Sub ShowRecord(Record As RecordPosition)
    Dim rg As Range
    Dim strRecord As String
    Dim lngFirstRec As Long
```

```
    Dim lngLastRec As Long
    Dim bolMoveRecord As Boolean

    lngFirstRec = mws.Range(mstrdbSavedRecords).Row + 1
    lngLastRec = mws.Range(mstrdbSavedRecords).Row + mws.Range(mstrdbSavedRecords).Rows.
    Count - 1

    Select Case Record
        Case FirstRec, LastRec, NewRec
            Set rg = mws.Range(mstrdbSavedRecords)
            Select Case Record
                Case FirstRec
                    Set rg = rg.Cells(2)
                Case LastRec
                    Set rg = rg.Cells(rg.Rows.Count)
                Case NewRec
                    Set rg = rg.Cells(1)
            End Select
            bolMoveRecord = True
        Case PreviousRec, NextRec
            strRecord = mws.Range(mstrdbDataValidationList)
            Set rg = mws.Range(mstrdbSavedRecords).Find(strRecord, , , xlWhole)
            If Record = NextRec And rg.Row < lngLastRec Then
                Set rg = rg.Offset(1)
                bolMoveRecord = True
            ElseIf Record = PreviousRec And rg.Row > lngFirstRec Then
                Set rg = rg.Offset(-1)
                bolMoveRecord = True
            End If
    End Select

    If bolMoveRecord Then
        'Move to selected record!
        mws.Range(mstrdbDataValidationList) = rg.Value
    End If
End Sub
```

Three of the possible types of records have a fixed position on the database: first record, last record, and new record. The other two, the previous and next records, have a relative position regarding the current record, with two restrictions:

- If `CurrentRecord` is the first record, `ShowRecord( )` cannot move backward.
- If `CurrentRecord` is the last record, `ShowRecord( )` cannot move forward.

The procedure begins by defining the `mstrdbSavedRercord` range row boundaries: the first and last range rows.

```
lngFirstRec = mws.Range(mstrdbSavedRecords).Row + 1
lngLastRec = mws.Range(mstrdbSavedRecords).Row + mws.Range(mstrdbSavedRecords).Rows.Count - 1
```

An outer Select Case statement verifies whether the Record argument received a reference to move to the first, last, or new record, according to the RecordPosition enumerator values.

```
Select Case Record
    Case FirstRec, LastRec, NewRec
```

If the move is to be made to the first, last, or new record, it is quite simple to do: it uses the rg object variable to set a reference to the mstrdbSavedRecords range and uses an inner Select Case statement to set the record position, using the Range.Cells( ) property. The first record is at the second range row (rg.Cells(2)), the last record is at the last range row (rg.Cells(Rows.Count)), and a new record is the first range row (rg.Cells(1)). The bolMoveRecord variable indicates an allowable move.

```
Set rg = mws.Range(mstrdbSavedRecords)
Select Case Record
    Case FirstRec
        Set rg = rg.Cells(2)
    Case LastRec
        Set rg = rg.Cells(rg.Rows.Count)
    Case NewRec
        Set rg = rg.Cells(1)
End Select
bolMoveRecord = True
```

Otherwise, it must make a relative move regarding the current record to the previous or next record. This time it uses Range.Find to first select the cell where the record resides.

```
Case PreviousRec, NextRec
    strRecord = mws.Range(mstrdbDataValidationList)
    Set rg = mws.Range(mstrdbSavedRecords).Find(strRecord, , , xlWhole)
```

Once the record is located, it first makes a double test to verify whether the move is to the next record *and* the current record is not the last record. If these two conditions are met, it sets a reference to the next record using the Range.Offset(1) method and uses bolMoveRecord to indicate an allowable move.

```
            If Record = NextRec And rg.Row < lngLastRec Then
                Set rg = rg.Offset(1)
                bolMoveRecord = True
```

Otherwise, the move must be to the previous record, so it does a double-check if the current record is not the first record before moving to the previous record using Range.Offset(-1).

```
        ElseIf Record = PreviousRec And rg.Row > lngFirstRec Then
            Set rg = rg.Offset(-1)
            bolMoveRecord = True
        End If
End Select
```

When the outer Select Case statement ends, it verifies if bolMoveRecord = True, and if it is, it uses the rg object variable to change the data validation list record, which will cascade-fire Sub Load( ), exhibiting the desired record.

```
    If bolMoveRecord Then
        'Move to selected record!
        mws.Range(mstrdbDataValidationList) = rg.Value
    End If
```

The CopyRecord and PasteRecord Methods

To allow copying record data between worksheet databases, the SheetDBEngine interface also offers the CopyRecord and PasteRecord methods that must be used in sequence (CopyRecord must be used before PasteRecord).

Both methods are implemented as Public Function procedures that return a Boolean value (True/False) indicating the operation success. The CopyRecord method implements this code:

```
Public Function CopyRecord(strRecord As String, rgOneSide As Range, rgManySide As Range) As
Boolean
    Dim rg As Range

    Set rg = mws.Range(mstrdbSavedRecords).Find(strRecord, , , xlWhole)
    If Not rg Is Nothing Then
        If Len(mstrdbOneSide) Then
            Set rgOneSide = Range(rg.Offset(0, mintdbRangeOffset), rg.Offset(0,
            mintdbRangeOffset + _
            mintdbOneSideColumnsCount - 1))
        End If

        If Len(mstrdbManySide1) Then
            Set rg = mws.Range(mstrdbManySidePrefix & FixName(strRecord))
            Set rgManySide = Range(rg.Offset(0, 0), rg.Offset
            (mintdbManySideRowsCount - 2, _
            mintdbManySideColumnsCount - 1))
        End If
        CopyRecord = True
    End If
End Function
```

Note that the CopyRecord( ) method expects to receive three arguments passed by reference. They are the record name whose data it must copy (strRecord) and two Range object variables—one to represent the one-side record data (if any) and another to represent the many-side record data (if any).

It begins searching the mstrSavedRecords range name for the desired record to be copied.

```
Set rg = mws.Range(mstrdbSavedRecords).Find(strRecord, , , xlWhole)
```

If the record is found, it verifies whether the database has the one-side record range name. If it does, it uses the Range property to attribute to the rgOneSide object variable the entire one-side record cells. It does this using the Range.Offset method and the database properties mintdbRangeOffset and mintdbOneSideColumnsCount to define the one-row rectangle that contains all the one-side record cells.

```
If Not rg Is Nothing Then
    If Len(mstrdbOneSide) Then
        Set rgOneSide = Range(rg.Offset(0, mintdbRangeOffset), rg.Offset(0,
        mintdbRangeOffset + _
                  mintdbOneSideColumnsCount - 1))
```

Once the one-side record was processed, it is time to verify whether the database also has a many-side record range name. If it does, it first finds the place where the many-side records are stored on the database worksheet.

```
If Len(mstrdbManySide1) Then
    Set rg = mws.Range(mstrdbManySidePrefix & FixName(strRecord))
```

Using again the Range property and the `Range.Offset` method with database properties `mintdbManySideRowsCount` and `mintdbManySideColumnsCount`, it defines a continuous range that encompasses all many-side worksheet cells, setting it to the `rgManySide` argument.

```
Set rgManySide = Range(rg.Offset(0, 0), rg.Offset(mintdbManySideRowsCount - 2,  _
                  mintdbManySideColumnsCount - 1))
```

And when all record cells have been set to the appropriate Range object variable arguments, the `CopyRecord` method returns `True` to indicate that it succeeds.

```
        CopyRecord = True
    End If
End Function
```

Once a record had been copied with the `CopyRecord` method, you can use the `PasteRecord` method to paste the record into the database, which follow these rules:

1. The record must be first copied with the `CopyRecord` method.
2. It supposes that the source and destination databases have the same record structure; it *will not* verify if there are record structure differences between the source and destination databases.
3. The `PasteRecord` method will search the database for the record being pasted. If it finds it, this data will be overwritten.
4. If `PasteAsNewRecord=True`, the record will be pasted as a new record, adding a counter suffix to its name.
5. Since `PasteRecord` directly manipulates the database structure, no database event will be triggered by the paste operation.
6. The `PasteRecord` method pastes new records at the bottom of the database, without sorting it. Use the `SheetDBEngine.Sort` method to sort the database records after one or more successive `PasteRecord` operations.

Now that you have an idea of the rules that the `PasteRecord` method follows, take a look at the code:

```
Public Function PasteRecord(strRecord As String, _
        rgOneSide As Range, _
        rgManySide As Range, _
        Optional PasteAsNewRecord As Boolean) As Boolean
```

```
Dim rg As Range
Dim strRangeName As String
Dim lngRow As Long
Dim intI As Integer
Dim bolProtect As Boolean
Dim bolWorksheetIsFull As Boolean
Dim bolRecordPaste As Boolean

Set rg = mws.Range(mstrdbSavedRecords).Find(strRecord, , , xlWhole)
If Not rg Is Nothing And PasteAsNewRecord Then
    'Add a name count suffix to paste existing record as new one
    Do Until rg Is Nothing
        'Find a new record name
        intI = intI + 1
        Set rg = mws.Range(mstrdbSavedRecords).Find(strRecord & intI, , , xlWhole)
    Loop
    strRecord = strRecord & intI
End If

Call Echo(False)
bolProtect = mws.ProtectContents
mws.Unprotect
strRangeName = mstrdbManySidePrefix & FixName(strRecord)

If rg Is Nothing Then
    'strRecord does not exist. Createt it!
    'Define sheet row where next Record data will be stored
    lngRow = NextEntryRow(bolWorksheetIsFull)

    'Verify if sheet is full
    If bolWorksheetIsFull Then
        'No more room to save data
        MsgBox "There is no more room to paste records", vbCritical, "Workdhseet
        database is full"
        Exit Function
    End If

    'Verify if mstrSavedRecords last rows is a empty cell
    Set rg = mws.Range(mstrdbSavedRecords)
    If Not rg.Cells(rg.Rows.Count, 1) = "" Then
        'Insert a new row at bottom of SavedRecords range name and update rg object
        rg.Resize(rg.Rows.Count + 1).Name = "'" & mws.Name & "'!" & mstrdbSavedRecords
        Set rg = mws.Range(mstrdbSavedRecords)
    End If

    'Position on new cell of SavedRecords range and save New Record name
    Set rg = rg.Cells(rg.Rows.Count, 1)
    rg = strRecord

    If Len(mstrdbManySide1) Then
        'Define Record name as 'rec_<strRecord>' and create it range name
```

```
            mws.Names.Add strRangeName, "='" & mws.Name & "'!" & mstrdbManySideFirstColumn &
            lngRow
            mws.Names(strRangeName).Visible = False
        End If
    End If

    If Len(mstrdbOneSide) Then
        'Paste the one side record
        Set rg = rg.Offset(0, mintdbRangeOffset)
        rgOneSide.Copy
        rg.PasteSpecial xlPasteValues
        bolRecordPaste = True
    End If

    If Len(mstrdbManySide1) Then
        'Paste the Many side records
        Set rg = mws.Range(strRangeName)
        rgManySide.Copy
        rg.PasteSpecial xlPasteValues
        bolRecordPaste = True
    End If

    If bolProtect Then mws.Protect
    Call Echo(True)
    PasteRecord = bolRecordPaste
End Function
```

Since the worksheet database structure does not allow record name duplication, the PasteRecord begins using the Range.Find method on the mstrdbSavedRecords range name to search the database for the record being pasted (strRecord argument), returning a reference to the cell to the rg object variable. If the record already exists on the database (Not rg Is Nothing) *and* PasteAsNewRecord = True, a new record name must be created.

```
Set rg = mws.Range(mstrdbSavedRecords).Find(strRecord, , , xlWhole)
If Not rg Is Nothing And PasteAsNewRecord Then
```

To create a new record name, a Do...Loop operation is conducted to add a counter suffix to the record name, using a Range.Find method until a new name is found.

```
    'Add a name count suffix to paste existing record as new one
    Do
        'Find a new record name
        intI = intI + 1
        Set rg = mws.Range(mstrdbSavedRecords).Find(strRecord & intI, , , xlWhole)
    Loop Until rg Is Nothing
    strRecord = strRecord & intI
End If
```

Once the record name is correctly set, it is time to set the stage: disable screen updates, recalculation, and events firing calling the Echo(False) procedure; store the worksheet protection state on the bolProtect variable; unprotect the worksheet (if protected, so the code can paste data in the worksheet); and define the many-side records range name (if any).

```
Call Echo(False)
bolProtect = mws.ProtectContents
mws.Unprotect
strRangeName = mstrdbManySidePrefix & FixName(strRecord)
```

Now verify again if the record already exists (`rg Is Nothing`). If it is a new record, use the `NextEntryRow( )` procedure to define whether there is still room in the worksheet to paste it. If the worksheet is full, issue a warning message and exit the `PasteRecord` method, returning `False`.

```
If rg Is Nothing Then
    'strRecord does not exist. Createt it!
    'Define sheet row where next Record data will be stored
    lngRow = NextEntryRow(bolWorksheetIsFull)

    'Verify if sheet is full
    If bolWorksheetIsFull Then
        'No more room to save data
        MsgBox "There is no more room to paste records", vbCritical, "Workdhseet database is
        full"
        Exit Function
    End If
```

If there is still room to paste the new record into the worksheet rows, the `lngRow Long` variable will hold a reference to the worksheet row where the many-side records must be saved. It is time to verify if the `mstrSavedRecords` range name last row, where the new record will be added, is empty.

```
'Verify if mstrSavedRecords last rows is a empty cell
Set rg = mws.Range(mstrdbSavedRecords)
If Not rg.Cells(rg.Rows.Count, 1) = "" Then
```

If the last `mstrdbSavedRecords` range is not empty, you need to add a new, empty row at the bottom of the range. You do this the traditional way: using the `Range.Resize` method along with the `Range.Name` property to resize it.

```
'Insert a new row at bottom of SavedRecords range name and update rg object
rg.Resize(rg.Rows.Count + 1).Name = "'" & mws.Name & "'!" & mstrdbSavedRecords
```

Since the range was resized, you need to update the rg object variable to reflect this row insertion.

```
Set rg = mws.Range(mstrdbSavedRecords)
```

And the new record is added on the last `mstrdbSavedRecords` range name row.

```
'Position on new cell of SavedRecords range and save New Record name
Set rg = rg.Cells(rg.Rows.Count, 1)
rg = strRecord
```

And since the record was added, the code checks whether there is a many-side record cells to save testing the length of the `mstrdbManySide1` database property. If it exists, the many-side record range name is created in the worksheet using the `Names` collection's `Add` method in the `lngRow` position determined earlier.

```
    If Len(mstrdbManySide1) Then
        'Define Record name as 'rec_<strRecord>' and create it range name
        mws.Names.Add strRangeName, "='" & mws.Name & "'!" & mstrdbManySideFirstColumn &
        lngRow
        mws.Names(strRangeName).Visible = False
    End If
End If
```

The code is now in the position to save the record data, either because the record already exists or because it has been created. So, it verifies whether the database record has its `mstrdbOneSide` property set. If it does, the entire range is pasted to the right of the `mstrdbSavedRecords` record position, using `Range.Copy` to copy the `rgOneSide` argument value on the clipboard and using `Range.PasteSpecial xlPasteValue` to paste it on the worksheet. To signal that this operation was performed, `bolRecordPaste` becomes `True`.

```
If Len(mstrdbOneSide) Then
    'Paste the one side record
    Set rg = rg.Offset(0, mintdbRangeOffset)
    rgOneSide.Copy
    rg.PasteSpecial xlPasteValues
    bolRecordPaste = True
End If
```

The same operations are repeated for the `mstrdbManySide1` database property to copy the `rgManySide` argument value to the clipboard and paste it in the worksheet on the range name represented by the `strRangeName` variable (and `bolRecordPaste = True` to signal the paste operation).

```
If Len(mstrdbManySide1) Then
    'Paste the Many side records
    Set rg = mws.Range(strRangeName)
    rgManySide.Copy
    rg.PasteSpecial xlPasteValues
    bolRecordPaste = True
End If
```

To end the operation, the worksheet protection is returned to its default mode; the screen updating, calculation, and events firing are turned on again; and `PasteRecord` returns the `bolRecordPaste` variable value, indicating if either the one-side record, the many-side records, or both have been pasted on the worksheet.

```
    If bolProtect Then mws.Protect
    Call Echo(True)
    PasteRecord = bolRecordPaste
End Function
```

Using CopyRecord/PasteRecord

To use the `CopyRecord` and `PasteRecord` methods, you must declare two Range object variables to hold references to the one-side and many-side record range and pass them first to the `CopyRecord` method (by reference) and then to the `PasteRecord` method on the VBA code, using a procedure like this:

```
Public Function CopyTest(strRecord as string)
    Dim rgOne As Range
```

```
    Dim rgMany As Range
    Dim intI As Integer

    If mdb Is Nothing then
        Set mdb = New SheetDBEngine
    End If

    For intI = 1 To 10
       If mdb.CopyRecord(strRecord, rgOne, rgMany) Then
           mdb.PasteRecord strRecord, rgOne, rgMany, True
       End If
    Next
    mdb.Sort
End Function
```

The procedure shown here receives an `strRecord` variable with the desired record name, creates an instance of the database engine class (if needed), and uses the `CopyRecord` method to try to copy the `strRecord` record. If the `CopyRecord` operation succeeds (`CopyRecord = True`), it uses a `For...Next` loop to paste ten new copies of the same record into the database.

■ **Attention** See Chapter 9 for a good example of how you can use the `CopyRecord` and `PasteRecord` methods to copy records from one worksheet database to another.

The Sort Method

The `Sort` method sorts the `mstrdbSavedRecords` range name in ascending order. It must be used after executing the `PasteRecord` method one or more times because there is a bug in Microsoft Excel VBA that does not sort hidden cells. The `Range.Sort` method fails to operate on range names located in hidden rows. To make the `PasteRecord` method execute faster, not unhide the `mstrSavedRecords` row, sort its cells, and hide its rows again, you must manually call the `SheetDBEngine.Sort` method after it calls the `SheetDBEngine.PasteRecord` one or more times.

This is the code executed by the `SheetDBEngine.Sort` method:

```
Public Sub Sort()
    Dim rg As Range
    Dim bolProtect As Boolean

    Call Echo(False)
    bolProtect = mws.ProtectContents
    mws.Unprotect
        Set rg = mws.Range(mstrdbSavedRecords)
        'Sort SavedRecords and find strRecord position
        Set rg = mws.Range(mws.Cells(rg.Row + 1, rg.Column), _
        mws.Cells(rg.Row + rg.Rows.Count - 1, rg.Column + mintdbRangeOffset +
        mintdbOneSideColumnsCount - 1))
        'Unhide range rows because Sort does not works well on hidden rows
        rg.EntireRow.Hidden = False
            rg.Sort rg.Cells(, 1)
        rg.EntireRow.Hidden = True
```

```
    If bolProtect Then mws.Protect
    Call Echo(True)
End Sub
```

To sort mstrdbSavedRecords, the code first makes a call to Echo(False) to disable screen updating, events firing, and calculation; the active sheet protect state is saved, and it is unprotected to allow sorting. The mstrSavedRecords is then attributed to the rg object variable and resized to not include the first entry ("New record") using the Cells property to define a range that encompasses all the mstrdbSavedRecords and one-side record cell columns.

```
  Call Echo(False)
  bolProtect = mws.ProtectContents
  mws.Unprotect
      Set rg = mws.Range(mstrdbSavedRecords)
      'Sort SavedRecords and find strRecord position
      Set rg = mws.Range(mws.Cells(rg.Row + 1, rg.Column), _
                  mws.Cells(rg.Row + rg.Rows.Count - 1, rg.Column + mintdbRangeOffset +
mintdbOneSideColumnsCount - 1))
```

The code uses the Range.EntireRows.Hidden property to show all hidden worksheet rows, sort the mstrdbSaveRecords range, and hide its rows again.

```
        rg.EntireRow.Hidden = False
            rg.Sort rg.Cells(, 1)
        rg.EntireRow.Hidden = True
```

The operation is finished and restores the worksheet protect state, screen updating, enable events, and calculation to their default states (Echo(True)).

```
    If bolProtect Then mws.Protect
    Call Echo(True)
End Sub
```

■ **Attention** All other code procedures you may find inside the SheetDBEngine class were already commented on in Chapter 7.

Using the SheetDBEngine Class

You can see two good examples of the SheetDBEngine class in action by extracting the BMI Companion Chart_SheetDBEngine.xlsm (Figure 8-11) and USDA Food Composer_SheetDBEngine.xlsm (Figure 8-12) macro-enabled workbooks from Chapter08.zip.

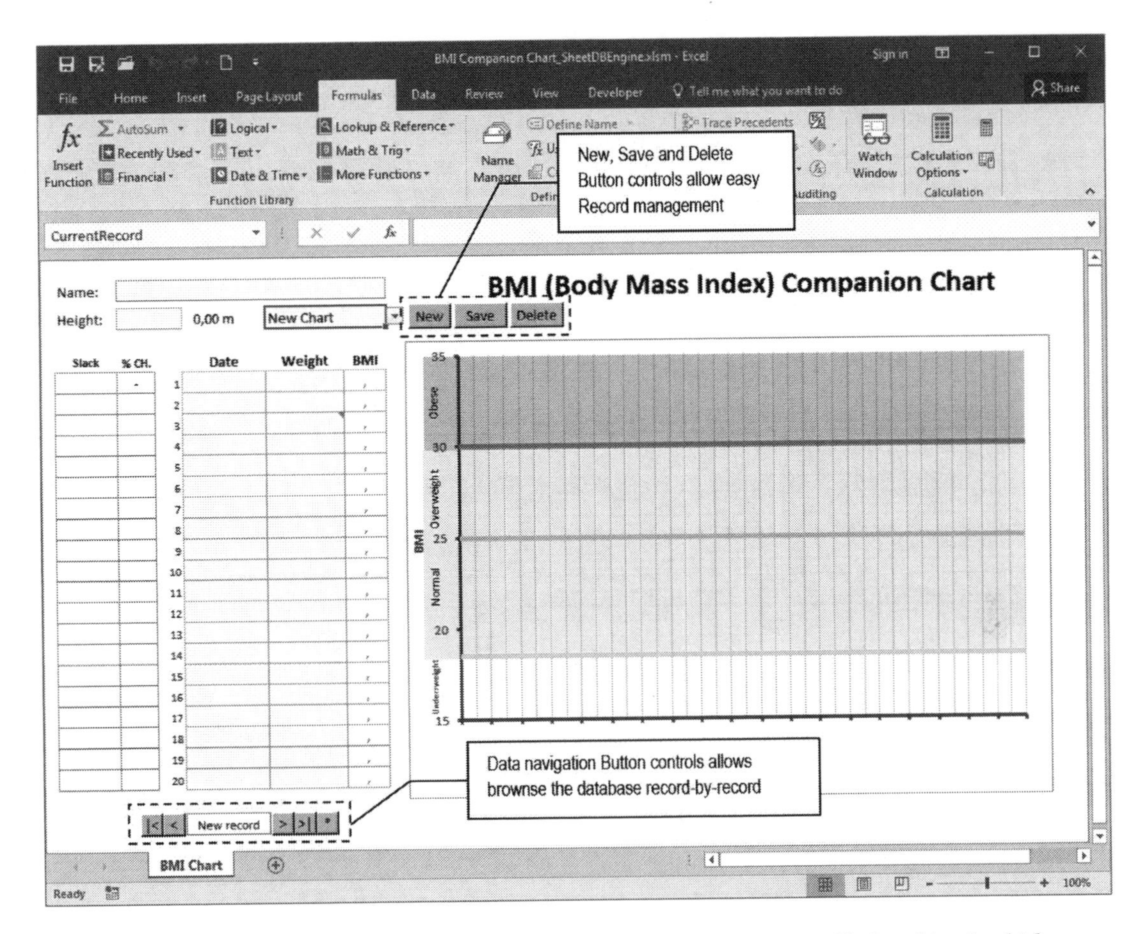

Figure 8-11. *This is the BMI Companion Chart_SheetDBEngine.xlsm macro-enabled workbook, which uses the SheetDBEngine class to implement the database storage system. Using the class ShowRecord, it can expose data navigation controls like Microsoft Access Forms do*

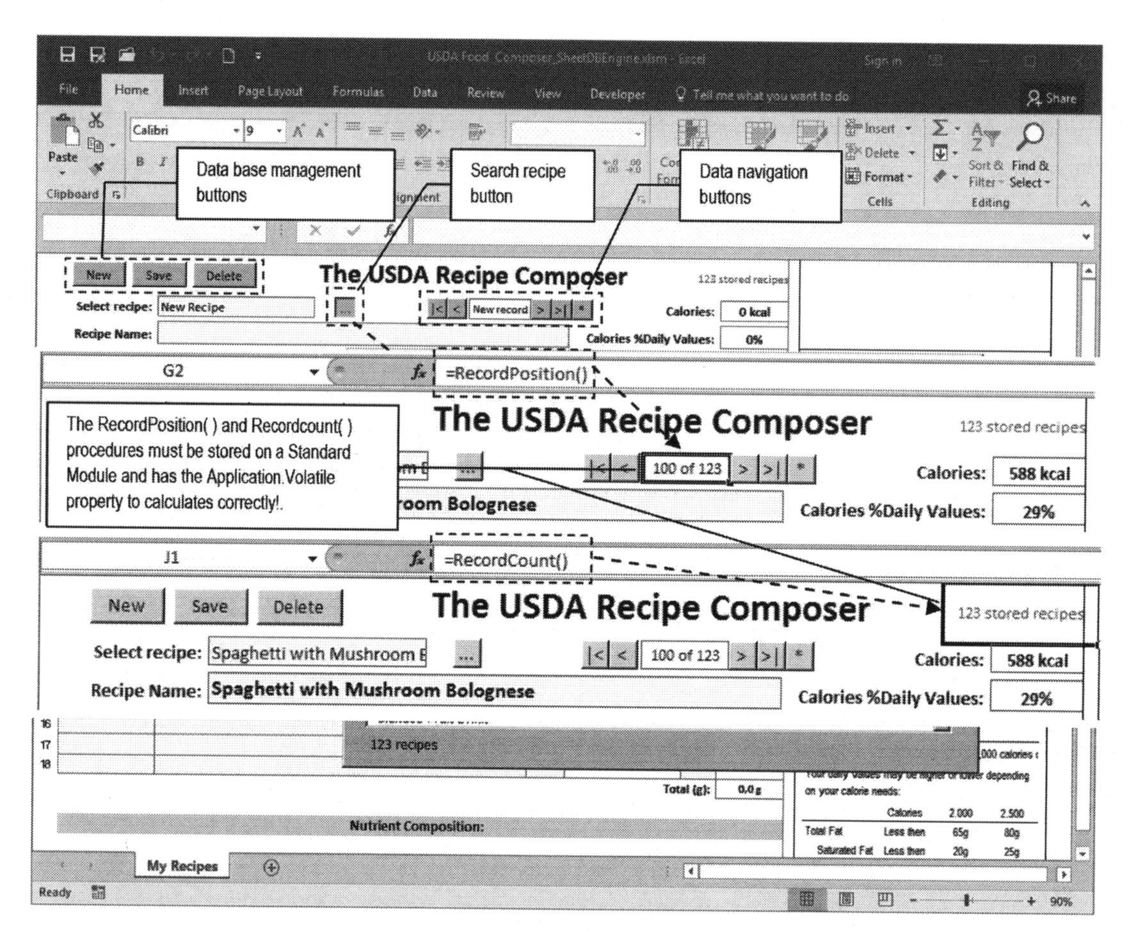

Figure 8-12. *This is USDA Food Composer_SheetDBEngine.xlsm macro-enabled workbook, which also uses the SheetDBEngine class to implement its database storage system and its ShowRecord method to expose data navigation controls*

Thanks to the SheetDBEngine.ShowRecord method, both workbooks present a kind of "navigation button," like Microsoft Access forms do, to allow you to do step-by-step navigation through all database records.

Try to use the USDA Food Composer_SheetDBEngine.xslm macro-enabled workbook to go to the first, next, previous, last, or new record. Click the "Search recipe" Button control, scroll the ListBox, select a recipe in the middle of the list, and double-click it. Notice how the data navigation indicator updates to indicate the record position. Add and delete a recipe and observe it again.

Cool, isn't it?

The code behind both worksheet applications is similar. Let's explore the USDA Food Composer_SheetDBEngine.xlsm My Recipes worksheet code. Since the SheetDBEngine class has events, you must use the VBA WithEvents statement to declare it and use the Worksheet object's Activate( ) and SelectionChange( ) events to create a new instance of the database class.

```
Option Explicit

'This variable receive frmSearchFoodItems selected item
Public SelectedFoodITem As Variant
```

```
Dim WithEvents mdb As SheetDBEngine

Public Sub Worksheet_Activate()
    If mdb Is Nothing Then
        Set mdb = New SheetDBEngine
    End If
End Sub

Private Sub Worksheet_SelectionChange(ByVal Target As Range)
    If mdb Is Nothing Then
        Set mdb = New SheetDBEngine
    End If
End Sub
```

As explained in Chapter 7, every time a recipe is saved (or deleted), the recipe information must be also saved (or deleted) from the `My_Recipes` range name located in the hidden USDA worksheet. Now that the `SheetDBEngine` class exposes the `AfterUpdate( )` and `AfterDelete( )` records, you can use them to execute such tasks, whenever one of those operations happens. Note how this is done when a record is saved.

```
Public Sub Save()
    mdb.SaveAs
End Sub

Private Sub mdb_AfterUpdate(Record As String)
    Dim strRecord As String
    'Update USDA My_Recipes range name with recipe nutritional data
    strRecord = Range("CurrentRecord")
    Call SaveInMyRecipes(strRecord)
End Sub
```

Know that whenever a record is saved by `Public Sub Save( )` and associated to the Save `Button` control, the `SheetDBEngine.SaveAs` method is called to save the record, and when it is done, the class raises the `AfterUpdate( )` event, which is then used to call the `Sub SaveInMyRecipes( )` procedure to also save the recipe information inside the USDA worksheet.

Now look at what happens when a record is deleted. `Public Sub DeleteRecord( )` calls the `iSheetDBEngine.Delete` method, and when it is done, the class raises the `AfterDelete( )` event, which is used to effectively delete the recipe entry from the USDA worksheet's `My_Recipes` range name.

```
Public Sub DeleteRecord()
    mdb.Delete
End Sub

Private Sub mdb_AfterDelete(Record As String)
    Dim rg As Range
    Dim rgRecipe As Range
    Dim strRecord As String

    'Delete recipe from USDA My_Recipes range name
    strRecord = Range("CurrentRecord")
    Set rg = Worksheets("USDA").Range("My_Recipes")
    Set rgRecipe = rg.Find(strRecord, , , xlWhole)
```

```
    If Not rgRecipe Is Nothing Then
        rgRecipe.EntireRow.Delete
    End If
End Sub
```

Producing Data Navigation Controls

Figures 8-10 and 8-11 show two worksheet applications that use data navigation `Button` controls to browse the database record set, moving to the first, previous, next, last, or new record with a button click. Every time a move is made, the database record position is automatically updated on the worksheet, the same way Microsoft Access tables, queries, and forms do.

Each `Button` control used to "data navigate" to the desired record is associated to a standard `Public` Sub procedure stored in the `basButton` control's standard module (to make them appear in the Excel Assign Macro dialog box). These are the move record procedures stored in the `basButton controls` module:

```
Public Sub MoveFirst()
    Dim obj As Object

    Set obj = ActiveSheet
    obj.MoveFirst
End Sub

Public Sub MoveLast()
    Dim obj As Object

    Set obj = ActiveSheet
    obj.MoveLast
End Sub

Public Sub MovePrevious()
    Dim obj As Object

    Set obj = ActiveSheet
    obj.MovePrevious
End Sub

Public Sub MoveNext()
    Dim obj As Object

    Set obj = ActiveSheet
    obj.MoveNext
End Sub

Public Sub MoveNew()
    Dim obj As Object

    Set obj = ActiveSheet
    obj.MoveNew
End Sub
```

All `Move...` procedures use the same technique described in Chapter 7. They declare a standard `obj As Object` variable to set a reference to the `ActiveSheet` object and then use *late binding* to compile correctly. The VBA code expects that the object exposes the desired method; if the method doesn't exist in the active sheet, VBA will raise a runtime error when the procedure is executed.

Inside the `MyRecipes` code module, you will find the expected `Public Function` procedures. Note that all of them use the `SheetDBEngine.ShowRecord` method, with the appropriate record constant.

```
Public Function MoveFirst()
    mdb.ShowRecord (FirstRec)
End Function

Public Function MoveLast()
    mdb.ShowRecord (LastRec)
End Function

Public Function MovePrevious()
    mdb.ShowRecord (PreviousRec)
End Function

Public Function MoveNext()
    mdb.ShowRecord (NextRec)
End Function

Public Function MoveNew()
    mdb.ShowRecord (NewRec)
End Function
```

Right-click each navigation `Button` control and select the Assign Macro context menu command to verify that all the `basButton controls Public Sub` procedures appear in the dialog box list and that the procedure associated to the control is not prefixed by any object or module name, which is a clear indication that they come from a standard module (Figure 8-13).

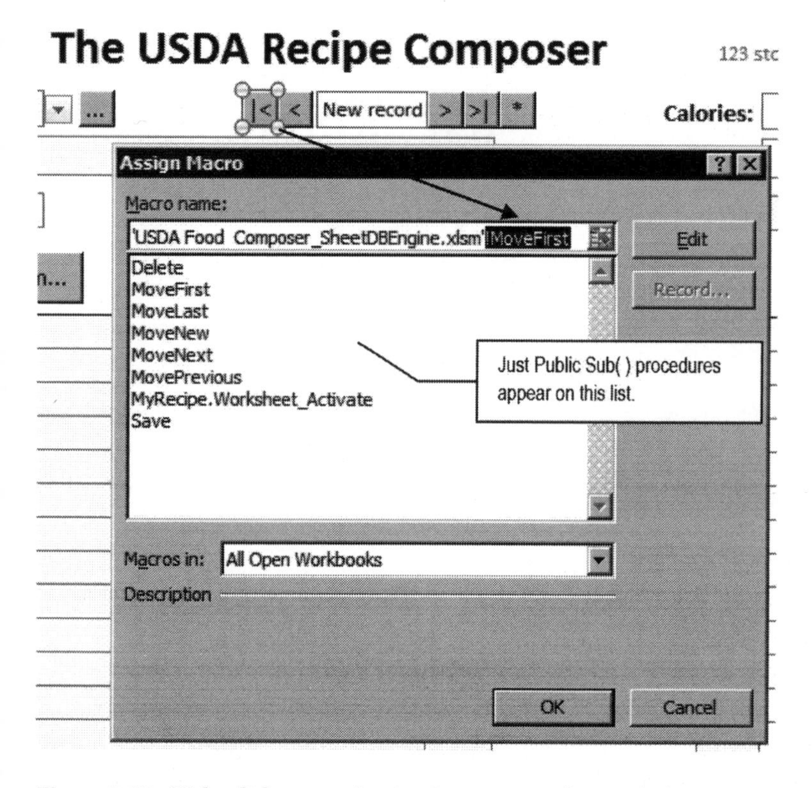

Figure 8-13. *Right-click any navigation Button control to verify that it is associated to a Public Sub procedure from a standard module. Each procedure must have its counterpart on the active sheet code module*

To return the total recipe number or the record position on the navigation buttons, the `My Recipes` worksheet code module has two other procedures: `Public Function RecordCount( )` and `Public Function Record Position( )`.

```
Public Function RecordCount() As Long
    RecordCount = mdb.RecordCount
End Function

Public Function RecordPosition() As String
    Dim strPosition As String

    If mdb.AbsolutePosition = 0 Then
        strPosition = "New record"
    Else
        strPosition = mdb.AbsolutePosition & " of " & mdb.RecordCount
    End If
    RecordPosition = strPosition
End Function
```

`Public Function RecordCount( )` just returns the `SheetDBEngine.RecordCount` property, while `Public Function RecordPosition( )` uses `SheetDBEngine.AbsolutePosition` to return "New Record" (when `AbsolutePosition = 0`) or the record absolute position regarding the total record count (`mdb.AbsolutePosition & " of " & mdb.RecordCount`).

To make these two functions work in a cell formula, like any other Excel function, you must do the following:

- Create a `Standard Module Function` procedure that calls the worksheet procedure
- Use the `Application.Volatile` property to force these `Standard Module` procedures to automatically recalculate
- Use these `Standard` module procedures from any cell formula

The `USDA Food Composer_SheetDBEngine.xlsm` workbook has the `basButton` control's `Standard` module, which is used to define the default `Button` control's procedures. All are declared as `Public Sub` procedures, so they can appear in the Excel Assign Macro dialog box. This is the code used by the `basButton` control's `RecordCount( )` and `RecordPosition( )` procedures:

```
Public Function RecordCount() As Long
    Dim obj As Object

    Set obj = ActiveSheet
    On Error Resume Next
    Application.Volatile
    RecordCount = obj.RecordCount
End Function

Public Function RecordPosition() As String
    Dim obj As Object

    Set obj = ActiveSheet
    On Error Resume Next
    Application.Volatile
    RecordPosition = obj.RecordPosition
End Function
```

After declaring the `obj As Object` object variable to set a reference to the `ActiveSheet` object, the `Application.Volatile` property is used to force the function to calculate whenever Excel calculates and then call the `obj.RecordCount` or `obj.RecordPosition` method using *late binding*, which means the code compiles normally, expecting that the `obj` variable has these methods. Figure 8-14 shows the formula used by the `My Recipes` worksheet cells `G2` and `J1`.

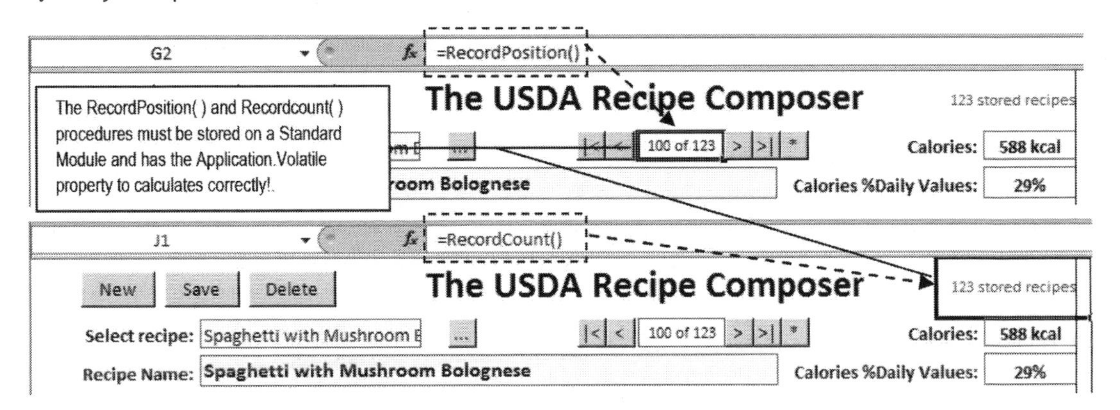

Figure 8-14. *The USDA Food Composer_SheetDBEngine.xlsm My Recipes worksheet uses the basButton control's Public Function RecordPosition( ) to return the current record position inside the database record set. This Standard Module procedure uses the Application.Volatile property to be updated whenever Excel recalculates*

■ **Attention** Be careful when you use the `Application.Volatile` property inside any `Public` procedure. When you do it, every time Excel calculates, the procedure also calculates, which may fire the code successively. That is why the `SheetDBEngine` class module defines `Application.Calculation = xlManual` on the `Sub Echo( )` procedure. When `Application.Calculation` is changed to `xlAutomatic`, every procedure that has the `Volatile` property is calculated.

Since the `My Recipes` worksheet of the `USDA Food Composes_SheetDBEngine.xlsm` macro-enabled workbook's `Button` controls are associated to standard module procedures that call worksheet module procedures, you can make as many copies as you want from `My Recipes`. Each copy will manage its own database record set. To guarantee that each worksheet's navigation buttons show the appropriate worksheet record count, the `Worksheet_Activate( )` event now calls the `Application.Calculate` method, forcing all volatile functions to calculate whenever the sheet tab is selected.

```
Public Sub Worksheet_Activate()
    If mdb Is Nothing Then
        Set mdb = New SheetDBEngine
    End If
    Application.Calculate
End Sub
```

Setting the Worksheet Database Class

The `SheetDBEngine` class examples cited in the previous section work quite well to manage different types of worksheet records that have a one-to-many database relationship, but it is still very difficult to set up. The user must find a safe worksheet area to store the database parameters, create the associated worksheet scope range names, appropriately define the values, and set up the worksheet code necessary to run the database class.

You need a simple way to do all these tasks. You need to build a "worksheet database wizard" that automates all the steps needed to implement the `SheetDBEngine` database class so it can be easily used to produce the desired results, which is to manage worksheet database records.

To create the worksheet database wizard, you will use a VBA `UserForm` to calculate and collect all the necessary information, create the worksheet range names, produce the worksheet code, and create all necessary controls needed to maintain the worksheet database. The data validation list cell fills with the database records and all the necessary `Button` controls to manage them.

Implementing the Worksheet Database Wizard

To implement a `UserForm` wizard for the database class module, you must use a `UserForm` with a `Multipage` control, where each control page is associated to a wizard step. To see a simple example about how you can implement such a wizard interface, open the `MultiPage.xlsm` macro-enabled workbook (that you can extract from the `Chapter08.zip` file), press Alt+F11 to show the Visual Basic IDE, and double-click the `frmMultiPage UserForm` in the Project Explorer tree (Figure 8-15).

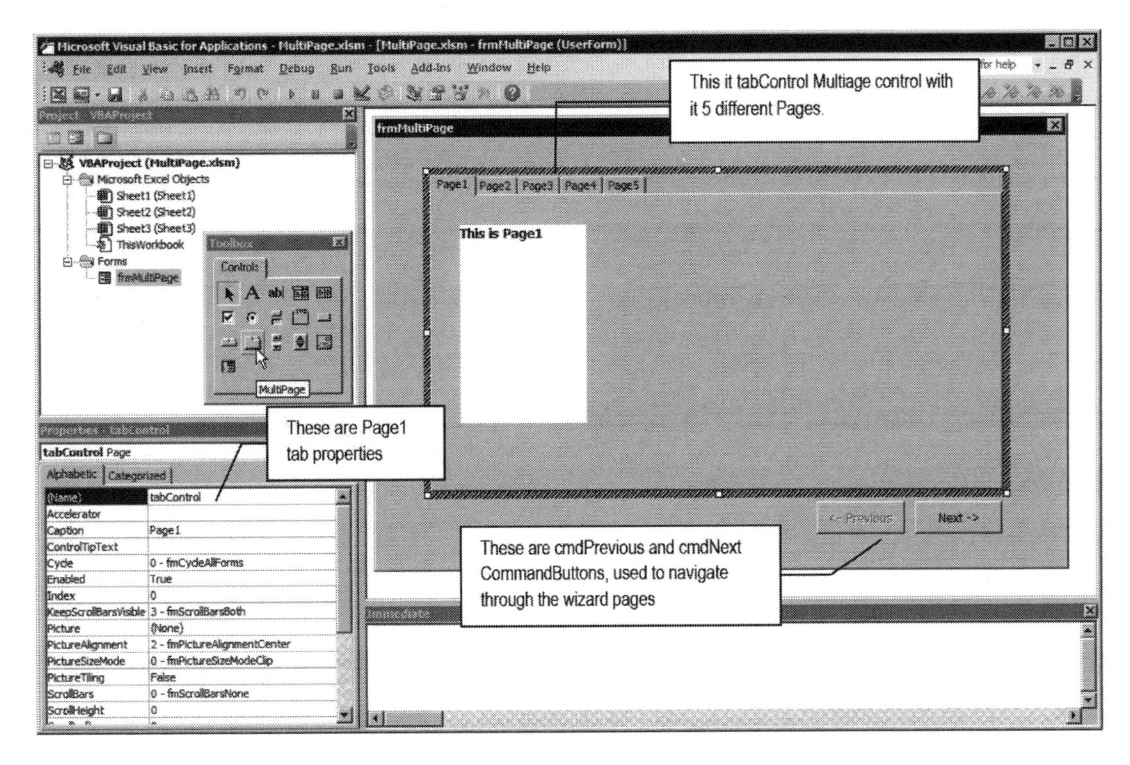

Figure 8-15. *This is the frmMultiPage UserForm from the MultiPage.xlsm workbook. It has a Multipage control with five tabs (Page1 to Page5). Use each page (tab) of the Multipage control to represent the steps that the wizard must follow, and use a pair of CommandButtons (cmdPrevious and cmdNext) to navigate through its pages*

Note that `frmMultiPage` has one `Multipage` control named `tabControl` (the `tab` prefix is commonly used to name the `Multipage` control according to the Reddick naming conventions, covered in Chapter 10). To select the `Multipage` control instead of one of its pages, click the control border.

Each control `Page` can be selected by clicking its associated tab, which is an independent control container where you can lay down other VBA controls to compose the `UserForm` wizard: one `Page` to each wizard step. Whenever a `Page` is selected, its controls are shown to the user.

To manage the `Multipage` control pages, right-click any `Page` tab caption and use the context menu to add, delete, rename, or move pages. To rename the `Tab` caption, use the `Page` control's `Caption` property.

Note that each `frmMultiPage` `Page` tab has its own `Label` control (using a different caption, background, and foreground color to call your attention as each tab is selected). The `Multipage` control is 0-based, meaning that the first `Page` (or tab) has `Index = 0` and the last `Page` has `Index = n-1`. To programmatically select the desired page, attribute to the `Multipage` control's `Value` property the desired `Page` index. The next instruction will always select the first `tabControl Multipage` control's `Page`:

```
tabControl.Value = 0   'This command will select the first tabControl page
```

The `Multipage` control can change the appearance and position of its `Page` tabs using two different properties: `Style` and `TabOrientation` (click the Multipage control border to set its properties).

- `Style`: Use this property to change the `Page` tab's appearance to
    - `0 - frmStyleTabs`: The default appearance

    - 1 - fmStyeButtons: To change the tab appearance to button controls
    - 1 - fmStyleNone: To remove the tab captions and control border

- TabOrientation: Use to change the position of the tabs. The default is 0 - fmTabOrientationTop, but you can show the tabs at the left, right, or bottom of the Multipage control.

To make the Multipage control work as a Microsoft Office wizard, use the UserForm_Initialize event to set MultiPage.Style = fmStyleNone and MultiPage.Value = 0 to show the first Page by default. The frmMultiPage UserForm has this code on its Initialize( ) event:

```
Private Sub UserForm_Initialize()
    Me.tabControl.Style = fmTabStyleNone
    Me.tabControl.Value = 0
End Sub
```

Navigating Through the UserForm Wizard

To see the frmMultiPage in action, double-click it in the VBA Explorer tree to show the UserForm design mode and press F5 to load it. Note that now it is showing its Page1 Page, and it has no tabs to select another Page control; it has just the cmdPrevious and cmdNext CommandButtons (Figure 8-16).

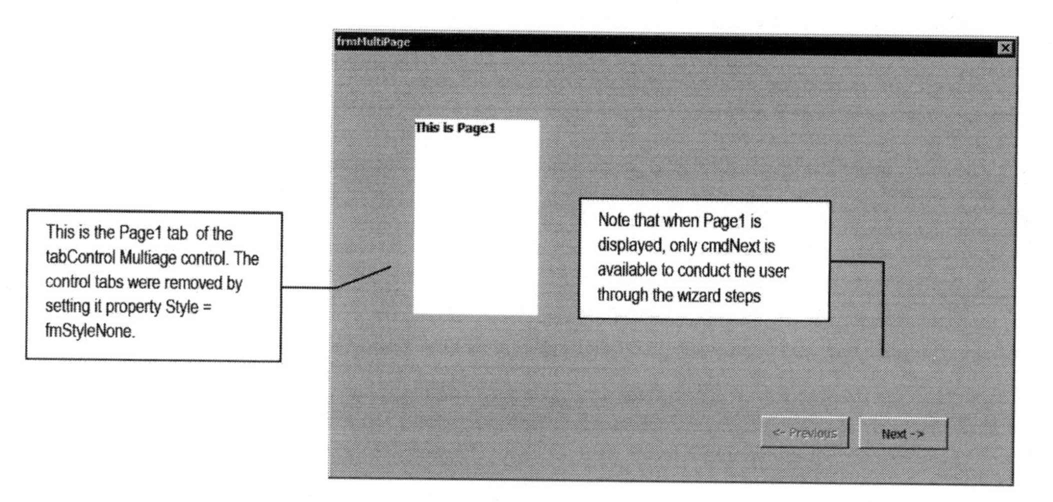

Figure 8-16. *When you press F5 in the Visual Basic IDE to load frmMultiPage UserForm, its Initialize( ) event fires, changing tabControl.Stye to fmStyleNone, removing the Multipage control tabs*

When you click cmdNext, the next tabControl Multipage Page is shown to the user. Note that now both cmdPrevious and cmdNext become enabled, so it can go back to the previous step. And if you keep click cmdNext until you reach the last Multipage Page, cmdNext will be disabled, like any wizard would do (Figure 8-17)!

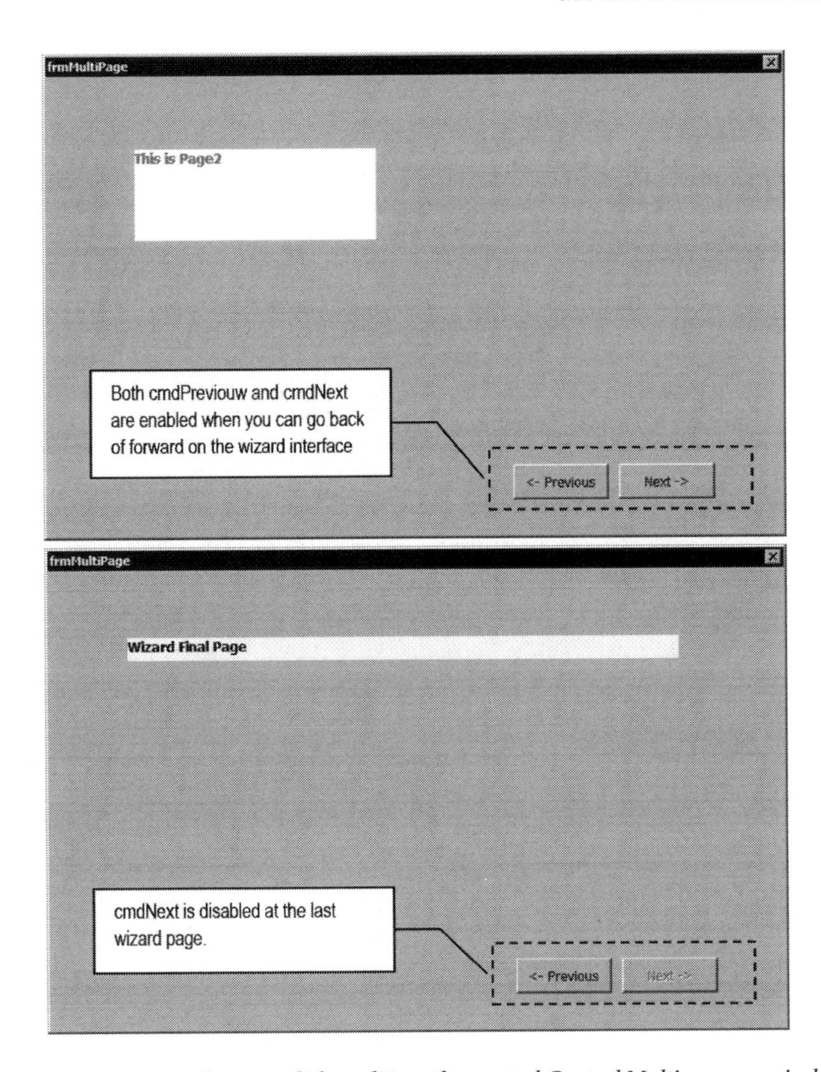

***Figure 8-17.** When you click cmdNext, the next tabControl Multipage page is shown to the user, until you reach the last Multipage control page, where cmdNext will be disabled*

Both the `cmdNext` and `cmdPrevious` command buttons use a single procedure called `ShowPage( )` to show the next/previous page of the `Multipage` control. To manage the `Multipage` control tab navigation, the `UserForm` declares the `ShowTab` enumerator, which is then called by each `CommandButton Click( )` event.

```
Private Enum ShowTab
    PreviousTab = -1
    NextTab = 1
End Enum

Private Sub cmdNext_Click()
    Call ShowPage(NextTab)
End Sub
```

```
Private Sub cmdPrevious_Click()
    Call ShowPage(PreviousTab)
End Sub
```

The ShowPage( ) code navigates through the Multipage control pages, setting the enabled property of the cmdNext and cmdPrevious CommandButtons:

```
Private Sub ShowPage(Action As ShowTab)
    Static sintPage As Integer
    Dim intMaxPages As Integer

    sintPage = sintPage + Action
    intMaxPages = Me.tabControl.Pages.Count - 1

    If sintPage < 0 Then sintPage = 0
    If sintPage > intMaxPages Then sintPage = intMaxPages
    Me.tabControl.Value = sintPage
End Sub
```

Note that the Static sintPage variable holds the position of the last-selected Page of the tabControl Multipage control, while intMaxPages holds the total page number.

```
    sintPage = sintPage + Action
    intMaxPages = Me.tabControl.Pages.Count - 1
```

The sintPage variable is always added by the Action argument value (which can be 1 or -1, according to the ShowTab enumerator received), and its value is used to verify which tab must be selected. If sintPage = 0, the first tab must be selected. If sintPage becomes negative (<0), it is set to the first Page index again.

```
    If sintPage < 0 Then sintPage = 0
```

If sintPage > intMaxPages, it must be set to the last Page tab.

```
    If sintPage > intMaxPages Then sintPage = intMaxPages
```

The tabControl is then moved to the desired tab.

```
    Me.tabControl.Value = sintPage
End Sub
```

■ **Attention** If you want to validate each wizard Page control value before moving to the next tab, use a centralized procedure, like Function ValidatePage( ), which receives the current Page index and validates all its controls returning True, before moving to the next tab, as follows:

```
If Action = NextTab Then
    If Not ValidatePage(Me.tabControl) then
        Exit Sub
```

```
        End If
    End If
    ...
    Me.tabControl.Value = sintPage
```

When the tabControl.Value property changes, VBA cascade-fires the tabControl_Change( ) event, which executes this simple code to enable/disable the cmdPrevious and cmdNext CommandButtons according to the tab selected. Note that cmdPrevious is disabled on tabControl to show the first page, while cmdNext is disabled on tabControl to show the last page.

```
Private Sub tabControl_Change()
    Me.cmdPrevious.Enabled = (Me.tabControl.Value > 0)
    Me.cmdNext.Enabled = (Me.tabControl.Value < Me.tabControl.Pages.Count - 1)
End Sub
```

■ **Attention** It is important to note that by using the tabControl_Change( ) event to control the enabled state of cmdPrevious and cmdNext, they will be always correctly synchronized when the tabControl.Value property changes, which can be made by user action or by the VBA code.

This is all you need to know about how to implement a wizard-like UserForm interface using a Multipage control that has one Page tab to each wizard step.

Required Database Properties

Among the 15 proposed database properties cited in section "Improving the Database Class Capacity" for the SheetDBEngine database engine class, just two of them are required to be defined so the database class can work properly.

- dbDataValitionList: The cell or range name where the data validation list will be created
- Either dbOneSide or dbManySide1 addresses or range names that indicate which worksheet cells have values that need to be saved by the database

All other properties will be either automatically defined by default values or derived from the one-side and/or many-side range addresses. And since not all worksheet database designs use the one-to-many record relationship (it can use just the one-side, many-side, or both), there is no need to define both dbOneSide and dbManySide1 properties; one of them is enough.

Using frmDBProperties UserForm

To help the user implement the database storage system, the database wizard must be capable of the following:

- Defining the first worksheet unused row where the database properties and the first database record will begin to be stored
- Showing all defined worksheet range names using ComboBox controls, so the user can select them to define the data validation list cell, the one-side record, and the many-side record values

- Creating the data validation list to select records saved in the database
- Allowing the user to change some database default values, like the record name (the default is `Record`) and the prefix used to identify the many-side record range names created by the database system (default is `rec_`)
- Creating the New, Save, and Delete control buttons in the worksheet, associated to default sheet code module procedures and its associated code

The `frmDBProperties_SheetDBEngine.xlsm` Excel macro-enabled workbook, which you can extract from the `Chapter08.zip` file, offers on its `Sheet1`, `Sheet2`, and `Sheet3` worksheets some simple worksheet input cell layouts that depict a generic one-to-many record relationship. Each sheet tab has a merged cell to place the data validation list from where the database saved records can be selected, a one-side record area, and a many-side records area (Figure 8-18).

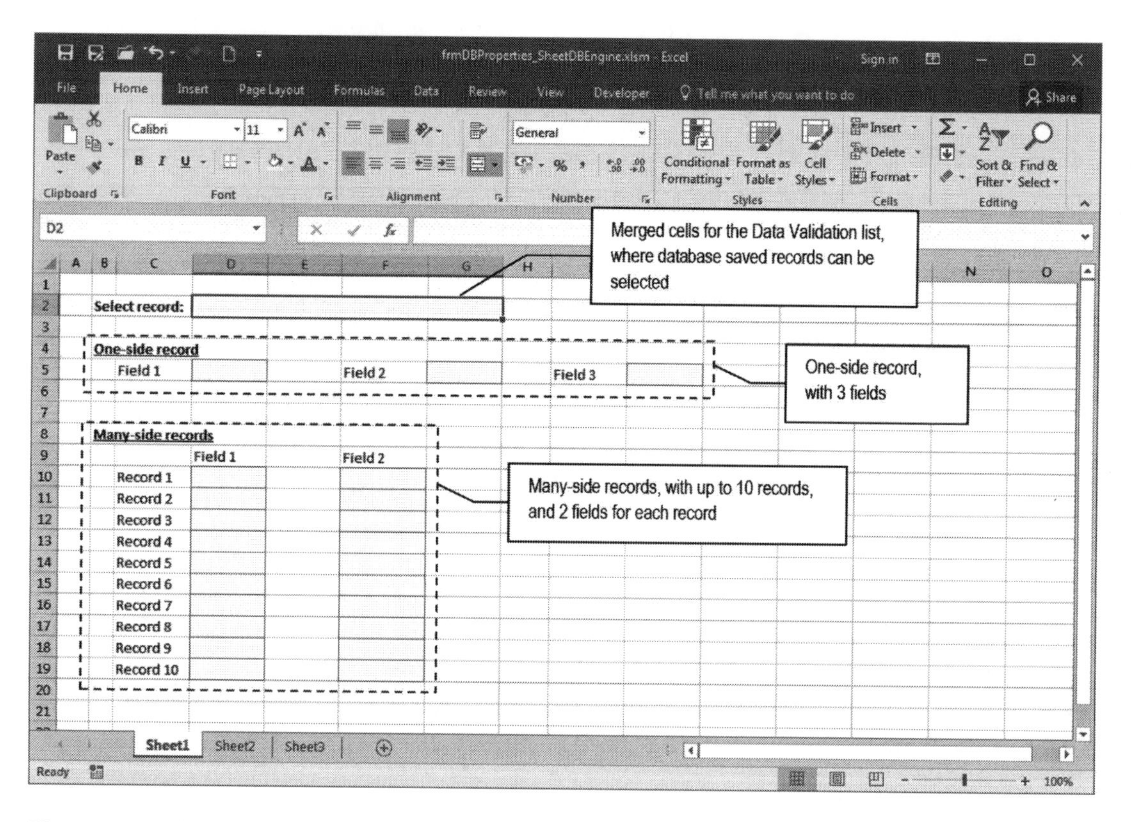

Figure 8-18. *This is the Sheet1 worksheet interface from the frmDBProperties_SheetDBEngine.xlsm Excel macro-enabled worksheet, which offers a default one-to-many database record relationship to save its data*

Press Alt+F11 to show the VBA interface and note that this workbook has the `frmDBProperties` `UserForm`, which uses a `Multipage` control (`tabControl`) and four `CommandButtons`: two in the top-left corner (`cmdDefine` and `cmdCancel`) to allow control the `UserForm` and two at the bottom-right corner (`cmdPrevious` and `cmdNext`) to navigate between the `tabControl Pages`. The `tabControl Multipage` control has six `Pages` that can be selected using the same programming technique described earlier in the section "Navigating Through the UserForm Wizard" (Figure 8-19).

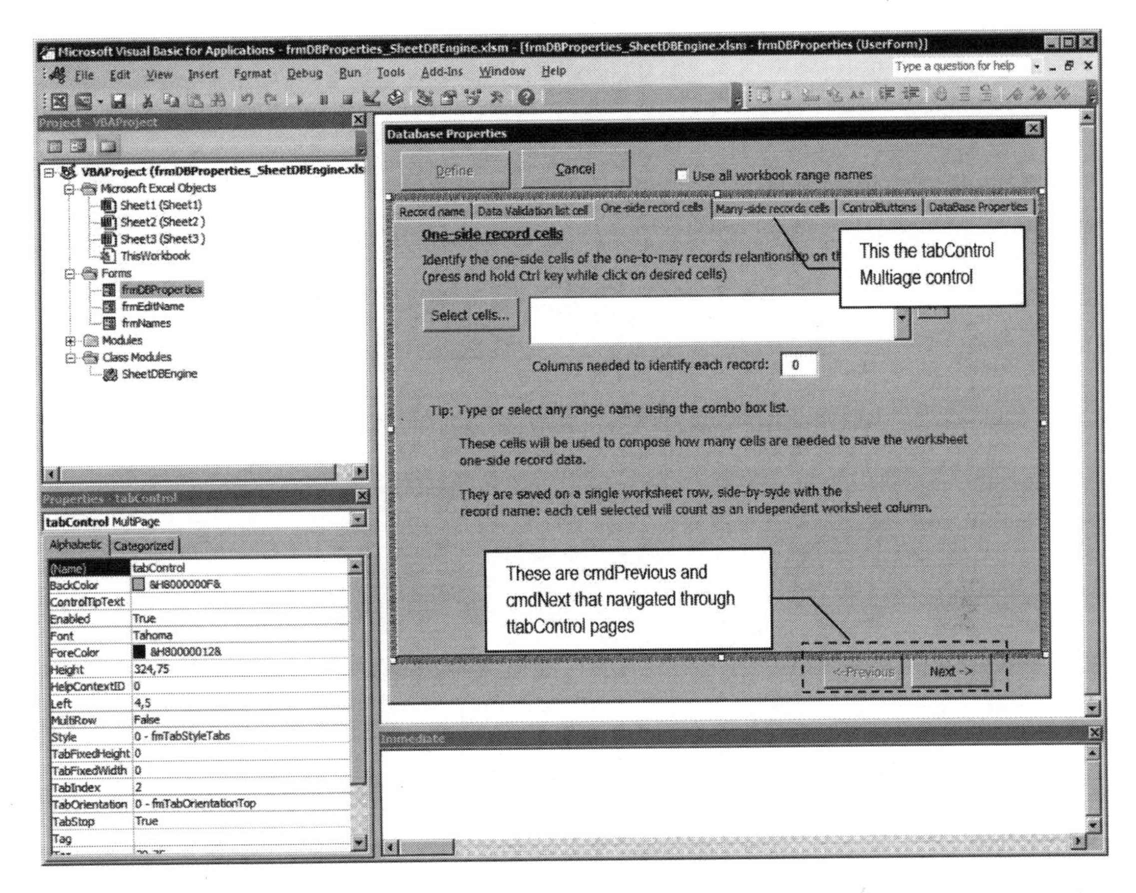

Figure 8-19. *This is frmDBProperties UserForm, with a Multipage control that uses five different pages and two CommandButtons at the bottom (cmdPrevious and cmdNext) to navigate through its pages*

Each `Multipage Page` tab has a defined usefulness.

- The first `tabControl Page` (record name) asks for a "record identification": a single substantive to be used to treat the record by the database system. The default value is `Record`.
- The second `tabControl Page` (data validation list cell) defines the data validation list cell.
- The third and fourth `tabControl Pages` define the cells that receives either the one-side or the many-side records cells that may have values to be saved by the database.
- The fifth `tabControl Page` defines the worksheet `Button` controls.
- The sixth `tabControl Page` shows a resumed view of all defined database properties.

For every database property defined in "Step 6: Save Database Properties as Range Names," there is at least one control to represent it inserted in one of the `tabControl Multipage Pages` (for some properties, a second control with the same name suffixed by 1 is offered on the sixth `tabControl Page` that resumes the database properties). So, to represent the `dbRecordName` property, the first `tabControl Page` (record name) has the `txtdbRecordName` text box, whose value is propagated in the `txtdbRecordName1` text box on the sixth `tabControl Page` (database properties). To represent the `dbDataValidationList` property, the second `tabControl Page` (data validation list cell) has the `txtdbDataValidationList` text box and also the `txtdbDataValidationList1` text box on the sixth `tabControl Page`, and so on.

Loading frmDBProperties

The wizard was built to verify on its `Initialize( )` event if the database system was already created on the active sheet. If it is, just the sixth `tabControl Page` (Database Properties) is shown in read-only mode. If not, the database must be set, and the wizard will show its first `tabControl Page`, like Figure 8-20 does.

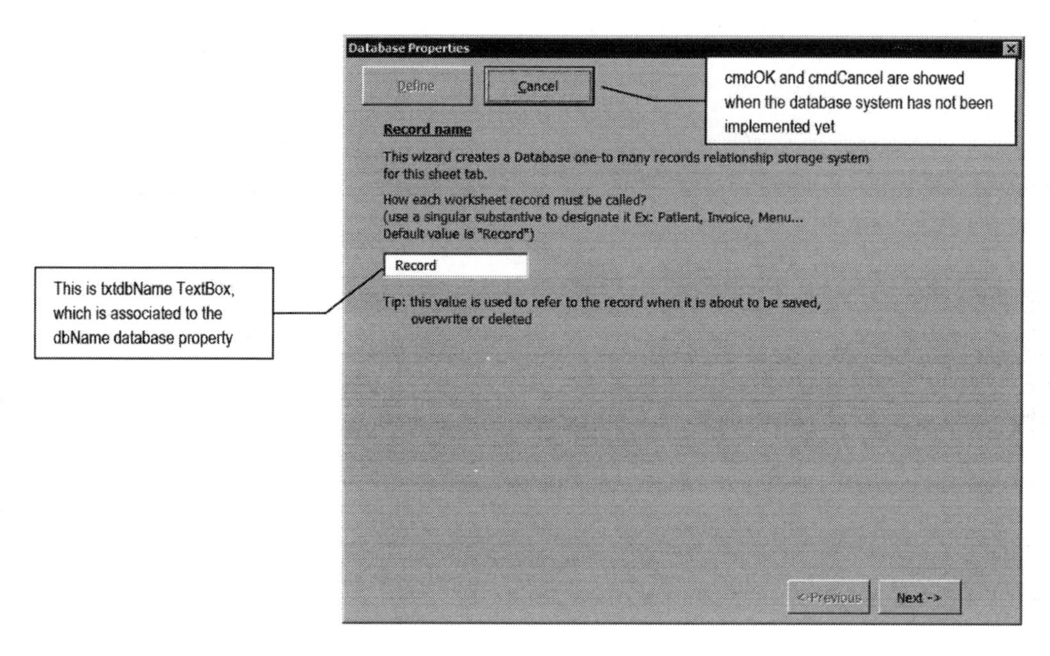

Figure 8-20. *When frmDBProperties is loaded, the tabControl Multipage control has its Page tabs hidden (property Style=frmTabStyleNone), showing a "wizard-like" interface to the user*

Since the aim of the `frmDBProperties` wizard is to walk the user through the process creation of the database property structure, its code must define a place in the worksheet where these database properties will be stored, using appropriate range names to identify each property name. So, the `UserForm_Initialize( )` event verifies whether one of these properties (`dbDataValidationList` range name) already exists to define what interface must be shown to the user. Take a look at the code:

```
Private Sub UserForm_Initialize()
    Dim ws As Worksheet
    Dim rg As Range
    Dim strName As String
    Dim strNameScope As String
    Dim intI As Integer
    Const conNormalWidth = 473
    Const conWhite = &HFFFFFF

    On Error Resume Next
```

```
    Me.Width = conNormalWidth
    Application.EnableEvents = False
    Set ws = Application.ActiveSheet
    strNameScope = "'" & ws.Name & "'!"
    Me.tabControl.Style = fmTabStyleNone

    Set rg = Range(strNameScope & "dbDataValidationList")
    If rg Is Nothing Then
        Me.tabControl.Value = 0
        Me.cmdPrevious.Visible = True
        Me.cmdNext.Visible = True
        Me.txtdbRecordsFirstRow = ActiveSheet.UsedRange.Row + ActiveSheet.UsedRange.Rows.
        Count + 3
        Me.txtRecordPosition = SetRecordPosition()
        Call LoadNames
    Else
        Me.cmdDefine.Caption = "Close"
        Me.cmdDefine.Accelerator = "C"
        Me.cmdDefine.Enabled = True
        Me.cmdCancel.Caption = "Remove"
        Me.cmdCancel.Accelerator = "R"
        Me.cmdPrevious.Visible = False
        Me.cmdNext.Visible = False
        Me.txtdbRecordName1.Locked = True
        Me.txtdbRecordName1.BackColor = conWhite
        Me.txtdbManySidePrefix.Locked = True
        Me.txtdbManySidePrefix.BackColor = conWhite

        'Update UserForm TextBoxes
        For intI = 1 To 15
            strName = Choose(intI, "dbRecordName", _
                             "dbDataValidationList", _
                             "dbSavedRecords", _
                             "dbRecordsFirstRow", _
                             "dbOneSide", _
                             "dbOneSideColumnsCount", _
                             "dbManySide1", _
                             "dbManySide2", _
                             "dbManySide3", _
                             "dbManySide4", _
                             "dbManySideFirstColumn", _
                             "dbManySideColumnsCount", _
                             "dbManySideRowsCount", _
                             "dbManySidePrefix", _
                             "dbRangeOffset")
            Me("txt" & strName) = ws.Range(strNameScope & strName)
        Next

        Call CalculateManySideRecords
        Me.tabControl.Value = Me.tabControl.Pages.Count - 1
    End If
End Sub
```

When the UserForm_Intialize( ) event fires, all necessary variables are declared, VBA unexpected errors are suppressed by an On Error Resume Next instruction, the UserForm Width property is set, Excel events firing is disabled, a reference to the current worksheet is set, and the tabControl Pages are hidden.

```
Private Sub UserForm_Initialize()
    ...
    On Error Resume Next

    Me.Width = conNormalWidth
    Application.EnableEvents = False
    Set ws = Application.ActiveSheet
    strNameScope = "'" & ws.Name & "'!"
    Me.tabControl.Style = fmTabStyleNone
```

These instructions prepare the stage. The code then tries to recover the dbDataValidationList database property, reading the dbDataValidationList range name value, as an indication that the active worksheet already has a database system implemented.

```
    Set rg = Range("'" & ws.Name & "'!dbDataValidationList")
```

Defining the Database

If the dbDataValidationList range name doesn't exist as a local sheet name, VBA will raise an error that will be ignored by the On Error Resume Next instruction, and the rg object variable will remain with its default value (Nothing), meaning that the frmDBProperties UserForm must walk the user step-by-step through implementing the database system beginning on the first tabControl Page.

```
    If rg Is Nothing Then
        Me.tabControl.Value = 0
```

The cmdPrevious and cmdNext CommandButtons become visible, and txtdbSaveRow (a TextBox positioned at the fifth tabControl Page) receives the row number where the database can be safely stored: three rows below the last used sheet row.

```
        Me.cmdPrevious.Visible = True
        Me.cmdNext.Visible = True
        Me. txtdbRecordsFirstRow = ActiveSheet.UsedRange.Row + ActiveSheet.UsedRange.Rows.
Count + 3
        Me.txtRecordPosition = SetRecordPosition()
```

Attention The Worksheet.UsedRange property returns a range address with the first and last used cell. That is why the code adds to the first used row (UsedRange,Row) the total number of rows used three extra rows (UsedRange.Rows.Count + 3).

To allow the user select any worksheet range name to define the cells associated to the one-side or the many-side worksheet records, it calls the LoadNames( ) procedure, which will produce a list of local (or workbook) range names that will be used to fill all on-side and many-side UserForm ComboBoxes.

```
        Call LoadNames
```

Loading Worksheet Names

The Sub LoadNames( ) procedure simply loops through the Names collection of the active worksheet or the entire workbook according to the value of the bolAllRangeNames optional argument (which is False by default), fills the varNames( ) array variable with the desired range names, and uses this array to define every desired ComboBox List property.

```
Private Sub LoadNames(Optional bolAllRangeNames As Boolean)
    Dim obj As Object
    Dim nm As Name
    Dim varNames() As Variant
    Dim intI As Integer

    On Error Resume Next

    'Load desired names on varNames() array
    If bolAllRangeNames Then
        Set obj = ThisWorkbook
    Else
        Set obj = ActiveSheet
    End If

    ReDim varNames(obj.Names.Count - 1)

    For Each nm In obj.Names
        varNames(intI) = nm.Name
        intI = intI + 1
    Next

    'Populate ComboBoxes
    Me.cbodbOneSide.List = varNames()
    Me.cbodbManySide1.List = varNames()
    Me.cbodbManySide2.List = varNames()
    Me.cbodbManySide3.List = varNames()
    Me.cbodbManySide4.List = varNames()
End Sub
```

Note the trick to select which range name scope must be used: the obj as Object variable is set to either the ActiveSheet or the ThisWorkbook (two different object types) according to the bolAllRangeNames as Boolean argument.

```
If bolAllRangeNames Then
    Set obj = ThisWorkbook
Else
    Set obj = ActiveSheet
End If
```

The varNames( ) array is then dimensioned according to the number of range names contained in the Names collection of the object represented by the obj variable (remember that array variables are 0-based, so you must use Names.Count - 1).

```
ReDim varNames(obj.Names.Count - 1)
```

And a For Each...Next loop is used to loop through all names of the object represented by the obj variable, filling the varNames( ) array with every nm.Name property associated to the scope of the active worksheet or the entire workbook (the intI variable is used to point to the desired array index).

```
    For Each nm In obj.Names
        varNames(intI) = nm.Name
        intI = intI + 1
    Next
```

Once the varNames( ) array is filled with the desired names, it is used to fulfill each wizard ComboBox List property.

```
    'Populate ComboBoxes
    Me.cbodbOneSide.List = varNames()
    Me.cbodbManySide1.List = varNames()
   ...
End Sub
```

And once LoadNames( ) is finished, the frmDBProperties UserForm is shown to the user, positioned at the first tabControl Page, and ready to walk through all the steps needed to implement the database properties (Figure 8-19).

Stepping Through frmDBProperties Wizard Pages

The frmDBProperties wizard needs to conduct the user using just five pages: one page to define an appropriate substantive to "treat" the worksheet records, a second page to define the cells for the record data validation list, a third page to define the address or range name that indicates the one-side record cells (if any), a fourth page to define the address or range name that indicats the many-side records cells (if any), and a fifth page to create the Button controls to manage the worksheet database system.

Step 1: Defining the Expression Used to Identify Worksheet Records

The first frmDBProperties wizard page is used to define how each database record will be treated, offering "Record" as the default substantive. Step 1 must do the following:

1. Allow the user to type a record name
2. Use the ValidatePage( ) procedure to check whether a name has been typed (this procedure is called from ShowPage( ) when the Next -> CommandButton is clicked).

The user is asked to type a single substantive to identify each worksheet one-side record, and the value typed is validated when it tries to go to the next wizard step, clicking the cmdNext CommandButton. You cannot leave the first page if the record name is empty (Figure 8-21).

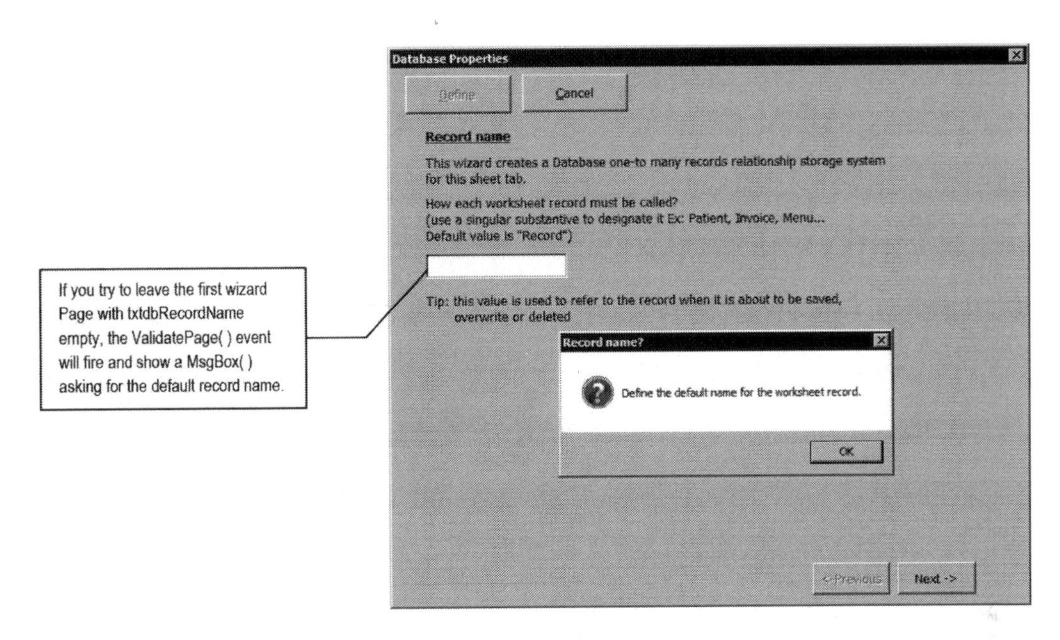

Figure 8-21. *frmDBProperties validates each wizard page as the user tries to advance to the next page. The first wizard page requires that a record name be typed*

This happens because each time the `cmdNext_Click( )` event fires, it calls the `Sub ShowPage( )` procedure passing to the `Action` argument the `NextTab` enumerator. The `ShowPage( )` procedure then calls the `ValidatePage( )` procedure, passing to the `intPage` argument the `sintPage` Static variable, which holds the last `tabControl.Value` property.

```
Private Sub cmdNext_Click()
    Call ShowPage(NextTab)
End Sub

Private Sub ShowPage(Action As ShowTab)
    Static sintPage As Integer
    Dim intMaxPages As Integer

    If Action = NextTab Then
        If Not ValidatePage(sintPage) Then Exit Sub
    End If

    sintPage = sintPage + Action
    intMaxPages = Me.tabControl.Pages.Count - 1

    If sintPage < 0 Then sintPage = 0
    If sintPage > intMaxPages Then sintPage = intMaxPages
    Me.tabControl.Value = sintPage
    Me.cmdDefine.Enabled = (sintPage = Me.tabControl.Pages.Count - 1)
    Me.chkWorkbookNames.Visible = (sintPage = 2 Or sintPage = 3)
End Sub
```

Note that this code is similar to the code in "Navigating Through the UserForm Wizard" earlier in this chapter, except for the validation code and its last two instructions, which change the cmdDefine.Enabled and chkWorkbookNames.Visible properties, according to the tabControl page selected.

When the ValidatePage( ) procedure receives the intPage argument, it uses a Select Case statement to validate the current tabControl page.

```
Function ValidatePage(intPage As Integer) As Boolean
    Dim strMsg As String
    Dim strTitle As String
    Dim bolValidateFail As Boolean

    Select Case intPage
        Case 0
            'Validata record name
            If Len(Me.txtdbRecordName) = 0 Then
                strMsg = "Define the default name for the worksheet record."
                strTitle = "Record name?"
                bolValidateFail = True
            End If
        Case 1
            'Validata Data Validation list
            If Len(Me.txtdbDataValidationList) = 0 Then
                strMsg = "Select a cell for the Records Data Validation list and try again."
                strTitle = "Data Validation list cell?"
                bolValidateFail = True
            End If
        Case 3
            'Validata OneSide and ManySide records
            If Me.txtdbOneSideColumnsCount = 0 And Me.txtdbManySideRowsCount = 0 Then
                strMsg = "Select the One-Side and/or the Many-Side cells that define the
                worksheet records ranges!"
                strTitle = "Select cells to be saved as worksheet records"
                bolValidateFail = True
            End If
    End Select

    If bolValidateFail Then
        MsgBox strMsg, vbQuestion, strTitle
    Else
        ValidatePage = True
    End If
End Function
```

Note that when intPage = 0 (meaning the first wizard page), ValidatePage( ) will verify whether anything has been typed inside the txtdbName TextBox. If it is empty, bolValidateFail = True, a MsgBox( ) function warns the user, and ValidatePage( ) returns False to the ShowPage( ) procedure.

```
Function ValidatePage(intPage As Integer) As Boolean
    ...
    Select Case intPage
        Case 0
```

```
            'Validata record name
            If Len(Me. txtdbRecordName) = 0 Then
                strMsg = "Define the default name for the worksheet record."
                strTitle = "Record name?"
                bolValidateFail = True
            End If
            ...
    End Select

    If bolValidateFail Then
        MsgBox strMsg, vbQuestion, strTitle
    Else
        ValidatePage = True
    End If
End Function
```

Step 2: Defining the Record's Data Validation List Cell

After defining how each database record must be treated on the first wizard page and clicking the `cmdNext` command button, the second `tabControl Page` will be shown. It asks you to define the record's data validation list cell. Note that the `txtdbDataValidationList` text box is locked and you must click the "Select cell" command button (`cmdDataValidationList`) to define it. Step 2 must do the following:

1. Allow the selection of a worksheet cell where the data validation list must be created by calling the `GetRange( )` procedure
2. Create a data validation list on the selected cell, pointing to the `SavedRecords` range name
3. Use the `ValidatePage( )` procedure to check whether a cell was selected for the record data validation list when the Next ➤ `CommandButton` is pressed.

■ Attention The `ValidatePage( )` procedure will not allow you to advance to the next page while `txtdbDataValidationList` is empty.

When you click the `cmdDataValidationList CommandButton`, the `frmDBProperties` will be hidden and the `Application.Inputbox` will be shown over the `Sheet1` worksheet to allow select the cell where the data validation list will be created (Figure 8-22).

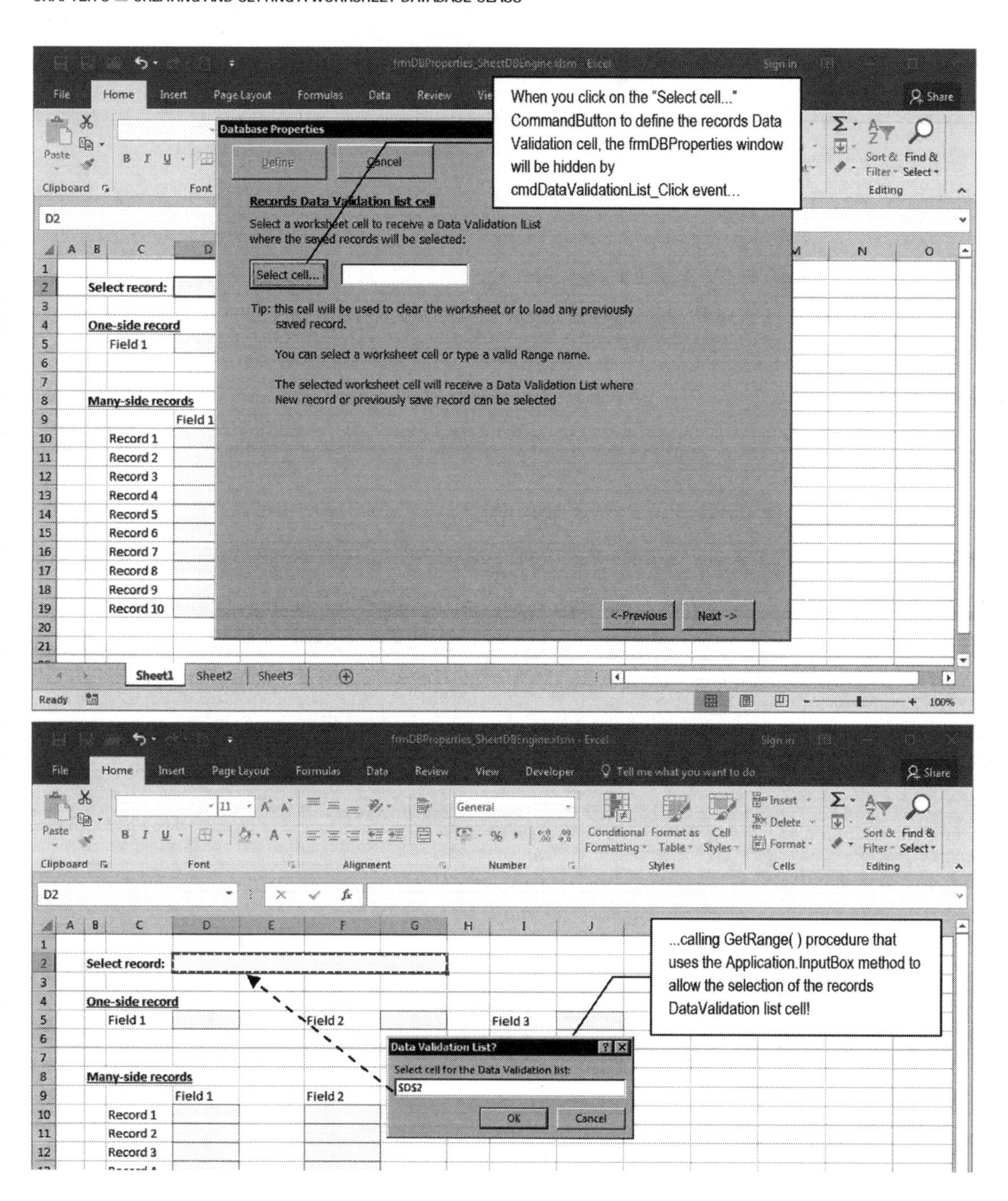

Figure 8-22. *The second tabControl Page asks the user to define the record data validation list cell. It has a "Select cell..." CommandButton that uses the Application.InputBox method to allow you to select the cell address*

This is the cmdDataValidationList_Click( ) event code:

```
Private Sub cmdDataValidationList_Click()
    Dim varFormula As Variant
    Dim varName As Variant
    Dim strListRange As String
    Dim strRange As String

    On Error Resume Next

    Me.Hide
    strRange = GetRange("Select cell for the Data Validation list:", "Data Validation
    List?", Me.txtdbDataValidationList)
    If Len(strRange) Then
        varName = Range(strRange).Name.Name
        If Len(varName) Then
            strRange = varName
        End If
        Me.txtdbDataValidationList = strRange
        Range(strRange).Merge

        'Verify if selected range has a data validation list
        strListRange = Range(strRange).Validation.Formula1
        If Len(strListRange) Then
            Me.txtdbSavedRecords = Mid(strListRange, 2)
        Else
            Me.txtdbSavedRecords = "SavedRecords"
        End If
    End If
    Me.Show
End Sub
```

Allowing the User to Select a Cell Range

The cmdDataValidationList_Click( ) event begins by hiding the frmDBProperties window by calling the UserForm.Hide method and then calls the Function GetRange( ) procedure, passing as an argument the message, title, and default value that must be displayed by the Application.InptuBox method.

```
    Me.Hide
    strRange = GetRange("Select cell for the Data Validation list:", "Data Validation
    List?", Me.txtdbDataValidationList)
```

The Function GetRange( ) procedure is responsible for allowing you to select any worksheet cells and returning the selection as a string of cell addresses. It has this code:

```
Private Function GetRange(strMsg As String, strTitle As String, Optional Default As Variant)
As String
    Dim varRg As Variant
    Dim rgArea As Range
    Dim strAddress As String
    Dim bolInvalidSelection As Boolean
```

```
    Const conRange = 8

    On Error Resume Next

    Set varRg = Application.InputBox(strMsg, strTitle, Default, , , , , conRange)
    If IsObject(varRg) Then
        For Each rgArea In varRg.Areas
            If Len(strAddress) Then
                strAddress = strAddress & ","
            End If
            strAddress = strAddress & rgArea.Address
        Next
        GetRange = strAddress
    End If
End Function
```

It calls the `Application.InputBox` method and sets the argument `Type=conRange` (`conRange=8`), meaning that it must return a `Range` object representing the cell(s) selected.

```
        Set varRg = Application.InputBox(strMsg, strTitle, Default, , , , , conRange)
```

■ **Attention** This is necessary because the `InputBox` string returns just the first 256 characters of the selected cell's address. By using `Type=conRange` to return a `Range` object, this limitation will no longer happen.

If the user selects the `Application.InputBox` Cancel button, nothing will be returned to the `varRg` variable, which will remain with its default `Empty` value. But if any cell has been selected, `varRg` will contain a `Range` object. The code will then step through the `Range.Areas` collection, using each selected `Area` to compose the selected cell's addresses string, which will be returned by the `GetRange( )` procedure (note that the code adds a comma between each selected range).

```
        If IsObject(varRg) Then
            For Each rgArea In varRg.Areas
                If Len(strAddress) Then
                    strAddress = strAddress & ","
                End If
                strAddress = strAddress & rgArea.Address
            Next
            GetRange = strAddress
        End If
End Function
```

■ **Attention** The `Range.Address` property returns just the first 256 characters of what is currently selected by it. It is necessary to run through the `Range.Areas` collection to return all selected cells when their textual length surpasses 256 characters.

Creating the Record's Data Validation List

When the GetRange( ) procedure ends, it will return either an empty string or a string containing the selected cell address where the record's data validation list must be created to the cmdDataValidationList_Click( ) event.

```
strRange = GetRange("Select cell for the Data Validation list:", "Data Validation List?",
Me.txtdbDataValidationList)
If Len(strRange) Then
```

If a cell address was selected, the procedure verifies whether the address is associated to a range name by checking whether the Range.Name property returns a valid Name object. If it does, use the Name object's .Name property to return the range name using the odd syntax Range(strRange).Name.Name. That is why the code begins disabling VBA errors with an On Error Resume Next instruction. If the range was not named, it has no Name object, and Range.Name will return an error.

```
On Error Resume Next
...
    varName = Range(strRange).Name.Name
```

▒ **Attention** The Range.Name property returns by default the name address.

Now the varName variable holds the range name or its default Empty value. So, the code verifies whether varName has any name inside it and attributes it to the strRange variable, strRange (that contains either the selected cell address or its range name) is stored in the txtdbDataValidationList TextBox, and the returned range is merged using the Range.Merge method.

```
If Len(varName) Then
    strRange = varName
End If
Me.txtdbDataValidationList = strRange
Range(strRange).Merge
```

Once the desired cell had been selected, it is time to create the record's data validation list on it. Since Validation is an object of the Microsoft Excel interface, the code verifies whether the selected cell has a data validation list by inspecting the Validation object's Formula1 property, storing it in the strListRange string variable.

```
'Verify if selected range has a data validation list
strListRange = Range(strRange).Validation.Formula1
```

If the selected range already has a data validation list, the strListRange variable now contains the formula used to fill the list. Since a formula always begin with a = character, it uses the VBA Mid( ) function to remove it from strListRange and store just the list address in the txtdbSavedRecords TextBox (positioned at the fifth tabControl Page).

```
If Len(strListRange) Then
    Me.txtdbSavedRecords = Mid(strListRange, 2)
```

If the selected range does not have a data validation list, the default SavedRecords name will be stored in the txtdbSavedRecords TextBox. Note that before the cmdDataValidationList_Click( ) event ends, it makes the frmDBProperties visible again by calling the UserForm.Show method.

```
        Else
            Me. txtdbSavedRecords = "SavedRecords"
        End If
    End If
    Me.Show
End Sub
```

Now the txtdbDataValidationList TextBox contains the cell address where the data validation list must be created, and the user can advance to the third wizard step.

Step 3: Defining the One-Side Record Cells

Having selected the record's data validation list cell, it is time to define which are the one-side record cells (if any). Step 3 must do the following:

1. Allow the selection of a valid range name or cell address (or cell addresses), by either selecting it in the cbodbOneSide ComboBox list or using the cmddbOneSide CommandButton
2. Select in cbodbOneSide either just worksheet names or all workbook names

---

■ **Attention** Since not all worksheets need to implement a one-side record, this page is not validated, and you can advance to the next page without any selection.

This is the first tabControl Page with a ComboBox (cbodbOneSide) that has been filled with all worksheet scope range names by the LoadNames( ) procedure (called in the UserForm_Initialize( ) event). It offers the chkWorkbookNames check box right above the tabControl Multipage control, allowing you to change the name scope used to fill this ComboBox list (Figure 8-23).

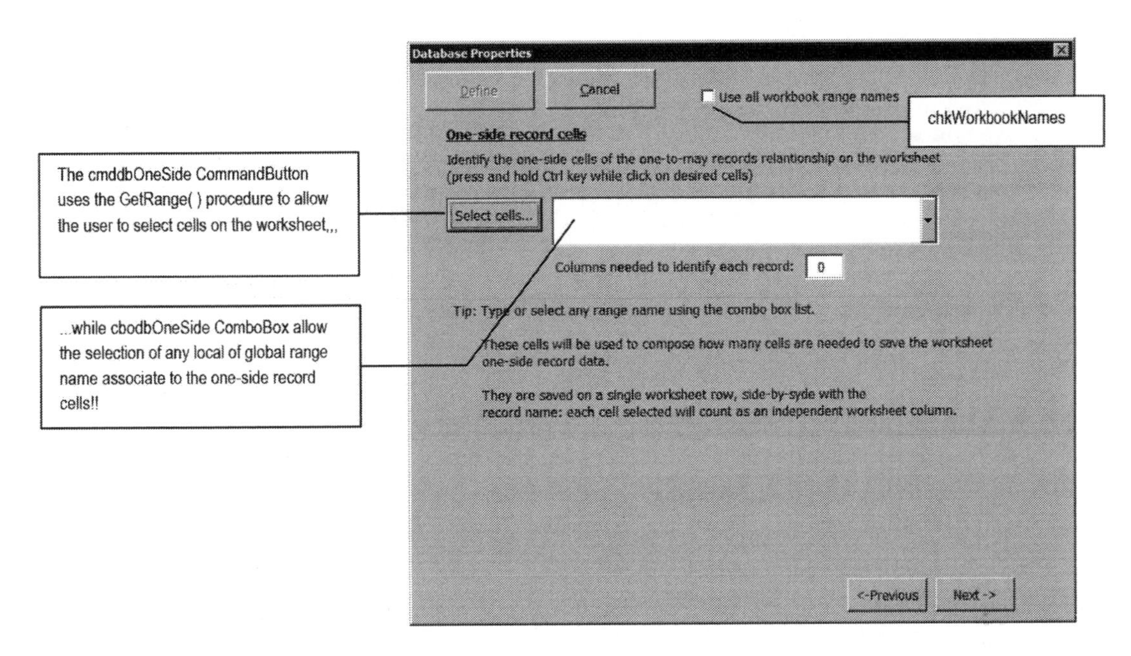

Figure 8-23. *This is the third wizard page, which asks the user to select the cells that must be used to compose the one-side worksheet record*

You cannot type a value in the `cbodbOneSide ComboBox`. You can just select a valid range name from its list or click the "Select cells" `CommandButton` to select the desired one-record worksheet cells.

If you click the "Select cells" `CommandButton`, the `cmddbOneSide_Click( )` event will fire, executing this code (note that it does not change the `cbodbOneSide ComboBox` value if the user clicks the `Application.Inputbox` Cancel button):

```
Private Sub cmddbOneSide_Click()
    Dim strRange As String
    Dim strMsg As String

    On Error Resume Next

    Me.Hide
    strMsg = "Select all cells that belongs to the 'one side' of the worksheet record." &
vbCrLf
    strRange = GetRange(strMsg, "One-side record sheet cells", Me.cbodbOneSide)
    If Len(strRange) Then
       Me.cbodbOneSide = strRange
    End If
    Me.Show
End Sub
```

The cmddbOneSide_Click( ) event calls GetRange( ) to allow the user to select all one-side record worksheet cells and attribute the selection (if any) to the cbodbOneSide ComboBox. But since by default the cbodbOneSide list is filled with all worksheet scope range names, you can both expand its list and select the appropriate range name that represents the one-side record cells. In both cases, this will fire the cbodbOneSide_Change( ) event, executing this code:

```
Private Sub cbodbOneSide_Change()
    Dim intCells As Integer
    Dim strAddress As String
    Const conColsDatabase = 6

    If IsRange(Me.cbodbOneSide) Then
        'Count cells selected
        intCells = CalculateOneSideColumns()
        Me.txtdbOneSideColumnsCount = intCells
        'Define save column for many-side records
        strAddress = Cells(1, intCells + conColsDatabase).Address
        strAddress = Left(strAddress, InStrRev(strAddress, "$"))
        Me.txtdbManySideFirstColumn = strAddress
    Else
        Me.cbodbOneSide = ""
        Me.txtdbOneSideColumnsCount = 0
        Me.txtdbManySideFirstColumn = ""
    End If

    If Left(Me.cbodbOneSide, 1) = "'" Then
        Me.txtdbOneSide = "'" & Me.cbodbOneSide
    Else
        Me.txtdbOneSide = Me.cbodbOneSide
    End If
    Me.cmdClearcbodbOneSide.Enabled = (Len(Me.cbodbOneSide) > 0)
End Sub
```

Whenever the cbodbOneSide ComboBox receives a value, either by the VBA code or by user typing, the code use the IsRange( ) procedure to verify whether the ComboBox value is a valid range name.

```
Private Function IsRange(strRange As String) As Boolean
    Dim rg As Range

    On Error Resume Next
    Set rg = Range(strRange)
    IsRange = (Err = 0)
End Function
```

■ **Attention** The IsRange( ) procedure tries to recover a Range object reference associated to text received by its strRange String argument. It will not raise errors when strRange strings refer to a valid Range reference.

If cbodbOneSide has a valid range name, it will call the CalculateOneSideColumns( ) procedure to count how many cells have the selected range and determine the worksheet column number needed to save the one-side record cells side by side right next to the record name.

```
'Count cells selected
intCells = CalculateOneSideColumns()
```

Calculating the Column Number Needed to Save the One-Side Record Cells

The function CalculateOneSideColumns( ) uses the cell's addresses or range name defined by the cbodbOneSide ComboBox to count how many cells exist on the selected one-side record cells and executes this code:

```
Private Function CalculateOneSideColumns() As Integer
    Dim rg As Range
    Dim rgArea As Range
    Dim strAddress As String
    Dim intNumCols As Integer
    Dim intI As Integer
    Dim intJ As Integer

    Set rg = Range(Me.cbodbOneSide)
    For Each rgArea In rg.Areas
        For intI = 1 To rgArea.Rows.Count
            For intJ = 1 To rgArea.Columns.Count
                If rgArea.Cells(intI, intJ).MergeCells Then
                    intI = intI + rgArea.Cells(intI, intJ).MergeArea.Rows.Count - 1
                    intJ = intJ + rgArea.Cells(intI, intJ).MergeArea.Columns.Count - 1
                End If
                intNumCols = intNumCols + 1
            Next
        Next
    Next
    CalculateOneSideColumns = intNumCols
End Function
```

The cells selected for the one-side record can be composed of unique or merged cells representing a single cell address (designated by the top-left corner cell of the merged range). All cells that comprise the one-side record can be isolated or grouped in individual ranges, so the code uses the Range.Areas collection to loop through all selected ranges.

```
Set rg = Range(Me.cbodbOneSide)
For Each rgArea In rg.Areas
```

Since an individual Range.Area can have just one single cell (or merged cell) or be comprised of many independent contiguous cells, the code needs to loop through the rows and columns using two nested For...Next loops. The outer loop will use the intI Integer variable to run through all Range.Area rows, while the inner loop will use the intJ Integer variable to run through all Range.Area columns.

```
For intI = 1 To rgArea.Rows.Count
    For intJ = 1 To rgArea.Columns.Count
```

To select each individual cell, the code uses the Range.Cells collection, using intI and intJ for each Cells(row, column) coordinate. Since any cell can be merged, it uses the Range.MergeCells property to verify whether the selected cell is merged. If it is, it will increment the intI and intJ counters by the number of rows or columns merged to count these merged cells as one unique cell. Note that it uses MergeArea.Rows.Count -1 and MergeArea.Columns.Count - 1 to sum the correct value for each counter.

```
                If rgArea.Cells(intI, intJ).MergeCells Then
                    intI = intI + rgArea.Cells(intI, intJ).MergeArea.Rows.Count - 1
                    intJ = intJ + rgArea.Cells(intI, intJ).MergeArea.Columns.Count - 1
                End If
```

If the cell is not merged, MergeCells = False, counting just one cell without incrementing the intI and intJ counters. The intNumCols Integer variable will hold the cell counting at the end of the For Each...Next loop, when all Range.Areas have been processed.

```
                intNumCols = intNumCols + 1
            Next intJ
        Next intI
    Next
    CalculateOneSideColumns = intNumCols
End Function
```

When the CalculateOneSideColumns( ) procedure ends, code control returns to the cmddbOneSide_Click( ) event procedure, and the value returned by the CalculatedOneSideColumns( ) procedure is stored in the intCells Integer variable, which is used to define the txtdbOneSideColumnsCount TextBox on the third wizard page.

```
        intCells = CalculateOneSideColumns()
        Me.txtdbOneSideColumnsCount = intCells
```

■ **Attention** Whenever a UserForm TextBox value has a duplicate control on the last wizard page, its TextBox.Change( ) event is used to update its value on the last wizard page. This is the code for the txtdbOneSideColumnsCount_Change( ) event (which duplicates its value using txtdbOneSideColumnsCount1).

```
Private Sub txtdbOneSideColumnsCount_Change()
    Me.txtdbOneSideColumnsCount1 = Me.txtdbOneSideColumnsCount
End Sub
```

Determining the Column to Save the Many-Side Records

To define the column letter where the many-side records will be saved, the Sub cbodbOneSide_Change() event procedure uses the constant conColsDatabase = 6 to define the first possible worksheet column where the database many-side record can be created. The database structure begins on column A (Figure 8-5) using two columns for the database properties (property name and value, one blank column, at least one column for the SavedRecords range name, and one extra blank column to separate the one-side record from the many-side records storage area). This constant value is added to the intCell Integer variable, which now holds how many columns will be needed to save the one-side record cells to generate the cell address for the first many-side record column.

```
Private Sub cbodbOneSide_Change()
    Const conColsDatabase = 6
    ...
        'Define save column for many-side records
        strAddress = Cells(1, intCells + conColsDatabase).Address
```

Since the Range.Address property returns an absolute address, by default the code uses the VBA InStrRev( ) function to search in reverse order the last $ character inside the strAddress String variable and extract the column letter using the VBA Left( ) function, which is saved in the txtdbManySideFirstColumn TextBox on the fifth wizard page.

```
        strAddress = Left(strAddress, InStrRev(strAddress, "$"))
        Me.txtdbManySideFirstColumn = strAddress
```

If IsRange( ) returns False, cbodbOneSide has no valid range name (the user tried to type a character), so cbodbOneSide and txtSaveCol must be cleaned, and txtdbOneSideColumnsCount must be set to zero.

```
    If IsRange(Me.cbodbOneSide) Then
    ...
    Else
        Me.cbodbOneSide = ""
        Me.txtdbOneSideColumnsCount = 0
        Me.txtdbManySideFirstColumn  = ""
    End If
```

The cbodbOneSide_Change( ) event then synchronizes txtdbOneSide (located on the fifth wizard page) with the cbodbOneSide value. Here is another trick to perform: if the current sheet tab name has a space (like "BMI Chart" or "Sheet 1"), the selected address will begin with a single quotation mark. But since Excel uses single quotation marks to insert cell text values, when this address is saved on the appropriate cell, it will lose the quote. The code must check for this character's existence and add an extra one before the proposed range name.

```
    If Left(Me.cbodbOneSide, 1) = "'" Then
        Me.txtdbOneSide = "'" & Me.cbodbOneSide
    Else
        Me.txtdbOneSide = Me.cbodbOneSide
    End If
```

The code finishes by changing the cmdClearcbodbOneSide CommandButton's Enabled property (the small button at the right of the cbodbOneSide ComboBox) according to the existence of any content in the cbodbOneSide ComboBox.

```
    Me.cmdClearcbodbOneSide.Enabled = (Len(Me.cbodbOneSide) > 0)
End Sub
```

Figure 8-24 shows the third wizard page after the frmDBProperties.xlsm workbook's Sheet1 OneSideRecord range was selected using the cbodbOneSide ComboBox list. Note how txtdbOneSideColumnsCount correctly shows the number of cells selected (three cells).

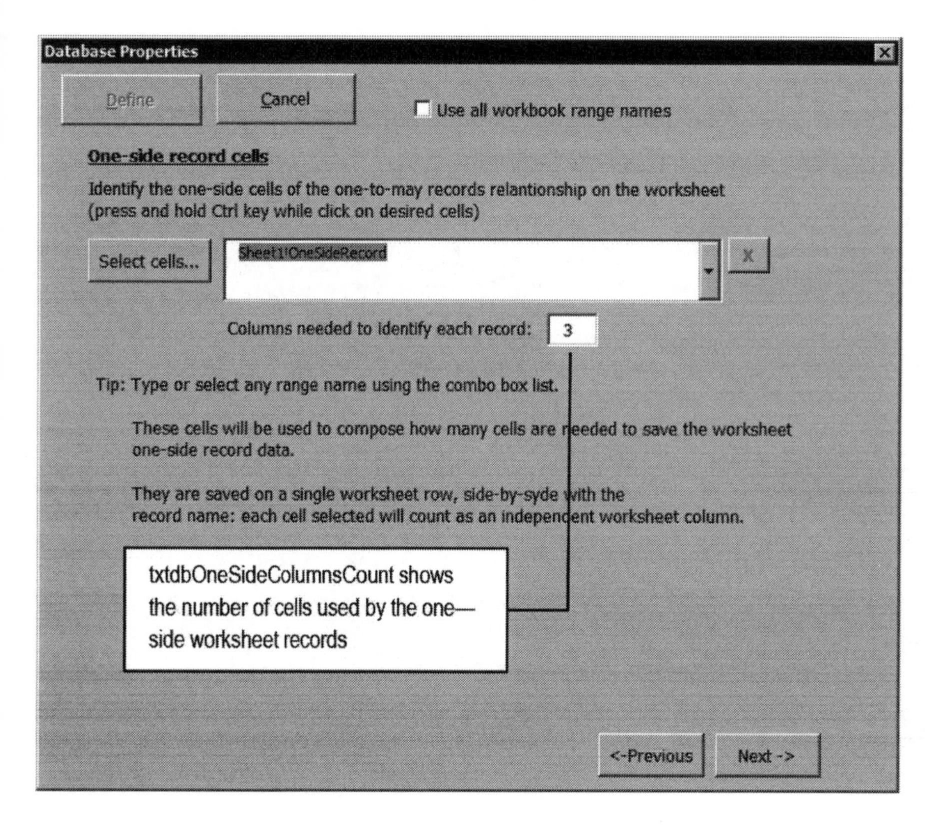

Figure 8-24. *This is the third wizard Page counting the number of cells inside the OneSideRecord range name. After the range is selected in the ComboBox list, the button at the right becomes enabled to clear the user selection*

Changing the Range Name Scope

At the top of the `UserForm` you find the `chkWorkbookNames CheckBox` that when checked shows every workbook name in the `cbodbOneSide ComboBox`. Whenever you click it, the `chkWorkbookNames_Change( )` event fires, executing this simple code:

```
Private Sub chkWorkbookNames_Click()
    Call LoadNames(Me.chkWorkbookNames)
End Sub
```

As you can see, it just calls the `LoadNames( )` procedure, passing as an argument the selection state of the `chkWorkbookNames CheckBox` to refill all `UserForm ComboBoxes` with the desired range names.

■ **Attention** The `LoadNames( )` procedure was analyzed in the section "Loading Worksheet Names" earlier in this chapter.

Clearing the One-Side Record Cell Selection

To clear the cbodbOneSide ComboBox, you can either select the control and press the Delete key or click the cmdClearcbodbOneSide CommandButton (the one with a red *X* character, right next the ComboBox). Note that this button becomes enabled whenever something has been selected for the one-side record cells, executing some simple code to clean it up.

```
Private Sub cmdClearcbodbOneSide_Click()
    Me.cbodbOneSide = ""
End Sub
```

■ **Attention** No matter the method you use, the cbodbOneSide_Change( ) event will cascade-fire to synchronize the UserForm interface.

Step 4: Defining the Many-Side Record Cells

On the fourth wizard page, you must define the many-side record worksheet cells to be saved by the database engine (Figure 8-25).

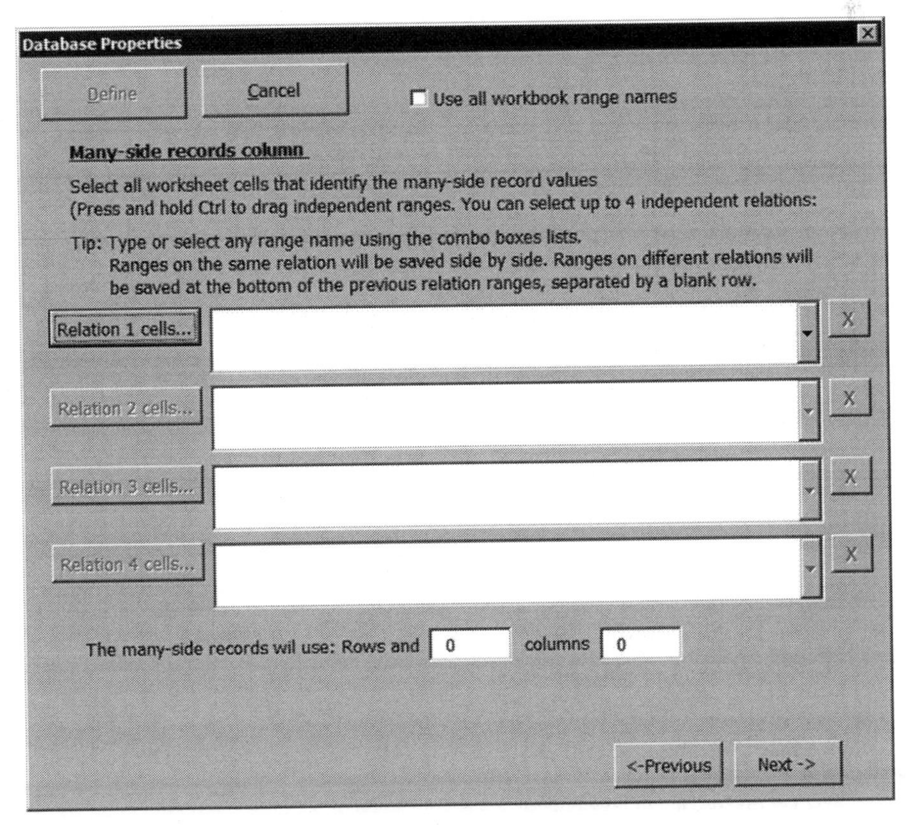

Figure 8-25. *This is the fourth wizard page that asks for database information from the user. It allows the selection of four different one-to-many records relations for the worksheet database. As explained by the page's "Tip," cells in the same relation will be saved side by side, using the same worksheet rows; cells in different relations will be saved behind the precedent relation using other worksheet rows*

There are no secrets here. Step 4 must do the following:

- Allow the user to define up to four different one-to-many record relationships on the worksheet
- Count the total number of rows and columns used by the many-side worksheet records
- Call ValidatePage( ) when the Next -> CommandButton is clicked to verify if either the one-side or the many-side record range name was defined

There are four different ComboBoxes, one for each many-side record range relation, named as cbodbManySide1 to cbodbManySide4. Each ComboBox has two CommandButtons: one on the left to allow the selection of worksheet cells (named as cmddbManySide1 to cmddbManySide4) and another on the right to clean up the ComboBox value (cmdClear1 to cmdClear4). Their values are defined either by selecting a cell range on the worksheet using the left CommandButton or by selecting a desired worksheet range name on the associated ComboBox list. This is the code executed by the cmddbManySide1_Click( ) event (the "Relation 1 cells" button at the right of cbodbManySide1):

```
Private Sub cmddbManySide1_Click()
    Call GetdbManySide(1)
End Sub
```

The GetdbManySide(1) procedure, which centralizes the code for the selection of each many-side records cell, executes this code:

```
Private Function GetdbManySide(intRelation As Integer) As String
    Dim strMsg As String
    Dim strRange As String

    Me.Hide
    strMsg = "Select all column cells that belongs to the  " & intRelation & " of the 'many
    side' worksheet record." & vbCrLf
    strRange = GetRange(strMsg, "Many-side record cells:  " & intRelation,
    Me("cbodbManySide" & intRelation))
    If Len(strRange) Then
        Me("cbodbManySide" & intRelation) = strRange
    End If
    Me.Show
End Function
```

As noted, it calls the GetRange( ) procedure, analyzed in the section "Allowing the User to Select a Cell Range" earlier in this chapter, to allow the user to select cells in the worksheet. Whichever the method is used to change the ComboBox value, it will cascade-fire the associated Change( ) event. The cbodbManySide_Change( ) event executes this code:

```
Private Sub cbodbManySide1_Change()
    If IsRange(Me.cbodbManySide1) Then
        If Left(Me.cbodbManySide1, 1) = "'" Then
            Me.txtdbManySide1 = "'" & Me.cbodbManySide1
        Else
            Me.txtdbManySide1 = Me.cbodbManySide1
        End If
```

```
    Else
        Me.cbodbManySide1 = ""
        Me.txtdbManySide1 = ""
    End If

    Me.cmdClear1.Enabled = (Len(Me.cbodbManySide1) > 0)
    Me.cbodbManySide2.Enabled = (Len(Me.cbodbManySide1) > 0)
    Me.cmddbManySide2.Enabled = (Len(Me.cbodbManySide1) > 0)
    Call CalculateManySideRecords
End Sub
```

You may note that it calls the `IsRange( )` procedure, analyzed in section "Defining the One-Side Record Cells" earlier in this chapter, to validate what is selected in the `cbodbManySide1` `ComboBox`, ensuring that just valid range names can be selected. And when it happens, it will set the value of `txtdbManySide1` (located on the last wizard page); otherwise, it will clear the `cbodbManySide1` `ComboBox` and the associated `txtdbManySide1` text box.

The code then synchronizes the enabled state of three other controls, according to the contents of `cbodbManySide1`: `cmdClear1` `CommandButton` (which is responsible for clearing the Relation 1 selection); `cbodbManySide2` `ComboBox`; and `cmddbManySide2` (which allows the selection of Relation 2's one-to-many record cells).

Counting Many-Side Record Rows and Columns

The `Sub cbodbManySide1_Change()` event procedure finishes by calling the `CalculateManySideRecords( )` procedure to calculate how many worksheet columns and rows are needed to save all four possible relations of many-side record cells selected.

Before you analyze the code, I need to call your attention to the "Tip" exposed by the fourth wizard page, which states "Ranges on the same relation will be saved side by side. Ranges on different relations will be saved at the bottom of the previous relation ranges, separated by a blank row."

This statement means that whenever more than one many-side record cell relation is selected, more worksheet rows will be needed to store these cells. The database storage system was defined this way to save the one-to-many record cells using the same worksheet presentation cell order, allowing you to easily recover records in case of a database storage failure.

The `CalculateManySideRecords( )` procedure works this way: it uses the `Range.Areas` collection to run through each selected cells relation, finding the maximum number of rows and columns each relation has, including a blank row to separate different relations rows (using the same technique described in Chapter 7).

At the end of the procedure, a rectangle with a given row and column count will define the total number of rows and columns needed to save the desired values.

```
Private Function CalculateManySideRecords() As Integer
    Dim rg As Range
    Dim rgArea As Range
    Dim strCtl As String
    Dim strRange As String
    Dim intI As Integer
    Dim intJ As Integer
    Dim intMaxRows As Integer
    Dim intNumRows As Integer
    Dim intMaxCols As Integer
    Dim intNumCols As Integer
```

```
    Dim intPos As Integer
    Dim intPos2 As Integer
    Dim nm As Name

    'Count how many rows are needed to save all range relations
    For intI = 1 To 4
        strCtl = Choose(intI, "cbodbManySide1", _
                        "cbodbManySide2", _
                        "cbodbManySide3", _
                        "cbodbManySide4")

        strRange = Me(strCtl)
        If Len(strRange) Then
            Set rg = Range(Me(strCtl))
            For Each rgArea In rg.Areas
                If rgArea.Rows.Count > intMaxRows Then
                    intMaxRows = rgArea.Rows.Count
                End If
                For intJ = 1 To rgArea.Columns.Count
                    If rgArea.Cells(1, intJ).MergeCells Then
                        intJ = intJ + rgArea.Cells(1, intJ).MergeArea.Columns.Count - 1
                    End If
                    intMaxCols = intMaxCols + 1
                Next intJ
            Next
            'Add an extra row to separate each "many-side" relation
            intNumRows = intNumRows + intMaxRows + 1
            'Update columns count
            If intMaxCols > intNumCols Then
                intNumCols = intMaxCols
            End If
            intMaxRows = 0
            intMaxCols = 0
        End If
    Next intI
    Me.txtdbManySideRowsCount = intNumRows
    Me.txtdbManySideColumnsCount = intNumCols
End Function
```

After declaring the variables, the procedure uses an outer For...Next loop to operate on each of the four possible one-to-many cell range relations. It uses the VBA Choose( ) function to recover each relation contents and analyze if it was selected in the UserForm wizard interface.

```
'Count how many rows are needed to save all range relations
For intI = 1 To 4
    strCtl = Choose(intI, "cbodbManySide1", _
                    "cbodbManySide2", _
                    "cbodbManySide3", _
                    "cbodbManySide4")

    strRange = Me(strCtl)
```

```
    If Len(strRange) Then
    ...
    End If
Next intI
```

For each relation selected, the code uses an outer `For Each...Next` loop to run through all ranges of the `Range.Areas` collection.

```
Set rg = Range(Me(strCtl))
For Each rgArea In rg.Areas
    ...
Next
```

The `intMaxRows Integer` variable will hold the maximum number of rows each relation has, which is determined by comparing the current value with the processed `rgArea.Rows.Count` property.

```
For Each rgArea In rg.Areas
    If rgArea.Rows.Count > intMaxRows Then
        intMaxRows = rgArea.Rows.Count
    End If
    Next intJ
Next
```

Attention Note that this code has no provision to worksheet designs where the many-side record cells use merged cells composed of more than one row. It is supposed that the many-side records cells *never* use a row's merged cells (you are invited to change the code to approach such complex designs).

To get the selected relation column count, the procedure will take into account that each `Range.Area` has the same formatting option for all cells on each of its columns. If the first range cell merges three column cells, it implies that all other rows of the same column are comprised by identical merged cells.

So, it runs a `For intJ...Next` loop through all range columns, using the `Cells` collection to get the first cell row of each `Area` column, verifying if it is merged. If it is, the `intJ` counter is increased by the `MergeArea.Columns.Count -1`, counting the merged cell only once. The `intMaxCols Integer` variable will hold the total number of columns needed for the selected relation.

```
For intJ = 1 To rgArea.Columns.Count
    If rgArea.Cells(1, intJ).MergeCells Then
        intJ = intJ + rgArea.Cells(1, intJ).MergeArea.Columns.Count - 1
    End If
    intMaxCols = intMaxCols + 1
Next intJ
```

The `For intJ...Next` loop ends when the `Range.Areas` collection is analyzed, adding to `intMaxRows` and an extra blank row to the `intNumRows Integer` variable, which will hold the total number of rows for every relation selected (on the first loop passage, `intMaxRows = 0`).

```
'Add an extra row to separate each "many-side" relation
intNumRows = intNumRows + intMaxRows + 1
```

The code verifies whether intMaxCols >= intNumCols, since intNumCols holds the maximum number of columns needed to save every relation selected and then clears intMaxRows and intMaxCols to process another cell's relation (it any).

```
        'Update columns count
        If intMaxCols > intNumCols Then
            intNumCols = intMaxCols
        End If
        intMaxRows = 0
        intMaxCols = 0
    End If
Next intI
```

When the outer For intI...Next relation loop ends, all relations have been processed, and the two text boxes located at the bottom of the fourth wizard page (txtdbManySideRowsCount and txtdbManySideColumnsCount) will receive the row and column counting, contained in the intNumRows and intNumCols Integer variables.

```
    Next intI
    Me.txtdbManySideRowsCount = intNumRows
    Me.txtdbManySideColumnsCount = intNumCols
End Function
```

The same code used by the Sub cbodbManySide1_Change() event is repeated for every other ComboBox associated to relations 2, 3, and 4's many-side record cells (if any).

Clearing Any Selected Relation

The last code that deserves to be mentioned is the one used by the cmdClear1_Click( ) event, used to clear the first many-side record cells relation. The fourth wizard page interface was built in such a way that the user must select each relation in sequence. If Relation 1, Relation 2, Relation 3, and Relation 4 ranges are selected and then Relation 1 (the first many-side relation) is cleared, the Relation 2 range must be transferred to Relation 1, Relation 3 must be transferred to Relation 2, Relation 4 must be transferred to Relation 3, and Relation 4 must be cleared (Figure 8-26).

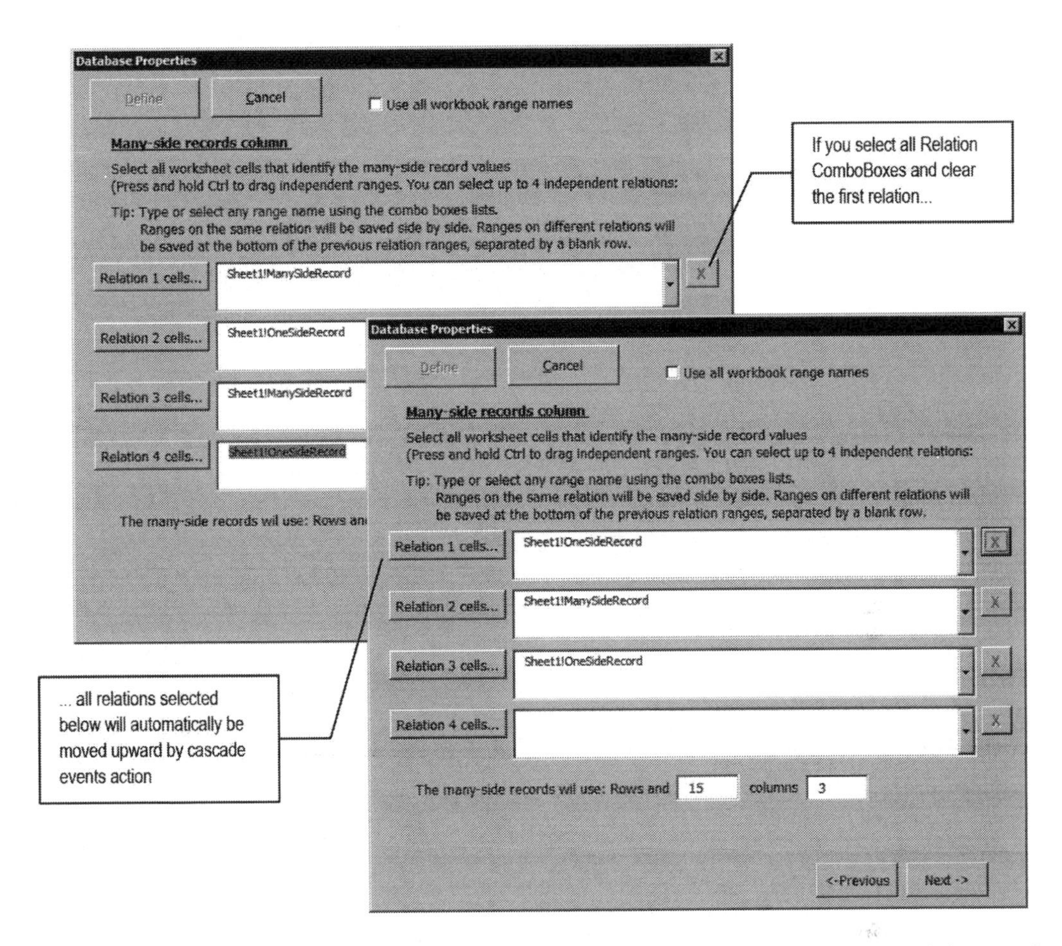

Figure 8-26. *Try to select any range name for all four one-to-many relations; then click cmdClear1 to clear the Relation 1 contents. This will cascade-fire events that will gracefully change relation values upward*

Whenever you click the `cmdClear1 CommandButton`, the `cmdClear1_Click( )` event fires, executing this code (which is quite similar to every other clear relation button):

```
Private Sub cmdClear1_Click()
    If Len(Me.cbodbManySide2) Then
        Me.cbodbManySide1 = Me.cbodbManySide2
        Call cmdClear2_Click
    Else
        Me.cbodbManySide1 = ""
    End If
End Sub
```

You may note that before Relation 1 is cleared, it checks whether the `cbodbManySide2 ComboBox` has any range selected. If it does, it attributes its selection to `cbodbManySide1`, cascade-firing the `cbodbManySide1_Change( )` event.

Then it calls the cmdClear2_Click( ) event, cascade-firing the cbodbManySide2_Change( ) event. And since the cmdClear2_Click event has similar code (just changing control references), it will check the cbodbManySide3 contents and eventually call cmdClear3, which will cascade-fire the cbodbManySide3_Change( ) event and synchronize the interface:

```
    If Len(Me.cbodbManySide2) Then
        Me.cbodbManySide1 = Me.cbodbManySide2
        Call cmdClear2_Click
```

If cbodbManySide2 ComboBox is empty, it just clears the associated ComboBox, cascading-fire the cbodbManySide1_Change( ) event, which will synchronize the interface.

```
    Else
        Me.cbodbManySide1 = ""
    End If
End Sub
```

■ **Attention** The ValidatePage( ) procedure validates if either the one-side record or the many-side record was selected by inspecting the txtdbOneSideColumnsCount And txtdbManySideRowsCount TextBox values. If both have a zero value, the user cannot advance to the next wizard page.

Step 5: Asking to Create Worksheet Database Button Controls

If either the one-side record or the many-side records have been selected, the wizard will allow the user to reach the next wizard page to confirm the creation of database Button controls (to go to a new record and save and delete records) and the creation of database "navigation buttons" (to easily move to the first, previous, next, last, and new records), and to hide the database rows and activate worksheet protection (Figure 8-27).

The fifth wizard page indicates that the database Button controls (New, Save, and Delete) will be created on the right of the data validation list cell, while the navigation buttons will be created on a specific cell (cell $C$21), which is defined by Function SetRecordPosition( ).

```
Public Function SetRecordPosition() As String
    Dim rg As Range
    Dim sngWidth As Single
    Const conNavigationButtonWidth = 17.3
    Const conRecordPosiciontCellWidth = 50.3

    Set rg = Range("$B$" & Me.txtdbRecordsFirstRow - 2)
    sngWidth = rg.Offset(0, -1).Width
    Do While (sngWidth < conNavigationButtonWidth * 2) And _
            (rg.Width < conRecordPosiciontCellWidth)
        sngWidth = sngWidth + rg.Width
        Set rg = rg.Offset(, 1)
    Loop
    SetRecordPosition = rg.Address(True, True)
End Function
```

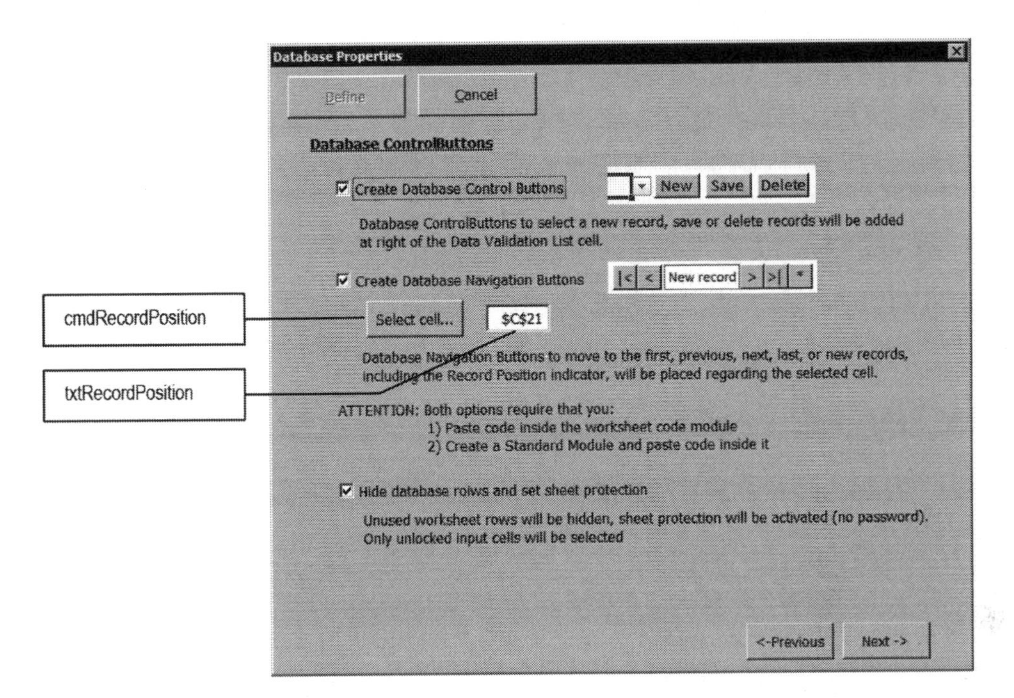

Figure 8-27. *The fifth wizard page asks for user confirmation to create database Button controls for record management and navigation and to hide the worksheet database rows and activate worksheet protection*

Each data navigation button has a specific `Width` property (`conNavigationButtonWidh = 17.3` points), while the record position cell needs a minimum `Width` property value (`conRecordPosiciontCellWidth = 50.3` points). The code tries to create the record position cell on column `B`, two rows above the first database row (determined by `txtRecordsFirstRow` value).

```
Set rg = Range("$B$" & Me.txtdbRecordsFirstRow - 2)
```

To verify whether this cell is appropriate for the data navigation buttons, it must have the following:

- The cell at the left is wide enough to contain two `Button` controls, each one 17.3 points wide (a 34.6-points width).
- The current cell (associated to the `rg` object variable) must be wide enough to exhibit the `=RecordPosition( )` function value: a minimum of 50.3 points.

This is done by first getting the width of the cell located on the left of the `rg` cell (`sngWidth`) and a `Do While...Loop` instruction to verify whether these two conditions are met.

```
sngWidth = rg.Offset(0, -1).Width
Do While (sngWidth < conNavigationButtonWidth * 2) And _
         (rg.Width < conRecordPosiciontCellWidth)
```

If the cell on the left of the `rg` cell is not wide enough to contain two navigation button widths (`rg.Offset(0,-1).Width < conNavigationButtonWidth * 2`) or the `rg` cell can't exhibit a record position expression (`rg.Width < conRecordPosiciontCellWidth`), the code adds the current cell width to the left cell width and uses `rg.OffSet(0,1)` to displace the proposed range one cell to the right and loop again, until an adequate cell is found.

```
        sngWidth = sngWidth + rg.Width
        Set rg = rg.Offset(, 1)
    Loop
```

When the loop ends, the cell selected to receive the record position information is returned as an absolute reference value.

```
    SetRecordPosition = rg.Address(True, True)
End Function
```

The user can change the proposed cell by clicking the cmdRecordPosition CommandButton, which executes this code:

```
Private Sub cmdRecordPosition_Click()
    Dim strRange As String
    Dim strMsg As String
    Dim strRecordPosition As String

    On Error Resume Next

    strRecordPosition = SetRecordPosition()
    strMsg = "Select cell to receive Record Position indicator:"
    strRange = GetRange(strMsg, "Record Position cell?", Me.txtRecordPosition)
    If Len(strRange) Then
        If Range(strRange).Column < Range(strRecordPosition).Column Then
            MsgBox "There is no room to create data navigation controls on selected cell.",
            vbCritical, "Invalid selection!"
        Else
            Me.txtRecordPosition = strRange
        End If
    End If
End Sub
```

It stores on the strRecordPosition variable the cell proposed by the SetRecordPosition( ) function and then uses the GetRange( ) procedure to allow the user to select another worksheet cell where the data navigation buttons must be set.

```
    strRecordPosition = SetRecordPosition()
    strMsg = "Select cell to receive Record Position indicator:"
    strRange = GetRange(strMsg, "Record Position cell?", Me.txtRecordPosition)
```

If the selected cell is on the left of the proposed range, it will be refused by the code, and the user will receive a MsgBox( ) warning. Otherwise, the selected cell is accepted and stored inside the txtRecordPosition text box.

```
        If Range(strRange).Column < Range(strRecordPosition).Column Then
            MsgBox "There is no room to create data navigation controls on selected cell.",
            vbCritical, "Invalid selection!"
        Else
            Me.txtRecordPosition = strRange
```

Attention The fifth wizard page doesn't need to be validated.

Step 6: Showing the Database Properties Page

If either the one-side record or the many-side records have been selected, the wizard will allow the user to reach the last page where a database property resumes showing what has been selected and how many records can be saved on the unused worksheet rows.

Figure 8-28 shows a resumed view for the `Sheet1` tab of the `fmDBProperties_SheetDBEngine.xlsm` workbook, after selecting the `OneSideRecord` range name to represent the one-side record and the `ManySideRecords` range name to represent the Relation 1 many-side records. Using such a one-to-many cell record structure, the `Sheet1` tab can save up to 95.323 records on its unused rows.

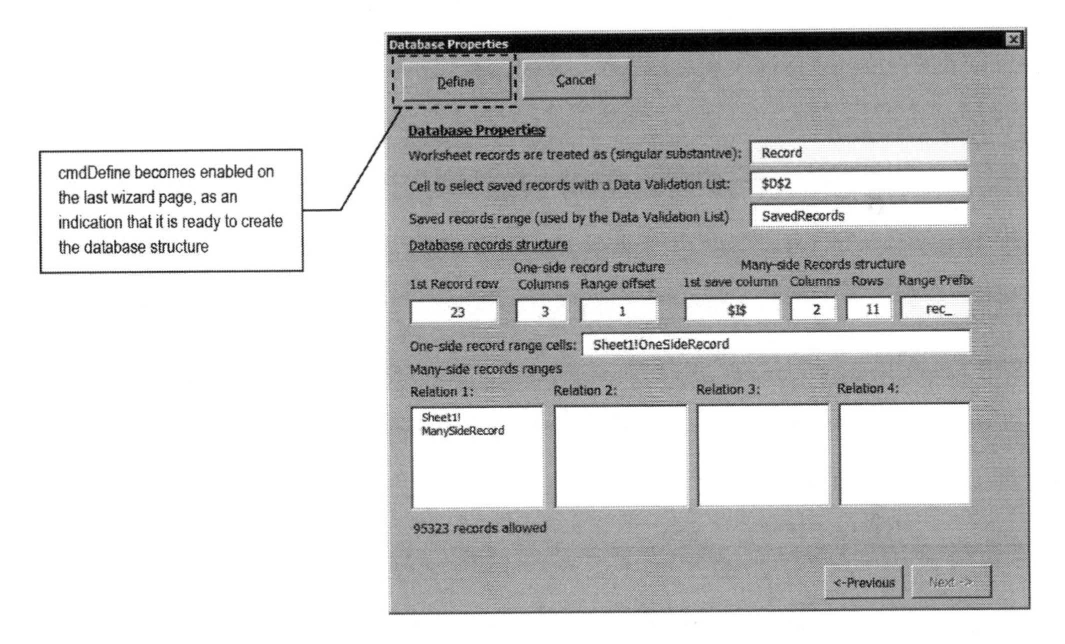

Figure 8-28. The last wizard page shows a resumed view of the database properties, including how many records can be saved in the worksheet

The total records allowed are calculated in the `tabControl_Change( )` event, which fires every time a different wizard page is selected, executing this code:

```
Private Sub tabControl_Change()
    Dim lngRec As Long

    On Error Resume Next

    If Me.txtdbManySideRowsCount = 0 Then
        lngRec = (ActiveSheet.Rows.Count - Me.txtdbRecordsFirstRow)
    Else
        lngRec = (ActiveSheet.Rows.Count - Me.txtdbRecordsFirstRow) / Me.txtdbManySideRowsCount
    End If
```

```
    Me.lblNumRecords.Caption = lngRec & " records allowed"
    Me.cmdPrevious.Enabled = (Me.tabControl.Value > 0)
    Me.cmdNext.Enabled = (Me.tabControl.Value < Me.tabControl.Pages.Count - 1)
End Sub
```

Note that it is quite similar to the code in section "Navigating Through the UserForm Wizard" earlier in this chapter. It calculates how many records are allowed by first verifying whether `txtdbManySideRowsCount = 0`. If it is, no selection is made to represent the many-side records; just the one-side record range was selected, and each worksheet record will use just one worksheet row. The number of records allowed must be calculated by the difference between all possible worksheet rows and the first used database row.

```
    If Me. txtdbManySideRowsCount = 0 Then
        lngRec = (ActiveSheet.Rows.Count - Me. txtdbRecordsFirstRow)
```

But if any selection was made to represent the many-side worksheet records, the number of records allowed is calculated by dividing the difference between all possible worksheet rows and the first used database row by `txtdbManySideRowsCount` (which contains the row count needed to save each record).

```
    Else
        lngRec = (ActiveSheet.Rows.Count - Me.txtdbRecordsFirstRow) / Me.
        txtdbManySideRowsCount
    End If
```

The number of records allowed is shown by changing the `lblNumRecords Label` control's `Caption` property, located at the bottom of the last wizard page.

```
    Me.lblNumRecords.Caption = lngRec & " records allowed"
```

▒ **Attention** When `frmDBProperties` is about to create the database structure on the worksheet, the database properties will allow the user to change the record name and the many-side record prefix. That is why these controls are formatted using a light yellow background.

Creating the Database Structure

When all database information has been collected and the last wizard page is reached, the `cmdDefine CommandButton` becomes enabled, ready to create the database, executing this code:

```
Private Sub cmdDefine_Click()
    Dim rg As Range
    Dim strRange As String
    Dim intI As Integer
    Dim intRow As Integer
    Dim intCol As Integer

    ActiveSheet.Unprotect
    If Me.cmdDefine.Caption = "Define" Then
        'Define database structure
        Call SetDataBase(CreateDatabase)
    End If
```

```
    'Hide or show database rows
    ActiveSheet.Range(Cells(Me.txtdbRecordsFirstRow, 1), _
                      Cells(ActiveSheet.Rows.Count, 1)).EntireRow.Hidden = Me.
                      chkHideDatabaseRows
    If Me.chkHideDatabaseRows Then
        'Unlock worksheet record cells
        Range(Me.txtdbDataValidationList).MergeArea.Locked = False
        For intI = 1 To 5
            strRange = Choose(intI, "cbodbOneSide", _
                                    "cbodbManySide1", _
                                    "cbodbManySide2", _
                                    "cbodbManySide3", _
                                    "cbodbManySide4")
            If Len(Me(strRange)) Then
                For Each rg In Range(Me(strRange)).Areas
                    For intRow = 1 To rg.Rows.Count
                        For intCol = 1 To rg.Columns.Count
                            rg.Cells(intRow, intCol).MergeArea.Locked = False
                        Next
                    Next
                Next
            End If
        Next
        'Active worksheet protection, selecting just unlocked cells
        ActiveSheet.Protect
        ActiveSheet.EnableSelection = xlUnlockedCells
    End If

    Unload Me
End Sub
```

The code first checks the cmdDefine.Caption property. If it is set to Define, it means that the database structure must be created, calling the SetDatabase(CreateDatabase) procedure.

```
ActiveSheet.Unprotect
If Me.cmdDefine.Caption = "Define" Then
    'Define database structure
    Call SetDataBase(CreateDatabase)
```

■ **Attention** As you will see later, when the database is already set on the active worksheet, cmdDefine will show "Close," while cmdCancel will show "Remove" in its Caption properties.

Using SetDatabase(CreateDatabase) to Define the Database Structure

The SetDatabase( ) procedure can be used to either create or remove the database properties from the active worksheet. When it receives the CreateDatabase argument, it will save all properties described on the Database Properties wizard page, beginning on the row defined by txtdbRecordsFirstRow, using column A to save the property name and column B to save the property value (similar to what is shown in Figure 8-5, which uses range A30:B44). It executes this code:

```
Private Sub SetDataBase(Operation As DataBaseOperation)
    Dim nm As Name
    Dim strNameScope As String
    Dim strName As String
    Dim intRow As Integer
    Dim intI As Integer
    Const conCol = "=$B$"
    Const conColD = 4

    Application.ScreenUpdating = False
    intRow = Me.txtdbRecordsFirstRow
    strNameScope = "'" & ActiveSheet.Name & "'!"
    'Create database range names on columns A:B
    For intI = 0 To 14
        strName = Choose(intI + 1, "dbRecordName", _
                                   "dbDataValidationList", _
                                   "dbSavedRecords", _
                                   "dbRecordsFirstRow", _
                                   "dbOneSide", _
                                   "dbOneSideColumnsCount", _
                                   "dbManySide1", _
                                   "dbManySide2", _
                                   "dbManySide3", _
                                   "dbManySide4", _
                                   "dbManySideFirstColumn", _
                                   "dbManySideColumnsCount", _
                                   "dbManySideRowsCount", _
                                   "dbManySidePrefix", _
                                   "dbRangeOffset")
        If Operation = CreateDatabase Then
            Set nm = Names.Add(strNameScope & strName, conCol & intRow + intI, False)
            Cells(intRow + intI, 1) = strName
            Cells(intRow + intI, 2) = Me("txt" & strName)
        Else
            Set nm = Names(strNameScope & strName)
            nm.Delete
            Cells(intRow + intI, 1).ClearContents
            Cells(intRow + intI, 2).ClearContents
        End If
    Next

    If Operation = CreateDatabase Then
        'Define SavedRecords range name on column D
        Set nm = Names.Add(strNameScope & Me.txtdbSavedRecords, "=" & Cells(intRow,
        conColD).Address, False)
        'Define SavedRecords data validation list
        Range(strNameScope & Me.txtdbSavedRecords) = "New " & Me.txtdbRecordName
        Range(Me.txtdbDataValidationList).Validation.Delete
        Range(Me.txtdbDataValidationList).Validation.Add xlValidateList, , , "=" & Me.
        txtdbSavedRecords
        Range(Me.txtdbDataValidationList).HorizontalAlignment = xlLeft
```

```
        Range(Me.txtdbDataValidationList) = "New " & Me.txtdbRecordName
        Call CreateDatabaseButtons
    Else
        Set nm = Names(strNameScope & Me.txtdbSavedRecords)
        nm.Delete
        Range(Me.txtdbDataValidationList).Validation.Delete
        Call DeleteDatabaseButtons
    End If
    Application.ScreenUpdating = True
End Sub
```

To define the database properties as worksheet scope range names, SetDatabase( ) begins by defining the row where the first property must be saved and the scope of the range names.

```
    intRow = Me. txtdbRecordsFirstRow
    strNameScope = "'" & ActiveSheet.Name & "'!"
```

To create the range names, SetDataBase( ) uses a For...Next loop to increment the intI integer variable and uses the VBA Choose( ) Function first argument, storing in the strName String variable a different range name at each loop passage.

```
    For intI = 0 To 13
        strName = Choose(intI + 1, "dbRecordName", _
                                   ...
                                   "dbRangeOffset")
       ...
    Next
```

Once strName has a property name, it verifies whether the Operation argument is defined to CreateDabase. If it is, the associated range name is created using the Names.Add method. Note that it uses strNameScope & strName to define a worksheet scope range name, and it uses the constant conCol="B" concatenated to intRow + intI to define an address in column B where the next range name must be created. Each name is created with the property Visible = False (the second argument of the Names.Add method).

```
        If Operation = CreateDatabase Then
            Set nm = Names.Add(strNameScope & strName, conCol & intRow + intI, False)
```

■ **Attention** All database properties are saved in TextBoxes whose name has a txt prefix in one of the tabControl pages.

Since all database property range names are hidden, you will not see them using the Excel Name Manager. To do so, use the frmNames UserForm.

Once the range name is created, it uses the Cells(row, column) property to save the database property name in column A (column argument = 1) and the database property value in column B (column argument = 2).

```
            Cells(intRow + intI, 1) = strName
            Cells(intRow + intI, 2) = Me("txt" & strName)
```

If Operation = RemoveDatabase, the Else clause will be executed to remove the range name and clear these worksheet cells.

```
        Else
            Set nm = Names(strNameScope & strName)
            nm.Delete
            Cells(intRow + intI, 1).ClearContents
            Cells(intRow + intI, 2).ClearContents
        End If
    Next
```

When the For...Next loop ends, all database property range names have been created (or deleted) on the active worksheet. Now it is time to create the range name defined by the txtSavedRecords TextBox in column D to store database record names, adding as the first item “New Record” (or whatever is the record name typed in txtdbRecordName at the first wizard page).

```
    If Operation = CreateDatabase Then
        'Define SavedRecords range name on column D
        Set nm = Names.Add(strNameScope & Me. txtdbSavedRecords, "=" & Cells(intRow,
conColD).Address, False)
        Range(strNameScope & Me.txtdbSavedRecords) = "New " & Me.txtdbRecordName
```

To end the database creation process, it sets the data validation list in the txtdbDataValidationList cell (defined on the second wizard page), using the txtSavedRecords range name to define the list range, with the aid of Validation object’s Delete and Add methods.

The first operation deletes any existing data validation lists that should exist in the txtdbDataValidationList cell using the Validation.Delete method.

```
        'Define SavedRecords data validation list
        Range(Me.txtdbDataValidationList).Validation.Delete
```

To create any data validation using VBA, you must use the Validation.Add method, which has this syntax:

```
Expression.Add(Type, AlertStyle, Operator, Formula1, Formula2
```

In this code:

> Expression: This is required; it is an object variable that represents a Validation object.
>
> Type: This is required; it a constant of xlDVType that can be set to xlValidateCustom, xlInputOnly, xlValidateList, lValidateWholeNumber, xlValidateDate, xlValidateDecimal, xlValidateTextLength, or xlValidateTime.
>
> AlertStyle: This is optional; it is a constant of XlDVAlertStyle that can be set to xlValidAlertInformation, xlValidAlertStop, or xlValidAlertWarning.

Operator: This is optional; it a constant of the XlFormatConditionOperator type that can be set to xlBetween, xlEqual, xlGreater, xlGreaterEqual, xlLess, xlLessEqual, xlNotBetween, or xlNotEqual.

Formula1: This is optional; it the first part of the data validation equation.

Formula2: This is optional; it is the second part of the data validation when Operator is xlBetween or xlNotBetween (otherwise, this argument is ignored).

So, to create a data validation list, the code uses the Validation.Add method with the xlValidateList constant for its first argument, specifying for the Formula1 argument a formula that points to the range name specified by txtdbSavedRecords.

```
        Range(Me.txtdbDataValidationList).Validation.Add xlValidateList, , , "=" & Me.
txtdbSavedRecords
```

The data validation list cell is left aligned using the Range.HorizontalAlignment method with the xlLeft constant, and its value is defined to New <txtdbRecordName>.

```
        Range(Me.txtdbDataValidationList).HorizontalAlignment = xlLeft
        Range(Me.txtdbDataValidationList) = "New " & Me.txtdbRecordName
        Call CreateDatabaseButtons
    Else
      ...
    End If
End Sub
```

■ **Attention** Note that when SetDatabase( ) receives RemoveDatabase on its Operation argument, the Else clause is executed, and both the range name used to save the worksheet records and the data validation list are removed, using the Delete method of the Name and Validate objects.

```
  Else
        Set nm = Names(strNameScope & Me.txtdbSavedRecords)
        nm.Delete
        Range(Me.txtdbDataValidationList).Validation.Delete
        Call DeleteDatabaseButtons
    End If
End Sub
```

Creating Sheet Tab Button Controls

The last frmDBProperties action is the creation of worksheet Button controls to perform three basic database operations (show a new record and save or delete an existing record) and the data navigation buttons, if both options have been selected on the fifth wizard page. This operation is made by the CreateDatabaseButtons( ) procedure, which executes this code:

```
Private Sub CreateDatabaseButtons()
    Dim ws As Worksheet
    Dim shp As Shape
    Dim rg As Range
    Dim dobjClipboard As New DataObject
    Dim strMsg As String
    Dim lngLeft As Long
    Const conColorLighBlue = 12419407

    Set ws = Application.ActiveSheet

    If Me.chkCreateButton controls Then
        'Create Database Button controls at right of Data Validation list
        '-----------------------------------------------------------
        Set rg = Range(Me.txtdbDataValidationList)
        If rg.MergeCells Then
            'Range has merged cells. Position on last right cell
            Set rg = Cells(rg.Row, rg.Column + rg.MergeArea.Columns.Count - 1)
        End If

        'Create New button
        lngLeft = rg.Left + rg.Width + 16
        Set shp = ws.Shapes.AddFormControl(xlButtonControl, lngLeft, rg.Top, 30, rg.Height)
        shp.OnAction = "MoveNew"
        shp.OLEFormat.Object.Text = "New"

        'Create Save button
        lngLeft = shp.Left + shp.Width + 5
        Set shp = ws.Shapes.AddFormControl(xlButtonControl, lngLeft, rg.Top, 30, rg.Height)
        shp.OnAction = "Save"
        shp.OLEFormat.Object.Text = "Save"

        'Create Delete button
        lngLeft = shp.Left + shp.Width + 5
        Set shp = ws.Shapes.AddFormControl(xlButtonControl, lngLeft, rg.Top, 35, rg.Height)
        shp.OnAction = "Delete"
        shp.OLEFormat.Object.Text = "Delete"
    End If

    If Me.chkCreateNavigationButtons Then
        'Create Data Navigation buttons
        '---------------------------------------
        Set rg = Range(Me.txtRecordPosition)
        rg.Formula = "=RecordPosition()"
        rg.HorizontalAlignment = xlCenter
        rg.Font.Size = 9
        rg.Borders.LineStyle = xlContinuous
        rg.Borders.Color = conColorLighBlue

        'Create MoveFirst button
        lngLeft = rg.Left - 2 * conMoveButtonWidth
```

```
        Set shp = ws.Shapes.AddFormControl(xlButtonControl, lngLeft, rg.Top,
        conMoveButtonWidth, rg.Height)
        shp.OnAction = "MoveFirst"
        'shp.OnAction = ws.CodeName & ".MoveFirst"
        shp.OLEFormat.Object.Text = "|<"

        'Create MovePreviousFirst button
        lngLeft = rg.Left - conMoveButtonWidth
        Set shp = ws.Shapes.AddFormControl(xlButtonControl, lngLeft, rg.Top,
        conMoveButtonWidth, rg.Height)
        shp.OnAction = "MovePrevious"
        shp.OLEFormat.Object.Text = "<"

        'Create MoveNext button
        lngLeft = rg.Left + rg.Width
        Set shp = ws.Shapes.AddFormControl(xlButtonControl, lngLeft, rg.Top,
        conMoveButtonWidth, rg.Height)
        shp.OnAction = "MoveNext"
        shp.OLEFormat.Object.Text = ">"

        'Create MoveLast button
        lngLeft = rg.Left + rg.Width + conMoveButtonWidth
        Set shp = ws.Shapes.AddFormControl(xlButtonControl, lngLeft, rg.Top,
        conMoveButtonWidth, rg.Height)
        shp.OnAction = "MoveLast"
        shp.OLEFormat.Object.Text = ">|"

        'Create MoveNew button
        lngLeft = rg.Left + rg.Width + 2 * conMoveButtonWidth
        Set shp = ws.Shapes.AddFormControl(xlButtonControl, lngLeft, rg.Top,
        conMoveButtonWidth, rg.Height)
        shp.OnAction = "MoveNew"
        shp.OLEFormat.Object.Text = "*"
    End If

    If Me.chkCreateButton controls Or Me.chkCreateNavigationButtons Then
        'Copy sheet modulce code and basButton controls code
        With dobjClipboard
            .SetText Me.txtButtonsCode.Text
            .PutInClipboard
            'Warn the user how to paste button codes on sheet module
            strMsg = "To create the database buttons code, select the worksheet code module"
            strMsg = strMsg & "place the text cursor behind the 'Option Explicit'
            instruction"
            strMsg = strMsg & "and press Ctrl+V to paste!"
            MsgBox strMsg, vbInformation, "WARNING: How to create buttons code!"
        End With
    End If
End Sub
```

Creating Database Button Controls

If chkCreateButton controls were kept selected on the fifth wizard page, the CreateDatabaseButtons( ) procedure will create three CommandButtons right next to the data validation list cell (pointed at by txtdbDataValidationList TextBox). To define the position of the first Button control, it first sets a Range object variable (rg) to the data validation list cell and then verifies if it is a merged cell using the Range. MergeCells property. If it is, it uses the Cells( ) property to set the rg object variable to its farthest-right merged cell. It does this by adding to the first data validation list column the total number of merged columns.

```
Private Sub CreateDatabaseButtons()
    ...
    If Me.chkCreateButton controls Then
        'Create Database Button controls at right of Data Validation list
        '-----------------------------------------------------------------
        Set rg = Range(Me.txtdbDataValidationList)
        If rg.MergeCells Then
            'Range has merged cells. Position on last right cell
            Set rg = Cells(rg.Row, rg.Column + rg.MergeArea.Columns.Count - 1)
        End If    Set ws = Application.ActiveSheet
```

Now that the desired cell is selected, it determines the position regarding worksheet cell A1 by adding the Range.Left property with the Range.Width property (both in points) plus 16 points (determined by trial-and-error experimentation), so it sits at the right of the data validation list control arrow.

```
        'Create New button
        lngLeft = rg.Left + rg.Width + 16
```

■ **Attention** The Range.Left property returns the screen position (in points) of the upper-left range cell corner relative to worksheet cell A1's upper-left corner.

To create each Button control, the code uses the Shapes Collection AddFormControl method, which has this syntax:

```
Expression.AddFormControl(Type, Left, Top, Width, Height)
```

In this code:

Expression: This is required; it is object variable that represents a Shapes object.

Type: This is required; it is a constant of XlFormControl type that can be set to the following:

xlButtonControl=0, for a Button

xlCheckBox=1, for a CheckBox

XlDropDown=2, for a ComboBox

XlEditBox=3, for a TextBox

XlGroupBox=4, for a GroupBox

XlLabel=5, for a Label

XlListBox=6, for a ListBox

XlOptionButton=7, for an OptionButton

XlScrollBar=8, for a Scrollbar

XlSpinner=9, for a Spinner

Left: This is required; it a Long value that sets the initial coordinates of the new object (in points) relative to the upper-left corner of cell A1 on a worksheet or to the upper-left corner of a chart.

Top: This is required; it is a Long value that sets the initial coordinates of the new object (in points) relative to the upper-left corner of cell A1 on a worksheet or to the upper-left corner of a chart.

Width: This is required; it is a Long value that sets the initial size of the new object, in points.

Height: This is required; it a Long value that sets the initial size of the new object, in points.

The Shapes.AddFormControl method returns a Shape object, which is an independent beast with a rich object model (everything you put in worksheet cells belongs to the Shapes collection, including a chart). Among them you use the following:

- The Shape.OnAction property to set the procedure name that is executed if Shape.Type is set to xlButton control and the Button control is clicked by the user action
- The Shape.OLEFormat property, which is another wild beast with its own object interface, from which you use derived OLEFormat.Object.Text to define the Button control caption

That is why CreateDatabaseButtons( ) declares the shp as Shape object variable: to set a reference of each created Button control. The first created button is New, which is associated to the MoveNew procedure. It is positioned 16 points to the right border of the data validation list cell, with its top-left corner touching the cell's top border (rg.Top), with a 30-point width (determined by trial and error) and the same row height (rg.Height):

```
Dim shp As Shape
...
    'Create New button
    lngLeft = rg.Left + rg.Width + 16
    Set shp = ws.Shapes.AddFormControl(xlButtonControl, lngLeft, rg.Top, 30, rg.Height)
    shp.OnAction = "MoveNew"
    shp.OLEFormat.Object.Text = "New"
```

Attention Since the MoveNew origin is not specified, it is supposed that it is stored on a Standard module as a Public Sub or Function procedure, available through all VBA project code.

The Save button is positioned 5 points to the right of the New button (note that the lngLeft variable now receives the shp.Left + shp.Width + 5 value to set the next button position), being associated to the Save procedure.

```
'Create Save button
lngLeft = shp.Left + shp.Width + 5
Set shp = ws.Shapes.AddFormControl(xlButtonControl, lngLeft, rg.Top, 30, rg.Height)
shp.OnAction = "Save"
shp.OLEFormat.Object.Text = "Save"
```

■ **Attention** The Delete button is then positioned 5 points to the right of the Save button using this same programming technique, being associated to the `Public Sub Delete( )` procedure.

Creating Database Navigation Buttons

If `chkCreateNavigationButtons` was kept selected on the fifth wizard page, the `CreateDatabaseButtons( )` procedure will now create the record position formula in the cell defined by the `txtRecordPosition TextBox`. It sets a reference to the desired cell and defines these Range object properties: `Formula`, `HorizontalAlignment`, `FontSize`, `Borders.LineStyle`, and `Borders.Color`. The result will be a cell with the text centered, a white background, and a blue border.

```
If Me.chkCreateNavigationButtons Then
    'Create Data Navigation buttons
    '---------------------------------------
    Set rg = Range(Me.txtRecordPosition)
    rg.Formula = "=RecordPosition()"
    rg.HorizontalAlignment = xlCenter
    rg.Font.Size = 9
    rg.Borders.LineStyle = xlContinuous
    rg.Borders.Color = conColorLighBlue
```

Each navigation button is created, beginning with the first `MoveFirst` button, and located by subtracting the `2 * conMoveButtonWidth` constant, on the left of the record position cell having the same cell height, being associated to `MoveFirst` procedure.

```
'Create MoveFirst button
lngLeft = rg.Left - 2 * conMoveButtonWidth
Set shp = ws.Shapes.AddFormControl(xlButtonControl, lngLeft, rg.Top,
conMoveButtonWidth, rg.Height)
shp.OnAction = "MoveFirst"
'shp.OnAction = ws.CodeName & ".MoveFirst"
shp.OLEFormat.Object.Text = "|<"
```

All other buttons are created with the same technique, at the desired position at each cell side. When the navigation buttons have been created, the `CreateDatabaseButtons( )` procedure does the last trick: it uses the `dobjClipboard as DataObject` variable to copy the `txtButtonsCode TextBox Text` property to the clipboard, using the `DataObject SetText` and `PutInClipboard` methods and warning the user with the VBA `MsgBox( ) Function` to paste the desired code on the active sheet code module.

```
With dobjClipboard
    .SetText Me.txtButtonsCode.Text
    .PutInClipboard
```

```
        'Warn the user how to paste button codes on sheet module
        strMsg = "To create the database buttons code, select the worksheet code module "
        strMsg = strMsg & "place the text cursor behind the 'Option Explicit' instruction "
        strMsg = strMsg & "and press Ctrl+V to paste!"
        MsgBox strMsg, vbInformation, "WARNING: How to create buttons code!"
    End With
End Sub
```

To understand this trick, you must know the following:

- The DataObject object has a rich object model, offering different methods to deal with different clipboard data (such as text, images, sound, and so on). Google it on the MSDN site to inspect its object interface.
- The frmDBProperties UserForm has the (not so) hidden txtButtonsCode TextBox on the right of its Multipage control. While the UserForm is in design mode, drag its right border to the right to show it (Figure 8-29). The txtButtonsCode TextBox has its Multiline property set to True, and its Value property contains all the code needed to create a new instance of the SheetDBEngine class module and execute all the code needed by the worksheet-created Button controls.

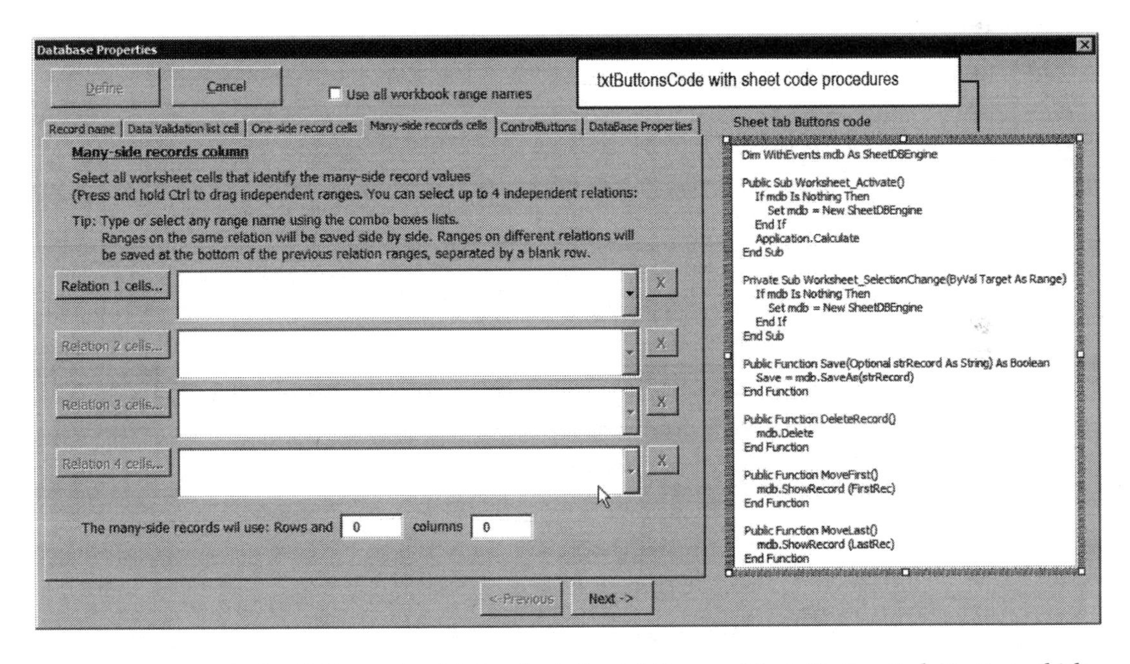

Figure 8-29. *Drag the frmDBProperties right border to the right to reveal the txtButtonsCode TextBox, which has its Multiline property set to True and its Value property defined to the code needed to implement the SheetDBEngine class and execute basic database procedures: save, delete, and show records*

■ **Attention** VBA calls *extensibility* the ability to modify its projects by writing code inside its IDE. This is done by setting a reference to the Microsoft Visual Basic for Applications Extensibility 5.3 (or higher) library (using the VBA Tools ➤ Reference menu option). Since many VBA-based viruses propagate and execute themselves by creating or modifying VBA code, most virus scanners may automatically and without confirmation delete Microsoft Excel `.xlsm` macro-enabled workbooks, which reference the `VBAProject` object. The method described here (storing the code inside a text box, copying it to the clipboard, and informing the user to paste it inside the code module) is secure, does not need to use extensibility, and will not be caught by virus scanner.

That is why `frmDBPRoperties` changes its `Width` property on the `Initialize( )` event (to hide `txtButtonsCode`) and asks the user to click the worksheet code module and press Ctr+V to create the code that will set the database engine and make the three control buttons work, when the wizard finishes its activities (Figure 8-30).

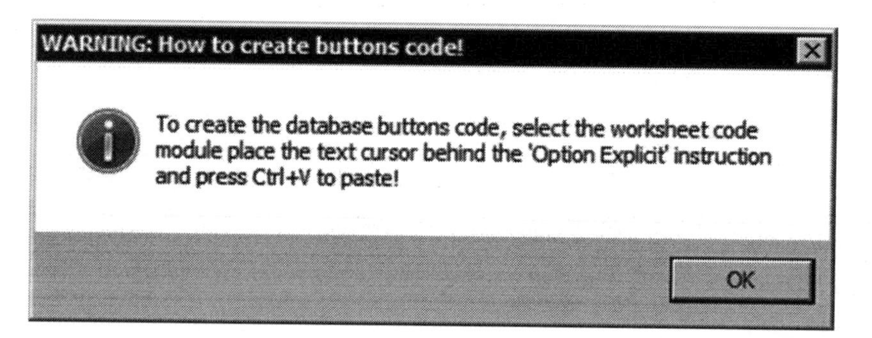

Figure 8-30. *This is the warning message sent by the frmDBProperties UserForm when it finishes implementing the database storage system inside the active sheet. It asks the user to select the active sheet code module and press Ctrl+V to create the code that will start and implement the worksheet record control*

Supposing that you have used the `Sheet1` worksheet from the `frmDBProperties_SheetDBEngine.xlsm` macro-enabled workbook and defined the `frmDBProperties UserForm` using the same values exhibited in Figure 8-28, the database will be created at row 23 and will look like Figure 8-31.

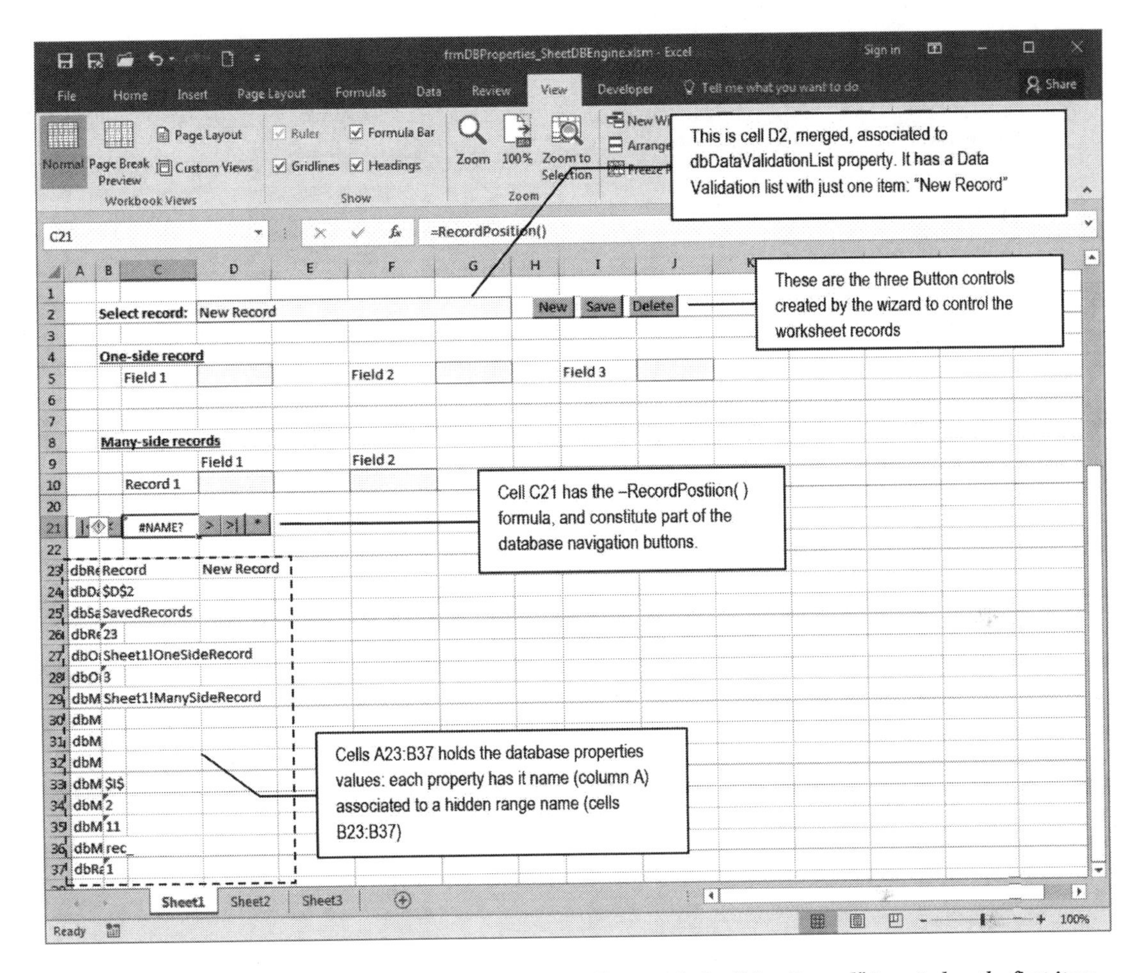

Figure 8-31. *This is the data validation list created at cell D2 with the "New Record" inserted as the first item and the database properties stored in cells A23:B36. Note that the "New Record" name is stored in cell D23, where the SavedRecords range name begins*

Pasting the Database Code

To finish the worksheet database implementation, press Alt+F11 to show the Visual Basic IDE, select the `Sheet1` code module in the Project Explorer tree, position the cursor below the `Option Explicit` statement, and press Ctrl+V. The code needed to run the `SheetDBEngine` class will be pasted in the worksheet module (Figure 8-32).

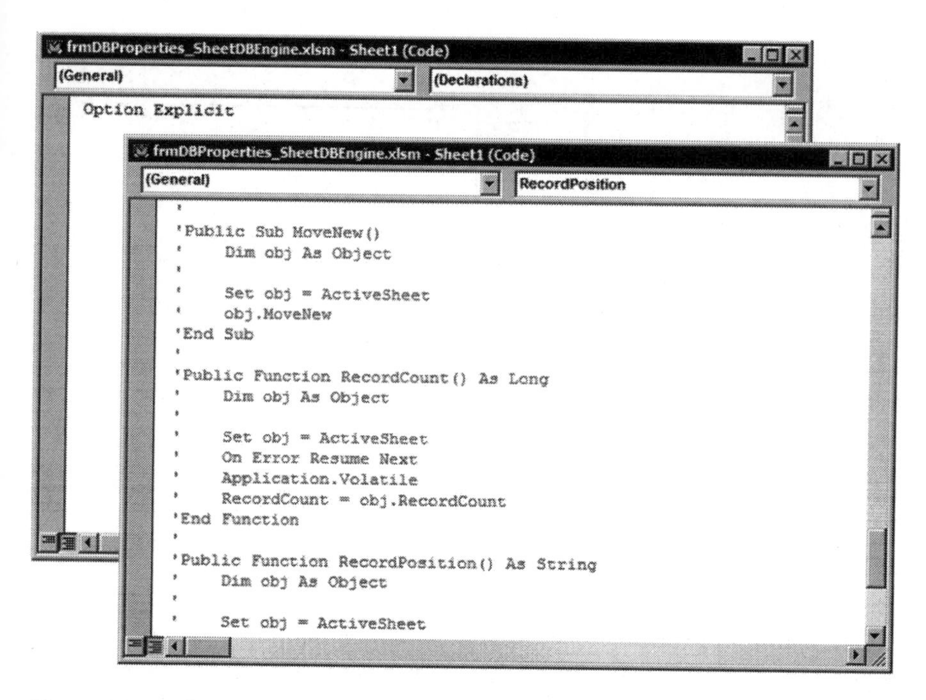

Figure 8-32. *Select the Sheet1 code module and press Ctrl+V to paste all the code needed to start and run the SheetDBEngine class module*

Understanding the Worksheet Database Code

The database code proposed by `frmDBProperties` has two parts.

- Normal code that belongs to the sheet code module
- Commented code that must be selected, cut, and pasted into a standard module

The worksheet normal code is the first code module part (press Ctrl+Home in the code module to go to its beginning, as shown in Figure 8-33).

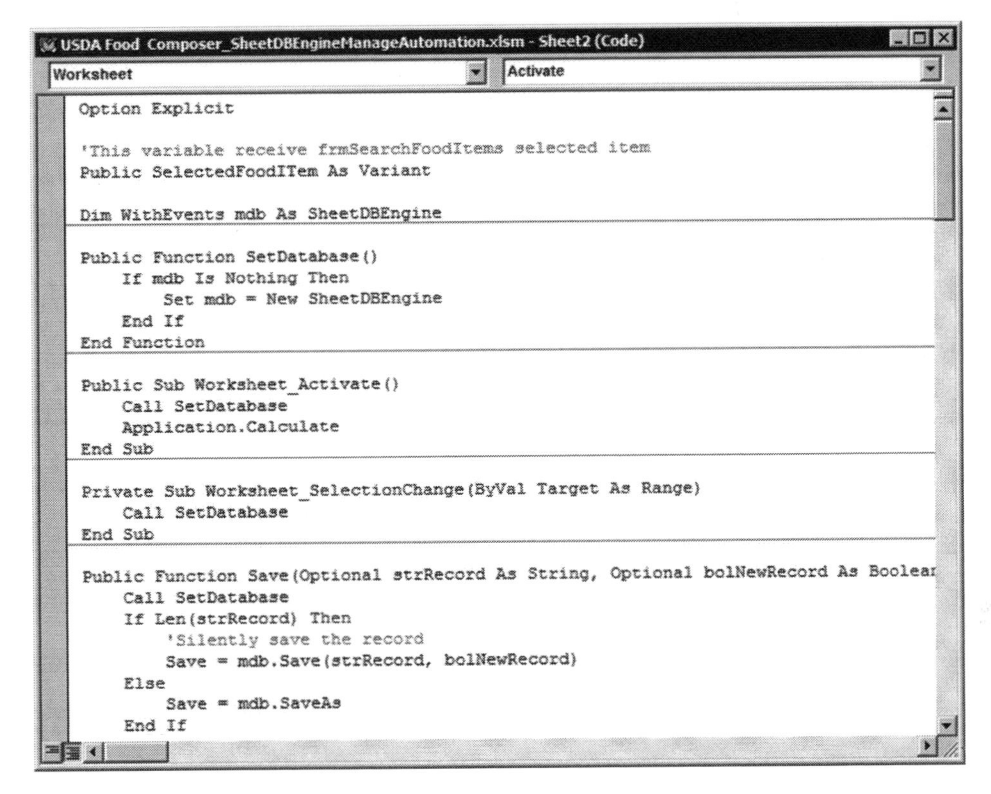

Figure 8-33. *This is part of the worksheet code that frmDBProperties copied to the clipboard. It has all the code needed to start the database engine and execute its basic procedures: show a new record, save and delete records, show the record count, record the position, and copy and paste record data between worksheet databases*

It declares a `WithEvents mdb as SheetDBEngine` object variable, allowing you to capture all database class events fired. To instantiate the class on any worksheet procedure, the code uses `Function SetDatabase( )`, which is called from two worksheet events: `Worksheet_Activate( )` and `Worksheet_SelectionChange( )`.

As explained in Chapter 4, the `Worksheet_Activate( )` event fires whenever the sheet tab is selected. It does not fire when the workbook opens and the sheet tab has the focus. The `Worksheet_SelectionChange( )` event fires whenever another worksheet cell is selected.

```
Dim WithEvents mdb As SheetDBEngine

Public Function SetDatabase()
    If mdb Is Nothing Then
        Set mdb = New SheetDBEngine
    End If
End Function

Public Sub Worksheet_Activate()
    Call SetDatabase
    Application.Calculate
End Sub
```

```
Private Sub Worksheet_SelectionChange(ByVal Target As Range)
    Call SetDatabase
End Sub
```

These instructions will guarantee that a new SheetDBEngine instance will be set to the mdb object variable whenever the sheet tab is selected or when any of its input cells receive the focus. Since the Worksheet_Activate( ) event does not fire when the workbook is opened for the worksheet that has the focus, there is a chance that the user will click any worksheet control button (like New, Save and so on). So, every other procedure calls SetDatabase( ) to instantiate the database class.

▪ **Attention** Note that the SetDatabase( ) code first verifies whether the object variable mdb is Nothing to not destroy it and instantiate it again. This will guarantee that it will be set only once.

The other three procedures that show a new record or save or delete the current record call SheetDBEngine class methods that allow these operations. For example, Sub Save( ) calls the SheetDBEngine.SaveAs( ) method, and Sub DeleteRecord( ) calls the SheetDBEngine.Delete method.

```
Public Function Save(Optional strRecord As String, Optional bolNewRecord As Boolean) As
Boolean
    Call SetDatabase
    If Len(strRecord) Then
        'Silently save the record
        Save = mdb.Save(strRecord, bolNewRecord)
    Else
        Save = mdb.SaveAs
    End If
End Function

Public Function DeleteRecord(Optional strRecord As String, Optional bolKeepInMyRecipes As
Boolean)
    Call SetDatabase
    If Len(strRecord) Then
        'Silently delete the record
        mdb.DeleteRecord (strRecord)
    Else
        mdb.Delete
    End If
End Function
```

There are also five other "Move" procedures to show a desired record, using the New Button control or the data navigation Button controls. To show the first record, Sub MoveFirst( ) calls SheetDBEngine.ShowRecord(FirstRec), while Sub MoveNext( ) calls SheetDBEngine.ShowRecord(NewRec) to show a new record.

```
Public Function MoveFirst()
    Call SetDatabase
    mdb.ShowRecord (FirstRec)
End Function
...
Sub MoveNew()
```

```
    Call SetDatabase
    mdb.ShowRecord (NewRec)
End Sub
```

Below the last "Move" method you will find the RecordCount( ) and RecordPosition( ) procedures, which use the SheetDBEngine AbsolutePosition and RecordCount properties.

```
Public Function RecordCount() As Long
    RecordCount = mdb.RecordCount
End Function

Public Function RecordPosition() As String
    Dim strPosition As String
    Dim lngPosition As Long

    lngPosition = mdb.AbsolutePosition

    If lngPosition = 0 Then
        strPosition = "New record"
    Else
        strPosition = lngPosition & " of " & mdb.RecordCount
    End If
    RecordPosition = strPosition
End Function
```

Then you find the CopyRecord( ), PasteRecord( ), and SortDatabase( ) procedures that implement the associated SheetDBEngine methods.

```
Public Function CopyRecord(strRecord As String, rgOneSide As Range, rgManySide As Range) As
Boolean
    Call SetDatabase
    CopyRecord = mdb.CopyRecord(strRecord, rgOneSide, rgManySide)
End Function

Public Function PasteRecord(strRecord As String, rgOneSide As Range, rgManySide As Range, _
Optional PasteAsNewRecord As Boolean) As Boolean
    Call SetDatabase
    PasteRecord = mdb.PasteRecord(strRecord, rgOneSide, rgManySide, PasteAsNewRecord)
End Function

Public Function SortDatabase()
    Call SetDatabase
    mdb.Sort
End Function
```

Creating the Standard Module

The commented code part begins after the RecordPosition( ) procedure, with commented directions explaining that you must create a standard module and cut, paste, and uncomment the code.

```
''---------------------------------------------------------------------------
 *** W A R N I N G ***
' Create a Standard Module (Insert, Module), copy all commented code bellow,
' paste inside the new module and uncomment the code
' Tip: To uncomment, use VBA Edit Toots (right click VBA Toolbar and choose Edit)
'---------------------------------------------------------------------------

'Public Sub Save()
'    Dim obj As Object
'
'    Set obj = ActiveSheet
'    obj.Save
'End Sub
...
'Public Function RecordPosition() As String
'    Dim obj As Object
'
'    Set obj = ActiveSheet
'    On Error Resume Next
'    Application.Volatile
'    RecordPosition = obj.RecordPosition
'End Function
```

Follow these instructions:

1. Put the mouse cursor to the left of the `'Public Sub Save( )` instruction (including the comment quote) and press Ctrl+Shift+End to select all commented code.
2. Press Ctrl+X to cut it from the sheet code module.
3. Use the VBA Insert ➤ Module menu command to insert a new standard module (`Module 1`).
4. Press Ctrl+V to paste the commented code.

To easily uncomment the code inside the new standard module, use the VBA Edit tools:

1. Right-click the VBA toolbar and choose Edit. The Edit toolbar will appear.
2. Select all commented code and click the Uncomment Block tool.
3. Use VBA Compile VBAProject menu command to compile the code.
4. Rename `Module 1` to `basButton` controls.

Now select the `Sheet1` worksheet and force the `SheetDBEngine` to start using one of these two methods:

- Change the input cell focus (this will fire the `Worksheet_SelectionChange( )` event)
- Select the `Sheet2` tab and reselect the `Sheet1` tab (this will fire `Worksheet_Activate( )` event)

The `Sheet1` database engine is already working. Try inserting, selecting, and deleting records!

Exhibiting Database Properties

Once you have defined the worksheet database properties, the next time you load the frmDBProperties UserForm, it will show its sixth tabControl Page with the Database Properties wizard resumed, offering just two CommandButtons: one to close the UserForm and another to remove the database structure (Figure 8-34).

■ **Attention** To reopen the frmDBProperties interface, click its code or the UserForm design and press the F5 function key.

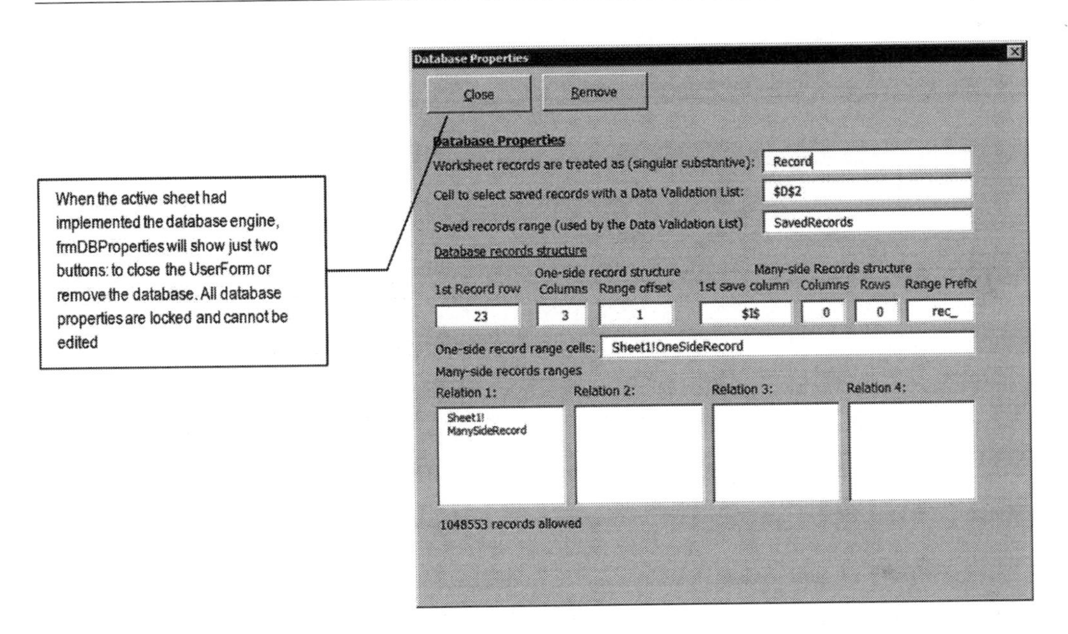

Figure 8-34. *This is how frmDBProperties UserForm is presented to the user after the active worksheet has implemented the database properties system*

All database properties are locked and cannot be edited. This happens on the frmDBProperties_Initialize( ) event, which verifies the active sheet has a local scope dbDataValidationList range name and, if it does, loads each range name property into the appropriate UserForm controls.

```
Private Sub UserForm_Initialize()
    Const conWhite = &HFFFFFF
    ...
    Set rg = Range(strNameScope & "dbDataValidationList")
    If rg Is Nothing Then
        ...
        Me.cmdDefine.Caption = "Close"
        Me.cmdDefine.Accelerator = "C"
        Me.cmdDefine.Enabled = True
        Me.cmdCancel.Caption = "Remove"
        Me.cmdCancel.Accelerator = "R"
        Me.cmdPrevious.Visible = False
```

```
        Me.cmdNext.Visible = False
        Me.txtdbRecordName1.Locked = True
        Me.txtdbRecordName1.BackColor = conWhite
        Me.txtdbManySidePrefix.Locked = True
        Me.txtdbManySidePrefix.BackColor = conWhite

        'Update UserForm TextBoxes
        For intI = 1 To 15
            strName = Choose(intI, "dbRecordName", _
                                   "dbDataValidationList", _
                                   "dbSavedRecords", _
                                   "dbRecordsFirstRow", _
                                   "dbOneSide", _
                                   "dbOneSideColumnsCount", _
                                   "dbManySide1", _
                                   "dbManySide2", _
                                   "dbManySide3", _
                                   "dbManySide4", _
                                   "dbManySideFirstColumn", _
                                   "dbManySideColumnsCount", _
                                   "dbManySideRowsCount", _
                                   "dbManySidePrefix", _
                                   "dbRangeOffset")
            Me("txt" & strName) = ws.Range(strNameScope & strName)
        Next

        Call CalculateManySideRecords
        Me.tabControl.Value = Me.tabControl.Pages.Count - 1
    End If
End Sub
```

The first instructions just change cmdDefine and cmdCancel Caption and Accelerator properties (the Accelerator property defines the Alt+Character key that fires each CommandButton Click( ) event; VBA automatically underlines it in the control's Caption property) and set cmdPrevious and cmdNext Visible = False, so the tabControl Page cannot be changed.

```
        Me.cmdDefine.Caption = "Close"
        Me.cmdDefine.Accelerator = "C"
        Me.cmdDefine.Enabled = True
        Me.cmdCancel.Caption = "Remove"
        Me.cmdCancel.Accelerator = "R"
        Me.cmdPrevious.Visible = False
        Me.cmdNext.Visible = False
```

The next instructions change the txtdbRecordName1 and txtdbManySidePrefix TextBox's Locked and BackColor properties, so the user cannot change its contents (Figure 8-27 shows both controls with a light yellow background to call attention to the fact that they can change their contents before creating the database system). Note that the code declares Constant conWhite = &HFFFFFF (which is the hexadecimal representation for white) to avoid magic numbers appearing in the code.

```
        Me.txtdbRecordName1.Locked = True
        Me.txtdbRecordName1.BackColor = conWhite
        Me.txtdbManySidePrefix.Locked = True
        Me.txtdbManySidePrefix.BackColor = conWhite
```

A For...Next loop is then used to generate the VBA Choose( ) Function argument, storing in the strName String variable a different range name at each loop passage, which is used to update the TextBox that has the same range name prefixed by txt.

```
        'Update UserForm TextBoxes
        For intI = 1 To 15
            strName = Choose(intI, "dbRecordName", _
                                   ...
                                   "dbRangeOffset")
            Me("txt" & strName) = ws.Range(strNameScope & strName)
        Next
```

Once all range names have been updated, it calls the CalculateManySideRecords( ) procedure to show how many records can be saved in worksheet rows and finally selects the last tabControl Page, changing the Value property to Pages.Count - 1 (since the Page value is 0-based).

```
Call CalculateManySideRecords
Me.tabControl.Value = Me.tabControl.Pages.Count - 1
```

Removing Database Properties

Note in Figure 8-33 that once a database system has been defined on the active worksheet, whenever you open frmDBProperties, it will change the cmdCancel.Caption property to Remove to allow the user to remove the database storage system from the active sheet. This is the cmdCancel_Click( ) event:

```
Private Sub cmdCancel_Click()
    Dim strMsg As String
    Dim strTitle As String

    If Me.cmdCancel.Caption = "Remove" Then
        strMsg = "Do you really want to remove this Database structure?" & vbCrLf & vbCrLf
        strMsg = strMsg & "    Just database properties will be removed. " & vbCrLf
        strMsg = strMsg & "    Existing records will remain on the worksheet." & vbCrLf &
vbCrLf
        strMsg = strMsg & "(This operation can be undone if close the workbook without
        saving it!)"
        strTitle = "Delete Database Properties?"
        If MsgBox(strMsg, vbYesNo + vbDefaultButton2 + vbCritical, strTitle) = vbYes Then
            'Remove Database properties
            Call SetDataBase(RemoveDatabase)
        End If
    End If
    Unload Me
End Sub
```

Note that if `cmdCancel.Caption = "Remove"`, it will warn the user that all database properties will be removed, and the only way to undo this operation is by closing the workbook without saving it (Figure 8-35). And once the user clicks Yes to confirm the database properties' deletion, it will call the `SetDataBase( )` procedure, passing as an argument the `RemoveDatabase DataOperation` enumerator.

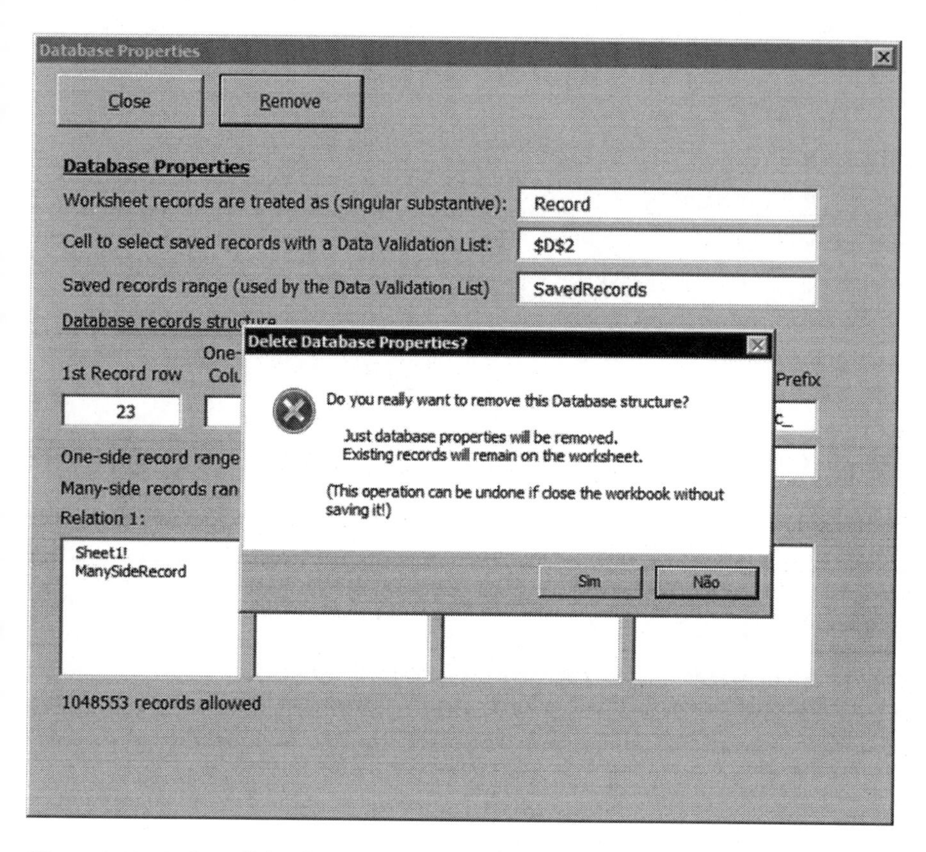

Figure 8-35. *Before all database properties are removed from the active worksheet, frmDBProperties will issue a MsgBox( ) warning to the user. As the message explains, once it is removed, the only way to undo the operation is to close the workbook without saving it*

■ **Attention** When `frmDBProperties` deletes the database structure, as explained by the `MsgBox( )` function, all database properties and associated range names will be removed from the worksheet, but the `SavedRecords` range name containing all records names, including the one-side and the many-side records, will be preserved, allowing the user to recover its records in case of database failure.

Note that although the database engine structure has been removed, the sheet code still operates. You must remove it manually to unload the `SheetDBEngine` class from memory.

Using the SheetDBEngine Class and frmDBProperties

The `frmDBProperties_SheetDBEngine.xlsm` Excel macro-enabled workbook offers on its `Sheet1`, `Sheet2`, and `Sheet3` sheet tabs different one-to-many database relationships so you can practice how to use the `frmDBProperties UserForm`. Each sheet tab has its own set of range names to allow an easier selection of what cells must be defined as the one-side and many-side records. Remember that you can use many-side records' Relation 1, Relation 2, Relation 3, to Relation 4 to save data in the same way it is presented to the user.

To use the `SheetDBEngine` class along with the `frmDBProperties` in other worksheet applications that need to define a database storage system, you will need to copy insert both files into the worksheet application VBA project, using one of these two methods:

- Extract from the `Chapter08.zip` files the `SheetDBEngine.cls`, `frmDBProperties.frm`, and `frmDBProperties.frx` files, and use the VBA File ➤ Import menu command.
- Open the `frmDBProperties_SheetDBEngine.xlsm` macro-enabled workbook, open the desired worksheet application, and using the VBA Explorer tree, drag the `frmDBProperties UserForm` and `SheetDBEngine` classes from one VBA project to another.

Once the worksheet application has both objects (the `SheetDBEngine` class and the `frmDBProperties UserForm`), double-click the `frmDBProperties` in the VBA Project Explorer and press the F5 function key to start it in the desired worksheet application where the database system must be implemented.

Conclusion

In this chapter, you learned how to use a VBA class module to encapsulate complex code that can be easily reused by other VBA projects. Such worksheet applications can be compared to an *n*-tiers database application, where each tier does a specific task.

- The first tier is the Excel worksheet that is responsible for doing the interface calculations (producing the Nutrition Facts label and nutrient analysis of any recipe).
- The second tier is the worksheet code module, which interacts with the class module to produce the desired functionality to deal with recipe records.
- The third tier is the class database code, which is responsible for managing the record set.

You can improve the class object interface using a set of properties, methods, and events carefully planned to make it resemble other popular database systems, hence improving its usability regarding its VBA implementation. You can also learn how to use a `UserForm` with a `Multipage` control to create a worksheet database wizard to set up the database class properties so it can be easily implemented in any worksheet application that need to store records on its unused rows.

Chapter Summary

In this chapter, you learned about the following:

- That using a class module you can easily encapsulate complex, lengthy code into a single, reusable object
- How to instantiate the class module so it can be reused by many worksheet tabs on the same workbook

- How to improve the database class using a set of `Public` variables and procedures to create its properties and methods and how to expose events that make the class more useful to the user needs
- How to use a set of `Button` controls to interact with the database class, using a `Standard` module with `Public Sub` procedures that call generic worksheet procedures to execute database saves, deletions, and move record operations
- How to create a database wizard to set up the database class interface, using a `Multipage` control

The next chapter will show how you can create a `UserForm` interface to manage your worksheet database records such as how to delete, save, export, and import them between different worksheet applications that share the same database engine structure.

CHAPTER 9

Exchanging Data Between Excel Applications

Many worksheet applications base your analysis on external data. This is the case with `USDA Food Composer.xlsm` and any other nutritional worksheet application solution that uses the ARS-USDA nutrient tables as primary data source. They all used some version of the Standard Reference (SR) file at the time they were built, which must be updated when another SR version becomes available.

On the other hand, as you begin to use the `SheetDBEngine` worksheet database class to store worksheet data as records, improving the use of your solution, chances are that you will need to create a way to exchange worksheet database records between different worksheet applications that reside in the same or different workbook file.

In this chapter, you will learn how to use VBA to exchange worksheet data between different applications, using code to update worksheet data sources, and how to exchange worksheet database records managed by the `SheetDBEngine` class between different sheet tabs. You can obtain all files and procedure code in this chapter by downloading the `Chapter 09.zip` and `Chapter09-1.zip` files from the book's Apress.com product page, located at `www.apress.com/9781484222041`, or from `http://ProgrammingExcelWithVBA.4shared.com`.

Updating the USDA Worksheet

Every year the United States Department of Agriculture (USDA) Agricultural Research Services (ARS) releases a new version of the USDA National Nutrient Database for Standard Reference file, with full versions using ASCII files or a single Microsoft Access database and abbreviated versions using either an ASCII file or a Microsoft Excel worksheet.

Besides this information, the ARS home page publishes the USDA Database for the Flavonoid Content of Selected Foods, which in November 2015 was on Release 3.2, also available as a Microsoft Access database file.

F. Morgado, *Programming Excel with VBA*, DOI 10.1007/978-1-4842-2205-8_9

Using the USDA Food List Creator Application

To deal with these two different sources of nutrient data, I developed a Microsoft Access application called `USDA Food List Creator.accdb`, which joins both databases to produce a complete USDA-ARS worksheet food table. It presents items by 100g of each food item, by the weight of its first available common measure, or by all available food item common measures, also producing an independent worksheet with all food item common measures available in each `USDA-ARS SRxx.accdb` and `Flav_Rxx-x.accbd` file release (where `xx` indicates the release version).

■ **Attention** You can download a copy of the `USDA Food List Creator.accdb` file by extracting it from `Chapter09-1.zip`, along with a copy of `SR28.accbd` (the latest SR file for nutrient content published in 2015) and `FLAV_R03-2.ACCDB` (the latest Flavonoid content for selected food items, also published in 2015). The application expects to find the nutrient files in folder `C\Dietary Guide to Excel Applications\SR28`, which is the default extraction folder.

To open the `USDA Food Item Creator.accdb` file, you will need Microsoft Access 2007 or newer. The application will show a splash screen and the Create USDA Food Tables - All Nutrients (Figure 9-1) form.

Dietary Guide to Excel Applications

USDA Food List Creator

by Flavio Morgado

Version 1.2, 2014
© Copyright 2010, Flavio Morgado.
All rights reserved

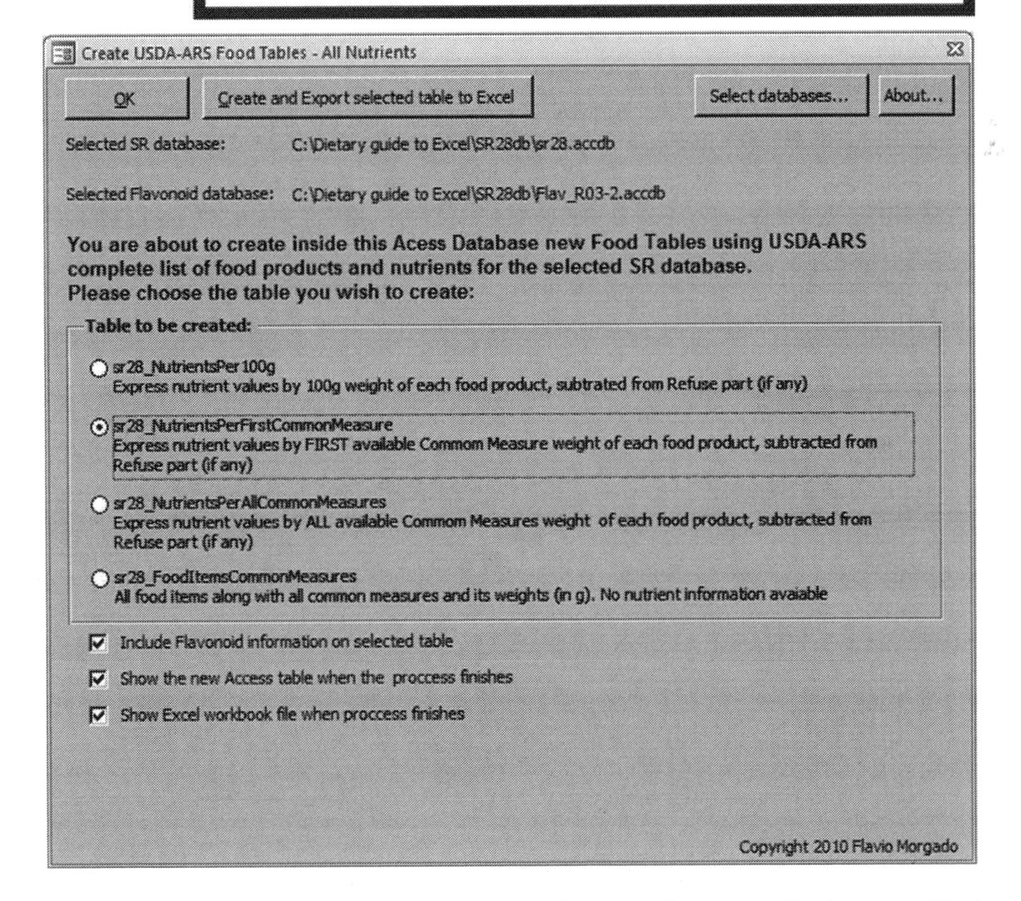

Figure 9-1. *This is the USDA Food Item Creator accdb Microsoft Access application screen. Use it to create new Microsoft Excel worksheet files for every new SRxx.accdb and Flav_Rxx-x.accbd file release*

The USDA Food List Creator.accdb Microsoft Access application allows you to do the following:

- Click the "Select databases" button to select the SRxx.accdb and Flav_Rxx-x.accdb files that must be used to generate the desired USDA worksheet.
- Select the USDA worksheet nutrient table that must be created as an Excel 2003 .xls file based on three different food item weights for the nutrient content.
    - By 100g of each food item
    - By the weight of the first food item common measure (default)
    - By the weight of all food item common measures
- Create the USDACommonMeasures worksheet that offers all food item common measures and their weights, without any nutrient information
- Use its three CheckBox controls to allow the following:
    - Include the Flavonoids nutrients information in the target USDA worksheet
    - Show the created Microsoft Access table with all food items and nutrient information available, used to export the results to Microsoft Excel format
    - Open Microsoft Excel to show the USDA food table (or USDACommonMeasures table) created by the application

The USDA worksheet generation uses a complex Microsoft Access query that is manipulated by a VBA code procedure that may take some minutes to complete. Please relax and wait for the application to finish. The workbook created by the application will be stored in the same folder where the selected SRxx.accdb file resides.

■ **Attention** You can use the USDA Food List Creator.accdb application to create USDA food tables based on earlier SRxx.mdb files that use the Microsoft Access 2003 database file format.

Inside the Chapter09-1.zip file you will also find copies of the sr28_NutrientsPerFirstCommonMeasure.xls and sr28_FoodItemsCommonMeasures.xls files, if you do not have Microsoft Access 2007 or newer versions installed on your computer.

The USDA Worksheet Updating Method

The USDA worksheet generated by the USDA Food List Creator.accdb Microsoft Access application creates a food table with an identical nutrient column sequence between SRxx releases; new nutrients that may appear in new SR versions (if any) will be added as new USDA worksheet table columns.

To manually change the current USDA worksheet version of any worksheet application to another USDA version, you must follow these steps:

1. Open the Excel workbook that has the USDA worksheet with the SR version you want to update (the current SR version workbook).
2. Open the Excel workbook that has the new USDA worksheet associated to the desired SR release (the new SR version workbook).
3. Copy all food items from the current SR version workbook `My_Recipes` range name to the empty `My_Recipes` range name of the new SR version workbook.
4. In the new SR version workbook USDA worksheet, resize both the `My_Recipes` and USDA range names to reflect the new `My_Recipes` food category items.
5. Delete in the current SR version workbook the USDA worksheet with the old SR version.
6. Delete in the current SR version workbook all Name objects that now show a `#REF!` error on the `Name.RefersTo` property (since the USDA worksheet was deleted).
7. Move from the new SR version workbook to current the SR version workbook the new, updated USDA worksheet, containing all `My_Recipes` range name food items.
8. If necessary, delete from the current SR version workbook the `USDACommonMeasure` worksheet and update it for a new SR version.
9. Save the application with the USDA worksheet associated to the new SR release.

All these steps must be executed *in this order* to produce the desired USDA SR updating. The good news is that you can surely implement them using VBA code.

Using the USDA Food Composer_SheetDBEnginebasUSDA.xlsm Application

The `USDA Food Composer_SheetDBEnginebasUSDA.xlsm` macro-enabled workbook is associated to USDA version SR27 released in August 2014 and shows the current USDA SR version in the bottom-left corner (Figure 9-2).

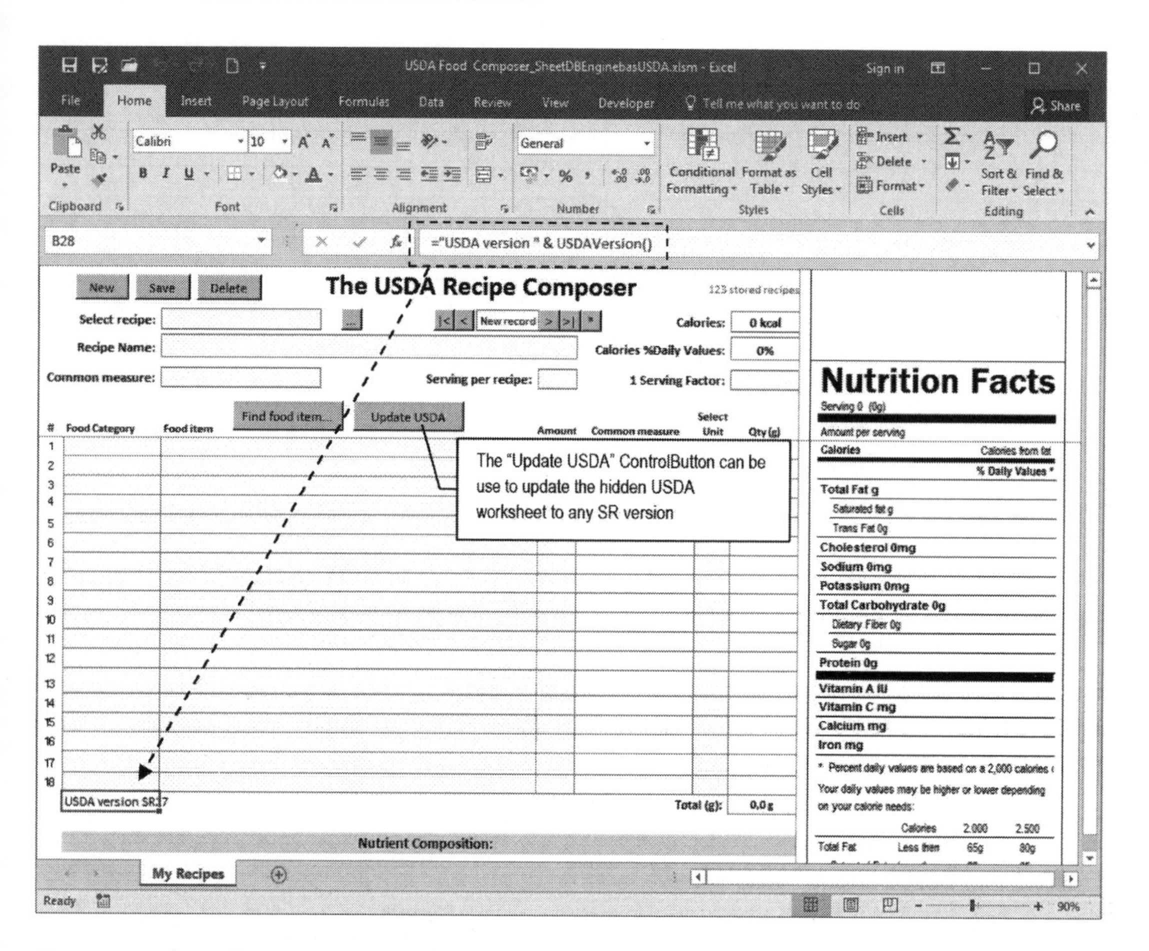

Figure 9-2. This is the USDA Food Composer_SheetDBEnginebasUSDA.xlsm workbook, which shows the current USDA SR version in cell B28, using Function USDAVersion( ) from the basUSDA standard module

---

■ **Attention** Right-click the `My Recipes` sheet tab, select Unhide to unhide the `USDA` and/or `USDACommonMeasures` worksheets, and see that both have an SR27 version.

To show the current USDA SR version in the user interface, you use `Public Function USDAVersion( )` from the `basUSDA` standard module, which executes this simple code:

```
Public Function USDAVersion() As String
    Application.Volatile
    USDAVersion = Left(Thisworkbook.Worksheets("USDA").Range("A1"), 4)
End Function
```

Note that function `USDAVersion( )` was tagged as `Application.Volatile`, meaning that it will be evaluated whenever the workbook calculates and that it just concatenates the words `USDA version` with the first four characters from the `ThisWorkbook USDA` worksheet's cell `A1`.

Assuming that you are extracting files `sr28_NutrientsPerFirstCommonMeasure.xls` and `sr28_FoodItemsCommonMeasures.xls` to the same folder, to update the USDA worksheet to the SR28 nutrients version, click the UpdateUSDA `ControlButton`, which will do the following:

1. Show an Open File dialog to select the folder workbook that has the USDA worksheet nutrient table with the desired SRxx update source (only `.xls` Excel 2003 files that begin with *SR* will be shown, in other words, the ones created with the USDA `Food List Creator.accbd` Microsoft Access application, as shown in Figure 9-3).

■ **Attention** You must use an `srxx_NutrientsPerFirstCommonMeasure.xls` workbook, which shows nutrient data based on the weight of the first food item common measure.

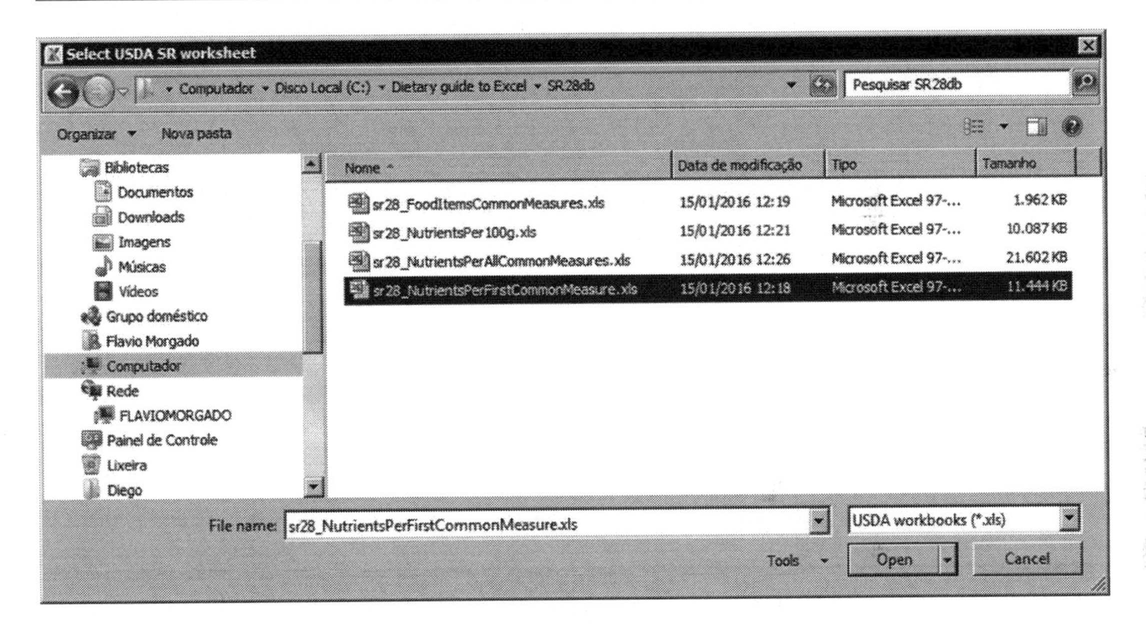

Figure 9-3. When you click the My Recipes Update USDA control button, the code asks you to select an .xls Excel 2003 file that begins with SR

2. A `MsgBox( )` warning will ask you to confirm that you want to update the current USDA worksheet to the selected SRxx version (Figure 9-4).

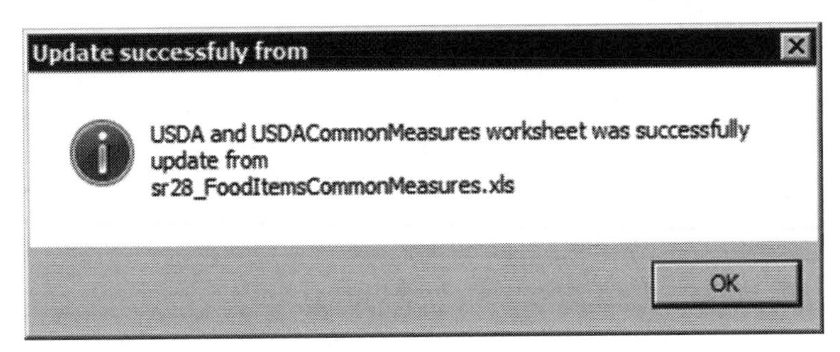

Figure 9-4. After selecting the desired srXX_NutrientsPerFirstCommonMeasure.xls workbook, confirm that you want to update the current USDA worksheet to the desired SR version

3. If the `SRxx_FoodItemsCommonMeasures.xls` file was not found in the selected folder, the procedure will show again the Open File dialog so you can try to select it.

4. If the operation succeeds, a `MsgBox( )` message will confirm whether `USDA` and/or the `USDACommonMeasures` worksheets were updated to the desired SR version (and cell `B28` will display the current `SRxx` version used by the application (Figure 9-5). After you dismiss it, the workbook is automatically saved.

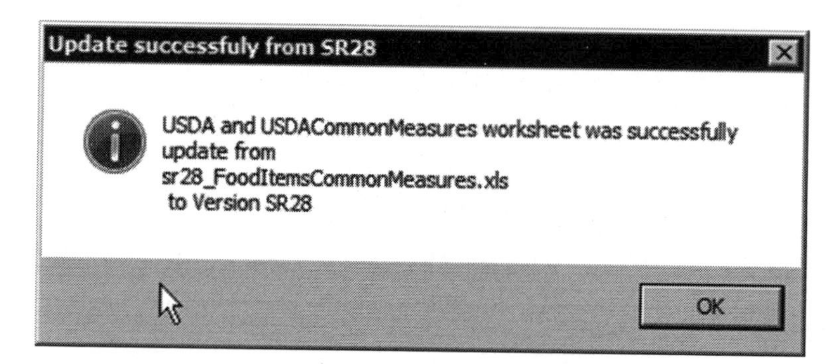

Figure 9-5. *When the updating process ends, a* `MsgBox( )` *will confirm what has been updated (USDA worksheet, USDACommonMeasures worksheet, or both)*

Figure 9-6 shows this process in action. Note that while the operation is conducted, the Excel status bar shows the updating USDA worksheet along with the percentage accomplished and a progress bar.

Figure 9-6. *While the USDA worksheet is updated, the Excel status bar shows the "Updating USDA worksheet" text, followed by the percent completed and a progress bar*

The `My Recipes` worksheet Update USDA control button is associated to `Public Function UpdateUSDA( )` from the basUSDA standard module, which executes this fully commented code:

```
Public Function UpdateUSDA()
    Dim wb As Workbook
    Dim wbUSDA As Workbook
```

```
Dim ws As Worksheet
Dim wsUSDA As Worksheet
Dim rgMyRecipes As Range
Dim rgUSDA As Range
Dim nm As Name
Dim strNewSRVersion As String
Dim strFileUSDA As String
Dim strFileCommonMsrs As String
Dim strMsg As String
Dim lngRows As Long

strFileUSDA = SelectUSDAFile(strNewSRVersion)
If Len(strFileUSDA) Then
    SetScreenEventsRecalc (False)
        Call UpdateStatusBar(0.33, , "Updating USDA worksheet")
        Set wb = ThisWorkbook
        Set ws = Worksheets("USDA")
        Set wbUSDA = Application.Workbooks.Open(strFileUSDA, False)
        Set wsUSDA = wbUSDA.Worksheets("USDA")

        'Copy and paste current My_recipes to new USDA worksheet
        ws.Range("My_Recipes").CurrentRegion.Copy
        wsUSDA.Range("My_Recipes").PasteSpecial xlPasteValues

        'Rebuild "My_Recipes" and "USDA" range names on new USDA Worksheet
        Set rgMyRecipes = wsUSDA.Range("My_Recipes")
        wsUSDA.Range("My_Recipes").Resize(rgMyRecipes.Rows.Count + 1, 1).Name = "My_
        Recipes"

        Set rgUSDA = wsUSDA.Range("USDA")
        lngRows = rgUSDA.Rows.Count + rgMyRecipes.Rows.Count - 1
        rgUSDA.Resize(lngRows, rgUSDA.Columns.Count).Name = "USDA"

        'Silently delete worksheets from this workbook
        Application.DisplayAlerts = False
            'Make worksheet visible before delete it to avoid Excel bug when save
            workbook
            ws.Visible = True
            wb.Worksheets("USDA").Delete
            'Search and delete invalid range names from this workbook
            For Each nm In wb.Names
                If InStr(nm.RefersTo, "#REF!") > 0 Then
                    nm.Delete
                End If
            Next

            'Move new USDA SR worksheet before USDACommonMeasures and hide it
            wbUSDA.Worksheets("USDA").Move wb.Worksheets("USDACommonMeasures")
            wb.Worksheets("USDA").Visible = False

            strMsg = "USDA worksheet updated to Version " & strNewSRVersion & vbCrLf
```

```
            strMsg = strMsg & "from file '" & Mid(strFileUSDA, InStrRev(strFileUSDA,
            "\") + 1) & "';"
            strMsg = strMsg & vbCrLf & vbCrLf
            Call UpdateStatusBar(0.66, , "Updating USDA worksheet")

            'Now try to update USDACommonMeasures worksheet: search it on same path
            strFileCommonMsrs = Dir(Left(strFileUSDA, InStrRev(strFileUSDA, "\")) &
            "SR??_FoodItemsCommonMeasures.xls")
            If Len(strFileCommonMsrs) = 0 Then
                'SR??_FoodItemCommonMeasures not found. Ask to select it!
                strFileCommonMsrs = Application.GetOpenFilename("USDA workbooks (*.xls),
                SR*.xls", ,
              "Select USDACommonMeasures SR worksheet", , False)
            End If

            If InStr(1, strFileCommonMsrs, "FoodItemsCommonMeasures") Then
                'SRxx_USDACommonMeasures.xls found or selected. Update
                USDACommonMeasures!
                wb.Worksheets("USDACommonMeasures").Visible = True
                wb.Worksheets("USDACommonMeasures").Delete
                Set wbUSDA = Application.Workbooks.Open(strFileCommonMsrs, False)
                wbUSDA.Worksheets("USDACommonMeasures").Move , wb.Worksheets("USDA")
                wb.Worksheets("USDACommonMeasures").Visible = False

                strFileCommonMsrs = Mid(strFileCommonMsrs, InStrRev(strFileCommonMsrs,
                "\") + 1)
                strMsg = strMsg & "USDACommonMeasures worksheet updated to Version " &
                Left(strFileCommonMsrs, 4) & vbCrLf
                strMsg = strMsg & "from file '" & strFileCommonMsrs & "'."
            End If
            Call UpdateStatusBar(1, , "Updating USDA worksheet")
        Application.DisplayAlerts = True
    SetScreenEventsRecalc (True)

    MsgBox strMsg, vbInformation, "Update successfully to Version " & strNewSRVersion
   Call UpdateStatusBar(0)
   ThisWorkbook.Save
  End If
End Function
```

Getting the SR Workbook File Name and Validating the SR Update

After declaring the variables, `Function UpdateUSDA( )` calls `SelectUSDAFile( )` to allow the selection of the new USDA SR version workbook. Note that it passes the `strNewVersion String` variable as a function argument.

```
Public Function UpdateUSDA()
    ...
    strFile = SelectUSDAFile(strNewSRVersion)
```

The Private Function SelectUSDAFile( ) procedure from basUSDA executes this code:

```
Private Function SelectUSDAFile(strNewSRVersion As String) As String
    Dim strFile As String
    Dim strSRVersion As String
    Dim strMsg As String

    'Get current USDA SR version on USDA A1 cell
    strSRVersion = Left(Worksheets("USDA").Range("A1").Value, 4)

    'Select USDA workbook
    strFile = Application.GetOpenFilename("USDA workbooks (*.xls), SR*.xls", , "Select USDA
    SR worksheet", , False)
    If strFile <> "False" Then
        'Get USDA version of selected file from file name
        strNewSRVersion = Mid(strFile, InStr(1, strFile, "SR"), 4)

        If strNewSRVersion = strSRVersion Then
            'Same USDA version
            strMsg = "The SR version you are trying to update (" & strNewSRVersion & _
                     ") is the same version already in use." & vbCrLf
            strMsg = strMsg & "Update anyway?"
        ElseIf Mid(strNewSRVersion, 3, 2) < Mid(strSRVersion, 3, 2) Then
            'Old SR version
            strMsg = "The SR version you are trying to update (" & strNewSRVersion & _
                     ") is older than current version (" & strSRVersion & ")."
            strMsg = strMsg & "Update anyway?"
        Else
            'New SR version
            strMsg = "Update current " & strSRVersion & " food table to USDA " &
            strNewSRVersion & " food table?"
        End If

        If MsgBox(strMsg, vbYesNo + vbDefaultButton2 + vbQuestion, "Update USDA worksheet?")
        = vbYes Then
            SelectUSDAFile = strFile
        End If
    End If
End Function
```

After getting the current USDA worksheet version from the USDA worksheet's cell A1, it uses the Application.GetOpenFileName method to show an Open File dialog and allow the selection of any Excel .xls workbook that begins with *SR* (note the filter SR*.xls used on the fourth method argument).

```
Private Function SelectUSDAFile(strFileSRVersion As String) As String
    ...
    'Get current USDA SR version on USDA A1 cell
    strSRVersion = Left(Worksheets("USDA").Range("A1"), 4)

    'Select USDA workbook
    strFile = Application.GetOpenFilename("USDA workbooks (*.xls), SR*.xls", , "Select USDA
SR worksheet", , False)
```

If the Open File dialog is closed without any file selected, `strFile = False`, and the procedure ends, returning an empty string to `Function UpdateUSDA( )`, which also ends doing nothing.

```
    Public Function UpdateUSDA()
        ...
        strFile = SelectUSDAFile(strNewSRVersion) -----------------------------,
,--->   If Len(strFile) Then----,                                              |
|       ...                     |                                              |
|       End If <----------------'                                              |
|   End Function                                                               |
|                                                                              |
|   Private Function SelectUSDAFile(strFileSRVersion As String) As String <----'
|       ...
|       strFile = Application.GetOpenFilename("USDA workbooks (*.xls), SR*.xls", ,
|       "Select USDA SR worksheet", , False)
|       If strFile <> "False" Then----,
|          ,,,                        |
|       End If <----------------------'
'---End Function
```

But if any `.xls` file beginning with the *SR* characters was selected (the ones created by the `USDA Food List Creator.accbd` Microsoft Access application), chances are that a new USDA worksheet of the desired SR version was selected. So, process it, getting the file SR version to verify whether the selected file has the same application SR current version, and produce a string message.

```
'Get USDA version of selected file from file name
strNewSRVersion = Mid(strFile, InStr(1, strFile, "SR"), 4)

If strNewSRVersion = strSRVersion Then
    'Same USDA version
    strMsg = "The SR version you are trying to update (" & strNewSRVersion & _")
                   is the same version already in use." & vbCrLf
    strMsg = strMsg & "Update anyway?"
```

Verify with the VBA `Mid( )` function if the new SR version number is smaller than the current one, configuring a downgrade update.

```
ElseIf Mid(strNewSRVersion, 3, 2) < Mid(strSRVersion, 3, 2) Then
    'Old SR version
    strMsg = "The SR version you are trying to update (" & strNewSRVersion & _")
                   is older than current version (" & strSRVersion & ")."
    strMsg = strMsg & "Update anyway?"
```

If not, the update is an upgrade. Issue a `MsgBox( )` to the user and exit `Function SelectUSDAFile( )`, which must return the path to the selected file (if any).

```
Else
    'New SR version
    strMsg = "Update current " & strSRVersion & " food table to USDA " & strNewSRVersion & "
    food table?"
End If
```

```
        If MsgBox(strMsg, vbYesNo + vbDefaultButton2 + vbQuestion, "Update USDA worksheet?")
        = vbYes Then
            SelectUSDAFile = strFile
        End If
    End If
End Function
```

Processing the Selected USDA SR Version

Once the desired file has been selected, it is time to set the stage to process it. This is made by first calling `SetScreenEventsRecalc(False)` to disable screen updates, events firing, and worksheet recalculations. The Excel status bar is then updated by calling the `Private Sub UpdateStatusBar( )` procedure, from the `basStatusBar` module.

```
Public Function UpdateUSDA()

    strFile = SelectUSDAFile(strNewSRVersion)
    If Len(strFile) Then
            SetScreenEventsRecalc (False)
                Call UpdateStatusBar(0.33, , "Updating USDA worksheet")
```

■ Attention Note that the `UpdateStatusBar( )` procedure receives 0.33 on its first argument (33%) and "Updating USDA worksheet" on its third argument, producing the results you see in Figure 9-6.

Updating the Excel Status Bar

To show a message and a progress bar in the Excel status bar and `Sub UpdateStatusBar( )` from the `basStatusbar` module, execute this code:

```
Public Sub UpdateStatusBar(sngValue As Single, Optional sngTotal As Single = 1, Optional
strText As String)
    Dim strStatusBar As String
    Dim sngPercent As Single
    Const conNumChars = 50
    Const conFillChar = 9608 'try 9609 for spaced char
    Const conEmptyChar = 9620

    If Abs(sngValue) > Abs(sngTotal) Then sngTotal = Abs(sngValue)
    If sngValue > 0 Then
        sngPercent = Abs(sngValue / sngTotal)
        strStatusBar = IIf(Len(strText), strText & " ", "Processing ")& Format(sngPercent,
        "0.0%") & "  "
        strStatusBar = strStatusBar & String(Int(conNumChars * sngPercent),
        ChrW(conFillChar))
        strStatusBar = strStatusBar & String(conNumChars - Int(conNumChars * sngPercent),
        ChrW(conEmptyChar))
```

```
    End If
    Application.StatusBar = strStatusBar
End Sub
```

The procedure receives three arguments:

- sngValue: The current progress bar value
- sngTotal = 1: The default maximum value for sngTotal
- strText: The text string that must be shown on the Excel status bar

It declares three constants: conNumChars = 50 to determine the progress bar length, and conFillChar = 9608 and conEmptyChars = 9620, which are Unicode code characters used to fill the progress bar with a black or empty block character of same width.

After avoiding that sngValue is greater than sngTotal, the procedure verifies whether sngValue = 0. If it is, the Excel status bar is cleaned up with an empty string (use UpdateStatusBar(0) to clean up the Excel status bar).

```
Public Sub UpdateStatusBar(sngValue As Single, Optional sngTotal As Single = 1, Optional
strText As String)
    ...
    If Abs(sngValue) > Abs(sngTotal) Then sngTotal = Abs(sngValue)
    If sngValue > 0 Then
    ...
    End If
    Application.StatusBar = strStatusBar
End Sub
```

Whenever sngValue > 0, it calculates the percentage to be shown by the status bar and uses it to define the status bar text. Note that if the strText optional argument is not used, the default text will be processing #.0% (the percentage amount is shown with one decimal).

```
sngPercent = Abs(sngValue / sngTotal)
strStatusBar = IIf(Len(strText), strText & " ", "Processing ")& Format(sngPercent,
"0.0%") & "  "
```

After the text is produced, it is concatenated with the number of conFillChars black block characters that must be used to produce a progress bar of fixed length. To produce such a number of characters, it uses the VBA String( ) function, which receives on its first argument the total number of characters (based on the Int(conNumChars * sngPercent) value) and the character to be used to indicate the percentage amount already processed (using the VBA ChrW( ) function).

```
        strStatusBar = strStatusBar & String(Int(conNumChars * sngPercent),
ChrW(conFillChar))
```

Since part of the conNumChars = 50 characters has been already filled with black blocks, fill the remaining characters (conNumChars - Int(conNumChars * sngPercent)) with conSpaceChars characters to produce the progress bar and update the Application.StatusBar property.

```
        strStatusBar = strStatusBar & String(conNumChars - Int(conNumChars * sngPercent),
ChrW(conEmptyChar))
    End If
    Application.StatusBar = strStatusBar
```

■ **Attention** Use Function TestStatusBar( ) from basStatusBar in the VBA Immediate window to test Sub UpdateStatusBar( ).

Note that the procedure comment gives a tip for using conFillChar = 9609, which produces a black block that has a width smaller than conEmptyChar = 9620. Try it with TestStatusBar( ) and note that the continue status bar growth shown in Figure 9-6 will be changed by a black block sequence status bar.

Updating the USDA Worksheet

After the Excel status bar is defined, four object variables are instantiated to reference the Thisworkbook and current USDA worksheet (wb and ws object variables) and the selected external workbook and USDA worksheet (wbUSDA and wsUSDA object variables) that has the desired updated data.

```
Set wb = ThisWorkbook
Set ws = Worksheets("USDA")
Set wbUSDA = Application.Workbooks.Open(strFile, False)
Set wsUSDA = wbUSDA.Worksheets("USDA")
```

Now it is time to copy all current My_Recipes range names between the current and the new USDA worksheets. This is done by selecting all the My_Recipes data using the Range.CurrentRegion property, followed by the Range.Copy method to copy the original data, and the Range.PasteSpecial xlPasteValues method to paste it in the new USDA worksheet.

```
'Copy and paste current My_recipes to new USDA worksheet
ws.Range("My_Recipes").CurrentRegion.Copy
wsUSDA.Range("My_Recipes").PasteSpecial xlPasteValues
```

Once the new USDA worksheet receives all current My_Recipes data, you need to update its My_Recipes and USDA range names to include all these newly inserted rows. This is done using Range.CurrentRegion to define the new range dimensions based on all pasted rows and then applying Range.Resize method. The My_Recipes range name has just one column, so just its row number is resized.

```
'Rebuild "My_Recipes" and "USDA" range names on new USDA Worksheet
Set rgMyRecipes = wsUSDA.Range("My_Recipes").CurrentRegion
wsUSDA.Range("My_Recipes").Resize(rgMyRecipes.Rows.Count + 1, 1).Name = "My_Recipes"
```

Now update the new USDA worksheet USDA range name to include all My_Recipes range name rows.

```
Set rgUSDA = wsUSDA.Range("USDA")
lngRows = rgUSDA.Rows.Count + rgMyRecipes.Rows.Count - 1
rgUSDA.Resize(lngRows, rgUSDA.Columns.Count).Name = "USDA"
```

With the new USDA worksheet correctly updated, delete the current USDA worksheet and move the new USDA worksheet to the workbook application. To avoid receiving Excel warning messages, the code sets Application.DisplayAlerts = False and then turns the current USDA worksheet visible before deleting it from the Worksheet.Delete method.

```
'Silently delete worksheets from this workbook
Application.DisplayAlerts = False
    'Make worksheet visible before delete it to avoid Excel bug when save workbook
    ws.Visible = True
    wb.Worksheets("USDA").Delete
```

■ **Attention** There is a bug on Microsoft Excel that will fire whenever you delete a hidden sheet tab and try to save the workbook. To avoid such a bug, always show a hidden worksheet before deleting it from the workbook.

At this code point, the current workbook has lost its USDA worksheet, so all the range names that have the `Name.RefersTo` property pointing to the USDA worksheet will show the `Name.RefersTo` = `#REF!` constant error. So, the code removes them all by looping through the `Workbook.Names` collection. Note that the code uses the VBA `InStr( )` function to verify whether the `Name.RefersTo` property contains the `#REF!` constant error.

```
'Search and delete invalid range names from this workbook
For Each nm In wb.Names
    If InStr(nm.RefersTo, "#REF!") > 0 Then
        nm.Delete
    End If
Next
```

The new, updated USDA worksheet is then moved from the external workbook to the current workbook with the `Worksheet.Move` method, using the `Before` argument to insert it before the hidden `USDACommonMeasures` worksheet, and the moved USDA worksheet is hidden.

```
'Move new USDA SR worksheet before USDACommonMeasures and hide it
wbUSDA.Worksheets("USDA").Move wb.Worksheets("USDACommonMeasures")
wb.Worksheets("USDA").Visible = False
```

■ **Attention** Since all `SR*.xls` files created by the `USDA Food List Creator.accdb` Microsoft Access application have just one USDA worksheet inside them, when you move the USDA worksheet from the external workbook to the current workbook application, the external workbook is automatically closed because it can't exist without at least one worksheet!

Since the USDA worksheet has been correctly updated, the procedure creates a text string on the `strMsg` string variable to indicate the updating made and then updates the Excel status bar using the `UpdateStatusBar( )` procedure.

```
strMsg = "USDA worksheet updated to Version " & strNewSRVersion & vbCrLf
strMsg = strMsg & "from file '" & Mid(strFileUSDA, InStrRev(strFileUSDA, "\") + 1) & "';"
strMsg = strMsg & vbCrLf & vbCrLf
Call UpdateStatusBar(0.66, , "Updating USDA worksheet")
```

Update the USDACommonMeasures Worksheet

With the USDA worksheet already updated to the desired SRxx version, it is time to also update the USDACommonMeasures worksheet to the same version. The code tries to find it on the same path of the selected file using the VBA Dir( ) function.

```
'Now try to update USDACommonMeasures worksheet: search it on same path
strFileCommonMsrs = Dir(Left(strFileUSDA, InStrRev(strFileUSDA, "\")) &
"SR??_FoodItemsCommonMeasures.xls")
```

The VBA Dir( ) function has this syntax:

```
Dir[(Pathname[, Attributes])]
```

In this code:

Pathname: This is optional; it is a string expression that specifies a file name, which may include its folder and drive. A zero-length string ("") is returned if the path name is not found.

Attributes: This is optional; it is a constant or numeric expression whose sum specifies the desired file attributes. If omitted, it returns files that match the path name but have no attributes. The values allowed are as follows:

VbNormal = 0: Files with no attributes (the default)

VbReadOnly = 1: Read-only files in addition to files with no attributes

VbHidden = 2: Hidden files in addition to files with no attributes

VbSystem = 4: System files or files with no attributes

VbVolume = 8: Volume label; ignored if any other attribute is specified

VbDirectory = 16: Folders and files with no attributes

VbAlias = 64: File name is an alias; available only on the Macintosh

Note that to use Dir( ) to find another file on the same file path, it extracts the strFileUSDA path using the VBA Left( ) and InsStrRev( ) functions to extract the path from strFileUSDA and then concatenates the SR??_FoodItemsCommonMeasures.xls string, which uses the ? local wildcard character to find any SR??_FoodItemCommonMeasures.xls file on the same path.

If the file *was not found*, the code uses the Application.GetOpenFilenName method to show an Open Dialog asking the user to select the USDACommonMeasures workbook.

```
If Len(strFileCommonMsrs) = 0 Then
    'SR??_FoodItemCommonMeasures not found. Ask to select it!
    strFileCommonMsrs = Application.GetOpenFilename("USDA workbooks (*.xls), SR*.xls", ,
       "Select USDACommonMeasures SR workbook", , False)
End If
```

The code uses InStr( ) to check whether strFileCommonMsrs has the FoodItemsCommonMeasures text on the file name, as an indication that the desired file was selected. If it is true, it follows the same process used to update the USDA worksheet: it turns the worksheet visible, deletes it, opens the new workbook, moves the new USDACommonMeasures worksheet to the application workbook, and hides it.

```
If InStr(1, strFileCommonMsrs, "FoodItemsCommonMeasures") Then
    'SRxx_USDACommonMeasures.xls found or selected. Update USDACommonMeasures!
    wb.Worksheets("USDACommonMeasures").Visible = True
    wb.Worksheets("USDACommonMeasures").Delete
    Set wbUSDA = Application.Workbooks.Open(strFileCommonMsrs, False)
    wbUSDA.Worksheets("USDACommonMeasures").Move , wb.Worksheets("USDA")
    wb.Worksheets("USDACommonMeasures").Visible = False
```

Once the USDACommonMeasures worksheet is updated, the workbook file name is extracted from inside the strFileCommonMsrs String variable using the VBA Mid( ) and InStrRev( ) functions and is used to update the strMsg String variable indicating the worksheet update.

```
     strFileCommonMsrs = Mid(strFileCommonMsrs, InStrRev(strFileCommonMsrs, "\") + 1)
     strMsg = strMsg & "USDACommonMeasures worksheet updated to Version " &
Left(strFileCommonMsrs, 4) & vbCrLf
     strMsg = strMsg & "from file '" & strFileCommonMsrs & "'."
 End If
```

The procedure ends by updating the Excel status bar to 100 percent; enabling Application.DisplayAlerts, screen updating, events firing, and calculation; and issuing a MsgBox( ) message indicating the update performed. When the MsgBox( ) is closed, it clears the Excel status bar using UpdateStatusBar(0) and saves the workbook.

```
            Call UpdateStatusBar(1, , "Updating USDA worksheet")
        Application.DisplayAlerts = True
    SetScreenEventsRecalc (True)

    MsgBox strMsg, vbInformation, "Update successfuly to Version " & strNewSRVersion
    Call UpdateStatusBar(0)
    ThisWorkbook.Save
  End If
End Function
```

Warning About USDA Worksheet Updates

Every time the USDA-ARS releases a new SRxx version, besides the possibility of the appearance of one or more new nutrient data columns, some food items are removed from the new version, and others are inserted into the new database. But for specific reasons that I can't anticipate, some food item names will suffer subtle changes.

If the Excel data validation lists used to select food items have multiple columns (like the VBA ComboBox control), you could fill the list with two USDA worksheet columns: the food item NDB_No (food item primary key) and the food item name. By hiding the first column (NDB_No) and setting it as the default value, you can select food items by name while the data validation list cell stores the NDB_No on the associated cell.

But Excel doesn't work this way. You must select food items by name and store the name value on the data validation list cell. The corollary to this behavior is if just one character is changed on any food item name, the worksheet application will not find it anymore in the USDA worksheet using the Excel VlookUp( ) function.

So, be aware of this problem when you update the USDA food tables: some of your worksheet calculations may fail.

This is the case with the `USDA Food Composer_SheetDBEnginebasUSDA.xlsm` application. After you update the `USDA` worksheet from version SR27 to SR28, some food items will change the name, and others will just disappear from the new nutrient data table, making some recipes stop calculating correctly because of the `#N/A!` constant error propagation.

Figure 9-7 shows what happened to Record 1 of the `My Recipes` worksheet ("Apple Custard Tart recipe"): the Calories, Calories %Daily Values, the Nutrition Facts food label values and all the Nutrient Composition areas stop calculating because one food item ("Water, tap, drinking," formatted in red) does not belong anymore to SR28 food table.

■ **Attention** All food items that cannot be found in the `USDA` worksheet by the Excel `VLOOKUP( )` function will be formatted in red in the food item data validation lists column (column C), because of a conditional formatting rule applied to these cells that changes the text color to red whenever this formula is `True` (returning the `#N/A!` constant error):

```
=ISNA(VLOOKUP(C10;USDA;1;FALSE))
```

Another conditional formatting rule hides all `#N/A!` (Not Available) constant errors in the `My Recipes` worksheet, changing the text color to white using this formula:

```
=ISNA($K$2)
```

Unprotect the `My Recipes` sheet tab and select the worksheet cells where the calculation fails to see the `#N/A!` constant errors hidden by this conditional formatting rule.

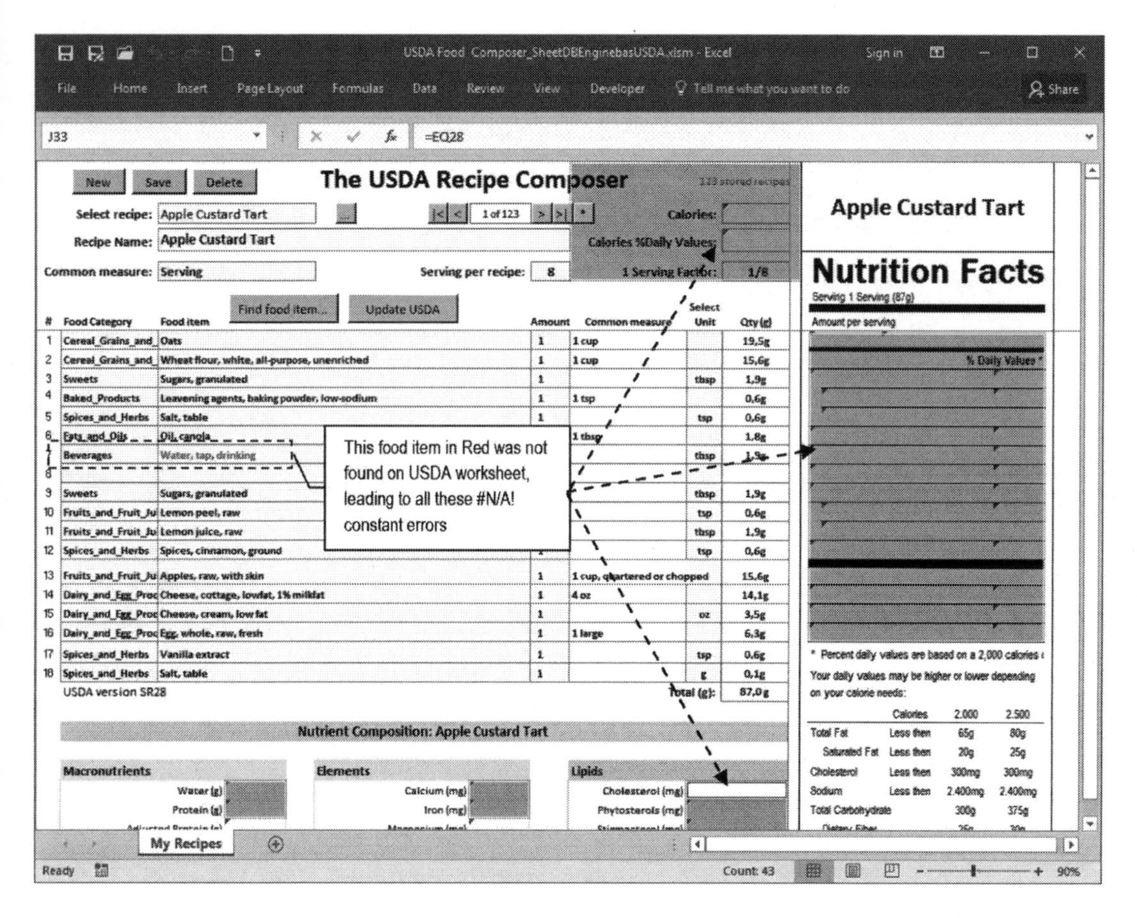

Figure 9-7. *When the USDA worksheet is updated, some food items may change name or be removed from the database, which will impact the worksheet application calculation. The recipe "Apple Custard Tart" stops to calculate because "Water, tap, drinking" could not be found in the USDA worksheet SR28 version*

■ **Attention** Earlier versions of Excel show the `#N/A!` and other Excel constant errors hidden with conditional formatting whenever you select cells. Figure 5-7 shows that this behavior does not happen in Excel 2016.

And if you *walk* along the `My Recipes` worksheet recipe records using the application's data navigation buttons, you will note that many other food items have not been found in the USDA SR28 food table, leading its recipes to fail to update its nutrient composition. To resolve this kind of problem during the `USDA` worksheet updating process, you need to use a VBA `UserForm`.

Using the USDA Food Composer_SheetDBEnginefrmUSDA.xlsm Application

The `USDA Food Composer_SheetDBEnginefrmUSDA.xlsm` macro-enabled workbook that you can extract from the `Chapter09.zip` file has the `frmUSDA` `UserForm`, which does the same updating operations made by `Sub UpdateUSDA( )` from `basUSDA` (which is also present in the workbook's VBA code), allowing you to resolve food item name conflicts between the current `USDA` worksheet and the new `SRxx` version selected for the update.

Press Alt+F11 to show the Visual Basic IDE and double-click `frmUSDA` in the Project Explorer tree to show its interface in design mode (Figure 9-8). Note that it has a progress bar in its top-right corner and a `ListBox` control at the bottom.

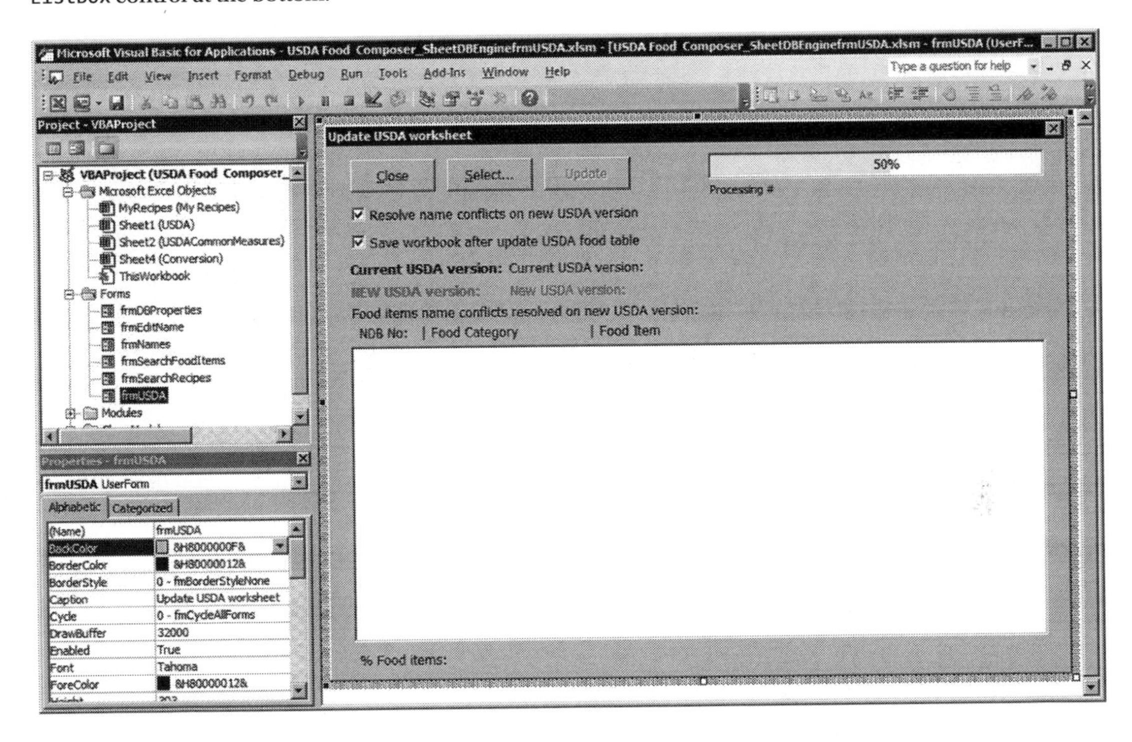

Figure 9-8. *This is the frmUSDA interface in design mode in the Visual Basic IDE*

Figure 9-9 shows how `frmUSDA` appears when you click the Update USDA `ControlButton` of the `My Recipes` worksheet from the `USDA Food Composer_SheetDBEnginefrmUSDA.xlsm` macro-enabled workbook. The `frmUSDA` `UserForm` offers a quite simple interface, with two `CommandButtons` (to close it or select the new SRxx file), two check boxes (to auto-resolve food item name conflicts and auto-save the workbook), and some details of the current USDA worksheet (SR version, total food items, and nutrient count).

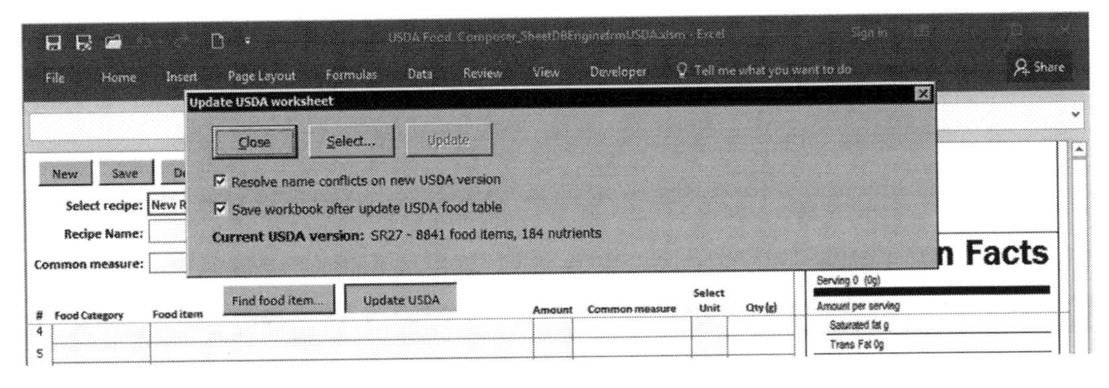

Figure 9-9. *This is frmUSDA from the USDA Food Composer_SheetDBEnginefrmUSDA.xlsm macro-enabled workbook. When you click the Update USDA ControlButton, frmUSDA is loaded and shows the current USDA version with some details of its nutrient database (total food items and nutrient count)*

The frmUSDA interface decreases its height to offer a simple interface when the UserForm_Initialize( ) event fires, executing this code:

```
Option Explicit
...
Dim mintNewVersion As Integer
...
Const mconSizeSmall = 131
Const mconSizeLarge = 393
...
Private Sub UserForm_Initialize()
    Me.Height = mconSizeSmall
    Me.Left = Application.Left + Application.Width / 2 - (Me.Width / 2)
    Me.Top = Application.Top + Application.Height / 2 - (mconSizeLarge) / 2
    Me.lblCurrentVersion.Caption = USDAVersion(Sheets("USDA"), mintCurrentVersion)
    Call UpdateProgressBar
End Sub
```

Note that frmUSDA appears centered inside the Microsoft Excel interface because its StartUpPosition property was set to 0 - Manual (other options are 1 - Center Owner; 2 - Center Screen; 3 - Windows Default), having its Left property being set by subtracting the Excel horizontal center (Application.Left +Application.Width / 2) from half of its width (-(Me.Width / 2)).

```
    Me.Left = Application.Left + Application.Width / 2 - (Me.Width / 2)
```

Its Top property is set by subtracting the Excel vertical center (Application.Top + Application.Height / 2) from half of its expanded height (- (mconSizeLarge) / 2).

```
    Me.Top =(Application.Top + Application.Height / 2 - (mconSizeLarge) / 2
```

■ **Attention** The mconSizeSmall = 132 points were determined by dragging up the UserForm bottom border in design mode to the desired size and inspecting the UserForm Height property in the VBA Properties window.

Showing USDA Worksheet Version Information

After the frmUserForm is correctly sized and positioned on the screen, decreasing the height of the lblCurrentVersion Label control receives the Function USDAVersion( ) return value.

```
    Me.lblCurrentVersion.Caption = USDAVersion(Sheets("USDA"), mintCurrentVersion)
```

This is the code executed by the Function USDAVersion( ) procedure code:

```
Private Function USDAVersion(ws As Worksheet, Optional intVersion As Integer) As String
    Dim wb As Workbook
    Dim rgUSDA As Range
    Dim rgMyRecipes As Range
    Dim strVersion As String
```

```
    Set wb = ws.Parent
    Set rgUSDA = wb.Worksheets("USDA").Range("USDA")
    Set rgMyRecipes = wb.Worksheets("USDA").Range("My_Recipes")
    strVersion = Left(ws.Range("A1"), 4)
    intVersion = Right(strVersion, 2)
    strVersion = strVersion & " - " & rgUSDA.Rows.Count - rgMyRecipes.Rows.Count & " food
    items, "
    strVersion = strVersion & rgUSDA.Columns.Count & " nutrients"
    USDAVersion = strVersion
End Function
```

The Private Function USDAVersion( ) receives two arguments: ws as Worksheet (to represent the desired USDA worksheet) and intVersion (an integer variable received by reference that is used to return the numerical SRxx version value).

It declares the wb as Workbook object variable and uses the ws.Parent property to set a reference to the workbook object where the ws worksheet resides.

```
    Set wb = ws.Parent
```

Now that a reference to the workbook was set, it uses the rgUSDA and rgMyRecieps object variables to set references to the USDA and My_Recipes range names of the ws worksheet.

```
    Set rgUSDA = wb.Worksheets("USDA").Range("USDA")
    Set rgMyRecipes = wb.Worksheets("USDA").Range("My_Recipes")
```

Then it uses those object references with the VBA Left( ) function to extract the first four characters (the SRxx value) from the USDA worksheet cell A1.

```
    strVersion = Left(ws.Range("A1"), 4)
```

The VBA Right( ) function is then used to extract the last two characters from the strVersion value (the integer part of the SRxx value, which has the numerical SR version) and attribute it to the intVersion argument.

```
    intVersion = Right(strVersion, 2)
```

To return how many food items the USDA worksheet associated to the ws object variable has, it subtracts the My_Recipes range rows from the USDA range rows, using the Range.Rows.Count property of each range name, which is concatenated to strVersion.

```
    strVersion = strVersion & " - " & rgUSDA.Rows.Count - rgMyRecipes.Rows.Count & " food
    items, "
```

And using the Range.Columns.Count, it returns how many nutrient columns the USDA range name has, using the strVersion String variable as the procedure return value.

```
    strVersion = strVersion & rgUSDA.Columns.Count & " nutrients"
    USDAVersion = strVersion
End Function
```

■ **Attention** For the current USDA worksheet, the result is shown in Figure 9-9 ("SR27 – 8841 food items, 184 nutrients").

Managing the UserForm Progress Bar

The progress bar you see in frmUSDA design mode in Figure 9-8 becomes invisible by calling Private Sub UpdateProgressBar(), without passing any arguments to this procedure as the last UserrForm_Initialize event instruction.

```
    Call UpdateProgressBar
End Sub
```

The frmUSDA progress bar is composed of three Label controls: lblTotal, lblPercent, and lblValue (Figure 9-10).

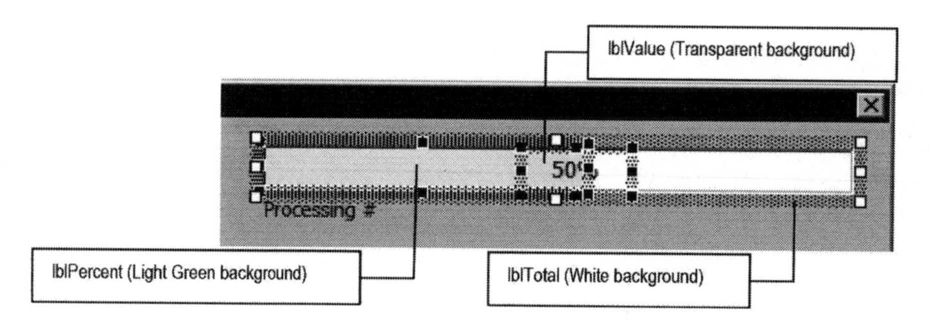

Figure 9-10. *These are the three Label controls used to compose the frmUSDA progress bar*

It works quite the same way as Sub UpdateStatusBar( ) from basUSDA. It calculates a percentage accomplished from any process and sets the lblPercent.Width property (with a light green background) as a percentage of the lblTotal.Width property, giving the illusion that the bar grows as the process progress. The Private Sub UpdateProgressBar( ) that manages it may receive up to three optional arguments (bolShow as Boolean, intValue as Integer, and intTotal as Integer = 1) and executes this code:

```
Private Sub UpdateProgressBar(Optional bolShow As Boolean,
                Optional sngValue As Integer, _
                Optional sngTotal As Integer = 1)
    Dim sngPercent As Single

    Me.lblTotal.Visible = bolShow
    Me.lblPercent.Visible = bolShow
    Me.lblValue.Visible = bolShow
    Me.lblProcessing.Visible = bolShow
    sngPercent = sngValue / sngTotal
    Me.lblPercent.Width = (Me.lblTotal.Width - 2) * sngPercent
    Me.lblValue.Caption = Format(sngPercent, "0.0%")
End Sub
```

Note that when Sub UpdateProgressBar( ) is called with no arguments, bolShow = False (default Boolean value), and all three Label controls become invisible, hiding the progress bar. Also note that both sngValue and sngTotal are single arguments, meaning that they can receive any real number.

Since lblPercent (light green background) must grow inside lblTotal (white background), to guarantee that lblPercent will not surpass the lblTotal right border, the code subtracts two points from lblTotal.Width to correctly set lblPercent.Width.

To make the control appear and draw correctly, its first argument must be True, as follows (supposing you need to reflect the percent accomplished on the 20th step of a total 1,250 steps):

```
Call UpdateProgressBar(True, 20, 1250)
```

Selecting the External SRxx Update Version

After frmUSDA has been loaded showing the current USDA SR version details (Figure 9-9), click the Select CommandButton to fire the cmdSelect_Click( ) event and select the desired SRxx_FoodItemsPerFirstCommonMeasure.xls file (Figure 9-11).

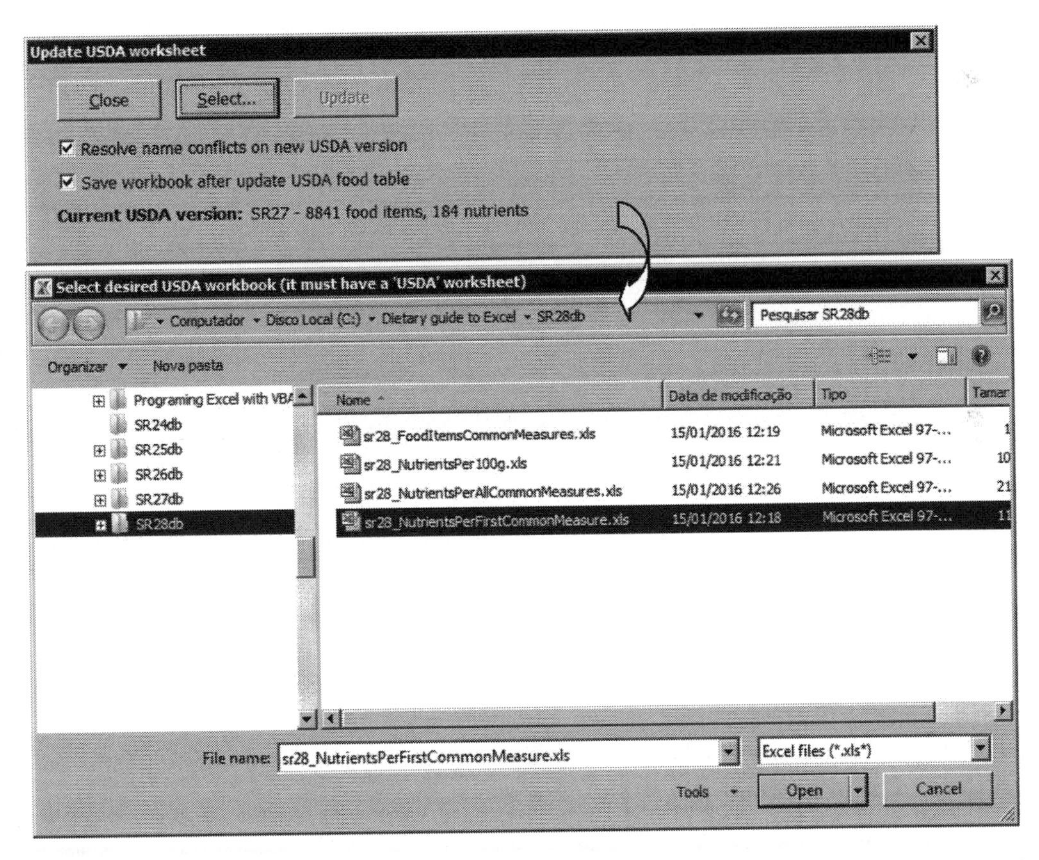

Figure 9-11. Click the frmUSDA Select CommandButton to select the desired SRxx_FoodItemsPerFirstCommonMeasure.xls file. In this figure, the current SR27 version will be updated to SR28

And once the `SRxx...xls` update file is selected, supposing that the "Resolve name conflicts on new USDA version" check box is selected, the code immediately begins to process it, showing the selected USDA version and its details in blue (food items and nutrient count) as an indication that a upgrade is in process, while the progress bar indicates how many food items have already been checked for name inconsistencies (food item name changes between versions). Note that the `cmdSelect CommandButton` control changes its `Caption` property to Cancel, allowing you to cancel the operation in progress (Figure 9-12).

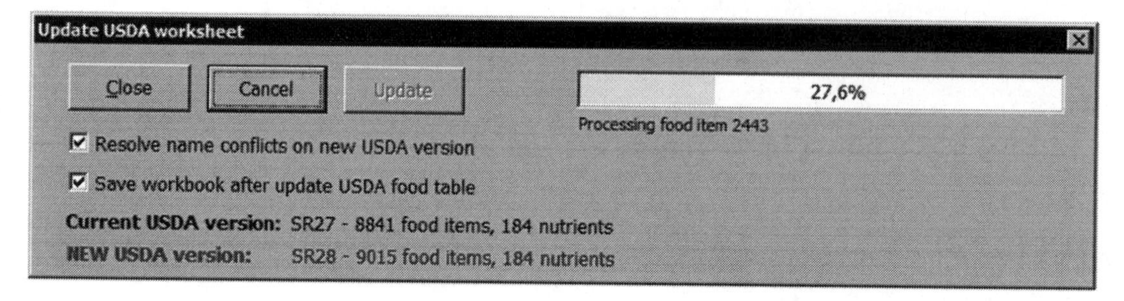

Figure 9-12. *After selecting the desired SRVersion, if chkResolveNames is selected ("Resolve name conflicts on new USDA version" check box), frmUSDA will immediately begin to process the new file, searching for food item name changes between SR versions. The progress bar indicates the process state*

When the name inconsistencies process ends, the `frmUSDA` interface expands to show the `lstFoodITems ListBox` control with all food items—and its food categories—whose names have changed between the two SR versions, and `cmdUpdate CommandButton` will become enabled, allowing you to update the current `USDA` worksheet to the desired SR version (Figure 9-13).

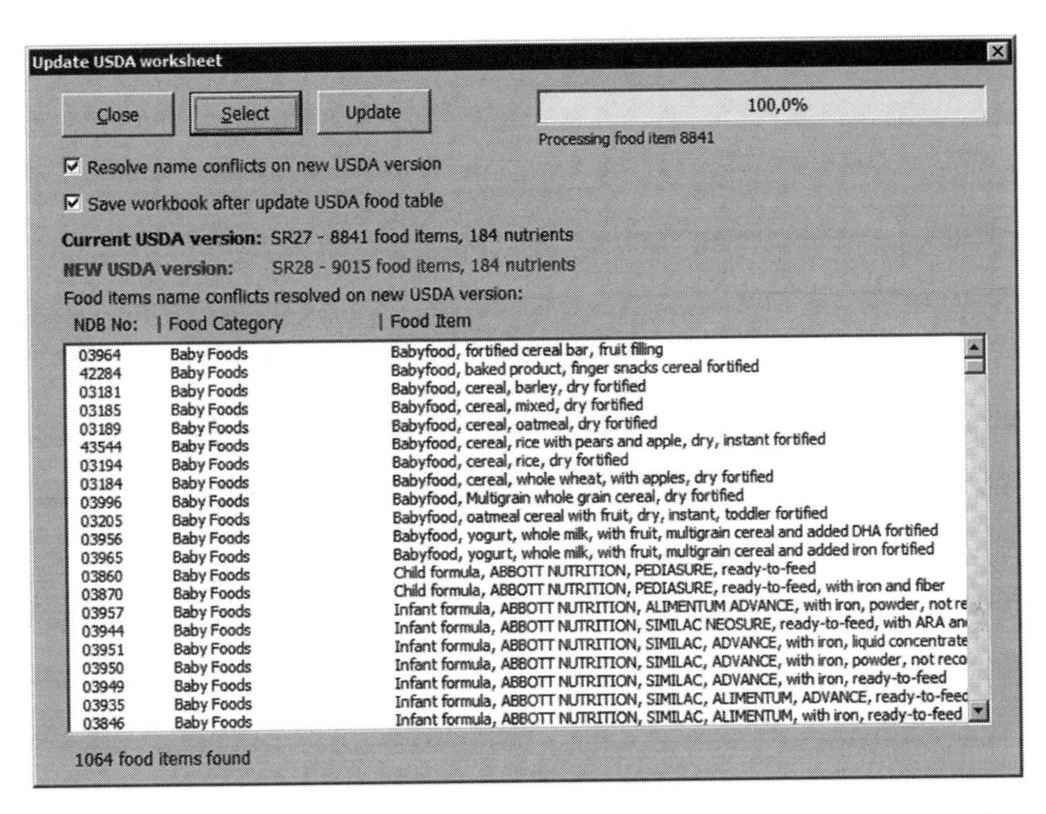

Figure 9-13. *When frmUSDA ends the food item name checking, it expands the interface to show the lstFoodItems ListBox control with all food item names, food categories, and which names changed between the two SR versions (1,064 food items found)*

■ **Attention** Note in Figure 9-13 that after an `SRxx...xls` file is selected, `cmdSelect` is still activated, allowing you to select another file.

To allow select the external USDA workbook with the SRxx version that must be used on the updated `cmdSelect_Click( )` event, execute this code:

```
Private Sub cmdSelect_Click()
    If Me.cmdSelect.Caption = "Cancel" Then
        mbolCancel = True
    Else
        Me.cmdUpdate.Enabled = False
        Call CloseUSDAWorkbook
        Call ShowCtls(False)
        If OpenUSDAWorkbook(mstrFile) Then
            Me.lblNewVersion.Caption = USDAVersion(mwsUSDA, mintNewVersion)
            Me.lblNew.ForeColor = IIf(mintNewVersion < mintCurrentVersion, mconRed,
            mconBlue)
            Me.lblNewVersion.ForeColor = IIf(mintNewVersion < mintCurrentVersion, mconRed,
            mconBlue)
```

```
            Me.lblNew.Visible = True
            Me.lblNewVersion.Visible = True
            If Me.chkResolveNames Then
                If ResolveNames() Then
                    Call ShowCtls(True)
                End If
            End If
            Me.cmdUpdate.Enabled = True
        End If
    End If
End Sub
```

The cmdSelects_Click( ) event code begins checking the cmdSelect.Caption property value. If it is Cancel, the mbolCancel module-level variable will be set to true—allowing the cancellation of the Private Sub ResolveNames( ) procedure that checks name inconsistencies between the two SR versions, as you will see later in this chapter. Otherwise, it first calls Private Sub CloseUSDAWorkbook( ) to close any open SRxx...xls workbook before opening a new one.

```
    If Me.cmdSelect.Caption = "Cancel" Then
        mbolCancel = True
    Else
        Me.cmdUpdate.Enabled = False
        Call CloseUSDAWorkbook

Private Sub CloseUSDAWorkbook()
    Dim strName As String

    On Error Resume Next
    strName = mwbUSDA.Name
    If Err = 0 Then
        mwbUSDA.Close False
        Set mwbUSDA = Nothing
    End If
    Call UpdateProgressBar
End Sub
```

Sub CloseUSDAWorkbook( ) disables VBA errors with an On Error Resume Next statement and tries to get the current SR external Workbook.Name property associated to the mwbUSDA module-level variable. If no error is generated (Err=0), this is an indication that the workbook is open, so it calls the Workbook.Close False method to close it without saving and hides the progress bar controls calling Sub UpdateProgressBar( ) with no arguments.

And once it is guaranteed that mwbUSDA points to nothing (no external workbook opened), the procedure calls Sub ShowCtls(False) to shrink frmUSDA and hide all controls that have their Tag property set to -1, executing a For...Next loop through the UserForm.Controls collection.

```
Private Sub ShowCtls(bolShow As Boolean)
    Dim intI As Integer

    Me.Height = IIf(bolShow, mconSizeLarge, mconSizeSmall)
    For intI = 0 To Me.Controls.Count - 1
        If Me.Controls(intI).Tag = "-1" Then
```

```
            Me.Controls(intI).Visible = bolShow
        End If
    Next
End Sub
```

When SubShowCtls( ) and code control and cmdSelect_Click calls the Private Sub OpenUSDAWorkbook(strFile) procedure to allow the selection of the desired SRxx...xls version's external workbook.

```
        If OpenUSDAWorkbook(mstrFile) Then
```

Opening the Desired SRxx...xls Version

The Private Sub OpenUSDAWorkbook( ) procedure, called from the cmdSelect_Click( ) event, is responsible for initializing the module-level variables mwbUSDA and mwsUSDA, which will point to the external SR version workbook and external USDA worksheet, respectively, executing this code:

```
Private Function OpenUSDAWorkbook(Optional USDAFile As String) As Boolean
    Dim ws As Worksheet
    Dim strFile As String
    Dim strFilter As String
    Dim strTitle As String
    Dim strMsg As String
    Dim bolFound As Boolean

    strFilter = "Excel files, *.xls*, All Files, *.*"
    strTitle = "Select desired USDA workbook (it must have a 'USDA' worksheet)"
    'Get workbook file name
    strFile = ShowDialogBox(OpenFile, , strTitle, , strFilter, 1)

    If Len(strFile) Then
        If strFile = ThisWorkbook.FullName Then
            MsgBox "Can´t use current workbook as source for itself", _
                   vbCritical, _
                   "You've selected the current Workbook!"
            Exit Function
        ElseIf IsWorkbookOpen(strFile) Then
            strMsg = "The selected file is already opened:" & vbCrLf
            strMsg = strMsg & strFile & vbCrLf
            strMsg = strMsg & "Close it and try again!"
            MsgBox strMsg, vbCritical, "Invalid Workbook!"
            Exit Function
        End If

        'Open USDA workbook
        Set mwbUSDA = Application.Workbooks.Open(strFile)

        'Verify if opened workbook has a "USDA" sheet tab
        For Each ws In mwbUSDA.Worksheets
            If (ws.Name = "USDA") And (Left(ws.Range("A1"), 2) = "SR") Then
                Set mwsUSDA = ws
```

```
                bolFound = True
                Exit For
            End If
        Next

        If bolFound Then
            USDAFile = strFile
            OpenUSDAWorkbook = True
            'Return the focus to current workbook
            ThisWorkbook.Activate
        Else
            MsgBox "The selected workbook doesn't has a 'USDA' worksheet to update food
            items information", _
                vbInformation, _
                "'USDA' sheet tab not found!"
            mwbUSDA.Close
            Set mwbUSDA = Nothing
        End If
    End If
End Function
```

Sub OpenUSDAWorkboook( ) begins by calling our old friend the ShowDialogBox( ) procedure (presented in Chapter 3). If the Open Dialog is canceled, the strFile variable will be an empty string, ending the procedure, which will also end the cmdSelect_Click( ) event and do nothing.

```
Private Function OpenUSDAWorkbook(Optional USDAFile As String) As Boolean
    ...
    strFilter = "Excel files, *.xls*, All Files, *.*"
    strTitle = "Select desired USDA workbook (it must have a 'USDA' worksheet)"
    'Get workbook file name
    strFile = ShowDialogBox(OpenFile, , strTitle, , strFilter, 1)

    If Len(strFile) Then
    ...
    End If
End Function
```

But if a file is selected, you must first validate it to verify whether it has the correct SR update. The first check verifies whether the selected file is the same application that you are trying to update, comparing strFile with the ThisWorkbook.FullName property.

```
If strFile = ThisWorkbook.FullName Then
    MsgBox "Can´t use current workbook as source for itself", _
           vbCritical, _
           "You've selected the current Workbook!"
    Exit Function
```

Being another workbook, the second check verifies whether the selected file is already opened, by calling the Private Function IsWorkbookOpen( ) procedure. If it returns True (meaning the workbook is opened), a MsgBox( ) will warn the user, and the procedure will end doing nothing.

```
ElseIf IsWorkbookOpen(strFile) Then
    strMsg = "The selected file is already opened:" & vbCrLf
    strMsg = strMsg & strFile & vbCrLf
    strMsg = strMsg & "Close it and try again!"
    MsgBox strMsg, vbCritical, "Invalid Workbook!"
    Exit Function
End If
```

Verify Whether a Workbook Is Open

The simplest way to verify whether a file is opened is to try opening it with exclusive file access, which can be made using the VBA file access Open statement, which has this syntax:

```
Open pathname For mode [Access] [lock] As [#]filenumber [Len=reclength]
```

In this code:

`pathname`: This is required; it is a string expression that specifies a file name, which may include its folder and drive.

`mode`: This is required; it is a keyword specifying the file mode: `Append`, `Binary`, `Input`, `Output`, or `Random`. If unspecified, the file is opened for `Random` access.

`access`:: This is optional; it is a keyword specifying the operations permitted on the open file: `Read`, `Write`, or `Read Write`.

`lock`: This is optional; it is a keyword specifying the operations restricted on the open file by other processes: `Shared`, `Lock Read`, `Lock Write`, and `Lock Read Write`.

`filenumber` : This is required; it is a valid file number in the range 1 to 511, inclusive. Use the VBA `FreeFile` function to obtain the next available file number.

`reclength`: This is optional; it is a number less than or equal to 32,767 (bytes). For files opened for random access, this value is the record length. For sequential files, this value is the number of characters buffered.

To close a file opened with the VBA `Open` statement, you must use the VBA `Close` statement, which has this syntax:

```
Close [filenumberlist]
```

In this code:

`[[#]filenumber] [, [#]filenumber]...`: This is optional; it can be one or more file numbers opened by the `Open` statement. If it's omitted, all active files are closed, and the association of a file with its file number ends.

To open a file with the VBA integer function, you reference the file with a number between 1 and 511 (that can be returned with VBA `FreeFile( )` function), and to close it, you use the VBA `Close` statement, optionally referencing the desired file number, as follows:

```
intFile = FreFile
Open strFile For Bynare Access Write Lock Read As #intFile
Close #intFile
```

The Private Function IsWorkbookOpen( ) verifies whether a workbook is opened in this way.

```
Private Function IsWorkbookOpen(strFile As String) As Boolean
    Dim intFreeFile As Long

    On Error Resume Next

    'Try to open the workbook with exclusive access
    intFreeFile = FreeFile
    Open strFile For Input Lock Read As #lintFreeFile
    Close #lintFreeFile
    IsWorkbookOpen = (Err > 0)
End Function
```

The procedure begins using On Error Resume Next to disable VBA errors and uses the VBA FreeFile( ) function to get the next available file number to open the file (yes, you can open up to 511 files using the VBA Open statement), attributing it to the intFreeFile Integer variable, which returns the next integer available (between 1 and 511) to represent the file. It then tries to open strFile using Input Lock Read access mode. If the file is already opened, VBA will not be able to lock the file generating an error (otherwise the file will be opened, so the next statement tries to close the file with the VBA Close statement). IsWorkbookOpen( ) returns the logical test Err>0, which will be true if an error is generated to try to lock an already open file.

Verify Whether the Workbook has an USDA Worksheet

If the selected strFile is still not opened, it will be opened by using the Application.Workbooks.Open method, which will set a reference to the mwbUSDA module-level variable.

```
'Open USDA workbook
Set mwbUSDA = Application.Workbooks.Open(strFile)
```

To verify whether the opened workbook has a valid USDA worksheet, it executes a For Each...Next loop through all workbook worksheets searching for the one whose Name property is USDA and whose cell A1 value begins with the SR characters. If such a worksheet is found, it sets bolFound = True and exits the loop.

```
For Each ws In mwbUSDA.Worksheets
    If (ws.Name = "USDA") And (Left(ws.Range("A1"), 2) = "SR") Then
        Set mwsUSDA = ws
        bolFound = True
        Exit For
    End If
Next
```

Having found a valid USDA worksheet, it sets strFile to the USDAFile procedure argument, and Function OpenUSDAWorkbook( ) ends returning True while returning the focus to the Thisworkbook.

```
If bolFound Then
    USDAFile = strFile
    OpenUSDAWorkbook = True
    'Return the focus to current workbook
    ThisWorkbook.Activate
```

Otherwise, the workbook is invalid, and it sends a VBA MsgBox( ) warning indicating that the file doesn't have a USDA worksheet to operate.

```
        Else
            MsgBox "The selected workbook doesn't has a 'USDA' worksheet to update food
            items information", _
                vbInformation, _
                "'USDA' sheet tab not found!"
            mwbUSDA..Close
            Set mwbUSDA = Nothing
        End If
    End If
End Function
```

Returning New USDA SR Version Details

Let's get back to the cmdSelect_Click( ) event after the desired SRxx...xls file was opened with OpenUSDAWorkbook( ). The opened file was stored in the mstrFile String module-level variable (passed by reference as a procedure argument), and the frmUSDA interface is updated again, using the USDAVersion(mintVersion) procedure to return information about the external, opened USDA worksheet (SR version, food items, and nutrient count), which are used to define lblNew and lblNewVersion.Caption properties. Note that both lblNew and lblNewVersion.ForeColor properties (text color) will be set to mconRed or mconBlue according to the new SRxx version being downgraded or upgraded, respectively (which is determined using the mintNewVersion < mintCurrentVersion expression).

```
Private Sub cmdSelect_Click()
    ...
        If OpenUSDAWorkbook(mstrFile) Then
            Me.lblNewVersion.Caption = USDAVersion(mwsUSDA, mintNewVersion)
            Me.lblNew.ForeColor = IIf(mintNewVersion < mintCurrentVersion, mconRed,
            mconBlue)
            Me.lblNewVersion.ForeColor = IIf(mintNewVersion < mintCurrentVersion, mconRed,
            mconBlue)
            Me.lblNew.Visible = True
            Me.lblNewVersion.Visible = True
```

Once information about the new selected SR version is displayed, if the chkResolveNames check box is selected (resolving name conflicts in the new USDA version), it calls the Private Function ResolveNames( ) procedure to check for food item name changes between the two SR versions. If ResolveNames( ) returns True, as an indication that name changes were found, it will call ShowCtls(True) to show the lstFoodItems ListBox with the food item whose names need to be updated. To allow updating of the current USDA worksheet, cmdUpdate becomes enabled.

```
            If Me.chkResolveNames Then
                If ResolveNames() Then
                    Call ShowCtls(True)
                End If
            End If
            Me.cmdUpdate.Enabled = True
        End If
    End If
End Sub
```

Finding Food Item Name Inconsistencies Between USDA Versions

To search and solve any food item name inconsistencies between the SR files, it is necessary to get each current USDA worksheet's food item, search it with NDB_No (the food item primary key) on the new USDA worksheet, and compare both food item names. If they differ, the food item name on the new USDA worksheet must be updated to the current USDA food item, which will avoid the calculation problem shown in Figure 9-7, earlier in this chapter, after the updated process is completed.

This operation is made by Private Function ResolveNames( ), called from cmdSelect_Click, which executes this code:

```
Private Function ResolveNames() As Boolean
    Dim wsUSDA As Worksheet
    Dim rgUSDA As Range
    Dim rgUSDANew As Range
    Dim rgItem As Range
    Dim rgNew As Range
    Dim intTotal As Integer
    Dim intI As Integer

    Set wsUSDA = Worksheets("USDA")
    Set rgUSDA = Range("USDA")
    'Exclude My_Recipes range from rgUSDA
    intTotal = rgUSDA.Rows.Count - Range("My_Recipes").Rows.Count
    Set rgUSDA = Range(wsUSDA.Cells(rgUSDA.Row, 1), wsUSDA.Cells(rgUSDA.Row + intTotal
    - 1, 1))
    Set rgUSDANew = mwsUSDA.Range("USDA")
    Set rgUSDANew = mwsUSDA.Range(mwsUSDA.Cells(rgUSDANew.Row, 1), mwsUSDA.Cells(rgUSDANew.
    Row + _
                    rgUSDANew.Rows.Count - 1, 1))
    Me.lblProcessing.Visible = True
    Me.cmdSelect.Caption = "Cancel"
    Application.ScreenUpdating = False
        For Each rgItem In rgUSDA
            'Allow loop cancellation
            DoEvents
            If mbolCancel Then
                mbolCancel = False
                If MsgBox("Do you want to cancel USDA Food items names update process?", _
                          vbQuestion + vbYesNo + vbDefaultButton2, _
                          "Cancel USDA update?") = vbYes Then
                    Me.cmdSelect.Caption = "Select"
                    Call UpdateProgressBar(False)
                    Me.lstFoodItems.Clear
                    Exit Function
                End If
            End If

            intI = intI + 1
            Me.lblProcessing.Caption = "Processing food item " & intI
            Set rgNew = rgUSDANew.Find(rgItem, , , xlWhole)
            If Not rgNew Is Nothing Then
```

```
                If rgNew.Offset(0, 1) <> rgItem.Offset(0, 1) Then
                    Me.lstFoodItems.AddItem rgNew
                    Me.lstFoodItems.Column(1, Me.lstFoodItems.ListCount - 1) = rgNew.
                   Offset(0, 2)
                    Me.lstFoodItems.Column(2, Me.lstFoodItems.ListCount - 1) = rgNew.
                    Offset(0, 1)
                    Me.lblFoodItems.Caption = Me.lstFoodItems.ListCount & " food items
                    found"
                    rgNew.Offset(0, 1) = rgItem.Offset(0, 1)
                End If
            End If
            Call UpdateProgressBar(True, intI, intTotal)
        Next
    Application.ScreenUpdating = True
    Me.cmdSelect.Caption = "Select"
    ResolveNames = True
End Function
```

Any USDA worksheet SR version produced with the USDA Food List Creator.accbd Microsoft Access application has in column A the food item NDB_No identification—an integer value that is the USDA food item primary key—and on column B, the food item name, which is used to select food items on all data validation lists of the USDA Food Composer...xlsm worksheet applications presented in this book. To allow such food item selection and calculate its nutrient data (using Excel Vlookup( ) functions), the USDA range name from the USDA worksheet begins in Column B, going through all other 184 nutrient columns of the SR27 and SR28 USDA versions.

So, the first ResolveNames( ) operation is to build two range names (rgUSDA and rgUSDANew, for current and new USDA worksheets) using just the first worksheet column and all its food items, excluding all My_Recipes range name food items from both worksheets (if any), using the worksheet Cells( ) collection to point to the desired range in column A.

```
Private Function ResolveNames() As Boolean
    ...
    Set wsUSDA = Worksheets("USDA")
    Set rgUSDA = Range("USDA")
    'Exclude My_Recipes range from rgUSDA
    intTotal = rgUSDA.Rows.Count - Range("My_Recipes").Rows.Count
    Set rgUSDA = Range(wsUSDA.Cells(rgUSDA.Row, 1), wsUSDA.Cells(rgUSDA.Row + intTotal
    - 1, 1))
    Set rgUSDANew = mwsUSDA.Range("USDA")
    Set rgUSDANew = mwsUSDA.Range(mwsUSDA.Cells(rgUSDANew.Row, 1), mwsUSDA.Cells(rgUSDANew.
    Row + _
            rgUSDANew.Rows.Count - 1, 1))
```

Now that both rgUSDA and rgUSDANew point to all NDB_No column data in both USDA worksheets, the stage is set: lblProcessing (the small Label control under the progress bar) is turned visible, cmdSelect.Caption exhibits Cancel to allow the loop cancelation, screen updates are disabled, and a For Each...Next loop begins on all rgUSDA column A cells.

```
    Me.lblProcessing.Visible = True
    Me.cmdSelect.Caption = "Cancel"
    Application.ScreenUpdating = False
        For Each rgItem In rgUSDA
```

The first loop operation uses a DoEvents statement to allow clicking cmdSelect (that now show Cancel) to cancel the loop. If this is made, the mbolCancel module-level variable will be set to true on the cmdSelect_Click( ) event, and a MsgBox( ) warning will ask the user to confirm the cancellation. By canceling the process, cmdSelect.Caption is updated, the progress bar is hidden (UpdateProgressBar(False)), lstFoodItems is cleared, and the procedure ends, allowing you to update the USDA worksheet partially according to food item name changes (all food items already processed will be updated).

```
'Allow loop cancellation
DoEvents
If mbolCancel Then
    mbolCancel = False
    If MsgBox("Do you want to cancel USDA Food items names update process?", _
              vbQuestion + vbYesNo + vbDefaultButton2, _
              "Cancel USDA update?") = vbYes Then
        Me.cmdSelect.Caption = "Select"
        Call UpdateProgressBar(False)
        Me.lstFoodItems.Clear
        Exit Function
    End If
End If
```

■ **Attention** Every VBA loop that has a DoEvents function takes longer to complete because VBA needs to look to see whether there are any events to process in the events queue before executing the next instruction.

As the loop continues to execute, the intI Integer variable and lblProcessing Label control are updated, and the current rgItem cell value, which points to an NDB_No cell on the current USDA worksheet, is searched on the new USDA worksheet using the Range.Find method.

```
For Each rgItem In rgUSDA
    ...
    intI = intI + 1
    Me.lblProcessing.Caption = "Processing food item " & intI
    Set rgNew = rgUSDANew.Find(rgItem, , , xlWhole)
    If Not rgNew Is Nothing Then
```

If the rgItem value is found in the new USDA worksheet (Not rgItem is Nothing), both worksheets have the same NDB_No food item and the procedure checks whether their names differ. If this is also true, the food item name and category are added to the lstFoodItems ListBox, and the food item name in the new USDA worksheet is updated to the food item name of the current USDA worksheet, using the Range.Offset() method. A call is made to UpdateProgressBar(True, intI, intTotal) to update the UserForm progress bar.

```
    If Not rgNew Is Nothing Then
        If rgNew.Offset(0, 1) <> rgItem.Offset(0, 1) Then
            Me.lstFoodItems.AddItem rgNew
            Me.lstFoodItems.Column(1, Me.lstFoodItems.ListCount - 1) = rgNew.Offset(0, 2)
            Me.lstFoodItems.Column(2, Me.lstFoodItems.ListCount - 1) = rgNew.Offset(0, 1)
```

```
            Me.lblFoodItems.Caption = Me.lstFoodItems.ListCount & " food items found"
            rgNew.Offset(0, 1) = rgItem.Offset(0, 1)
        End If
    End If
    Call UpdateProgressBar(True, intI, intTotal)
Next
```

When the For Each... Loop ends, the code returns to the cmdSelect_Click( ) event, which will call ShowCtls(True) to expand the frmUSDA interface and show all food item names changed, allowing you to update the current USDA worksheet to the selected SR version, as shown in Figure 9-13.

Updating the USDA Worksheet

Once all food items have been processed, the cmdUpdate becomes enabled, ready to process the cmdUSDA_Click( ) event to perform the USDA worksheet updating, which executes almost the same code made by the UpdateUSDA( ) procedure from the basUSDA standard module and deserves no more consideration (except that it uses object module-level variables to reference the external USDA worksheet and verify the chkSave CheckBox control state before saving the workbook).

```
Private Sub cmdUpdate_Click()
    Dim wb As Workbook
    Dim wbUSDA As Workbook
    Dim ws As Worksheet
    Dim wsUSDA As Worksheet
    Dim rgMyRecipes As Range
    Dim rgUSDA As Range
    Dim nm As Name
    Dim strNewSRVersion As String
    Dim strFileUSDA As String
    Dim strFileCommonMsrs As String
    Dim strMsg As String
    Dim lngRows As Long

    Call SetScreenEventsRecalc(False)
        Call UpdateProgressBar(True, 1, 3)
        Set wb = ThisWorkbook
        Set ws = Worksheets("USDA")

        'Copy and paste current My_recipes to new USDA worksheet
        ws.Range("My_Recipes").CurrentRegion.Copy
        mwsUSDA.Range("My_Recipes").PasteSpecial xlPasteValues

        'Rebuild "My_Recipes" and "USDA" range names on new USDA Worksheet
        Set rgMyRecipes = mwsUSDA.Range("My_Recipes").CurrentRegion
        mwsUSDA.Range("My_Recipes").Resize(rgMyRecipes.Rows.Count + 1, 1).Name = "My_
        Recipes"

        Set rgUSDA = mwsUSDA.Range("USDA")
        lngRows = rgUSDA.Rows.Count + rgMyRecipes.Rows.Count - 1
        rgUSDA.Resize(lngRows, rgUSDA.Columns.Count).Name = "USDA"
```

```
        'Silently delete worksheets from this workbook
        Application.DisplayAlerts = False
            'Make worksheet visible before delete it to avoid Excel bug when save workbook
            mwsUSDA.Visible = True
            wb.Worksheets("USDA").Delete
            'Search and delete invalid range names from this workbook
            For Each nm In wb.Names
                If InStr(nm.RefersTo, "#REF!") > 0 Then
                    nm.Delete
                End If
            Next

            'Move new USDA SR worksheet before USDACommonMeasures and hide it
            mwbUSDA.Worksheets("USDA").Move wb.Worksheets("USDACommonMeasures")
            wb.Worksheets("USDA").Visible = False

            strMsg = "USDA worksheet updated to Version SR" & mintNewVersion & vbCrLf
            strMsg = strMsg & "from file '" & Mid(mstrFile, InStrRev(mstrFile, "\") + 1) &
             "';"
            strMsg = strMsg & vbCrLf & vbCrLf
            Call UpdateProgressBar(True, 2, 3)

            'Now try to update USDACommonMeasures worksheet: search it on same path
            strFileCommonMsrs = Dir(Left(mstrFile, InStrRev(mstrFile, "\")) & "SR??_
            FoodItemsCommonMeasures.xls")
            If Len(strFileCommonMsrs) = 0 Then
                'SR??_FoodItemCommonMeasures not found. Ask to select it!
                strFileCommonMsrs = Application.GetOpenFilename("USDA workbooks (*.xls),
                SR*.xls", , _
                  "Select USDACommonMeasures SR workbook", , False)
            End If

            If InStr(1, strFileCommonMsrs, "FoodItemsCommonMeasures") Then
                'SRxx_USDACommonMeasures.xls found or selected. Update USDACommonMeasures!
                wb.Worksheets("USDACommonMeasures").Visible = True
                wb.Worksheets("USDACommonMeasures").Delete
                Set mwbUSDA = Application.Workbooks.Open(strFileCommonMsrs, False)
                mwbUSDA.Worksheets("USDACommonMeasures").Move , wb.Worksheets("USDA")
                wb.Worksheets("USDACommonMeasures").Visible = False

                strFileCommonMsrs = Mid(strFileCommonMsrs, InStrRev(strFileCommonMsrs, "\")
                + 1)
                strMsg = strMsg & "USDACommonMeasures worksheet updated to Version " & _
                       Left(strFileCommonMsrs, 4) & vbCrLf
                strMsg = strMsg & "from file '" & strFileCommonMsrs & "'."
            End If
            Call UpdateProgressBar(True, 3, 3)
        Application.DisplayAlerts = True
    SetScreenEventsRecalc (True)
```

```
    MsgBox strMsg, vbInformation, "Update successfuly to Version " & strNewSRVersion
    If Me.chkSave Then
        ThisWorkbook.Save
    End If
    Unload Me
End Sub
```

■ **Attention** Try to run frmUSDA to update the USDA Food Composer_SheetDBEnginefrmUSDA.xlsm macro-enabled workbook from SR27 to SR28 and note that now the food item name change problem pointed at by Figure 9-7 doesn't happen anymore. Anyway, all recipes whose food item names selected to compose recipes changed between USDA table SR27 and SR28 versions must be manually changed.

Managing Worksheet Application Data

Let's suppose for a moment that the USDA Food Composer_SheetDBEnginefrmUSDA.xslm worksheet application becomes a huge success among users who need to generate Nutrition Facts food labels and nutrient profiles for the many recipes and food products. Chances are that they will try to do the following:

- Create copies of the My Recipes worksheet inside the workbook to better manage its recipes
- Update each recipe nutrient data saved in the My_Recipes range using the new USDA worksheet
- Copy recipes between two My Recipes worksheets on the same or different workbooks
- Delete some or all recipes from any My Recipes sheet tab

The USDA Food Composer_SheetDBEngineManageAutomation.xlsm macro-enabled workbook has the frmManageRecipes UserForm that allows selection of the operation to be performed, the workbook and worksheet target, and the recipes that will be affected by the selected operation (Figure 9-14).

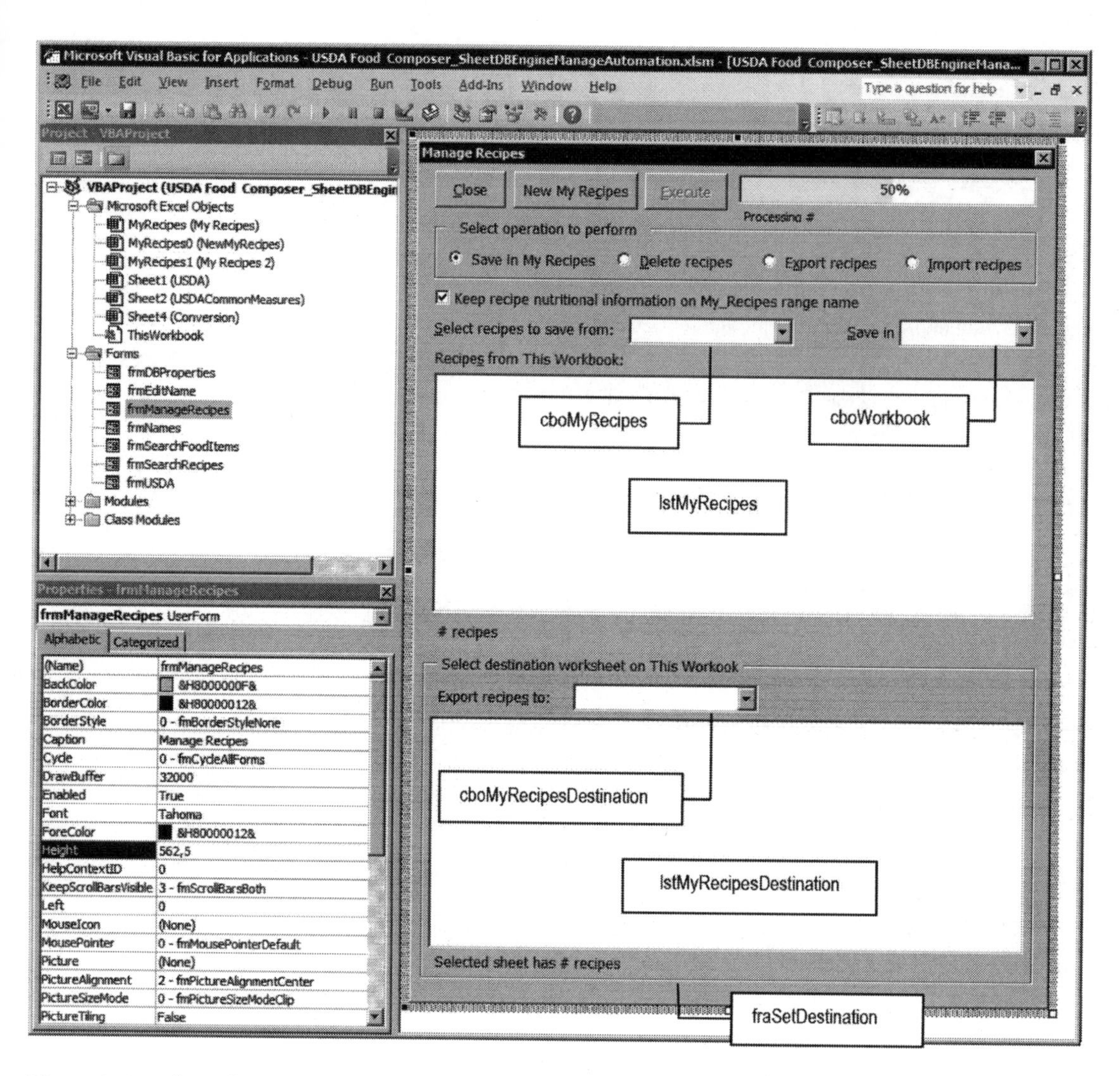

Figure 9-14. *This is frmManageRecipes from USDA Food Composer_SheetDBEngineManageAutomation.xlsm that allows you to perform multiple operations on the My_Recipes worksheet using VBA automation*

■ **Attention** The `frmManageRecipes UserForm` interface was built while keeping in mind that the `fraSetDestination Frame` control will be displaced, hiding `cboMyRecipes`, `cboWorkbook`, and `lstMyRecipes` whenever necessary to select the destination worksheet while importing or exporting recipes.

When you click the Manage `ControlButton` of the `My Recipes` worksheet to show `frmManageRecipes` in a reduced view, allowing you to select any of all recipes stored on the active sheet (Figure 9-15).

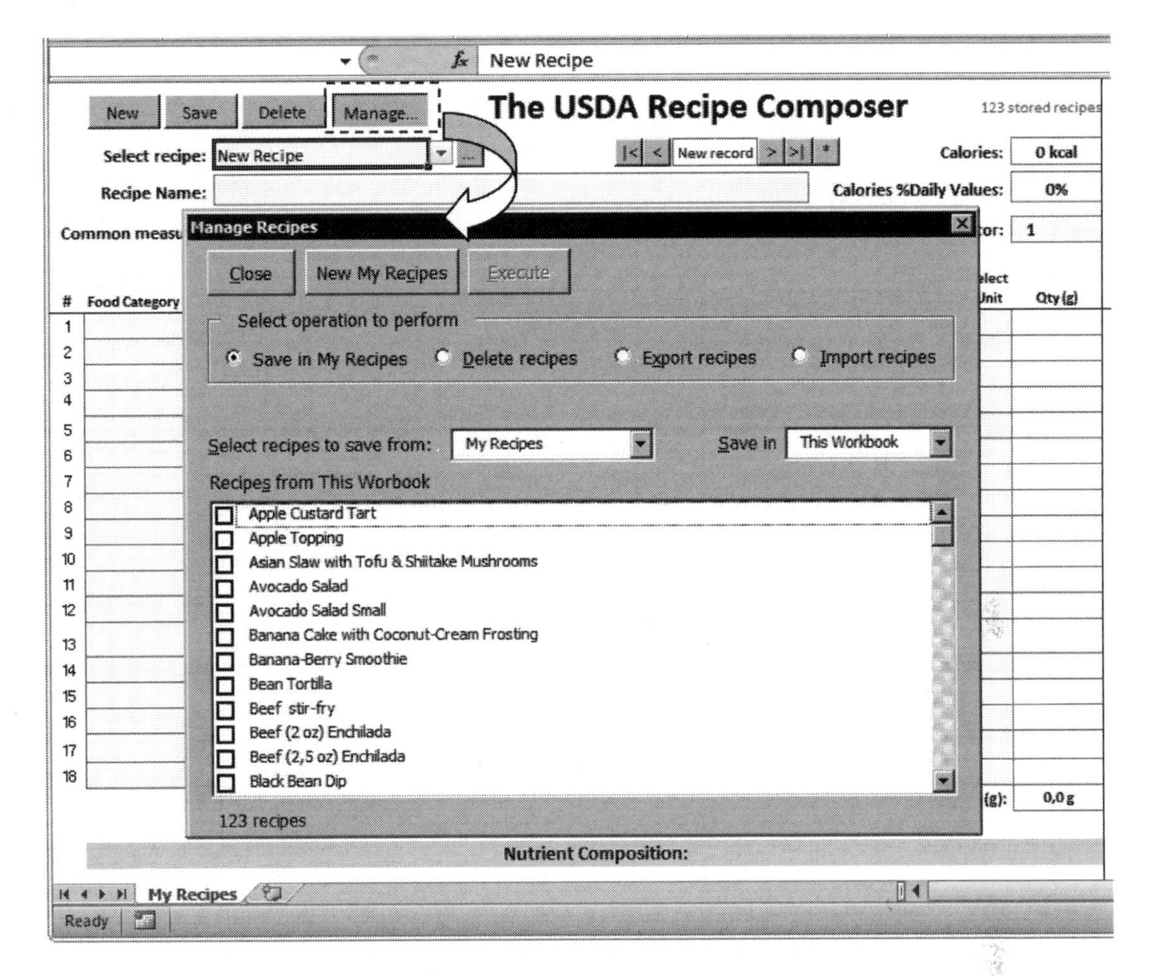

Figure 9-15. This is how the frmManageRecipes UserForm appears to the application user, allowing you to perform different operations on any number of selected recipes

The frmManageRecipes_Initialize( ) Event

When you click the Manage `ControlButton`, `Public Function Manage( )` from the `MyRecipes` worksheet module is executed, calling `SetDatabase( )` to guarantee that the `SheetDBEngine` class is running and allowing you to automate its database services before `frmManageRecipes` is loaded into memory.

```
Public Function Manage()
    Call SetDatabase
    frmManageRecipes.Show
End Function
```

This will fire the `frmManageRecipes UserForm_Initialize( )` event, which will execute this code:

```
Private Sub UserForm_Initialize()
    Dim rg As Range
    Const conInitialHeight = 329
```

```
    Me.Height = conInitialHeight
    mintWorkbookHasMyRecipes = MyRecipesCount()
    Call LoadcboMyRecipes
    Me.cboWorkbook.AddItem "This Workbook"
    Me.cboWorkbook.AddItem "External Workbook"
    Me.cboWorkbook.ListIndex = 0
End Sub
```

Although the UserForm_Initialize( ) event has just a few lines of code, it generates an intense code activity to set the initial appearance of the frmManageRecipes UserForm. It declares constant conInitialHeight = 329 to shrink the user form to the desired height and then calls Private Function MyRecipesCount( ) to count how many copies of the My Recipes sheet tab exist on ThisWorkbook.

```
    Me.Height = conInitialHeight
    mintWorkbookHasMyRecipes = MyRecipesCount()
```

Counting My Recipes Copies

The Private Function MyRecipesCount( ) executes this code:

```
Private Function MyRecipesCount() As Integer
    Dim ws As Worksheet
    Dim intI As Integer

    For Each ws In Worksheets
        If Left(ws.CodeName, 9) = "MyRecipes" Then
            intI = intI + 1
        End If
    Next
    MyRecipesCount = intI
End Function
```

It uses a For Each...Next loop to run through the Worksheets collection using the VBA Left( ) function to select worksheets whose first nine characters of the CodeName property begin with MyRecipes, updating the intI Integer variable, which is used to return the function value.

Filling ComboBox Lists with the LoadcboMyRecipes( ) Procedure

When Function MyRecipesCount( ) ends with returning the code control to the UserForm_Initialize( ) event, it calls Private Sub LoadcboMyRecipes( ) to either fill the cboMyRecipes ComboBox (the "Select recipes to save from" control) or cboMyRecipesDestination with all the My Recipes copies that may exist on this or an external workbook.

```
Private Sub LoadcboMyRecipes(Optional wb As Workbook, Optional bolSetDestination As Boolean)
    Dim ws As Worksheet
    Dim cbo As ComboBox
    Dim intI As Integer
    Const conBlack = 0
    Const conRed = 255
```

```
    If wb Is Nothing Then
        Set wb = ThisWorkbook
    End If
    Set cbo = IIf(bolSetDestination, Me.cboMyRecipesDestination, Me.cboMyRecipes)

    cbo.Clear
    For Each ws In wb.Worksheets
        If (Left(ws.CodeName, 9) = "MyRecipes") And (ws.Visible = xlSheetVisible) _
            And (Me.optImport Or ws.Name <> Me.cboMyRecipes) Then
            cbo.AddItem ws.Name
            If ws.Name = ActiveSheet.Name Then
                intI = Me.cboMyRecipes.ListCount - 1
            End If
        End If
    Next
    cbo.ListIndex = intI
    If Not bolSetDestination Then
        Me.lblRecipesFrom.Caption = "Recipes from " & IIf(wb.Name = ThisWorkbook.Name, "This
        Workbook", wb.Name)
        Me.lblRecipesFrom.ForeColor = IIf(wb.Name = ThisWorkbook.Name, conBlack, conRed)
    End If
End Sub
```

This procedure receives two optional arguments: the wb as Workbook argument and bolSetDestination (used to indicate which ComboBox control must be filled). When the wb argument points to Nothing (a condition that will always happen when the argument is missing), it receives a reference to the ThisWorkbook object.

```
If wb Is Nothing Then
    Set wb = ThisWorkbook
End If
```

And the VBA IIF( ) function is used to verify the bolSetDestination Boolean variable and set a pointer to the desired ComboBox control, using the cbo as ComboBox object variable to reference it.

```
Set cbo = IIf(bolSetDestination, Me.cboMyRecipesDestination, Me.cboMyRecipes)
```

The desired ComboBox is cleared, and a For Each...Next loop is performed through the Worksheets collection, verifying which ones must be used to fill the ComboBox list. They must have the first nine characters equal to MyRecipes and must be visible to an "Import recipes" operation but must also be different from the one chosen on cboMyRecipes for an Export operation to this workbook.

```
cbo.Clear
For Each ws In wb.Worksheets
    If Left(ws.CodeName = "MyRecipes", 9) Then And (ws.Visible = xlSheetVisible) _
        And (Me.optImport Or ws.Name <> Me.cboMyRecipes) Then
```

When this is true, the worksheet name is added to desired ComboBox list, while the intI integer variable takes care to select the list position of the active sheet (if the workbook has more than one My Recipes copy).

```
            Me.cbo.AddItem ws.Name
            If ws.Name = ActiveSheet.Name Then
                intI = Me.cbo.ListCount - 1
            End If
        End If
    Next
    Me.cbo.ListIndex = intI
```

This instruction will cascade fire ComboBox_Click( ) event

Selecting Worksheets with the cboMyRecipes_Click( ) Event

When the loop ends, the active sheet is selected in cboMyRecipes by setting the ListIndex property to the intI Integer variable, which will cascade-fire the cboMyRecipes_Click( ) or cboMyRecipesDestination_Click event, according to the bolSetDestination argument. The cboMyRecipes_Click( ) event executes this code:

```
Private Sub cboMyRecipes_Click()
    Dim rg As Range
    Dim wb As Workbook

    If mwb Is Nothing Then
        Worksheets((cboMyRecipes)).Activate
        Set mobjMyRecipes = ActiveSheet
        Set rg = mobjMyRecipes.Range("SavedRecords")
    Else
        mwb.Worksheets((cboMyRecipes)).Activate
        Set rg = mwb.Worksheets((cboMyRecipes)).Range("SavedRecords")
    End If
    Call LoadCurrentRecipes(rg)
    Call UpdateProgressBar(False)
End Sub
```

The Sub cboMyRecipes_Click( ) event first checks that the mwb as Workbook module-level variable is set to Nothing (which always happens on the UserForm_Initialize( ) event), and if it is, it activates the worksheet, selects cboMyRecipes, sets the mobjMyRecipes as Object module-level variable to the active sheet, and sets the rg as Range object variable to point to the active sheet SavedRecords range name (the one that has all recipes stored on current worksheet database).

```
If mwb Is Nothing Then
    Worksheets((cboMyRecipes)).Activate
    Set mobjMyRecipes = ActiveSheet
    Set rg = mobjMyRecipes.Range("SavedRecords")
```

Loading lstRecipes with LoadCurrentRecipes( )

Since the rg object variable now points to the active sheet recipes, the procedure calls LoadCurrentRecipes(rg) to fill the frmManageRecipes lstRecipes ListBox.

```
Private Sub LoadCurrentRecipes(rg As Range)
    Dim intFirstRecipe As Integer
    Dim intI As Integer
```

```
    Me.lstRecipes.Clear
    intFirstRecipe = IIf(mintWorkbookHasMyRecipes > 0, 2, 1)
    For intI = intFirstRecipe To rg.Rows.Count
        If Len(rg.Cells(intI)) > 0 Then
            Me.lstRecipes.AddItem rg.Cells(intI)
        End If
    Next

    Me.lblSelected.Caption = Me.lstRecipes.ListCount & " recipes"
    Me.cmdExecute.Enabled = False
End Sub
```

The frmManageRecipes UserForm can exchange recipe information between any USDA Food Composer...xlsm application or recipe nutrient data with any Excel workbook that has a USDA worksheet with a My_Recipes range name, so the code clears the lstRecipes ListBox and uses a VBA IIF( ) instruction to set the intFirstRecipe Integer variable according to the mintWorkbookHasMyRecipes module-level variable. If mintWorkbookHasMyRecipes >0, it is an indication that it has at least one My_Recipes worksheet, making intFirstRecipe = 2 to exclude the first SavedRecords range name item (usually New Recipe).

```
Me.lstRecipes.Clear
intFirstRecipe = IIf(mintWorkbookHasMyRecipes > 0, 2, 1)
```

Then it uses a For...Next loop to rung from intFirstRecipe to rg.Rows.Count, using the Range.Cells property to add all cells with some value to the lstRecipes ListBox list.

```
For intI = intFirstRecipe To rg.Rows.Count
    If Len(rg.Cells(intI)) > 0 Then
        Me.lstRecipes.AddItem rg.Cells(intI)
    End If
Next
```

When the For...Next loops ends, the lblSelected.Caption property is updated to indicate how many recipes were found, and cmdExecute is disabled, because no recipe is still selected.

```
    Me.lblSelected.Caption = Me.lstRecipes.ListCount & " recipes"
    Me.cmdExecute.Enabled = False
End Sub
```

Finishing LoadcboMyRecipes( )

When LoadCurrentRecipes( ) ends, it returns the code control to cboMyRecipes_Click( ), which will hide the UserForm progress bar (call UpdateProgressBar(False)) and also end, returning the code control to the LoadcboMyRecipes( ) procedure, which updates the frmManageRecipes lblRecipesFrom.Caption Label control (the label that sits right above the lstRecipes ListBox), indicating from where the recipes comes.

```
    Me.lblRecipesFrom.Caption = "Recipes from " & IIf(wb.Name = ThisWorkbook.Name, "This
Workbook", wb.Name)
    Me.lblRecipesFrom.ForeColor = IIf(wb.Name = ThisWorkbook.Name, conBlack, conRed)
End Sub
```

■ **Attention** Note that when the recipes come from an external workbook, the wb workbook variable Name property *is not* the ThisWorkbook.Name property, its text color (Foreground property) will be red.

Finishing the frmManageRecipes_Initialize( ) Event

When the LoadcboMyRecipes( ) procedure ends, it returns the code control to the UserForm_Initialize( ) event, which finishes by adding two items to the cboWorkbooks ComboBox (ThisWorkbook and External Workbook) and setting the default value to This Workbook.

```
    Me.cboWorkbook.AddItem "This Workbook"
    Me.cboWorkbook.AddItem "External Workbook"
    Me.cboWorkbook.ListIndex = 0
End Sub
```

Inserting Copies of the My Recipes Sheet Tab

It is more than probable that as the users become acquainted with your worksheet database application, they will need to create copies of the sheet tab application inside the workbook to better manage its records. Thinking about the My Recipes worksheet of the USDA Food Composer.xlsm application, wouldn't be nice if the user could use a sheet tab to store dessert recipes, another tab to store vegetarian recipes, another one for pasta recipes, and so on?

Although it is quite straightforward to create such worksheet copies using the Excel interface (right-click the desired sheet tab and choose the Move or Copy menu command), when you do this in a sheet tab that already has some records inside it, the new copy will receive all the existing records, and you will also need to delete the copied records before beginning to store the records you want.

The frmManageRecipes UserForm does this for you. Whenever you click its New My Recipes CommandButton (cmdNewMyRecipes), a new, empty copy of the My Recipes worksheet is created and begins to appear in the UserForm cboMyRecipes ComboBox (Figure 9-16).

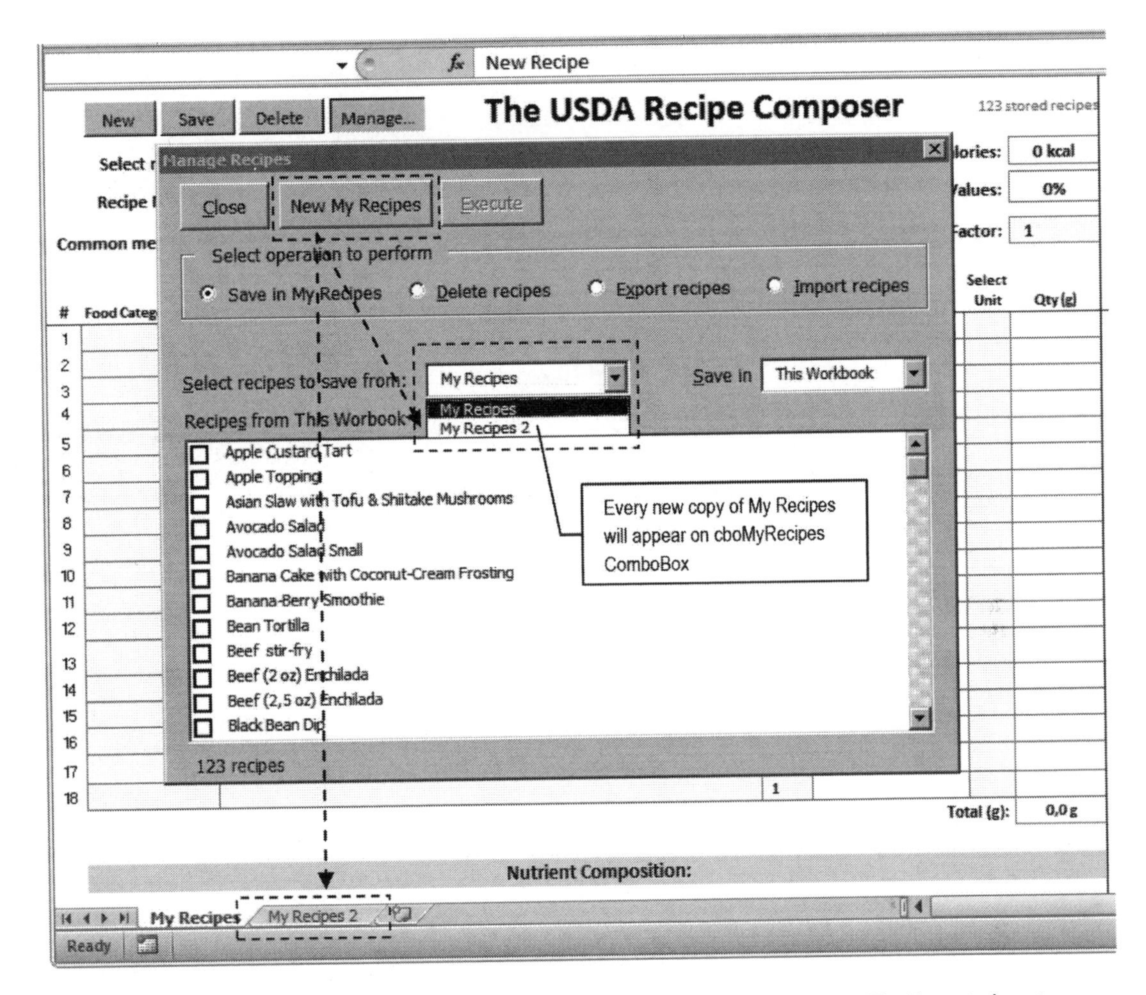

Figure 9-16. *Click the New My Recipes CommandButton of the frmManageRecipes UserForm to insert new, empty copies of the My Recipes sheet tab. Each copy will be named with a counter suffix*

There is a simple trick here:

- The `USDA Food Composer_SheetDBEngineManageAutomation` has a very hidden copy of the `My Recipes` sheet tab with no records and three changed properties: `Name = NewMyRecipes`, `CodeName = MyRecipes0`, and `Visible = 2 - SheetVeryHidden` (which hides the sheet tab from the Move or Copy dialog box, as shown in Figure 9-17).

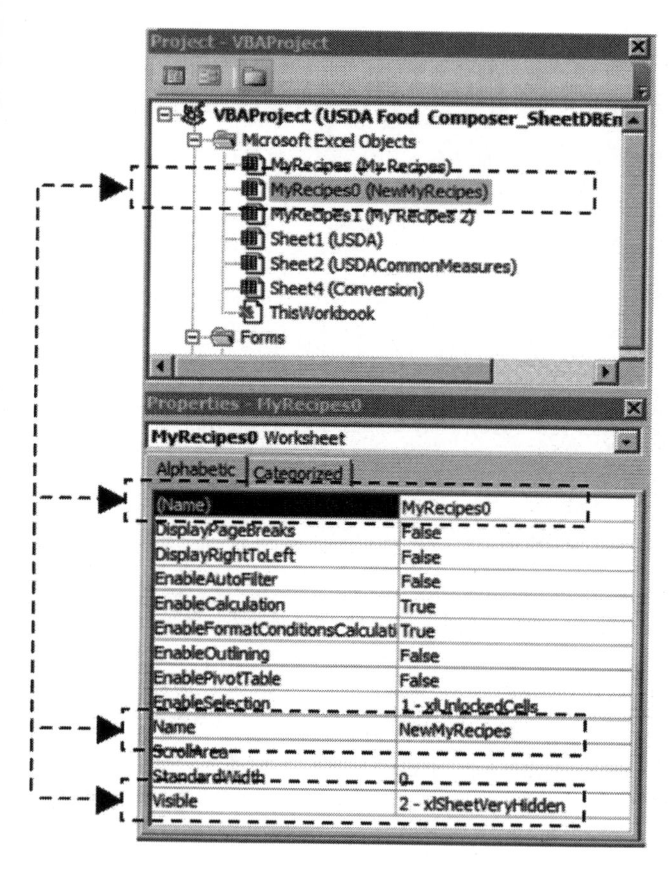

Figure 9-17. *The MyRecipes0 (NewMyRecipes) sheet tab is hidden in the Excel interface from user eyes because the Visible property was changed to 2 – xlSheetVeryHidden in the VBA Properties window*

- The `cmdNewMyRecipes_Click( )` event makes a new copy of the `NewMyRecipes` worksheet, changes its name to My Recipes # (where # is the suffix counter), and sets its `Visible` property to `1 - xlSheetVisible`.

The new `My Recipes #` copy will be placed on the right of the last visible copy. Take a look at the `cmdNewMyRecipes_Click( )` event code in the next listing.

```
Private Sub cmdNewMyRecipes_Click()
    Dim wsCurrent As Worksheet
    Dim ws As Worksheet
    Dim wsNew As Worksheet
    Dim lngErr as Long
    Dim intI As Integer
    Dim intAfter As Integer

    Call SetScreenEventsRecalc(False)
        'Save the active sheet
        Set wsCurrent = ActiveSheet
        'Find last visible MyRecipes
        For Each ws In Worksheets
```

```
            'Set ws = Worksheets(intI)
            If ws.Visible = xlSheetVisible And Left(ws.CodeName, 9) = "MyRecipes" Then
                intAfter = ws.Index
                intI = intI + 1
            End If
        Next
        intI = intI + 1

        'Make a copy of NewMyRecipes sheet very hidden
        Set ws = Worksheets("NewMyRecipes")
        ws.Visible = xlSheetHidden
        ws.Copy , Worksheets(intAfter)
        ws.Visible = xlSheetVeryHidden

        'Get new NewMyRecipes and change it name
        Set wsNew = Worksheets(intAfter + 1)
        Do
            On Error Resume Next
                'Try to change new MyRecipes name
                wsNew.Name = "My Recipes " & intI
                lngErr = Err
                intI = intI + 1
            On Error GoTo 0
        Loop While lngErr <> 0

        'Make sheet visible
        wsNew.Visible = xlSheetVisible
        'Activate the new sheet to start it database services
        wsNew.Activate
        'Restore the activesheet
        wsCurrent.Activate
        'Update cboMyRecipes
        Call LoadcboMyRecipes
    Call SetScreenEventsRecalc(True)
End Sub
```

To allow the cmdNewMyRecipes_Click( ) event code to run smoother in the user interface, it begins by disabling screen updates, events firing, and calculation while setting a reference to the ActiveSheet object.

```
Call SetScreenEventsRecalc(False)
    'Save the active sheet
    Set wsCurrent = ActiveSheet
```

Now it is time to determine the last visible My Recipes sheet tab position (Index property) to insert the new worksheet on the right. This is made using a For Each...Next loop through all visible My Recipes copies (which is determined by checking its first nine CodeName property characters).

```
'Find last visible MyRecipes
For Each ws In Worksheets
    'Set ws = Worksheets(intI)
    If ws.Visible = xlSheetVisible And Left(ws.CodeName, 9) = "MyRecipes" Then
```

```
        intAfter = ws.Index
        intI = intI + 1
    End If
Next
intI = intI + 1
```

Note that the intAfter Integer variable is updated to the Worksheet.Index property every time a visible copy of My Recipes is found. The intI Integer variable is used as a counter to indicate how many copies were found. When the loop ends, intI is incremented again to generate the most probable counter to the new sheet tab name.

The code sets a reference to the very hidden NewMyRecipes worksheet and changes its Visible property to xlSheetHidden because the Worksheet.Copy method fails to work on very hidden worksheets.

```
'Make a copy of NewMyRecipes sheet very hidden
Set ws = Worksheets("NewMyRecipes")
ws.Visible = xlSheetHidden
```

Now the Worksheet.Copy method is applied using its After argument's Worksheets(intAfter) to set a reference to the worksheet that must be on the left of this new worksheet copy (which will be named as NewMyRecipes (2)). And once the copy is made, the NewMyRecipes worksheet is turned very hidden again.

```
ws.Copy , Worksheets(intAfter)
ws.Visible = xlSheetVeryHidden
```

After the copy is made, it is time to change the name to My Recipes #, where # means a counter that is still not in use. Since the worksheet application user can change sheet names wherever, you must try to change the worksheet name inside a Do...Loop instruction until a valid name is found.

To do this, the code first sets a reference to the new worksheet (Index = intAfter+1), initiates the loop, and disables VBA error handling with an On Error Resume Next statement. It then tries to change the worksheet name. If the name already exists, an error will be raised, and the error number is stored on the lngErr Long variable. The intI Integer variable is incremented, and the error handler is reset with an On Error GoTo 0 statement. The Do...Loop will continue until lngErr = 0 (no error found after trying to change the sheet name).

```
'Get new NewMyRecipes and change it name
Set wsNew = Worksheets(intAfter + 1)
Do
    On Error Resume Next
        'Try to change new MyRecipes name
        wsNew.Name = "My Recipes " & intI
        lngErr = Err
        intI = intI + 1
     On Error GoTo 0
Loop While lngErr <> 0
```

When the loop ends, the new My Recipes # sheet tab needs to be turned visible (because its Visible property is xlSheetVeryVisible), and the new sheet is activated to start the SheetDBEngine class with the database services (the wsNew.Activate command will fire the Worksheet_Activate( ) event).

```
'Make sheet visible
wsNew.Visible = xlSheetVisible
```

```
'Activate the new sheet to start it database services
wsNew.Activate
```

The code finishes restoring the active sheet, updating the cboMyRecipes list with a call to LoadcboMyRecipes( ), and restoring screen updating, events firing, and calculation to its default values.

```
    'Restore the activesheet
    wsCurrent.Activate
    'Update cboMyRecipes
    Call LoadcboMyRecipes
Call SetScreenEventsRecalc(True)
```

■ Attention You may want to create an undefined number of the My Recipes copy. If you have trouble removing them (by right-clicking the sheet tab and choosing Delete), press Alt+F11 to open the Visual Basic IDE, double-click the worksheet copy you want to delete in the VBA Explorer tree to show its code module, press Ctrl+A to select all its code, and delete it before deleting the worksheet.

After creating the My Recipes 2 sheet tab, I suggest saving this workbook with a new name, to use it on other examples of this chapter (save it as `USDA Food Composer_SheetDBEngineManageAutomation1`).

You can extract a copy of the `USDA Food Composer_SheetDBEngineManageAutomation1.xlsm` macro-enabled workbook from the `Chapter09.zip` file.

Selecting Desired Recipes

You may note in Figures 9-15 and 9-16 that whenever frmManageRecipes has recipes to select on its lstRecipes ListBox, it offers a check box to the right of each recipe name, which is possible because the lstRecipes.ListStyle property was set to 1 - fmListStyleOption. It also has the MultiSelect property set to 2 - fmMultiSelectExtended, which allows you to select as many list items as you want: dragging the mouse to select successive list items and pressing Ctrl+click to select by random.

Whenever one or more recipes are selected, the lstRecipes_Change( ) event fires, executing this code:

```
Private Sub lstRecipes_Change()
    Dim intI As Integer

    If Not mbolCancelEvent Then
        Call UpdateProgressBar(False)
        Set mcolSelected = New Collection
        For intI = 0 To Me.lstRecipes.ListCount - 1
            If Me.lstRecipes.Selected(intI) Then
                mcolSelected.Add intI, Me.lstRecipes.Column(0, intI)
            End If
        Next

        Me.cmdExecute.Enabled = (mcolSelected.Count > 0)
        Me.lblSelected.Caption = mcolSelected.Count & " recipe(s) selected"
    End If
End Sub
```

The `lstRecipes_Change( )` event does the trick already mentioned in this book: it uses the `mbolCancelEvent` module-level variable to verify whether the event code must be executed (the event code will not be executed when `mbolCancelEvent = True`).

`mbolCancelEvent = False` whenever the `UserForm` is interacting with the user, allowing the user to select the `ListBox` items. It hides the progress bar case it is exhibiting (`UpdateProgressBar(False)`) using the same technique described in the section "Managing the `UserForm` Progress Bar" earlier in this chapter. To hold a reference of every item selected in the `ListBox` control, it uses the `mcolSelected as Collection` module-level variable, which is cleared by setting it to a `New Collection` whenever a `ListBox` item is selected.

```
If Not mbolCancelEvent Then
    Call UpdateProgressBar(False)
    Set mcolSelected = New Collection
```

The `mcolSelected` collection is then filled by a `For...Next` loop through all `lstRecipes` list items, adding each selected item (`lstRecipes.Selected`) using the `Collection.Add` method and using the item list position (`intI`) as the `Collection.Item` value and the recipe name as the `Collection.Key` value.

```
For intI = 0 To Me.lstRecipes.ListCount - 1
    If Me.lstRecipes.Selected(intI) Then
        mcolSelected.Add intI, Me.lstRecipes.Column(0, intI)
    End If
Next
```

When the loop ends, the `cmdExecute.Enabled` property enables/disables the `CommandButton` if the `mcolSelected Collection` variable has at least one item, and the `lblSelected.Caption Label` control is updated to reflect how many recipes are selected in the list (Figure 9-18).

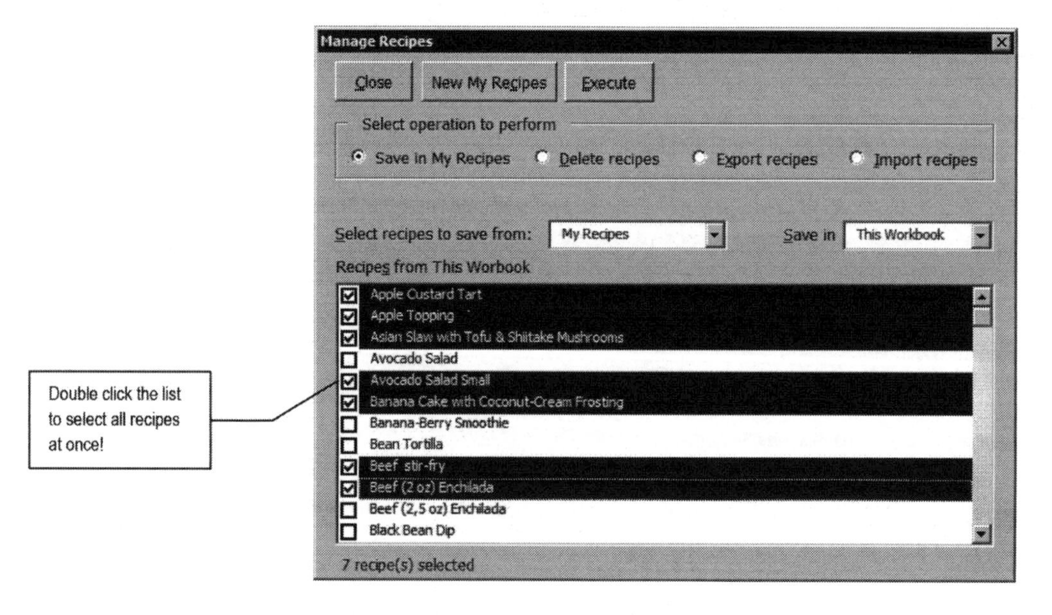

Figure 9-18. *Whenever one or more items are selected in the lstRecipes ListBox, the lstRecipes_Change( ) event fires, fulfilling the mcolSelected Collection variable with all selected items, and the UserForm interface is updated to reflect the selection, enabling the cmdExecute CommandButton and the number of selected recipes*

```
        Me.cmdExecute.Enabled = (mcolSelected.Count > 0)
        Me.lblSelected.Caption = mcolSelected.Count & " recipe(s) selected"
    End If
End Sub
```

Selecting All Recipes at Once

To select all recipes at once, double-click the `lstRecipes` `ListBox`, which will fire the `lstRecipes_DblClick( )` event, executing this code:

```
Private Sub lstRecipes_DblClick(ByVal Cancel As MSForms.ReturnBoolean)
    Dim intI As Integer

    For intI = 0 To Me.lstRecipes.ListCount - 1
        Me.lstRecipes.Selected(intI) = True
    Next
End Sub
```

This procedure uses a `For...Next` loop to run through all `lstRecipes` items, selecting one recipe at a time. And if you are wondering if the `Me.lstRecipes.Selected(intI) = True` instruction cascade-fires the `lstRecipes_Change( )` event at each selection, you are absolutely right.

■ **Attention** You could set the module-level variable `mbolCancelEvent = True` before executing the loop and set it to `False` again after the loop ends, avoiding the cascade event, but the code runs so fast that I felt it unnecessary to do so. Feel free to try for yourself.

And once the desired recipes have been selected, it is time to select the desired operation to be performed (Save in My Recipes, Delete recipes, Export recipes, or Import recipes) before click the `cmdExecute` `CommandButton`.

Saving Recipe Nutritional Information in the My_Recipes Range Name

The Save in My Recipes option allows you to select the desired recipes (double-click to select them all) and save an updated version of the nutritional information in the `My_Recipes` range name of the `USDA` worksheet. You may want to conduct such an operation after a new USDA SR version is used to update the `USDA Food Composer...xlsm` application.

It is important to note that any worksheet application that implements database services using the `SheetDBEngine` class knows nothing about the database itself, while the database engine knows nothing about what the worksheet application does with the records. For example, the `My Recipes` worksheet of the `USDA Food Composer_SheetDBEngine.xlsm` macro-enabled workbook knows nothing about if, how, or where its data is saved, while the `SheetDBEngine` class knows nothing about the meaning of its records data or if and how they are used to calculate the nutritional value of each recipe.

Here is where VBA automation enters into action. To save an updated version of the nutritional information of any recipe in the My_Recipes range name, you need to do the following:

1. Load the recipe.
2. Wait for the worksheet application to calculate its nutritional value (which is immediate).
3. Copy its nutritional data from the My Recipes worksheet to the clipboard.
4. Paste it in the appropriate My_Recipes range name food item of the USDA worksheet.

To do such operations, you may first check that the Save in My Recipes option is selected (optSaveInMyRecipes OptionButton), select the desired recipes to be processed (Figure 9-18), and click the Execute CommandButton (cmdExecute) to perform the operation. The frmManageRecipes UserForm will show its progress bar running and a visual indication of which recipe is being processed, unselecting the recipe name in the lstRecipes ListBox as its nutritional data has been already copied between the My Recipes and USDA worksheets (Figure 9-19).

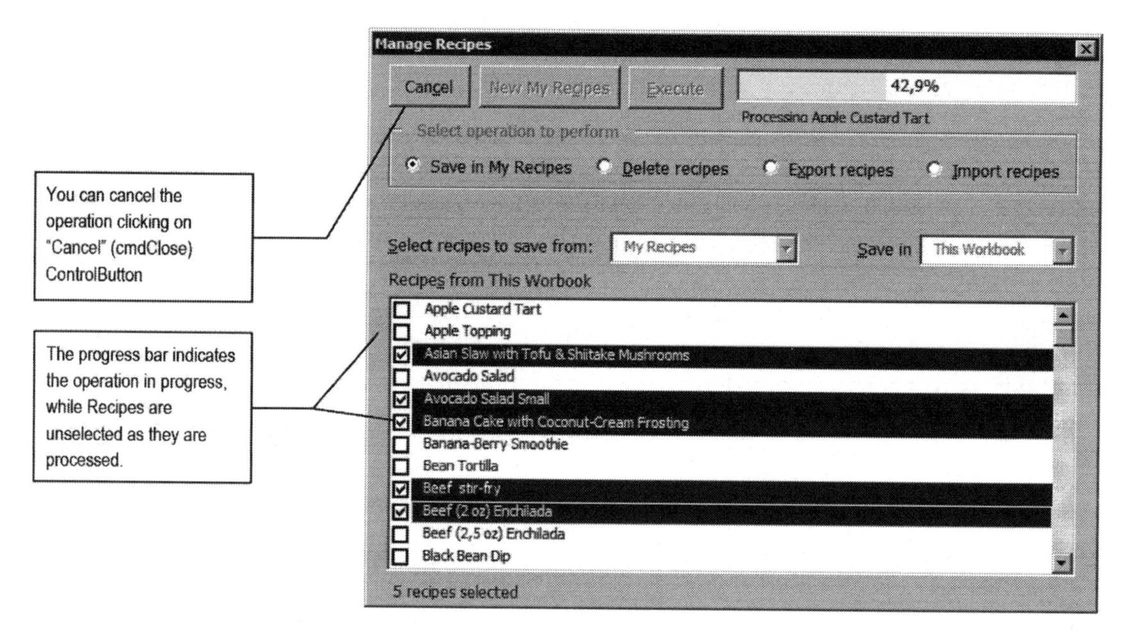

Figure 9-19. *When you click cmdExecute, frmManageRecipes begins to process each selected recipe, showing its progress bar and unchecking the last processed recipe*

Attention Note that the My Recipes worksheet loads each recipe as it is processed by the frmManageRecipes UserForm.

You can cancel the operation while it is running by clicking the Cancel (cmdClose) CommandButton and clicking the Execute CommandButton to start it again with the remaining selected recipes, until all recipes are processed (and unselected) in the lstRecipes ListBox.

When all recipes have been processed, the frmManageRecipes ListBox will have no item selected, the progress bar will indicate 100%, cmdClose.Caption = Close, cmdExecute will be disabled, and the My Recipes worksheet will show the last processed recipe automatically (Beef (20z) Enchilada), as shown in Figure 9-20).

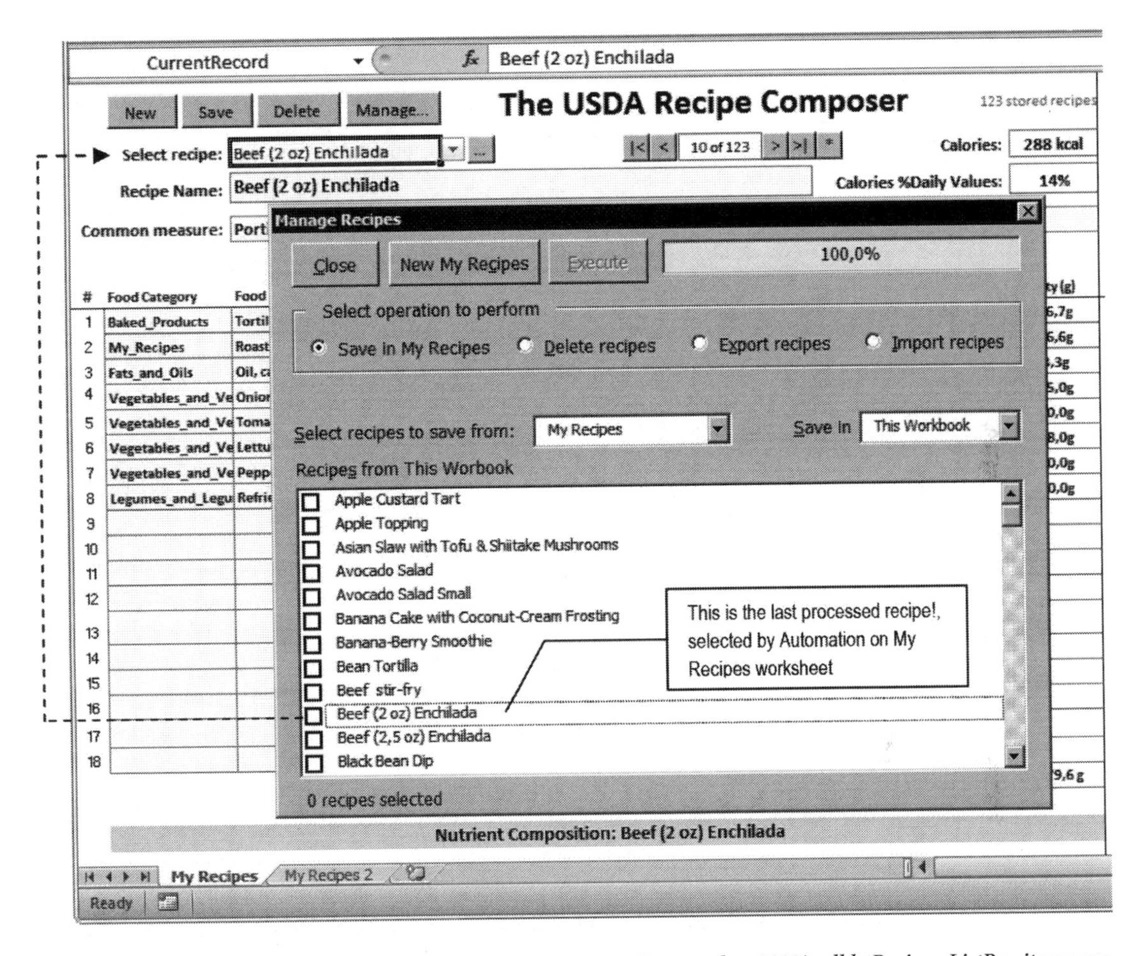

Figure 9-20. *When the selected process finishes, the progress bar reaches 100%, all lstRecipes ListBox items are unselected, and the My Recipes worksheet shows the last processed recipe automatically*

Note that whenever the Save in My Recipes tionButton is selected (optSaveInMyRecipes), the frmManageRecipes UserForm updates the UserForm interface by locking the cboWorkbook value to the This Workbook option by executing the Sub optSaveInMyRecipes_Click( ) event.

```
Private Sub optSaveInMyRecipes_Click()
    Call CloseExternalWorkbook

    If mbolUpdateInterface Then
        Call LoadcboMyRecipes
        mbolUpdateInterface = False
    End If
```

```
    Call UpdateProgressBar(False)
    Me.lblRecipes.Caption = "Select recipes to Save from:"
    Me.lblWorkbook.Caption = "Save in:"
    Me.cboWorkbook.ListIndex = 0
    Me.cboWorkbook.Locked = True
    Me.chkOption.Visible = False
End Sub
```

This code calls CloseExternalWorkbook( ) to close any workbook already opened by the UserForm code (we will see more about this procedure later in this chapter), and if the module-level variable mbolUpdateInterface = True, it calls LoadcboMyRecipes( ) to update the cboMyRecipes list. It also hides the UserForm progress bar with a call to UpdateProgressBar(False) and sets the default interface options: lblRecipes.Caption = Selec recipes to Save from, lblWorkbook.Caption = Save in, cboWorkbook is set to its first list item (This Workbook) and locked, and the chkOption check box is hidden (chkOption.Visible = False).

Supposing that the Save in My Recipes option (optSaveInMyRecipes) is selected and all desired recipes have been selected, when you click the Execute CommandButton, it will run the next code fragment of the cmdExecute_Click( ) event.

■ **Attention** Since cmdExecute_Click( ) is quite extensive, we will analyze just its relevant code parts for each OptionButton operation performed by frmManageRecipes UserForm.

```
Private Sub cmdExecute_Click()
    Dim wb As Workbook
    Dim ws As Object
    ...
    If Me.cmdExecute.Caption = "Continue" Then
        ...
    Else
        Set wb = ThisWorkbook
        Set ws = ActiveSheet
        Call EnableControls(False)

        If Me.optDelete Then         ...
           ...
        End If

        Call ProcessRecipes(wb, ws)
    End If
End Sub
```

As you can see in the cmdExecute_Click( ) event code fragment, it just sets the wb = ThisWorkbook and ws = Activesheet object variables and calls EnableControls(False) to disable the frmManageRecipes controls.

Updating the frmManateRecipes Interface with EnableControls( )

The Private Sub EnableControls( ) receives the bolEnable as Boolean argument, executing this code:

```
Private Sub EnableControls(bolEnabled As Boolean)
    Me.cmdExecute.Enabled = bolEnabled And (mcolSelected.Count > 0)
    Me.cmdClose.Caption = IIf(bolEnabled, "Close", "Cancel")
    Me.cmdNewMyRecipes.Enabled = bolEnabled
    Me.fraOperation.Enabled = bolEnabled
    Me.cboMyRecipes.Enabled = bolEnabled
    Me.cboWorkbook.Enabled = bolEnabled
    Me.chkOption.Enabled = (bolEnabled Or optExport Or optImport)
End Sub
```

As you can see, Sub EnableControls( ) is responsible for enabling/disabling controls on the UserForm according to the bolEnabled argument. Note that cmdExecute will be enabled just when bolEnabled = True *and* there is at least one recipe selected in the lstRecipes ListBox (mcolSelected.Count > 0). The cmdClose.Caption text alternates between Close and Cancel according to the bolEnabled value, and chkOption will always be enabled for import/export operations.

Processing Selected Recipes with ProcessRecipes( )

After synchronizing the frmManageInterface controls, cmdExecute_Click( ) calls Sub ProcessRecipes(wb, ws) to effectively process the selected recipes, which is the heart of the frmManageRecipes UserForm, being responsible for the following:

- Performing a loop through all selected recipes
- Updating the UserForm progress bar
- Executing the desired operation in each recipe (save in the My_Recipes range name, delete, export, or import recipes)
- Unchecking the already processed recipes in lstRecipes and removing them from the mcolSelected Collection module-level variable

It receives two arguments (the workbook and worksheet where it needs to act), executing this code:

```
Private Sub ProcessRecipes(wb As Workbook, ws As Worksheet)
    Dim varItem As Variant
    Dim strRecipe As String
    Dim intTotal As Integer
    Dim intI As Integer
    Dim intJ As Integer

    mbolCancel = False
    mbolCancelEvent = True

    'Freeze MyRecipes updating
    mobjMyRecipes.ScreenUpdating (False)
    Application.Cursor = xlWait
        'Clear and show the Progress Bar
        Call UpdateProgressBar(True)
```

```
        intTotal = mcolSelected.Count
        For Each varItem In mcolSelected
            If mbolCancel Then
                If Me.optDelete Then
                    Call RebuildCollection
                End If
                Exit For
            End If

            'Updade Progress Bar (intI = recipe count, intJ = lstRecipes.Index)
            intI = intI + 1
            intJ = varItem + IIf(optDelete, 1 - intI, 0)
            strRecipe = Me.lstRecipes.Column(0, intJ)
            Me.lblProcessing.Caption = "Processing " & strRecipe
            Call UpdateProgressBar(True, intI, intTotal)
            DoEvents

            If Me.optDelete Then
                mobjMyRecipes.DeleteRecord strRecipe, Me.chkOption
                Me.lstRecipes.RemoveItem (intJ)
            Else
                If (Me.optExport Or Me.optImport) And mintWorkbookHasMyRecipes Then
                    Call TransferRecipe(strRecipe)
                Else
                    'Save just recipe nutritional information on 'USDA'!My_Recipes
                    Call SaveInMyRecipes(strRecipe)
                End If
                Me.lstRecipes.Selected(varItem) = False
            End If
            mcolSelected.Remove (1)
            Me.lblSelected.Caption = mcolSelected.Count & " recipes selected"
        Next
    mbolCancelEvent = False
    Me.lblProcessing.Visible = False

    If Me.optExport Or Me.optSaveInMyRecipes Then
        wb.Save
        If Me.optExport Then
            If (mxl.Hwnd <> Application.Hwnd) Then
                wb.Close True
                mxl.Quit
            End If
            Set mxl = Nothing
            Set mwb = Nothing
            Set mws = Nothing
        End If
    End If

    Me.cmdExecute.Enabled = (mcolSelected.Count > 0)
    Me.cmdClose.Caption = "Close"
    Application.Cursor = xlDefault
    mobjMyRecipes.ScreenUpdating (True)
End Sub
```

Before beginning to process the selected recipes, Sub ProcessRecipes( ) sets the stage for operating in the frmManageRecipes interface: mbolCancel = False to allow canceling the operation when it becomes True, and mbolCancelEvent = True to avoid fire-cascading events when the lstRecipes ListBox is manipulated by the code.

```
Private Sub ProcessRecipes(wb As Workbook, ws As Worksheet)
    ...
    mbolCancel = False
    mbolCancelEvent = True
```

▒ Attention The mobjMyRecipes as Object module-level variable, which has a reference to the source My Recipes worksheet, has its ScreenUpdating property set to False. That is why it is declared as Object. If it was declared as Worksheet, VBA would raise an error here since the Worksheet object does not have a ScreenUpdating method (just the My Recipes worksheet has it!).

The UserForm progress bar is turned visible with 0%, and intTotal receives the total number of selected recipes in the lstRecipes ListBox (mcolSelected.Count property).

```
'Freeze MyRecipes updating
mobjMyRecipes.ScreenUpdating (False)
Application.Cursor = xlWait
    'Clear and show the Progress Bar
    Call UpdateProgressBar(True)
    intTotal = mcolSelected.Count
```

Looping Through All Selected Recipes

Since ProcessRecipes( ) is used for all four operations (saved in my recipes, delete, export, or import recipes), let's see how the For Each...Next loop is used to update the frmManageRecipes UserForm interface as recipes are processed.

The loop uses the varItem as Variant variable to get each mcolSelected collection's Item value: the lstRecipes.Index value of each selected recipe. It then verifies the mbolCancel module-level variable value. If mbolCancel = True (the user clicks Cancel), it checks whether the operation in progress is deleting recipes, and if it is, it calls the Sub RebuildCollection( ) procedure (which will be analyzed later in this chapter) and exits the loop.

```
For Each varItem In mcolSelected
    If mbolCancel Then
        If Me.optDelete Then
            Call RebuildCollection
        End If
        Exit For
    End If
    ...
Next
```

If the For Each...Next loop is not canceled, the intI Integer variable is incremented to reflect the number of recipes already processed, and the intJ Integer variable receives the varItem value, which is the lstRecipes.Index value of the recipe being processed (note that if a "Delete recipes" operation is in progress, varItem is added to 1 - intI).

```
For Each varItem In mcolSelected
    ...
    'Updade Progress Bar (intI = recipe count, intJ = lstRecipes.Index)
    intI = intI + 1
    intJ = varItem + IIf(optDelete, 1 - intI, 0)
```

The recipe name in process is selected with the ListBox Column property, using the intJ value for its Row argument (recipe position on the list). lblProcessing.Caption receives the name of the recipe being processed, and the UserForm progress bar is updated to show the percentage already accomplished, using intI (current recipe processed) and intTotal (total recipes to be processed) as references to the UpdateProgressBar( ) procedure. The DoEvents function allows other events to be processed (like canceling the operation).

```
strRecipe = Me.lstRecipes.Column(0, intJ)
Me.lblProcessing.Caption = "Processing " & strRecipe
Call UpdateProgressBar(True, intI, intTotal)
DoEvents
```

The desired operation is then processed. If it is not a "Delete recipes" operation, the recipe is processed, and the varItem value (recipe position on the list) is used to indicate which recipe must be unselected in the lstRecipes ListBox. Since the loop is running through all mcolSelected as Collection items, the item processed is always the first item, so it is removed from the collection using the mcolSelected.Remove(1) method (the next item to be processed will become the first item inside the collection) and the lblSelected. The Label control (at the bottom of the lstRecipes ListBox) has its Caption property updated with an mcolSelected.Count value to indicate how many recipes are still selected for processing.

```
    If Me.optDelete Then
        ...
    Else
        ...
        Me.lstRecipes.Selected(varItem) = False
    End If
    mcolSelected.Remove (1)
    Me.lblSelected.Caption = mcolSelected.Count & " recipes selected"
Next
```

If the operation in progress is an export or import of a *recipe record* to a worksheet that has at least one My Recipes worksheet (mintWorkbookHasMyRecipes > 0), the loop calls TransferRecipe(strRecipe). Otherwise, the operation in progress is to save just the recipe nutritional information in the My_Recipes range name of the USDA worksheet, which can be on this or an external workbook, calling the Sub SaveInMyRecipes( ) procedure.

```
'Give a time to load recipe information
DoEvents
If (Me.optExport Or Me.optImport) And mintWorkbookHasMyRecipes Then
    ...Call TransferRecipe(strRecipe)
```

```
Else
    'Save just recipe nutritional information on 'USDA'!My_Recipes
    Call SaveInMyRecipes(strRecipe)
End If
```

Saving Recipe Nutritional Information with SaveInMyRecipes( )

The frmManageRecipes UserForm allows you to get recipe nutritional information from two different sources.

- By selecting the desired recipe as a food item in the My_Recipes range name of some USDA worksheet, from this or another workbook
- By loading the desired recipe record on the My Recipes worksheet and getting its nutritional information by selecting the NewRecipe range name

If you do not have the curiosity to inspect how the My Recipes worksheet or any of its copies produce the recipe nutritional information, you must be aware that in its hidden columns it uses the cell range Y10:GW27 to calculate each food item nutrient value based on the food item weight used by the recipe with the aid of Excel VlookUp( ) functions. The cell V28:GW28 range (the NewRecipe range name) is used to sum each nutrient values for all possible recipe food items (Figure 9-21).

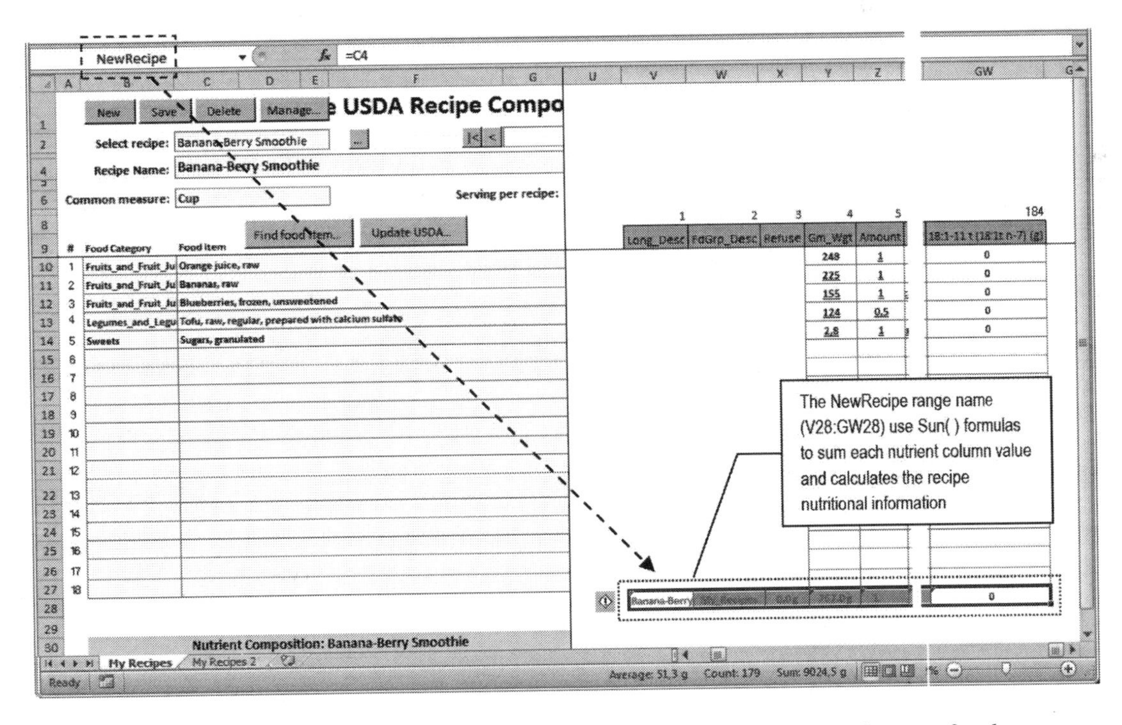

Figure 9-21. *The My Recipes worksheet has the NewRecipe range name that returns the sum of each calculated nutrient for the selected recipe from range V28:GW28*

To save any recipe amount, you just need to load the recipe and copy the NewRecipe range name to the correct row of the My_Recipes range name. Or just exchange food items of the My_Recipes range names between two different USDA worksheets.

One of these two operations is conducted by the `Private Sub SaveInMyRecipes( )` procedure, which receives the recipe name on its `strRecipe as String` argument and executes this code:

```
Private Sub SaveInMyRecipes(strRecipe As String)
    Dim rgSource As Range
    Dim rgDestination As Range
    Dim rgSourceRecipe As Range
    Dim rgDestinationRecipe As Range
    Dim rgUSDA As Range
    Dim bolNewRecipe As Boolean

    Set rgUSDA = Range("USDA")
    Set rgDestination = Range("My_Recipes")
    Range("CurrentRecord") = strRecipe

   If Me.optSaveInMyRecipes Or Me.optExport Then
        Set rgSourceRecipe = Range("NewRecipe")
        If Me.optExport Then
            Set rgDestination = mws.Range("My_Recipes")
            Set rgUSDA = mws.Range("USDA")
        End If
    Else
        'Import operation
        Set rgSource = mws.Range("My_Recipes")
        'Find recipe on Source range
        Set rgSourceRecipe = rgSource.Find(strRecipe, , , xlWhole)
        'Resize rgSourceRecipe
        Set rgSourceRecipe = rgSourceRecipe.Resize(1, Range("NewRecipe").Columns.Count)
    End If

    'Find rgRecipe on Desination range
    Set rgDestinationRecipe = rgDestination.Find(strRecipe, , , xlWhole)
    If rgDestinationRecipe Is Nothing Then
        'Insert new recipe on destination range
        rgDestination.Resize(rgDestination.Rows.Count + 1).Name = "My_Recipes"
        'Update rgDestination object variable to contain My_Recipes new row
        If Me.optExport Then
            Set rgDestination = mws.Range("My_Recipes")
        Else
            Set rgDestination = Range("My_Recipes")
        End If
        'Position on new cell of My_Recipes range
        Set rgDestinationRecipe = rgDestination.Cells(rgDestination.Rows.Count, 1)
        bolNewRecipe = True
    End If

    'Copy rgSourceRecipe content
    rgSourceRecipe.Copy
    rgDestinationRecipe.PasteSpecial xlPasteValues

    If bolNewRecipe Then
```

```
        'A New Recipe was inserted on USDA My_Recipes. Resize and sort it!
        rgDestination.Resize( , rgUSDA.Columns.Count).Sort rgDestination.Cells(1, 1)
        rgUSDA.Resize(rgUSDA.Rows.Count + 1).Name = "USDA"
    End If
End Sub
```

Sub SaveInMyRecipes( ) declares object variables to reference the source and destination workbooks and worksheets and initialize them according to the operation to perform: save in My_Recipes, export, or import. By default, rgUSDA, rgDestination, and rgSourceRecipe (for an export operation) are set to these ThisWorkbook range names:

```
Set rgUSDA = Range("USDA")
Set rgDestination = Range("My_Recipes")
If Me.optSaveInMyRecipes Or Me.optExport Then
    Set rgSourceRecipe = Range("NewRecipe")
```

It then loads the recipe being processed on the active sheet by changing the Range(CurrentRecord) value, effectively automating the process (the Worksheet_Change( ) event will fire).

```
Range("CurrentRecord") = strRecipe
```

If the selected operation is to export data, both rgUSDA and rgDestination are redefined to the external USDA range name (associated to the mws as Worksheet module-level variable).

```
If Me.optExport Then
    Set rgUSDA = mws.Range("USDA")
    Set rgDestination = mws.Range("My_Recipes")
End If
```

But if the operation is to import data, the rgSource object variable is set to the external My_Recipes range name (associated to the mws as Worksheet module-level variable), while rgSourceRecipe is defined by searching the selected recipe inside rgSource, using the Range.Find method.

```
Else
    'Import operation
    Set rgSource = mws.Range("My_Recipes")
    'Find recipe on Source range
    Set rgSourceRecipe = rgSource.Find(strRecipe, , , xlWhole)
```

Once the recipe was found in the My_Recipes range name, the rgSourceRange object variable is resized to encompass all NewRecipes range name nutrient columns, using the Range.Resize method.

```
    'Resize rgSourceRecipe
    Set rgSourceRecipe = rgSourceRecipe.Resize(1, Range("NewRecipe").Columns.Count)
End If
```

After rgUSDA, rgDestination, and rgSourceRecipe have been correctly defined, the code uses the Range.Find method to verify whether the selected recipe already has a record in the destination My_Recipes range.

```
'Find rgRecipe on Desination range
Set rgDestinationRecipe = rgDestination.Find(strRecipe, , , xlWhole)
```

If the record is not found, it must insert it at the bottom of the My_Recipes range, so rgDestination is increased by one row using the Range.Resize method and updated with the Range.Name property.

```
'Insert new recipe on destination range
rgDestination.Resize(rgDestination.Rows.Count + 1).Name = "My_Recipes"
```

And once rgDestination is resized, its references are updated to reflect the new inserted row.

```
If Me.optExport Then
    Set rgDestination = mws.Range("My_Recipes")
Else
    Set rgDestination = Range("My_Recipes")
End If
```

Now that rgDestination already has a pointer to the resized range, it is used to set rgDestinationRecipe to point to the new inserted row, using the rgDestination.Cells( ) property and bolNewRecipe = True to signal a new record.

```
    Set rgDestinationRecipe = rgDestination.Cells(rgDestination.Rows.Count, 1)
    bolNewRecipe = True
End If
```

Now the recipe nutritional information is transferred between worksheets using the rgSource.Copy and rgDestinationRecipe.PasteSpecial methods.

```
'Copy rgSourceRecipe content
rgSourceRecipe.Copy
rgDestinationRecipe.PasteSpecial xlPasteValues
```

If a new recipe is inserted on the My_Recipes range name, it needs to be sorted to put the new food item in its correct order. This is made by first applying the Range.Resize method to include all USDA range nutrient columns (note that it uses just the Range.Resize Columns argument) and then uses Range.Sort to sort it, using the first rgDestination column (rgDestination.Cells(1,1)) for its Key1 argument.

```
If bolNewRecipe Then
    'A New Recipe was inserted on USDA My_Recipes. Resize and sort it!
    rgDestination.Resize( , rgUSDA.Columns.Count).Sort rgDestination.Cells(1, 1)
```

Since a new record was inserted, to finish the operation, the code also needs to resize the USDA range name to include this new My_Recipes row (note that it uses just the Range.Resize Rows argument), using the Range.Name property to update the range name.

```
        rgUSDA.Resize(rgUSDA.Rows.Count + 1).Name = "USDA"
    End If
End Sub
```

The strRecipe nutritional information was saved on the destination My_Recipes range name, and the code control returns to Sub ProcessRecipes( ), which deselects the item on lstRecipes and removes it from the mcolSelected variable using the Collection.Remove method, updating the lblSelected.Caption property to indicate how many recipes are still selected to process using the Collection.Count property.

```
            Call SaveInMyRecipes(strRecipe)
        End If
        Me.lstRecipes.Selected(varItem) = False
    End If
    mcolSelected.Remove (1)
    Me.lblSelected.Caption = mcolSelected.Count & " recipes selected"
Next
```

Finishing the Save in My Recipes Operation

When all selected recipes have been processed, Sub ProcessRecipes( ) verifies the operation in progress is "Export recipes" or "Save in My Recipes," and if this is true, it calls the Workbook.Save method to save the workbook. To finish the operation, it synchronizes the frmManageRecipes interface by changing the cmdExecute.Enable property, returns the mouse cursor to its default state, and uses automation to execute the active sheet to redefine the database's mobjMyRecipes.ScreenUpdating property to True, forcing database controls to update on the worksheet application interface.

```
        If Me.optExport Or Me.optSaveInMyRecipes Then
            wb.Save
            ...
        End If

        Me.cmdExecute.Enabled = (mcolSelected.Count > 0)
        Me.cmdClose.Caption = "Close"
    Application.Cursor = xlDefault
    mobjMyRecipes.ScreenUpdating (True)
End Sub
```

Exporting and Importing Recipe Data

▒ **Attention** To understand this section, I hope you have created at least one copy of the My Recipes worksheet using the New My Recipes CommandButton of the frmManageRecipes UserForm, as indicated in section "Inserting Copies of My Recipes Sheet Tab" earlier in this chapter.

The frmManageRecipes UserForm allows you to export recipe nutritional data to any USDA worksheet as a new food item to the My_Recipes range name or as recipe records to any other USDA Food Composer...xlsm macro-enabled worksheet that has a My Recipes worksheet database application.

Whenever you click the "Export recipes" OptonButton, three small changes take place in the frmManageRecipes interface: the chkOption CheckBox is shown, and the two Label controls on the left of cboMyRecipes and cboWorkbook ComboBoxes update their Caption property to indicate the appropriate operation (Figure 9-22).

▒ **Attention** If the current workbook has just one My Recipes worksheet, the cboWorkbook ComboBox will be changed to "External Workbook" and locked.

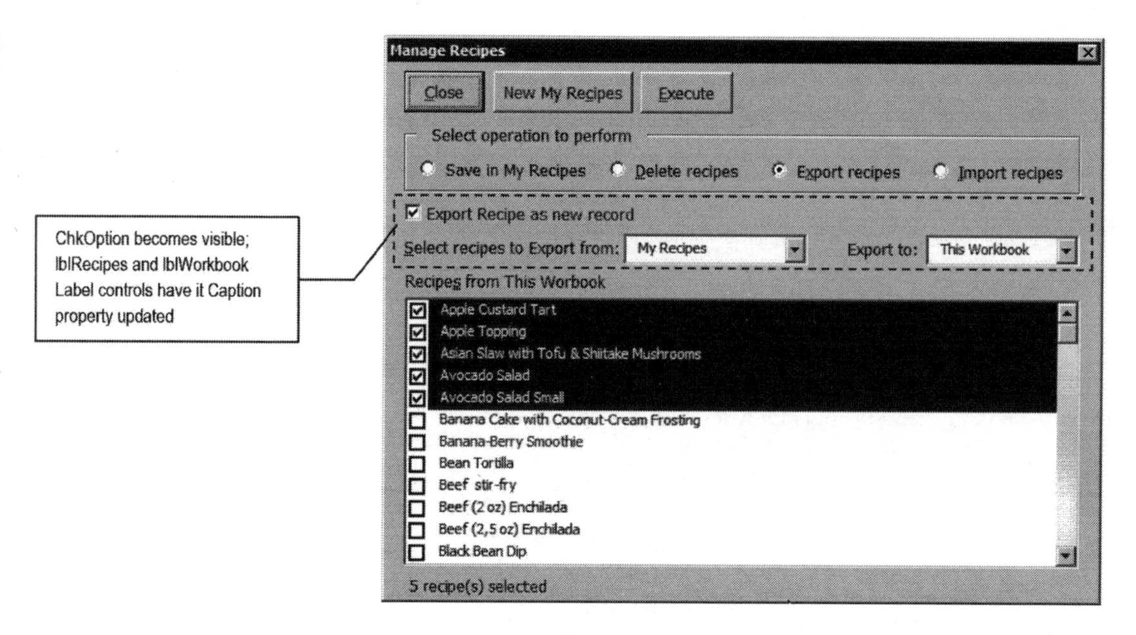

Figure 9-22. *Whenever you select "Export recipes," the frmManageRecipes interface shows some cosmetic changes on its interface*

These small interface changes are made by the `Sub optExport_Click( )` event, which executes this code:

```
Private Sub optExport_Click()
    Call CloseExternalWorkbook

    If mbolUpdateInterface Then
        Call LoadcboMyRecipes
        mbolUpdateInterface = False
    End If
    Call UpdateProgressBar(False)
    Me.lblRecipes.Caption = "Select recipes to Export from:"
    Me.chkOption.Caption = "Export Recipe as new record"
    Me.chkOption.Visible = True
    Me.cboWorkbook.Locked = (MyRecipesCount() = 1)
    Me.cboWorkbook.ListIndex = IIf(Me.cboWorkbook.Locked, 1, 0)
    Me.lblWorkbook.Caption = "Export to:"
End Sub
```

This code is similar to the code used by `optSaveInMyRecipes`: it calls `CloseExternalWorkbook( )` to close any possible external workbook, checks whether `mbolUpdateInterface = True`, and if it is, calls `LoadcboMyRecipes( )`, hides the `UserForm` progress bar with `UpdateProgressBar(False)`, and updates interface controls to better reflect the nature of the export operation.

Note, however, that `cboWorkbook` will have its property `Locked = True` only if this workbook has just one `My Recipes` worksheet. If this is true, it will also set `cboWorkbook = Export Recipes (ListIndex = 1)`, which is the unique export operation allowed. Otherwise, `cboWorkbook` will be unlocked and set to "Import recipes."

```
Me.cboWorkbook.Locked = (MyRecipesCount() = 1)
Me.cboWorkbook.ListIndex = IIf(Me.cboWorkbook.Locked, 1, 0)
```

Exporting Recipes to This Workbook

Let's suppose you want to export the first five recipes from My Recipes to another worksheet of this workbook like Figure 9-22 suggests. After clicking the Execute CommandButton, the frmManageRecipes will change its interface to allow the selection of the destination worksheet inside this workbook (Figure 9-23).

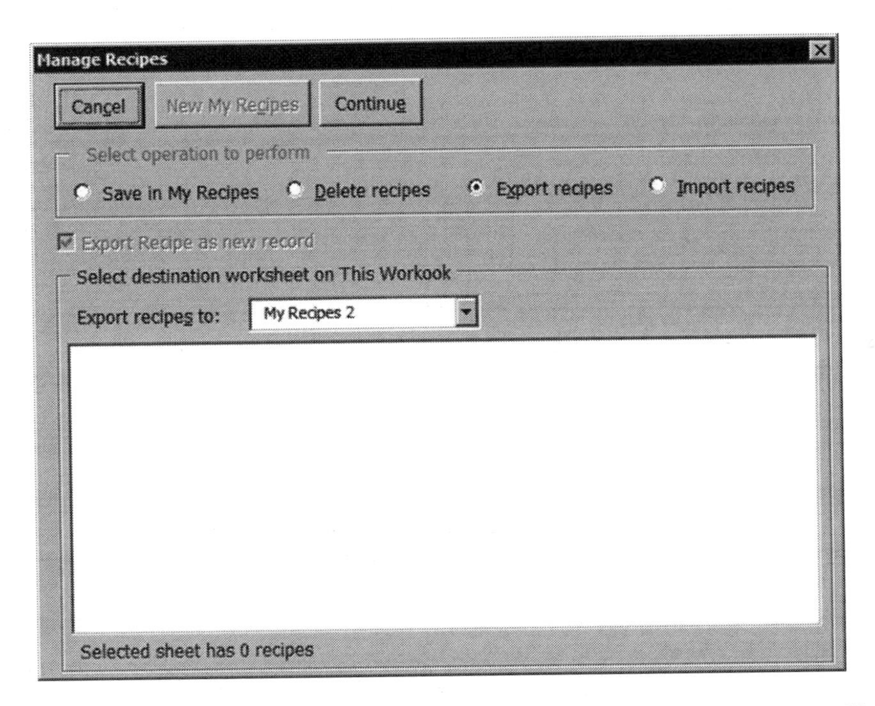

Figure 9-23. *Whenever you need to export recipes to another worksheet of this workbook (or to an external workbook that has more than one My Recipes worksheet), frmManageRecipes will change its interface, repositioning the fraSetDestination Frame control, to allow the selection of the destination worksheet*

All copies of My Recipes made with the New My Recipes CommandButton (but the selected cboMyRecipes source worksheet) will appear in cboMyRecipesDestination (the "Export recipes to" ComboBox). Note that My Recipes 2 is selected by default, showing no recipes at all in the lstExternalRecipes ListBox and that just the Cancel (cmdClose) and Continue (cmdExecute) CommandButtons are enabled, allowing you to cancel or continue the operation after selecting the desired destination worksheet.

■ **Attention** At this point, you can click Cancel to return to the previous UserForm interface (Figure 9-22), where you can select again the desired recipes or just click Execute to return to this interface to select the destination worksheet.

This interface change is made by the cmdExecute_Click( ) event, which executes the next code fragment of the cmdExecute_Click( ) event.

```
Option Explicit

Dim mxl As Excel.Application
Dim mwb As Workbook
Dim mws As Worksheet
...
Private Sub cmdExecute_Click()
    ...
    ElseIf Not Me.optSaveInMyRecipes Then
        If Me.optExport And Me.cboWorkbook = "External Workbook" Then
            'Export operation to external workbook
            If GetExternalWorkbook() Then
                'Load recipes of first external MyRecipes
                Me.cboMyRecipesDestination.ListIndex = 0
                bolSetDestination = True '(mintWorkbookHasMyRecipes > 1)
            Else
                Call EnableControls(True)
                Exit Sub
            End If
        Else
            'Operation is Import or Export to ThisWokbook
            'Select destination My Recipes?
            bolSetDestination = (MyRecipesCount() > 1)
            If Me.optExport Then
                'Export to other sheet tab of ThisWorkbook
                Set mxl = Application
                Set mwb = ThisWorkbook
            End If
        End If
        If bolSetDestination Then
            If optImport Or Me.cboWorkbook = "This Workbook" Then
                'ThisWorkbook has more than one possible destination
                Call LoadcboMyRecipes(wb, True)
            End If
            Call SelectMyRecipesDestination(True)
            If Me.optExport Then
                Set mws = mwb.Worksheets((cboMyRecipesDestination))
            End If
            Exit Sub
        End If
    End If
```

To export recipes to another worksheet of the same workbook, cmdExecute_Click( ) first checks if this is an “Export recipes” or “Import recipes” operation (Not optSaveInMyRecipes), and if this is true, it checks whether this is not an “Export recipes” operation *to an external workbook*.

```
    ElseIf Not Me.optSaveInMyRecipes Then
        If Me.optExport And Me.cboWorkbook = "External Workbook" Then
        ...
        Else
```

The procedure executes the Else clause when this is an "Import recipes" or "Export recipes" *to this workbook*, so verify whether there is more than one possible destination worksheet using the MyRecipesCount( ) procedure.

```
Else
    'Operation is Import or Export to ThisWokbook
    'Select destination My Recipes?
    bolSetDestination = (MyRecipesCount() > 1)
```

If this is an "Export recipes" operation to this workbook, it sets the module-level variables mxl as Application and mwb as Workbook to point to the current Excel application and the ThisWorkbook object.

```
    If Me.optExport Then
        'Export to other sheet tab of ThisWorkbook
        Set mxl = Application
        Set mwb = ThisWorkbook
    End If
End If
```

For any kind of export or import operation, it verifies whether bolSetDestination = True as an indication that the UserForm interface must be changed to allow you to select the destination worksheet. If this is an import or export operation to ThisWorkbook, it calls LoadcboMyRecipes(wb, True) to load all possible destination worksheets in the cboMyRecipesDestination ComboBox.

```
If bolSetDestination Then
    If optImport Or Me.cboWorkbook = "This Workbook" Then
        'ThisWorkbook has more than one possible destination
        Call LoadcboMyRecipes(wb, True)
    End If
```

■ **Attention** For a description about how LoadcboMyRecipes( ) uses its arguments to load the cboMyRecipesDestination ComboBox list, see the section "Filling ComboBox Lists with LoadcboMyRecipes( )" earlier in this chapter.

When the second argument (bolSetDestination) of the LoadcboMyRecipes( ) procedure is True, it will load the cboMyRecipesDestination list and set its value to the first list item, which will cascade-fire the cboMyRecipesDestination_Click( ) event (the same event fired when another destination worksheet is selected in this ComboBox).

Showing Destination Worksheet Recipes on lstRecipesDestination

Whenever a destination worksheet is selected in the cboMyRecipesDestination ComboBox, the Click( ) event fires, showing all recipes stored in the selected worksheet database and executing this code:

```
Private Sub cboMyRecipesDestination_Click()
    Dim wb As Workbook
    Dim ws As Worksheet
    Dim rg As Range
    Dim intI As Integer
```

```
    If Me.cboMyRecipesDestination.ListCount > 0 Then
        Set wb = IIf(Me.optExport, mwb, ThisWorkbook)
        Set ws = wb.Worksheets((cboMyRecipesDestination))
        Set mws = ws
        'Provision to start Database services on destination sheet
        ws.Activate
        Me.lstRecipesDestination.Clear
        Set rg = ws.Range("SavedRecords")
        For intI = 2 To rg.Rows.Count
            If Len(rg.Cells(intI)) > 0 Then
                Me.lstRecipesDestination.AddItem rg.Cells(intI)
            End If
        Next
        Me.lblRecipesDestination.Caption = "Selected sheet has " & Me.lstRecipesDestination.
ListCount & " recipes"

        If mwb.Name = ThisWorkbook.Name Then
            'Keep focus on activesheet if operation will be performed on ThisWorkbook
            Worksheets((Me.cboMyRecipes)).Activate
        End If
    End If
```

The frmManageRecipes UserForm can export recipes to this or another external workbook, so the cmdExecute_Click( ) event calls the cboMyRecipesDestionation.Clear method before loading it with all the possible worksheet export targets. This action fires the cboMyRecipesDestination_Change( ) event with no list items, so it first checks whether the ListCount property is > 0 to execute the operation. If it is empty (ListCount = 0), the event code ends, doing nothing.

```
Private Sub cboMyRecipesDestination_Change()
    ...
    If Me.cboMyRecipesDestination.ListCount > 0 Then
      ...
    End If
End Sub
```

If cboMyRecipesDestination.ListCount > 0, it checks whether the action in progress is an “Export recipes” operation. If it is, the wb as Workbook object variable is set to the wmb as Workbook module-level variable already set (which will point to the desirable workbook target). Otherwise, it is an “Import recipes” operation, which will always be made to the ThisWorkbook.

```
Set wb = IIf(Me.optExport, mwb, ThisWorkbook)
```

Having selected the desired destination workbook, the code sets a reference to the destination worksheet using a double pair of parentheses to automatically convert the cboMyRecipesDestination value to the string argument of the Worksheets collection, attributing the destination worksheet to the mws module-level variable. The destination worksheet’s Activate method is executed to guarantee that its database services have been started allowing you to automate the export/import operation between the two worksheet applications.

```
Set ws = wb.Worksheets((cboMyRecipesDestination))
Set mws = ws
```

```
'Provision to start Database services on destination sheet
ws.Activate
```

■ **Attention** Always use double parentheses to surround a procedure argument whenever you want VBA to make an automatic conversion from number to string, and vice versa.

The lstRecipesDestination ListBox is cleared, and a range object reference is set to the destination worksheet's SavedRecords range name (the database records list). A For int=2 to...Next loop is then used to fill its list with all recipes on its SavedRecords range name, discarding the first recipe ("New recipe").

```
Me.lstRecipesDestination.Clear
Set rg = ws.Range("SavedRecords")
For intI = 2 To rg.Rows.Count
    If Len(rg.Cells(intI)) > 0 Then
        Me.lstRecipesDestination.AddItem rg.Cells(intI)
    End If
Next
```

When the loop ends, the lblExternalRecipes.Caption property is updated with the recipe count, and it checks whether the mwb as Workbook module-level variable points to ThisWorkbook (the workbook that executes the code). If it does, the source worksheet, selected on cboMyRecipes, is activated again.

```
        Me.lblRecipesDestination.Caption = "Selected sheet has " & Me.
lstRecipesDestination.ListCount & " recipes"


        If mwb.Name = ThisWorkbook.Name Then
            'Keep focus on activesheet if operation will be performed on ThisWorkbook
            Worksheets((Me.cboMyRecipes)).Activate
        End If
    End If
End Sub
```

Changing the frmManageRecipes Interface with SelectMyRecipesDestination( )

When the cboMyRecipesDestination_Change( ) event finishes, the code control returns to the cmdExecute_Click( ) event, which calls SelectMyRecipesDestination(True) to change the frmManageRecipes interface to allow the user to select the destination worksheet.

```
Call SelectMyRecipesDestination (True)
```

The Private Sub SelectMyRecipesDestination( ) procedure executes this code:

```
Private Sub SelectMyRecipesDestination(bolSelect As Boolean)
    Dim strCaption As String
    Dim bolExternalWorkbook As Boolean
    Const conBlack = 0
    Const conRed = 255
```

```
    If (Me.cboWorkbook = "External Workbook") And (Not mwb Is Nothing) Then
        bolExternalWorkbook = True
        strCaption = "Select destination on " & IIf(optImport, "This Workbook", mwb.Name)
    Else
        strCaption = "Select destination on This Workbook"
    End If
    Me.fraSetDestination.Caption = strCaption
    Me.fraSetDestination.ForeColor = IIf(bolExternalWorkbook And optExport, conRed,
    conBlack)
    Me.fraSetDestination.Top = Me.cboMyRecipes.Top
    Me.fraSetDestination.Left = Me.fraOperation.Left
    Me.fraSetDestination.Visible = bolSelect
    Me.lblMyRecipesDestination.Caption = IIf(Me.optImport, "Import recipes from:", "Export
    recipes to:")
    Me.cboMyRecipes.Visible = Not bolSelect
    Me.lstRecipes.Visible = Not bolSelect
    Me.cmdExecute.Enabled = bolSelect
    Me.cmdExecute.Caption = IIf(bolSelect, "Continue", "Execute")
End Sub
```

The procedure verifies whether an operation with an external workbook is in progress to set the bolExternalWorkook and strCaption variables and uses the bolSelect argument to change the frmManageRecipes interface, manipulating some controls' Visible properties (also changing fraSetDestination Frame control's Top and Left properties to put them right over cboMyRecipes, cboWorkbook, and lstRecipes). To finish the operation, it changes the cmdExecute Enabled and Caption properties and updates the UserForm interface.

When bolSelect =False, the destination recipe controls are hidden (Figure 9-22); otherwise, they become visible (Figure 9-23), also changing the cmdExecute Caption property from Execute to Continue.

And once again, when the SelectMyRecipesDestination( ) procedure ends, the code control returns to the cmdExecute_Click( ) event, which verifies whether this is an "Export recipes" operation and finishes the code by setting the mws as Worksheet module-level variable to reference the worksheet selected in the cboMyRecipesDestination ComboBox (note the double parentheses used to make an automatic variable-type conversion to String).

```
        If Me.optExport Then
            Set mws = mwb.Worksheets((cboMyRecipesDestination))
        End If
        Exit Sub
    End If
End If
```

Executing the Export Recipes Operation

At this point, frmManageRecipes must be like Figure 9-23, ready to cancel or continue the "Export recipes" operation. If you click the cmdExecute CommandButton that now shows "Continue" on its Caption property, the cmdExecute_Click( ) event will fire again, executing this code:

```
Private Sub cmdExecute_Click()
    ...
    If Me.cmdExecute.Caption = "Continue" Then
```

```
        Call SelectMyRecipesDestination(False)
        Call ProcessRecipes(mwb, mws)
    Else
        ...
    End If
End Sub
```

As you can see, it calls again `SelectMyRecipesDestination(False)` to return `frmManageRecipes` to its default interface and then calls `ProcessRecipes(mwb, mws)` to export the selected recipes to the destination worksheet. Figure 9-24 shows the `UserForm` progress bar indicating that this operation is in progress. Each recipe is loaded in the `My Recipes` worksheet interface automatically and saved in the `My Recipes 2` worksheet interface (which is not shown). When the operation finishes, select My Recipes 2 in `cboMyRecipes` and note that it now shows that it has received all the selected recipes (Figure 9-25).

■ **Attention** When the export recipes process ends, the `My Recipes` source worksheet must be showing the last exported recipe record because of the automation process.

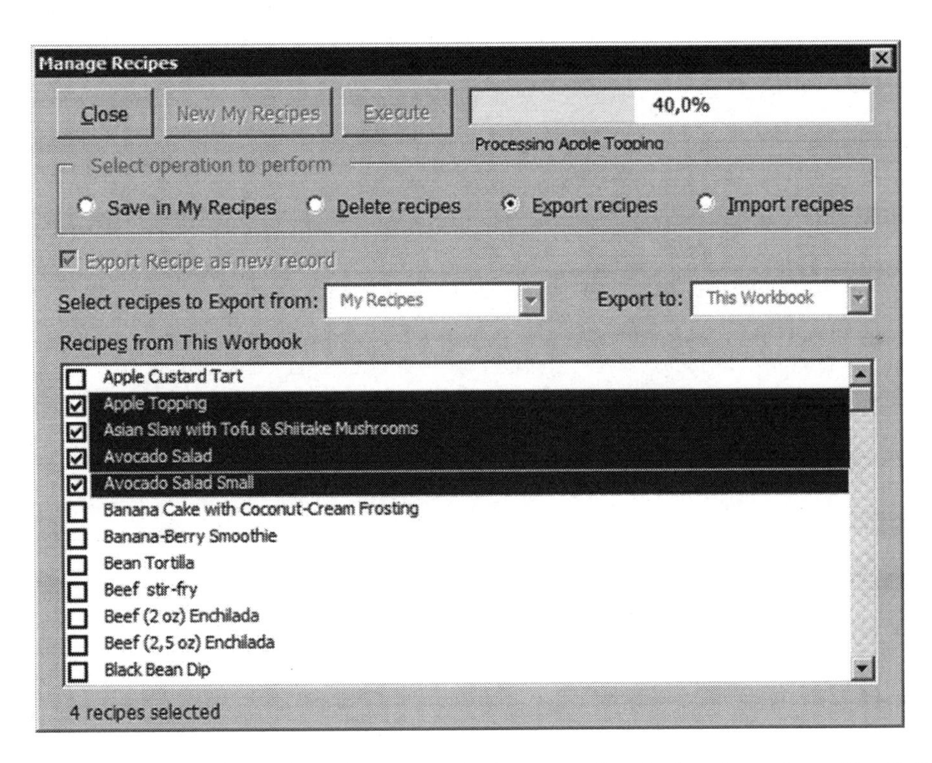

Figure 9-24. *This figure shows frmManageRecipes exporting the first five recipes of the My Recipes worksheet to the My Recipes 2 worksheet*

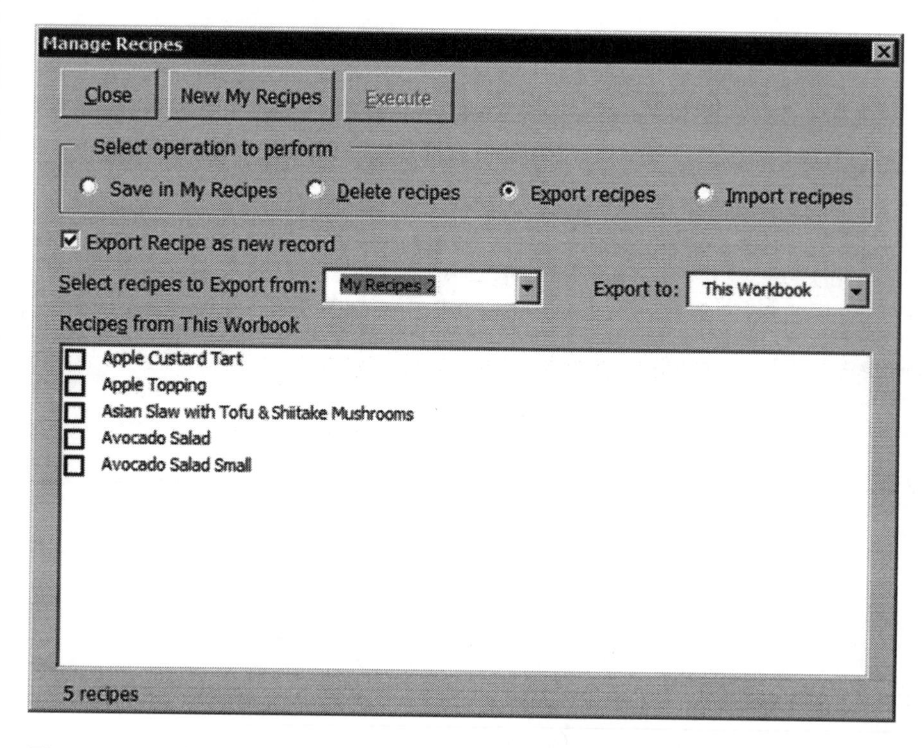

Figure 9-25. *When the "Export recipes" operation ends, select the My Recipes 2 worksheet in cboMyRecipes and note that it now has the five exported recipes!*

Looping Through All Selected Recipes to Import/Export Records

The next code fragment points to what happens inside `ProcessRecipes( )` to export/import the selected record. It calls `Sub TransferRecipe(strRecipe)` to import/export the selected records if `mintWorkbookHasMyRecipes > 0` (an indication that the destination workbook has a `My Recipes` worksheet to be automated).

```
Private Sub ProcessRecipes(wb As Workbook, ws As Worksheet)
    ...
        For Each varItem In mcolSelected
            ...
                If (Me.optExport Or Me.optImport) And mintWorkbookHasMyRecipes Then
                    Call TransferRecipe(strRecipe)
                    ...
                End If
                ...
        Next
```

Exchanging Recipe Records Automatically with TransferRecipe( )

`Private Sub TransferRecipe( )` is responsible for effectively exchanging recipe records between two selected worksheets. One worksheet is the record source, while other is the record destination. They just change places regarding the operation in progress (either export or import), following a single rule: it can just import records from an external workbook that has another `My Recipes` worksheet and execute this code:

```
Private Sub TransferRecipe(strRecipe As String)
    Dim xl As Application
    Dim wsSource As Worksheet
    Dim wsDestination As Object
    Dim rgSource As Range
    Dim rgDestination As Range
    Dim rgMyRecipes As Range
    Dim rgArea As Range
    Dim strNewRecipe As String
    Dim strMsg As String
    Dim strTitle As String
    Dim intI As Integer
    Dim bolNewRecipe As Boolean

    If Me.optExport Then
        Set xl = mxl
        Set wsSource = ActiveSheet
        Set wsDestination = mws
    Else
        Set xl = Application
        Set wsSource = mws
        Set wsDestination = ActiveSheet
    End If
    Set rgSource = wsSource.Range("RecipeData")
    Set rgDestination = wsDestination.Range("RecipeData")

    'Verify if recipe already exist on wsDestination
    strNewRecipe = strRecipe
    Set rgMyRecipes = wsDestination.Range("SavedRecords").Find(strNewRecipe, , , xlWhole)
    If rgMyRecipes Is Nothing Then
        'strRecipe das not exist on destination
        bolNewRecipe = True
        wsDestination.Range("CurrentRecord") = "New Recipe"
    Else
        'Recipe already exist on desination. Transfer recipe as new recipe?
        If Me.chkOption Then
            'Add a name count suffix to paste existing record as new one
            Do
                'Find a new record name
                intI = intI + 1
                strNewRecipe = strRecipe & intI
                Set rgMyRecipes = mws.Range("SavedRecords").Find(strNewRecipe, , , xlWhole)
            Loop Until rgMyRecipes Is Nothing
```

```
            bolNewRecipe = True
        Else
            'strRecipe exist! Ask to overwrite it.
            strMsg = "Recipe '" & strNewRecipe & "' already existe on '" & wsDestination.
            Parent.Name & "." & vbCrLf
            strMsg = strMsg & "Overwrite it?"
            strTitle = "Overwrite recipe '" & strNewRecipe & "'?"
            If MsgBox(strMsg, vbYesNoCancel + vbDefaultButton2 + vbQuestion, strTitle) =
            vbYes Then
                wsDestination.Range("CurrentRecord") = strNewRecipe
            Else
                Exit Sub
            End If
        End If
    End If

    wsSource.Range("CurrentRecord") = strRecipe
    DoEvents
    xl.Calculation = xlCalculationManual
    For Each rgArea In rgSource.Areas
        wsDestination.Range(rgArea.Address).Value = rgArea.Value
    Next
    xl.Calculation = xlCalculationAutomatic
    Application.DisplayAlerts = False
    wsDestination.Save strNewRecipe, bolNewRecipe
    DoEvents
End Sub
```

After declaring all the variables it needs, Sub TransferRecipes( ) determines what the source and destination workbook and worksheets are according to the operation being processed using the mxl, wsSource, and wsDestination object variables. For an "Export recipes" operation, wsSource is set to the active sheet (the one that is automated with the recipe being processed), while wsDestination is set to the mws module-level object variable (if it is an import operation, they simply exchange places).

```
If Me.optExport Then
    Set xl = mxl
    Set wsSource = ActiveSheet
    Set wsDestination = mws
Else
    Set xl = Application
    Set wsSource = mws
    Set wsDestination = ActiveSheet
End If
```

And once the source and destination workbook and worksheets are set, it is time to set the rgSource and rgDestination as Range variables to point to the desired RecipeData range.

```
Set rgSource = wsSource.Range("RecipeData")
Set rgDestination = wsDestination.Range("RecipeData")
```

■ **Attention** The `RecipeData` range name includes all input cells of the `My Recipes` worksheet: the one-side and the many-side record cells. You can select it in the Excel Name box and see what cells it selects. Since it selects noncontiguous cells, its name does not appear in the Excel Name box after being selected in the range name list (Figure 9-26).

Before exporting/importing the selected recipes, `TransferRecipe( )` uses the `Range.Find` method to verify whether the recipe being processed already exists on the destination worksheet.

```
'Verify if recipe already exist on wsDestination
strNewRecipe = strRecipe
Set rgMyRecipes = wsDestination.Range("SavedRecords").Find(strNewRecipe, , , xlWhole)
```

If the recipe record being processed does not exist in the destination worksheet, it is automated by defining the `CurrentRecord` range name (the database data validation list) as New Recipe, and `bolNewRecipe` receives a `True` indication that this is a new recipe record.

```
If rgMyRecipes Is Nothing Then
    'strRecipe das not exist on destination
    wsDestination.Range("CurrentRecord") = "New Recipe"
    bolNewRecipe = True
```

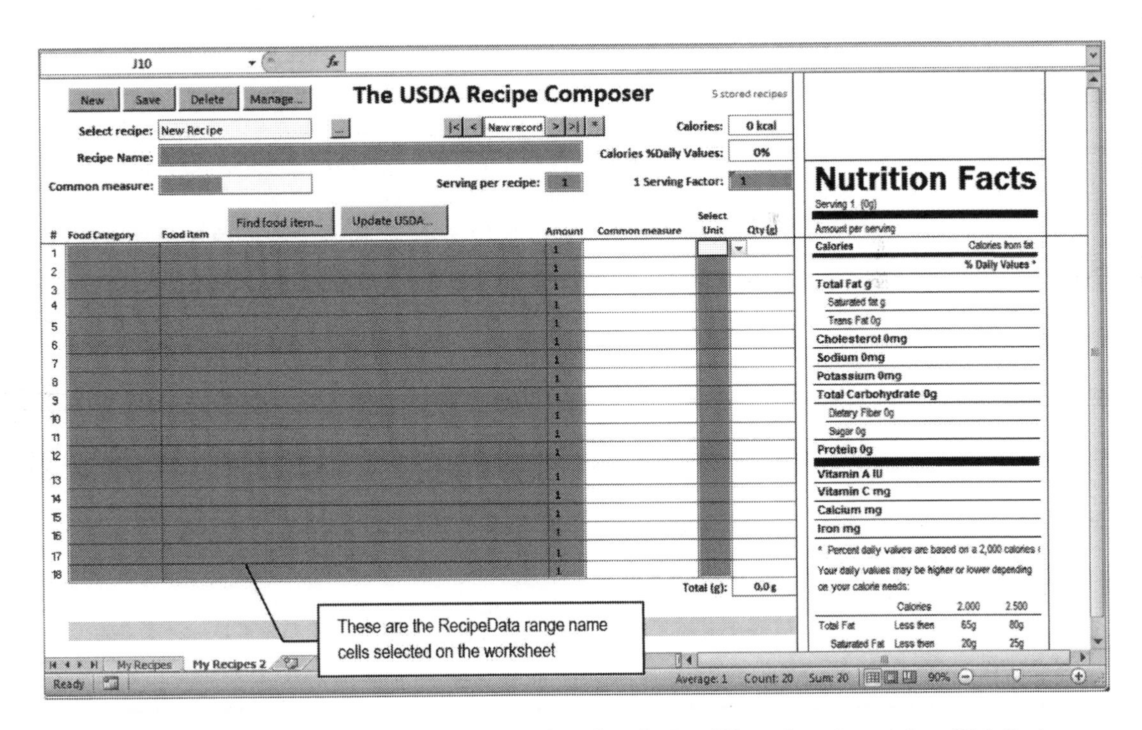

Figure 9-26. This is the RecipeData range name, selected in the Excel Name box. It contains all My Recipes input cells, including both the one-side and many-side record cells

If the recipe already exists on the destination worksheet, a decision must be made according to the chkOption CheckBox value ("Export Recipe as new record" option) that you can see in Figures 9-22 to 9-25. If the option is checked, the recipe must be saved as a new record, changing the recipe name to one that does not exist in the destination database. This is done using a Do...Loop instruction that uses the strNewRecipe String variable to successively concatenate a counter suffix (intI) to the original recipe name (strRecipe), using Range(SaveRecords).Find to search the new name in the destination database. The loop ends when the new strNewRecipe name does not exist in the destination worksheet.

```
'Recipe already exist on desination. Transfer recipe as new recipe?
If Me.chkOption Then
    'Add a name count suffix to paste existing record as new one
    Do
        'Find a new record name
        intI = intI + 1
        strNewRecipe = strRecipe & intI
        Set rgMyRecipes = mws.Range("SavedRecords").Find(strNewRecipe, , , xlWhole)
    Loop Until rgMyRecipes Is Nothing
    bolNewRecipe = True
```

But if the user unchecks the "Export Recipe as new record" option *and* the record already exists in the destination database, a VBA MsgBox( ) function will warn the user before overwriting the recipe. If the user decides to overwrite the existing recipe, the code will automate the destination worksheet by loading the recipe being processed so its recipe data can be changed.

```
Else
    'strRecipe exist! Ask to overwrite it.
    strMsg = "Recipe '" & strNewRecipe & "' already existe on '" & wsDestination.
    Parent.Name & "." & vbCrLf
    strMsg = strMsg & "Overwrite it?"
    strTitle = "Overwrite recipe '" & strNewRecipe & "'?"
    If MsgBox(strMsg, vbYesNoCancel + vbDefaultButton2 + vbQuestion, strTitle) = vbYes Then
        wsDestination.Range("CurrentRecord") = strNewRecipe
```

If it decides to not overwrite the existing recipe, an Exit Sub instruction will exit Sub TransferRecipe( ), returning the code control to Sub ProcessRecipes( ) to process the next selected recipe (if any).

```
        Else
            Exit Sub
        End If
    End If
End If
```

When the Sub TransferRecipe( ) procedure reaches this code point, it is ready to exchange the records between the two selected worksheets, so it is time to automate the source worksheet by loading the recipe being processed. A DoEvents instruction is used here to allow any further database events to execute before the recipe is loaded.

```
wsSource.Range("CurrentRecord") = strRecipe
DoEvents
```

Now it is time to really exchange recipe data between the two worksheet application interfaces, which is done using a `For Each...Next` loop through the `Range.Areas` collection of the `rgSource` range. Note that before the loop is executed, the destination workbook (represented by the `xl as Application` object variable) is put in manual calculation mode to avoid any `Volatile` procedure from running as each cell value is changed by the loop.

```
xl.Calculation = xlCalculationManual
For Each rgArea In rgSource.Areas
    wsDestination.Range(rgArea.Address).Value = rgArea.Value
Next
```

When the loop ends, all values from the source worksheet record have been transferred to the equivalent cells of the destination worksheet record, and the destination workbook is returned to automatic calculation. The code then uses `Application.DisplayAlerts = False` to disable any OLE messages issued by the destination application, which may fire when exporting records to an external workbook.

```
xl.Calculation = xlCalculationAutomatic
Application.DisplayAlerts = False
```

To effectively save the record on the destination workbook, the code calls the `wsDestination.Save` procedure on the destination worksheet, using the destination database engine to automate the task to save the `strNewRecipe` recipe record on its record structure. After the record is saved, another `DoEvents` function is issued to allow time for any database events to be processed, and the code control returns to the `Sub ProcessRecipes( )` procedure.

```
    wsDestination.Save strNewRecipe, bolNewRecipe
    DoEvents
End Sub
```

■ **Attention** This is why `wsDestination` is declared as `Object` instead of as `Worksheet`. Since the Excel `Worksheet` object does not have a `Save` method, it avoids that `wsDestination.Save` instruction code generates a VBA compile error.

Finishing the Export Recipes Operation

When the code control returns to `Sub ProcessRecipes( )` and all selected recipes have been processed, the code verifies whether the operation processed was an "Export recipes" or "Save in My Recipes" operation. If this is true, it first saves the database, which is especially important when recipes have been exported to an external workbook.

```
If Me.optExport Or  Me.optSaveInMyRecipes Then
    wb.Save
```

If the selected recipes have been exported to an external workbook (represented by the `mxl as Application` module-level variable), this workbook must be saved *and* closed. The easy way to determine with VBA code if two different opened workbooks are the same is to compare the `Hwnd` properties (read as *H*andle to a *wind*ow).

```
If Me.optExport Then
    If (mxl.Hwnd <> Application.Hwnd) Then
```

If mxl.Hwnd and Application.Hwnd differ, it means that they are different applications, and the external workbook must be closed, silently saving itself using the Workbook.Close method and passing True to its save argument, and the external Excel application must be finished, using its Application.Quit method.

```
    wb.Close True
    mxl.Quit
End If
```

As good programming practice, all object variables used to represent the external Application, Workbook, and Worksheet objects are set to nothing, finishing the Sub ProcessRecipes( ) procedure.

```
            Set mxl = Nothing
            Set mwb = Nothing
            Set mws = Nothing
        End If
    End If
    ...
End Sub
```

When Sub ProcessRecipes( ) ends, it returns the code control to the cmdExecute_Click( ) event, which will update the frmManageRecipes interface by calling EnableControls(True) and executing the next code fragment:

```
    Call ProcessRecipes(mwb, mws)
    Call EnableControls(True)
End Sub
```

Exporting/Importing Recipe Records to/from an External Workbook

Let's see how frmManageRecipes implements an "Export recipes" operation when cboWorkbook is defined to "External workbook." After selecting the desired recipes to export, when the user clicks Execute, they will be asked to select the external worksheet where the recipes must be exported (Figure 9-27).

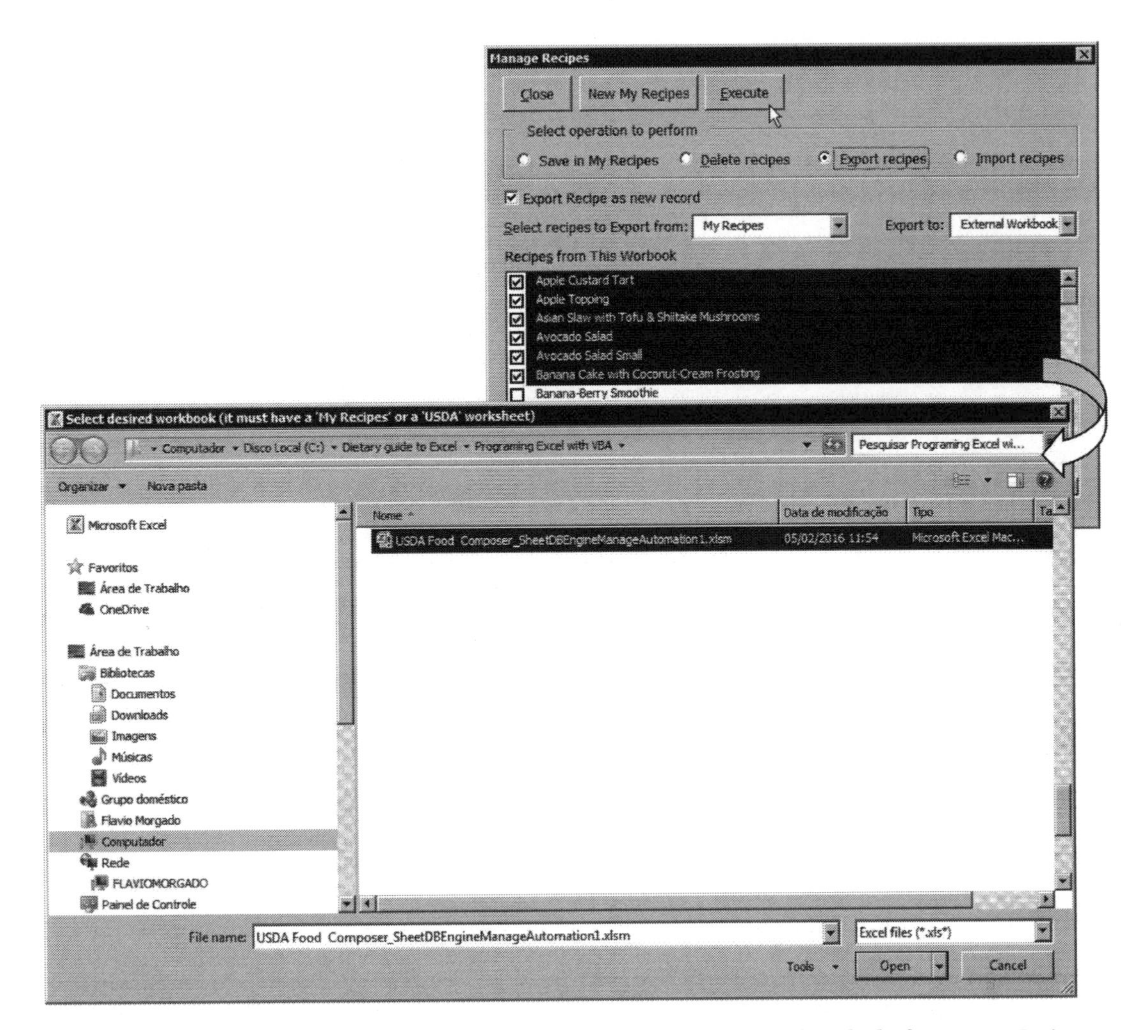

Figure 9-27. *When the "Export recipes" operation is made to an external workbook, the frmManageRecipes UserForm will show the Open dialog box to select the desired workbook to where the selected recipes must be exported*

Note how the cmdExecute_Click( ) event code deals with this situation:

```
Private Sub cmdExecute_Click()
    ....
        If Me.optExport And Me.cboWorkbook = "External Workbook" Then
            'Export operation to external workbook
            If GetExternalWorkbook() Then
                'Load recipes of first external MyRecipes
                Me.cboMyRecipesDestination.ListIndex = 0
                bolSetDestination = True
            Else
                Call EnableControls(True)
                Exit Sub
            End If
        Else
```

As you can see, the code calls `GetExternalWorkbook( )` to allow the selection of the destination workbook to where the recipes must be exported.

Selecting the Destination Workbook with GetExternalWorkbook( )

The `Private Function GetExternalWorkbook( )` must do the following:

1. Allow the selection of an Excel workbook to be used as destination
2. Return `True` if the selected workbook has either a `My Recipes` or a `USDA` worksheet, using the `mintWorkbookHasMyRecipes` module-level variable to indicate how many copies of the `My Recipes` worksheet it has
3. Use `mxl as Application` and `mwb as Workbook` to set a reference to the `Excel.Application` and `Workbook` objects selected
4. Use the `mws as Worksheet` module-level variable to set a reference to either the `My_Recipes` or `USDA` worksheet found in the destination workbook

---

■ **Attention** By default, `GetExternalWorkbook( )` will give precedence to a `My Recipes` worksheet over `USDA` worksheet whenever the destination workbook has both of them.

```
Private Function GetExternalWorkbook() As Boolean
    Dim wb As Workbook
    Dim ws As Worksheet
    Dim wsMyRecipes As Worksheet
    Dim wsUSDA As Worksheet
    Dim rg As Range
    Dim strFile As String
    Dim strFilter As String
    Dim strTitle As String
    Dim strMsg As String
    Dim bolWorkbookOpen As Boolean
    Dim bolWorkbookOpenInThisApp As Boolean
    Dim bolFound As Boolean

    strFilter = "Excel files, *.xls*, All Files, *.*"
    strTitle = "Select desired workbook (it must have a 'My Recipes' or a 'USDA' worksheet)"
    'Get workbook file name
    strFile = ShowDialogBox(OpenFile, , strTitle, , strFilter, 1)

    If Len(strFile) Then
        bolWorkbookOpen = IsWorkbookOpen(strFile)
        If bolWorkbookOpen Then
            'Verify if workbook is already open on THIS Excel.Application
            For Each wb In Application.Workbooks
                If wb.FullName = strFile Then
                    bolWorkbookOpenInThisApp = True
                    Set mxl = Application
                    Set mwb = wb
                    Exit For
```

```
            End If
        Next
    End If

    If (Not bolWorkbookOpen) Or bolWorkbookOpenInThisApp Then
        Application.Cursor = xlWait
            If Not bolWorkbookOpenInThisApp Then
                'Open workbook
                Set mxl = New Excel.Application
                mxl.DisplayAlerts = False
                Set mwb = mxl.Workbooks.Open(strFile)
            End If

            'Verify if the opened workbook has a "My Recipes" or "USDA" sheet tab
            mintWorkbookHasMyRecipes = 0
            mbolCancelEvent = True
            Me.cboMyRecipesDestination.Clear
            For Each ws In mwb.Worksheets
                'Confirm by it CodeName if it is the desired sheet tab
                If Left(ws.CodeName, 9) = "MyRecipes" And ws.Visible = xlSheetVisible
                Then
                    If wsMyRecipes Is Nothing Then
                        Set wsMyRecipes = ws
                    End If
                    mintWorkbookHasMyRecipes = mintWorkbookHasMyRecipes + 1
                    Me.cboMyRecipesDestination.AddItem ws.Name
                    bolFound = True
                ElseIf (ws.Name = "USDA") And (Left(ws.Range("A1"), 2) = "SR") Then
                    'Verify if "USDA" sheet tab has as My_Recipes range name
                    On Error Resume Next
                    Set rg = ws.Range("My_Recipes")
                    If Not rg Is Nothing Then
                        'wsUSDA has a pointer to "USDA"
                        Set wsUSDA = ws
                        bolFound = True
                    End If
                End If
            Next
            mbolCancelEvent = False
        Application.Cursor = xlDefault

        If bolFound Then
            Set mws = IIf(mintWorkbookHasMyRecipes > 0, wsMyRecipes, wsUSDA)
        Else
            Call CloseExternalWorkbook
            MsgBox "The selected workbook doesn't has a 'My Recipe' or 'USDA' worksheet
            to manage recipes data", _
                    vbInformation, _
                    "'My Recipes' or 'USDA' sheet tab not found!"
        End If
    End If
```

```
    End If
    GetExternalWorkbook = bolFound
End Function
```

After declaring all its variables, GetExternalWorkbook( ) uses the ShowDialogBox( ) procedure to issue the Windows Open dialog box and allows the destination workbook to be selected.

```
strFilter = "Excel files, *.xls*, All Files, *.*"
strTitle = "Select desired workbook (it must have a 'My Recipes' or a 'USDA' worksheet)"
'Get workbook file name
strFile = ShowDialogBox(OpenFile, , strTitle, , strFilter, 1)
```

If a workbook has been selected, it uses Function IsWorkbookOpen( ) (analyzed in section "Verify Whether a Workbook Is Open" earlier in this chapter) to verify whether the selected workbook is already opened by the user.

```
If Len(strFile) Then
    bolWorkbookOpen = IsWorkbookOpen(strFile)
```

If the selected file is already opened, it first checks whether it is a member of this Application.Workbooks Collection object, executing a For Each...Next loop that compares each opened Workbook.FullName property with the strFile argument. If it finds a match, it defines bolWorkbookOpenInThisApp = True, sets the module-level variables mxl and mwb to point to the appropriate objects, and exits the loop.

```
If bolWorkbookOpen Then
    'Verify if workbook is already open on THIS Excel.Application
    For Each wb In Application.Workbooks
        If wb.FullName = strFile Then
            bolWorkbookOpenInThisApp = True
            Set mxl = Application
            Set mwb = wb
            Exit For
        End If
    Next
End If
```

The next text verifies whether the selected workbook file is not open in another Excel windows *or* if it is already opened in this Application object.

```
If (Not bolWorkbookOpen) Or bolWorkbookOpenInThisApp Then
```

If one of these two conditions is met, the code changes the Application.Cursor property and tests whether the selected file *is not* opened in this Application interface. If the selected file is not open, it sets the mxl module-level variable to a new, hidden Microsoft Excel Application object, sets DisplayAlerts = False to avoid OLE messages when it is automated, and uses the Workbooks.Open method to load the desired workbook file, using the mwb module-level variable to reference it.

```
Application.Cursor = xlWait
    If Not bolWorkbookOpenInThisApp Then

    'Open workbook
```

```
        Set mxl = New Excel.Application
        mxl.DisplayAlerts = False
        Set mwb = mxl.Workbooks.Open(strFile)
    End If
```

Now that it already has the mwb module-level variable pointing to the external workbook file, it is time to set the stage to verify that the external workbook object has a My Recipes or USDA worksheet to use as a destination. It sets mintWorkbookHasMyRecipes = 0 and mbolCancelEvent = True to avoid cascade events inside the loop, while also clearing the cboMyRecipesDestination ComboBox.

```
'Verify if the opened workbook has a "My Recipes" or "USDA" sheet tab
mintWorkbookHasMyRecipes = 0
mbolCancelEvent = True
Me.cboMyRecipesDestination.Clear
```

A For Each... Next loop is made through the mwb.Worksheets collection to search for the desired worksheet references. The first test searches for any My Recipes worksheet (one that has its CodeName property's first nine characters = MyRecipes and not hidden inside the destination workbook).

```
For Each ws In mwb.Worksheets
    'Confirm by it CodeName if it is the desired sheet tab
    If Left(ws.CodeName, 9) = "MyRecipes" And ws.Visible = xlSheetVisible Then
```

The first worksheet match is set to the wsMyRecipes object variable, while mintWorkbookHasMyRecipes is incremented, and the cboMyRecipesDestination list is filled with the worksheet name. The bolFound Boolean variable signals the finding.

```
If wsMyRecipes Is Nothing Then
    Set wsMyRecipes = ws
End If
mintWorkbookHasMyRecipes = mintWorkbookHasMyRecipes + 1
Me.cboMyRecipesDestination.AddItem ws.Name
bolFound = True
```

The second test verifies whether each workbook name is USDA and the first two characters of its A1 cell are SR, as a clear indication that it is a USDA worksheet produced by the USDA Food List Creator.accdb Microsoft Access application.

```
ElseIf (ws.Name = "USDA") And (Left(ws.Range("A1"), 2) = "SR") Then
```

If such a worksheet is found, the code disables VBA error messages with an On Error Resume Next instruction and tries to set a reference to the My_Recipes range name.

```
'Verify if "USDA" sheet tab has as My_Recipes range name
On Error Resume Next
Set rg = ws.Range("My_Recipes")
```

If the range is found, it uses the wsUSDA object variable to set a reference to this worksheet, and once more bolFound =True to confirm the finding.

```
        If Not rg Is Nothing Then
            'wsUSDA has a pointer to "USDA"
            Set wsUSDA = ws
            bolFound = True
        End If
    End If
Next
```

When this second loop ends, the code sets `mbolCancelEvent = False` and resets the `Application.Cursor` to it default mouse pointer.

```
    mbolCancelEvent = False
Application.Cursor = xlDefault
```

If `bolFound = True`, a destination worksheet was found inside the selected workbook file, and a VBA `IIF( )` function is used to test the `mintWorkbookHasMyRecipes` variable and set the `mws` module-level variable with a reference to either `wsMyRecipes` or `wsUSDA` (note that there is a precedence to the `My Recipes` worksheet over the USDA worksheet):

```
If bolFound Then
    Set mws = IIf(mintWorkbookHasMyRecipes > 0, wsMyRecipes, wsUSDA)
```

Otherwise, no destination worksheet is found on a selected file, and a call is made to the `CloseExternalWorkbook( )` procedure.

```
Else
    Call CloseExternalWorkbook
```

Closing Any Opened Workbook with CloseExternalWorkbook( )

The `Private Sub CloseExternalWorkbook( )` procedure is called by many events of `frmManageRecipes`, which is responsible for closing any references to an external workbook and avoiding keeping a hidden Excel `Application` object open when it is no longer necessary. The procedure executes this code:

```
Private Sub CloseExternalWorkbook()
    If Not mxl Is Nothing Then
        If mxl.Hwnd <> Application.Hwnd Then
            mxl.DisplayAlerts = False
            mwb.Close
            mxl.Quit
            Set mws = Nothing
            Set mwb = Nothing
            Set mxl = Nothing
        End If
    End If
End Sub
```

There is a trick already mentioned here: if the `mxl` module-level variable is pointing to an `Excel.Application` object, the code compares both `mxl.Hwnd` and `Application.Hwnd` properties: the application handles that indicate its `mxl` and the current `Application` object that points to the same object.

```
Private Sub CloseExternalWorkbook()
    If Not mxl Is Nothing Then
        If mxl.Hwnd <> Application.Hwnd Then
```

If they are not the same, the code sets again the external `mxl.DisplayAlerts = False` properties to avoid the external application displaying OLE messages and then closes the external workbook without saving it. As good programming practice, all three object variables used to point to the destination application, workbook, and worksheet (`mxl`, `mwb`, and mws) are set to `Nothing`.

Finishing GetExternalWorkbook( )

After `CloseExternalWorkbook( )` ends and there is no external workbook opened by the `UserForm`, the code control returns to the `GetExternalWorkbook( )` procedure, which uses a VBA `MsgBox( )` function to warn the user that the selected workbook file does not fit to export or import recipe operations, and the procedure ends indicating if it has opened an external workbook file.

```
                MsgBox "The selected workbook doesn't has a 'My Recipe' or 'USDA' worksheet
                to manage recipes data", _
                         vbInformation, _
                         "'My Recipes' or 'USDA' sheet tab not found!"
            End If
        End If
    End If
    GetExternalWorkbook = bolFound
End Function
```

Finishing the cmdExecute_Click( ) Event

When `Private Function GetExternalWorkbook( )` finishes, the code control returns to the `cmdExecute_Click( )` event, which will text the function result, and if it is `True` (an external workbook was opened to export the selected recipes), it will set `cboMyRecipesDestination` to the first list item and define `bolSetDestination = True`. Otherwise, no external workbook was opened, and the code will just call `EnableControls(True)` and exit the event code with an `Exit Sub` instruction.

```
If GetExternalWorkbook() Then
    'Load recipes of first external MyRecipes
    Me.cboMyRecipesDestination.ListIndex = 0
    bolSetDestination = True
Else
    Call EnableControls(True)
    Exit Sub
End If
```

Supposing that an external workbook was opened, `cmdExecute_Click( )` verifies whether `bolSetDestination = True`, and if it is, it calls `SelectMyRecipesDestination(True)` to change the `UserForm` interface to allow select/inspect all possible destination worksheet recipes. If this is an export operation, it uses the `cboMyRecipesDestionation` value to set a reference to the `mws as Worksheet` module-level variable, exiting the code with an `Exit Sub` instruction.

```
If bolSetDestination Then
    ...
    Call SelectMyRecipesDestination(True)
    If Me.optExport Then
        Set mws = mwb.Worksheets((cboMyRecipesDestination))
    End If
    Exit Sub
End if
```

Supposing that you had selected as the destination workbook USDA Food Composer_SheetDBEngineManageAutomation1.xlsm, as suggested by Figure 9-27, when the cmdExecute_Click( ) event finishes, the frmManageRecipes UserForm must be similar to Figure 9-28, (showing all recipes from its My Recipes worksheet, while also exposing its My Recipes 2 worksheet, with no recipes at all).

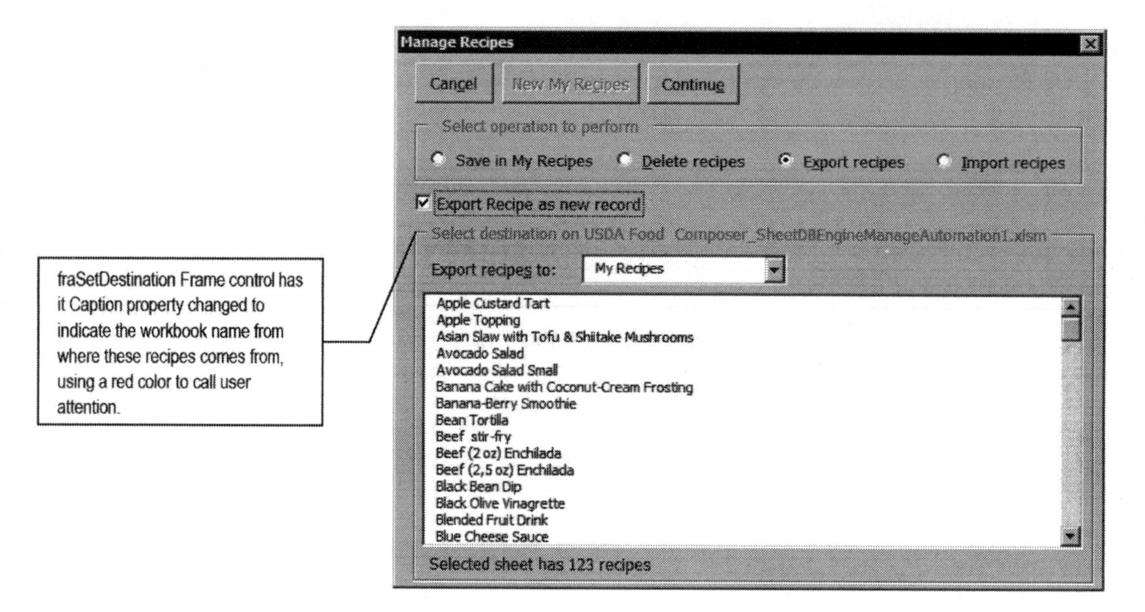

Figure 9-28. *The frmManageRecipes UserForm changes its interface, showing the fraSetDestination Frame control over cboMyRecipes, cboWorkbook, and lstRecipes, to allow you to select/inspect the destination worksheet that will receive the selected recipes*

Once again, there are just two possibilities: cancel the operation or continue with the exporting process. By clicking the Cancel CommandButton (cmdClose), the external workbook is closed, and the UserForm interface is updated, allowing you to select another workbook to export to.

```
Private Sub cmdClose_Click()
    Select Case Me.cmdClose.Caption
        ...
        Case "Cancel"
            If Me.fraSetDestination.Visible = True Then
                mintWorkbookHasMyRecipes = MyRecipesCount()
                Call CloseExternalWorkbook
                Call SelectMyRecipesDestination(False)
            Else
```

```
                mbolCancel = True
            End If
            Call EnableControls(True)
    End Select
```

By clicking the Continue CommandButton (cmdExecute), the process continues by first updating the UserForm interface and then calling ProcessRecipes(mwb, mws).

```
Private Sub cmdExecute_Click()
    ...
    If Me.cmdExecute.Caption = "Continue" Then
        Call SelectMyRecipesDestination(False)
        Call ProcessRecipes(mwb, mws)
    Else
        ...
    End If
End Sub
```

Supposing that you click Continue, the selected recipes can be exported in two different ways, according to the chkOption CheckBox value ("Export Recipe as new record" option) and whether the destination worksheet already has any of the selected recipes:

- If chkOption is checked, each existing recipe will be renamed with a suffix counter and added to the destination database.
- If chkOption is unchecked, a VBA MsgBox( ) warning will be sent to the application user so it can make a decision: cancel the recipe transfer (No or Cancel option) or overwrite the destination recipe (Yes option) (Figure 9-29).

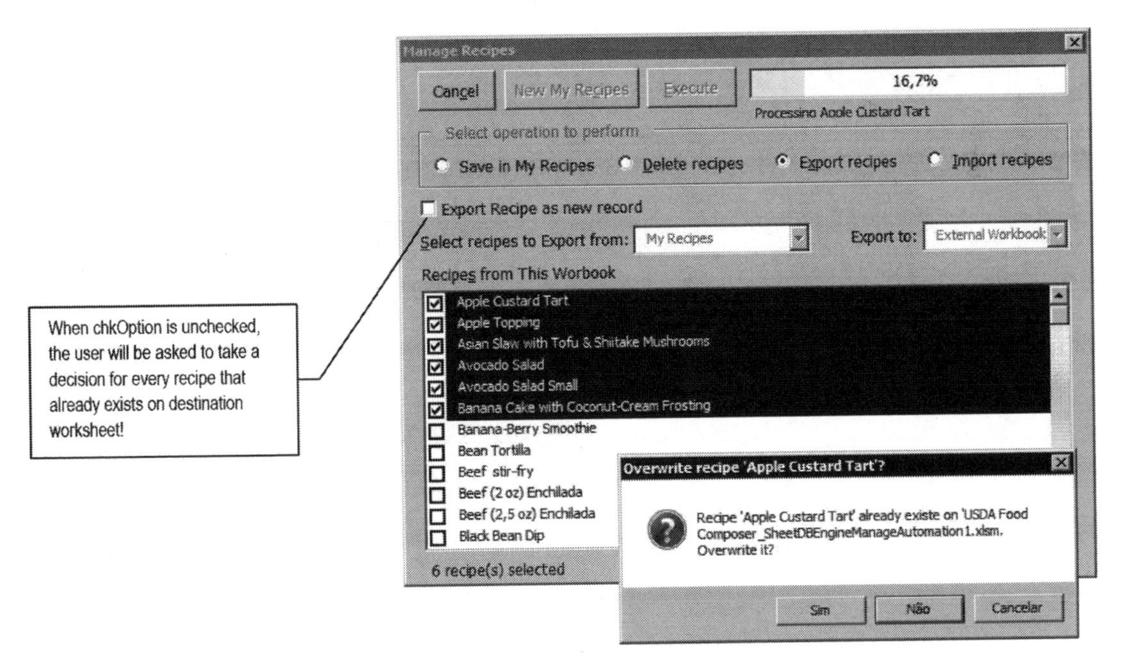

Figure 9-29. *If chkOption is checked, any recipe already existing in the destination worksheet will be saved as a new recipe record, with a suffix counter added to its name. Otherwise, the interface will ask the user to make a decision: cancel the operation (for this recipe) or overwrite the destination recipe*

The next code fragment of Sub TransferRecipes( ) shows how the VBA MsgBox( ) function deals with such situations. Note that it offers two options to cancel the overwriting (No and Cancel), keeping No as the default option. The procedure will end whenever No or Cancel is selected, returning the code control to Sub ProcessRecipes( ), as explained earlier in this chapter.

```
Private Sub TransferRecipe(strRecipe As String)
        ...
        If Me.chkOption Then
            ...
        Else
            'strRecipe exist! Ask to overwrite it.
            strMsg = "Recipe '" & strNewRecipe & "' already existe on '" & wsDestination.
            Parent.Name & "." & vbCrLf
            strMsg = strMsg & "Overwrite it?"
            strTitle = "Overwrite recipe '" & strNewRecipe & "'?"
            If MsgBox(strMsg, vbYesNoCancel + vbDefaultButton2 + vbQuestion, strTitle) =
            vbYes Then
                wsDestination.Range("CurrentRecord") = strNewRecipe
            Else
                Exit Sub
            End If
        End If
        ...
End Sub
```

When Sub ProcessRecipes( ) ends to process all recipes on an export operation, it calls the Workbook object's Save and Close methods to close the external workbook after saving it.

Importing Recipes from an External Workbook

Whenever you click the “Import recipes” OptonButton, a Windows Open dialog box is immediately shown to allow select the source workbook from where the recipes must be imported (Figure 9-30).

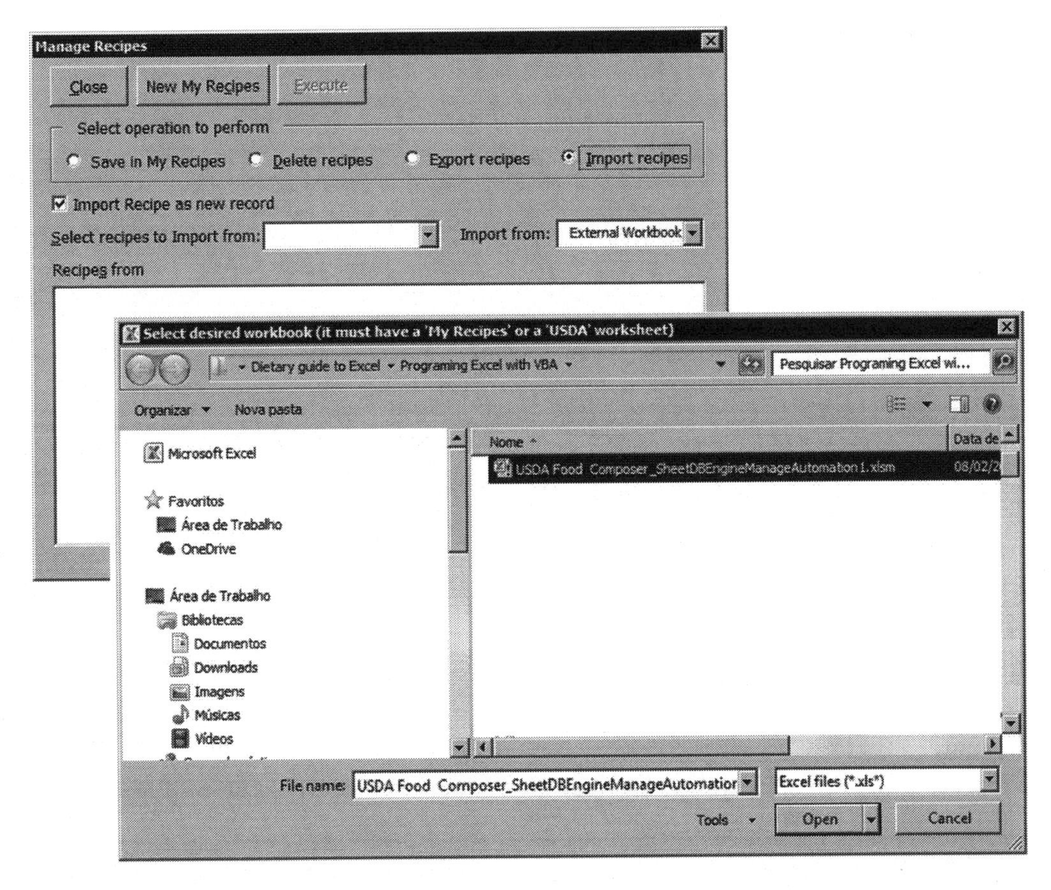

***Figure 9-30.** The "Import recipes" option also makes small updates in frmManageRecipes interface labels and controls, which immediately shows the Windows Open dialog box to allow you to select the source workbook from where the recipes must be imported*

Such unpredictable behavior is because of the original `frmManageRecipes` interface project, which reserved `cboRecipes` and `lstRecipes` to always be the recipe's source: the primary local where recipes must be selected to conduct an operation. Note in Figure 9-30 that `lblRecipes` shows "Import recipes from," `lblWorkbook` shows "Import from," `cboWorbook` shows "External workbook," `chkOption` shows "Import recipes as new recipe," and `cboRecipes` and `lstRecipes` have their lists cleared. These changes happened on the `optImport_Click( )` event, which executes this code:

```
Private Sub optImport_Click()
    Dim rg As Range

    Call CloseExternalWorkbook
    Call UpdateProgressBar(False)
    mbolUpdateInterface = True
    Me.lblRecipes.Caption = "Select recipes to Import from:"
    Me.lblRecipesFrom.Caption = "Recipes from "
    Me.chkOption.Caption = "Import Recipe as new record"
    Me.chkOption.Visible = True
    Me.cboWorkbook.Locked = True
```

```
    Me.lblWorkbook.Caption = "Import from:"
    mbolCancelEvent = True
        Me.cboMyRecipes.Clear
        Me.lstRecipes.Clear
        Me.cboWorkbook.ListIndex = 1
    mbolCancelEvent = False
    If GetExternalWorkbook() Then
        Call LoadcboMyRecipes(mwb)
    End If
End Sub
```

There is no surprise here. Any open workbook is closed with a call to Sub CloseExternalWorkbook( ), and the progress bar is hidden with a call to Sub UpdateProgressBar(False). Next, the lblRecipes, lblRecipesFrom, chkOption, and lblWorkbook Caption properties are updated, while cboWorkbook is locked to reflect the import operation, and, after setting mboCancelEvent = True, both cboMyRecipes and lstRecipes are cleared, avoiding firing cascade events. The Windows Open dialog box is shown by making a call to Function GetExternalWorkbook( ), which leaves the interface in the state depicted by Figure 9-30.

If the user closes the Open dialog box by clicking its Close or Cancel button, no file will be selected, GetExternalWorkbook( ) = False, and the optImport_Click( ) event ends doing nothing. The frmManageRecipes interface will look like Figure 9-31.

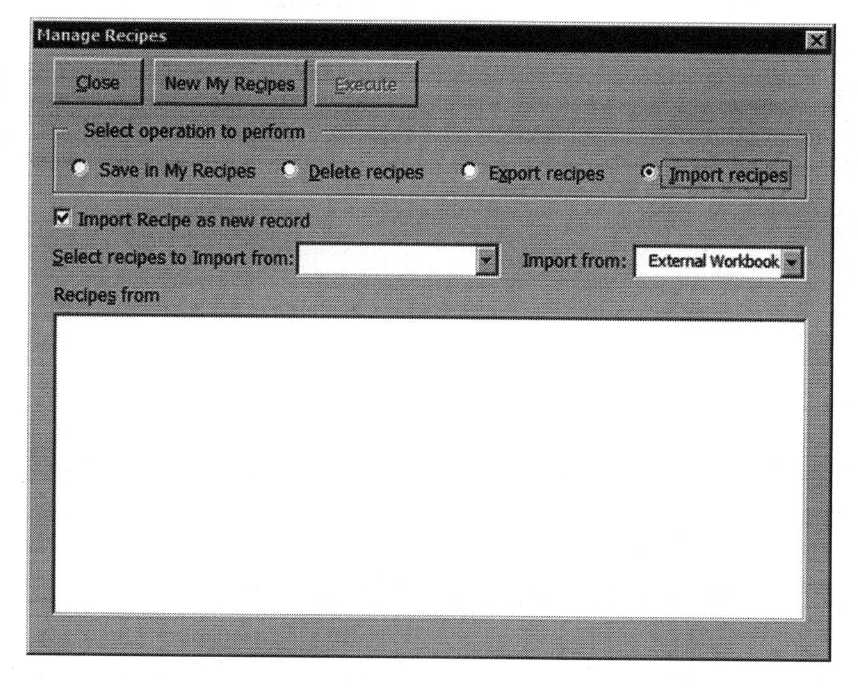

Figure 9-31. *This is the frmManageRecipes interface after selecting the optImport ("Import recipes") option and canceling the Open dialog box without selecting any destination workbook*

To show again the Windows Open dialog box needed to select the source workbook for an "Import recipes" operation, the user must either select any other option button and reselect optImport to fire again the optImport_Click( ) event or just double-click the optImport Option button, which will fire the optImport_DblClick( ) event, executing this code:

```
Private Sub optImport_DblClick(ByVal Cancel As MSForms.ReturnBoolean)
    Call optImport_Click
End Sub
```

Easy, huh?

But supposing that you had selected optImport and chose as the source workbook for the import operation USDA Food Composer_SheetDBEngineManageAutomation1.xlsm (as suggested by Figure 9-30), when you close the Windows Open dialog and return the code control to the optImport_Click( ) event, it calls Sub LoadcboMyRecipes(mwb) to load all recipes of the first My Recipes worksheet found in the selected, source workbook, changing the frmManageRecipes interface to something like Figure 9-32.

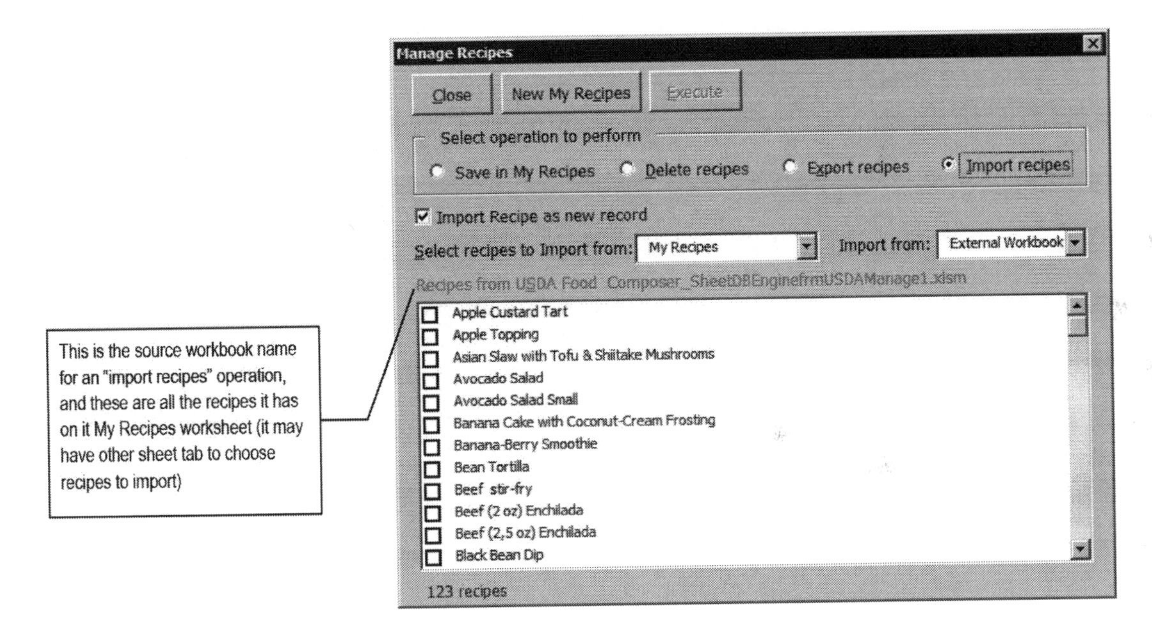

Figure 9-32. *This is the frmManageRecipes interface after selecting USDA Food Composer_ SheetDBEngineManageAuto-mation1.xlsm as the workbook source for an "Import recipes" operation. Note that the lblRecipesFrom Label control shows the source workbook name in red to call attention to the fact that those recipes belong to an external source*

frmManageRecipes is waiting for the user to select the recipes to be imported by enabling the cmdExecute CommandButton to allow execution of this operation. Supposing that you had selected the first five recipes and clicked cmdExecute, if the current application has just one My Recipes worksheet, the import operation will begin immediately. Otherwise, the frmManageRecipes interface will be changed to allow selection of the destination worksheet to where the recipes must be imported (Figure 9-33).

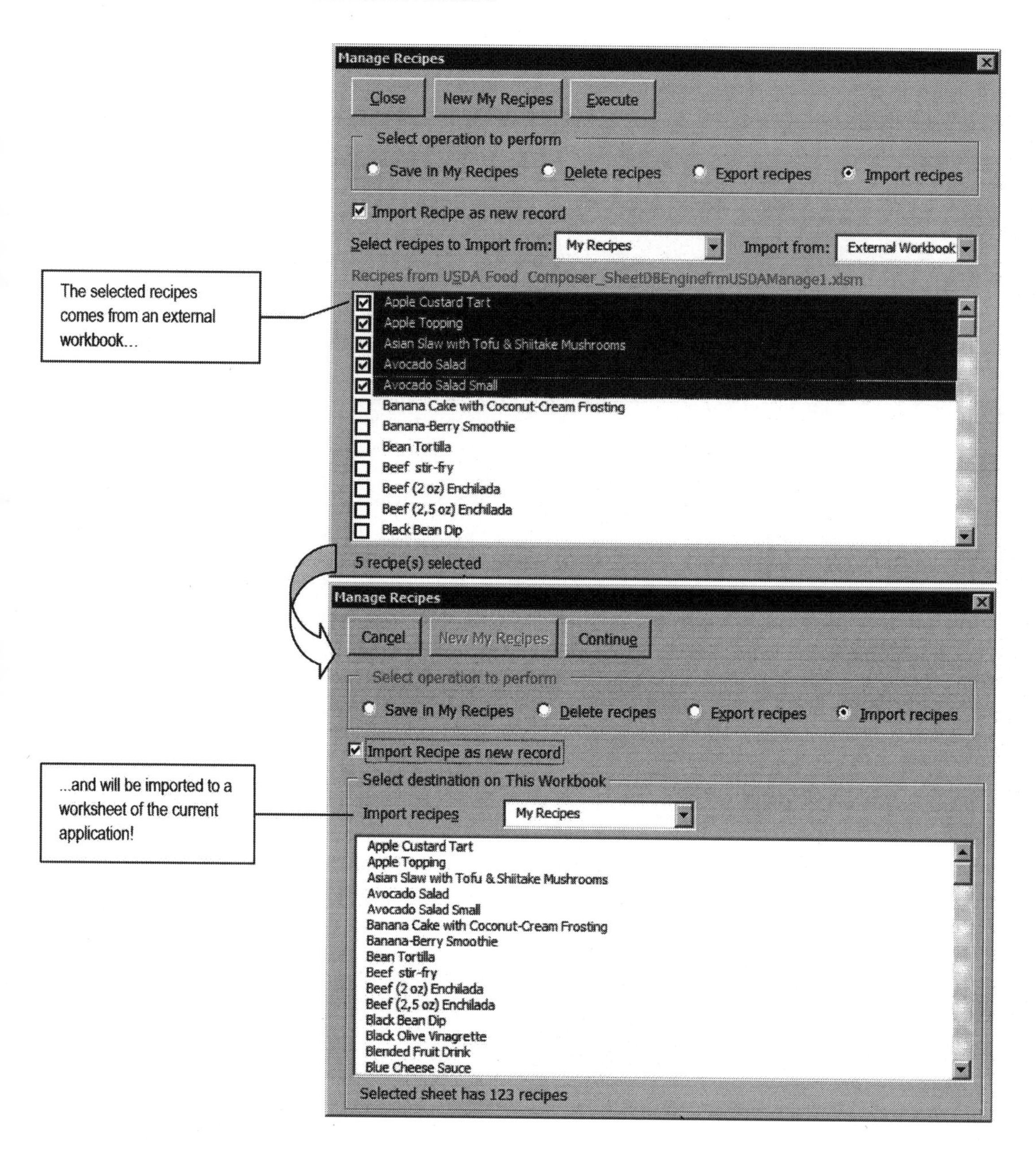

Figure 9-33. *When an "Import recipes" operation is in progress, the user selects recipes from a source My Recipes worksheet on an external workbook and eventually selects the destination worksheet in the current application*

The next code fragment shows what happens in cmdExecute_Click when an "Import recipes" operation is in progress:

```
Private Sub cmdExecute_Click()
    ...
```

```
        Else
            'Operation is Import or Export to ThisWokbook
            'Select destination My Recipes?
            bolSetDestination = (MyRecipesCount() > 1)
            ...
        End If

        If bolSetDestination Then
            If optImport Or Me.cboWorkbook = "This Workbook" Then
                'ThisWorkbook has more than one possible destination
                Call LoadcboMyRecipes(wb, True)
            End If
            Call SelectMyRecipesDestination(True)
            ...
             Exit Sub
```

As you can see, it uses `Function MyRecipesCount( )` to indicate whether the current application has more than one `My Recipes` to choose as a destination worksheet, and if this is `True`, it calls `Sub LoadcboMyRecipes(wb, True)` and `Sub SelectMyRecipesDestination(True)` to allow you to select the destination worksheet inside `ThisWorkbook`.

This will change `cmdContinue Caption` to Continue, which will execute this code to begin importing the selected recipes:

```
Private Sub cmdExecute_Click()
    ...
    If Me.cmdExecute.Caption = "Continue" Then
        Call SelectMyRecipesDestination(False)
        Call ProcessRecipes(mwb, mws)
```

Once again, a call to `Sub SelectMyRecipesDestination(False)` restores the `frmManageRecipes` to the default interface, and the selected recipes are processed by calling `Sub ProcessRecipes(mwb, mws)`—the external object variables that now point to the external workbook and selected worksheet (Figure 9-34)!

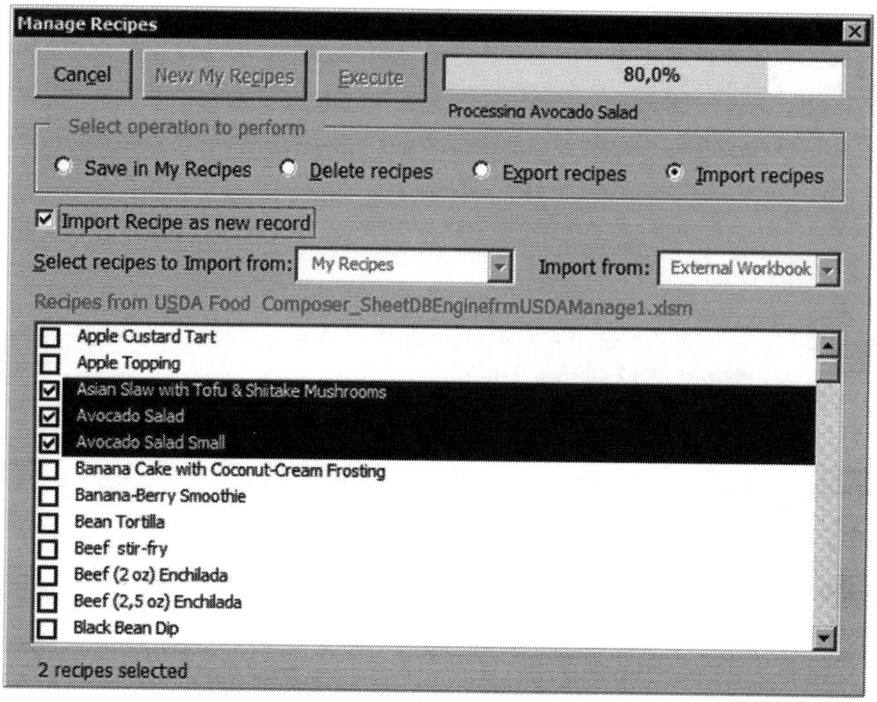

Figure 9-34. *When an "Import recipes" operation is in progress, the frmManageRecipes UserForm processes recipes on an external workbook, with the interface indicating how many recipes still remain unprocessed*

Note, however, that `Sub ProcessRecipes( )` does not close the external workbook after processing all the selected recipes. It will remain open, allowing you to select other worksheets and recipes to be imported, until another operation option is selected or the `cmdClose` Command button is selected to close the `UserForm`.

Supposing that you chose to import the first five recipes, keeping `chkOption` checked ("Import recipes as new record"), as suggested by Figures 9-33 and 9-34, since these recipes already exist in the destination worksheet, they will be saved with the same name concatenated by a suffix counter (Figure 9-35).

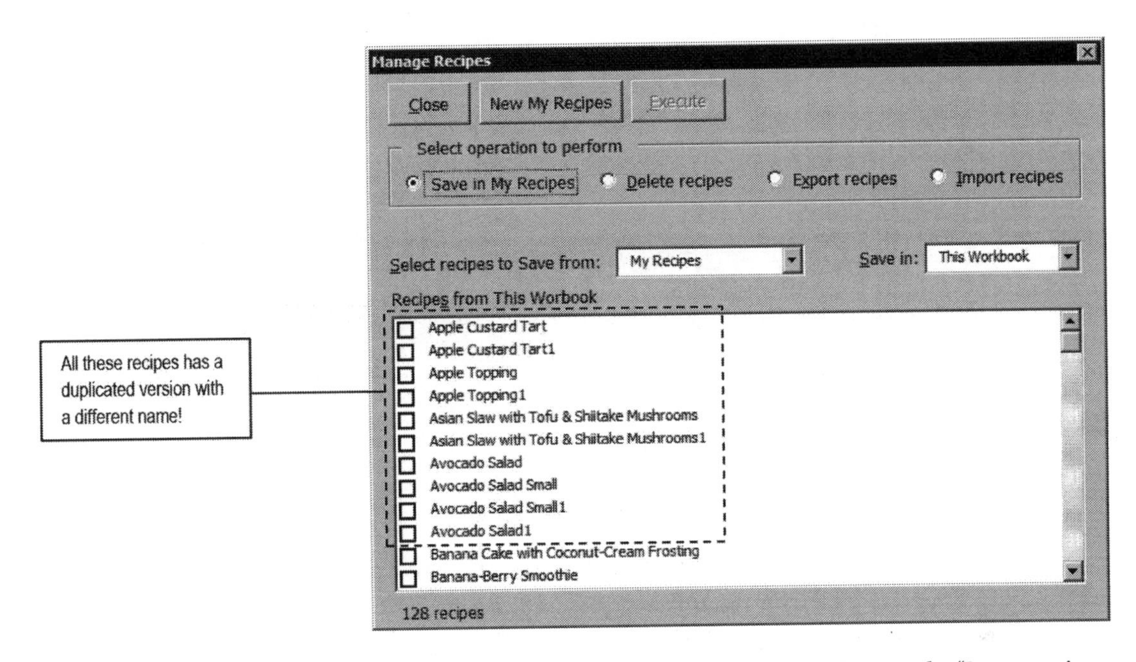

Figure 9-35. *If you import recipes that already exist in the destination worksheet, keeping the "Import recipes as new record" option checked, the imported recipes will be saved with the same name concatenated by a suffix counter*

Deleting Recipes from This Workbook

Using Figure 9-35 as metaphor, when you have a situation where recipes have been duplicated by an "Import recipes" operation, let's suppose that the user wants to delete all duplicated recipes. Just select them in the frmManageRecipes interface and choose the "Delete recipes" option, firing the optDelete_Click( ) event.

```
Private Sub optDelete_Click()
    Call CloseExternalWorkbook

    If mbolUpdateInterface Then
        Call LoadcboMyRecipes
        mbolUpdateInterface = True
    End If
    Call UpdateProgressBar(False)
    Me.lblRecipes.Caption = "Select recipes to Delete from:"
    Me.chkOption.Caption = "Keep Recipe nutritional information on My_Recipes"
    Me.chkOption.Visible = True
    Me.cboWorkbook.ListIndex = 0
    Me.cboWorkbook.Locked = True
    Me.lblWorkbook.Caption = "Delete from:"
End Sub
```

As you can see, when another operation is selected, it first calls Sub CloseExternalWorkbook( ) to close any workbook opened by the UserForm. If mbolUpdateInterface = True (which is always true after an "Import recipes" operation), it calls Sub LoadcboMyRecipes( ) with no argument to update the UserForm interface to ThisWorkbook. The progress bar is hidden and the label controls of the user form are updated to reflect the "Delete recipes" operation (note that chkOption now indicates "Keep recipe nutritional information" on My_Recipes).

After selecting the recipes to be deleted, cmdExecute will be enabled, and if the user clicks it, it will receive a warning message before beginning the deletion process (Figure 9-36).

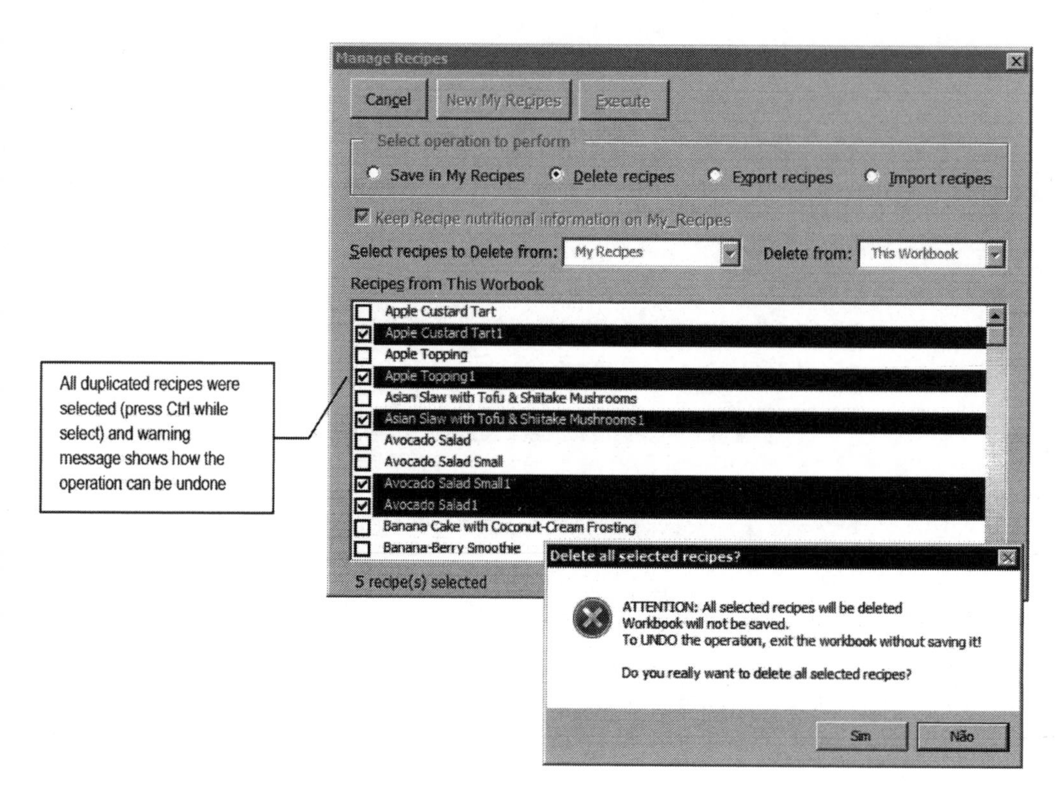

Figure 9-36. *The frmManageRecipes uses a VBA MsgBox( ) to warn the user before beginning a "Delete recipes" operation. Note that the code does not save the workbook, allowing you to undo the delete operation by closing it without saving*

The next code fragment shows how this happens inside the cmdExecute_Click( ) event:

```
Private Sub cmdExecute_Click()
    ...
    Else
        Set wb = ThisWorkbook
        Set ws = ActiveSheet
        Call EnableControls(False)
        If Me.optDelete Then
            strMsg = "ATTENTION: All selected recipes will be deleted" & vbCrLf
            strMsg = strMsg & "Workbook will not be saved." & vbCrLf
            strMsg = strMsg & "To UNDO the operation, exit the workbook without saving it!"
            & vbCrLf & vbCrLf
            strMsg = strMsg & "Do you really want to delete all selected recipes?"
            strTitle = "Delete all selected recipes?"
```

```
            If MsgBox(strMsg, vbYesNo + vbCritical + vbDefaultButton2, strTitle) = vbNo Then
                EnableControls (True)
                Exit Sub
            End If
            ...
        End If
        Call ProcessRecipes(wb, ws)
    End If
End Sub
```

Note that if the No option of the MsgBox( ) warning is selected, the procedure calls EnableControls(True) to update the UserForm interface and exit doing nothing; otherwise, it just calls Sub ProcessRecipes(wb, ws) to delete the selected recipes, which now will execute the next code fragment:

```
Private Sub ProcessRecipes(wb As Workbook, ws As Worksheet)
    ...
        For Each varItem In mcolSelected
            ...
            'Updade Progress Bar (intI = recipe count, intJ = lstRecipes.Index)
            intI = intI + 1
            intJ = varItem + IIf(optDelete, 1 - intI, 0)
           ...
            If Me.optDelete Then
                mobjMyRecipes.DeleteRecord strRecipe, Me.chkOption
                Me.lstRecipes.RemoveItem (intJ)
                ...
            End If
            mcolSelected.Remove (1)
            Me.lblSelected.Caption = mcolSelected.Count & " recipes selected"
        Next
    ...
End Sub
```

It is quite interesting that to delete the selected recipes, the code uses pure database automation, calling the mbojMyRecipes.DeleteRecord( ) method and passing the recipe name and chkOption state to indicate whether the recipe nutritional information must be kept in the USDA worksheet's My_Recipes range name. This operation deserves these considerations:

- An object reference to the worksheet used as the recipe source to be deleted is stored inside the mobjMyRecipes as Object variable. It can't be declared as Worksheet because the Worksheet object does not have a Delete method, which will raise a VBA compile error.
- Whenever the current application has more than one My Recipes worksheet, chances are that the user may duplicate recipes on different sheet tabs. So, when the user decides to delete any duplicated copy, the recipe nutritional information will also be deleted from the My_Recipes range name, and if the deleted recipe is used as a food item of any other recipe, the recipe nutritional value will fail to calculate. That is why the option "Keep recipe nutritional information on My_Recipes" is shown.

■ **Attention** The user can always rebuild the My_Recipes range name using the frmManageRecipes "Saving in My Recipes" option.

Also note something quite interesting: Sub ProcessRecipes( ) deletes recipes from the top to the bottom, removing them from both the lstRecipes ListBox and the mcolSelected Collection variable.

Since the mcolSelected Collection.Index value is associated to the recipe position in lstRecipes, as each recipe is deleted, all other recipes have their new ListIndex property decreased by 1. That is why the varItem value (mcolSelected.Index value) used by the For Each...Next loop is decreased by 1 minus the number of items already deleted from the list (intI counter), which produces the desired visual effect (keep items selected as they are deleted, from the top down (Figure 9-37).

```
intI = intI + 1
intJ = varItem + IIf(optDelete, 1 - intI, 0)
...
    Me.lstRecipes.RemoveItem (intJ)
```

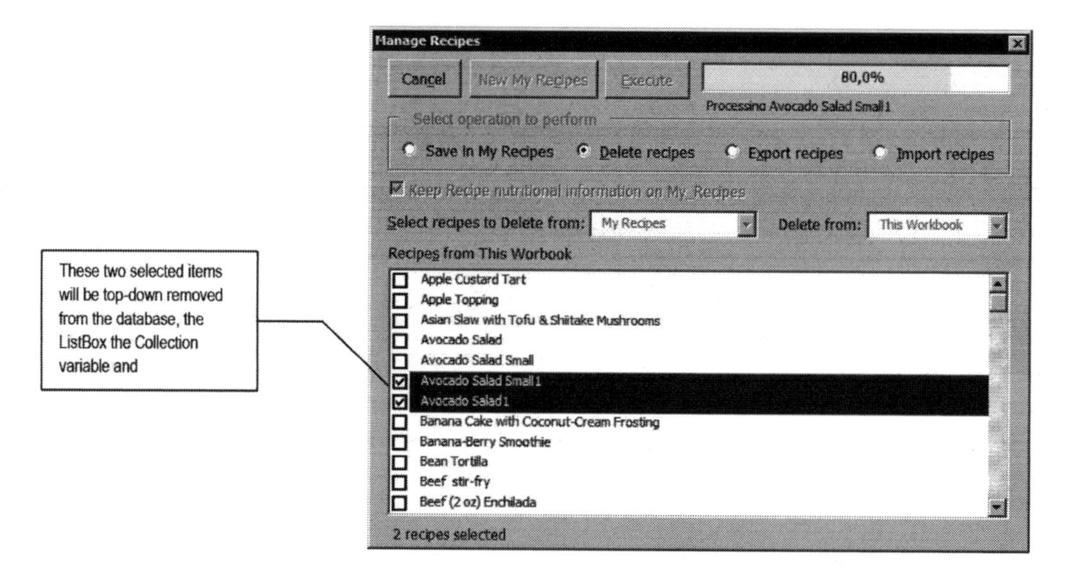

Figure 9-37. *When recipes are deleted from the selected My Recipes worksheet, the frmManageRecipes removes them from the lstRecipes ListBox as they are deleted in a top-down operation*

■ **Attention** At any moment the user can cancel and restart the "Delete recipes" operation again from the remaining recipes.

The "Delete recipes" operation is faster because it just automates the database, having no need to load each recipe before deleting it.

Exporting/Importing Recipes with Database Copy/Paste Methods

You may notice that both export and import recipe operations are quite slow, because of the double-sided automation that takes place in these processes. The code needs to load the recipe in the source worksheet and save it in the destination worksheet using the `SheetDBEngine` class to automate the database engine of each worksheet application.

You can speed up this process by using the `SheetDBEngine CopyRecord` and `PasteRecord` methods, implemented as the `My Recipe` worksheet methods with the same name, and a good example of how much it can improve the process speed can be appreciated in the `USDA Food Compose_SheetDBEngineManageCopyPasteRecords.xlsm` macro-enabled workbook that you can also extract from the `Chapter09.zip` file, which has a `My Recipes 2` worksheet copy, with no recipe records.

Figure 9-38 shows the process running to copy the first 11 recipes from the `My Recipes` worksheet to the `My Recipes 2` worksheet. Note that the figure shows a "New recipe" record behind the `frmManageRecipesCopyPaste UserForm`, which is evidence that no automation is required from the worksheet side (the code doesn't need to load each recipe to save it on the destination worksheet). Also note that the `UserForm` progress bar now shows the time elapsed before the process begins, and if you try the example, you will notice a considerable speed increase during the export process to another worksheet of the same workbook. (To export the first 11 recipes to the empty `My Recipes 2` worksheet, the entire process takes 3.8 seconds on my computer. To copy all 123 recipes, it takes about 61 seconds—you may experience a different time on your PC.)

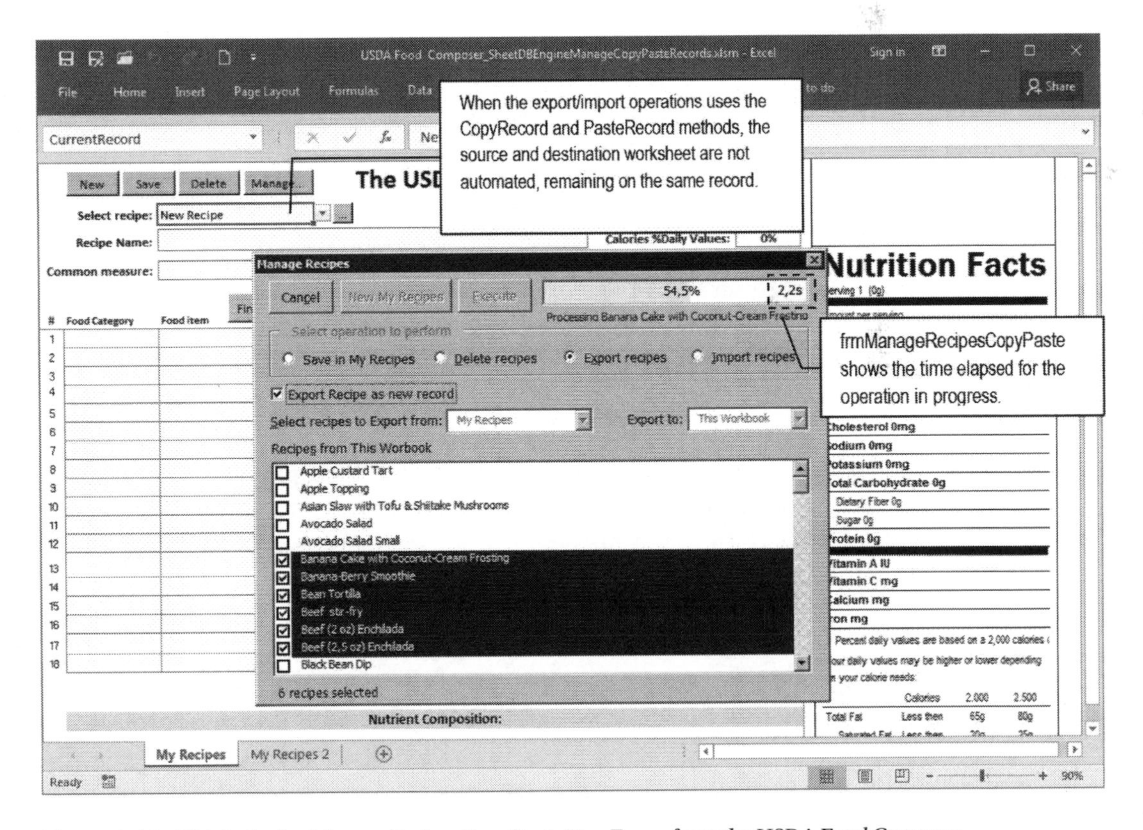

Figure 9-38. This is the frmManageRecipesCopyPaste UserForm, from the USDA Food Compose_SheetDBEngineManageCopyPasteRecords.xlsm macro-enabled workbook, that uses the SheetDBEngine CopyRecord and PasteRecord methods to speed up the export and import recipe operations between worksheets

■ **Attention** I will leave it as exercise for you to verify how `Sub UpdateProgressBar( )` implements the time elapse control, which may also be implemented on `frmManageRecipes` to compare both export/import recipe methods.

■ **Tip** The code uses the VBA `Timer( )` function, which returns the number of seconds past since midnight.

To understand what happens in this code, you must remember the following:

- The `CopyRecord` and `PasteRecord` methods don't fire `SheetDBEngine` class events.
- The `SheetDBEngine` class knows nothing about its records. It doesn't know that each recipe recalculates its nutritional information using the `NewRecipe` range name and that this information is needed to save each recipe's nutrient values in `USDA` worksheet's `My_Recipes` range name.

Since the `SheetDBEngine` class's `CopyRecord` and `PasteRecord` methods work by exchanging the one-side and many-side record range values, you must change both methods in the `My Recipes` worksheet (and also in the very hidden `NewMyRecipes` worksheet) to allow them to also exchange the nutrient profile of each copy/paste recipe.

Enhancing the MyRecipes.CopyRecord Method

If you look in the `USDA Food Compose_SheetDBEngineManageCopyPasteRecords.xlsm My Recipes` worksheet's `CopyRecord` method, you will realize that it executes this code to copy the desired recipe record data from the source worksheet:

```
Public Function CopyRecord(strRecord As String, rgOneSide As Range, rgManySide As Range, _
    Optional rgNutrients As Range) As Boolean Dim ws As Worksheet

    CopyRecord = mdb.CopyRecord(strRecord, rgOneSide, rgManySide)
    'Copy nutrient information from USDA

    Set ws = Worksheets("USDA")
    Set rgNutrients = ws.Range("My_Recipes").Find(strRecord, , , xlWhole)
    If Not rgNutrients Is Nothing Then
        'Recipe found. Extend it to all it nutrients
        Set rgNutrients = rgNutrients.Resize(, ws.Range("USDA").Columns.Count)
    End If
End Function
```

The `MyRecipes.CopyRecord` method now has an extra argument: `Optional rgNutrients as Range`. Note that the code receives the `strRecord` argument to indicate which recipe record data must be copied and calls the `SheetDBEngine.CopyRecord` method to recover it, passing by reference its `rgOneSide` and `rgManySide` object variables.

```
CopyRecord = mdb.CopyRecord(strRecord, rgOneSide, rgManySide)
```

The CopyRecord( ) operation returns true when the record is found, and since the record is not loaded in the worksheet application, to recover its nutrient data information, it searches inside the My_Recipes range name using the Range.Find method.

```
Set rgNutrients = ws.Range("My_Recipes").Find(strRecord, , , xlWhole)
```

If the recipe record was found, rgNutrients is not nothing, so the code resizes the range to get all USDA range columns, effectively returning on the rgNutrient argument all recipe nutrient data information.

```
If Not rgNutrients Is Nothing Then
    'Recipe found. Extend it to all it nutrients
    Set rgNutrients = rgNutrients.Resize(, ws.Range("USDA").Columns.Count)
```

Enhancing the MyRecipes.PasteRecord Method

Now look at how the My Recipes worksheet's PasteRecord method was implemented to allow you to save each recipe's nutrient data in the destination worksheet and in the My_Recipes range name of the destination workbook.

```
Public Function PasteRecord(strRecord As String, _
                rgOneSide As Range, _
                rgManySide As Range, _
                Optional PasteAsNewRecord As Boolean, _
                Optional rgNutrients As Range) As Boolean
    Dim ws As Worksheet
    Dim rgMyRecipes As Range
    Dim rgRecipe As Range
    Dim rgUSDA As Range
    Dim bolRecordPaste As Boolean

    bolRecordPaste = mdb.PasteRecord(strRecord, rgOneSide, rgManySide, True)
    If bolRecordPaste Then
        'Paste rgNutrients on USDA
        Set ws = Worksheets("USDA")
        Set rgUSDA = ws.Range("USDA")
        Set rgMyRecipes = ws.Range("My_Recipes")
        rgMyRecipes.Resize(rgMyRecipes.Rows.Count + 1).Name = "My_Recipes"
        'update rg object variable to contain My_Recipes new row
        Set rgMyRecipes = ws.Range("My_Recipes")
        'Position on new cell of My_Recipes range
        Set rgRecipe = rgMyRecipes.Cells(rgMyRecipes.Rows.Count, 1)
        'Copy and paste nutrient information
        rgNutrients.Copy
        rgRecipe.Resize(1, rgNutrients.Columns.Count).PasteSpecial xlPasteValues
        rgRecipe = strRecord
        'Sort My_Recipes to include the new pasted recipe
        rgMyRecipes.Resize(, rgUSDA.Columns.Count).Sort rgMyRecipes.Cells(1, 1)
```

```
        'Resize USDA range name to encompass this new recipe
        rgUSDA.Resize(rgUSDA.Rows.Count + 1).Name = "USDA"
    End If
    PasteRecord = bolRecordPaste
End Function
```

Note that the MyRecipes.PasteRecord method also declares the Optional rgNutrient as Range argument to exchange the recipe nutrient data profile. It delegates to the SheetDBEngine.PasteRecord method of the destination worksheet the task of pasting the record in the worksheet database (the fifth method argument was set to True to paste the record as a new record).

```
bolRecordPaste = mdb.PasteRecord(strRecord, rgOneSide, rgManySide, True)
```

PasteRecord = True when the record is correctly pasted on the destination worksheet by the SheetDBEngine class, and now it needs to paste the record nutrient data inside the USDA worksheet My_Recipes range name of the destination workbook, so it sets object variables to the appropriate range names.

```
If bolRecordPaste Then
    'Paste rgNutrients on USDA
    Set ws = Worksheets("USDA")
    Set rgUSDA = ws.Range("USDA")
    Set rgMyRecipes = ws.Range("My_Recipes")
```

Then it resizes the My_Recipes destination range to allow it to receive an extra recipe row (note that it uses the Range.Resize method to resize it and uses Range.Name to rebuild the range and updates the object variable to reflect this size change).

```
rgMyRecipes.Resize(rgMyRecipes.Rows.Count + 1).Name = "My_Recipes"
'update rg object variable to contain My_Recipes new row
Set rgMyRecipes = ws.Range("My_Recipes")
```

To copy the recipe nutrient data, the code sets rgRecipe to the first column of the new range row using the Range.Cells collection.

```
'Position on new cell of My_Recipes range
Set rgRecipe = rgMyRecipes.Cells(rgMyRecipes.Rows.Count, 1)
```

It then copies the rgNutrients argument to the clipboard, using the Range.Copy method, resizes rgRecipe to have the same column count of rgNutrients, and uses the Range.PasteSpecial xlPasteValues to paste the recipe nutrient data, using just one procedure row.

```
'Copy and paste nutrient information
rgNutrients.Copy
rgRecipe.Resize(1, rgNutrients.Columns.Count).PasteSpecial xlPasteValues
```

To keep the recipe data with its original name (which may differ from the record name that may be changed by a suffix counter), the code updates the recipe name in the My_Recipes first column, and the entire My_Recipes range name is sorted to correctly position the new recipe in ascending order (note that the rgMyRecipes is first resized to include all recipe nutrient columns, before applying the Range.Sort method).

```
rgRecipe = strRecord
'Sort My_Recipes to include the new pasted recipe
rgMyRecipes.Resize(, rgUSDA.Columns.Count).Sort rgMyRecipes.Cells(1, 1)
```

And the PasteRecord( ) operation ends by resizing the USDA range name to encompass the new recipe inserted on its My_Recipes food category.

```
        'Resize USDA range name to encompass this new recipe
        rgUSDA.Resize(rgUSDA.Rows.Count + 1).Name = "USDA"
    End If
    PasteRecord = bolRecordPaste
End Function
```

Updating the frmManageRecipesCopyPaste Code

The frmManageRecipeCopyPaste suffers just two updates from its older brother frmManageRecipes studied in the previous sections.

The first change is in Sub TransferRecipes( ), which now doesn't need to deal with any operation regarding saving the recipe nutrient data on the My_Recipes range name, which considerably simplifies the procedure code, both in terms of variable declaration and in instruction number.

```
Private Sub TransferRecipe(strRecipe As String)
    Dim xl As Application
    Dim wsSource As Object
    Dim wsDestination As Object
    Dim rgOneSide As Range
    Dim rgManySide As Range
    Dim rgNutrients As Range

    If Me.optExport Then
        Set xl = mxl
        Set wsSource = ActiveSheet
        Set wsDestination = mws
    Else
        Set xl = Application
        Set wsSource = mws
        Set wsDestination = ActiveSheet
    End If

    xl.Calculation = xlCalculationManual
    Application.DisplayAlerts = False
        If wsSource.CopyRecord(strRecipe, rgOneSide, rgManySide, rgNutrients) Then
            Call wsDestination.PasteRecord(strRecipe, rgOneSide, rgManySide, True,
rgNutrients)
        End If
    xl.Calculation = xlCalculationAutomatic
    Application.DisplayAlerts = False
End Sub
```

As you can see, after setting the appropriate references to the source and destination worksheets according to an export or import recipe operation, the code just disables automatic calculation on the destination worksheet (represented by the `xl As Application` object variable), turns off `Application.DisplayAlerts` (to disable any OLE messages between applications), and calls the source worksheet `CopyRecord` method for the selected recipe. Note that it passes the recipe name and the `rgNutrients as Range` object variable to receive the recipe nutrient data (if any).

```
    xl.Calculation = xlCalculationManual
    Application.DisplayAlerts = False
        If wsSource.CopyRecord(strRecipe, rgOneSide, rgManySide, rgNutrients) Then
```

■ **Attention** The source worksheet calls the `SheetDBEngine.CopyRecord` method to automate the copy record data task.

If the `ws.Source.CopyRecord` method succeeds (returns `True`), it calls the destination worksheet's `PasteRecord` method, passing by reference all object variables it needs to paste the record as a new record inside its database structure.

```
    Call wsDestination.PasteRecord(strRecipe, rgOneSide, rgManySide, True, rgNutrients)
End If
```

■ **Attention** The destination worksheet will also call the `SheetDBEngine.PasteRecord` method to automate the paste record data task.

The `Sub TransferRecipe( )` procedure ends by reactivating the destination workbook calculation and the Excel `DisplayAlerts` property.

```
    xl.Calculation = xlCalculationAutomatic
    Application.DisplayAlerts = False
End Sub
```

The second change suffered by `frmManageRecipesCopyPaste UserForm` comes from the fact that the `SheetDBEngine.PasteRecord` method always pastes records on the bottom of the database `SavedRecords` range name, meaning that after all records are exported/imported to the destination worksheet, the database needs to be sorted.

That is why `Sub ProcessRecipes` now has a call to the destination worksheet's `SortDatabase` method after its `For Each...Next` loop ends.

```
Private Sub ProcessRecipes(wb As Workbook, ws As Object)
    ...
        For Each varItem In mcolSelected
          ...
        Next
        ...
        If Me.optExport Or Me.optImport Then
            'Sort the database
```

```
            ws.SortDatabase
        End If
        ...
End Sub
```

And this is all you need to know about the frmManageRecipesCopyPaste UserForm!

Attention The workbooks that use the frmManageRecipes and frmManageRecipesCopyPaste methods studied in this chapter are incompatible because the first one (USDA Food Composer_SheetDBEngineManageAutomation.xlsm) doesn't implement in the My Recipes worksheet (and all its copies) the rgNutrients as Range argument on its CopyRecord and PasteRecord methods. To make them talk with their new brother, these methods must be updated to execute the same code used by USDA Food Composer_SheetDBEngineManageCopyPasteRecords.

Conclusion

This chapter showed how to exchange data between two different Excel workbooks or applications. The first case studied how to update a source database worksheet (such as the USDA worksheet, with thousands of food items and 184 nutrient columns) to a new version inside a dietary application using both pure code and a UserForm.

You also learned how two well-built worksheet applications can exchange database records using automation either on the source and destination worksheets.

All these operations were conducted using simple VBA UserForm interfaces that implement different private procedures to execute their tasks. You had the opportunity to verify how a well-built class module, such as SheetDBEngine, can be useful to automate the worksheet records database, by either automatically loading, saving, copying, or pasting records between different worksheets that can be on the same or different workbooks, as well as how to synchronize a UserForm interface regarding the operation it is executing, using a simple set of controls, like a ListBox associated to a progress bar (that can also implemented using the Excel status bar) to allow follow any task progress until it is finished.

Chapter Summary

In this chapter, you learned about the following:

- How to use the USDA Food List Creator.accb Microsoft Access application to open and process any SRxx.mdb or SRxx.accdb Microsoft Access nutrient database to generate a USDA or USDACommonMeasures worksheet to use as a database nutrient search for dietary worksheet applications
- How to update the USDA worksheet used on a worksheet application using VBA code
- How to use the Excel status bar to produce a progress bar that indicates progress
- How the USDA Food Composer...xlsm worksheet application may fail when a food item changes it name between different SRxx updates
- How to use a VBA UserForm to produce a better user interface to update the USDA worksheet, searching for food items whose Ndb_No code is the same but food item name changed between two or more SRxx versions

- How to produce a `UserForm` progress bar using three `Label` controls and a centralized procedure code
- How to exchange data between two different worksheets using VBA code
- How to copy or move worksheets between two different workbooks using VBA code
- How to verify if a workbook is already opened to use it in VBA code
- How to manage worksheet application data, exchanging it between two different worksheet applications using VBA automation
- How to use the `SheetDBEngine` class's `CopyRecord` and `PasteRecord` to exchange worksheet records between two different worksheets databases

In the next chapter, you will learn how to take VBA to the next programmable level by using the Windows API to enhance some aspects of the presentation of your worksheet applications.

CHAPTER 10

Using the Windows API

Now that you have a good understanding of VBA and how to use it to produce good interfaces to your worksheet applications, it is time to expand this knowledge to the unknown, incomprehensible world of the Windows application programming interface (API) that makes Microsoft Windows work.

Why should you care about it? For one simple reason: by using the Windows API, you can extend VBA capabilities to the next level, doing programming tricks that you see in other software interfaces and that you can't do using just the regular VBA language.

In this chapter, I will give you some guidance about the Windows API: what it is, how to declare them, how to call DLL API procedures, and how you can use the enormous amount of VBA/DLL code available on the Internet.

You can obtain all the files and procedure code in this chapter by downloading the `Chapter10.zip` file from the book's Apress.com product page, located at `www.apress.com/9781484222041`, or from `http://ProgrammingExcelWithVBA.4shared.com`.

The Microsoft Windows API

In the old DOS days, every time an executable program was created, all the code it needed to perform its many functions was statically linked into the executable code. So, if 100 programs were doing string operations (like using the `Left( )` and `Mid( )` functions), all of them needed to compile these string functions inside the executable file, reproducing the same static code in every executable file that needed it.

Microsoft Windows changed this by exposing an operating system based on an API uses dynamic link libraries (DLLs) to offer all the code a program needs to exist on the Windows operating system. Instead of storing the code functionally inside each executable file, programmers have the ability to use small code *declares* that call the functionality they need from the desired DLLs, leaving to the operating system the task of keeping the code available. These commonly used DLLs are system files with the `.dll` extension normally stored in the `\Windows\System` folder, unless they were specifically created for the program that needs them, in which case they probably reside in the program folder.

Besides many other things, these DLLs functions are responsible for creating a window, changing the window properties, interacting with the many protocols available (like TCP/IP, HTTP, MailTo, and so on), playing multimedia files, printing, and saving files—all the operations you graciously perform on the Microsoft Windows system using the same dialog box in every program.

There is an important detail to know about DLLs: they are all written in the C or C++ language, which is quite different from VBA, from the number of variable types it can receive to the way they work. To use a DLL library of functions, you need to know which function you need to call and the DLL file where it resides, the arguments it needs, its presentation order, how they are manipulated by the DLL function, and what value it returns.

As you can see, when it comes to Windows API programming, there is a steep learning curve for VBA programmers.

F. Morgado, *Programming Excel with VBA*, DOI 10.1007/978-1-4842-2205-8_10

In the next sections, I will give you some information about DLLs and the code you can grab from the Internet to create some special effects. This is just a primer of the possibilities to teach you simple tricks that may enhance your worksheet applications with a professional touch.

Using Declare Statements

By definition, a `Declare` statement is a way to grant access to a DLL function inside a VBA module. It must be made in the declaration section of the module as a private or public procedure, using one of these syntaxes, if it is a `Function` or `Sub` procedure:

```
Declare [Function][Sub] PublicName Lib "LibName" [Alias "alias"] [([[ByVal] variable [As
type] ...])] [As Type]
```

In this code:

`Function, Sub`: This indicates whether it is a call for a `Function` or `Sub` procedure.

`PublicName`: This is the procedure name in your VBA project.

`LibName`: This is the DDL file where the procedure resides.

`Alias "alias"`: This is the original procedure name inside the DLL file.

`Variable [As type]`: These are the procedure arguments and expected type values.

The declare statement for each DLL function is well-documented on the Internet, and most of the time you just have to copy/paste it into your code, paying attention to the following:

- DLLs declared on standard modules are always public by default.
- DLLs declared on `UserForm` or `Class` modules are private by default and must be preceded by the `Private` keyword.
- To avoid name conflicts on your code modules, you must use the optional `Alias` clause to give an alias name to each DLL function declared in your code.

These are the basics about DLLs declares. The next code instructions show how the `SetTimer( )` function of the `User32.dll` library can be declared as a public procedure on any standard module:

```
Declare Function SetTimer Lib "user32" (ByVal hwnd As Long, ByVal nIDEvent As Long, ByVal
uElapse As Long, ByVal lpTimerFunc As Long) As Long
```

To avoid conflict with other possible declarations in other modules, it is usual to give the function call an alias by adding a personal prefix to the procedure name, as it must be used by the VBA project (I personally use the FM_ prefix, from `Flavio Morgado`, when necessary).

```
Declare Function FM_SetTimer Lib "user32" Alias SetTimer (ByVal hwnd As Long, ByVal nIDEvent
As Long, ByVal uElapse As Long, ByVal lpTimerFunc As Long) As Long
```

If you declare it inside a `UserForm` or a `Class` module, it will be private to the module, with no need to alias the procedure name, but it is imperative that you prefix it with the VBA `Private` keyword.

```
'DLL declaration inside a UserForm or Class module
Private Declare Function SetTimer Lib "user32" (ByVal hwnd As Long, ByVal nIDEvent As Long,
                                    ByVal uElapse As Long, ByVal lpTimerFunc As Long) As Long
```

Besides those basic instructions, you do not need to know how a DLL procedure must be declared. Just copy and paste the declaration code from the Internet to the desired code module, turn it private or alias it, and it is ready to be used by your VBA code.

Constants Declaration

Many API procedures are based on predefined values that you must pass to their arguments so they work properly. These values are always documented along the API declaration and must be declared in the module declaration section where the procedure is declared.

These constant values are mainly bit flags, commonly declared as decimal or hexadecimal values (a value that begins with &H characters that define the constant using another number scale). The next instruction declares the GWL_STYLE constant using a decimal value:

```
Private Const GWL_STYLE = -16
```

The next instruction declares the WS_CAPTION constant using a hexadecimal value:

```
Private Const WS_CAPTION = &HC00000
```

Since a single API procedure can use many different flags, alone or combined, to give the desired effect, any code module that uses API declarations ends up with a lot of constant declarations, and many of them are not always used in the code (as you will see in the next sections).

Window Handles

Have you ever wondered how the mechanism behind Microsoft Windows allows its *window programs* to react to mouse clicks? If you have multiple windows opened, one in front of the other, as soon as you click any part of a window that is underneath the window pile, that window immediately comes to the front, receiving the system focus and activating the selected window and its command.

This is possible because every window has a unique identifier called a *window handle*. You can think of this like a suitcase handle—it allows you to grab and take the suitcase anywhere you want.

The window handle is a long integer that uniquely identifies each window and allows the Microsoft Windows operating system to control it. In fact, in Microsoft Windows, not just an application window has a handle; everything that can react to mouse events has its own window handle. So, besides the application handle, every other window control has a handle. The borders and the close, minimize, restore, and maximize buttons have their own handles. And every other application part like menus, toolbar controls, and so on, has its own window handle.

To control such an immense number of window handles, Microsoft Windows has what is called a Windows *handle tree*, where each main application handle behaves like a basic folder, and all handles inside the application behave like subfolders. When you close an application handle, all the handles are also closed, releasing system resources.

This concept is important, because in Chapter 9, you used the Excel.Application.Hwnd property (which returns any Excel window application handle value) as a way to differentiate the Application object (the one where the code is running) from other possible open Excel.Application objects opened by the code by just comparing their Hwnd property values.

Be aware that not all window structures are equal! In fact, they belong to different window classes, according to the type of data they can contain. For example, applications that can open multiple documents—called *multiple document interfaces*—and use an `MDIForm` to be built (such as Excel, Word, Access, and so on) are associated with the `Omain` window classes, while the document opened within them comes from a different window class called `MDIClient` (like each workbook window inside Excel or document window inside Word), which comes from the `Form` class. The `UserForm` window you use from Visual Basic is still another different beast, coming from the `ThunderFrame` class. Each one has its own handle.

This is a pure concept, because when you come to use Windows DLLs, it is often necessary to grab the window handle of the object you want to manipulate in code so you can obtain the desired result.

Class Instance Handle

All objects you create as instances of a `Class` module (like the `SheetDBEngine` class) have their own handle, so they can react to system messages. To recover any class object instance handle, you must use the undocumented VBA `ObjPtr( )` function (read as "object pointer"), which returns the pointer to the interface referenced by an object variable, with this syntax:

```
ObjPtr(<ObjectVariableName>)
```

In this code:

> `ObjectVariableName`: This is the class object variable instance whose pointer (handle) you want to recover.

The next example shows how you can recover a class object instance variable handle in your code:

```
Dim cClass1 as New Class1
Dim hWnd as long

HWnd = ObjPtr(cClass1)
```

▒ **Attention** Note in the previous example that the `cClass1` object variable was declared `As New Class1`, meaning that it is created the first time it is referenced by the code.

Creating a Timer Class

The first project using VBA and the Windows API is one that creates a timer class, where the user can set the timer interval, enable or disable the Timer, and fire a `Timer( )` event whenever the timer interval expires. This is made using the `SetTimer( )` and `KillTimer( )` functions of `User32.dll`:

- `SetTimer( )` starts the timer and defines a VBA function to be called when the timer expires, beginning another timer. It returns a `Long` integer indicating the timer ID.
- `KillTimer( )` kills a timer already set using the long integer that represents it.

Investigating the `SetTimer( )` function on MSDN web site, you will notice that its syntax is quite complex to understand from a novice VBA programmer's perspective. It can be translated to the following:

```
Declare Function SetTimer Lib "user32" (ByVal hwnd As Long, ByVal nIDEvent As Long, ByVal
uElapse As Long, ByVal lpTimerFunc As Long) As Long
```

In this code:

`hWnd`: This is the window handle to be associated with the timer. Use the `Application.Hwnd` property to define it.

`nIDEvent`: This is a handle to an object that will receive the timer event (`UserForm` or `Class` handle).

`uElapse`: This is a long integer for the timer interval in milliseconds (maximum interval is $2^{31} = 2.147.483.648$ ms $\cong 596$ hours, or 24.8 days).

`lpTimerFunc`: This is the address of the callback procedure to be called when the timer expires. This procedure must exist on the object represented by the `nIDEvent`.

The `SetTimer( )` function return value is explained this way:

- If the function succeeds and the `hWnd` parameter is `NULL`, the return value is an integer identifying the new timer. An application can pass this value to the `KillTimer` function to destroy the timer.
- If the function succeeds and the `hWnd` parameter is not `NULL`, then the return value is a nonzero integer. An application can pass the value of the `nIDEvent` parameter to the `KillTimer` function to destroy the timer.
- If the function fails to create a timer, the return value is zero. To get extended error information, call the `GetLastError` API.

Although the documentation does not clearly explain it, most DLLs that need a callback `Sub` procedure to call require that it be declared this way (where `publicname` can be any name you want):

```
Sub PublicName(ByVal hwnd As Long, ByVal uMsg As Long, ByVal idEvent As Long, ByVal dwTime
As Long)
    ...
End Sub
```

The arguments can be translated as follows:

`PublicName`: This is the procedure name as declared in the VBA code module.

`hWnd`: This is the window handle associated with the timer.

`uMsg`: This is the timer message sent.

`idEvent`: This is a long integer that identifies the object handle that will receive the timer message when it fires (specified by the `nIDEvent` of the `SetTimer( )` API function).

`dwTime`: This is the number of milliseconds that have elapsed since the system was started. This is the value returned by the `GetTickCount` API function.

▒ **Attention** `GetTickCount( )` is a DLL function from `Kernel32.dll` that retrieves the number of milliseconds that have elapsed since the system was started, up to 49.7 days using a 10 to 16 milisecond precision. It is a high-performance timer counter that can be used instead of the VBA `Time( )` and `Timer( )` functions.

Once a timer is set, you can reset it by calling again SetTimer( ) with another timer interval or you can stop it by calling the KillTimer( ) API function, which has this syntax:

```
Declare Function KillTimer Lib "user32" (ByVal hwnd As Long, ByVal nIDEvent As Long) As Long
```

In this code:

hWnd: This is the window handle associated with the timer.

nIDEvent: This is a handle to the object that received the timer event.

To create a useful VBA timer code using the SetTimer( ) and KillTimer( ) Windows APIs, you must use a class module so you can create as many timer instances as needed using a single, centralized code.

Extract the Timer Class.xlsm macro-enabled workbook from the Chapter10.zip file, press Alt+F11 to show the VBA IDE, and double-click the Timer Class module to show its code. You will see that it declares three module-level variables (one to hold the timer ID and two others to hold the class Interval and Enabled property values), both SetTimer( ) and KillTimer( ) DLL procedures, and the Timer( ) event in the class module declaration section.

```
Option Explicit

Dim mlngTimer As Long
Dim mlngInterval As Long
Dim mbolEnabled As Boolean

Private Declare Function SetTimer Lib "user32" (ByVal hwnd As Long, ByVal nIDEvent As Long,
ByVal uElapse As Long, ByVal lpTimerFunc As Long) As Long
Private Declare Function KillTimer Lib "user32" (ByVal hwnd As Long, ByVal nIDEvent As Long)
As Long

Event Timer()
```

To set the timer interval, the Timer Class uses a pair of Property Let/Get procedures to implement the Interval property. Note that it just accepts values greater than zero, using the mlngInterval module-level variable to store it.

```
Public Property Get Interval() As Long
    Interval = mlngInterval
End Property

Public Property Let Interval(ByVal lngInterval As Long)
    If lngInterval > 0 Then
        mlngInterval = lngInterval
    End If
End Property
```

The timer is enabled/disabled using another pair of Property Let/Get procedures to implement its Enabled property.

```
Public Property Get Enabled() As Boolean
    Enabled = mbolEnabled
End Property
```

```
Public Property Let Enabled(ByVal bolEnabled As Boolean)
    If bolEnabled And mlngInterval > 0 Then
        'ObjPtr(Me) returns the class object handle
        mlngTimer = SetTimer(Application.hwnd, ObjPtr(Me), mlngInterval, AddressOf
TimerProc)
    Else
        'mlngTimer is the timer id for this class
        Call KillTimer(Application.hwnd, mlngTimer)
        mlngTimer = 0
    End If
    mbolEnabled = bolEnabled
End Property
```

As you can see, the `Property Let Enabled( )` procedure is used to set the timer. It begins verifying whether the argument `bolEnabled = True` and whether the class module-level variable `mlngInterval > 0`, indicating that the `Timer.Interval` property has been set. If this is true, it calls the `SetTimer( )` API to set the timer; otherwise, it calls `KillTimer( )` and resets the `mlngTimer` pointer.

```
If bolEnabled And mlngInterval > 0 Then
    'ObjPtr(Me) returns the class object handle
    mlngTimer = SetTimer(Application.hwnd, ObjPtr(Me), mlngInterval, AddressOf TimerProc)
Else
    'mlngTimer is the timer id for this class
    Call KillTimer(Application.hwnd, mlngTimer)
    mlngTimer = 0
End If
```

Note how it uses the `SetTimer( )` function arguments:

- The `mWnd` argument is set to the `Application.hWnd` property (the handle of the Excel application window).
- The `nIDEvent` argument uses the VBA `ObjPr(Me)` function to return the class instance handle, as explained in the section "Class Handle" earlier in this chapter.
- The `uElapse` argument is set to the `mlngInterval` module-level variable.
- The `lpTimerFunc` argument uses the VBA `AddressOf` statement to return the address of the `TimerProc( )` procedure, from the `basTimer` standard module.

```
mlngTimer = SetTimer(Application.hwnd, ObjPtr(Me), mlngInterval, AddressOf TimerProc)
```

The `SetTimer( )` API will return a long integer to the `mlngTimer` module-level variable, indicating the ID of the timer associated to this instance of the `Timer` class.

To raise the `Timer( )` event, the `Timer` class also declares the `RaiseTimer( )` method (as a `Public Sub` procedure of the class module).

```
Public Sub RaiseTimer()
    RaiseEvent Timer
End Sub
```

And whenever the class is destroyed, its `Class_Terminate( )` event fires, calling the `KillTimer( )` API to stop the timer associated to this instance of the class module (if any). Note that it uses the `Application.Hwnd` property and the `mlngTimer` ID to stop the timer.

```
Private Sub Class_Terminate()
    Call KillTimer(Application.hwnd, mlngTimer)
End Sub
```

The TimerProc( ) Procedure

Since a call to the `Timer` class's `Let Enable(True)` property procedure sets a new timer, passing to the `SetTimer( )` API the memory address pointer of the `TimerProc( )` procedure, if you inspect it in the `basTimer` module, you will note that it executes this code:

```
Sub TimerProc(ByVal hwnd As Long, ByVal uMsg As Long, ByVal clsTimer As Timer, ByVal dwTime
As Long)
    clsTimer.RaiseTimer
End Sub
```

Note that it changes the original `IDEvent as Long` argument to the `ByVal clsTimer as Timer` object, which is evidence that object interfaces are in fact long integers. Since `SetTimer( )` uses the `ObjPtr(Me)` value (the handle to the `Timer` class instance) to the `nIDEvent` argument, this value is internally passed by `SetTimer( )` to the `clsTimer` argument, effectively identifying the class module. The code just calls the `clsTimer.RaiseTimer` event to raise the timer event.

Using the Timer Class

To use the `Timer` class, you just need to declare an object variable as `Private WithEvents ... as Timer` on the declaration section of the `UserForm` module.

```
Private WithEvents mTimer1 as Timer
```

You then use the `Timer.Interval` (in milliseconds) and `Timer.Enabled` properties to activate the timer.

```
mTimer1.Interval = 1000
mTimer1.Enabled = True
```

You use the `mTimer1_Timer( )` event to do whatever you want in the code.

```
Sub mTimer1_Timer( )
    'Code goes here!
End Sub
```

The `frmTimer UserForm` from the `Timer Class.xlsm` macro-enabled workbook uses this technique to implement four different timers, which are set to 1000 ms (1 second), 500 ms (0.5s), 250 ms (0.25s), and 125 ms (0.125s). Just click the Enabled check box of each timer, and it will begin to fire using the defined time. Change the timer interval to any value to see whether it runs faster or slower (Figure 10-1).

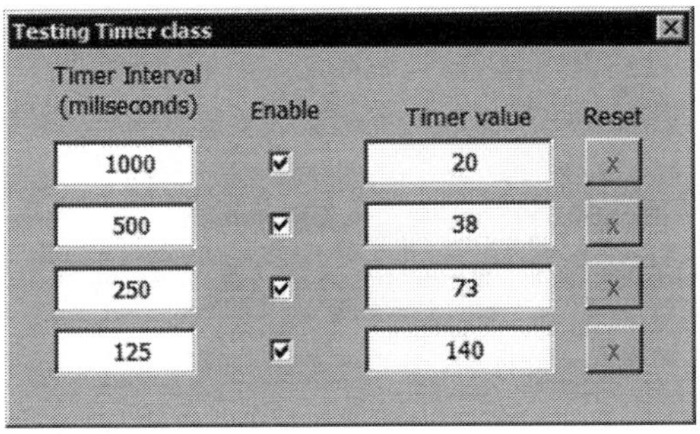

Figure 10-1. *The frmTimer implements four timers (mTimer1 to mTimer4) with default Interval values of 1000, 500, 250, and 125 milliseconds. Click the Enabled CheckBox to begin each timer. Change the timer interval to see it run at another speed*

The frmTimer UserForm declares four Timer module-level variables: mTimer1 to mTimer4.

```
Option Explicit

Dim WithEvents mTimer1 As Timer
Dim WithEvents mTimer2 As Timer
Dim WithEvents mTimer3 As Timer
Dim WithEvents mTimer4 As Timer
```

When you click each timer's Enable CheckBox, it instantiates the associated module-level variable, sets the timer interval, and enables/disables the timer. This is the chkInterval1_Click( ) event.

```
Private Sub chkInterval1_Click()
    Set mTimer1 = New Timer
    mTimer1.Interval = Me.txtInterval1
    mTimer1.Enabled = Me.chkInterval1
End Sub
```

To see the mTimer1 object variable work, the UserForm code uses the mTimer1_Timer( ) event to set the txtTimer1 TextBox value.

```
Private Sub mTimer1_Timer()
    Me.txtTimer1 = Me.txtTimer1 + 1
End Sub
```

To change the mTimer1.Interval while the timer is running, the code uses the txtInterval1 TextBox's Change( ) event.

```
Private Sub txtInterval1_Change()
    If Not IsNumeric(Me.txtInterval1) Then
        Me.txtInterval1 = 0
    ElseIf Me.txtInterval1 < 0 Then
        Me.txtInterval1 = 0
    End If
    mTimer1.Interval = Me.txtInterval1
    Call chkInterval1_Click
End Sub
```

Quite simple code, uh? It just accepts numeric, greater-than-zero `Timer.Interval` properties and calls again the `chkInterval1_Click( )` event to reset the timer. (Figure 10-2 shows a diagram view of the entire process.)

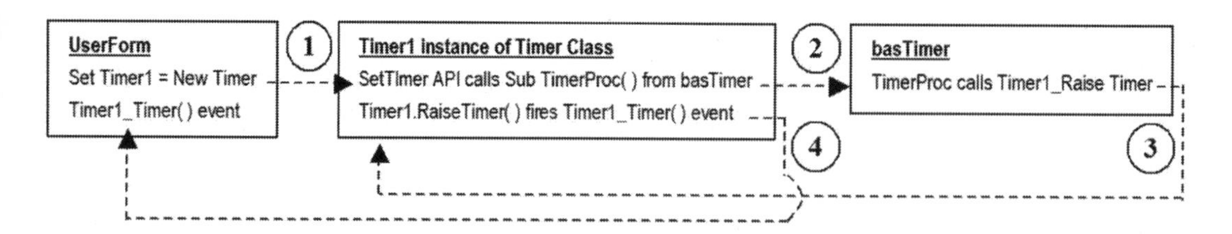

Figure 10-2. *This diagram explains what happens with the Timer class that uses the SetTimer( ) API to create a timer object*

■ **Attention** You can make a more robust `Timer` class by validating the `Interval` value in the `Property Let Interval( )` procedure.

UserForm Handle

It is surprising that the VBA `UserForm` object doesn't expose a `Hwnd` property, like the `Application` object does. Although the `UserForm` object doesn't have a handle in design mode, it *must have* a handle as soon as it is loaded so it can react to system and mouse events. This is a runtime where only the `Hwnd` property is available.

The only way you can get any `UserForm` handle is to use the `FindWindowA( )` API function from `User32.dll`, which can be declared in this way:

```
Private Declare Function fm_FindWindow Lib "user32" Alias "FindWindowA" (ByVal lpClassName
As String, ByVal lpWindowName As String) As Long
```

In this code:

> `lpClassName`: This is the class name of the object whose handle you want to find. If `lpClassName` is `NULL`, it finds any window whose title matches the `lpWindowName` parameter.
>
> `lpWindowName`: This is the window name (window caption text).

The `FindWindow( )` API function returns a `Long Integer` indicating the window handle. Note that this API procedure declaration gives the `fm_FindWindow` alias to the `FindWindowA( )` function, which expects to receive two arguments: the VBA `UserForm` class name (`ThunderFrame`) or `NULL` and the `UserForm` caption property. This procedure returns to the `UserForm` handle if it finds the window by its caption property. It returns `NULL` if it fails to find any window match.

▒ **Attention** Many `UserForm` handle procedures you find on Internet will temporarily store the current `UserForm` caption in a local variable, change the `UserForm.Caption` property to an improbable value (such as `Me.Caption` and `Timer`, where `Timer` returns the number of seconds past since midnight), and call the `FindWindow( )` API to get the `UserForm` handle, restoring the `UserForm` caption to its original value before the procedure finishes.

```
Public Function Hwnd(frm As Object) As Long
    Dim varHwnd As Variant
    varHwnd = fm_FindWindow("ThunderDFrame", frm.Caption)
    If Not IsNull(varHwnd) Then
        Hwnd = CLng(varHwnd)
    End If
End Function
```

The technique is quite simple. It declares the `varHwnd as Variant` variable (since `Variant` is the only variable type that can receive a `null` value) to receive the `fm_FindWindow( )` return value, and if this value is not `null`, it uses the VBA `CLng(varHwnd)` function to convert it to a `Long Integer` that is used as the `Hwnd( )` procedure returned value.

Setting Bit Values

The Windows system stores object values using a `Long` integer, because a `Long` integer has up to 32 bits that can be associated to 32 different `Boolean` options, which can be turned on/off by just changing any bit value from 0 (to disable) to 1 (to enable) and vice versa, using the `OR`, `AND`, and `NOT` VBA operators.

Let's suppose that a given set of properties is stored using an 8-byte value (a value that has just 8 bits). All these properties are represented by the integer number 231 (using a decimal representation). By using the Windows Calculator applet with the Programming option set (Show ➤ Programming menu), you can easily see that the 231 decimal values is represented as the binary value 11100111.

Counting from right to left, the 1, 2, 3, 6, 7, and 8 bits are set (value = 1), while the 4 and 5 bits are not set (value = 0).

If the property you want to set is on the 4th bit, you can turn it on by using the `OR` operator to combine the 11100111 binary representations (231 in decimal) with 00001000 (8 in decimal), as follows:

```
      11100111 (231)
OR
      00001000 (008)
     ----------
      11101111 (239)
```

The `OR` operator combined all 8 bits from both binary numbers, setting just the fourth bit to 1 (or `True`, in programming language) and keeping all other bits on their default states. It `OR`*ed* two 8-byte numbers, meaning that each bit from these two numbers must be set if either one *or* the other is set!

Mathematically speaking, the `OR` operator is equal to the + operator, giving the same result as 231 + 8 = 239.

To unset the fourth bit again and disable the property associated with it, you use the AND NOT operator, meaning that it will *negate* the second bit entirely (all the 0s will become 1s, and all the 1s will become 0s) before applying the AND operator to set just bits that are set in both numbers, as follows:

```
         11101111 (239)                  11101111
AND NOT                   =    AND
         00001000 (008)                  11110111  <--[NOT]
         --------                        --------
         11100111 (231)                  11100111
```

To verify whether a given bit is set, just use the AND operator to combine the desired number with the bit to test. If the tested bit is set, the result will be the number with the bit to set or, in logical terms, will be considered True (anything different than zero).

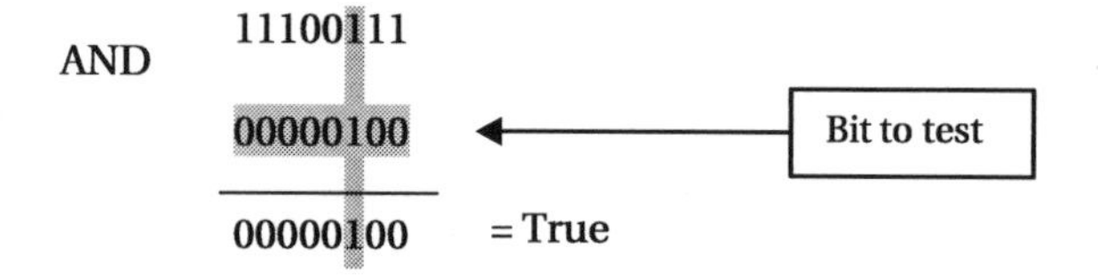

■ **Attention** Although any number can be represented using a different number scale (like octal, decimal, or hexadecimal formats), it is important to note that the leftmost bit (the 8th bit for an 8-bit byte, or the 32nd bit for a 32-byte number), also known as the *most significant bit*, is reserved to the number sign on the decimal scale. So, it is not uncommon to face constant flags using decimal negative numbers, meaning that the most significant bit was set (like GWL_STYLE = -16).

Animating the UserForm Window

You can change the way a UserForm loads and unloads by using the Function AnimateWindow( ) API from User32.dll, which has this syntax:

```
Declare Function AnimateWindow Lib "user32" (ByVal Hwnd As Long, ByVal dwTime As Long,
ByVal dwFlags As Long) As Boolean
```

In this code:

- Hwnd: This is the handle of the UserForm.
- dwTime: This is the time in milliseconds the animation takes to play.
- dwFlags: This is the type of animation associated to these constants.
  - AW_ACTIVATE = &H20000: This activates the window (do not use it with AW_HIDE).
  - AW_BLEND = &H80000: This uses a fade effect to show the window (or hides it if used with AW_HIDE).
  - AW_CENTER = &H10: This expands the window outward.

AW_HIDE = &H10000: This hides the window if used with AW_BLEND (the windows is shown).

AW_HOR_POSITIVE = &H1: This animates the window from left to right.

AW_HOR_NEGATIVE = &H2: This animates the window from right to left.

AW_SLIDE = &H40000: This uses slide animation.

AW_VER_POSITIVE = &H4: This animates the window from top to bottom.

AW_VER_NEGATIVE = &H8: This animates the window from bottom to top.

Except when using animation to hide a window (using dwFlags = AW_BLEND or AW_HIDE), you can use the OR operator to combine AW_HOR_POSITIVE or AW_HOR_NEGATIVE with AW_VER_POSITIVE or AW_VER_NEGATIVE alone or with the AW_SLIDE flag to produce a diagonal animation. The AW_ACTIVATE flag must always be used to show the animation.

The UserForm_APIs.xlsm macro-enabled workbook (that you can extract from the Chapter10.zip file) has the basUserFormAPIs standard module, which declares in its Declaration section thefm_FindWindow( ) (to find the UserForm handle) and fm_AnimateWindow( ) (to animate a UserForm window) aliased API procedures, along with all animation constants needed.

```
Option Explicit

'DLL declarations to change UserForm animation or appearance
Private Declare Function fm_FindWindow Lib "user32" _
Alias "FindWindowA" (ByVal lpClassName As String, ByVal lpWindowName As String) As Long
Private Declare Function fm_AnimateWindow Lib "user32" _
Alias "AnimateWindow" (ByVal Hwnd As Long, ByVal dwTime As Long, ByVal dwFlags As Long)
As Boolean
...
'Window animation constants
Const AW_ACTIVATE = &H20000         'Activates the window.
Const AW_BLEND = &H80000            'Window has a fade in of fade out effect (if used with
                                    AW_HIDE)
Const AW_CENTER = &H10              'Window expand from center
Const AW_HIDE = &H10000             'Hide the window when used with AW_BLEND
Const AW_HOR_POSITIVE = &H1         'Window animates from left to right
Const AW_HOR_NEGATIVE = &H2         'Window animates from right to left
Const AW_SLIDE = &H40000            'Windows use slide animation (specify direction horizontal
                                    or vertical)
Const AW_VER_POSITIVE = &H4         'Window animates from top to bottom
Const AW_VER_NEGATIVE = &H8         'Window animates from bottom to top
```

To deal with the AnimateWindow( ) API animation constants, basUserFormAPIs also declares the Animation enumerator to combine these animation flags using more significant names.

```
Public Enum Animation
    Appear = AW_BLEND Or AW_ACTIVATE
    DiagonalToBottomLeft = AW_HOR_NEGATIVE Or AW_VER_POSITIVE Or AW_ACTIVATE
    DiagonalToBottomRight = AW_HOR_POSITIVE Or AW_VER_POSITIVE Or AW_ACTIVATE
    DiagonalToTopLeft = AW_HOR_NEGATIVE Or AW_VER_NEGATIVE Or AW_ACTIVATE
    DiagonalToTopRight = AW_HOR_POSITIVE Or AW_VER_NEGATIVE Or AW_ACTIVATE
    Disappear = AW_BLEND Or AW_HIDE
    Expand = AW_CENTER Or AW_ACTIVATE
```

```
    SlideToBotton = AW_SLIDE Or AW_VER_POSITIVE Or AW_ACTIVATE
    SlideToLeft = AW_SLIDE Or AW_HOR_NEGATIVE Or AW_ACTIVATE
    SlideToRight = AW_SLIDE Or AW_HOR_POSITIVE Or AW_ACTIVATE
    SlideToTop = AW_SLIDE Or AW_VER_POSITIVE Or AW_ACTIVATE
    SlideDiagonalToBottomLeft = AW_SLIDE Or AW_HOR_NEGATIVE Or AW_VER_POSITIVE Or AW_ACTIVATE
    SlideDiagonalToBottomRight = AW_SLIDE Or AW_HOR_POSITIVE Or AW_VER_POSITIVE Or AW_ACTIVATE
    SlideDiagonalToTopLeft = AW_SLIDE Or AW_HOR_NEGATIVE Or AW_VER_NEGATIVE Or AW_ACTIVATE
    SlideDiagonalToTopRight = AW_SLIDE Or AW_HOR_POSITIVE Or AW_VER_NEGATIVE Or AW_ACTIVATE
    ToBotton = AW_VER_POSITIVE Or AW_ACTIVATE
    ToLeft = AW_HOR_NEGATIVE Or AW_ACTIVATE
    ToRight = AW_HOR_POSITIVE Or AW_ACTIVATE
    ToTop = AW_VER_NEGATIVE Or AW_ACTIVATE
End Enum
```

Note that most Enum Animation declarations use the desired animation OR*ed* with the AW_ACTIVATE flag (except the Disappear enumerator), and to produce a diagonal animation effect, the code uses more than one constant flag (like DiagonalToBottomLeft = AW_HOR_NEGATIVE Or AW_VER_POSITIVE Or AW_ACTIVATE).

■ **Attention** Although Microsoft MSDN documentation for the WindowAnimate( ) function states that you can combine different flags with AW_HIDE to produce different closing effects, just the AW_BLEND constant works with AW_HIDE to produce a fade-out effect.

The Animate( ) Procedure

To produce the UserForm animation, use the basUserFormAPIs Public Sub Animate( ) procedure, which executes this code:

```
Public Sub Animate(frm As Object, Animation As Animation, Optional Duration As Long = 1500)
    Dim lngHwnd As Long

    'Get frm UserForm Handle
    lngHwnd = Hwnd(frm)

    'Center UserForm on Application window
    With frm
        .Top = (Application.Top + Application.Height / 2) - .Height / 2
        .Left = (Application.Left + Application.Width / 2) - .Width / 2
    End With

    'Animate the UserForm
    fm_AnimateWindow lngHwnd, Duration, Animation
End Sub
```

The function Animate( ) receives three arguments: frm as Object (a reference to a loaded UserForm), Animation as Animation (the desired animation enumerator), and the Optional Duration as Long = 1500 argument (animation duration, with 1,500 milliseconds—1.5 s—as default value).

It then uses the Hwnd(frm) function (as cited in the section "UserForm Handle") to attribute the frm as Object UserForm handle to the lngHwnd variable.

```
'Get frm UserForm Handle
lngHwnd = Hwnd(frm)
```

The frm UserForm is then centralized inside the Excel window. (Note that it uses Application.Top + Application.Height / 2 to find the Excel window's vertical center point and subtracts .Height / 2 = half the UserForm height; the same is done to find the horizontal center point.)

```
With frm
    .Top = (Application.Top + Application.Height / 2) - .Height / 2
    .Left = (Application.Left + Application.Width / 2) - .Width / 2
End With
```

The UserForm associated to the lngHwnd Long Integer is animated as desired by calling the fm_AnimateWindow( ) aliased API.

```
fm_AnimateWindow lngHwnd, Duration, Animation
```

To animate any UserForm when it is loaded, verify whether property ShowModal = False and make a call to Function Animate( ) on the UserForm_Initialize( ) event, as follows (note that the code uses the Appear enumerator in the Animation argument, accepting the default duration of 1500 ms):

```
Private Sub UserForm_Initialize()
    Animate Me, Appear
End Sub
```

Since Function Animate( ) declares the Animation as Animation enumerator argument, you can easily select the desired effect from the VBA constant list (Figure 10-3).

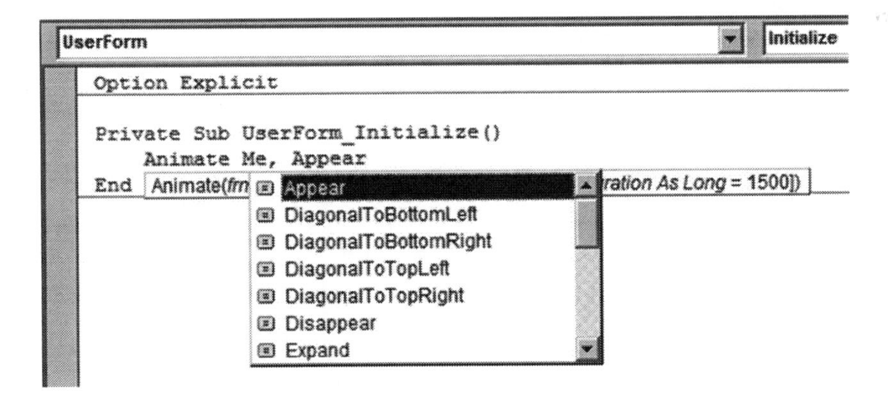

Figure 10-3. *The Animation as Animation enumerator argument of Function Animate( ) allows the user to easily select the desired animation to be applied when loading a UserForm*

■ **Attention** If you do not set the UserForm property's ShowModal = False, Visual Basic will raise error 400, "Form already displayed; can't show modally," when the loading animation finishes.

To animate any UserForm by an external procedure, you must first load the desired UserForm using the Load method and then call Function Animate( ) to animate it, as follows:

```
Public Function AnimateUserForm( )
    Load UserForm1
    Animate UserForm1, Appear
End Function
```

To animate any UserForm when it is unloaded, make a call to Function Animate( ) on the UserForm _QueryClose( ) event (the last event fired before the UserForm is terminated), using the Disappear enumerator on the Animation argument, like this (this code also uses a default duration of 1500 ms):

```
Private Sub UserForm_QueryClose(Cancel As Integer, CloseMode As Integer)
    Animate Me, Disappear
End Sub
```

The Sheet1 worksheet from the UserForm_APIs.xlsm macro-enabled workbook defines in the Animation range (merged cells B3:D3) a data validation list filled with all possible Animation enumerators (defined in the range M2:M20) to easily test each possible animation applied to the frmAnimate UserForm (which has no code). Click the list and select the desired animation to see it running, and once frmAnimation is shown, select the Disappear list option to see it fade at the desired duration time (change the Duration range name, cell F3, to apply the selected animation effect at a different speed, as shown in Figure 10-4).

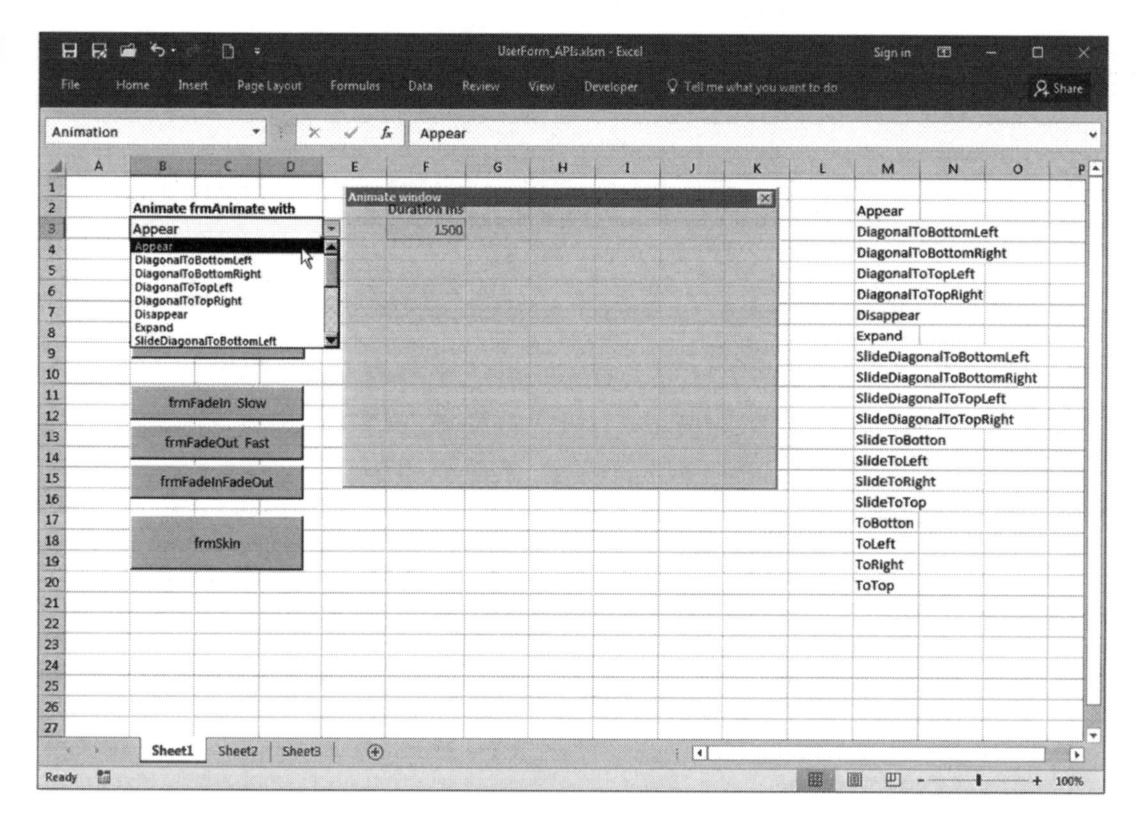

Figure 10-4. *Select the desired animation on cell B3's data validation list (the Animation range name) to apply it to frmAnimate. Change the Duration range name to reapply the effect*

To apply successive animations to the frmAnimate UserForm, for every animation effect but Discard, the code needs to unload the UserForm, update the Excel interface, and load it again. This is made in the Sheet1_Change( ) event, which fires every time any Sheet1 cell value changes, executing this code:

```
Private Sub Worksheet_Change(ByVal Target As Range)
    Dim Animation As Animation
    Dim bolDisappear As Boolean

    If Target.Address = Range("Animation").Address Or Target.Address =
Range("Duration").Address Then
        Select Case Trim(Range("Animation"))
            Case "Appear"
                Animation = Appear
            Case "DiagonalToBottomLeft"
                Animation = DiagonalToBottomLeft
            Case "DiagonalToBottomRight"
                Animation = DiagonalToBottomRight
            Case "DiagonalToTopLeft"
                Animation = DiagonalToTopLeft
            Case "DiagonalToTopRight"
                Animation = DiagonalToTopRight
            Case "Disappear"
                Animation = Disappear
                bolDisappear = True
            Case "Expand"
                Animation = Expand
            Case "SlideDiagonalToBottomLeft"
                Animation = SlideDiagonalToBottomLeft
            Case "SlideDiagonalToBottomRight"
                Animation = SlideDiagonalToBottomRight
            Case "SlideDiagonalToTopLeft"
                Animation = SlideDiagonalToTopLeft
            Case "SlideDiagonalToTopRight"
                Animation = SlideDiagonalToTopRight
            Case "SlideToBotton"
                Animation = SlideToBotton
            Case "SlideToLeft"
                Animation = SlideToLeft
            Case "SlideToRight"
                Animation = SlideToRight
            Case "SlideToTop"
                Animation = SlideToTop
            Case "ToBotton"
                Animation = ToBotton
            Case "ToLeft"
                Animation = ToLeft
            Case "ToRight"
                Animation = ToRight
            Case "ToTop"
                Animation = ToTop
        End Select
```

```
        If Not bolDisappear Then
            Unload frmAnimate
            Application.ScreenUpdating = True
            Load frmAnimate
        End If

        Animate frmAnimate, Animation, Range("Duration")

        If bolDisappear Then Unload frmAnimate
    End If
End Sub
```

The code declares the Animation as Animation enumerator value and verifies whether the change happens in the Animation or Duration range name.

```
Private Sub Worksheet_Change(ByVal Target As Range)
    Dim Animation As Animation
    Dim bolDisappear As Boolean

    If Target.Address = Range("Animation").Address Or Target.Address =
Range("Duration").Address Then
```

If this is true, it uses a Select Case statement to set the desired animation enumerator constant according to the value selected in the Animation range name (note that it uses the VBA Trim( ) function to remove undesired spaces in the Animation range value, and if Disappear is selected, bolDisappear = True).

```
Select Case Trim(Range("Animation"))
    Case "Appear"
        Animation = Appear
    ,,,
    Case "Disappear"
        Animation = Disappear
        bolDisappear = True
    ...
End Select
```

■ **Attention** There is no way to programmatically iterate through VBA Enumerator items. The Selected Case statement is an alternative way to do this.

Then the code verifies whether Disappear *was not* selected by testing not bolDisappear. If this is true, it unloads frmAnimate, updates the Excel interface, and loads it again before applying the selected effect.

```
If Not bolDisappear Then
    Unload frmAnimate
    Application.ScreenUpdating = True
    Load frmAnimate
End If
```

The animation effect (even Disappear) is then applied by calling `Function Animate( )` to `frmAnimate`, using the `Animation` variable value and the duration defined by the `Duration` range name.

```
        Animate frmAnimate, Animation, Range("Duration")
```

If Disappear was selected, the `UserForm` is now hidden and must be unloaded.

```
        If bolDisappear Then Unload frmAnimate
    End If
End Sub
```

Manipulating the UserForm Window

Suffice it to say by now that the VBA `UserForm` window belongs to the `ThunderFrame` class and that every time a `UserForm` is loaded into your PC memory, the `UserForm_Initialize( )` event fires and loads all its properties, which are used to draw the `UserForm` window when the form is shown, immediately after the `UserForm_Activate( )` event fires (Figure 10-5).

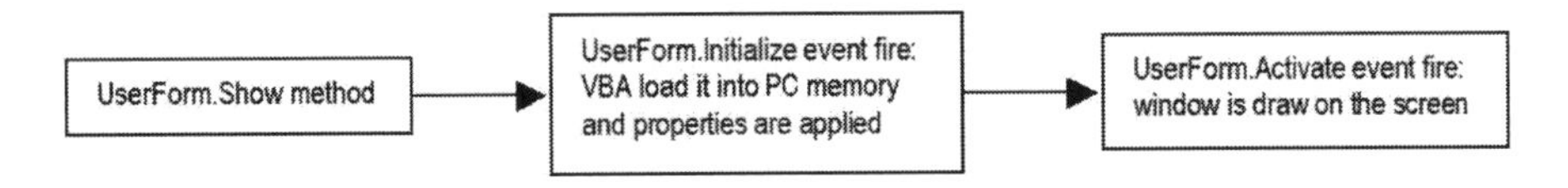

Figure 10-5. *When a UserForm is loaded into memory, its visual properties are defined in the UserForm.Initialize( ) event, but its window is created immediately before the UserForm_Activate( ) event fires*

Thinking in API terms, when a `UserForm.Show` method is called, the Windows operating system gets from the `UserForm` the `ThunderFrame` class definition a `Long` integer that identifies many window properties and sets them accordingly.

In most form classes, these basic properties are stored in address `-16` (`FFFFFFFFFFFFFFF0` in hexadecimal) inside the form structure definition. If you check on the Internet for API code to manipulate the `UserForm` window, you will immediately note that this address is normally attributed to a constant: `GWL_STYLE = -16`.

This means that if you want to manipulate the `UserForm` window to change some of its properties, you must put code in the `UserForm_Initialize( )` event or call the `UserForm.Hide` method to hide it, change its properties, and call the `UserForm.Show` method to draw it again.

The API procedure that grabs these `UserForm` property values as a `Long Integer` number is called `GetWindowLong( )` and has this syntax:

```
Declare Function GetWindowLongA Lib "user32" (ByVal Hwnd As Long, ByVal nIndex As Long)
As Long
```

In this code:

> `Hwnd`: This is the handle of the `UserForm`.
>
> `Nindex`: This is the position inside the `UserForm` class from where it must extract the `Long` integer.

The GetWindowLong( ) API function returns a long integer with all desired UserForm properties from the Nindex address position inside the class structure. Once you get it, you just need to know the bit position that you want to manipulate, change its value, and set it again to the UserForm structure, using the API procedure SetWindowLong( ), which has this syntax:

```
Declare Function SetWindowLongA Lib "user32" (ByVal hwnd As Long, ByVal nIndex As Long, _
                                              ByVal dwNewLong As Long) As Long
```

In this code:

Hwnd: This is the UserForm handle.

NIndex: This is the position inside the UserForm class to where the Long integer must be set.

DwNewLong: This is the Long value that must be set.

Most Internet API code uses dedicated procedures to set a given UserForm property (read as "set the bit inside the ThunderClass frame") and another procedure to verify whether the bit is set.

The UserForm Title Bar

The bit associated with the presence of the UserForm title bar inside the Long integer that represents the window properties is normally attributed to constant WS_CAPTION = &HC00000 (12582912 in decimal, 23rd and 24th bits set), although it can also be removed by setting the constant WS_DLGFRAME = &H400000 (4194304 in decimal, 23rd bit set).

■ **Attention** It seems like it is the 23rd bit that sets/removes the UserForm title bar. You can achieve the same result using both the WS_CAPTION and WS_DLGFRAME constants.

To add/remove a UserForm title bar, you need to set/unset this bit on the Long integer property byte and call the DrawMenuBar( ) API to change the UserForm appearance, which is declared in this way:

```
Declare Function DrawMenuBar Lib "user32" (ByVal Hwnd As Long) As Long
```

In this code:

Hwnd: This is the UserForm handle.

So, to remove the UserForm caption, you can use code like the following (supposing that the fm_FindWindow( ), fm_GetWindowLong( ), fm_SetWindowLong( ), and fm_DrawMenuBar( ) aliased APIs were declared):

```
Const GWL_STYLE = (-16)
Const WS_CAPTION = &HC00000

Sub RemoveTitleBar (frm as Object)
    Dim lngHwnd as Long
    Dim lngWinInfo as Long

    lngHwnd = Hwnd(frm)
```

```
    If lngHwnd > 0 then
        lngWinInfo  = fm_GetWindowLong(lngHwnd, GWL_STYLE)
        'clear frmUserForm SysMenu bit
        lngWinInfo = lngWinInfo And (Not WS_CAPTION)
        fm_SetWindowLong lngHwnd, GWL_STYLE, lngWinInfo
        fm_DrawMenuBar lngHwnd
    End If
End Function
```

Do you get it? After using the Hwnd( ) function to get the UserForm handle (associated with the frm as Object argument), you use the fm_GetWindowLong( ) aliased API to get the UserForm Long integer properties (using the GWL_STYLE constant to indicate from where this value must be retrieved), attributing it to the lngWinInfo variable.

```
lngHwnd = Hwnd(frm)
If lngHwnd > 0 then
    lngWinInfo  = fm_GetWindowLong(lngHwnd, GWL_STYLE)
```

To unset the WS_CAPTION bit from the lngWinInfo value, you use the And (Not WS CAPTION) operators, as explained in section "Setting Bit Values" earlier in this chapter.

```
lngWinInfo = lngWinInfo And (Not WS_CAPTION)
```

Once the desired bit is unset, you call fm_SetWindowLong to update the UserForm Long Integer, effectively disabling the title bar bit, and call the fm_DrawMenuBar lngHwnd API to set/remove the UserForm title bar.

```
fm_SetWindowLong lngHwnd, GWL_STYLE, lngWinInfo
fm_DrawMenuBar lngHwnd
```

▒ **Attention** This code will remove the UserForm title bar if called from the UserForm_Initialize( ) event, because at this point the UserForm window is not still drawn by the Windows system. To call it from a Command button, you must call the UserForm Hide and Show methods to update the window, adding/removing the title bar.

By making a small code change, you can declare a bolEnabled as a Boolean argument and allow the procedure to either set or remove the UserForm title bar, as follows:

```
Sub RemoveTitleBar (frm as Object, bolEnabled as Boolean)
    Dim lngHwnd as Long
    Dim lngWinInfo as Long

    lngHwnd = Hwnd(frm)
    If lngHwnd > 0 then
        lngWinInfo  = fm_GetWindowLong(lngHwnd, GWL_STYLE)
        'clear frmUserForm SysMenu bit
        If bolEnabled then
```

```
            lngWinInfo = lngWinInfo Or WS_CAPTION
        Else
            lngWinInfo = lngWinInfo And (Not WS_CAPTION)
        End If
        fm_SetWindowLong lngHwnd, GWL_STYLE, lngWinInfo
      fm_DrawMenuBar lngHwnd
    End If
    frm.Hide
    frm.Show
End Function
```

Note in the previous code that now you use the OR operator to set the desired bit (bolEnabled=True; UserForm has a title bar) or the AND NOT operators to remove the bit (bolEnabled = False). Also note that now the procedure calls the frm.Hide and frm.Show methods to allow the UserForm window to redraw, with or without a title bar.

To add/remove the UserForm Close button (the "X" button in the top-right corner), add/remove the maximize or minimize buttons, or add/remove a resizable border, you use the same code, changing the constant used to set/unset the lngWinInfo bit associated with these properties, as declared here:

```
Private Const GWL_STYLE = (-16)
Private Const WS_CAPTION = &HC00000
Private Const WS_MAXIMIZEBOX = &H10000
Private Const WS_MINIMIZEBOX = &H20000
Private Const WS_SYSMENU = &H80000
Private Const WS_THICKFRAME = &H40000
```

The Appearance( ) Procedure

Instead of making one procedure to manipulate each property, you can write a single, centralized procedure that manipulates any one of them, according to the argument it receives.

This is exactly what the frmAppearance UserForm, from the UserForm_APIs.xlsm macro-enabled workbook (that can be extracted from the Chapter10.zip file) does: it uses a single centralized procedure to change the UserForm appearance regarding the presence of a title bar; close, maximize, minimize buttons; resizable border; and transparency (Figure 10-6).

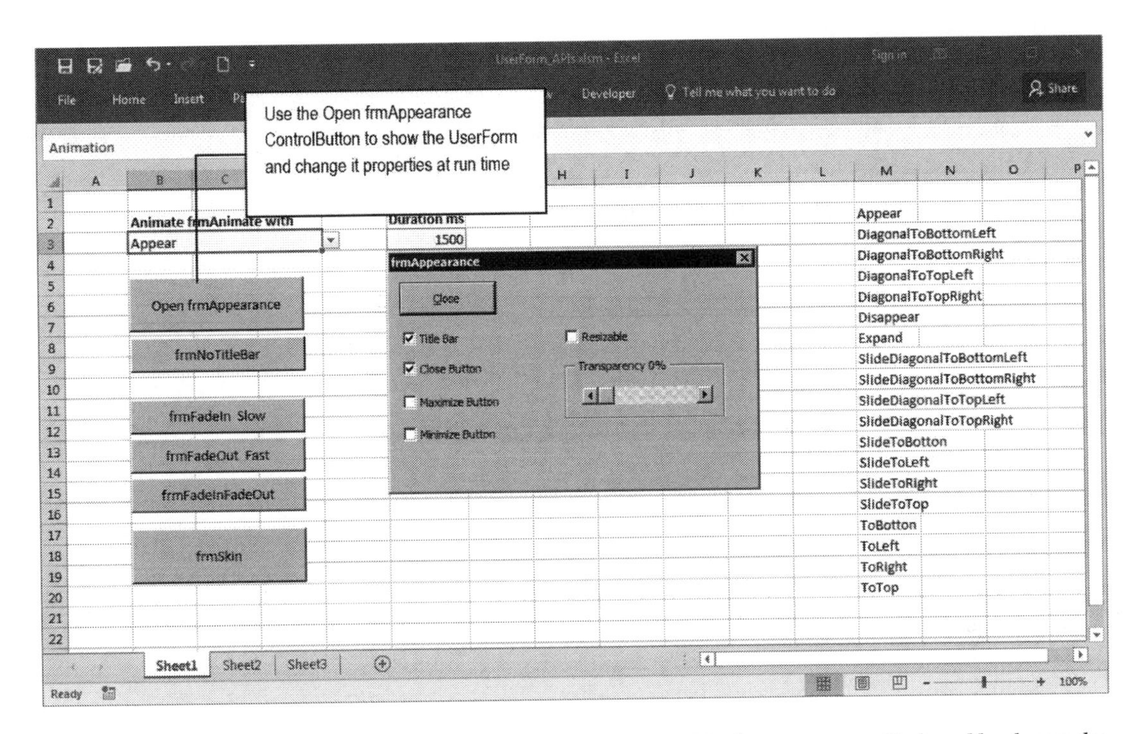

***Figure 10-6.** The frmAppearance UserForm from the UserForm_APIs.xlsm macro-enabled workbook uses the Windows APIs to manipulate the UserForm title bar; close, maximize and minimize buttons; resizable border; and transparency*

By inspecting the `basUserFormAPI` standard module, you will notice that it declares all API functions and constants needed to manipulate the `UserForm` properties in the `Declaration` section. Note that it also declares `Public Enum FormStyle` to give more significant names to the property that must be set.

```
Option Explicit
...
Private Declare Function fm_GetWindowLong Lib "user32" _
Alias "GetWindowLongA" (ByVal Hwnd As Long, ByVal nIndex As Long) As Long
Private Declare Function fm_SetWindowLong Lib "user32" _
Alias "SetWindowLongA" (ByVal Hwnd As Long, ByVal nIndex As Long, _
ByVal dwNewLong As Long) As Long
Private Declare Function fm_DrawMenuBar Lib "user32" Alias "DrawMenuBar" (ByVal Hwnd As
Long) As Long
Private Declare Function fm_SetLayeredWindowAttributes Lib "user32" _
Alias "SetLayeredWindowAttributes" (ByVal Hwnd As Long, ByVal crKey As Byte, _
ByVal bAlpha As Byte, ByVal dwFlags As Long) As Long

Public Enum FormStyle
    TitleBar = WS_CAPTION
    ResizableBorder = WS_THICKFRAME
    MaximizeButton = WS_MAXIMIZEBOX
    MinimizeButton = WS_MINIMIZEBOX
    CloseButton = WS_SYSMENU
End Enum
```

To manipulate the UserForm properties, it uses the Public Sub Appearance( ) procedure, which executes this code:

```
Public Sub Appearance(frm As Object, Style As FormStyle, bolEnabled As Boolean, Optional
bolRepaint As Boolean = True)
    Dim lngHwnd As Long
    Dim lngWinInfo As Long
    Dim lngHeight As Long

    'Get frm UserForm Handle
    lngHwnd = Hwnd(frm)
    'Get frm USerForm window style
    lngWinInfo = fm_GetWindowLong(lngHwnd, GWL_STYLE)

    If bolEnabled Then
        'Set frm UserForm bits
        lngWinInfo = lngWinInfo Or Style
    Else
        'Clear frm UserForm bits
        lngWinInfo = lngWinInfo And (Not Style)
    End If
    fm_SetWindowLong lngHwnd, GWL_STYLE, lngWinInfo

    If ((Style And TitleBar) = TitleBar) Then
        fm_DrawMenuBar (lngHwnd)
    End If

    If bolRepaint Then
        frm.Hide
        frm.Show
    End If
End Sub
```

As you can see, Sub Appearance( ) receives four arguments: frm as Object (the UserForm to manipulate), Style as FormStyle (the attribute or attributes to be set), bolEnabled (to turn on/off the desired attribute or attributes), and the optional bolRepaint (to automatically repaint the frm UserForm).

■ **Attention** As you will see later, sometimes it is not desirable that the UserForm automatically repaints as its properties are changed by the Sub Appearance( ) procedure.

The code uses Public Function Hwnd( ) to retrieve the frm UserForm handle and uses the fm_GetWindowLong( ) aliased API to retrieve the frm UserForm Long integer properties. According to the bolEnabled argument, it sets/unsets the desired bits represented by the Style argument and uses the fm_SetWindowLong( ) aliased API to set again the UserForm Long integer properties and eventually calls frm_DrawMenyBar, if the Style argument has the WSCAPTION flag set.

To effectively change the UserForm appearance regarding what it receives on the Style argument, the Hide and Show methods are called to force the UserForm to redraw, showing the appropriate properties.

▒ **Attention** The last paragraph uses "desired bits," allowing more than one bit at a time, because you can call `Appearance( )` with some of the `Style` constants to set more than one property at once, if possible.

In the `frmAppearance UserForm` module you can see how some operations are easily made to change the `UserForm` window appearance, sometimes needing a small adjustment. This is the case with the `chkTitleBar_Click( )` event, which fires whenever `frmAppearance chkTitleBar` changes its value.

```
Option Explicit

Dim mlngHeight As Long

Private Sub UserForm_Initialize()
    mlngHeight = Me.InsideHeight
End Sub
...
Private Sub chkTitleBar_Click()
    Appearance Me, TitleBar, Me.chkTitleBar
    Me.Height = mlngHeight
End Sub
```

When `frmAppearance` loads, the `Initialize( )` event stores the `UserForm.Height` property on the `mlngHeight` module-level variable and uses this value to return it to the default size after a call is made to the `Sub Appearance Me, TitleBar, Me.chkTitleBar` procedure, adding/removing the `UserForm` title bar as the `chkTitleBar` CheckBox control changes its value. This is necessary because after the `UserForm` title bar is removed and set again, the form grows by the size of the title bar (comment the `Me.Height = mlngHeight` instruction to see this happen).

The UserForm Close, Maximize, and Minimize Buttons, and Resizable Border

To set the `UserForm` Close, Maximize, and Minimize buttons or a resizable border, you just need to make a call to the `Sub Appearance( )` procedure, passing it the appropriate `Style` constant. The next procedure code shows what happens when you click the `chkCloseButton`, `chkMaximize`, `chkMinimize`, and `chkResizable` check boxes to add/remove the `UserForm` properties:

```
Private Sub cmdClose_Click()
    Unload Me
End Sub

Private Sub chkTitleBar_Click()
    Appearance Me, TitleBar, Me.chkTitleBar
    Me.Height = mlngHeight
End Sub

Private Sub chkCloseButton_Click()
    Appearance Me, CloseButton, Me.chkCloseButton
End Sub

Private Sub chkMaximize_Click()
    Appearance Me, MaximizeButton, Me.chkMaximize
End Sub
```

```
Private Sub chkMinimize_Click()
    Appearance Me, MinimizeButton, Me.chkMinimize
End Sub

Private Sub chkResizable_Click()
    Appearance Me, ResizableBorder, Me.chkResizable
End Sub
```

Changing More Than One UserForm Property at Once

As explained, you can set the UserForm properties on the UserForm_Initialize( ) event or change more than one property at any time by making a single call to the Appearance( ) procedure.

The frmNoTitleBar UserForm from the UserForm_APIs.xlsm macro-enabled workbook does this! It removes the title bar using the UserForm_Initialize( ) event and adds/removes the Close and Maximize buttons on the UserForm_Click( ) event (which fires when you click inside it), running this simple code:

```
Option Explicit

Private Sub UserForm_Initialize()
    Appearance Me, TitleBar + , False
End Sub

Private Sub UserForm_Click()
    Static sbolEnabled As Boolean

    sbolEnabled = Not sbolEnabled
    Appearance Me, TitleBar + ResizableBorder + MaximizeButton + CloseButton, sbolEnabled
End Sub
```

Note that the UserForm_Click( ) event passes TitleBar + + ResizableBorder + MaximizeButton + CloseButton to the Sub Appearance( ) Style argument, which allows you to make the UserForm border resizable while also enabling its Maximize and Close buttons. It uses the Static sbolEnabled as Boolean variable to alternate the on/off state of these properties.

To see them in action, double-click the frmNoTitleBar object in the VBA Object Explorer tree and press F5 to load it. You will notice that it appears without a title bar, but whenever you click inside it, the title bar automatically appears, showing the Minimize, Restore, Maximize, and Close buttons, with a resizable border.

■ **Attention** When you set the bit that shows the Maximize and Minimize buttons, both buttons will become visible, but just the bit set will work. By showing just the Maximize button, the Minimize button will appear, but disabled, and vice versa.

Any UserForm that calls the Sub Appearance( ) procedure from the Initialize( ) event must set the ShowModal property to False. Otherwise, the code will stop on the last procedure instruction (frm.Show) until the UserForm is closed, and a VBA runtime error 91 ("Object variable or With block not set") could appear when the UserForm is closed, after setting its properties using the Windows APIs (change the frmNoTitleBar ShowModal property to True, open it, click inside it, and close it to see for yourself).

The UserForm Transparency

At the UserForm ThunderFrame class structure's -20 position (FFFFFFFFFFFFFFEC in hexadecimal, normally associated with constant GWL_EXSTYLE = -20), you can recover extended window properties associated, for example, to the UserForm transparency: the value of the alpha channel or opacity value.

To create a transparent effect, you need to use the GetWindowLong( ) API to get this Long integer value, manipulate its 20th bit using constant WS_EX_LAYERED = &H80000 (10000000000000000000 in binary, 20th bit), and use the SetLayredWindowAttributes( ) API function to change the window alpha channel opacity from 0 (totally transparent) to 255 (totally opaque), which is declared in this way:

```
Declare Function SetLayeredWindowAttributes Lib "user32" (ByVal Hwnd As Long, ByVal crKey As
Byte, ByVal bAlpha As Byte, ByVal dwFlags As Long) As Long
```

In this code:

Hwnd: This is the UserForm handle.

crKey: This is the chroma key, which is a color reference that will be used as the transparent color. Use 0 to specify every color.

bAlpha: This is the alpha value used to describe the window opacity (must be between 0 = totally transparent and 255 = totally opaque).

dwFlags: This is an action to be taken.

LWA_ALPHA = &H2: This determines the opacity of the layered window, if crKey = 0.

LWA_COLORKEY = &H1: This uses the crKey color value as the transparency color.

The Transparency( ) Procedure

To change the UserForm transparency, basUSerFormAPIs implements Public Function Transparency( ), which executes this code:

```
Public Sub Transparency(frm As Object, sngTransparency As Single) 'As Boolean
    Dim lngHwnd As Long
    Dim lngWinInfo As Long
    Dim intOpacity As Integer

    'Get frm UserForm Handle
    lngHwnd = Hwnd(frm)
    'Get frm USerForm window style
    lngWinInfo = fm_GetWindowLong(lngHwnd, GWL_EXSTYLE)
    'Set extended form bit
    fm_SetWindowLong lngHwnd, GWL_EXSTYLE, lngWinInfo Or WS_EX_LAYERED
    intOpacity = 255 - (255 * sngTransparency)
    fm_SetLayeredWindowAttributes lngHwnd, 0, intOpacity, LWA_ALPHA
End Sub
```

The code calls function Hwnd(frm) to retrieve the UserForm handle, uses fm_GetWindowLong( ) with constant GWL_EXSTYLE to retrieve the Long integer associated with the extended form properties, and sets the WS_EX_LAYERED bit value to indicate that the transparency effect will be changed.

```
lngWinInfo = fm_GetWindowLong(lngHwnd, GWL_EXSTYLE)
'Set extended form bit
fm_SetWindowLong lngHwnd, GWL_EXSTYLE, lngWinInfo Or WS_EX_LAYERED
```

Since it receives the sngTransparency argument as a percent value (from 0 to 1) to the desired transparency effect, it must change it into an opacity value that goes from 0 to 255 (opacity is the reciprocal of transparency; 100 percent transparent means 0 percent opaque).

```
intOpacity = 255 - (255 * sngTransparency)
```

Once the opacity is calculated, it calls the fm_SetLayeredWindowAttributes( ) aliased API to change the UserForm opacity, using the crKey = 0 argument, to manipulate all colors opacity at once and passes LWA_ALPHA = &H26 (100110 in binary) to the dwFlags argument.

```
fm_SetLayeredWindowAttributes lngHwnd, 0, intOpacity, LWA_ALPHA
```

On the UserForm side, the frmAppearance uses the fraTransparency Frame control to contain the scrTransparency ScrollBar control and change its own transparency. The scrTransparency control has set its properties Mini = 0, Max = 100, SmallChange = 5, and LargeChange = 10, to allow changing the transparency value from 0 to 100 (by 5 transparency points when click the scroll bar arrows and by 20 transparency points inside the scroll bar at the left or right of the scroll bar button). Whenever the scrTransparency ScrollBar control value changes, it fires the scrTransparency_Change( ) event, executing this code:

```
Private Sub scrTransparency_Change()
    Dim sngValue As Single

    sngValue = Me.scrTransparency / 100
    Me.fraTransparency.Caption = "Transparency" & Format(sngValue, "0%")
    Transparency Me, sngValue
End Sub
```

As you can see, whenever you drag the scrTransparency ScrollBar control to a new value, it calculates the percent transparency on the sngValue variable and sets the fraTransparency Frame control caption accordingly. Then it calls Function Transparency( ) from basUserFormAPIs, passing as an argument a reference to itself (Me) and the desired transparency percent value (Figure 10-7).

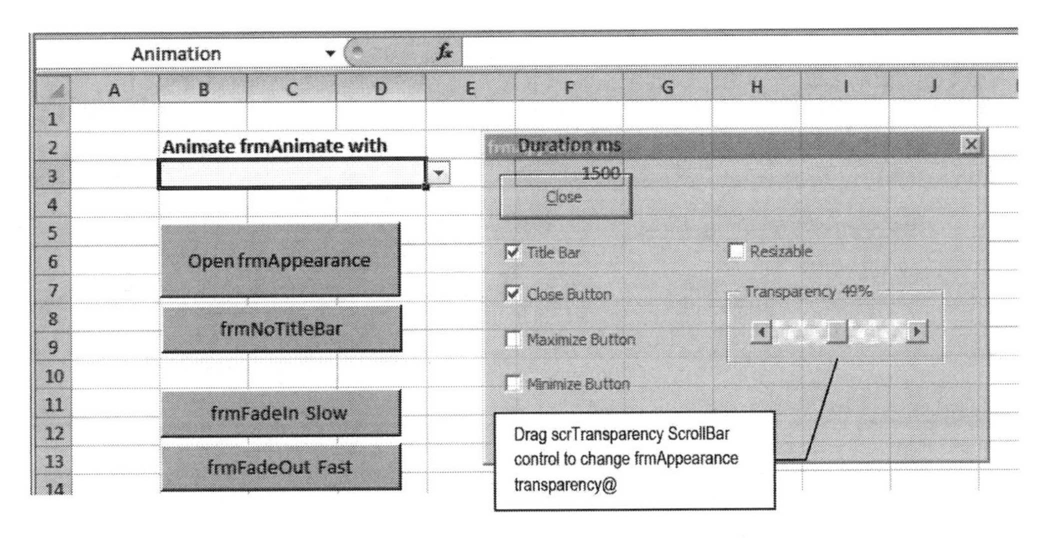

Figure 10-7. *Drag the scrTransparency ScrollBar control to the left or right to change the UserForm transparency. It calls Function Transparency( ) from basUserFormAPIs to change the UserForm opacity value (the reciprocal of transparency)*

■ **Attention** In `basUserFormAPIs`, change the `Function Transparency( )` instruction `lngHwnd = Hwnd(frm)` to `lngHwnd = Application.Hwnd` and then drag the `scrTransparency ScrollBar` control on `frmAppearance` to change the Excel window's transparency.

Note that if you drag the `scrTransparency ScrollBar` control to 100 percent transparency, the `frmAppearance UserForm` disappears from the Excel interface and cannot be selected anymore. It is still loaded but unreachable.

The Fade( ) Procedure

You can use `Sub Transparency( )` to create a "fade-in" or "fade-out" effect to animate any `UserForm` like the `AnimateWindow( )` API does using the `AW_Blend` constant.

You just need to use a `For...Next` loop to loop 100 times, changing the `UserForm` transparency from 100 percent to 0 percent to fade in and from 0 percent to 100 percent to fade out, taking care to use a specific delay in milliseconds at each loop step.

This was implemented in the `basUserFormAPI` standard module using two different procedures: `Public Sub Fade( )` is used to interact with the user regarding the fade effect, and `Private Sub FadeEffect( )` is used to apply the `UserForm` transparency in the desired direction (fade-in or fade-out).

To allow an easy selection of the fade effect and speed, two enumerators were declared in `basUserFormAPI`: `FadeMethod` and `FadeSpeed`.

```
Public Enum FadeMethod
    FadeIn = 1
    FadeOut = 2
    FadeInFadeOut = 3
End Enum

Public Enum FadeSpeed
```

```
    Slow = 1
    Fast = 2
End Enum
```

To apply the fade effect, use the Public Sub Fade( ) procedure, which executes this code:

```
Public Sub Fade(frm As Object, Fading As FadeMethod, Optional Speed As FadeSpeed = Slow, _
                Optional WaitSeconds As Integer)
    Dim sngTime As Single

    Call FadeEffect(frm, Fading, Speed)
    frm.Repaint
    If Fading = FadeInFadeOut Then
        If WaitSeconds > 0 Then
            sngTime = Time
            Do
            Loop While DateDiff("s", sngTime, Time) < WaitSeconds
        End If
        Call FadeEffect(frm, FadeOut, Speed)
    End If

    If Fading = FadeInFadeOut Or Fading = FadeOut Then
        Unload frm
    End If
End Sub
```

Note that it receives three arguments: frm as Object (the UserForm), Fading as FadeMethod (the fade direction), and the Optional Speed as FadeSpeed = Slow argument (the fade speed).

Independent of the fade method selected, it first calls FadeEffect(frm, Fadding, Speed) to apply the desired fade method at the desired speed, which executes this code:

```
Private Sub FadeEffect(frm As Object, Fading As FadeMethod, Optional Speed As
FadeSpeed = Slow)
    Dim sngTimer As Single
    Dim sngTransparency As Single
    Dim sngMaxTime As Single
    Dim intI As Integer
    Const conSlow = 0.04
    Const conFast = 0.01

    sngMaxTime = IIf(Speed = Fast, conFast, conSlow)
    For intI = 0 To 100
        sngTransparency = intI / 100
        If Fading = FadeIn Or Fading = FadeInFadeOut Then sngTransparency = 1 -
sngTransparency
        Transparency frm, sngTransparency
        sngTimer = Timer
        Do
        Loop While (Timer - sngTimer) < sngMaxTime
    Next
End Sub
```

Note that the Sub FadeEffect( ) declares the constants conSlow = 0;04 and conFast = 0.01, which relate to tenths of milliseconds, and attributes to sngMaxTime the desired constant value according to the Speed argument.

```
sngMaxTime = IIf(Speed = Fast, conFast, conSlow)
```

It then begins a For...Next loop that will execute 100 times, attributing to sngTransparency the desired transparency percentage.

```
For intI = 0 To 100
    sngTransparency = intI / 100
```

If Fading = FadeIn or Fading = FadeInFadeOut, sngTransparency must go from 100 percent to 05 transparency, so it changes sngTransparency accordingly.

```
If Fading = FadeIn Or Fading = FadeInFadeOut Then sngTransparency = 1 - sngTransparency
```

Apply the desired transparency effect to the UserForm associated to the frm as Object argument.

```
Transparency frm, sngTransparency
```

Now it is time to give a delay between each loop step. This is done by attributing the VBA function Timer( ) value (the number of seconds since midnight) to the sngTimer variable and executing a Do...Loop while the difference between sngTimer and Timer( ) is smaller than the sngMaxTime milliseconds value.

```
sngTimer = Timer
Do
Loop While (Timer - sngTimer) < sngMaxTime
```

When the loop ends, the UserForm changes the transparency from 0 to 100 (or vice versa), during from 100 * 0.01 ms ≅ 1 second (if sngMaxTime = conFast = 0.01) to 100 * 0.04 ≅ 4 seconds (if sngMaxTime = conSlow = 0.04).

The code returns the control to the Sub Fade( ) effect, which now verifies whether the argument Fading = FadInFadeOut, and if this is true, it verifies whether the WaitSeconds argument is greater than zero (indicating that the fade must do a stop before fading out).

```
If Fading = FadeInFadeOut Then
    If WaitSeconds > 0 Then
```

If WaitSeconds > 0, the desired delay is applied, using again a Do...Loop that uses the VBA DateDiff("s", sngTime, Time) function to define the difference between sngTime and another call to the VBA Time( ) function in seconds. The loop will last while this difference is smaller than WaitSeconds.

```
    sngTime = Time
    Do
    Loop While DateDiff("s", sngTime, Time) < WaitSeconds
End If
```

Having or not executing a delay, since argument Fading = FadeInFadeOut, it must now execute a fade-out effect at the same fade-in speed.

```
    Call FadeEffect(frm, FadeOut, Speed)
End If
```

And if the Fading argument is either FadeOut or FadeInFadeOut, the frm UserForm must be unloaded, ending the fade effect.

```
    If Fading = FadeInFadeOut Or Fading = FadeOut Then
        Unload frm
    End If
End Sub
```

Using frmFadeIn UserForm

You can use the frmFadeIn Slow ControlButton to open the frmFadeIn UserForm and watch it fade in using the default slow speed, because it executes this code on the UserForm_Initialize( ) and Activate( ) events.

```
Private Sub UserForm_Initialize()
    Transparency Me, 1
End Sub

Private Sub UserForm_Activate()
    Me.Repaint
    Fade Me, FadeIn
End Sub
```

Note that it first applies a 100 percent transparency to itself using Transparency Me, 1 on the UserForm_Initialize( ) event, and when the Activate( ) event fires, it uses the Repaint method to appear 100 percent transparent before calling Fade Me, FadeIn to slowly fade in to the user environment!

Using frmFadeOut UserForm

The frmFadeOut UserForm can be tested by clicking the frmFadeOut Fast ControlButton, which is waiting to be clicked to fade out, executing this code:

```
Private Sub UserForm_Activate()
    Transparency Me, 0
End Sub

Private Sub UserForm_Click()
    Unload Me
End Sub

Private Sub UserForm_QueryClose(Cancel As Integer, CloseMode As Integer)
    Fade Me, FadeOut, Fast
End Sub
```

Note that now it needs to add a 0 percent transparency using Transparency Me, 0 on the UserForm_Activate( ) event before fading out itself using a call to Fade Me, FadeOut, Fast on the UserForm_QueryClose( ) event (the last event fired before the UserForm_Terminate( ) event), when the UserForm is unloaded by any means.

Using frmFadeInFadeOut UserForm

To apply an automated fade-in/fade-out slow effect with a three-second wait between them, use the `frmFadeInFadeOut` `ControlButton`, which loads `frmFadeInFadeOut` that executes this codes:

```
Private Sub UserForm_Initialize()
    Transparency Me, 1
End Sub

Private Sub UserForm_Activate()
    Me.Repaint
    Fade Me, FadeInFadeOut, Slow, 3
End Sub
```

This time, the `UserForm` is first made 100 percent transparent by calling `Transparency Me, 1` on the `UserForm_Initialize( )` event, and when the `Activate( )` event fires, it uses the `Repaint` method to repaint itself totally transparent and calls `Fade Me, FadeInFadeOut, Slow, 3` to create a fade-in/fade-out effect with a three-second wait in between. This is the same thing most programs do to show the splash screen at startup (like Excel does).

■ **Attention** You must make the `UserForm` 100 percent transparent or 100 percent opaque before applying a `FadeIn` or `FadeOut` effect, respectively.

Applying a Skin to a UserForm

Since the advent of Windows XP, a new class of window appeared in some popular applications, such as Windows Media Player, Nero Burning Room, and so on. Instead of using a rectangular window, they all use a specially designed version with different shapes, colors, and positions of the menu bar and window controls (such as minimize, restore, maximize, and close buttons, if any).

This type of form shape is usually associated with the word *skin*, and you can use a bunch of API procedures to apply a skin to any `UserForm`.

These are the steps to apply a `UserForm` skin:

1. Produce a bitmap image that will shape the `UserForm` (reserve space to put controls, if necessary).
2. Surround the image with a chroma key background color that must be changed to transparent (usually white or black).
3. Use the same chroma key color in the `UserForm Background` property.
4. Attribute the bitmap image to the `UserForm Picture` property, setting `PictureSizeMode` to `3 - fmPictureSizeModeZoom`.
5. Use a code procedure to remove the `UserForm` title bar and change every pixel associated with the chroma key color to transparent.

The first four steps are design specific and do not require much knowledge, except that the bitmap image produced must be a BMP file (avoid using JPEG files because they add no white pixel artifacts that can surround the image), be surrounded by the chroma key color as best as possible, and have the smallest possible size in terms of pixel count so that the process of changing the chroma key pixels to transparent run as fast as your computer can. To apply the fifth step, you need to use the Windows API functions.

Device Contexts

Although a VBA UserForm does not offer a line or shape control, Microsoft Access forms offer both of them, and Visual Basic, since its first version, also offers a circle control. So, it is possible to draw on the UserForm surface.

This is possible because Windows offers what is called a *device context*, which is a programmable structure that can receive directly drawn color manipulation, like Paint does. The most famous device contexts are the display screen and the printer, although Windows supports many other devices, such as plotters, image acquisition devices, and so on.

Regarding the UserForm, the device context in design mode corresponds to the background area, which is the rectangular place where the layout controls produce the desired results. But when a UserForm is in running mode, the device context may be considered as the UserForm image, with everything you put on it.

To get the device context (DC) of any UserForm and manipulate the pixels, Windows offers the API Function GetDC( ) from User32.dll, which has this syntax:

```
Declare Function GetDC Lib "user32" (ByVal Hwnd As Long) As Long
```

In this code:

Hwnd: This is the window handle.

The GedDC( ) API function returns a Long Integer, representing the handle for the device context to the specified window.

And since the device context uses about 800 bytes of memory, once you finish manipulating it, you must release it from memory using the API's Function ReleaseDC( ), which has this syntax:

```
Declare Function ReleaseDC Lib "user32" (ByVal Hwnd As Long, ByVal hdc As Long) As Long
```

In this code:

Hwnd: This is the window handle.

hdc: This is the handle to the device context to be released.

Once a UserForm device context handle had been obtained, you can loop through the pixels using the Windows API's Function GetPixel( ), which retrieves the RGB Long Integer color value of any pixel, given its x, y coordinates. It has this syntax:

```
Declare Function GetPixel Lib "gdi32" (ByVal hdc As Long, ByVal X As Long, ByVal Y As Long)
As Long
```

In this code:

hdc: This is the device context handle.

X, Y: This is a zero-based point to check in the logical coordinates of the bitmap image.

The GetPixel( ) API returns a Long Integer indicating the pixel color value. If the pixel is outside the clipping region of the UserForm, it will return the CLR_INVALID = &HFFFFFFFF constant.

There is one warning regarding coordinate transformations between the UserForm device context pixel count: you need to multiply the UserForm internal dimensions by 4/3 (the basic image aspect ratio) to convert the UserForm InternalWidth and InternalHeight properties to the X, Y pixels count, respectively.

The next code fragment illustrates how you can iterate through all the pixels of a UserForm device context, getting one at a time:

```
Dim lngPixelsX As Long
Dim lngPixelsY As Long
Dim lngHwnd As Long
Dim lngHwndDC As Long
Dim lngPixelColor As Long
Const AspectRatio = 4 / 3

'Get UserForm and it DeviceContext handles
lngHwnd = Hwnd(UserForm1)
lngHwndDC = GetDC(lngHwnd)

lngPixelsX = UserForm1.InsideWidth * AspectRatio
lngPixelsY = UserForm1.InsideHeight * AspectRatio

For lngY = 0 To lngPixelsY - 1
    For lngX = 0 To lngPixelsX - 1
        lngPixelColor = GetPixel(lngHwndDC, lngX, lngY)
        ...
    Next
Next
```

That was easy, huh? There is no such complexity to step through each UserForm pixel and get its color.

Changing the UserForm Shape Using Windows Regions

Now that you can access the UserForm image and iterate through each of its pixels, you need to know that you will not change its colors to change the UserForm shape. Instead, you will analyze the UserForm picture, row by row, finding regions composed of continuous pixels where the colors are different than the one used as the chroma key (usually the white color).

These colored regions are then used to compose a new shape (representing the skin image), row by row, until all UserForm pixels are processed. When this new shape is created, it is applied as the new UserForm shape, effectively creating the "skin" effect.

These regions are graphic device interface (GDI) objects that describe an area in a device context object, having its own handle be manipulated by other Windows API functions. To create these pixel colored regions, you need to use the Windows API's Function CreateRectRgn( ) from gdi32.dll, which has this syntax:

```
Declare Function CreateRectRgn Lib "gdi32" (ByVal Left As Long, ByVal Top As Long, _
ByVal Right As Long, Bottom As Long) As Long
```

In this code:

Left, Top: These are coordinates that describe the upper-left corner.

Right, Bottom: These are coordinates that describe the lower-right corner. The Right and Bottom coordinates must be at least 1 pixel greater than the Left and Top coordinates, respectively, to create a rectangular region. They are not considered part of the region.

The CreateRectRgn( ) API returns a long integer associated to the graphic device interface's object handle. The next call to CreateRectRgn( ) creates an empty region, with no pixel count, and stores its handle on lngHwndRgn.

```
lngHwndRgn = CreateRectRgn(0, 0, 0, 0)
```

Since the Right and Bottom coordinates do not belong to the region, the next call creates a 1-pixel region associated to the pixel position (0,0).

```
lngHwndRgn = CreateRectRgn(0, 0, 1, 1)
```

Whenever you create a device context region with the GetDC( ) API, you must be sure to delete it to free the memory used by using Function DeleteObject( ) from gdi32.dll, which has this syntax:

```
Declare Function DeleteObject Lib "gdi32" (ByVal hObject As Long) As Long
```

In this code:

hObject: This is the handle to the graphic device interface object to be deleted.

To create a complex region, composed of different colored rectangular regions, you must use the Windows API's Function CombinRgn( ) from gdi32.dll, which has this syntax:

```
Declare Function CombineRgn Lib "gdi32" (ByVal hDestRgn As Long, ByVal hSrcRgn1 As Long, _
ByVal hSrcRgn2 As Long, ByVal nCombineMode As Long) As Long
```

In this code:

hDestRgn: This is the handle to the new combined region with dimensions defined by combining two other regions (this new region must exist before CombineRgn( ) is called).

hSrcRgn1: This is the handle to the first of two regions to be combined.

hSrcRgn2: This is the handle to the second of two regions to be combined.

nCombineMode: This is the mode indicating how the two regions will be combined.

RGN_AND: This creates the intersection of the two combined regions.

RGN_COPY: This creates a copy of the region identified by hSrcRgn1.

RGN_DIFF: This combines the parts of hSrcRgn1 that are not part of hSrcRgn2.

RGN_OR: This creates the union of two combined regions.

RGN_XOR: This creates the union of two combined regions except for any overlapping areas.

The CombineRgn( ) API function returns one of these constant values:

NULLREGION: This region is empty.

SIMPLEREGION: This region is a single rectangle.

COMPLEXREGION: This region is more than a single rectangle.

ERROR: No region is created.

The CombineRgn( ) API can be used to successively combine an empty region with different regions of a device context, using the RGN_XOR constant on the nCmbineMode argument and producing a complex, nonrectangular region (an irregular bitmap). Once you have such an irregular region, you can use the Function SetWindowRgn( ) API from user32.dll to apply it as the new UserForm appearance, which has this syntax:

```
Declare Function SetWindowRgn Lib "user32.dll" (ByVal Hwnd As Long, ByVal hRgn As Long,
ByVal bRedraw As Long) _As Long/
```

In this code:

Hwnd: This is the window handle.

hRgn: This is the region handle. It sets the window region of the window to this region. If hRgn is NULL, the function sets the window region to NULL.

bRedraw: This uses True to specify that the system must redraw the window after setting the window region.

The SetWindowRgn( ) API returns a nonzero value if it succeeds; it otherwise returns zero.

Let's see a practical example. Figure 10-8 shows a small bitmap composed of 10x10 pixels, representing a diamond shape composed of a black border and yellow background, entirely surrounded by white color pixels (the chroma key).

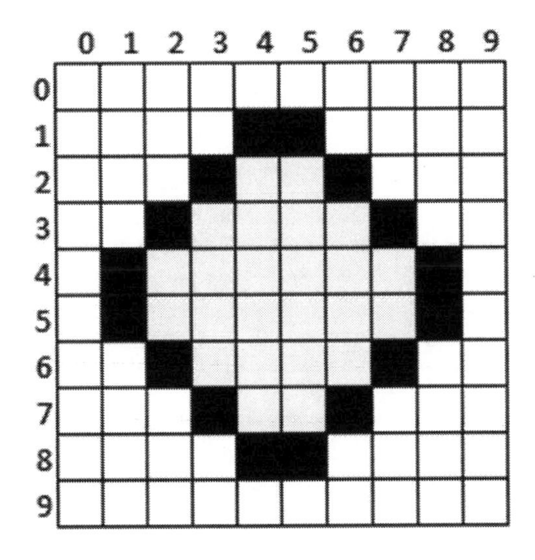

Figure 10-8. *A bitmap of 10x10 pixels, surrounding by white color pixels (chroma key). The pixels count and grid were used to indicate the pixel position*

Supposing that you want to use the Figure 10-8 bitmap as a UserForm skin using the white color as the chroma key (the transparent color), you must apply it to the UserForm.Picture property, get the UserForm device context, and use two nested loops to run through each image row, pixel-by-pixel, until a nonwhite pixel is found, creating successive one-row rectangular regions of every other nonwhite colored pixel on the same row.

Since the bitmap row 0 is entirely white, you can create a one-row bitmap region from row 1, pixels 4 and 5, using CreateRectRgn( ) in this way:

```
lngTempRegn = CreateRectRgn(4, 1, 2, 6)
```

(left = 4, top = 1) indicates the rectangle's (left, top) initial pixel coordinates, and Right = 6, Bottom = 2 indicates the (right, bottom) rectangle coordinates that do not belong to the first rectangular region.

You can create another one-row bitmap from row 2, pixels 3 to 6, using CreateRectRgn( ) in this way:

```
lngTempRegn = CreateRectRgn(3, 2, 3, 7)
```

(Left = 3, Top = 2) indicates the rectangle's (left, top) initial pixel, and Right = 7, Bottom = 3 indicates the (right, bottom) rectangle coordinates that do not belong to this second rectangular region.

You can use the CombineRgn( ) API to combine these two different one-row rectangular regions in this way:

```
Dim lngSkinRgn as Long
Dim lngTempRgn as Long

lngSkinRgn = CreateRectRgn(0, 0, 0, 0)  'Empty rectangular region
lngTempRgn = CreateRectRgn(4, 1, 2, 6) 'First one-row rectangular region
lngSkinRgn = CombineRgn(lngSkinRgn, lngSkingRgn, lngTempRgn, RGN_OR)
DeleteObject(lngTempRgn)

lngTempRgn = CreateRectRgn(3, 2, 3, 7) "Second one-row rectangular region
lngSkinRgn = CombineRgn(lngSkinRgn, lngSkingRgn, lngTempRgn, RGN_OR)
DeleteObject(lngTempRgn)
```

This last code fragment begins defining lngSkinRgn as an empty rectangle and then uses the CombineRgn( ) API to combine it with the first rectangular region comprising all of row 1's colored pixels. After deleting the lngTempRgn region, it combines lngSkinRgn again with the second rectangular region comprising all of row 2's colored pixels.

By continue to combine all rectangular regions of rows 3 to 8, lngSkinRgn will end up with a handle to the diamond shape bitmap that has an irregular border, using just its nonwhite pixels, which can be applied as the new UserForm window shape using the SetWindowRgn( ) API.

The Skin( ) Procedure

The basUserFormAPI module declares the GetDC( ), ReleaseDC( ), GetPixel( ), CreateRectRgn( ), CombineRgn( ), SetWindowRgn( ), and DeleteObject( ) aliased APIs.

```
Option Explicit

...
'DLL declaration to change UserForm skin
Private Declare Function fm_GetDC Lib "user32" Alias "GetDC" (ByVal Hwnd As Long) As Long
Private Declare Function fm_ReleaseDC Lib "user32" Alias "ReleaseDC" (ByVal Hwnd As Long,
ByVal hdc As Long) _As Long
Private Declare Function fm_GetPixel Lib "gdi32" Alias "GetPixel" (ByVal hdc As Long, ByVal
X As Long, ByVal Y As Long) As Long
```

```
Private Declare Function fm_CreateRectRgn Lib "gdi32" Alias "CreateRectRgn" (ByVal Left As
Long, ByVal Top As Long, ByVal Right As Long, ByVal Bottom As Long) As Long
Private Declare Function fm_CombineRgn Lib "gdi32" Alias "CombineRgn" (ByVal hDestRgn As
Long, ByVal hSrcRgn1 _
As Long, ByVal hSrcRgn2 As Long, ByVal nCombineMode As Long) As Long
Private Declare Function fm_SetWindowRgn Lib "user32" Alias "SetWindowRgn" (ByVal Hwnd As
Long, ByVal hRgn As Long, ByVal bRedraw As Long) As Long
Private Declare Function fm_DeleteObject Lib "gdi32" Alias "DeleteObject" (ByVal hObject As
Long) As Long
```

The code also has the TransparentColor enumerator and the fully commented Public Sub Skin( ), which uses these APIs to apply a UserForm skin:

```
Public Enum TransparentColor
    White = 16777215
    Black = 0
End Enum
...
Public Sub Skin(frm As Object, Optional TransparentColor As TransparentColor = White)
    'Apply a UserForm Skin using the UserForm Picture image
    'TransparentColor argument indicates the chroma-key color (White or Black)
    'Use this procedure on a UserForm 100% transparent with no Title bar
    Dim lngHwnd As Long
    Dim lngHwndDC As Long
    Dim lngPixelsX As Long
    Dim lngPixelsY As Long
    Dim lngSkinRgn  As Long
    Dim lngX As Long
    Dim lngY As Long
    Dim lngPixel As Long
    Dim lngLeft As Long
    Dim lngNewRgn  As Long
    Dim bolNewRgn  As Boolean
    Const Color_Invalid As Long = &HFFFFFFFF
    Const RGN_OR As Long = 2
    Const AspectRatio = 4 / 3

    'Get UserForm and it DeviceContext handles
    lngHwnd = hwnd(frm)
    lngHwndDC = fm_GetDC(lngHwnd)

    'Set UserForm BackColor and border
    frm.BackColor = TransparentColor
    frm.BorderStyle = fmBorderStyleNone

    'Get UserForm dimensions in pixels
    lngPixelsY = frm.InsideHeight * AspectRatio
    lngPixelsX = frm.InsideWidth * AspectRatio
```

```
    'Create a new, empty rectangular gdi region
    lngSkinRgn = fm_CreateRectRgn(0, 0, 0, 0)

    'Loop through all USerForm pixels rows
    For lngY = 0 To lngPixelsY - 1
        'Loop through all UserForm pixels columns on each row
        For lngX = 0 To lngPixelsX - 1
           lngPixel = fm_GetPixel(lngHwndDC, lngX, lngY)
           If bolNewRgn Then
               If lngPixel = TransparentColor Then
                   'Define the new region
                   lngNewRgn = fm_CreateRectRgn(lngLeft, lngY, lngX + 1, lngY + 1)
                   'Add the new region to existing regions
                   Call fm_CombineRgn(lngSkinRgn, lngSkinRgn, lngNewRgn, RGN_OR)
                   'Delete the new region
                   Call fm_DeleteObject(lngNewRgn)
                   bolNewRgn = False
               End If
           Else
               If lngPixel <> TransparentColor And lngPixel <> Color_Invalid Then
                   'Begin to define a new region
                   bolNewRgn = True
                   lngLeft = lngX
               End If
           End If
        Next lngX
        'Restart a new region for every pixel row
        lngLeft = 0
        bolNewRgn = False
    Next lngY

    'lngSkinRgn has now all no transparent regions. Apply the skin!
    fm_SetWindowRgn lngHwnd, lngSkinRgn, True
    'Release the new UserForm device context
    fm_DeleteObject lngSkinRgn
End Sub
```

The Skin( ) procedure is in fact quite small, but it's inflated a bit because of its long section declaration and its many comments. It receives two arguments: `frm as Object` (the UserForm reference) and `TransparentColor as TransparentColor = White` (the chroma key that will be discarded from the UserForm picture; the default is white).

As the first comment states, the code expects to receive a UserForm reference that has no title bar and is 100 percent transparent, so you need to call Sub Appearance( ) and Sub Transparency( ) before calling Sub Skin( ) to apply a skin effect to the UserForm.

After declaring the many variables, it gets a handle to the UserForm and its device context, using the Function Hwnd( ) and the fm_GetDC( ) aliased API, and effectively takes a picture of the UserForm appearance (the UserForm device context is independent of its alpha channel; in other words, it doesn't care about the UserForm transparency).

```
'Get UserForm and it DeviceContext handles
lngHwnd = hwnd(frm)
lngHwndDC = fm_GetDC(lngHwnd)
```

It then sets the UserForm BackColor = TransparentColor argument to guarantee that it has the same chroma key color and sets UserForm BorderStyle = fmBorderStyleNone to remove the border.

```
'Set UserForm BackColor and border
frm.BackColor = TransparentColor
frm.BorderStyle = fmBorderStyleNone
```

To determine how many pixel columns and rows the UserForm has, it uses the UserForm InternalWidth and InternalHeight properties multiplied by the AspectRatio = 4/3 constant (which gives a pretty good pixel count approximation to most screen resolutions).

```
'Get UserForm dimensions in pixels
lngPixelsX = frm.InsideWidth * AspectRatio
lngPixelsY = frm.InsideHeight * AspectRatio
```

And before looping through the UserForm device context pixels, it declares an empty rectangular region using the fm_CreateRectRgn( ) aliased API, attributing the region handle to the lngSkinRgn variable.

```
'Create a new, empty rectangular gdi region
lngSkinRgn = fm_CreateRectRgn(0, 0, 0, 0)
```

It then sets two nested For...Next loops. The outer loop runs lngY through all the device context pixels rows, while the inner one runs lngX through all its pixels columns.

```
'Loop through all USerForm pixels rows
For lngY = 0 To lngPixelsY - 1
    'Loop through all UserForm pixels columns on each row
    For lngX = 0 To lngPixelsX - 1
```

For each device context row, it takes each pixel column using the fm_GetPixel( ) aliased API and stores its color value on the lngPixel variable.

```
lngPixel = fm_GetPixel(lngHwndDC, lngX, lngY)
```

Then the code verifies whether the code has already begun to define a nonchroma key region (a bitmap region that has no transparent color), testing the bolNewRgn Boolean variable value. While bolNewRgn = False, the codes steps to the Else clause and verifies whether the lngPixel color is different from the selected TransparentColor argument (the chroma key). Note that the code tests it against the Color_Invalid constant, which may appear if the selected pixel is outside the device context dimensions because of the approximate pixel count provided by the AspectRatio approximation.

Whenever lngPixel <> TransparentColor And lngPixel <> Color_Invalid is true, it means that a colored bitmap pixel was found, such the diamond-shaped first black pixel color at point (4, 1) shown in Figure 10-9, and a new region must begin. So, the code sets bolNewRegion = True and stores the X (column) position of this first pixel.

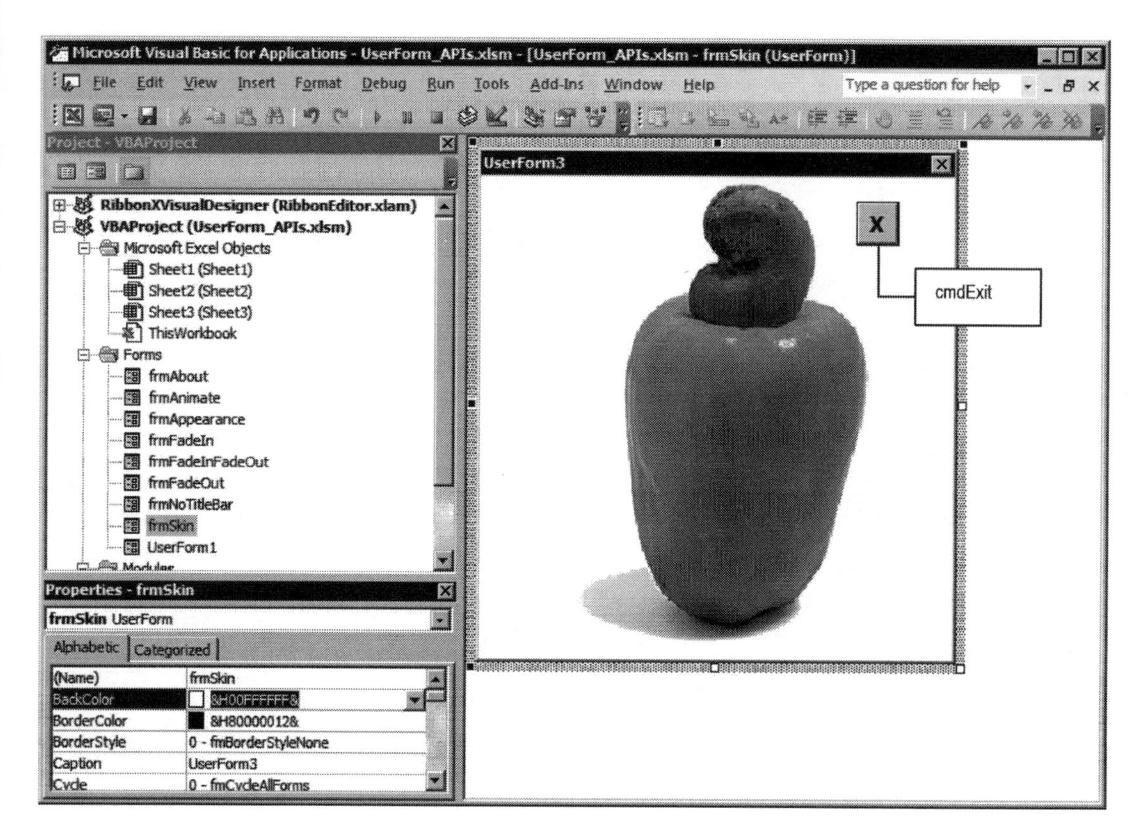

Figure 10-9. *This is the frmSkin UserForm, which uses the Caju.bmp image as the Picture property and sets BackColor = &H00FFFFFF&, which is the same bitmap background as the chroma key*

```
If bolNewRgn Then
    ...
Else
    If lngPixel <> TransparentColor And lngPixel <> Color_Invalid Then
        bolNewRgn = True
        lngLeft = lngX
    End If
End If
```

Now bolNewRgn = True, and the loop will find the next pixel of the chroma color on the same row (for the diamond shape of Figure 10-8, row 1, this pixel will be at column 6).

```
If bolNewRgn Then
    If lngPixel = TransparentColor Then
```

That is why it is so important that the bitmap be surrounded by the transparent color. When such a pixel is found, it is time to use the aliased API's `frm_CreateRctRgn( )` to create a new region that goes from (`lngLeft, lngY`) to (`lngX+1, lngY+1`), apply `fm_CombineRgn( )` to combine it with the empty `lngSkinRgn` region, and use `fm_DeleteObject( )` delete the new region. This will free its memory resources.

```
        'Define the new region
        lngNewRgn = fm_CreateRectRgn(lngLeft, lngY, lngX + 1, lngY + 1)
        'Add the new region to existing regions
        Call fm_CombineRgn(lngSkinRgn, lngSkinRgn, lngNewRgn, RGN_OR)
        'Delete the new region
        Call fm_DeleteObject(lngNewRgn)
        bolNewRgn = False
    End If
End If
```

After the new region was combined with lngSkinRgn, it makes bolNewRgn = False so the code can chase another colored region on the same or next bitmap rows. Note that whenever the loop reaches the end of a row (by processing all its lngX pixels columns), it makes lngLeft = 0 and bolNewRgn = False to guarantee that just one-row rectangular pixel regions will be combined.

When all row pixels are processed, lngSkinRgn will have an irregular region associated to the picture whose skin you want to apply to the UserForm, so it calls the fm_SetWindowRgn( ) aliased API to apply it as the new UserForm window and uses the fm_DeleteObject( ) aliased API to delete this device context and free the memory it uses.

```
    'lngSkinRgn has now all no transparent regions. Apply the skin!
    fm_SetWindowRgn lngHwnd, lngSkinRgn, True
    'Release the new UserForm device context
    fm_DeleteObject lngSkinRgn
End Sub
```

The frmSkin UserForm

The UserForm_APIs.xlsm macro-enabled workbook has the frmSkin ControlButton, which loads the frmSkin UserForm and applies its Picture property bitmap as a skin. If you inspect the UserForm in design mode, you will note that it uses the Caju.bmp bitmap (also found in the Chapter10.zip file) on the Picture property and has BackColor = &H00FFFFFF& (the hexadecimal value of the white background) to guarantee that the bitmap be totally surrounded by white, which is the chroma key (Figure 10-9).

■ **Attention** Caju is a northeast Brazilian exquisite fruit, also famous for its external nut, which is appreciated worldwide. The Caju.bmp file is 172 KB using 200x293 24-bit color pixels.

The VBA Properties window returns color values in the hexadecimal color string because this color mode allows you to define by intuition the RGB red, green, and blue color components of the selected color. From right to left of the &H00FFFFFF& color associated to the white color, the first two FF characters are associated to the Blue component (255 in decimal), the most intense blue; the next two FF characters are associated to the Green component, or the most intense green; and the leftmost two FF characters are associated to the Red component, or the most intense red. The VBA Function RGB(255, 255, 255) produces white, the most intense visual color. The leftmost 00 characters are the alpha channel value—the color transparency, where 00 means totally opaque.

Note that frmSkin also has the small cmdExit CommandButton with property Caption = X to allow it to be easily closed.

Before applying the `frmSkin` background image as a skin, it first needs to become 100 percent transparent and with no title bar, which is done in the `UserForm_Initialize( )` event, calling Sub `Appearance( )` and `Transparency( )`, which were already analyzed in this chapter.

```
Private Sub UserForm_Initialize()
    Appearance Me, TitleBar, False, False
    Transparency Me, 1
End Sub
```

Once the `UserForm` appearance is totally rendered with 100 percent transparency, the skin is applied on the `UserForm_Activate( )` event.

```
Private Sub UserForm_Activate()
    Me.Repaint
    Skin Me
    Transparency Me, 0
End Sub
```

Note that the code first calls the `UserForm.Repaint` method to force it to repaint and then calls the `Skin Me` procedure, accepting the default `TransparentColor = White = 16777215` (the white color in decimal). To force the `UserForm` to appear, it calls again `Transparency Me, 0`, making it 100 percent opaque.

The effect is applied quite fast because the `Caju.bmp` bitmap is a small `.bmp` file. It could even be smaller as a JPG file, but if you use this resource, it is probably to show some strange artifacts surrounding the `UserForm`, such as almost white pixels, that are really not 100 percent white.

Figure 10-10 shows `frmSkin` floating over `Sheet1` of the `frmUserForm_APIs.xlsm` macro-enabled workbook. Note that you can drag it to anywhere inside the Excel window using the left mouse button.

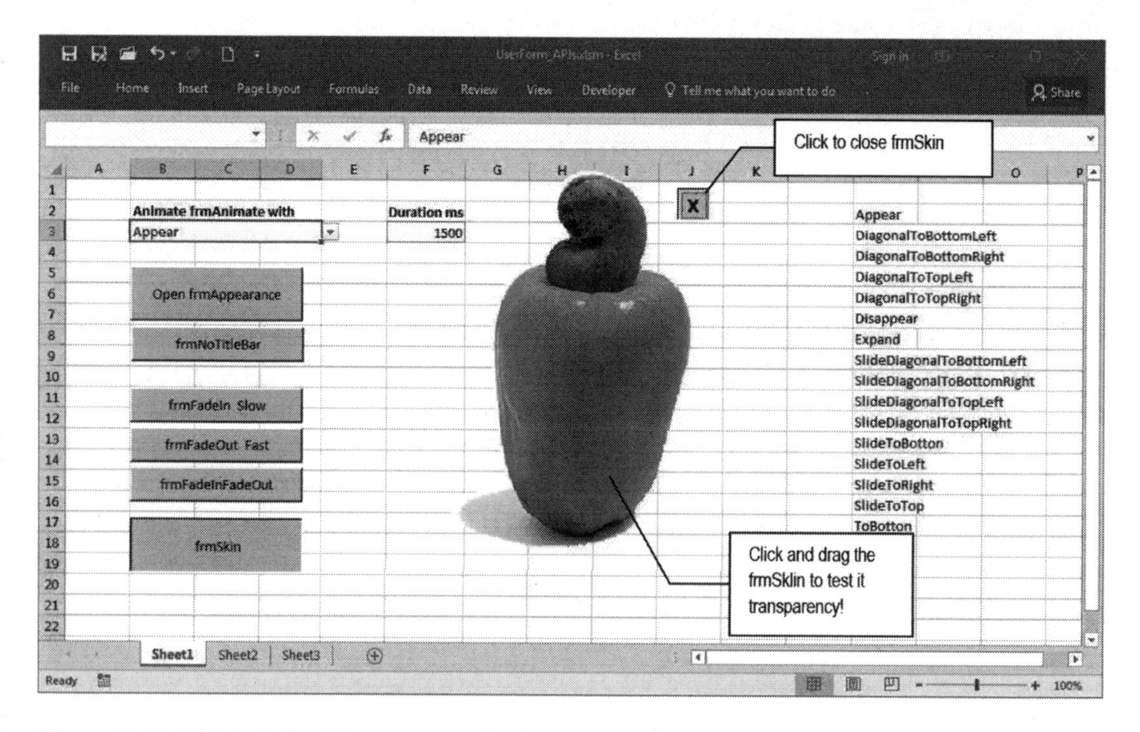

Figure 10-10. *This is the frmSkin UserForm after loading the Caju.bmp image as the skin. Click and drag the UserForm to anywhere in the Excel window to appreciate its selective transparency*

You can drag the frmSkin over the Sheet1 worksheet because it stores the UserForm position on the msngX and msngY module-level variables in the UserForm_MouseDown( ) event (which fires when you click any mouse button), just for the left mouse button (Button = 1).

```
Private Sub UserForm_MouseDown(ByVal Button As Integer, ByVal Shift As Integer, ByVal X As
Single, ByVal Y As Single)
    If Button = 1 Then
        msngX = X
        msngY = Y
    End If
End Sub
```

The code repositions frmSkin when you drag the mouse by adding (X-msngX) to the current Letf property and (Y-sngY) to the current Top property, where X, Y are the new move coordinates.

```
Private Sub UserForm_MouseMove(ByVal Button As Integer, ByVal Shift As Integer, ByVal X As
Single, ByVal Y As Single)
    If Button = 1 Then
        Me.Left = Me.Left + (X - msngX)
        Me.Top = Me.Top + (Y - msngY)
    End If
End Sub
```

When you close the frmSkin UserForm by either clicking cmdExit or double-clicking it, it disappears smoothly from the Sheet1 interface because of a call to Fade Me, FadeOut, Fast on the QueryClose ( ) event.

```
Private Sub UserForm_QueryClose(Cancel As Integer, CloseMode As Integer)
    Fade Me, FadeOut, Fast
End Sub
```

■ **Attention** You can change the frmSkin Picture property to any other bitmap to verify how it behaves as a skin. Inside the Chapter10.zip file you will also find the Apple.bmp and Donut.bmp bitmaps that can be used as good skin examples.

The USDA Food Composer_frmAbout.xlsm Application

A good first impression is achieved by using all these UserForm effects on the popular splash screen that appears when a professional application is started. Such an example can be obtained by opening the USDA Food Composer_frmAbout.xlsm macro-enabled workbook (that you can extract from the Chapter10.zip file), which shows frmAbout as the splash screen welcome window (note that it smoothly fades in, remains for three seconds on the screen, and then fades out until it completely disappears, as shown in Figure 10-11).

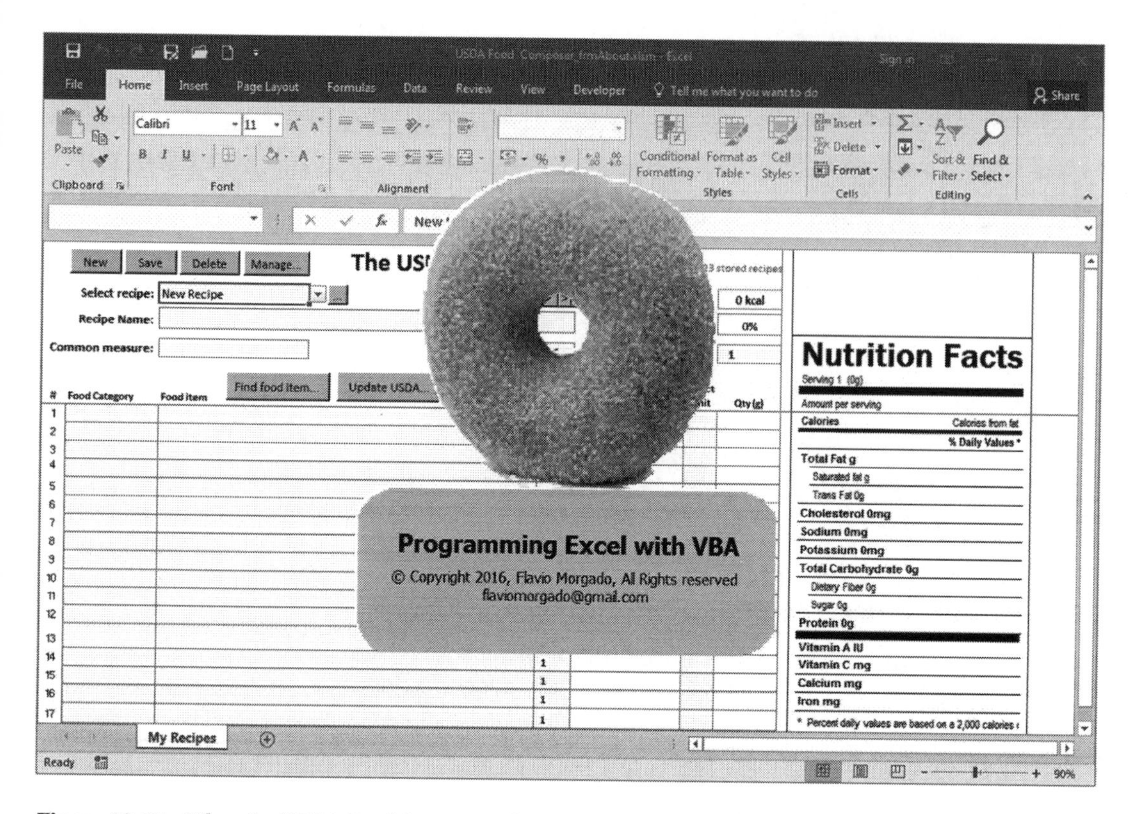

Figure 10-11. *When the USDA Food Composer_frmAbout.xlsm macro-enabled workbook is opened, it shows frmAbout, which smoothly fades in and remains for about three seconds on the screen before fading out*

Note that even the donut hole is transparent!

To make it appear, it uses the same technique employed by `frmSkin`, but to disappear after a three-second wait, it sets the `mTimer` object variable on the `UserForm_Initialize( )` event with `Interval = 3000` (3000 ms = 3 seconds).

```
Dim WithEvents mTimer As Timer
Dim msngX As Single
Dim msngY As Single
Dim mbolCancelUnload As Boolean

Private Sub UserForm_Initialize()
    Set mTimer = New Timer
    mTimer.Interval = 3000
    Appearance Me, TitleBar, False, False
    Transparency Me, 1
End Sub
```

On the `UserForm_ActivateEvent( )`, immediately after calling `Skin Me`, the code verifies whether `mbolCancelUnload = False` (default value) to activate the timer, setting `mTimer.Enabled = True`. When the `mTimer_Timer( )` event fires after a three-second delay, it unloads the `UserForm`, which will cascade-fire the `UserForm_QueryClose( )` event, where a call to `Fade Me, FadeOut, Fast` makes it smoothly disappear!

```
Private Sub UserForm_Activate()
    Me.Repaint
    Skin Me
    Fade Me, FadeIn, Fast
    If Not mbolCancelUnload Then
        mTimer.Enabled = True
    End If
End Sub

Private Sub mTimer_Timer()
    Unload Me
End Sub

Private Sub UserForm_QueryClose(Cancel As Integer, CloseMode As Integer)
    Fade Me, FadeOut, Fast
End Sub
```

The `frmAbout` `UserForm` uses the `Donut.bmp` bitmap file, which has a green round rectangle at its bottom, where you can lay out some `Label` controls to give application information. It also has a hidden `cmdExit` `CommandButton` control. Figure 10-12 shows `frmAbout` in design mode inside VBA IDE.

Figure 10-12. *This is the frmAbout UserForm from the USDA Food Composer_frmAbout.xlsm macro-enabled workbook, which uses the Donut.bmp bitmap image in the Picture property. It has some Label controls over the image's green round rectangle to give application information to the user when the UserForm appears on the screen*

After `frmAbout` is dismissed, you can click the About `ControlButton` on the `My Recipes` worksheet to force it to appear again over the application interface, but this time it does not automatically disappear. It shows the `imgCloseButton` `Image` control to be unloaded and can also be dragged onto the worksheet screen.

This is done with `Sub ShowfrmAbout( )`, from `basControlButtons`, which uses this code:

```
Public Sub ShowfrmAbout()
    With frmAbout
        .CancelUnload
        .Show
    End With
End Sub
```

This code loads `frmAbout` by referencing it on a `With frmAbout...End With` structure (cascade-firing the `UserForm_Initialize( )` event) and calling the `CancelUnload( )` method. Then the code shows `frmAbout` with the `UserForm.Show` method (cascade-firing the `UserForm_Activate( )` event). This is the code executed by the `frmAbout.CancelUnload( )` method:

```
Public Sub CancelUnload()
    Me.imgCloseButton.Visible = True
    mbolCancelUnload = True
End Sub
```

Quite simple, huh? It turns `imgCloseButton` visible and avoids the `mTimer` object to be activated by setting `mboCancelUnload = True` (the `imgCloseButton_Click( )` event unloads the `UserForm`).

Conclusion

This chapter gave you some guidance about how to use the Windows API's procedures inside VBA code to produce effects that are not allowed from the standard Visual Basic language.

It briefly discussed how to declare a DLL procedure, how and why you should care about aliasing procedures, why they use so many constant declarations, and how you set bits using the `OR`, `AND`, and `NOT` operators to produce the desired bit effect.

You were also introduced to the window `Handle` concept and how to obtain the `UserForm` handle to use the Windows APIs.

All the code inside this chapter was inspired from many Internet web sites that currently offer code without explaining what really happens when the code is executed. I hope that after reading the chapter you feel more comfortable searching for and copying and pasting other VBA DLLs procedures to use in your applications.

For a better understanding of Windows DLLs, I frequently base my research on these references:

- Daniel Appleman books such as *Visual Basic Programmer's Guide to the Win32 API*, *Win32 API Puzzle Book and Tutorial for Visual Basic Programmers*, and *Developing ActiveX Components With Visual Basic: A Guide to the Perplexed*
- Ken Gets books such as *VBA Developer's Handbook* and *Microsoft Access 2000 Developer's Handbook*

Chapter Summary

In this chapter, you learned about the following:

- That Microsoft Windows bases its inner workings on a set of DLL files, each one with its own set of procedures, comprising what is called the API
- How to declare a DLL procedure and its constant values
- That you can avoid code conflict by giving an alias to your DLL procedure declarations
- That most DLL procedures have a set of constant values that must also be declared
- That these constant values can appear either in decimal or hexadecimal notation
- The meaning of handle on the Microsoft Windows operating system
- How to get the handle for the `Application`, `UserForm`, and `Class` module instance objects
- That some API procedures need the address of a callback procedure, which is a VBA procedure that must be declared with specific arguments that will be called back by the API
- How to create a `Timer` object using some Windows APIs and a `Class` module
- How to change the `UserForm` window appearance using Windows APIs
- How to create transparency and fade effects on a `UserForm`
- How to implement a skin effect based on a bitmap image
- How to produce a splash screen to be presented to the user when your Excel application starts

In the next chapter, you will learn how to create a personalized Microsoft Excel ribbon using third-party applications developed using VBA to enhance the professional appeal of your worksheet applications.

CHAPTER 11

Producing a Personal Ribbon Using RibbonEditor.xlam

The VBA environment became part of Microsoft Office when version 4 was released in 1993, exposing the `CommandBar` object that offered the traditional VBA object model to manipulate Office's toolbars (allowing users to create, delete, edit, and personalize them). This programmable approach remained for 14 years, until the release of Microsoft Office 2007.

When Microsoft 2007 appeared, offering the new ribbon concept, at first glance I was uncomfortable, partly because my 15-inch monitor used a maximum 1024x768 resolution and obscured a significant part of the screen. It also repositioned the controls, giving me the same frustrating sensation that happens when I enter my favorite store and realize that everything has changed places. Both experiences cost me an enormous amount of time to do something that was so fast and easy to accomplish before.

The ribbon approach allows personalization in any Microsoft Office application, via the creation of new tabs, groups, and controls. Such changes become permanent in the application interface (like Excel), not just for your application.

That is not all: both the `CommandBar` object and the VBA object model used to interact with the Microsoft Office toolbars completely disappeared from the scene! The enlightened Microsoft programmers chose to offer XML programming as the only way to interact with the `Ribbon` object, renouncing the VBA language, which developers had used in the past to program in any Microsoft application. This was a clear indication that Microsoft had a big conflict inside its organization. It now had new Internet programmers that despised the VBA history and the entire VBA programmer community!

But Microsoft would be in permanent debt to the entire VBA programmer community if it had not offered a new `Ribbon` object model with properties, methods, and events to program Microsoft Office applications, like it always had before.

To interact with the `Ribbon` object to personalize your applications in Microsoft Office 2007 or newer, you must learn XML, use a set of different technologies (including the Visual Studio interface, which I personally refuse to do), or base your programming style on external solutions produced by brave people who felt uncomfortable like myself and envisioned a simpler way to allow a programmable ribbon interaction, like Ron de Bruin (with his CustomUIEditor solution, which is an XML editor that interacts more easily with ribbon XML programming) and Andy Pope (who produced the `RibbonEditor.xlam` solution, which is a VBA add-in that offers a high-level design interface to the awkward Office ribbon). Both authors have Internet presences to manage their solutions, and I recommend you to do a Google search to find them.

This last chapter is a small one, intended to give you some guidance about how you can create a personal ribbon for your application instead of using the default Excel tools. Since the ribbon cannot be easily programmed using just VBA objects, I will base the chapter on the `RibbonEditor.xlam` VBA application produced by Andy Pope, which you can download from `www.andypope.info/vba/ribboneditor.htm`.

F. Morgado, *Programming Excel with VBA*, DOI 10.1007/978-1-4842-2205-8_11

How Personal Ribbon Information Is Stored

Microsoft Excel uses a folder storage system for its `.xls*` file formats (`.xlsx`, `.xlsm`, and so on), which you can see by renaming the file extension to `.zip` and opening the file in any ZIP application like IZarc (freeware), WinZip, WinRar, and many others.

To produce a personal ribbon for your application, the XML code to this new `Ribbon` object must be stored inside the Microsoft Excel XLSM file using a file named `CustomUI.xml` for Microsoft Office 2007 or `CustomUI14.xml` for Microsoft Office 2010–2016 files that use new options like Backstage View. The Backstage View feature, offered by Microsoft 2010 or later, is the way Office manages its File tab options instead of the circular Office 2007 button.

■ **Attention** The Microsoft Office application will use the `CustomUI.xml` file in any Office version since version 2007if its personal ribbon XML file does not use any Backstage View option. But if your personal ribbon uses any Backstage View command and you open the application in Excel 2010 or later versions, it will use only the `CustomUI14.xml` file stored inside it.

Both Ron de Bruin's CustomUIEditor and Andy Pope's `RibbonEditor.xlam` add-in allow the creation of the `CustomUI.xml` and `CustomUI14.xml` files inside your XLSM macro-enabled workbook. To make your Excel application backward compatible with any Microsoft Excel version since 2007, you must always code the `CustomUI.xml` or `CustomUI14.xml` file inside your worksheet applications.

Using RibbonEditor.xlam

To produce and save the `CustomUI.xml` or `CustomUI14.xml` file responsible for showing a new `Ribbon` object in your Excel application, I recommend you download the `RibbonEditor.xlam` application from Andy Pope's AJT web site. It's a simple Microsoft Excel workbook that uses the `.xlam` extension to define it as an Excel add-in (the add-in will be downloaded as a `.zip` file; then you must extract the Excel add-in workbook).

■ **Attention** Since this add-in has VBA code that needs to access the VBA structure, it might be considered as a virus by your antivirus solution; you may need to disable your antivirus solution to allow the add-in to download.

Once you have downloaded the add-in to your hard drive, use these steps to allow the `RibbonEditor.xlam` add-in to access the VBA environment of your Excel applications:

1. Select Excel File ➤ Options to show the Excel Options dialog box, and select the Trust Center option in the right panel.
2. Click the Trust Center button to open the Excel Trust Center dialog box.
3. In the Trust Center dialog box, select Macro Settings in the vertical right panel to show the Trust Center dialog and select "Trust access to the VBA project object model" (Figure 11-1).

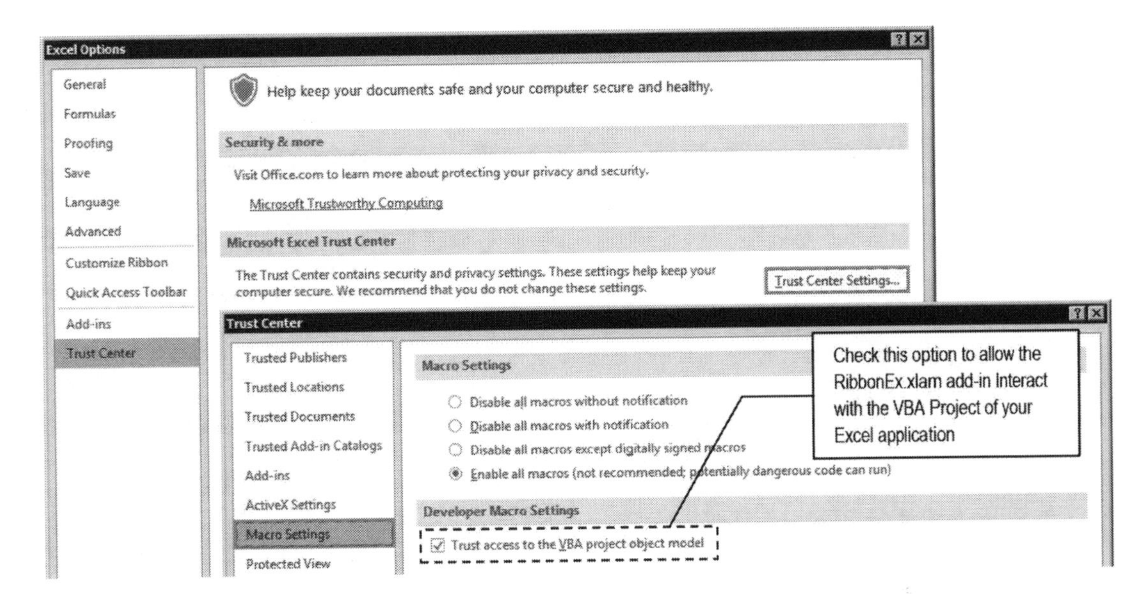

Figure 11-1. *To install the RibbonEditor.xlam add-in that you can download from Andy Pope's web site, first check the "Trust access to the VBA project object model" option in the Macro Setting option in the Excel Trust Center dialog*

■ **Attention:** Note in Figure 11-1 that since I am an experienced Excel user who is aware of the damage that any Excel `.xlsm` or `.xlam` macro-enabled workbook files can produce, my Excel environment also has the "Enable all macros (not recommended, potential dangerous code can run)" option selected.

To install the `RibbonEditor.xlam` add-in in Excel 2016, you also need to add it to the available Excel add-ins of your Excel environment, following these steps:

4. In the Excel Options dialog box, select the Add-ins option in the vertical right panel.

5. In the Manage combo box at the bottom of the Excel Options dialog box, select the Excel Add-ins option and click the Go button.

6. In the Excel Add-ins dialog box, click the Browse button and select the folder where you extracted the `RibbonEditor.xlam` add-in after downloading it.

7. Select the RibbonX Visual Designer option to install the `RibbonEditor.xlam` add-in in your Excel environment (Figure 11-2).

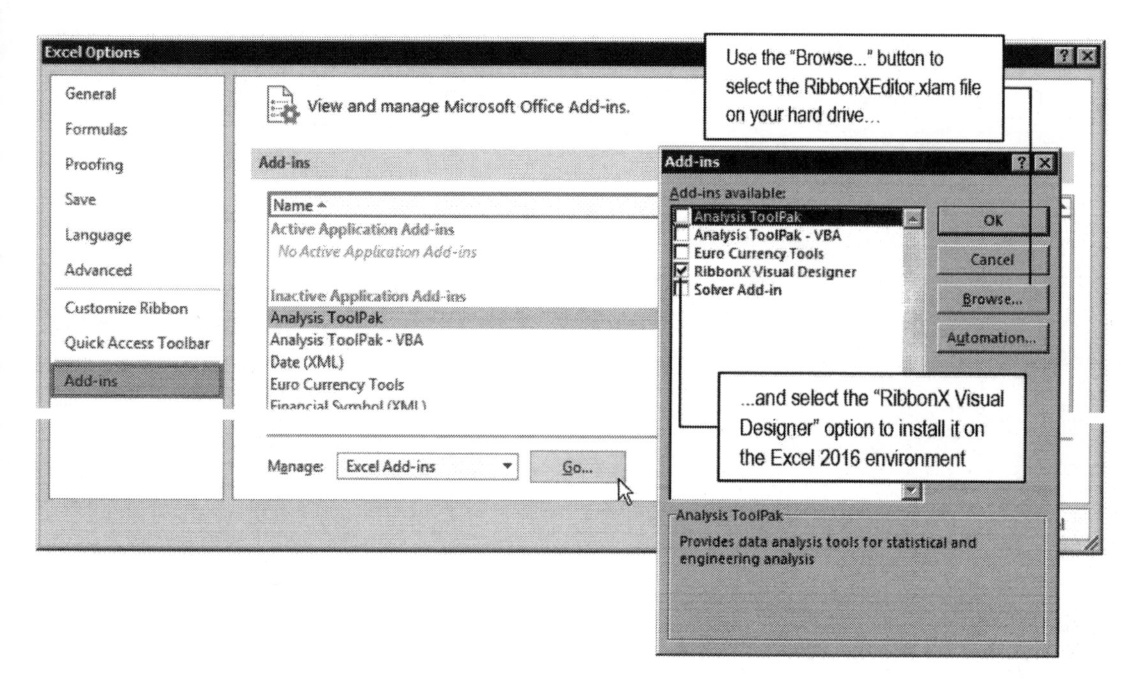

Figure 11-2. *Use the Excel Options dialog box's Add-ins option to show the Excel Add-ins dialog box. Use the Browse button to navigate to the folder where you extracted the RibbonEditor.xlam add-in and select the RibbonX Visual Designer add-in to install it in your Excel environment*

These steps will install the RibbonX Visual Designer Excel add-in that can open the `.xlsm` Excel macro-enabled workbook and store inside its file storage system the `CustomUI.xlm` or `CustomUI14.xml` file (used just by Office 2010 and all previous versions), which provides a graphic interface to manipulate the ribbon. Figure 11-3 shows how to open the RibbonX Visual Designer add-in application after installing it from Andy Pope's web site.

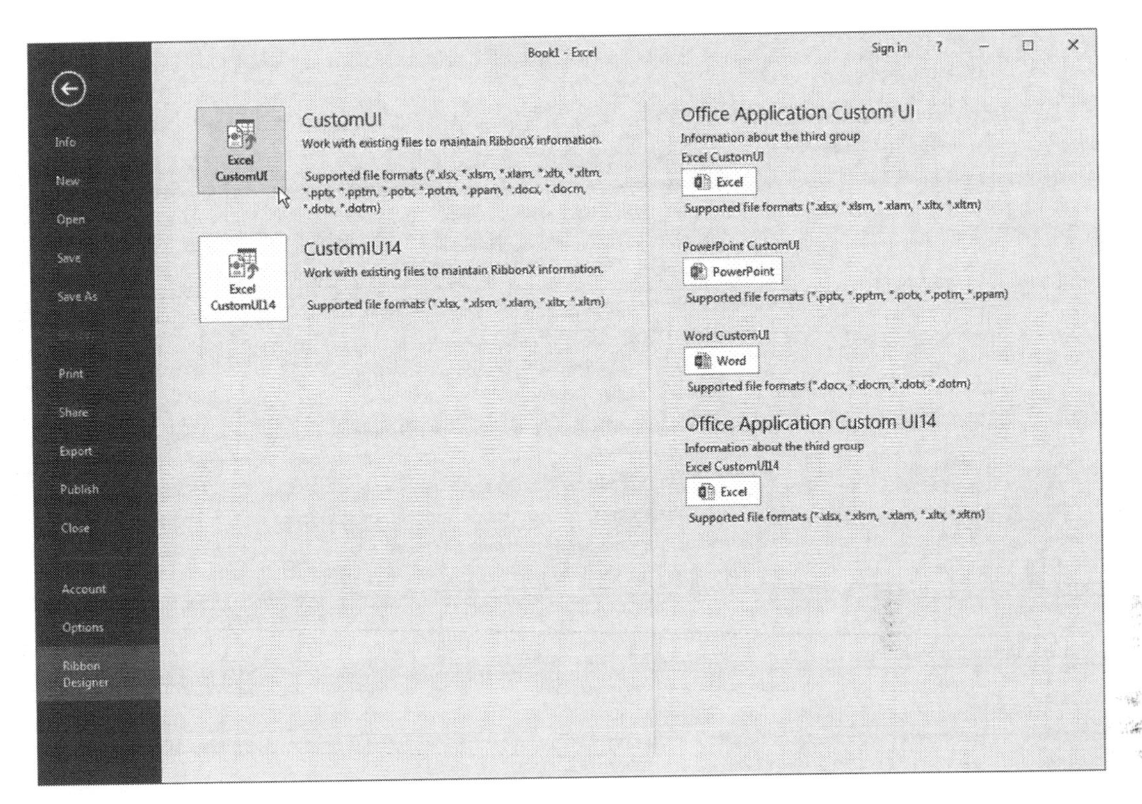

***Figure 11-3.** After installing the RibbonEditor.xlam application, load it using the File ➤ Add-In ➤ Ribbon Designer ➤ Load RibbonX menu command to manipulate CustomUI.xlm or CustonUI14.xlm (for Excel 2010 or later versions). The RibbonEditor.xlam Designer uses the BackStage view of Microsoft Excel 2016*

Attention Since the `CustomUI.xlm` file will work in any Microsoft Excel 2007 or later environment, this chapter will use this option to produce the personalized ribbon described in the next sections.

Once the `RibbonEditor.xlam` add-in CustomUI option is selected, it will ask you to select the desired Excel `.xls*` file where the `CustomUI14.xml` file will be created (Figure 11-4 shows the selection of the `USDA Food Composer_frmAbout.xlsm` file).

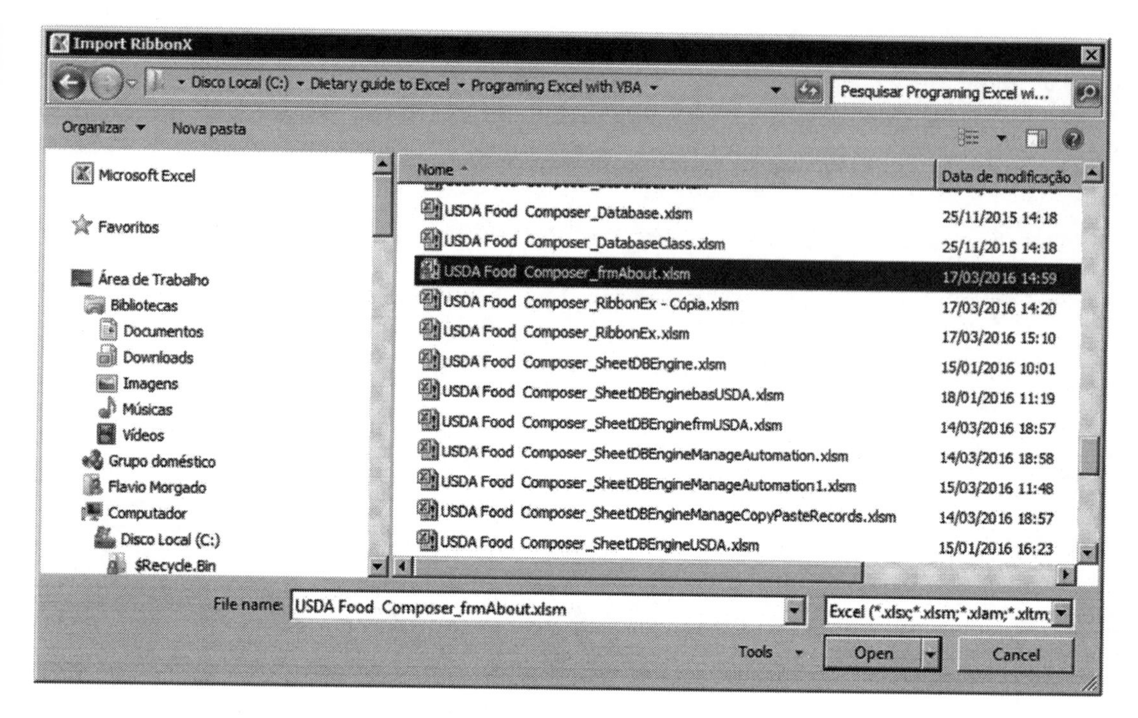

Figure 11-4. *Select the desired Excel .xlsm macro-enabled workbook where the personalized ribbon must be created. This figure selects the USDA Food Composer_frmAbout.xlsm file*

After the desired Excel file is opened, the `RibbonEditor.xlam` designer will show the current `CustomUI.xml` file with all its options (if any), showing all default ribbon commands (with blue text and light green icons for its tabs) and the personalized ones (if any, using black text and white tabs). Figure 11-5 shows the `USDA Food Composer_frmAbout.xlsm` file, which still has no ribbon customization.

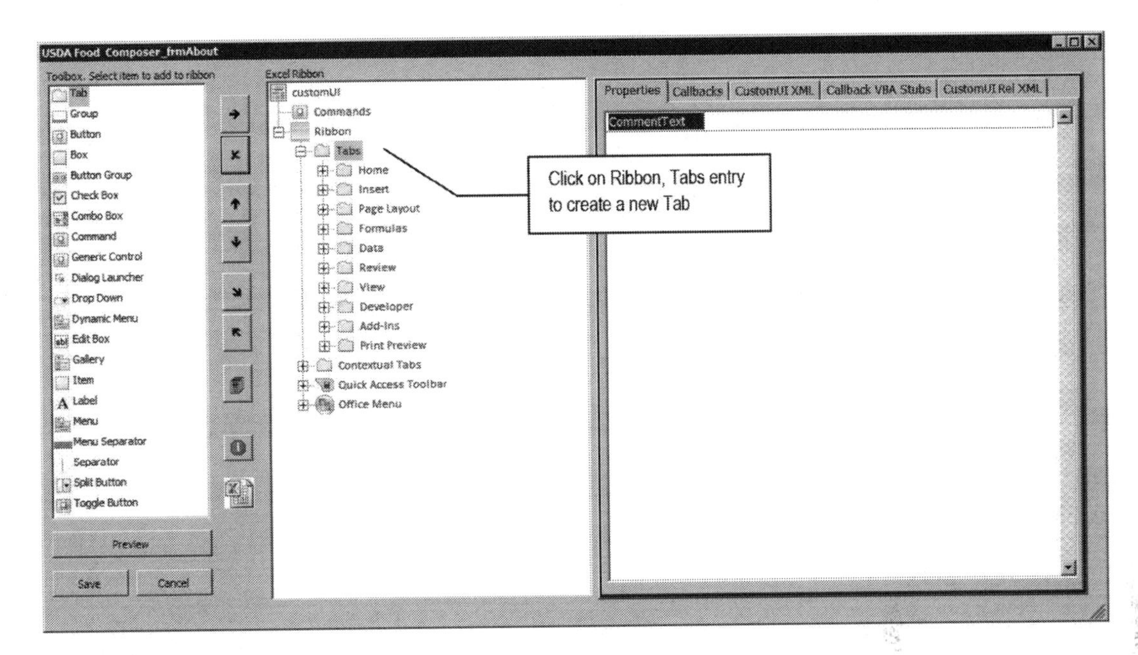

Figure 11-5. *After the RibbonEditor.xlam add-in opens the file, it will show a VBA UserForm with all the default ribbon commands and the current personal ribbon, if any*

■ **Attention** The file where the `CustomUI14.xml` file will be created must be closed before being opened by the `RibbonEditor.xlam` add-in.

Adding Tabs, Groups, and Buttons Using the RibbonX Add-In

I will not take too long to explain how to use the `RibbonEditor.xlam` add-in since the AJP web site is full of examples. The add-in interface has three different areas: the Toolbox on the left, the ribbon tree in the center, and the Properties area on the right.

To add a new ribbon to your application, select the Tab option in the Toolbox, click the `Ribbon` entry and then the `Tabs` entry in the `Ribbon` tree, and press the right arrow button. A new tab option will be inserted below the `Print Preview` default tab (Figures 11-6 and 11-7).

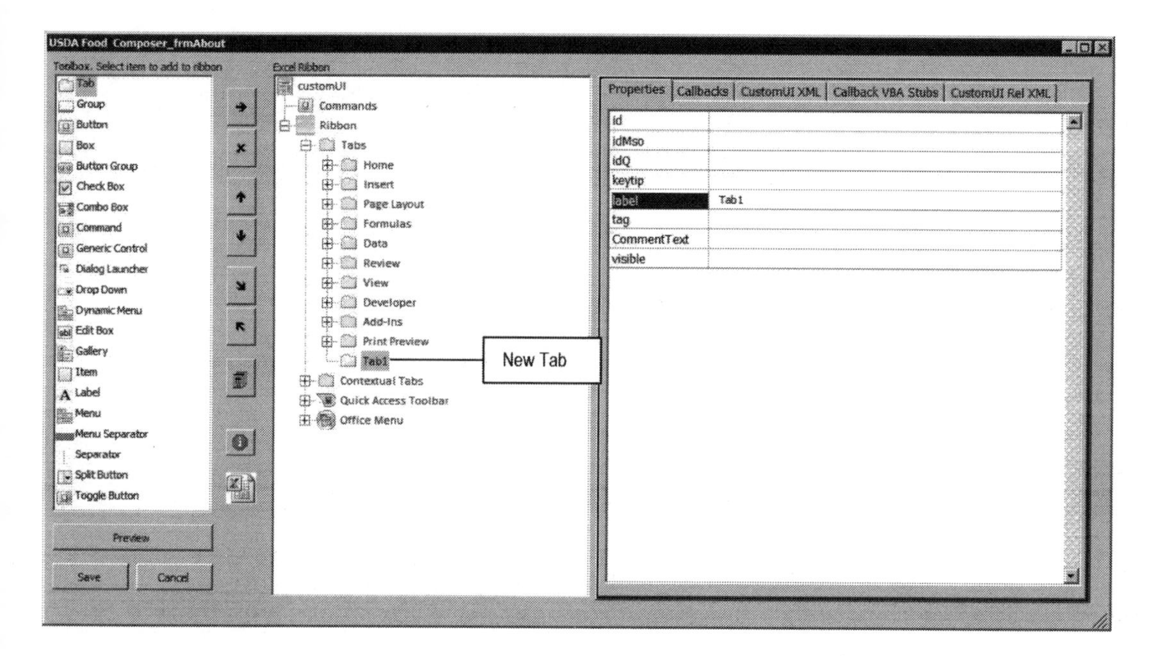

Figure 11-6. *Click the Ribbon ➤ Tabs option in the Ribbon tree area of the RibbonEditor.xlam interface to insert a new personalized tab in your worksheet application*

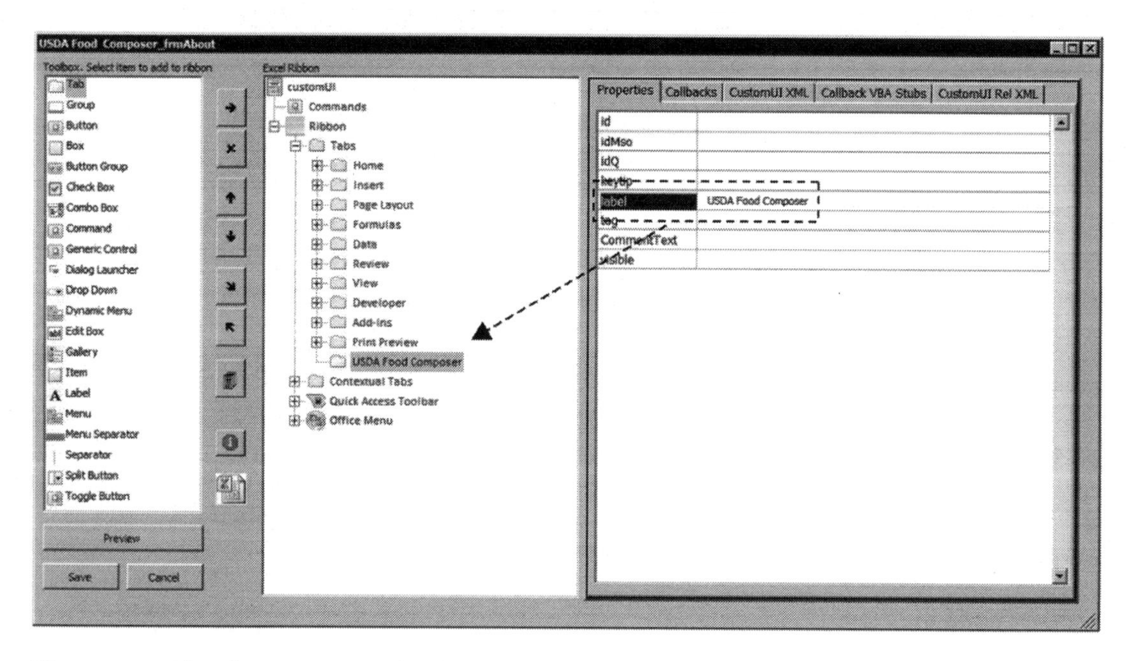

Figure 11-7. *After changing any item's Label property, click the item in the RibbonEditor.xlam Ribbon tree to update its Label property value*

To personalize the tab, change its `Label` property on the Properties area on the right side of `RibbonEditor.xlam` interface. After changing the tab's `Label` property, click the new Tab option again in the `RibbonEditor.xlam` Ribbon tree to update its value (Figure 11-8).

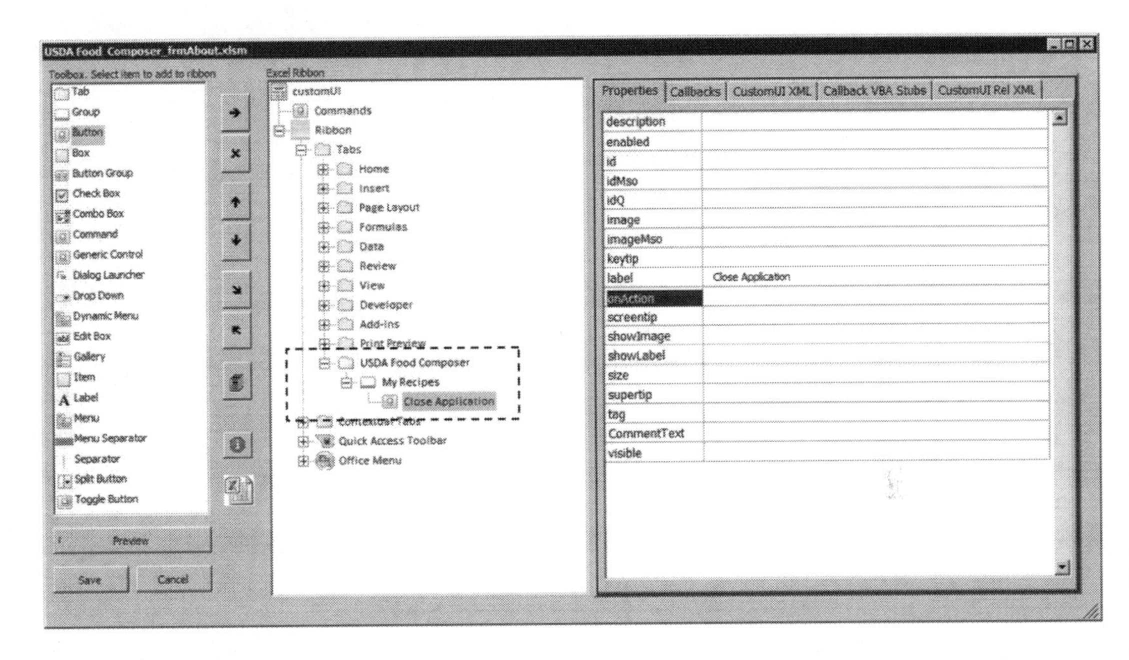

Figure 11-8. *This is the USDA Food Composer tab, which has a My Recipes group and a Close Application button*

To create a new group, select Tab in the `RibbonEditor.xlam` add-ins `Ribbon` tree, double-click the `Group` item of the toolbar, and change its `Label` property accordingly. To add a button to any tab group, select the desired group on the `Ribbon` tree and double-click the Button option on the toolbar.

Figure 11-8 shows the `USDA Food Composer` tab, with the `My Recipes` group that has the Close Application button.

To personalize any button, click the desired button in the `Ribbon` tree and change these properties to indicate what the button does:

- `Label`: This is the `Button` name.
- `ShowLabel`: Set this to `True` to make the label appear below the button.
- `ShowImage`: Set this to `True` to make the image selected on the `ImageMso` option be associated to the button.
- `Size`: Select `Large` or `Small` to set the button image size.
- `ImageMso`: Click the ... button to the right of the property to show the Icon Gallery dialog, where you can select the desired image. To see large icons, set `Size = Large`.
- `OnAction`: This is the VBA Sub procedure name that will be executed (must be on a standard module).

Figure 11-9 shows how the Close Application button was personalized (`Label = Close Application; ShowLabel = True; ShowImage = True; Size = Large; Visible = True; ImageMso = CancelRequest`).

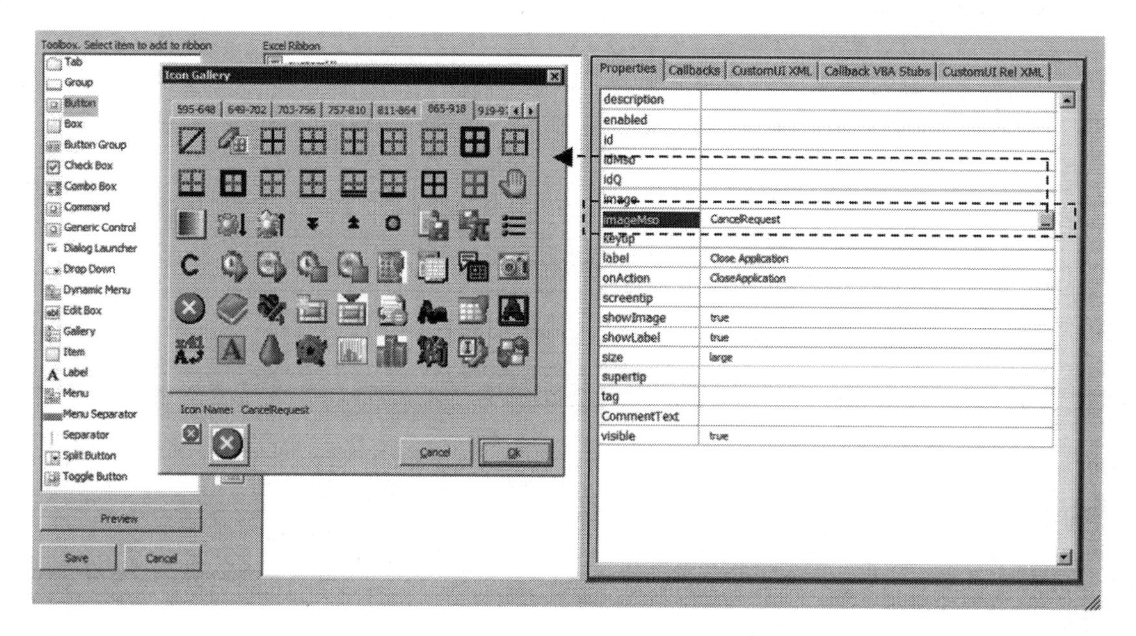

Figure 11-9. *Use the Property tab to define the Button control properties. Use the ImageMso property ellipses to show the Icon Gallery dialog to select the desired images. Use the OnAction property to define the VBA Public Sub procedure name that will be executed when the button is clicked*

■ **Attention** Set `Size = True` before clicking the ellipses button of the `ImageMso` property to select the desired images using large icons. The `CancelRequest` icon used by the Close Application button can be found on pages 865–918 of the Icon Gallery dialog.

The `Public Sub CloseApplication( )` procedure defined in the `OnAction` property of the Close Application button must be declared as stated by the `CallBack VBA Subs` tab (all procedure declarations for every `Button` control inserted on the `USDA Food Composer` tab will appear on this tab page, as shown in Figure 11-10).

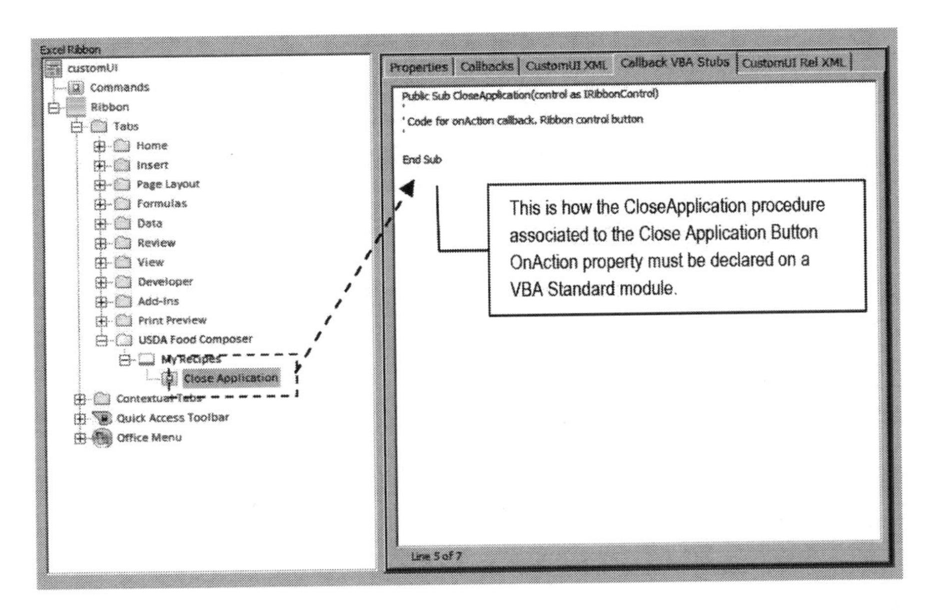

Figure 11-10. *Once all buttons are created on the desired USDA Food Composer tab, click the Callback VBA Subs page to verify how the procedure name associated to the Button OnAction property should be declared*

To associate the `USDA Ribbon Control` tab to the `.xlsm` Excel macro-enabled workbook opened by `RibbonEditor.xlam`, click its Save button. This will add a new ribbon tab as the last item of the default Excel ribbon. Minimize the add-in window, and open the workbook in Excel to see that it now has a USDA Food Composer tab, with a My Recipes group and the Close Application button (this tab will appear just for this document, as shown in Figure 11-11).

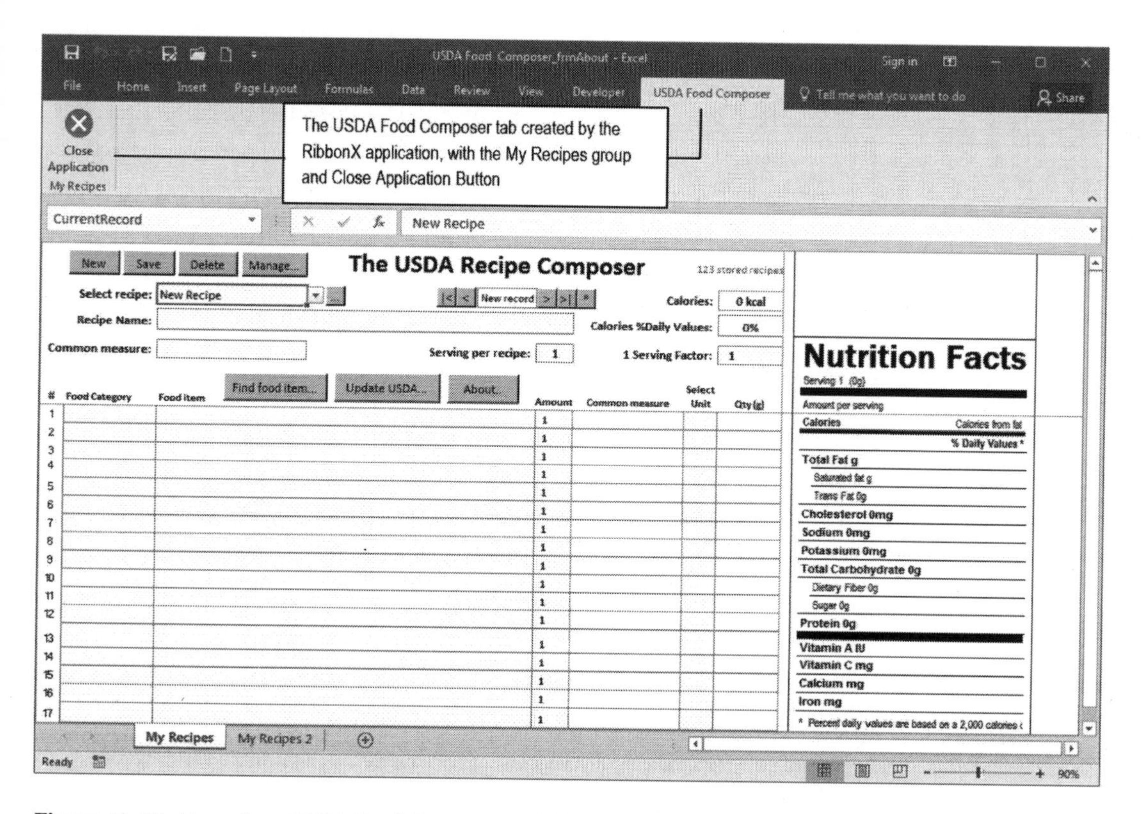

***Figure 11-11.** To see how USDA Food Composer Tab behaves in the .xslm macro-enabled application, open it in Excel and click the new tab (that will appear just for this workbook)*

Supposing that you had selected and copied the proposed code on the `Callback VBA Subs` tab, press Alt+F11 to show the VBA environment, insert a new module in the VBA project, and paste the code. Then insert the code instruction to make it execute the desired command.

Figure 11-12 shows the `basRibbonEditor.xlam` module inserted in the `USDA Food Composer_frmAbout.xlsm` macro-enabled workbook, which now has the `Public Sub CloseApplication( )` procedure (note that `RibbonEditor.xlam` specifies that the procedure declare the `control as iRibbonControl` object).

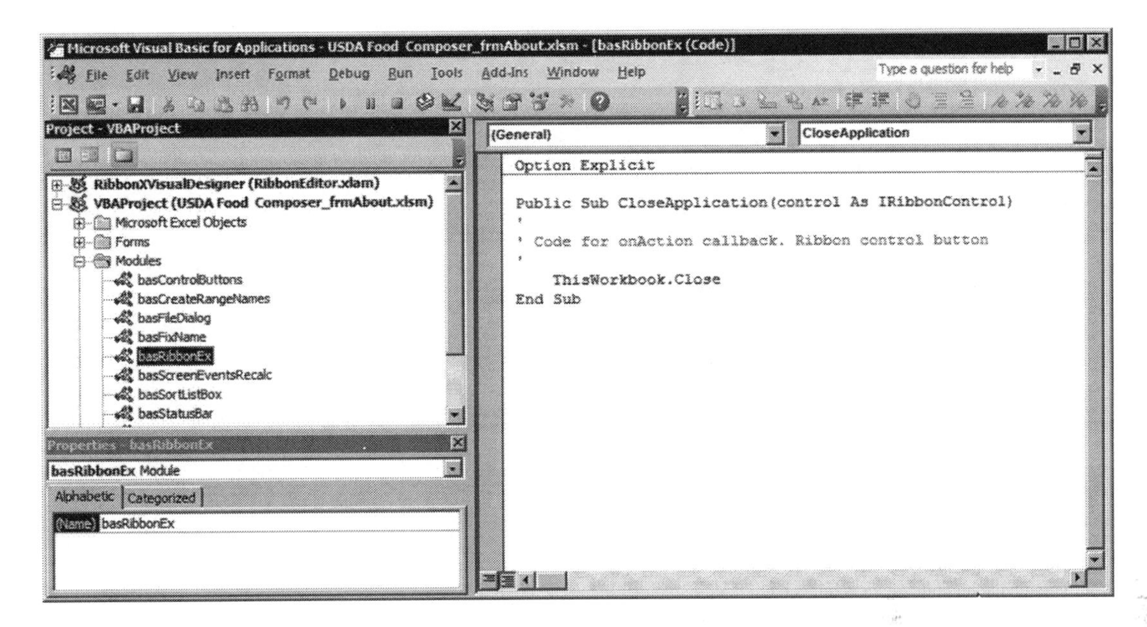

Figure 11-12. *To make the Close Application button work, paste the proposed RibbonEditor.xlam code in a standard VBA module and insert the desired instructions. The Sub CloseApplication( ) procedure just calls the ThisWorkbook.Close method to close the Excel application window*

Once you have defined the code for the Close Application button, save the workbook and try the button. The workbook must be closed for this to work!

To hide the default Excel ribbon using just the `USDA Food Composer` tab and its groups (or any other tabs) created by `RibbonEditor.xlam`, whenever the `USDA Food Composer_frmAbout.xlsm` application is loaded, restore the add-in window, click the `Ribbon` option of the `Ribbon` tree, and set its property as `StartFromScratch = True` (Figure 11-13).

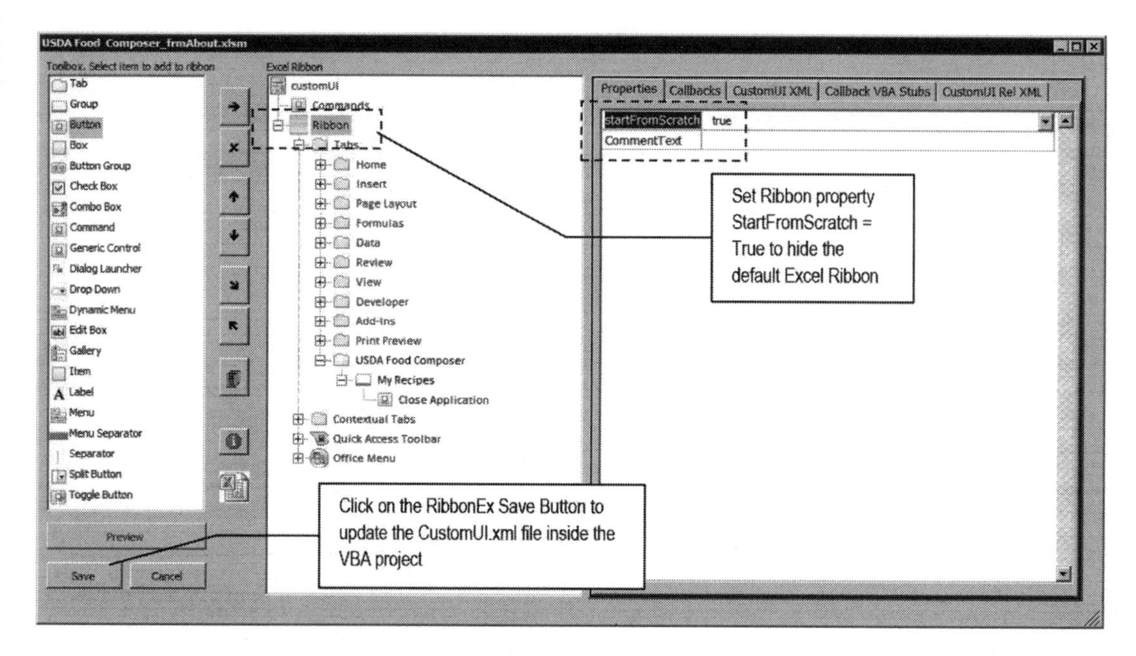

Figure 11-13. *Select the Ribbon object in the RibbonEditor.xlam Ribbon tree, and change the property StartFromScratch = True to totally hide the Microsoft Excel default ribbon. Now just the USDA Food Composer tab will appear for the selected workbook application*

Click the `RibbonEditor.xlam` Save button to update the `CustomUI.xml` file inside the workbook VBA project and reopen the application in Excel. Now it will show just the USDA Food Composer ribbon, created by this add-in (Figure 11-14).

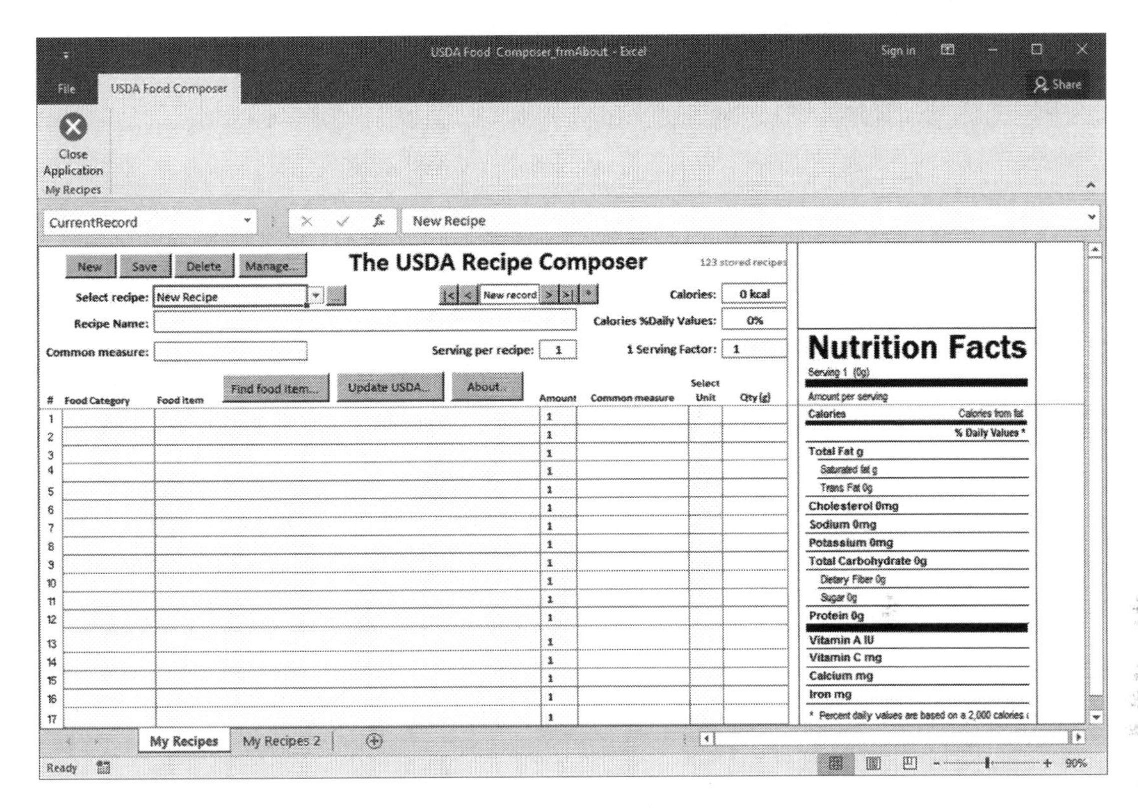

Figure 11-14. *This is the USDA Food Composer_frmAbout.xlsm macro-enabled workbook without the Excel ribbon, using just the USDA Food Composer tab created by RibbonEditor.xlam. This was possible after selecting the Ribbon object in the RibbonEditor.xlam Ribbon tree, setting the property StarFromScratch = True, and saving the CustomUI.xml project inside the workbook application*

Removing the CustomUI.xml File from a Workbook Application

You can easily remove the `CustomUI.xml` file from a workbook application in two ways:

- In the `RibbonEditor.xlam` interface, select all the created tabs, click their "X" delete item button (below the right arrow button), and click Save to update the workbook.
- Rename the `.xlsm` extension to `.zip`; open the project in WinZip, WinRar, Izarc, or any other `.zip` file; and manually delete the `CustomUI` folder from the project. Remember to rename the file to `.xlsm`.

Producing a Nice Ribbon with RibbonEditor.xlam

To produce a nice ribbon with `RibbonEditor.xlam`, you must use different groups, the Separator item to divide the groups, and different `Button` control sizes (set the property `Size = Small` before selecting the desired icon using the `ImageMso` property).

■ **Attention** It is easy to insert new ribbon controls by using the `RibbonEditor.xlam` Clone button. The new button will be inserted with most properties already set.

You can see a good example by extracting the `USDA Food Composer_RibbonX.xlsm` application from the `Chapter11.zip` file. It has all the `My Recipes` worksheet ControlButtons duplicated on the `USDA Food Composer` tab produced with `RibbonEditor.xlam` (Figure 11-15). To show how each of its buttons work, inspect the `basRibbonEditor.xlam` module in its VBA project. To see how the `USDA Food Composer` tab was produced, open the project inside the `RibbonEditor.xlam` add-in using the CustomUI editor.

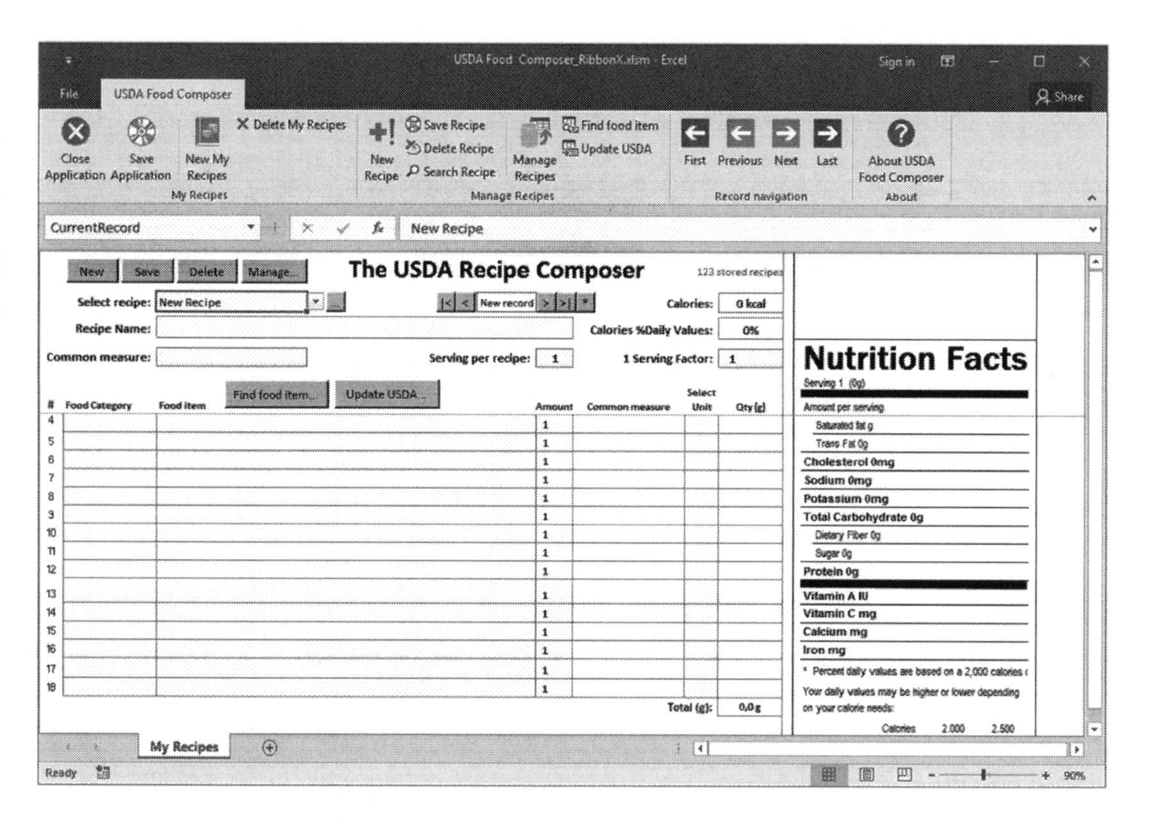

Figure 11-15. This is the USDA Food Composer_RibbonX.xlsm macro-enabled workbook that you can extract from the Chapter11.zip file. It hides Excel's default ribbon and duplicates all the My Recipes buttons on the USDA Food Composer ribbon, produced with the RibbonEditor.xlam add-in

---

■ **Attention** I think that the duplication of all My Recipes worksheet `ControlButtons` on the USDA Food Composer ribbon deserves a better discussion about the interface produced.

Conclusion

In this chapter, you learned that despite Microsoft offering a simple object interface to the ribbon, you can use free, third-party add-ins available on the Internet to produce a personalized ribbon for your Excel applications.

The entire chapter used the RibbonX add-in available from Andy Pope's AJP web site to produce a nice personalized ribbon for the `USDA Food Composer_RibbonX.xlsm` macro-enabled workbook, which you can extract from the `Chapter11.zip` file.

Chapter Summary

In this chapter, you learned about the following:

- That to create personalized ribbon tabs for your Excel application, you must create a `CustomUI.xml` folder inside the Microsoft Excel file, using XML code to produce it
- That the `RibbonEditor.xlam` application from Andy Pope's web site is a nice alternative for producing a personalized ribbon
- That you must select the ribbon's Tabs option in the `RibbonEditor.xlam Ribbon` tree to insert a new personalized tab
- That you can double-click the `RibbonEditor.xlam` Toolbox to insert new items inside a personalized tab or group
- How to use the `Button` control properties to make a button work
- That you must set the control `OnAction` property to the desired `Public Sub` procedure name of the application VBA project
- How to use the `CallBack VBA Subs` tab to copy the base code needed to run each `Ribbon` control
- How to remove a personalized ribbon from a workbook application by deleting all the personalized tabs using `RibbonEditor.xlam`'sDelete item button
- How to hide the Excel default ribbon by setting the `RibbonEditor.xlam Ribbon` object property as `StartFromScratch = True`

Afterword

Writing this book was a great challenge for me: I had to learn Excel VBA programming as I wrote it because my first intention was to try every Excel feature using the VBA Immediate window. If it works in the VBA Immediate window, it surely works inside any VBA code procedure.

I had a slight notion about how each book chapter should proceed. Chapter 1 should make a brief introduction to the VBA integrated development environment (IDE) and the VBA language, with some simple examples about variable declarations, loop structures, and so on. The next three chapters should touch on the Excel object model from the top down: first the `Application` object, then the `Workbook` object, then the `Worksheet` object, and finally the `Range` object.

But how to make the book new and interesting with an original approach was still a mystery for me.

I've always heard people talk about the old adage, "The path is made walking," and since I had made "the way" (to Santiago de Compostela, Spain), this truth began to become incomparably clear in my mind.

Using these basic principles, Chapter 2 was approached by covering some important `Application` object methods that allow Excel programmers to manipulate computer folders and files (the `FileDialog`, `GetOpenFileName`, and `GetSaveAsFileName` methods), how to interact with the user and the worksheet (the `InputBox` method), and how to create a timer (the `OnTime` method). I also used a `Class` module with the `Application` object events to create the example that can control when a sheet tab name changes while I wrote the chapter. And I liked it!

At that time, I still had no idea about how to approach the next chapters. To produce Chapter 3, the `UserForm` appeared suddenly by trial-and-error programming as a good approach to learn VBA, because it offered an excellent way to explore some important `Workbook` object events (like the Splash Screen dialog that implements a timer to unload itself). The `frmOpenWorkbooks UserForm`, which deals with different `Workbook` objects using pure VBA code, was especially important because it offered a solid approach to the `ListBox` control and many of its properties and methods, as well as allowed me to study interface synchronization techniques.

When I finished Chapter 3, I was so impressed with the results that I thought I would repeat the same approach with sheet tabs in Chapter 4 using the `frmWorksheets UserForm`. Once again, when I finished the chapter, I concluded, "This is astounding! I did it again!"

I am not bragging, OK? I was so surprised by what I had accomplished in Chapters 2, 3, and 4 that I decided to repeat the same `UserForm` approach to provide consistent VBA knowledge to the so-important `Range` object in Chapter 5.

This time I had to make a great effort to study and learn the `Range` object and its many properties and methods, like the `Address` and `Cells` properties; the `Areas` collection; and the `Resize`, `Protect`, and `Unprotect` methods. The final result, materialized in the `frmRange UserForm`, also quite impressed me.

Have you ever felt yourself touched by an improbable inspiration coming from nowhere? At the time I wrote these words, I was remembering how much I appreciated the work done so far. I sat and prayed for the undeserved heavenly inspiration that was pushing me forward, and immediately I felt my eyes shallows of water. Thank you again and forever, my Guardian Angel, for your permanent companion.

F. Morgado, *Programming Excel with VBA*, DOI 10.1007/978-1-4842-2205-8

Although the Range object programming had been touched on a convenient and practical way, it still had many other methods to be explained, so I dove into Chapter 6 to teach how you can deal with large worksheet data sets using a VBA code. This chapter's intention was to teach how to search and filter large data sets such as the USDA nutrition table worksheet using VBA, and once again I used the UserForm approach. I used frmRangeFind and frmRangeFilter to teach how to use the Range.Find and Range.AutoFilter methods. This chapter did not surprise me because I knew in advance what message it had to convey; I just didn't know how to approach it.

Then came Chapter 7. This chapter needed to be written because I wanted to teach how to use a programmable approach to store worksheet application data as database records in a worksheet's unused cells. The entire process should be based on a data validation list and range names stored inside constant values (at this time, the code had only four constants: the record name, the data validation list cell, and the one-side and many-side record range names). In my mind this was the perfect way to practice the knowledge built so far in all previous book chapters. The worksheet applications used (the BMI Companion Chart.xlsm and USDA Food Composer_Database.xlsm macro-enabled workbooks) successfully explained that the proposed database structure could work to save worksheet data as records.

After the success in implementing a worksheet database record structure based on a set of VBA procedures, I began to dream about a database class module. The production of Chapter 8 consumed me for an appreciable amount of time. Would I be able to encapsulate on a single Class module all the code needed to implement a generic worksheet database system? Would it work as expected?

I started in on the code. Trial and error for weeks.... When the SheetDBEngine class code stabilized, I realized that it was too complex to be implemented without the aid of a wizard. The original four constants were now fifteen, and I had to go back and rewrite Chapter 7 using these fifteen database constants so they could be explained in a single chapter. So, I decided to produce a UserForm wizard as a way to help implement this generic worksheet database system on any worksheet, and this was the moment when frmDatabaseProperties was born.

While I wrote about the SheetDBEngine class code, I found many, many bugs in the original database code provided in Chapter 7. So I went back and fixed them. I did it again and again, until Chapter 8 was concluded. During this process, I had to manually clean up the worksheet database records, one by one, and this tedious work showed me that I needed to offer an easy way to delete unwanted database records. This was the second reason I wrote Chapter 9. The first reason was because a new SRxx.mdb ARS nutritional table version had been launched, and when I manually updated the hidden USDA worksheet to this new version, some recipes failed to calculate because some of the food item names had been changed.

There was an urgent need to write about how to use VBA code to silently update a worksheet application data set, and that is why Chapter 9 begins by working with two different methods to update the hidden USDA worksheet: using simple VBA code and using a UserForm to verify data discrepancies that may appear in any new USDA food table version.

Chapter 9 also presented the frmManage UserForm to allow, delete, and save recipes' nutritional data and export and import recipe records between two different worksheet applications using automation. While I wrote it, I found again many other bugs that were still not seen in the SheetDBEngine class module code. And while I corrected them, I had also to correct the standard database code used in Chapter 7 and the SheetDBEngine code in Chapter 8, rewriting again both chapters while Chapter 9 was written. And once again, I did it again and again, as an extreme test of my patience and effort.

The frmManage UserForm was first written to delete records, using a simple interface with a single ListBox control that worked well for the job. This first interface was quite easily adapted to save recipe nutritional data in the USDA food table—an operation that I had to manually make for many recipes when the SRxx.accdb Access file changed from SR27 to SR28.

When it came time to make Export operations between different worksheets or workbooks, it also fit the job well, but when I tried to make an import operation, the interface failed. That is why frmManage suddenly shows the Open dialog box when the user selects the "Import recipes" option.

Since the interface worked well for three basic operations (delete, save nutritional data, and export) and I had gone too far to rewrite the entire chapter, I needed to make some adaptations. Looking now for `frmManage`, I think that it must be completely rewritten using a different approach. But the code example is solid and good for a code book.

I also felt that the automation code used to export and import records between workbooks or worksheets was quite slow, so it was up to me to create the `SheetDBEngine` class's `CopyRecord` and `PasteRecord` methods, which led me to rewrite again Chapters 7 and 8 (text and code), as if I was stuck in an endless loop....

When Chapter 9 was finished, I felt that the book could be considered done. But in my original plans, I always envisioned writing about Windows API programming with VBA and the indomitable ribbon. So, I needed to write two other new chapters.

Chapter 10 was challenging to write because of it being an extensive issue that should be written in a way to give some guidance, using simple and useful programming examples that would make sense to the Excel programmer. After searching the Internet for the most prevalent API issues, I decided to write about how to easily change the `UserForm` appearance.

The chapter begins by teaching you how to declare Windows APIs and call them from VBA code. The first API code teaches you how to create a `Timer` object using `Class` modules and API calls so it can be reused as many times as needed. The second issue was to define the Microsoft Windows handle concept and show how to obtain the `UserForm` handle. It then teaches you how to change some `UserForm` properties that cannot be done with just VBA: removing the title bar, making it resizable, and so on. This basic issue touches on the first principle of combining bits using OR and AND NOT VBA operators to produce the desired final product.

This leads me to the `UserForm` Skin code, found on a German web site, that needs to be explained since a lot of API code just exists to be copied/pasted without ever explaining why it works. Looking now to Chapter 10, I think that it works very well for my basic proposal.

Chapter 11 was written to show how you can produce a personalized ribbon interface to give the best appearance to your worksheet applications. Since it is hard to use VBA alone to produce such simple results, I base the chapter on Andy Pope's RibbonX VBA add-in, which I think is a good start while Microsoft refuses to offer a better programmable approach to such tasks.

All this "code talk" has a reason: it is to call your attention to the fact that it is hard to produce good, solid, concise, and bulletproof code. You need to begin, write, and rewrite it as many times as needed so it becomes trustable. To produce a simple, useful `UserForm`, you must spend a good amount of time just to define the interface, anticipating the user needs and making it intuitive and useful to the task it must accomplish. Then you code it!

After all the work I have done on this book, the code may still be full of bugs that you, the reader, will find. Please forgive me when you find them. I have made my best efforts to remove them one by one, but this is almost impossible.

I also want to testify that the best code quality control ever produced is to write about the code. As I wrote about the code I produced, I removed many, many errors and unnecessary instructions. The lesson is: if you want to produce better code, write about it.

I hope you like the book and that it is useful to you as a good first step into the world of VBA worksheet application programming.

Sincerely yours,
Flavio Morgado
March 28, 2016

Index

A

B

C

F. Morgado, *Programming Excel with VBA*, DOI 10.1007/978-1-4842-2205-8

D

E

F, G, H, I, J, K

L

M

N

O

P, Q

R

S, T

U

V

W, X, Y, Z

Apress®
THE EXPERT'S VOICE™